FOURTEENTH EDITION

Junqueira's
Basic Histology
TEXT AND ATLAS

Anthony L. Mescher, PhD

Professor of Anatomy and Cell Biology
Indiana University School of Medicine
Bloomington, Indiana

New York   Chicago   San Francisco   Athens   London   Madrid   Mexico City
Milan   New Delhi   Singapore   Sydney   Toronto

This book was set in Minion Pro by Cenveo® Publisher Services.
The editors were Michael Weitz and Brian Kearns.
The production supervisor was Catherine H. Saggese.
Project management was provided by Sonam Arora, Cenveo Publisher Services.
RR Donnelley was the printer and binder.

This book is printed on acid-free paper.

The Figure Credit section for this book begins on page 527 and is considered an extension of the copyright page.

Contents

Key Features of
Junqueira's Basic Histology,
Fourteenth Edition:

- **Recognized for more than three decades** as the most authoritative, comprehensive, and effective approach to understanding medical histology

- **Unmatched** in its ability to explain the relationship between cell and tissue structure and their function in the human body

- **Updated** to the latest research and developments on each topic

- **New and additional Medical Applications** throughout each chapter provide clinical relevance for every subject

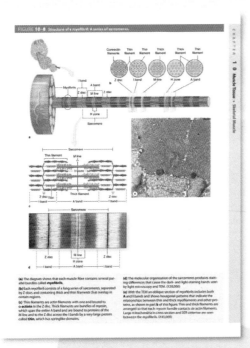

1,500+ illustrations bring important concepts to life

- **New design** incorporates full color for each figure and table in addition to color micrographs of each cell and tissue

- **Figure legends** summarize and provide easy access to key points of each topic

- **Author Anthony L. Mescher, Ph.D.** has over 30 years' experience teaching cell biology and histology to medical students

- **Summaries of Key Points** at the end of each chapter list main points concisely for very easy review

- **Many new tables** help integrate each histological topic with relevant physiological aspects of the system

- **New, detailed tables of contents** within each chapter.

Student-suggested detailed legends

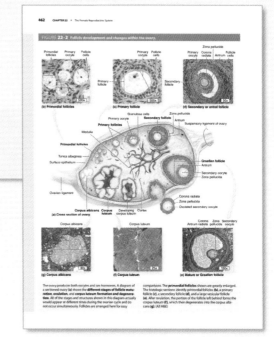

State-of-the art micrographs are the core of an all-new art program

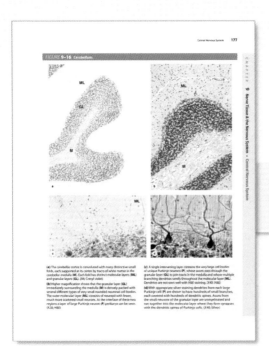

Medical Applications are incorporated into every chapter

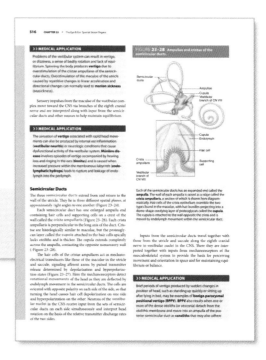

Preface

With this 14th edition, *Junqueira's Basic Histology* continues as the preeminent source of **concise yet thorough** information on human tissue structure and function. For nearly 45 years this educational resource has met the needs of learners for a well-organized and concise presentation of **cell biology and histology** that integrates the material with that of **biochemistry, immunology, endocrinology, and physiology** and provides an excellent foundation for subsequent studies in **pathology**. The text is prepared specifically for students of medicine and other health-related professions, as well as for advanced undergraduate courses in tissue biology. As a result of its value and appeal to students and instructors alike, *Junqueira's Basic Histology* has been translated into a dozen different languages and is used by medical students throughout the world.

This edition now includes with each chapter a set of multiple-choice **Self-Test Questions** that allow readers to assess their comprehension and knowledge of important material in that chapter. At least a few questions in each set utilize clinical vignettes or cases to provide context for framing the medical relevance of concepts in basic science, as recommended by the US National Board of Medical Examiners. As with the last edition, each chapter also includes a **Summary of Key Points** designed to guide the students concerning what is clearly important and what is less so. **Summary Tables** in each chapter organize and condense important information, further facilitating efficient learning.

Each chapter has been revised and shortened, while coverage of specific topics has been expanded as needed. Study is facilitated by modern page design. Inserted throughout each chapter are more numerous, short paragraphs that indicate how the information presented can be used medically and which emphasize the foundational relevance of the material learned.

The art and other figures are presented in each chapter, with the goal to simplify learning and integration with related material. The McGraw-Hill medical illustrations, now used throughout the text and supplemented by numerous animations in the electronic version of the text, are the most useful, thorough, and attractive of any similar medical textbook. Electron and light micrographs have been replaced throughout the book as needed, and again make up a complete atlas of cell, tissue, and organ structures fully compatible with the students' own collection of glass or digital slides. A virtual microscope with over 150 slides of all human tissues and organs is available: http://medsci.indiana.edu/junqueira/virtual/junqueira.htm.

As with the previous edition, the book facilitates learning by its **organization**:

- An opening chapter reviews the **histological techniques** that allow understanding of cell and tissue structure.
- Two chapters then summarize the structural and functional organization of **human cell biology**, presenting the cytoplasm and nucleus separately.
- The next seven chapters cover the **four basic tissues** that make up our organs: epithelia, connective tissue (and its major sub-types), nervous tissue, and muscle.
- Remaining chapters explain the organization and functional significance of these tissues in **each of the body's organ systems**, closing with up-to-date consideration of cells in the eye and ear.

For additional review of what's been learned or to assist rapid assimilation of the material in *Junqueira's Basic Histology*, McGraw-Hill has published a set of 200 full-color **Basic Histology Flash Cards**, Anthony Mescher author. Each card includes images of key structures to identify, a summary of important facts about those structures, and a clinical comment. This valuable learning aid is available as a set of actual cards from Amazon.com, or as an app for smart phones or tablets from the online App Store.

With its proven strengths and the addition of new features, I am confident that *Junqueira's Basic Histology* will continue as one of the most valuable and most widely read educational resources in histology. Users are invited to provide feedback to the author with regard to any aspect of the book's features.

Anthony L. Mescher
Indiana University School of Medicine
mescher@indiana.edu

Acknowledgments

I wish to thank the students at Indiana University School of Medicine and the undergraduates at Indiana University with whom I have studied histology and cell biology for over 30 years and from whom I have learned much about presenting basic concepts most effectively. Their input has greatly helped in the task of maintaining and updating the presentations in this classic textbook. As with the last edition the help of Sue Childress and Dr. Mark Braun was invaluable in slide preparation and the virtual microscope for human histology respectively.

A major change in this edition is the inclusion of self-assessment questions with each topic/chapter. Many of these questions were used in my courses, but others are taken or modified from a few of the many excellent review books published by McGraw-Hill/Lange for students preparing to take the U.S. Medical Licensing Examination. These include *Histology and Cell Biology: Examination and Board Review*, by Douglas Paulsen; *USMLE Road Map: Histology*, by Harold Sheedlo; and *Anatomy, Histology, & Cell Biology: PreTest Self-Assessment & Review*, by Robert Klein and George Enders. The use here of questions from these valuable resources is gratefully acknowledged. Students are referred to those review books for hundreds of additional self-assessment questions.

I am also grateful to my colleagues and reviewers from throughout the world who provided specialized expertise or original photographs, which are also acknowledged in figure captions. I thank those professors and students in the United States, as well as Argentina, Canada, Iran, Ireland, Italy, Pakistan, and Syria, who provided useful suggestions that have improved the new edition of *Junqueira's Basic Histology*. Finally, I am pleased to acknowledge the help and collegiality provided by the staff of McGraw-Hill, especially editors Michael Weitz and Brian Kearns, whose work made possible publication of this 14th edition of *Junqueira's Basic Histology*.

Anthony L. Mescher
Indiana University School of Medicine
mescher@indiana.edu

1 Histology & Its Methods of Study

Histology is the study of the tissues of the body and how these tissues are arranged to constitute organs. This subject involves all aspects of tissue biology, with the focus on how cells' structure and arrangement optimize functions specific to each organ.

Tissues have two interacting components: cells and extracellular matrix (ECM). The ECM consists of many kinds of macromolecules, most of which form complex structures, such as collagen fibrils. The ECM supports the cells and contains the fluid transporting nutrients to the cells, and carrying away their wastes and secretory products. Cells produce the ECM locally and are in turn strongly influenced by matrix molecules. Many matrix components bind to specific cell surface receptors that span the cell membranes and connect to structural components inside the cells, forming a continuum in which cells and the ECM function together in a well-coordinated manner.

During development, cells and their associated matrix become functionally specialized and give rise to fundamental types of tissues with characteristic structural features. Organs are formed by an orderly combination of these tissues, and their precise arrangement allows the functioning of each organ and of the organism as a whole.

The small size of cells and matrix components makes histology dependent on the use of microscopes and molecular methods of study. Advances in biochemistry, molecular biology, physiology, immunology, and pathology are essential for a better knowledge of tissue biology. Familiarity with the tools and methods of any branch of science is essential for a proper understanding of the subject. This chapter reviews common methods used to study cells and tissues, focusing on microscopic approaches.

› PREPARATION OF TISSUES FOR STUDY

The most common procedure used in histologic research is the preparation of tissue slices or "sections" that can be examined visually with transmitted light. Because most tissues and organs are too thick for light to pass through, thin translucent sections are cut from them and placed on glass slides for microscopic examination of the internal structures.

The ideal microscopic preparation is preserved so that the tissue on the slide has the same structural features it had in the body. However, this is often not feasible because the preparation process can remove cellular lipid, with slight distortions of cell structure. The basic steps used in tissue preparation for light microscopy are shown in Figure 1–1.

Fixation

To preserve tissue structure and prevent degradation by enzymes released from the cells or microorganisms, pieces of

FIGURE **1–1** Sectioning fixed and embedded tissue.

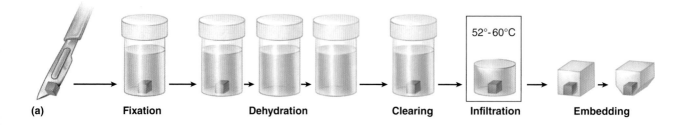

(a)   Fixation   Dehydration   Clearing   Infiltration   Embedding

52°-60°C

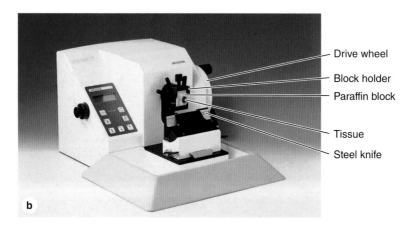

Drive wheel

Block holder
Paraffin block

Tissue
Steel knife

b

Most tissues studied histologically are prepared as shown, with this sequence of steps **(a)**:

- **Fixation:** Small pieces of tissue are placed in solutions of chemicals that cross-link proteins and inactivate degradative enzymes, which preserves cell and tissue structure.
- **Dehydration:** The tissue is transferred through a series of increasingly concentrated alcohol solutions, ending in 100%, which removes all water.
- **Clearing:** Alcohol is removed in organic solvents in which both alcohol and paraffin are miscible.
- **Infiltration:** The tissue is then placed in melted paraffin until it becomes completely infiltrated with this substance.
- **Embedding:** The paraffin-infiltrated tissue is placed in a small mold with melted paraffin and allowed to harden.
- **Trimming:** The resulting paraffin block is trimmed to expose the tissue for sectioning (slicing) on a microtome.

Similar steps are used in preparing tissue for transmission electron microscopy (TEM), except special fixatives and dehydrating solutions are used with smaller tissue samples and embedding involves epoxy resins which become harder than paraffin to allow very thin sectioning.

(b) A **microtome** is used for sectioning paraffin-embedded tissues for light microscopy. The trimmed tissue specimen is mounted in the paraffin block holder, and each turn of the drive wheel by the histologist advances the holder a controlled distance, generally from 1 to 10 μm. After each forward move, the tissue block passes over the steel knife edge and a section is cut at a thickness equal to the distance the block advanced. The paraffin sections are placed on glass slides and allowed to adhere, deparaffinized, and stained for light microscope study. For TEM, sections less than 1 μm thick are prepared from resin-embedded cells using an ultramicrotome with a glass or diamond knife.

organs are placed as soon as possible after removal from the body in solutions of stabilizing or cross-linking compounds called **fixatives**. Because a fixative must fully diffuse through the tissues to preserve all cells, tissues are usually cut into small fragments before fixation to facilitate penetration. To improve cell preservation in large organs fixatives are often introduced via blood vessels, with vascular perfusion allowing fixation rapidly throughout the tissues.

One widely used fixative for light microscopy is formalin, a buffered isotonic solution of 37% formaldehyde. Both this compound and glutaraldehyde, a fixative used for electron microscopy, react with the amine groups (NH_2) of proteins, preventing their degradation by common proteases. Glutaraldehyde also cross-links adjacent proteins, reinforcing cell and ECM structures.

Electron microscopy provides much greater magnification and resolution of very small cellular structures and fixation must be done very carefully to preserve additional "ultrastructural" detail. Typically in such studies glutaraldehyde-treated tissue is then immersed in buffered osmium tetroxide, which preserves (and stains) cellular lipids as well as proteins.

Embedding & Sectioning

To permit thin sectioning fixed tissues are infiltrated and embedded in a material that imparts a firm consistency. Embedding materials include paraffin, used routinely for light microscopy, and plastic resins, which are adapted for both light and electron microscopy.

Before infiltration with such media the fixed tissue must undergo **dehydration** by having its water extracted gradually by transfers through a series of increasing ethanol solutions, ending in 100% ethanol. The ethanol is then replaced by an organic solvent miscible with both alcohol and the embedding medium, a step referred to as **clearing** because infiltration with the reagents used here gives the tissue a translucent appearance.

The fully cleared tissue is then placed in melted paraffin in an oven at 52°-60°C, which evaporates the clearing solvent and promotes **infiltration** of the tissue with paraffin, and then **embedded** by allowing it to harden in a small container of paraffin at room temperature. Tissues to be embedded with plastic resin are also dehydrated in ethanol and then infiltrated with plastic solvents that harden when cross-linking polymerizers are added. Plastic embedding avoids the higher temperatures needed with paraffin, which helps avoid tissue distortion.

The hardened block with tissue and surrounding embedding medium is trimmed and placed for sectioning in an instrument called a **microtome** (Figure 1–1). Paraffin sections are typically cut at 3-10 μm thickness for light microscopy, but electron microscopy requires sections less than 1 μm thick. One micrometer (1 μm) equals 1/1000 of a millimeter (mm) or 10^{-6} m. Other spatial units commonly used in microscopy are the nanometer (1 nm = 0.001 μm = 10^{-6} mm = 10^{-9} m) and angstrom (1 Å = 0.1 nm or 10^{-4} μm). The sections are placed on glass slides and stained for light microscopy or on metal grids for electron microscopic staining and examination.

❯❯ MEDICAL APPLICATION

Biopsies are tissue samples removed during surgery or routine medical procedures. In the operating room, biopsies are fixed in vials of formalin for processing and microscopic analysis in a pathology laboratory. If results of such analyses are required before the medical procedure is completed, for example to know whether a growth is malignant before the patient is closed, a much more rapid processing method is used. The biopsy is rapidly frozen in liquid nitrogen, preserving cell structures and making the tissue hard and ready for sectioning. A microtome called a **cryostat** in a cabinet at subfreezing temperature is used to section the block with tissue, and the frozen sections are placed on slides for rapid staining and microscopic examination by a pathologist.

Freezing of tissues is also effective in histochemical studies of very sensitive enzymes or small molecules because freezing, unlike fixation, does not inactivate most enzymes. Finally, because clearing solvents often dissolve cell lipids in fixed tissues, frozen sections are also useful when structures containing lipids are to be studied histologically.

Staining

Most cells and extracellular material are completely colorless, and to be studied microscopically tissue sections must be stained (dyed). Methods of staining have been devised that make various tissue components not only conspicuous but also distinguishable from one another. Dyes stain material more or less selectively, often behaving like acidic or basic compounds and forming electrostatic (salt) linkages with ionizable radicals of macromolecules in tissues. Cell components such as nucleic acids with a net negative charge (anionic) have an affinity for basic dyes and are termed **basophilic**; cationic components, such as proteins with many ionized amino groups, stain more readily with acidic dyes and are termed **acidophilic**.

Examples of basic dyes include toluidine blue, alcian blue, and methylene blue. Hematoxylin behaves like a basic dye, staining basophilic tissue components. The main tissue components that ionize and react with basic dyes do so because of acids in their composition (DNA, RNA, and glycosaminoglycans). Acid dyes (eg, eosin, orange G, and acid fuchsin) stain the acidophilic components of tissues such as mitochondria, secretory granules, and collagen.

Of all staining methods, the simple combination of **hematoxylin and eosin (H&E)** is used most commonly. Hematoxylin stains DNA in the cell nucleus, RNA-rich portions of the cytoplasm, and the matrix of cartilage, producing a dark blue or purple color. In contrast, eosin stains other cytoplasmic structures and collagen pink (Figure 1–2a). Here eosin is considered a **counterstain**, which is usually a single dye applied separately to distinguish additional features of a tissue. More complex procedures, such as trichrome stains (eg, Masson trichrome), allow greater distinctions among various extracellular tissue components.

The **periodic acid-Schiff (PAS) reaction** utilizes the hexose rings of polysaccharides and other carbohydrate-rich tissue structures and stains such macromolecules distinctly purple or magenta. Figure 1–2b shows an example of cells with carbohydrate-rich areas well-stained by the PAS reaction. The DNA of cell nuclei can be specifically stained using a modification of the PAS procedure called the Feulgen reaction.

Basophilic or PAS-positive material can be further identified by enzyme digestion, pretreatment of a tissue section with an enzyme that specifically digests one substrate. For example, pretreatment with ribonuclease will greatly reduce cytoplasmic basophilia with little overall effect on the nucleus, indicating the importance of RNA for the cytoplasmic staining.

Lipid-rich structures of cells are revealed by avoiding the processing steps that remove lipids, such as treatment with heat and organic solvents, and staining with **lipid-soluble dyes** such as **Sudan black,** which can be useful in diagnosis of metabolic diseases that involve intracellular accumulations of cholesterol, phospholipids, or glycolipids. Less common methods of staining can employ **metal impregnation** techniques, typically using solutions of silver salts to visual certain ECM fibers and specific cellular elements in nervous tissue. The Appendix lists important staining procedures used for most of the light micrographs in this book.

FIGURE **1–2** Hematoxylin and eosin (H&E) and periodic acid-Schiff (PAS) staining.

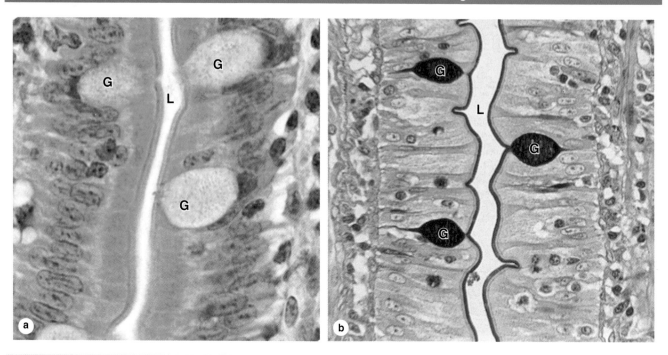

Micrographs of epithelium lining the small intestine, **(a)** stained with H&E, and **(b)** stained with the PAS reaction for glycoproteins. With H&E, basophilic cell nuclei are stained purple while cytoplasm stains pink. Cell regions with abundant oligosaccharides on glycoproteins, such as the ends of the cells at the lumen **(L)** or the scattered mucus-secreting goblet cells **(G)**, are poorly stained. With PAS, however, cell staining is most intense at the lumen, where projecting microvilli have a prominent layer of glycoproteins at the lumen (L) and in the mucin-rich secretory granules of goblet cells. Cell surface glycoproteins and mucin are PAS-positive because of their high content of oligosaccharides and polysaccharides respectively. The PAS-stained tissue was counterstained with hematoxylin to show the cell nuclei. (a. X400; b. X300)

Slide preparation, from tissue fixation to observation with a light microscope, may take from 12 hours to 2½ days, depending on the size of the tissue, the embedding medium, and the method of staining. The final step before microscopic observation is mounting a protective glass coverslip on the slide with clear adhesive.

› LIGHT MICROSCOPY

Conventional bright-field microscopy, as well as more specialized applications like fluorescence, phase-contrast, confocal, and polarizing microscopy, are all based on the interaction of light with tissue components and are used to reveal and study tissue features.

Bright-Field Microscopy

With the **bright-field microscope** stained tissue is examined with ordinary light passing through the preparation. As shown in Figure 1–3, the microscope includes an optical system and mechanisms to move and focus the specimen. The optical components are the **condenser** focusing light on the object to be studied; the **objective** lens enlarging and projecting the image of the object toward the observer; and the **eyepiece**

(or ocular lens) further magnifying this image and projecting it onto the viewer's retina or a charge-coupled device (CCD) highly sensitive to low light levels with a camera and monitor. The total magnification is obtained by multiplying the magnifying power of the objective and ocular lenses.

The critical factor in obtaining a crisp, detailed image with a light microscope is its **resolving power**, defined as the smallest distance between two structures at which they can be seen as separate objects. The maximal resolving power of the light microscope is approximately 0.2 μm, which can permit clear images magnified 1000-1500 times. Objects smaller or thinner than 0.2 μm (such as a single ribosome or cytoplasmic microfilament) cannot be distinguished with this instrument. Likewise, two structures such as mitochondria will be seen as only one object if they are separated by less than 0.2 μm. The microscope's resolving power determines the quality of the image, its clarity and richness of detail, and depends mainly on the quality of its objective lens. Magnification is of value only when accompanied by high resolution. Objective lenses providing higher magnification are designed to also have higher resolving power. The eyepiece lens only enlarges the image obtained by the objective and does not improve resolution.

Virtual microscopy, typically used for study of bright-field microscopic preparations, involves the conversion of a

FIGURE 1–3 Components and light path of a bright-field microscope.

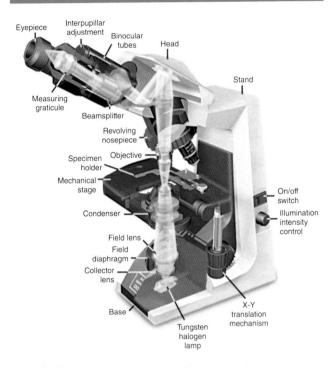

Photograph of a bright-field light microscope showing its mechanical components and the pathway of light from the substage lamp to the eye of the observer. The optical system has three sets of lenses:

- The **condenser** collects and focuses a cone of light that illuminates the tissue slide on the stage.
- **Objective** lenses enlarge and project the illuminated image of the object toward the eyepiece. Interchangeable objectives with different magnifications routinely used in histology include X4 for observing a large area (field) of the tissue at low magnification; X10 for medium magnification of a smaller field; and X40 for high magnification of more detailed areas.
- The two **eyepieces** or oculars magnify this image another X10 and project it to the viewer, yielding a total magnification of X40, X100, or X400.

(Used with permission from Nikon Instruments.)

stained tissue preparation to high-resolution digital images and permits study of tissues using a computer or other digital device, without an actual stained slide or a microscope. In this technique regions of a glass-mounted specimen are captured digitally in a grid-like pattern at multiple magnifications using a specialized slide-scanning microscope and saved as thousands of consecutive image files. Software then converts this dataset for storage on a server using a format that allows access, visualization, and navigation of the original slide with common web browsers or other devices. With advantages in cost and ease of use, virtual microscopy is rapidly replacing light microscopes and collections of glass slides in histology laboratories for students.

Fluorescence Microscopy

When certain cellular substances are irradiated by light of a proper wavelength, they emit light with a longer wavelength—a phenomenon called **fluorescence**. In **fluorescence microscopy**, tissue sections are usually irradiated with ultraviolet (UV) light and the emission is in the visible portion of the spectrum. The fluorescent substances appear bright on a dark background. For fluorescent microscopy the instrument has a source of UV or other light and filters that select rays of different wavelengths emitted by the substances to be visualized.

Fluorescent compounds with affinity for specific cell macromolecules may be used as fluorescent stains. Acridine orange, which binds both DNA and RNA, is an example. When observed in the fluorescence microscope, these nucleic acids emit slightly different fluorescence, allowing them to be localized separately in cells (Figure 1–4a). Other compounds such as DAPI and Hoechst stain specifically bind DNA and are used to stain cell nuclei, emitting a characteristic blue fluorescence under UV. Another important application of fluorescence microscopy is achieved by coupling compounds such as fluorescein to molecules that will specifically bind to certain cellular components and thus allow the identification of these structures under the microscope (Figure 1–4b). Antibodies labeled with fluorescent compounds are extremely important in immunohistologic staining. (See the section Visualizing Specific Molecules.)

Phase-Contrast Microscopy

Unstained cells and tissue sections, which are usually transparent and colorless, can be studied with these modified light microscopes. Cellular detail is normally difficult to see in unstained tissues because all parts of the specimen have roughly similar optical densities. **Phase-contrast microscopy**, however, uses a lens system that produces visible images from transparent objects and, importantly, can be used with living, cultured cells (Figure 1–5).

Phase-contrast microscopy is based on the principle that light changes its speed when passing through cellular and extracellular structures with different refractive indices. These changes are used by the phase-contrast system to cause the structures to appear lighter or darker in relation to each other. Because they allow the examination of cells without fixation or staining, phase-contrast microscopes are prominent tools in all cell culture laboratories. A modification of phase-contrast microscopy is **differential interference microscopy** with Nomarski optics, which produces an image of living cells with a more apparent three-dimensional (3D) aspect (Figure 1–5c).

Confocal Microscopy

With a regular bright-field microscope, the beam of light is relatively large and fills the specimen. Stray (excess) light reduces contrast within the image and compromises the resolving

FIGURE **1–4** **Appearance of cells with fluorescent microscopy.**

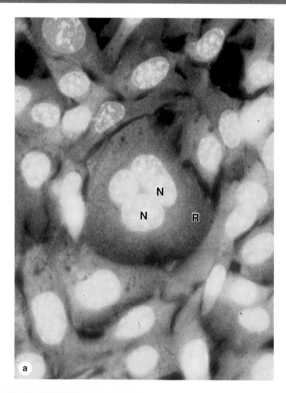
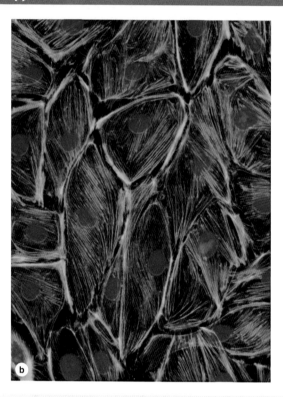

Components of cells are often stained with compounds visible by fluorescence microscopy.

(a) Acridine orange binds nucleic acids and causes DNA in cell nuclei **(N)** to emit yellow light and the RNA-rich cytoplasm **(R)** to appear orange in these cells of a kidney tubule.

(b) Cultured cells stained with DAPI (4′,6-diamino-2-phenylindole) that binds DNA and with fluorescein-phalloidin that binds actin

filaments show nuclei with blue fluorescence and actin filaments stained green. Important information such as the greater density of microfilaments at the cell periphery is readily apparent. (Both X500)

(*Figure 1–4b, used with permission from Drs Claire E. Walczak and Rania Rizk, Indiana University School of Medicine, Bloomington.*)

FIGURE **1–5** **Unstained cells' appearance in three types of light microscopy.**

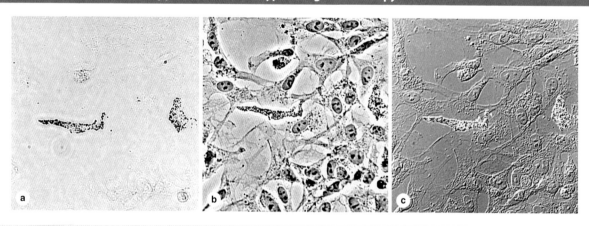
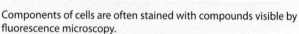

Living neural crest cells growing in culture appear differently with various techniques of light microscopy. Here the *same field* of unstained cells, including two differentiating pigment cells, is shown using three different methods (all X200):

(a) Bright-field microscopy: Without fixation and staining, only the two pigment cells can be seen.

(b) Phase-contrast microscopy: Cell boundaries, nuclei, and cytoplasmic structures with different refractive indices affect

in-phase light differently and produce an image of these features in *all* the cells.

(c) Differential interference microscopy: Cellular details are highlighted in a different manner using Nomarski optics. Phase-contrast microscopy, with or without differential interference, is widely used to observe live cells grown in tissue culture.

(*Used with permission from Dr Sherry Rogers, Department of Cell Biology and Physiology, University of New Mexico, Albuquerque, NM.*)

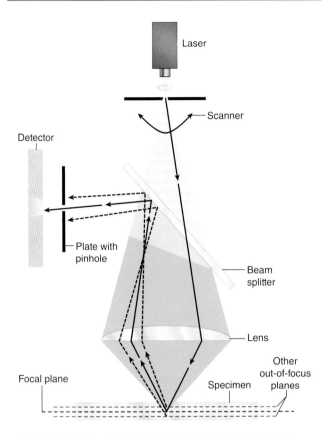

FIGURE **1–6** Principle of confocal microscopy.

Laser

Scanner

Detector

Plate with pinhole

Beam splitter

Lens

Other out-of-focus planes

Focal plane

Specimen

Although a very small spot of light originating from one plane of the section crosses the pinhole and reaches the detector, rays originating from other planes are blocked by the blind. Thus, only one very thin plane of the specimen is focused at a time. The diagram shows the practical arrangement of a confocal microscope. Light from a laser source hits the specimen and is reflected. A beam splitter directs the reflected light to a pinhole and a detector. Light from components of the specimen that are above or below the focused plane is blocked by the blind. The laser scans the specimen so that a larger area of the specimen can be observed.

power of the objective lens. Confocal microscopy (Figure 1–6) avoids these problems and achieves high resolution and sharp focus by using (1) a small point of high-intensity light, often from a laser, and (2) a plate with a pinhole aperture in front of the image detector. The point light source, the focal point of the lens, and the detector's pinpoint aperture are all optically conjugated or aligned to each other in the focal plane (confocal), and unfocused light does not pass through the pinhole. This greatly improves resolution of the object in focus and allows the localization of specimen components with much greater precision than with the bright-field microscope.

Confocal microscopes include a computer-driven mirror system (the beam splitter) to move the point of illumination across the specimen automatically and rapidly. Digital images captured at many individual spots in a very thin plane of focus are used to produce an "optical section" of that plane. Creating such optical sections at a series of focal planes through the specimen allows them to be digitally reconstructed into a 3D image.

Polarizing Microscopy

Polarizing microscopy allows the recognition of stained or unstained structures made of highly organized subunits. When normal light passes through a **polarizing** filter, it exits vibrating in only one direction. If a second filter is placed in the microscope above the first one, with its main axis perpendicular to the first filter, no light passes through. If, however, tissue structures containing oriented macromolecules are located between the two polarizing filters, their repetitive structure rotates the axis of the light emerging from the polarizer and they appear as bright structures against a dark background (Figure 1–7). The ability to rotate the direction of vibration of polarized light is called **birefringence** and is

FIGURE **1–7** Tissue appearance with bright-field and polarizing microscopy.

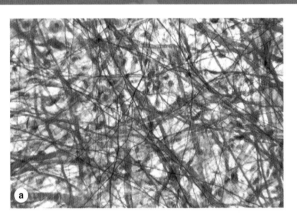

Polarizing light microscopy produces an image only of material having repetitive, periodic macromolecular structure; features without such structure are not seen. Pieces of thin, unsectioned mesentery were stained with red picrosirius, orcein, and hematoxylin, placed on slides and observed by bright-field **(a)** and polarizing **(b)** microscopy.

(a) With bright-field microscopy collagen fibers appear red, with thin elastic fibers and cell nuclei darker. (X40)

(b) With polarizing microscopy, only the collagen fibers are visible and these exhibit intense yellow or orange birefringence. (a: X40; b: X100)

a feature of crystalline substances or substances containing highly oriented molecules, such as cellulose, collagen, microtubules, and actin filaments.

The utility of all light microscopic methods is greatly extended through the use of digital cameras. Many features of digitized histological images can be analyzed quantitatively using appropriate software. Such images can also be enhanced to allow objects not directly visible through the eyepieces to be examined on a monitor.

❯ ELECTRON MICROSCOPY

Transmission and scanning electron microscopes are based on the interaction of tissue components with beams of electrons.

The wavelength in an electron beam is much shorter than that of light, allowing a 1000-fold increase in resolution.

Transmission Electron Microscopy

The **transmission electron microscope (TEM)** is an imaging system that permits resolution around 3 nm. This high resolution allows isolated particles magnified as much as 400,000 times to be viewed in detail. Very thin (40-90 nm), resin-embedded tissue sections are typically studied by TEM at magnifications up to approximately 120,000 times.

Figure 1–8a indicates the components of a TEM and the basic principles of its operation: a beam of electrons focused using electromagnetic "lenses" passes through the tissue section to produce an image with black, white, and intermediate

---

FIGURE 1–8 Electron microscopes.

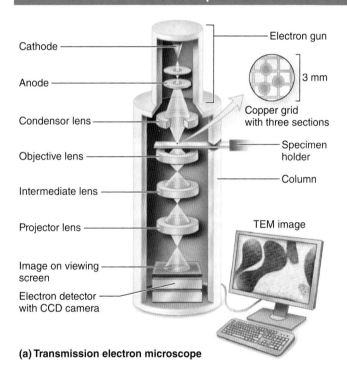

(a) Transmission electron microscope

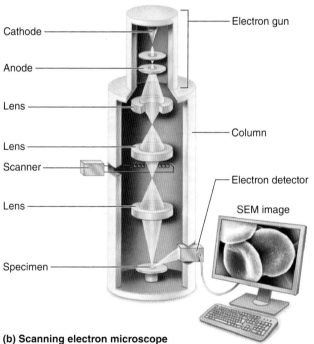

(b) Scanning electron microscope

Electron microscopes are large instruments generally housed in a specialized EM facility.

(a) Schematic view of the major components of a transmission electron microscope (TEM), which is configured rather like an upside-down light microscope. With the microscope column in a vacuum, a metallic (usually tungsten) filament (cathode) at the top emits electrons that travel to an anode with an accelerating voltage between 60 and 120 kV. Electrons passing through a hole in the anode form a beam that is **focused electromagnetically** by circular electric coils in a manner analogous to the effect of optical lenses on light.

The first lens is a condenser focusing the beam on the section. Some electrons interact with atoms in the section, being absorbed or scattered to different extents, while others are simply transmitted through the specimen with no interaction. Electrons reaching the objective lens form an image that is then magnified and finally projected on a fluorescent screen or a charge-coupled device (CCD) monitor and camera.

In a TEM image areas of the specimen through which electrons passed appear bright (electron lucent), while denser areas or those that bind heavy metal ions during specimen preparation absorb or deflect electrons and appear darker (electron dense). Such images are therefore always black, white, and shades of gray.

(b) The scanning electron microscope (SEM) has many similarities to a TEM. However, here the focused electron beam does not pass through the specimen, but rather is moved sequentially (scanned) from point to point across its surface similar to the way an electron beam is scanned across a television tube or screen. For SEM specimens are coated with metal atoms with which the electron beam interacts, producing reflected electrons and newly emitted secondary electrons. All of these are captured by a detector and transmitted to amplifiers and processed to produce a black-and-white image on the monitor. The SEM shows only surface views of the coated specimen but with a striking 3D, shadowed quality. The inside of organs or cells can be analyzed after sectioning to expose their internal surfaces.

shades of gray regions. These regions of an electron micrograph correspond to tissue areas through which electrons passed readily (appearing brighter or electron-lucent) and areas where electrons were absorbed or deflected (appearing darker or more electron-dense). To improve contrast and resolution in TEM, compounds with **heavy metal ions** are often added to the fixative or dehydrating solutions used for tissue preparation. These include osmium tetroxide, lead citrate, and uranyl compounds, which bind cellular macromolecules, increasing their electron density and visibility.

Cryofracture and **freeze etching** are techniques that allow TEM study of cells without fixation or embedding and have been particularly useful in the study of membrane structure. In these methods very small tissue specimens are rapidly frozen in liquid nitrogen and then cut or fractured with a knife. A replica of the frozen exposed surface is produced in a vacuum by applying thin coats of vaporized platinum or other metal atoms. After removal of the organic material, the replica of the cut surface can be examined by TEM. With membranes the random fracture planes often split the lipid bilayers, exposing protein components whose size, shape, and distribution are difficult to study by other methods.

Scanning Electron Microscopy

Scanning electron microscopy (SEM) provides a high-resolution view of the surfaces of cells, tissues, and organs. Like the TEM, this microscope produces and focuses a very narrow beam of electrons, but in this instrument the beam does not pass through the specimen (Figure 1–8b). Instead, the surface of the specimen is first dried and spray-coated with a very thin layer of heavy metal (often gold) which reflects electrons in a beam scanning the specimen. The reflected electrons are captured by a detector, producing signals that are processed to produce a black-and-white image. SEM images are usually easy to interpret because they present a three-dimensional view that appears to be illuminated in the same way that large objects are seen with highlights and shadows caused by light.

AUTORADIOGRAPHY

Microscopic **autoradiography** is a method of localizing newly synthesized macromolecules in cells or tissue sections. Radioactively labeled metabolites (nucleotides, amino acids, sugars) provided to the living cells are incorporated into specific macromolecules (DNA, RNA, protein, glycoproteins, and polysaccharides) and emit weak radiation that is restricted to those regions where the molecules are located. Slides with radiolabeled cells or tissue sections are coated in a darkroom with photographic emulsion in which silver bromide crystals act as microdetectors of the radiation in the same way that they respond to light in photographic film. After an adequate exposure time in lightproof boxes, the slides are developed photographically. Silver bromide crystals reduced by the radiation produce small black grains of metallic silver, which under either the light microscope or TEM indicate the locations of radiolabeled macromolecules in the tissue (Figure 1–9).

Much histological information becomes available by autoradiography. If a radioactive precursor of DNA (such as tritium-labeled thymidine) is used, it is possible to know which cells in a tissue (and how many) are replicating DNA

FIGURE **1–9** Microscopic autoradiography.

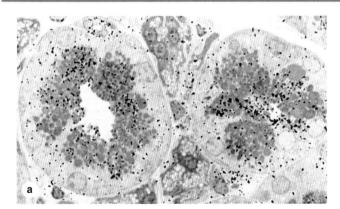

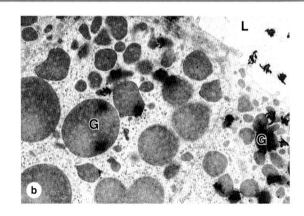

Autoradiographs are tissue preparations in which particles called **silver grains** indicate the cells or regions of cells in which specific macromolecules were synthesized just prior to fixation. Shown here are autoradiographs from the salivary gland of a mouse injected with ³H-fucose 8 hours before tissue fixation. Fucose was incorporated into oligosaccharides, and the free ³H-fucose was removed during fixation and sectioning of the gland. Autoradiographic processing and microscopy reveal locations of newly synthesized glycoproteins containing that sugar.

(a) Black grains of silver from the light-sensitive material coating the specimen are visible over cell regions with secretory granules and the duct indicating glycoprotein locations. (X1500)

(b) The same tissue prepared for TEM autoradiography shows silver grains with a coiled or amorphous appearance again localized mainly over the granules **(G)** and in the gland lumen **(L)**. (X7500)

(Figure 1–9b, used with permission from Drs Ticiano G. Lima and A. Antonio Haddad, School of Medicine, Ribeirão Preto, Brazil.)

and preparing to divide. Dynamic events may also be analyzed. For example, if one wishes to know where in the cell protein is produced, if it is secreted, and its path in the cell before being secreted, several animals are injected with a radioactive amino acid and tissues collected at different times after the injections. Autoradiography of the tissues from the sequential times will indicate the migration of the radioactive proteins.

› CELL & TISSUE CULTURE

Live cells and tissues can be maintained and studied outside the body in culture (in vitro). In the organism (in vivo), cells are bathed in fluid derived from blood plasma and containing many different molecules required for survival and growth. Cell culture allows the direct observation of cellular behavior under a phase-contrast microscope and many experiments technically impossible to perform in the intact animal can be accomplished in vitro.

The cells and tissues are grown in complex solutions of known composition (salts, amino acids, vitamins) to which serum or specific growth factors are added. Cells to be cultured are dispersed mechanically or enzymatically from a tissue or organ and placed with sterile procedures in a clear dish to which they adhere, usually as a single layer (Figure 1–5). Such preparations are called **primary cell cultures**. Some cells can be maintained in vitro for long periods because they become immortalized and constitute a permanent **cell line**. Most cells obtained from normal tissues have a finite, genetically programmed life span. However certain changes (some related to oncogenes; see Chapter 3) can promote cell immortality, a process called **transformation**, and are similar to the initial changes in a normal cell's becoming a cancer cell. Improvements in culture technology and use of specific growth factors now allow most cell types to be maintained in vitro.

As shown in Chapter 2, incubation of living cells in vitro with a variety of new fluorescent compounds that are sequestered and metabolized in specific compartments of the cell provides a new approach to understanding these compartments both structurally and physiologically. Other histologic techniques applied to cultured cells have been particularly important for understanding the locations and functions of microtubules, microfilaments, and other components of the cytoskeleton.

> ›› **MEDICAL APPLICATION**
>
> Cell culture is very widely used to study molecular changes that occur in cancer; to analyze infectious viruses, mycoplasma, and some protozoa; and for many routine genetic or chromosomal analyses. Cervical cancer cells from a patient later identified as Henrietta Lacks, who died from the disease in 1951, were used to establish one of the first cell lines, called **HeLa cells**, which are still used in research on cellular structure and function throughout the world.

› ENZYME HISTOCHEMISTRY

Enzyme histochemistry (or cytochemistry) is a method for localizing cellular structures using a specific enzymatic activity present in those structures. To preserve the endogenous enzymes histochemical procedures usually use unfixed or mildly fixed tissue, which is sectioned on a cryostat to avoid adverse effects of heat and organic solvents on enzymatic activity. For enzyme histochemistry (1) tissue sections are immersed in a solution containing the substrate of the enzyme to be localized; (2) the enzyme is allowed to act on its substrate; (3) the section is then put in contact with a marker compound that reacts with a product of the enzymatic action on the substrate; and (4) the final product from the marker, which must be insoluble and visible by light or electron microscopy, precipitates over the site of the enzymes, identifying their location.

Examples of enzymes that can be detected histochemically include the following:

- **Phosphatases**, which remove phosphate groups from macromolecules (Figure 1–10).
- **Dehydrogenases**, which transfer hydrogen ions from one substrate to another, such as many enzymes of the citric acid (Krebs) cycle, allowing histochemical identification of such enzymes in mitochondria.
- **Peroxidase**, which promotes the oxidation of substrates with the transfer of hydrogen ions to hydrogen peroxide.

> ›› **MEDICAL APPLICATION**
>
> Many enzyme histochemical procedures are used in the medical laboratory, including Perls' Prussian blue reaction for iron (used to diagnose the iron storage diseases, hemochromatosis and hemosiderosis), the PAS-amylase and alcian blue reactions for polysaccharides (to detect glycogenosis and mucopolysaccharidosis), and reactions for lipids and sphingolipids (to detect sphingolipidosis).

› VISUALIZING SPECIFIC MOLECULES

A specific macromolecule present in a tissue section may also be identified by using tagged compounds or macromolecules that bind *specifically* with the molecule of interest. The compounds that interact with the molecule must be visible with the light or electron microscope, often by being tagged with a detectible label. The most commonly used labels are fluorescent compounds, radioactive atoms that can be detected with autoradiography, molecules of peroxidase or other enzymes that can be detected with histochemistry, and metal (usually gold) particles that can be seen with light and electron microscopy. These methods can be used to detect and localize specific sugars, proteins, and nucleic acids.

FIGURE **1–10** Enzyme histochemistry.

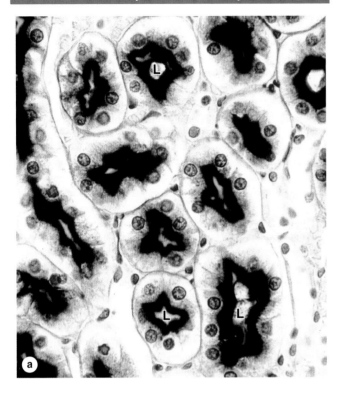

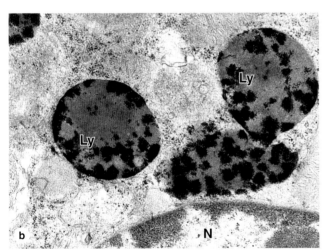

(a) Micrograph of cross sections of kidney tubules treated histochemically to demonstrate alkaline phosphatases (with maximum activity at an alkaline pH) showing strong activity of this enzyme at the apical surfaces of the cells at the lumens **(L)** of the tubules. (X200)

(b) TEM image of a kidney cell in which acid phosphatase has been localized histochemically in three lysosomes **(Ly)** near the nucleus **(N)**. The dark material within these structures is lead phosphate that precipitated in places with acid phosphatase activity. (X25,000)

(*Figure 1–10b, used with permission from Dr Eduardo Katchburian, Department of Morphology, Federal University of São Paulo, Brazil.*)

Examples of molecules that interact specifically with other molecules include the following:

- **Phalloidin**, a compound extracted from mushroom, *Amanita phalloides*, interacts strongly with the actin protein of microfilaments.
- **Protein A**, purified from *Staphylococcus aureus* bacteria, binds to the Fc region of antibody molecules, and can therefore be used to localize naturally occurring or applied antibodies bound to cell structures.
- **Lectins**, glycoproteins derived mainly from plant seeds, bind to carbohydrates with high affinity and specificity. Different lectins bind to specific sugars or sequences of sugar residues, allowing fluorescently labeled lectins to be used to stain specific glycoproteins or other macromolecules bearing specific sequences of sugar residues.

Immunohistochemistry

A highly specific interaction between macromolecules is that between an antigen and its antibody. For this reason labeled antibodies are routinely used in **immunohistochemistry** to identify and localize many specific proteins, not just those with enzymatic activity that can be demonstrated by histochemistry.

The body's immune cells interact with and produce **antibodies** against other macromolecules—called antigens—that are recognized as "foreign," not a normal part of the organism, and potentially dangerous. Antibodies belong to the **immunoglobulin** family of glycoproteins and are secreted by lymphocytes. These molecules normally bind specifically to their provoking antigens and help eliminate them.

Widely applied for both research and diagnostic purposes, every immunohistochemical technique requires an antibody against the protein that is to be detected. This means that the protein must have been previously purified using biochemical or molecular methods so that antibodies against it can be produced. To produce antibodies against protein x of a certain animal species (eg, a human or rat), the isolated protein is injected into an animal of another species (eg, a rabbit or a goat). If the protein's amino acid sequence is sufficiently different for this animal to recognize it as foreign—that is, as an antigen—the animal will produce antibodies against the protein.

Different groups (clones) of lymphocytes in the injected animal recognize different parts of protein x and each clone produces an antibody against that part. These antibodies are collected from the animal's plasma and constitute a mixture of **polyclonal antibodies**, each capable of binding a different region of protein x.

It is also possible, however, to inject protein x into a mouse and a few days later isolate the activated lymphocytes and place them into culture. Growth and activity of these cells can be prolonged indefinitely by fusing them with lymphocytic tumor cells to produce hybridoma cells. Different hybridoma clones produce different antibodies against the several parts of protein *x* and each clone can be isolated and cultured separately so that the different antibodies against protein *x* can be collected separately. Each of these antibodies is a **monoclonal antibody**. An advantage to using a monoclonal antibody rather than polyclonal antibodies is that it can be selected to be highly specific and to bind strongly to the protein to be detected, with less nonspecific binding to other proteins that are similar to the one of interest.

In immunohistochemistry a tissue section that one believes contains the protein of interest is incubated in a solution containing antibody (either monoclonal or polyclonal) against this protein. The antibody binds specifically to the protein and after a rinse the protein's location in the tissue or cells can be seen with either the light or electron microscope by visualizing the antibody. Antibodies are commonly tagged with fluorescent compounds, with peroxidase or alkaline phosphatase for histochemical detection, or with electron-dense gold particles for TEM.

As Figure 1–11 indicates, there are **direct and indirect methods of immunocytochemistry**. The direct method just involves a labeled antibody that binds the protein of interest.

Indirect immunohistochemistry involves sequential application of two antibodies and additional washing steps. The (primary) antibody specifically binding the protein of interest is not labeled. The detectible tag is conjugated to a **secondary antibody** made in an animal species different ("foreign") from that which made the primary antibody. For example, primary antibodies made by mouse lymphocytes (such as most monoclonal antibodies) are specifically recognized and bound by antibodies made in a rabbit or goat injected with mouse antibody immunoglobulin.

The indirect method is used more widely in research and pathologic tests because it is more sensitive, with the extra level of antibody binding serving to amplify the visible signal. Moreover, the same preparation of labeled secondary antibody can be used in studies with different primary antibodies (specific for different antigens) as long as all these are made in the same species. There are other indirect methods that involve the use of other intermediate molecules, such as the biotin-avidin technique, which are also used to amplify detection signals.

Examples of indirect immunocytochemistry are shown in Figure 1–12, demonstrating the use of this method with cells in culture or after tissue sectioning for both light microscopy and TEM.

›› MEDICAL APPLICATION

Because cells in some diseases, including many cancer cells, often produce proteins unique to their pathologic condition, immunohistochemistry can be used by pathologists to diagnose many diseases, including certain types of tumors and some virus-infected cells. Table 1-1 shows some applications of immunocytochemistry routinely used in clinical practice.

Hybridization Techniques

Hybridization usually implies the specific binding between two single strands of nucleic acid, which occurs under appropriate conditions if the strands are complementary. The greater the similarities of their nucleotide sequences, the more readily the complementary strands form "hybrid" double-strand molecules. Hybridization at stringent conditions allows the specific identification of sequences in genes or RNA. This can

FIGURE 1–11   Immunocytochemistry techniques.

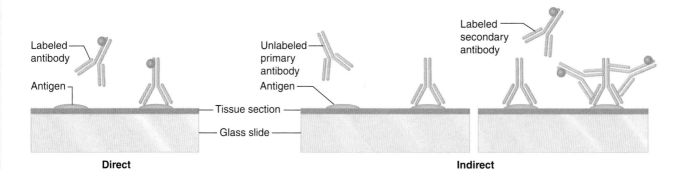

Direct   **Indirect**

Immunocytochemistry (or immunohistochemistry) can be direct or indirect. **Direct immunocytochemistry** (left) uses an antibody made against the tissue protein of interest and tagged directly with a label such as a fluorescent compound or peroxidase. When placed with the tissue section on a slide, these labeled antibodies bind specifically to the protein (antigen) against which they were produced and can be visualized by the appropriate method. **Indirect immunocytochemistry** (right) uses first a **primary antibody** made against the protein (antigen) of interest and applied to the tissue section to bind its specific antigen. Then a **labeled secondary antibody** is obtained that was (1) made in another species against immunoglobulin proteins (antibodies) from the species in which the primary antibodies were made and (2) labeled with a fluorescent compound or peroxidase. When the labeled secondary antibody is applied to the tissue section, it specifically binds the primary antibodies, indirectly labeling the protein of interest on the slide. Because more than one labeled secondary antibody can bind each primary antibody molecule, labeling of the protein of interest is amplified by the indirect method.

FIGURE **1-12** Cells and tissues stained by immunohistochemistry.

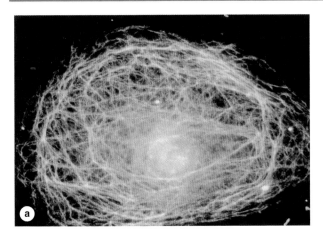

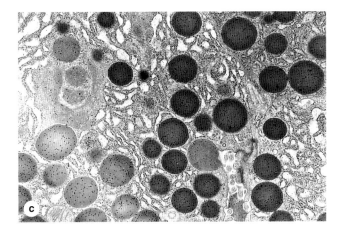

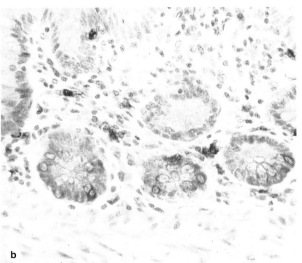

Immunocytochemical methods to localize specific proteins can be applied to either light microscopic or TEM preparations using a variety of labels.

(a) A single cultured uterine cell stained fluorescently to reveal a meshwork of intermediate filaments (green)

throughout the cytoplasm. Primary antibodies against the filament protein desmin and fluorescein isothiocyanate (FITC)–labeled secondary antibodies were used in the indirect staining technique, with the nucleus counterstained blue with DAPI. (X650)

(b) A section of small intestine treated with an antibody against the enzyme lysozyme. The secondary antibody labeled with peroxidase was then applied and the localized brown color produced histochemically with the peroxidase substrate 3,3′-diamino-azobenzidine (DAB). The method demonstrates lysozyme-containing structures in scattered macrophages and in the large clusters of cells. Nuclei were counterstained with hematoxylin. (X100)

(c) A section of pancreatic cells in a TEM preparation incubated with an antibody against the enzyme amylase and then with protein A coupled with gold particles. Protein A has high affinity toward antibody molecules and the resulting image reveals the presence of amylase with the gold particles localized as very small black dots over dense secretory granules and developing granules (left). With specificity for immunoglobulin molecules, labeled protein A can be used to localize any primary antibody. (X5000)

(*Figure 1–12c, used with permission from Dr Moise Bendayan, Departments of Pathology and Cell Biology, University of Montreal, Montreal, Canada.*)

TABLE **1-1** Examples of specific antigens with diagnostic importance.

Antigens	Diagnosis
Specific cytokeratins	Tumors of epithelial origin
Protein and polypeptide hormones	Certain endocrine tumors
Carcinoembryonic antigen (CEA)	Glandular tumors, mainly of the digestive tract and breast
Steroid hormone receptors	Breast duct cell tumors
Antigens produced by viruses	Specific virus infections

occur with cellular DNA or RNA when nucleic acid sequences in solution are applied directly to prepared cells and tissue sections, a procedure called **in situ hybridization** (ISH).

This technique is ideal for (1) determining if a cell has a specific sequence of DNA, such as a gene or part of a gene (Figure 1–13), (2) identifying the cells containing specific messenger RNAs (mRNAs) (in which the corresponding gene is being transcribed), or (3) determining the localization of a gene in a specific chromosome. DNA and RNA of the cells must be initially denatured by heat or other agents to become completely single-stranded and the nucleotide sequences of interest are detected with **probes** consisting of single-stranded complementary DNA (cDNA). The probe may be obtained by cloning, by polymerase chain reaction (PCR) amplification of the target sequence, or by chemical synthesis if the desired sequence is short. The probe is tagged with nucleotides containing a radioactive isotope (localized by autoradiography) or modified with a small compound such as digoxigenin (identified by immunocytochemistry). A solution containing the probe is placed over the specimen under conditions allowing hybridization and after the excess unbound probe is washed off, the localization of the hybridized probe is revealed through its label.

FIGURE **1–13** **In situ hybridization (ISH).**

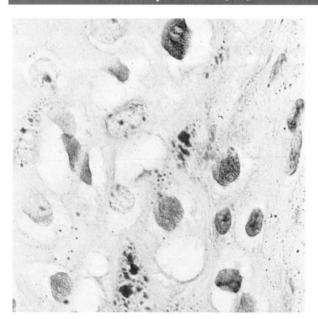

In situ hybridization of this tissue section with probes for the human papilloma virus (HPV) reveals the presence of many cells containing the virus. The section was incubated with a solution containing a digoxigenin-labeled complementary DNA (cDNA) probe for the HPV DNA. The probe was then visualized by direct immunohistochemistry using peroxidase-labeled antibodies against digoxigenin. This procedure stains brown only those cells containing HPV. (X400; H&E)

(Used with permission from Dr Jose E. Levi, Virology Lab, Institute of Tropical Medicine, University of São Paulo, Brazil.)

Warts on the skin of the genitals and elsewhere are due to infection with the human papilloma virus (HPV) which causes the characteristic benign proliferative growth. As shown in Figure 1–12 such virus-infected cells can often be demonstrated by ISH. Certain cancer cells with unique or elevated expression of specific genes are also localized in tumors and studied microscopically by ISH.

》 INTERPRETATION OF STRUCTURES IN TISSUE SECTIONS

In studying and interpreting stained tissue sections, it is important to remember that microscopic preparations are the end result of a series of processes that began with collecting the tissue and ended with mounting a coverslip on the slide. Certain steps in this procedure may distort the tissues slightly, producing minor structural abnormalities called **artifacts** not present in the living tissue.

One such distortion is minor shrinkage of cells or tissue regions produced by the fixative, by the ethanol, or by the heat needed for paraffin embedding. Shrinkage can create artificial spaces between cells and other tissue components. Such spaces can also result from the loss of lipids or low-molecular-weight substances not preserved by the fixative or removed by the dehydrating and clearing fluids. Slight cracks in sections may also appear as large spaces in the tissue.

Other artifacts may include small wrinkles in the section (which the novice may confuse with linear structures in tissue) and precipitates from the stain (which may be confused with cellular structures such as cytoplasmic granules). Students must be aware of the existence of artifacts and able to recognize them.

Another difficulty in the study of histologic sections is the impossibility of differentially staining all tissue components on one slide. A single stain can seldom demonstrate well nuclei, mitochondria, lysosomes, basement membranes, elastic fibers, etc. With the light microscope, it is necessary to examine preparations stained by different methods before an idea of the whole composition and structure of a cell or tissue can be obtained. The TEM allows the observation of cells with all its internal structures and surrounding ECM components, but only a few cells in a tissue can be conveniently studied in these very small samples.

Finally, when a structure's **three-dimensional** volume is cut into very thin sections, the sections appear microscopically to have only two dimensions: length and width. When examining a section under the microscope, the viewer must always keep in mind that components are missing in front of and behind what is being seen because many tissue structures are thicker than the section. Round structures seen microscopically may actually be portions of spheres or tubes. Because structures in a tissue have different orientations, their two-dimensional (2D) appearance will also vary depending on the plane of section. A single convoluted tube will appear in a tissue section as many separate rounded or oval structures (Figure 1–14).

FIGURE **1–14** Interpretation of 3D structures in 2D sections.

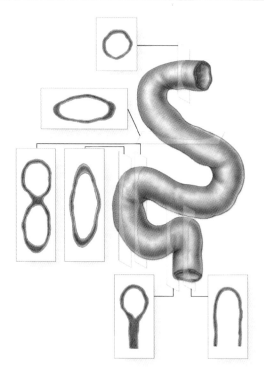

In thin sections 3D structures appear to have only two dimensions. Such images must be interpreted correctly to understand the actual structure of tissue and organ components. For example, blood vessels and other tubular structures appear in sections as round or oval shapes whose size and shape depend on the transverse or oblique angle of the cut. A highly coiled tube will appear as several round and oval structures. In TEM sections of cells, round structures may represent spherical organelles or transverse cuts through tubular organelles such as mitochondria. It is important to develop such interpretive skill to understand tissue and cell morphology in microscopic preparations.

Histology & Its Methods of Study SUMMARY OF KEY POINTS

Preparation of Tissues for Study

- Chemical fixatives such as formalin are used to preserve tissue structure by cross-linking and denaturing proteins, inactivating enzymes, and preventing cell autolysis or self-digestion.
- Dehydration of the fixed tissue in alcohol and clearing in organic solvents prepare it for embedding and sectioning.
- Embedding in paraffin wax or epoxy resin allows the tissue to be cut into very thin sections (slices) with a microtome.
- Sections are mounted on glass slides for staining, which is required to reveal specific cellular and tissue components with the microscope.
- The most commonly used staining method is a combination of the stains hematoxylin and eosin (H&E), which act as basic and acidic dyes, respectively.
- Cell substances with a net negative (anionic) charge, such as DNA and RNA, react strongly with hematoxylin and basic stains; such material is said to be "basophilic."
- Cationic substances, such as collagen and many cytoplasmic proteins react with eosin and other acidic stains and are said to be "acidophilic."

Light Microscopy

- **Bright-field microscopy**, the method most commonly used by both students and pathologists, uses ordinary light and the colors are imparted by tissue staining.
- **Fluorescence microscopy** uses ultraviolet light, under which only fluorescent molecules are visible, allowing localization of fluorescent probes which can be much more specific than routine stains.
- **Phase-contrast microscopy** uses the differences in refractive index of various natural cell and tissue components to produce an image without staining, allowing observation of living cells.
- **Confocal microscopy** involves scanning the specimen at successive focal planes with a focused light beam, often from a laser, and produces a 3D reconstruction from the images.

Autoradiography

- This process localizes cell components synthesized using **radioactive precursors** by detecting silver grains produced by weakly emitted radiation in a photographic emulsion coating the tissue section or cells.
- With either light microscopy or TEM, autoradiography permits unique studies of processes such as tissue growth (using radioactive DNA precursors) or cellular pathways of macromolecular synthesis.

Cell & Tissue Culture

- Cells can be grown in vitro from newly **explanted** tissues (primary cultures) or as long-established cell lines and can be examined in the living state by phase-contrast light microscopy.

Enzyme Histochemistry

- **Histochemical** (or **cytochemical**) **techniques** use specific enzymatic activities in lightly fixed or unfixed tissue sections to produce visible products in the specific enzyme locations.
- Fixation and paraffin embedding denatures most enzymes, so histochemistry usually uses **frozen tissue** sectioned with a **cryostat**.
- Enzyme classes for which histochemical study is useful include phosphatases, dehydrogenases, and peroxidases, with peroxidase often conjugated to antibodies used in immunohistochemistry.

Visualizing Specific Molecules

- Some substances specifically bind certain targets in cells.
- **Immunohistochemistry** is based on specific reactions between an antigen and antibodies labeled with visible markers, often fluorescent compounds or peroxidase for light microscopy and gold particles for TEM.
- If the cell or tissue antigen of interest is detected by directly binding a labeled **primary antibody** specific for that antigen, the process is considered **direct immunohistochemistry**.

- **Indirect immunohistochemistry** uses an unlabeled primary antibody that is detected bound to its antigen with labeled **secondary antibodies**.
- The indirect immunohistochemical method is more commonly used because the added level of antibody binding amplifies the signal detected and provides greater technical flexibility.
- Specific gene sequences or mRNAs of cells can be detected microscopically using labeled complementary DNA (cDNA) probes in a technique called **in situ hybridization (ISH)**.

Interpretation of Structures in Tissue Sections

- Many steps in tissue processing, slide preparation, and staining can introduce minor **artifacts** such as spaces and precipitates that are not normally present in the living tissue and must be recognized.
- Sections of cells or tissues are essentially 2D planes through 3D structures, and understanding this fact is important for their correct interpretation and study.

Histology & Its Methods of Study ASSESS YOUR KNOWLEDGE

1. In preparing tissue for routine light microscopic study, which procedure immediately precedes clearing the specimen with an organic solvent?
   a. Dehydration
   b. Fixation
   c. Staining
   d. Clearing
   e. Embedding

2. Which of the following staining procedures relies on the cationic and anionic properties of the material to be stained?
   a. Enzyme histochemistry
   b. Periodic acid-Schiff reaction
   c. Hematoxylin & eosin staining
   d. Immunohistochemistry
   e. Metal impregnation techniques

3. In a light microscope used for histology, resolution and magnification of cells are largely dependent on which component?
   a. Condenser
   b. Objective lens
   c. Eyepieces or ocular lenses
   d. Specimen slide
   e. The control for illumination intensity

4. Cellular storage deposits of glycogen, a free polysaccharide, could best be detected histologically using what procedure?
   a. Autoradiography
   b. Electron microscopy
   c. Enzyme histochemistry
   d. Hematoxylin & eosin staining
   e. Periodic acid-Schiff reaction

5. Adding heavy metal compounds to the fixative and ultrathin sectioning of the embedded tissue with a glass knife are techniques used for which histological procedure?
   a. Scanning electron microscopy
   b. Fluorescent microscopy
   c. Enzyme histochemistry
   d. Confocal microscopy
   e. Transmission electron microscopy

6. Resolution in electron microscopy greatly exceeds that of light microscopy due to which of the following?
   a. The wavelength of the electrons in the microscope beam is shorter than that of a beam of light.
   b. The lenses of an electron microscope are of greatly improved quality.
   c. For electron microscopy the tissue specimen does not require staining.
   d. The electron microscope allows much greater magnification of a projected image than a light microscope provides.
   e. An electron microscope can be much more finely controlled than a light microscope.

7. Microscopic autoradiography uses radioactivity and can be employed to study what features in a tissue section?
   a. The types of enzymes found in various cell locations
   b. Cellular sites where various macromolecules are synthesized
   c. The sequences of mRNA made in the cells
   d. The dimensions of structures within the cells
   e. The locations of genes transcribed for specific mRNA

8. To identify and localize a specific protein within cells or the extracellular matrix one would best use what approach?
   a. Autoradiography
   b. Enzyme histochemistry
   c. Immunohistochemistry
   d. Transmission electron microscopy
   e. Polarizing microscopy

9. In situ hybridization is a histological technique used to visualize what type of macromolecule?
   a. Proteins
   b. Carbohydrates
   c. Certain enzymes
   d. Nucleic acids
   e. Lipids

10. Hospital laboratories frequently use unfixed, frozen tissue specimens sectioned with a cryostat for rapid staining, microscopic examination, and diagnosis of pathological conditions. Besides saving much time by avoiding fixation and procedures required for paraffin embedding, frozen sections retain and allow study of what macromolecules normally lost in the paraffin procedure?
    a. Carbohydrates
    b. Small mRNA
    c. Basic proteins
    d. Acidic proteins
    e. Lipids

Answers: 1a, 2c, 3b, 4e, 5e, 6a, 7b, 8c, 9d, 10e

CHAPTER

2 The Cytoplasm

Cells and extracellular material together comprise all the tissues that make up the organs of multicellular animals. In all tissues, cells themselves are the basic structural and functional units, the smallest living parts of the body. Animal cells are eukaryotic, with distinct membrane-limited nuclei surrounded by **cytoplasm** which contains various membrane-limited organelles and the cytoskeleton. In contrast, the smaller prokaryotic cells of bacteria typically have a cell wall around the plasmalemma and lack nuclei and membranous cytoplasmic structures.

› CELL DIFFERENTIATION

The human organism consists of hundreds of different cell types, all derived from the zygote, the single cell formed by the merger of a spermatozoon with an oocyte at fertilization. The first zygotic cellular divisions produce cells called **blastomeres**, and as part of the early embryo's inner cell mass blastomeres give rise to all tissue types of the fetus. Explanted to tissue culture cells of the inner cell mass are called **embryonic stem cells**. Most cells of the fetus undergo a specialization process called **differentiation** in which they differentially express sets of genes that mediate specific cytoplasmic activities, becoming very efficient in specialized functions and usually changing their shape accordingly. For example, muscle cell precursors elongate into fiber-like cells containing large arrays of actin and myosin. All animal cells contain actin filaments and myosins, but muscle cells are specialized for using these proteins to convert chemical energy into forceful contractions.

Major cellular functions performed by specialized cells in the body are listed in Table 2–1. It is important to understand that the functions listed there can be performed by most cells of the body; specialized cells have greatly expanded their capacity for one or more of these functions during differentiation. Changes in cells' microenvironments under normal and pathologic conditions can cause the same cell type to have variable features and activities. Cells that appear similar structurally often have different families of receptors for signaling molecules such as hormones and extracellular matrix (ECM) components, causing them to behave differently. For example, because of their diverse arrays of receptors, breast fibroblasts and uterine smooth muscle cells are exceptionally sensitive to female sex hormones while most other fibroblasts and smooth muscle cells are insensitive.

› THE PLASMA MEMBRANE

The **plasma membrane** (**cell membrane** or plasmalemma) that envelops every eukaryotic cell consists of phospholipids, cholesterol, and proteins, with oligosaccharide chains covalently linked to many of the phospholipid and protein molecules. This limiting membrane functions as a selective barrier regulating the passage of materials into and out of the cell and facilitating the transport of specific molecules. One important role of the cell membrane is to keep constant the ion content of cytoplasm, which differs from that of the extracellular fluid. Membranes also carry out a number of specific recognition and signaling functions, playing a key role in the interactions of the cell with its environment.

TABLE 2–1	Differentiated cells typically specialize in one activity.
General Cellular Activity	**Specialized Cell(s)**
Movement	Muscle and other contractile cells
Form adhesive and tight junctions between cells	Epithelial cells
Synthesize and secrete components of the extracellular matrix	Fibroblasts, cells of bone and cartilage
Convert physical and chemical stimuli into action potentials	Neurons and sensory cells
Synthesis and secretion of degradative enzymes	Cells of digestive glands
Synthesis and secretion of glycoproteins	Cells of mucous glands
Synthesis and secretion of steroids	Certain cells of the adrenal gland, testis, and ovary
Ion transport	Cells of the kidney and salivary gland ducts
Intracellular digestion	Macrophages and neutrophils
Lipid storage	Fat cells
Metabolite absorption	Cells lining the intestine

Although the plasma membrane defines the outer limit of the cell, a continuum exists between the interior of the cell and extracellular macromolecules. Certain plasma membrane proteins, the **integrins**, are linked to both the cytoskeleton and ECM components and allow continuous exchange of influences, in both directions, between the cytoplasm and material in the ECM.

Membranes range from 7.5 to 10 nm in thickness and consequently are visible only in the electron microscope. The line between adjacent cells sometimes seen faintly with the light microscope consists of plasma membrane proteins plus extracellular material, which together can reach a dimension visible by light microscopy.

Membrane phospholipids are amphipathic, consisting of two nonpolar (hydrophobic or water-repelling) long-chain fatty acids linked to a charged polar (hydrophilic or water-attracting) head that bears a phosphate group (Figure 2–1a). Phospholipids are most stable when organized into a double layer (bilayer) with the hydrophobic fatty acid chains located in a middle region away from water and the hydrophilic polar head groups contacting the water (Figure 2–1b). Molecules of cholesterol, a sterol lipid, insert at varying densities among the closely-packed phospholipid fatty acids, restricting their movements and modulating the fluidity of all membrane components. The phospholipids in each half of the bilayer are different. For example, in the well-studied membranes of red blood cells phosphatidylcholine and sphingomyelin are more abundant in the outer half, while phosphatidylserine and phosphatidylethanolamine are more concentrated in the inner layer. Some of the outer layer's lipids, known as **glycolipids**, include oligosaccharide chains that extend outward from the cell surface and contribute to a delicate cell surface coating called the **glycocalyx** (Figures 2–1b and 2–2). With the transmission electron microscope (TEM) the cell membrane—as well as all cytoplasmic membranes—may exhibit a trilaminar appearance after fixation in osmium tetroxide; osmium binds the polar heads of the phospholipids and the oligosaccharide chains, producing the two dark outer lines that enclose the light band of osmium-free fatty acids (Figure 2–1b).

Proteins are major constituents of membranes (~50% by weight in the plasma membrane). **Integral proteins** are incorporated directly within the lipid bilayer, whereas **peripheral proteins** are bound to one of the two membrane surfaces, particularly on the cytoplasmic side (Figure 2–2). Peripheral proteins can be extracted from cell membranes with salt solutions, whereas integral proteins can be extracted only by using detergents to disrupt the lipids. The polypeptide chains of many integral proteins span the membrane, from one side to the other, several times and are accordingly called **multipass proteins**. Integration of the proteins within the lipid bilayer is mainly the result of hydrophobic interactions between the lipids and nonpolar amino acids of the proteins.

Freeze-fracture electron microscope studies of membranes show that parts of many integral proteins protrude from both the outer or inner membrane surface (Figure 2–2b). Like those of glycolipids, the carbohydrate moieties of glycoproteins project from the external surface of the plasma membrane and contribute to the glycocalyx (Figure 2–3). They are important components of proteins acting as **receptors**, which participate in important interactions such as cell adhesion, cell recognition, and the response to protein hormones. As with lipids, the distribution of membrane polypeptides is different in the two surfaces of the cell membranes. Therefore, all membranes in the cell are asymmetric.

Studies with labeled membrane proteins of cultured cells reveal that many such proteins are not bound rigidly in place and are able to move laterally (Figure 2–4). Such observations as well as data from biochemical, electron microscopic, and other studies showed that membrane proteins comprise a moveable mosaic within the fluid lipid bilayer, the well-established **fluid mosaic model** for membrane structure (Figure 2–2a). Unlike the lipids, however, lateral diffusion of many membrane proteins is often restricted by their cytoskeletal attachments. Moreover, in most epithelial cells tight junctions between the cells (see Chapter 4) also restrict lateral diffusion of unattached transmembrane proteins and outer layer lipids, producing different domains within the cell membranes.

Membrane proteins that are components of large enzyme complexes are also usually less mobile, especially those involved in the transduction of signals from outside the cell. Such protein complexes are located in specialized membrane patches termed **lipid rafts** with higher concentrations of

FIGURE **2–1** **Lipids in membrane structure.**

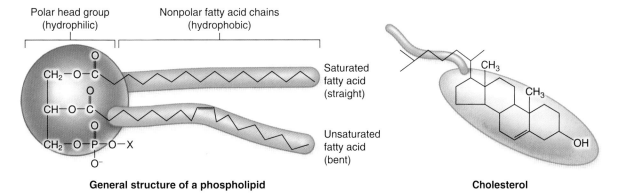

Polar head group
(hydrophilic)

Nonpolar fatty acid chains
(hydrophobic)

Saturated
fatty acid
(straight)

Unsaturated
fatty acid
(bent)

General structure of a phospholipid

Cholesterol

a

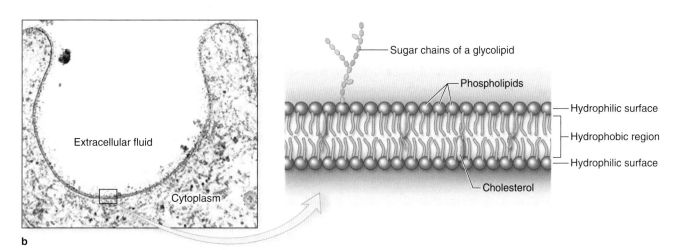

Extracellular fluid

Cytoplasm

Sugar chains of a glycolipid

Phospholipids

Hydrophilic surface

Hydrophobic region

Hydrophilic surface

Cholesterol

b

(a) Membranes of animal cells have as their major lipid components **phospholipids** and **cholesterol**. A phospholipid is amphipathic, with a phosphate group charge on the polar head and two long, nonpolar fatty acid chains, which can be straight (saturated) or kinked (at an unsaturated bond). Membrane cholesterol is present in about the same amount as phospholipid.

(b) The amphipathic nature of phospholipids produces the bilayer structure of membranes as the charged (hydrophilic) polar heads spontaneously form each membrane surface, in direct contact with water, and the hydrophobic nonpolar fatty acid chains are buried in the membrane's middle, away from water. Cholesterol molecules are also amphipathic and are interspersed less evenly

throughout the lipid bilayer; cholesterol affects the packing of the fatty acid chains, with a major effect on membrane fluidity. The outer layer of the cell membrane also contains **glycolipids** with extended carbohydrate chains.

Sectioned, osmium-fixed cell membrane may have a faint trilaminar appearance with the transmission electron microscope (TEM), showing two dark (electron-dense) lines enclosing a clear (electron-lucent) band. Reduced osmium is deposited on the hydrophilic phosphate groups present on each side of the internal region of fatty acid chains where osmium is not deposited. The "fuzzy" material on the outer surface of the membrane represents the **glycocalyx** of oligosaccharides of glycolipids and glycoproteins. (X100,000)

cholesterol and saturated fatty acids which reduce lipid fluidity. This together with the presence of scaffold proteins that maintain spatial relationships between enzymes and signaling proteins allows the proteins assembled within lipid rafts to remain in close proximity and interact more efficiently.

Transmembrane Proteins & Membrane Transport

The plasma membrane is the site where materials are exchanged between the cell and its environment. Most small

molecules cross the membrane by the general mechanisms shown schematically in Figure 2–5 and explained as follows:

- **Diffusion** transports small, nonpolar molecules directly through the lipid bilayer. Lipophilic (fat-soluble) molecules diffuse through membranes readily, water very slowly.
- **Channels** are multipass proteins forming transmembrane pores through which ions or small molecules pass selectively. Cells open and close specific channels for Na^+,

FIGURE **2–2** Proteins associated with the membrane lipid bilayer.

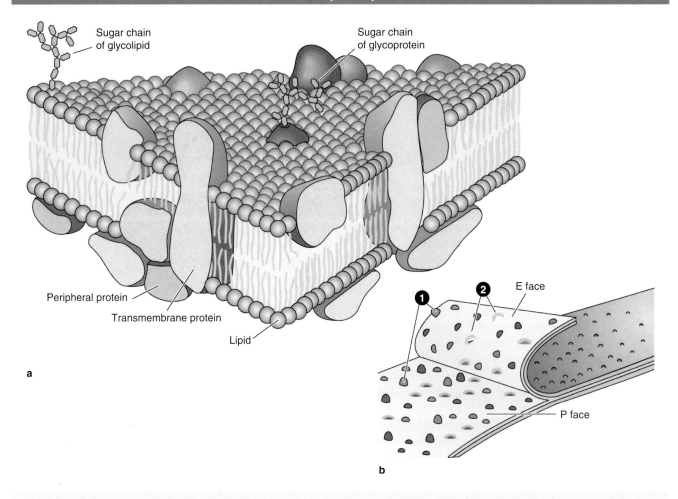

Sugar chain of glycolipid

Sugar chain of glycoprotein

Peripheral protein

Transmembrane protein

Lipid

a

E face

P face

b

(a) The fluid mosaic model of membrane structure emphasizes that the phospholipid bilayer of a membrane also contains proteins inserted in it or associated with its surface (peripheral proteins) and that many of these proteins move within the fluid lipid phase. **Integral proteins** are firmly embedded in the lipid layers; those that completely span the bilayer are called **transmembrane proteins**. Hydrophobic amino acids of these proteins interact with the hydrophobic fatty acid portions of the membrane lipids. Both the proteins and lipids may have externally exposed oligosaccharide chains.

(b) When cells are frozen and fractured (**cryofracture**), the lipid bilayer of membranes is often cleaved along the hydrophobic

center. Splitting occurs along the line of weakness formed by the fatty acid tails of phospholipids. Electron microscopy of such cryofracture preparation replicas provides a useful method for studying membrane structures. Most of the protruding membrane particles seen (1) are proteins or aggregates of proteins that remain attached to the half of the membrane adjacent to the cytoplasm (P or protoplasmic face). Fewer particles are found attached to the outer half of the membrane (E or extracellular face). Each protein bulging on one surface has a corresponding depression (2) on the opposite surface.

K^+, Ca^{2+} and other ions in response to various physiological stimuli. Water molecules usually cross the plasma membrane through channel proteins called **aquaporins**.

■ **Carriers** are transmembrane proteins that bind small molecules and translocate them across the membrane via conformational changes.

Diffusion, channels, and carrier proteins operate passively, allowing movement of substances across membranes

down a concentration gradient due to its kinetic energy. In contrast, membrane **pumps** are enzymes engaged in **active transport**, utilizing energy from the hydrolysis of adenosine triphosphate (ATP) to move ions and other solutes across membranes, against often steep concentration gradients. Because they consume ATP pumps they are often referred to as **ATPases**.

These transport mechanisms are summarized with additional detail in Table 2–2.

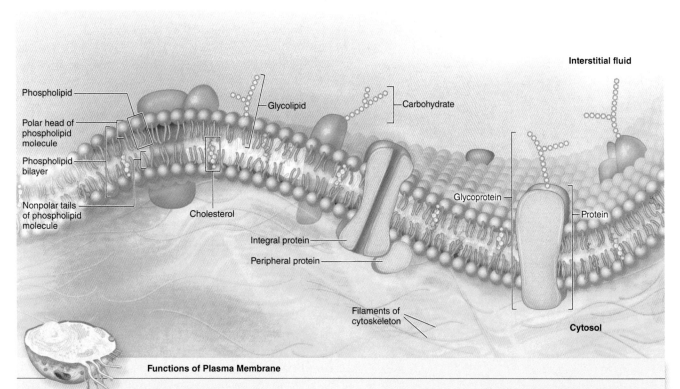

FIGURE **2–3** **Membrane proteins.**

Functions of Plasma Membrane

1. **Physical barrier:** Establishes a flexible boundary, protects cellular contents, and supports cell structure. Phospholipid bilayer separates substances inside and outside the cell.
2. **Selective permeability:** Regulates entry and exit of ions, nutrients, and waste molecules through the membrane.
3. **Electrochemical gradients:** Establishes and maintains an electrical charge difference across the plasma membrane.
4. **Communication:** Contains receptors that recognize and respond to molecular signals.

Both protein and lipid components often have covalently attached oligosaccharide chains exposed at the external membrane surface. These contribute to the cell's **glycocalyx**, which provides important antigenic and functional properties to the cell surface. Membrane proteins serve as receptors for various signals coming from outside cells, as parts of intercellular connections, and as selective gateways for molecules entering the cell.

Transmembrane proteins often have multiple hydrophobic regions buried within the lipid bilayer to produce a channel or other active site for specific transfer of substances through the membrane.

Vesicular Transport: Endocytosis & Exocytosis

Macromolecules normally enter cells by being enclosed within folds of plasma membrane (often after binding specific membrane receptors) which fuse and pinch off internally as cytoplasmic **vesicles** (or vacuoles) in a general process known as **endocytosis**. Three major types of endocytosis are recognized, as summarized in Table 2–2 and Figure 2–6.

1. **Phagocytosis** ("cell eating") is the ingestion of particles such as bacteria or dead cell remnants. Certain blood-derived cells, such as macrophages and neutrophils, are specialized for this activity. When a bacterium becomes bound to the surface of a neutrophil, it becomes surrounded by extensions of plasmalemma and cytoplasm which project from the cell in a process dependent on cytoskeletal changes. Fusion of the membranous folds encloses the bacterium in an intracellular vacuole called a **phagosome**, which then merges with a lysosome for degradation of its contents as discussed later in this chapter.

2. **Pinocytosis** ("cell drinking") involves smaller invaginations of the cell membrane which fuse and entrap extracellular fluid and its dissolved contents. The resulting **pinocytotic vesicles** (~80 nm in diameter) then pinch off inwardly from the cell surface and either fuse with lysosomes or move to the opposite cell surface where they fuse with the membrane and release their contents outside the cell. The latter process, called **transcytosis**,

FIGURE 2–4 Experiment demonstrating the fluidity of membrane proteins.

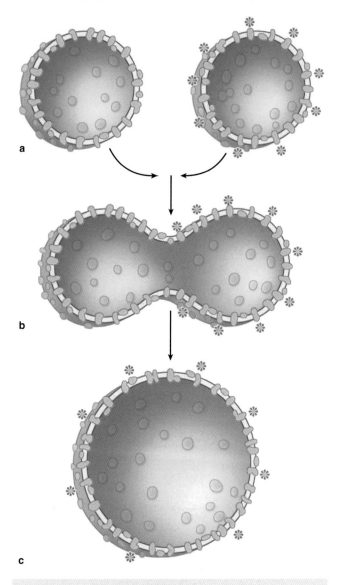

a

b

c

(a) Two types of cells were grown in tissue cultures, one with fluorescently labeled transmembrane proteins in the plasma-lemma (right) and one without.

(b) Cells of each type were fused together experimentally into hybrid cells.

(c) Minutes after the fusion of the cell membranes, the fluorescent proteins of the labeled cell spread to the entire surface of the hybrid cells. Such experiments provide important data supporting the fluid mosaic model. However, many membrane proteins show more restricted lateral movements, being anchored in place by links to the cytoskeleton.

accomplishes bulk transfer of dissolved substances across the cell.

3. **Receptor-mediated endocytosis:** Receptors for many substances, such as low-density lipoproteins and protein

hormones, are integral membrane proteins at the cell surface. High-affinity binding of such **ligands** to their receptors causes these proteins to aggregate in special membrane regions that then invaginate and pinch off internally as vesicles.

The formation and fate of vesicles in receptor-mediated endocytosis also often depend on specific peripheral proteins on the cytoplasmic side of the membrane (Figure 2–7). The occupied cell-surface receptors associate with these cytoplasmic proteins and begin invagination as **coated pits**. The electron-dense coating on the cytoplasmic surface of such pits contains several polypeptides, the major one being **clathrin**. Clathrin molecules interact like the struts of a geodesic dome, forming that region of cell membrane into a cage-like invagination that is pinched off in the cytoplasm as a **coated vesicle** (Figure 2–7b) with the receptor-bound ligands inside. Another type of receptor-mediated endocytosis very prominent in endothelial cells produces invaginations called **caveolae** (L. *caveolae*, little caves) that involve the membrane protein **caveolin**.

In all these endocytotic processes, the vesicles or vacuoles produced quickly enter and fuse with the **endosomal compartment**, a dynamic collection in the peripheral cytoplasm of membranous tubules and vacuoles (Figure 2–7). The clathrin molecules separate from the coated vesicles and recycle back to the cell membrane for the formation of new coated pits. Vesicle trafficking through the endosomal compartment is directed largely through peripheral membrane G proteins called **Rab proteins**, small GTPases that bind guanine nucleotides and associated proteins.

As shown in Figure 2–7, phagosomes and pinocytotic vesicles typically fuse with lysosomes within the endosomal compartment for digestion of their contents, while molecules entering by receptor-mediated endocytosis may be directed down other pathways. The membranes of many late endosomes have ATP-driven H^+ pumps that acidify their interior, activating the hydrolytic enzymes of lysosomes and in other endosomes causing ligands to uncouple from their receptors, after which the two molecules are sorted into separate endosomes. The receptors may be sorted into recycling endosomes and returned to the cell surface for reuse. Low-density lipoprotein receptors, for example, are recycled several times within cells. Other endosomes may release their entire contents at a separate domain of the cell membrane (transcytosis), which occurs in many epithelial cells.

Movement of large molecules from inside to outside the cell usually involves vesicular transport in the process of **exocytosis**. In exocytosis a cytoplasmic vesicle containing the molecules to be secreted fuses with the plasma membrane, resulting in the release of its contents into the extracellular space without compromising the integrity of the plasma membrane (see "Transcytosis" in Figure 2–7a). Exocytosis is triggered in many cells by transient increase in cytosolic Ca^{2+}. Membrane fusion during exocytosis is highly regulated, with selective interactions between several specific membrane proteins.

FIGURE **2–5** Major mechanisms by which molecules cross membranes.

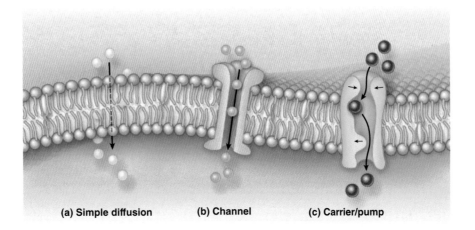

(a) Simple diffusion   **(b) Channel**   **(c) Carrier/pump**

Lipophilic and some small, uncharged molecules can cross membranes by simple diffusion **(a)**.

Most ions cross membranes in multipass proteins called channels **(b)** whose structures include transmembrane ion-specific pores.

Many other larger, water-soluble molecules require binding to sites on selective carrier proteins **(c)**, which then change their conformations and release the molecule to the other side of the membrane.

Diffusion, channels and most carrier proteins translocate substances across membranes using only kinetic energy. In contrast, **pumps** are carrier proteins for active transport of ions or other solutes and require energy derived from ATP.

Exocytosis of macromolecules made by cells occurs via either of two pathways:

- **Constitutive secretion** is used for products that are released from cells continuously, as soon as synthesis is complete, such as collagen subunits for the ECM.
- **Regulated secretion** occurs in response to signals coming to the cells, such as the release of digestive enzymes from pancreatic cells in response to specific stimuli. Regulated exocytosis of stored products from epithelial cells usually occurs specifically at the apical domains of cells, constituting a major mechanism of glandular secretion (see Chapter 4).

Portions of the cell membrane become part of the endocytotic vesicles or vacuoles during endocytosis; during exocytosis, membrane is returned to the cell surface. This process of membrane movement and recycling is called **membrane trafficking** (Figure 2-7a). Trafficking of membrane components occurs continuously in most cells and is not only crucial for maintaining the cell but also for physiologically important processes such as reducing blood lipid levels.

In many cells subpopulations of vacuoles and tubules within the endosomal compartment accumulate small vesicles *within* their lumens by further invaginations of their limiting membranes, becoming **multivesicular bodies**. While multivesicular bodies may merge with lysosomes for selective degradation of their content, this organelle may also fuse with the plasma membrane and release the intralumenal vesicles outside the cell. The small (<120 nm diameter) vesicles released are called **exosomes**, which can fuse with other cells transferring their contents and membranes.

Signal Reception & Transduction

Cells in a multicellular organism communicate with one another to regulate tissue and organ development, to control their growth and division, and to coordinate their functions. Many adjacent cells form communicating **gap junctions** that couple the cells and allow exchange of ions and small molecules (see Chapter 4).

Cells also use about 25 families of **receptors** to detect and respond to various extracellular molecules and physical stimuli. Each cell type in the body contains a distinctive set of cell surface and cytoplasmic receptor proteins that enable it to respond to a complementary set of signaling molecules in a specific, programmed way. Cells bearing receptors for a specific ligand are referred to as **target cells** for that molecule. The routes of signal molecules from source to target provide one way to categorize the signaling process:

- In **endocrine signaling**, the signal molecules (here called **hormones**) are carried in the blood from their sources to target cells throughout the body.
- In **paracrine signaling**, the chemical ligand diffuses in extracellular fluid but is rapidly metabolized so that its effect is only local on target cells near its source.
- In **synaptic signaling**, a special kind of paracrine interaction, neurotransmitters act on adjacent cells through special contact areas called **synapses** (see Chapter 9).

TABLE 2–2	Mechanisms of transport across the plasma membrane.	
Process	**Type of Movement**	**Example**
PASSIVE PROCESSES	Movement of substances down a concentration gradient due to the kinetic energy of the substance; no expenditure of cellular energy is required; continues until equilibrium is reached (if unopposed)	
Simple diffusion	Unassisted net movement of small, nonpolar substances down their concentration gradient across a selectively permeable membrane	Exchange of oxygen and carbon dioxide between blood and body tissues
Facilitated diffusion	Movement of ions and small, polar molecules down their concentration gradient; assisted across a selectively permeable membrane by a transport protein	
Channel-mediated	Movement of ion down its concentration gradient through a protein channel	Na^+ moves through Na^+ channel into cell
Carrier-mediated	Movement of small, polar molecule down its concentration gradient by a carrier protein	Transport of glucose into cells by glucose carrier
Osmosis	Diffusion of water across a selectively permeable membrane; direction is determined by relative solute concentrations; continues until equilibrium is reached	Solutes in blood in systemic capillaries "pulls" fluid from interstitial space back into the blood
ACTIVE PROCESSES	Movement of substances requires expenditure of cellular energy	
Active transport	Transport of ions or small molecules across the membrane against a concentration gradient by transmembrane protein pumps	
Primary	Movement of substance up its concentration gradient; powered directly by ATP	Ca^{2+} pumps transport Ca^{2+} out of the cell Na^+/K^+ pump moves Na^+ out of cell and K^+ into cell
Secondary	Movement of a substance up its concentration gradient is powered by harnessing the movement of a second substance (eg, Na^+) down its concentration gradient	
Symport	Movement of substance up its concentration gradient in the same direction as Na^+	Na^+/glucose transport
Antiport	Movement of substance up its concentration gradient in the opposite direction from Na^+	Na^+/H^+ transport
Vesicular transport	Vesicle formed or lost as material is brought into a cell or released from a cell	
Exocytosis	Bulk movement of substance out of the cell by fusion of secretory vesicles with the plasma membrane	Release of neurotransmitter by nerve cells
Endocytosis	Bulk movement of substances into the cell by vesicles forming at the plasma membrane	
Phagocytosis	Type of endocytosis in which vesicles are formed as particulate materials external to the cell are engulfed by pseudopodia	White blood cell engulfing a bacterium
Pinocytosis	Type of endocytosis in which vesicles are formed as interstital fluid is taken up by the cell	Formation of small vesicles in capillary wall to move substances
Receptor-mediated endocytosis	Type of endocytosis in which plasma membrane receptors first bind specific substances; receptor and bound substance then taken up by the cell	Uptake of cholesterol into cells

FIGURE **2–6** Three major forms of endocytosis.

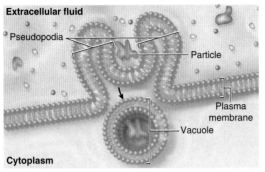

a Phagocytosis

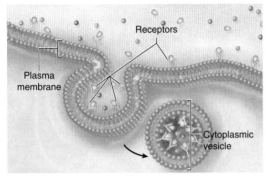

c Receptor-mediated endocytosis

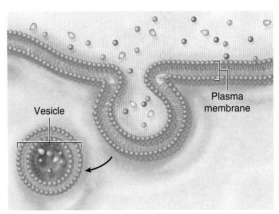

b Pinocytosis

There are three general types of endocytosis:

(a) Phagocytosis involves the extension from the cell of surface folds or pseudopodia which engulf particles such as bacteria,

and then internalize this material into a cytoplasmic vacuole or **phagosome**.

(b) In **pinocytosis** the cell membrane forms similar folds or invaginates (dimples inward) to create a pit containing a drop of extracellular fluid. The pit pinches off inside the cell when the cell membrane fuses and forms a pinocytotic vesicle containing the fluid.

(c) Receptor-mediated endocytosis includes membrane proteins called **receptors** that bind specific molecules (ligands). When many such receptors are bound by their ligands, they aggregate in one membrane region, which then invaginates and pinches off to create a vesicle or **endosome** containing both the receptors and the bound ligands.

- In **autocrine signaling**, signals bind receptors on the same cells that produced the messenger molecule.
- In **juxtacrine signaling**, important in early embryonic tissue interactions, the signaling molecules are cell membrane–bound proteins which bind surface receptors of the target cell when the two cells make direct physical contact.

Receptors for hydrophilic signaling molecules, including polypeptide hormones and neurotransmitters, are usually transmembrane proteins in the plasmalemma of target cells. Three important functional classes of such receptors are shown in Figure 2–8:

- **Channel-linked receptors** open associated channels upon ligand binding to promote transfer of molecules or ions across the membrane.
- **Enzymatic receptors**, in which ligand binding induces catalytic activity in associated peripheral proteins.
- **G protein–coupled receptors** upon ligand binding stimulate associated G proteins which then bind the guanine nucleotide GTP and are released to activate other cytoplasmic proteins.

>> MEDICAL APPLICATION

Many diseases are caused by defective receptors. For example, **pseudohypoparathyroidism** and one type of **dwarfism** are caused by nonfunctioning parathyroid and growth hormone receptors, respectively. In these two conditions the glands produce the respective hormones, but the target cells cannot respond because they lack normal receptors.

Ligands binding such receptors in a cell membrane can be considered first messengers, beginning a process of **signal transduction** by activating a series of intermediary enzymes downstream to produce changes in the cytoplasm, the nucleus, or both. Channel-mediated ion influx or activation of kinases can activate various cytoplasmic proteins, amplifying the signal. Activated G proteins target ion channels or other membrane-bound effectors that also propagate the signal further into the cell (Figure 2–8). One such effector protein is the enzyme adenyl cyclase which generates large quantities of second messenger molecules, such as cyclic adenosine

FIGURE **2–7**   Receptor-mediated endocytosis involves regulated membrane trafficking.

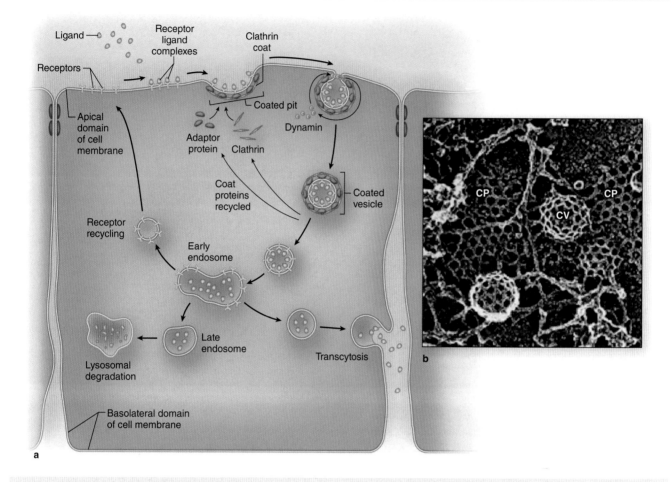

Major steps during and after endocytosis are indicated diagrammatically in part **a**. Ligands bind with high affinity to specific surface receptors, which then associate with specific cytoplasmic proteins, including clathrin and adaptor proteins, and aggregate in membrane regions to form **coated pits**. Clathrin facilitates invagination of the pits, and another peripheral membrane protein, **dynamin**, forms constricting loops around the developing neck of the pit, causing the invagination to pinch off as a **coated vesicle**. The clathrin lattice of coated pits **(CP)** and vesicles **(CV)** is shown ultrastructurally in part **b**.

Internalized vesicles lose their clathrin coats, which are recycled, and fuse with other endosomes that comprise the **endosomal compartment**. Ligands may have different fates within the endosomal compartment:

- Receptors and ligands may be carried to late endosomes and then to lysosomes for **degradation.**
- Ligands may be released from the receptors and the empty receptors sequestered into **recycling endosomes** and returned to the cell surface for reuse.
- Other endosomal vesicles containing ligands may move to and fuse with another cell surface, where the ligands are released again outside the cell in the process of **transcytosis.**

(*Figure 2–7b, used with permission from Dr John Heuser, Department of Cell Biology and Physiology, Washington University School of Medicine, St. Louis, MO.*)

monophosphate (cAMP). Other second messengers include 1,2-diacylglycerol (DAG), and inositol 1,4,5-triphosphate (IP_3). The ionic changes or second messengers amplify the first signal and trigger a cascade of enzymatic activity, usually including kinases, leading to changes in gene expression or cell behavior. Second messengers may diffuse through the cytoplasm or be retained locally by scaffold proteins for more focused amplification of activity.

Low molecular weight *hydrophobic* signaling molecules, such as steroids and thyroid hormones, bind reversibly to

carrier proteins in the plasma for transport through the body. Such hormones are lipophilic and pass by diffusion through cell membranes, binding to specific cytoplasmic receptor proteins in target cells. With many steroid hormones, receptor binding activates that receptor, enabling the complex to move into the nucleus and bind with high affinity to specific DNA sequences. This generally increases the level of transcription of those genes. Each steroid hormone is recognized by a different member of a family of homologous receptor proteins.

FIGURE **2–8** **Major types of membrane receptors.**

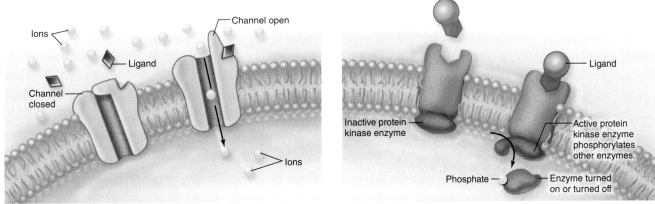

a **Channel-linked receptors**

b **Enzymatic receptors**

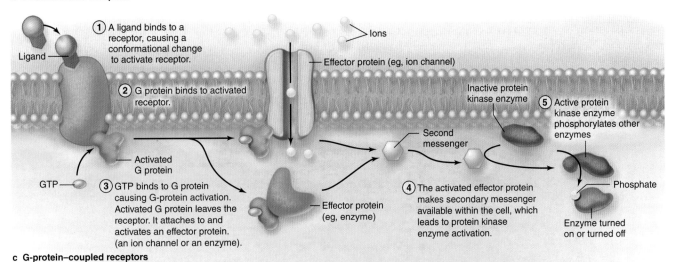

① A ligand binds to a receptor, causing a conformational change to activate receptor.

② G protein binds to activated receptor.

③ GTP binds to G protein causing G-protein activation. Activated G protein leaves the receptor. It attaches to and activates an effector protein. (an ion channel or an enzyme).

④ The activated effector protein makes secondary messenger available within the cell, which leads to protein kinase enzyme activation.

⑤ Active protein kinase enzyme phosphorylates other enzymes

c **G-protein–coupled receptors**

Protein and most small ligands are hydrophilic molecules that bind transmembrane protein receptors to initiate changes in the target cell.

(a) Channel-linked receptors bind ligands such as neurotransmitters and open to allow influx of specific ions. **(b) Enzymatic receptors** are usually protein kinases that are activated to phosphorylate (and usually activate) other proteins upon ligand binding. **(c) G-protein–coupled receptors** bind ligand, changing the conformation of its G-protein subunit, allowing it to bind GTP, and activating and releasing this protein to in turn activate other proteins such as ion channels and adenyl cyclase.

❱ CYTOPLASMIC ORGANELLES

Inside the cell membrane the fluid cytoplasm (or cytosol) bathes metabolically active structures called **organelles**, which may be membranous (such as mitochondria) or nonmembranous protein complexes (such as ribosomes and proteasomes). Most organelles are positioned in the cytoplasm by movements along the polymers of the **cytoskeleton**, which also determines a cell's shape and motility.

Cytosol also contains hundreds of enzymes, such as those of the glycolytic pathway, which produce building blocks for larger molecules and break down small molecules to liberate energy. Oxygen, CO_2, electrolytic ions, low-molecular-weight substrates, metabolites, and waste products all diffuse through cytoplasm, either freely or bound to proteins, entering or leaving organelles where they are used or produced.

Ribosomes

Ribosomes are macromolecular machines, about 20×30 nm in size, which assemble polypeptides from amino acids on molecules of transfer RNA (tRNA) in a sequence specified by mRNA. A functional ribosome has two subunits of different sizes bound to a strand of mRNA. The core of the small ribosomal subunit is a highly folded ribosomal RNA (rRNA) chain associated with more than 30 unique proteins; the core of the large subunit has three other rRNA molecules and nearly 50 other basic proteins.

The rRNA molecules in the ribosomal subunits not only provide structural support but also position transfer RNAs (tRNA) molecules bearing amino acids in the correct "reading frame" and catalyze the formation of the peptide bonds. The more peripheral proteins of the ribosome seem to function primarily to stabilize the catalytic RNA core.

These ribosomal proteins are themselves synthesized in cytoplasmic ribosomes, but are then imported to the nucleus where they associate with newly synthesized rRNA. The ribosomal subunits thus formed then move from the nucleus to the cytoplasm where they are reused many times, for translation of any mRNA strand.

During protein synthesis many ribosomes typically bind the same strand of mRNA to form larger complexes called **polyribosomes**, or **polysomes** (Figure 2–9). In stained preparations of cells polyribosomes are intensely basophilic because of the numerous phosphate groups of the constituent RNA molecules that act as polyanions. Thus, cytoplasmic regions that stain intensely with hematoxylin and basic dyes, such as methylene and toluidine blue, indicate sites of active protein synthesis.

Proper folding of new proteins is guided by protein chaperones. Denatured proteins or those that cannot be refolded properly are conjugated to the protein ubiquitin that targets them for breakdown by proteasomes (discussed below). As indicated in Figure 2–9, proteins synthesized for use within the cytosol (eg, glycolytic enzymes) or for import into the nucleus and certain other organelles are synthesized on polyribosomes existing as isolated cytoplasmic clusters. Polyribosomes attached to membranes of the endoplasmic reticulum (ER) translate mRNAs coding for membrane proteins of the ER, the Golgi apparatus, or the cell membrane; enzymes to be stored in lysosomes; and proteins to undergo exocytosis from secretory vesicles.

Endoplasmic Reticulum

The cytoplasm of most cells contains a convoluted membranous network called the **endoplasmic reticulum (ER)**. As shown in Figure 2–10 this network (reticulum) extends from the surface of the nucleus throughout most of the cytoplasm and encloses a series of intercommunicating channels called **cisternae** (L. *cisternae*, reservoirs). With a membrane surface up to 30 times that of the plasma membrane, the ER is a major site for vital cellular activities, including biosynthesis of proteins and lipids. Numerous polyribosomes attached to the membrane in some regions of ER allow two types of ER to be distinguished.

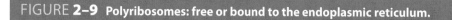

FIGURE **2–9** **Polyribosomes: free or bound to the endoplasmic reticulum.**

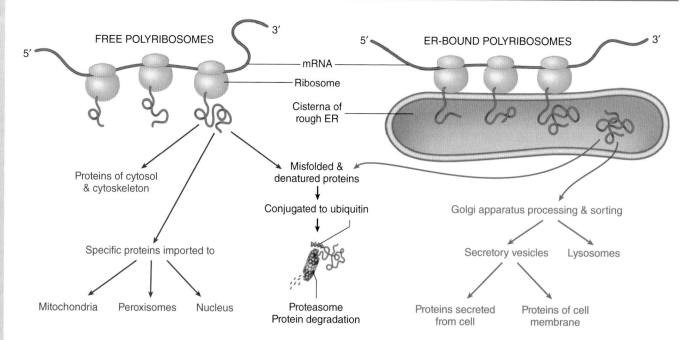

Free polyribosomes (not attached to the endoplasmic reticulum, or ER) synthesize cytosolic and cytoskeletal proteins and proteins for import into the nucleus, mitochondria, and peroxisomes.
  Proteins that are to be incorporated into membranes, stored in lysosomes, or eventually secreted from the cell are made on

polysomes attached to the membranes of ER. The proteins produced by these ribosomes are segregated during translation into the interior of the ER's membrane cisternae.
  In both pathways misfolded proteins are conjugated to ubiquitin and targeted for proteasomal degradation.

FIGURE **2–10** Rough and smooth endoplasmic reticulum.

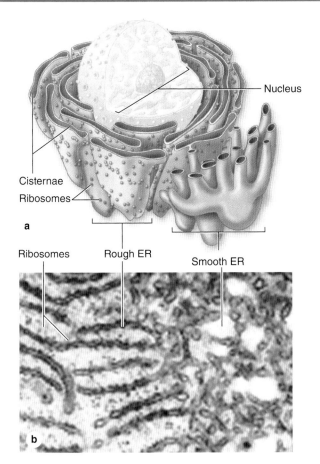

Nucleus

Cisternae

Ribosomes

a

Ribosomes   Rough ER   Smooth ER

b

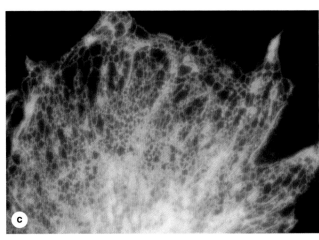

c

Functions of Endoplasmic Reticulum

1. **Synthesis:** Provides a place for chemical reactions
   a. Smooth ER is the site of lipid synthesis and carbohydrate metabolism
   b. Rough ER synthesizes proteins for secretion, incorporation into the plasma, membrane, and as enzymes within lysosomes
2. **Transport:** Moves molecules through cisternal space from one part of the cell to another, sequestered away from the cytoplasm
3. **Storage:** Stores newly synthesized molecules
4. **Detoxification:** Smooth ER detoxifies both drugs and alcohol

(a) The **endoplasmic reticulum** is an anastomosing network of intercommunicating channels or **cisternae** formed by a continuous membrane, with some regions that bear polysomes appearing rough and other regions appearing smooth. While RER is the site for synthesis of most membrane-bound proteins, three diverse activities are associated with smooth ER: (1) lipid biosynthesis, (2) detoxification of potentially harmful compounds, and (3) sequestration of Ca^{++} ions. Specific cell types with well-developed SER are usually specialized for one of these functions.

(b) By TEM cisternae of RER appear separated, but they actually form a continuous channel or compartment in the cytoplasm.

The interconnected membranous cisternae of RER are flattened, while those of SER are frequently tubular. (14,000X)

(c) In a very thin cultured endothelial cell, both ER (green) and mitochondria (orange) can be visualized with vital fluorescent dyes that are sequestered specifically into those organelles. This staining method with intact cells clearly reveals the continuous, lacelike ER present in all regions of the cytoplasm.

(Figure 2–10c, © 2015 Thermo Fisher Scientific, Inc. Used under permission.)

Rough Endoplasmic Reticulum

Rough endoplasmic reticulum (RER) is prominent in cells specialized for protein secretion, such as pancreatic acinar cells (making digestive enzymes), fibroblasts (collagen), and plasma cells (immunoglobulins). The RER consists of saclike as well as parallel stacks of flattened cisternae (Figure 2–10), each limited by membranes that are continuous with the outer membrane of the nuclear envelope. The presence of polyribosomes on the cytosolic surface of the RER confers basophilic staining properties on this organelle when viewed with the light microscope.

The major function of RER is production of membrane-associated proteins, proteins of many membranous organelles, and proteins to be secreted by exocytosis. Production here includes the initial (core) glycosylation of glycoproteins, certain other posttranslational modifications of newly formed polypeptides, and the assembly of multichain proteins. These activities are mediated by resident enzymes of the RER and by protein complexes that act as chaperones guiding the folding of nascent proteins, inhibiting aggregation, and generally monitoring protein quality within the ER.

Protein synthesis begins on polyribosomes in the cytosol. The 5′ ends of mRNAs for proteins destined to be segregated in the ER encode an N-terminal **signal sequence** of 15-40 amino acids that includes a series of six or more hydrophobic residues. As shown in Figure 2–11, the newly translated

FIGURE **2–11** Movement of polypeptides into the RER.

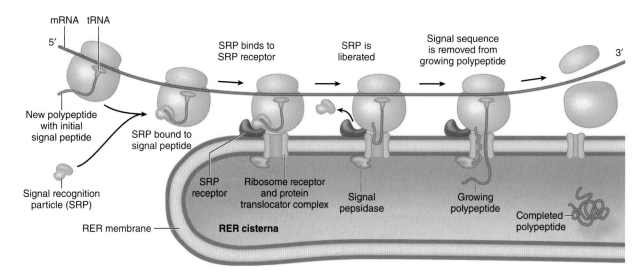

The newly translated amino terminus of a protein to be incorporated into membranes or sequestered into vesicles contains 15-40 amino acids that include a specific sequence of hydrophobic residues comprising the **signal sequence** or signal peptide. This sequence is bound by a signal-recognition particle (SRP), which then recognizes and binds a receptor on the ER. Another receptor in the ER membrane binds a structural protein of the large ribosomal subunit, more firmly attaching the ribosome to the ER. The hydrophobic signal peptide is translocated through a protein pore (translocon) in the ER membrane, and the SRP is freed for reuse. The signal peptide is removed from the growing protein by a peptidase and translocation of the growing polypeptide continues until it is completely segregated into the ER cisterna.

signal sequence is bound by a protein complex called the **signal-recognition particle (SRP)**, which inhibits further polypeptide elongation. The SRP-ribosome-nascent peptide complex binds to SRP receptors on the ER membrane. SRP then releases the signal sequence, allowing translation to continue with the nascent polypeptide chain transferred to a **translocator complex** (also called a translocon) through the ER membrane (Figure 2–11). Inside the lumen of the RER, the signal sequence is removed by an enzyme, signal peptidase. With the ribosome docked at the ER surface, translation continues with the growing polypeptide pushing itself while chaperones and other proteins serve to "pull" the nascent polypeptide through the translocator complex. Upon release from the ribosome, posttranslational modifications and proper folding of the polypeptide continue.

RER has a highly regulated system to prevent nonfunctional proteins being forwarded to the pathway for secretion or to other organelles. New proteins that cannot be folded or assembled properly by chaperones undergo ER-associated degradation (ERAD), in which unsalvageable proteins are translocated back into the cytosol, conjugated to ubiquitin, and then degraded by proteasomes.

As mentioned, proteins synthesized in the RER can have several destinations: intracellular storage (eg, in lysosomes and specific granules of leukocytes), provisional storage in cytoplasmic vesicles prior to exocytosis (eg, in the pancreas, some endocrine cells), and as integral membrane proteins. Diagrams in Figure 2–12 show a few cell types with distinct differences in the destinations of their major protein products and how these differences determine a cell's histologic features.

> ❯❯ **MEDICAL APPLICATION**
>
> Quality control during protein production in the RER and properly functioning ERAD to dispose of defective proteins are extremely important and several inherited diseases result from malfunctions in this system. For example, in some forms of osteogenesis imperfecta bone cells synthesize and secrete defective procollagen molecules which cannot assemble properly and produce very weak bone tissue.

Smooth Endoplasmic Reticulum

Regions of ER that lack bound polyribosomes make up the smooth endoplasmic reticulum (SER), which is continuous with RER but frequently less abundant (Figure 2–10). Lacking polyribosomes, SER is not basophilic and is best seen with the TEM. Unlike the cisternae of RER, SER cisternae are more tubular or saclike, with interconnected channels of various shapes and sizes rather than stacks of flattened cisternae.

SER has three main functions, which vary in importance in different cell types.

- Enzymes in the SER perform *synthesis of phospholipids and steroids*, major constituents of cellular membranes. These lipids are then transferred from the SER to other membranes by lateral diffusion into adjacent

FIGURE 2–12 Protein localization and cell morphology.

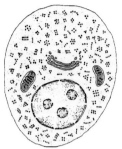
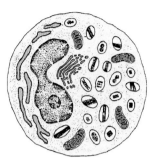
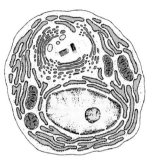
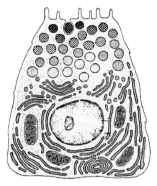

(a) Erythroblast   **(b)** Eosinophilic leukocyte   **(c)** Plasma cell   **(d)** Pancreatic acinar cell

The ultrastructure and general histologic appearance of a cell are determined by the nature of the most prominent proteins the cell is making.

(a) Cells that make few or no proteins for secretion have very little RER, with essentially all polyribosomes free in the cytoplasm.

(b) Cells that synthesize, segregate, and store various proteins in specific secretory granules or vesicles always have RER, a Golgi apparatus, and a supply of granules containing the proteins ready to be secreted.

(c) Cells with extensive RER and a well-developed Golgi apparatus show few secretory granules because the proteins undergo exocytosis immediately after Golgi processing is complete. Many cells, especially those of epithelia, are *polarized*, meaning that the distribution of RER and secretory vesicles is different in various regions or poles of the cell.

(d) Epithelial cells specialized for secretion have distinct polarity, with RER abundant at their basal ends and mature secretory granules at the apical poles undergoing exocytosis into an enclosed extracellular compartment, the lumen of a gland.

membranes, by phospholipid transfer proteins, or by vesicles which detach from the SER for movement along the cytoskeleton and fusion with other membranous organelles. In cells that secrete steroid hormones (eg, cells of the adrenal cortex), SER occupies a large portion of the cytoplasm.

- Other SER enzymes, including those of the cytochrome P450 family, allow *detoxification of potentially harmful exogenous molecules* such as alcohol, barbiturates, and other drugs. In liver cells these enzymes also process endogenous molecules such as the components of bile.

- SER vesicles are also responsible for *sequestration and controlled release of Ca²⁺*, which is part of the rapid response of cells to various stimuli. This function is particularly well developed in striated muscle cells, where the SER has an important role in the contraction process and assumes a specialized form called the **sarcoplasmic reticulum** (see Chapter 10).

›› MEDICAL APPLICATION

Jaundice denotes a yellowish discoloration of the skin and is caused by accumulation in extracellular fluid of bilirubin and other pigmented compounds, which are normally metabolized by SER enzymes in cells of the liver and excreted as bile. A frequent cause of jaundice in newborn infants is an underdeveloped state of SER in liver cells, with failure of bilirubin to be converted to a form that can be readily excreted.

Golgi Apparatus

The dynamic organelle called the **Golgi apparatus**, or Golgi complex, completes posttranslational modifications of proteins produced in the RER and then packages and addresses these proteins to their proper destinations. The organelle was named after histologist Camillo Golgi who discovered it in 1898. The Golgi apparatus consists of many smooth membranous saccules, some vesicular, others flattened, but all containing enzymes and proteins being processed (Figure 2–13). In most cells the small Golgi complexes are located near the nucleus.

As shown in Figure 2–13, the Golgi apparatus has two distinct functional sides or faces, formed by the complex traffic of vesicles within cells. Material moves from the RER cisternae to the Golgi apparatus in small, membrane-enclosed carriers called **transport vesicles** that are transported along cytoskeletal polymers by motor proteins. The transport vesicles merge with the Golgi-receiving region, or *cis face*. On the opposite side of the Golgi network, at its shipping or *trans face*, larger saccules or vacuoles accumulate, condense, and generate other vesicles that carry completed protein products to organelles away from the Golgi (Figure 2–13).

Formation of transport vesicles and secretory vesicles is driven by assembly of various coat proteins (including clathrin), which also regulate vesicular traffic to, through, and beyond the Golgi apparatus. Forward movement of vesicles in the *cis* Golgi network of saccules is promoted by the **coat protein COP-II**, while retrograde movements in that region involve **COP-I**. Other membrane proteins important

FIGURE 2–13 Golgi apparatus.

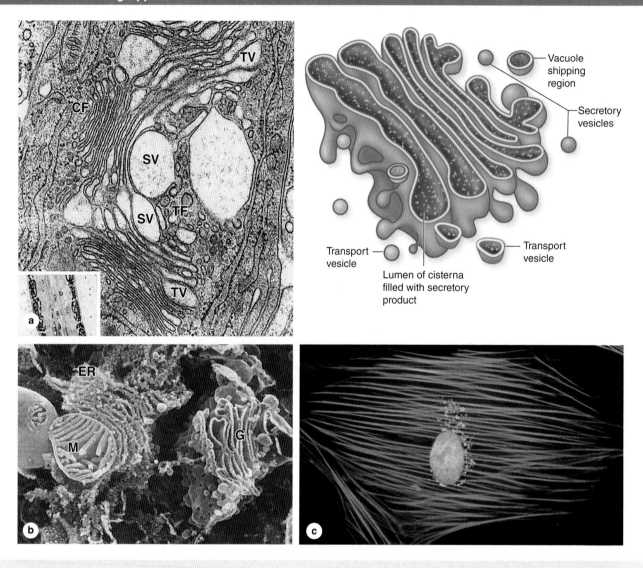

The **Golgi apparatus** is a highly plastic, morphologically complex system of membrane vesicles and cisternae in which proteins and other molecules made in the RER undergo further modification and sorting into specific vesicles destined for different roles in the cell.

(a) TEM of the Golgi apparatus provided early evidence about how this organelle functions. To the left is a cisterna of RER and close to it are many small vesicles at the *cis* face **(CF)**, or receiving face, of the Golgi apparatus, merging with the first of several flattened Golgi cisternae. In the center are the characteristic flattened, curved, and stacked medial cisternae of the complex. Cytological and molecular data suggest that other transport vesicles **(TV)** move proteins serially through the cisternae until at the *trans* face **(TF)**, or shipping region, larger condensing secretory vesicles **(SV)** and other vacuoles emerge to carry the modified proteins elsewhere in the cell. Formation and fusion of the vesicles through the Golgi apparatus is controlled by specific membrane proteins. (X30,000) **Inset:** A small region of a Golgi apparatus in a 1-µm section from a silver-stained cell, demonstrating abundant glycoproteins within cisternae.

(b) Morphological aspects of the Golgi apparatus are revealed more clearly by SEM, which shows a three-dimensional snapshot of the region between RER and the Golgi membrane compartments. Cells may have multiple Golgi apparatuses, each with the general organization shown here and typically situated near the cell nucleus. (X30,000)

(c) The Golgi apparatus location can be clearly seen in intact cultured cells processed by immunocytochemistry using an antibody against golgin-97 to show the many complexes of Golgi vesicles (green), all near the nucleus, against a background of microfilaments organized as stress fibers and stained with fluorescent phalloidin (violet). Because of the abundance of lipids in its many membranes, the Golgi apparatus is difficult to visualize in typical paraffin-embedded, H&E-stained sections. In developing white blood cells with active Golgi complexes, the organelle can sometimes be seen as a faint unstained juxtanuclear region (sometimes called a "Golgi ghost") surrounded by basophilic cytoplasm.

(*Figure 2–13b reproduced, with permission from Naguro T, Iino A. Prog Clin Biol Res. 1989;295:250; Figure 2–13c, © 2015 Thermo Fisher Scientific, Inc. Used under permission.*)

for directed vesicle fusion include various Rab proteins and other enzymes, receptors and specific binding proteins, and fusion-promoting proteins that organize and shape membranes. Depending on the activity of these proteins, vesicles are directed toward different Golgi regions and give rise to lysosomes or secretory vesicles for exocytosis.

As indicated in Figure 2–14, Golgi saccules at sequential locations contain different enzymes at different *cis*, *medial*, and *trans* levels. Enzymes of the Golgi apparatus are important for glycosylation, sulfation, phosphorylation, and limited proteolysis of proteins. Along with these activities, the Golgi apparatus initiates packing, concentration, and storage of secretory

products. Protein movements through the Golgi and the control of protein processing are subjects of active research.

Secretory Granules

Originating as condensing vesicles in the Golgi apparatus, **secretory granules** are found in cells that store a product until its release by exocytosis is signaled by a metabolic, hormonal, or neural message (regulated secretion). The granules are surrounded by membrane and contain a concentrated form of the secretory product (Figure 2–15). The contents of some secretory granules may be up to 200 times more

FIGURE **2–14** **Summary of functions within the Golgi apparatus.**

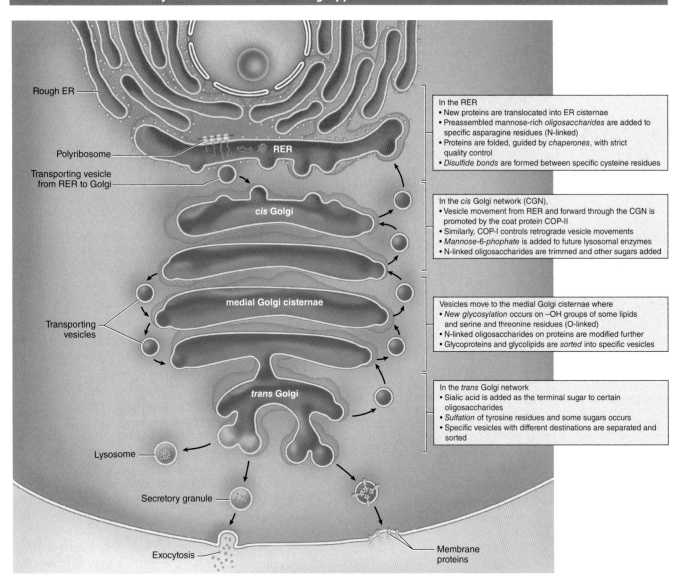

The main molecular processes are listed at the right, with the major compartments where they occur. In the *trans* Golgi network, the proteins and glycoproteins combine with specific receptors that guide them to the next stages toward their destinations.

FIGURE **2–15** Secretory granules.

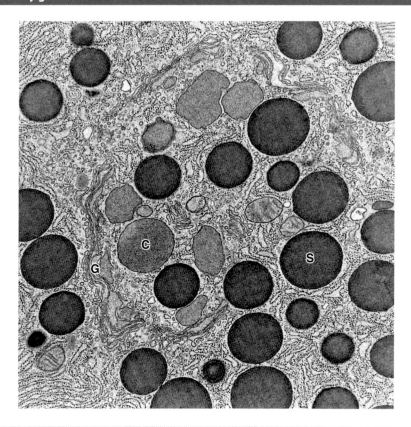

TEM of one area of a pancreatic acinar cell shows numerous mature, electron-dense **secretory granules (S)** in association with condensing vacuoles **(C)** of the Golgi apparatus **(G)**. Such granules form as the contents of the Golgi vacuoles become more condensed. In H&E-stained sections secretory granules are often shown as intensely eosinophilic structures, which in polarized epithelial cells are concentrated at the apical region prior to exocytosis. (X18,900)

concentrated than those in the cisternae of the RER. Secretory granules with dense contents of digestive enzymes are also referred to as **zymogen granules**.

Lysosomes

Lysosomes are sites of intracellular digestion and turnover of cellular components. Lysosomes (Gr. *lysis*, solution, + *soma*, body) are membrane-limited vesicles that contain about 40 different hydrolytic enzymes and are particularly abundant in cells with great phagocytic activity (eg, macrophages, neutrophils). Although the nature and activity of lysosomal enzymes vary depending on the cell type, the most common are acid hydrolyases such as proteases, nucleases, phosphatase, phospholipases, sulfatases, and β-glucuronidase. As can be seen from this list, lysosomal enzymes are capable of breaking down most macromolecules.

Lysosomes, which are usually spherical, range in diameter from 0.05 to 0.5 μm and present a uniformly granular, electron-dense appearance in the TEM (Figure 2–16). In macrophages and neutrophils, lysosomes are slightly larger and visible with the light microscope, especially after histochemical staining.

Cytosolic components are protected from these enzymes by the membrane surrounding lysosomes and because the enzymes have optimal activity at an acidic pH (~5.0). Any leaked lysosomal enzymes are practically inactive at the pH of cytosol (~7.2) and harmless to the cell.

Lysosomal hydrolases are synthesized and segregated in the RER and then transferred to the Golgi apparatus, where the enzymes are further modified and packaged in vacuoles that form lysosomes. The marker mannose-6-phosphate (M6P) is added by a phosphotransferase in the *cis* Golgi only to the N-linked oligosaccharides of the hydrolases destined for lysosomes. Membrane receptors for M6P-containing proteins in the *trans* Golgi network then bind these proteins and divert them from the secretory pathway for segregation into lysosomes.

Material taken from outside the cell by endocytosis is digested when the membrane of the phagosome or pinocytotic vesicle fuses with a lysosome. This mixes the endocytosed material with the lysosomal enzymes and activates proton pumps in the lysosomal membrane that acidify the contents, allowing digestion. The composite, active organelle is now termed a secondary or

FIGURE **2–16** Lysosomes.

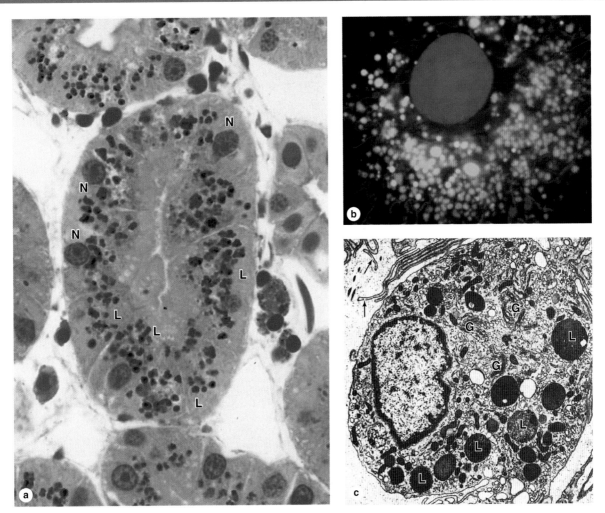

Lysosomes are spherical membrane-enclosed vesicles that function as sites of intracellular digestion and are particularly numerous in cells active after the various types of endocytosis. Lysosomes are not well shown on H&E-stained cells but can be visualized by light microscopy after staining with toluidine blue.

(a) Cells in a kidney tubule show numerous purple lysosomes **(L)** in the cytoplasmic area between the basally located nuclei **(N)** and apical ends of the cells at the center of the tubule. Using endocytosis, these cells actively take up small proteins in the lumen of the tubule, degrade the proteins in lysosomes, and then release the resulting amino acids for reuse. (X300)

(b) Lysosomes in cultured vascular endothelial cells can be specifically stained using fluorescent dyes sequestered into these organelles (green), which are abundant around the blue Hoechst-stained nucleus. Mitochondria (red) are scattered among the lysosomes.

(c) In the TEM lysosomes **(L)** have a characteristic very electron-dense appearance and are shown here near groups of Golgi cisternae **(G)**. The less electron-dense lysosomes represent heterolysosomes in which digestion of the contents is under way. The cell is a macrophage with numerous fine cytoplasmic extensions (arrows). (X15,000)

(Figure 2–16b, © 2015 Thermo Fisher Scientific, Inc. Used under permission.)

heterolysosome. Heterolysosomes are generally somewhat larger and have a more heterogeneous appearance in the TEM because of the wide variety of materials they may be digesting (Figure 2–16c).

During this digestion of macromolecules, released nutrients diffuse into the cytosol through the lysosomal membrane. Indigestible material is retained within a small vacuolar remnant called a **residual body** (Figure 2–17). In some long-lived

cells (eg, neurons, heart muscle), residual bodies can accumulate over time as granules of **lipofuscin**.

Besides degrading exogenous macromolecules, lysosomes also function in the removal of excess or nonfunctional organelles in a process called **autophagy** (Figures 2–17 and 2–18). A membrane from SER forms around the organelle or cytoplasmic portion to be removed, producing an **autophagosome** (Gr. *autos*, self, + *phagein*, to eat, + *soma*). These then

FIGURE **2–17** Lysosomal functions.

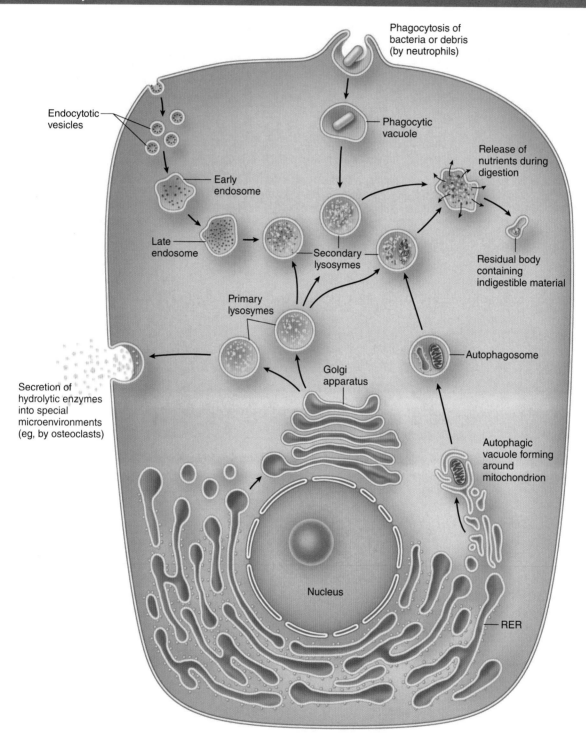

Synthesis of lysosomal enzymes occurs in the RER, with packaging in the Golgi apparatus. Endocytosis produces vesicles that fuse with **endosomes** before merging with **lysosomes**. Phagocytic vacuoles (or phagosomes) fuse with primary lysosomes to become **secondary lysosomes** (or heterolysosomes), in which ingested material is degraded. **Autophagosomes**, such as those depicted here with a mitochondrion in the process of digestion, are formed after nonfunctional or surplus organelles become enclosed with membrane and the resulting structure fuses with a lysosome. The products of lysosomal digestion are recycled to the cytoplasm, but indigestible molecules remain in a membrane-enclosed **residual body**, which may accumulate in long-lived cells as lipofuscin. In some cells, such as osteoclasts, the lysosomal enzymes are secreted into a restricted extracellular compartment.

FIGURE **2–18** Autophagy.

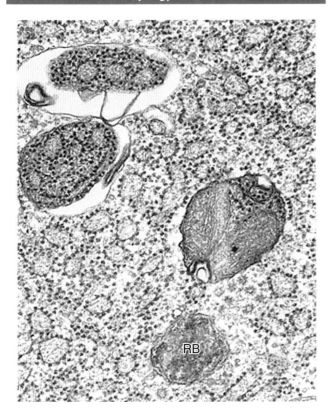

Autophagy is a process in which the cell uses lysosomes to dispose of excess or nonfunctioning organelles or membranes. Membrane that appears to emerge from the SER encloses the organelles to be destroyed, forming an autophagosome that then fuses with a lysosome for digestion of the contents. In this TEM the two autophagosomes at the **upper left** contain portions of RER more electron dense than the neighboring normal RER and one near the **center** contains what may be mitochondrial membranes plus RER. Also shown is a vesicle with features of a residual body **(RB)**. (30,000X)

fuse with lysosomes for digestion of the enclosed material. Autophagy is enhanced in secretory cells that have accumulated excess secretory granules and in times of nutrient stress, such as starvation. Digested products from autophagosomes are reused in the cytoplasm.

Diseases categorized as *lysosomal storage disorders* stem from defects in one or more of the digestive enzymes present in lysosomes, usually due to a mutation leading to a deficiency of one of the enzymes, or defects due to faulty posttranslational processing. In cells that must digest the substrate of the missing or defective enzyme following autophagy, the lysosomes cannot function properly. Such cells accumulate large secondary lysosomes or residual bodies filled with the indigestible macromolecule. The accumulation of these vacuoles may eventually interfere with normal cell or tissue function, producing symptoms of the disease. A few lysosomal storage diseases are listed in Table 2–3, with the enzyme involved for each and the tissue affected.

Proteasomes

Proteasomes are very small abundant protein complexes not associated with membrane, each approximately the size of the small ribosomal subunit. They function to degrade denatured or otherwise nonfunctional polypeptides. Proteasomes also remove proteins no longer needed by the cell and provide an important mechanism for restricting activity of a specific protein to a certain window of time. Whereas lysosomes digest organelles or membranes by autophagy, proteasomes deal primarily with free proteins as individual molecules.

As shown in Figure 2–9, the proteasome is a cylindrical structure made of four stacked rings, each composed of seven proteins including proteases. At each end of the cylinder is a regulatory particle that contains ATPase and recognizes proteins with attached molecules of **ubiquitin**, an abundant cytosolic 76-amino acid protein found in all cells. Misfolded or denatured proteins, or short-lived proteins with oxidized amino acids, are recognized by chaperones and targeted for destruction by other enzyme complexes that conjugate ubiquitin to lysine residues of the protein, followed by formation of a polyubiquitin chain. Ubiquinated proteins are recognized by the regulatory particles of proteasomes, unfolded by the ATPase using energy from ATP, and then translocated into the core of the cylindrical structure and degraded by endopeptidases. The ubiquitin molecules are released for reuse and the peptides produced may be broken down further to

TABLE **2–3** Examples of lysosomal storage diseases caused by defective lysosomal enzymes.

Disease	Faulty Enzyme	Main Organs Affected
Hurler syndrome (MPS I)	α-L-Iduronidase	Skeleton and nervous system
McArdle syndrome	Muscle phosphorylase	Skeletal muscles
Tay-Sachs	GM_2-gangliosidase	Nervous system
Gaucher	Glucocerebrosidase	Liver and spleen
I-cell disease	Phosphotransferase for M6P formation	Skeleton and nervous system

amino acids or they may have other specialized destinations, such as the antigen-presenting complexes of cells activating an immune response.

Mitochondria

Mitochondria (Gr. *mitos*, thread, + *chondros*, granule) are membrane-enclosed organelles with arrays of enzymes specialized for aerobic respiration and production of **adenosine triphosphate (ATP)**, with high-energy phosphate bonds, which supplies energy for most cellular activities. Glycolysis converts glucose anaerobically to pyruvate in the cytoplasm, releasing some energy. The rest of the energy is captured when pyruvate is imported into mitochondria and oxidized to CO_2 and H_2O. Mitochondrial enzymes yield 15 times more ATP than is produced by glycolysis alone. Some of the energy released in mitochondria is not stored in ATP but is dissipated as heat that maintains body temperature.

Mitochondria are usually elongated structures with diameters of 0.5-1 μm and lengths up to 10 times greater. They are highly plastic, rapidly changing shape, fusing with one another and dividing, and are moved through the cytoplasm along microtubules. The number of mitochondria is related to the cell's energy needs: cells with a high-energy metabolism (eg, cardiac muscle, cells of some kidney tubules) have abundant mitochondria, whereas cells with a low-energy metabolism have few mitochondria. Similarly, mitochondria of differentiated cells are concentrated in cytoplasmic regions where energy utilization is more intense.

Mitochondria are often large enough to be visible with the light microscope as numerous discrete organelles (Figure 2–19).

Under the TEM each mitochondrion is seen to have two separated and very different membranes that together create two compartments: the innermost **matrix** and a narrow **intermembrane space** (Figure 2–20a). Both mitochondrial membranes contain a higher density of protein molecules than other membranes in the cell and have reduced fluidity. The **outer membrane** is sieve-like, containing many transmembrane proteins called **porins** that form channels through which small molecules such as pyruvate and other metabolites readily pass from the cytoplasm to the intermembrane space.

The **inner membrane** has many long folds called **cristae**, which project into the matrix and greatly increase the membrane's surface area (Figure 2–20). The number of cristae in mitochondria also corresponds to the energy needs of the cell. The lipid bilayer of the inner membrane contains unusual phospholipids and is highly impermeable to ions (Figure 2–20). Integral proteins include various transport proteins that make the inner membrane selectively permeable to the small molecules required by enzymes in the matrix. Mitochondrial matrix enzymes include those that oxidize pyruvate and fatty acids to form acetyl coenzyme A (CoA) and those of the citric acid cycle that oxidize acetyl CoA, releasing CO_2 as waste and small energy-rich molecules that provide electrons for transport along the **electron-transport chain** (or respiratory chain). Enzymes and other components of this chain are embedded in the inner membrane and allow oxidative phosphorylation, which produces most of the ATP in animal cells.

Formation of ATP by oxidative phosphorylation enzymes occurs by a **chemiosmotic process**. Membrane proteins guide the small electron carrier molecules through closely packed enzyme complexes so that the electrons move sequentially along the chain. Electron transfer is coupled with oriented proton uptake and release, with protons accumulating in the intermembrane space (Figure 2–20) and producing an **electrochemical gradient** across the inner membrane. Membrane-associated proteins of the **ATP synthase** system form large (10-nm), multisubunit, globular complexes on stalk-like structures that project from the matrix side of the inner membrane (Figure 2–20). This enzyme complex contains a channel through which protons flow down the electrochemical gradient, crossing the membrane back into the matrix. Passage of protons through this channel causes rotation of specific polypeptides in the globular ATP synthase complex, converting the energy of proton flow into the mechanical energy of protein movement. Mechanical energy is stored in the new phosphate bond of ATP by other subunit polypeptides binding adenosine diphosphate (ADP) and inorganic phosphate (Figure 2–20). A steady torrent of protons along the gradient allows each of these remarkable synthase complexes to produce more than 100 molecules of ATP per second.

Another role for mitochondria occurs at times of cell stress, when the protein cytochrome c is released from the inner membrane's electron transport chain. In the cytoplasm

FIGURE **2-19** Mitochondria in the light microscope.

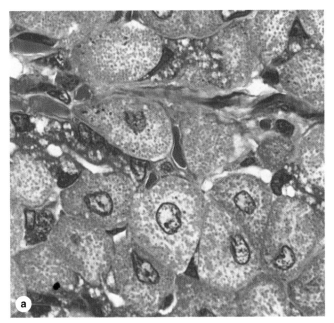

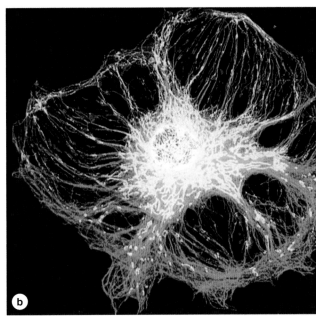

(a) In certain sectioned cells stained with H&E, mitochondria appear throughout the cytoplasm as numerous eosinophilic structures. The mitochondria usually appear round or slightly elongated and are more numerous in cytoplasmic regions with higher energy demands, such as near the cell membrane in cells undergoing much active transport. The central nuclei are also clearly seen in these cells.

(b) Entire mitochondria can be shown in cultured cells, such as the endothelial cells shown here, and often appear as the elongated structures (shown in yellow or orange here), usually arrayed in

parallel along microtubules. Such preparations also show that mitochondrial shape can be quite plastic and variable. Specific mitochondrial staining such as that shown here involves incubating living cells with specific fluorescent compounds that are specifically sequestered into these organelles, followed by fixation and immunocytochemical staining of the microtubules. In this preparation, microtubules are stained green and mitochondria appear yellow or orange, depending on their association with the green microtubules. The cell nucleus was stained with DAPI (4′,6-diamidino-2-phenylindole).

this protein activates sets of proteases that degrade all cellular components in a regulated process called **apoptosis** which results in rapid cell death (see Chapter 3).

New mitochondria originate by growth and division (fission) of preexisting mitochondria. During cell mitosis each daughter cell receives approximately half the mitochondria in the parent cell.

Unlike most organelles mitochondria are partly autonomous of nuclear genes and activities. The mitochondrial matrix contains a small circular chromosome of DNA, ribosomes, mRNA, and tRNA, all with similarities to the corresponding bacterial components. Protein synthesis occurs in mitochondria, but because of the reduced amount of mitochondrial DNA, only a small subset of mitochondrial proteins is produced locally. Most are encoded by nuclear DNA and synthesized on free polyribosomes of the cytosol. These proteins have short terminal amino acid sequences that serve as signals for their uptake across the mitochondrial membranes. The observation that mitochondria have

certain bacterial characteristics led with later work to the understanding that mitochondria evolved from an ancestral aerobic prokaryote that lived symbiotically within an ancestral eukaryotic host cell.

Peroxisomes

Peroxisomes are spherical organelles enclosed by a single membrane and named for their enzymes producing and degrading hydrogen peroxide, H_2O_2 (Figure 2–21). **Oxidases** located here oxidize substrates by removing hydrogen atoms that are transferred to molecular oxygen (O_2), producing H_2O_2. Peroxidases such as **catalase** immediately break down H_2O_2, which is potentially damaging to the cell. These enzymes also inactivate various potentially toxic molecules, including some prescription drugs, particularly in the large and abundant peroxisomes of liver and kidney cells.

Other diverse enzymes in peroxisomes complement certain functions of the SER and mitochondria in the metabolism

FIGURE **2–20** Mitochondrial structure and ATP formation (Legend Opposite).

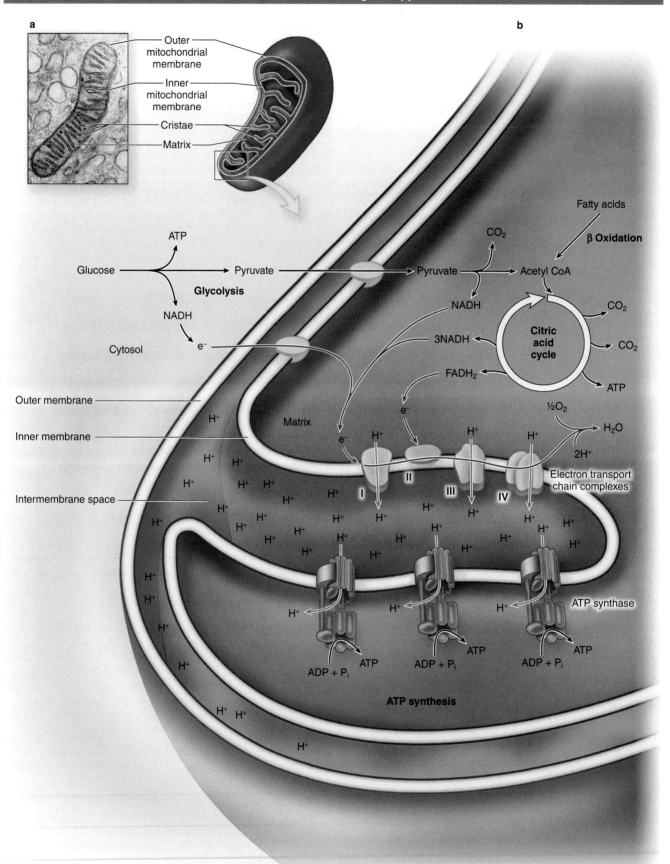

(a) The two mitochondrial membranes and the innermost matrix can be seen in the TEM and diagram. The **outer membrane** is smooth and the **inner membrane** has many sharp folds called **cristae** that increase its surface area greatly. The **matrix** is a gel with a high concentration of enzymes.

(b) Metabolites such as pyruvate and fatty acids enter mitochondria via membrane porins and are converted to acetyl CoA by matrix enzymes of the **citric acid cycle** (or Krebs cycle), yielding some ATP and NADH (nicotinamide adenine dinucleotide), a major source of electrons for the **electron-transport chain**. The movement of electrons through the protein complexes of the inner membrane's electron-transport system is accompanied by

the directed movement of protons (H+) from the matrix into the intermembranous space, producing an electrochemical gradient across the membrane. Other membrane-associated proteins make up the **ATP synthase** systems, each of which forms a globular complex on a stalk-like structure projecting from the matrix side of the inner membrane. A channel in this enzyme complex allows proton flow down the electrochemical gradient and across the membrane back into the matrix. The flow of protons causes rapid spinning of specific polypeptides in the globular ATP synthase complex, converting the energy of proton flow into mechanical energy, which other subunit proteins store in the new phosphate bond of ATP.

of lipids and other molecules. Thus, the β-oxidation of long-chain fatty acids (18 carbons and longer) is preferentially accomplished by peroxisomal enzymes that differ from their mitochondrial counterparts. Certain reactions leading to the formation of bile acids and cholesterol also occur in peroxisomes.

Peroxisomes form in two ways: budding of precursor vesicles from the ER or growth and division of preexisting peroxisomes. Peroxisomal proteins are synthesized on free polyribosomes and have targeting sequences of amino acids at either terminus recognized by receptors located in the peroxisomal membrane for import into the organelle.

FIGURE 2–21 Peroxisomes.

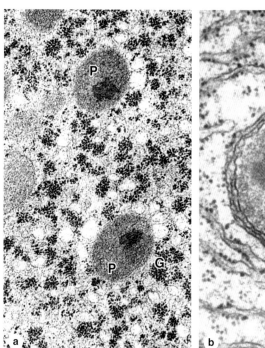 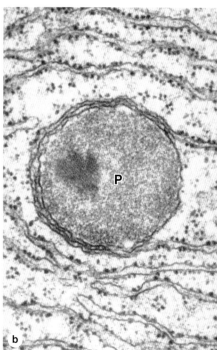 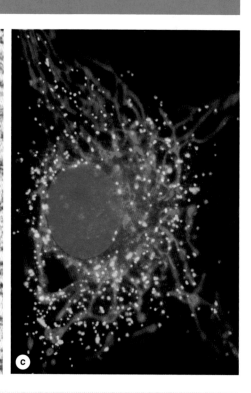

Peroxisomes are small spherical, membranous organelles, containing enzymes that use O_2 to remove hydrogen atoms from fatty acids, in a reaction that produces hydrogen peroxide (H_2O_2) that must be broken down to water and O_2 by another enzyme, **catalase.**

(a) By TEM peroxisomes **(P)** generally show a matrix of moderate electron density. Aggregated electron-dense particles represent glycogen (G). (X30,000)

(b) Peroxisomes **(P)** in most species are characterized by a central, more electron-dense crystalloid aggregate of constituent enzymes, as shown here. (X60,000)

(c) A cultured endothelial cell processed by immunocytochemistry shows many peroxisomes (green) distributed throughout the cytoplasm among the vitally stained elongate mitochondria (red) around the DAPI-stained nucleus (blue). Peroxisomes shown here were specifically stained using an antibody against the membrane protein PMP70.

(Figure 2–21c, © 2015 Thermo Fisher Scientific, Inc. Used under permission.)

>> MEDICAL APPLICATION

Several fairly rare disorders arise from defective peroxisomal proteins. **Neonatal adrenoleukodystrophy** is caused by a defective integral membrane protein needed for transport of very-long-chain fatty acids into the peroxisome for β-oxidation. Accumulation of these fatty acids in body fluids can disrupt the myelin sheaths in nerve tissue, causing severe neurologic symptoms. Deficiencies of peroxisomal enzymes cause **Zellweger syndrome** that affects the structure and functions of several organ systems.

> THE CYTOSKELETON

The cytoplasmic **cytoskeleton** is a complex array of (1) microtubules, (2) microfilaments (also called actin filaments), and (3) intermediate filaments. These protein polymers determine the shapes of cells, play an important role in the movements of organelles and cytoplasmic vesicles, and also allow the movement of entire cells. Important properties, functions, and locations of the cytoskeletal components are summarized in Table 2–4.

TABLE **2–4**	Properties of cytoskeletal components (microtubules, microfilaments, and intermediate filaments).

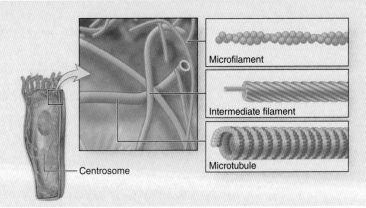

Microfilament

Intermediate filament

Microtubule

Centrosome

General Function of Cytoskeleton

1. **Structural:** Provides structural support to cell; stabilizes junctions between cells
2. **Movement:** Assists with cytosol streaming and cell motility; helps move organelles and materials throughout cell; helps move chromosomes during cell division

	Microtubules	**Microfilaments**	**Intermediate filaments**
Subunit	Heterodimers of αβ-tubulin	G-actin monomers	Antiparallel tetramers of 2 rod-like dimers
Overall structure	Hollow tube with a wall of 13 parallel protofilaments	2 intertwined filaments of F-actin	Cable of 4 intertwined protofibrils, each consisting of bundled tetramers associated end to end
Diameter	25 nm	5-7 nm	8-10 nm
Monomeric proteins	α and β tubulin (54 kDa)	Globular G-actin (42 kDa)	Various α-helical rod-like proteins (~55 kDa, Table 2–5)
Polarity	+ and − ends	+ and − ends	No apparent polarity
Relative stability	Dynamic in cytoplasm; stable in axonemes	Dynamic	Stable
General locations	Radiating through cytoplasm from concentration at centrosomes; axonemes	Concentrated beneath cell membrane; in cell extensions like microvilli	Arrayed throughout cytoplasm; at desmosomes; inside nuclear envelop
Key functions	Maintain cell's shape and polarity; provide tracks for organelle and chromosome movement; move cilia and flagella	Contract and move cells; change cell shape; cytokinesis; cytoplasmic transport and streaming	Strengthen cell and tissue structure; maintain cell shape; maintain nuclear shape (lamins)

Microtubules

Within the cytoplasm of all eukaryotic cells are fine tubular structures known as **microtubules** (Table 2–4; Figure 2–22), most of which are highly dynamic in length. Microtubules

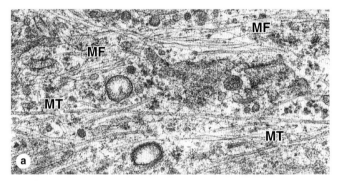

FIGURE **2–22** Microtubules and actin filaments in cytoplasm.

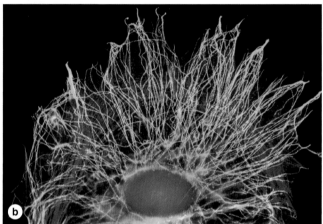

(a) Microtubules (MT) and actin microfilaments (MF) can both be clearly distinguished in this TEM of fibroblast cytoplasm, which provides a good comparison of the relative diameters of these two cytoskeletal components. (X60,000)

(b) Arrays of microfilaments and microtubules are easily demonstrated by immunocytochemistry using antibodies against their subunit proteins, as in this cultured cell. Actin filaments (red) are most concentrated at the cell periphery, forming prominent circumferential bundles from which finer filaments project into cellular extensions and push against the cell membrane. Actin filaments form a dynamic network important for cell shape changes such as those during cell division, locomotion, and formation of cellular processes, folds, pseudopodia, lamellipodia, microvilli, etc, which serve to change a cell's surface area or give direction to a cell's crawling movements.

Microtubules (green/yellow) are oriented in arrays that generally extend from the centrosome area near the nucleus into the most peripheral extensions. Besides serving to stabilize cell shape, microtubules form the tracks for kinesin-based transport of vesicles and organelles into the cell periphery and dynein-based transport toward the cell nucleus.

(*Figure 2–22b, used with permission from Dr Albert Tousson, University of Alabama—Birmingham High Resolution Imaging Facility, Birmingham.*)

are also organized into larger, more stable arrays called **axonemes** in the cytoplasmic extensions called cilia (discussed in Chapter 4) and flagella. Each microtubule is hollow, with an outer diameter of 25 nm and a wall 5-nm thick, a structure that confers significant rigidity to help maintain cell shape. Microtubules vary in length, but can become many micrometers long. Two or more microtubules are often linked side by side by protein arms or bridges, which are particularly important in the axonemes of cilia and flagella.

As indicated in Table 2–4, the protein subunit of a microtubule is a heterodimer of α and β **tubulin**, each with a molecular mass of about 50 kDa. Under appropriate conditions the tubulin heterodimers polymerize to form the microtubules, which have a slightly spiral organization. The tubulin subunits align lengthwise as protofilaments, with 13 parallel protofilaments forming the circumference of each microtubule wall.

Polymerization of tubulins is directed by **microtubule organizing centers (MTOCs)**, which contain short assemblies of tubulin that act as nucleating sites for further polymerization. Microtubules are polarized structures, with growth (polymerization) occurring more rapidly at the (+) end (Figure 2–23). Microtubules show **dynamic instability**, with continuous cycles of polymerization and depolymerization at steady-state conditions, which depend on concentrations of tubulin, Ca^{2+}, Mg^{2+}, and the presence of various **microtubule-associated proteins (MAPs)**. Energy for assembly is derived from GTP bound to incoming tubulin subunits. Individual microtubules shorten as depolymerization exceeds growth. Microtubule stability varies greatly with cellular location and function; microtubules of cilia are very stable, while those of the mitotic spindle are short-lived.

The dominant MTOC in most cells is the **centrosome**, which is organized around two cylindrical **centrioles**, each about 0.2 μm in diameter and 0.3-0.5 μm in length. Each centriole is composed of nine highly organized microtubule triplets (Figure 2–24). With their long axes at right angles, the paired centrioles organize nearby tubulin complexes and other proteins as a pericentriolar matrix found close to the nucleus of nondividing cells. Before cell division, more specifically during the period of DNA replication, each centrosome duplicates itself so that now each centrosome has two pairs of centrioles. During mitosis, the centrosome divides into halves, which move to opposite poles of the cell, and become organizing centers for the microtubules of the mitotic spindle.

Microtubules also form part of the system for intracellular transport of membranous vesicles, macromolecular complexes, and organelles. Well-studied examples include axoplasmic transport in neurons, melanin transport in pigment cells, chromosome movements by the mitotic spindle, and vesicle movements among different cell compartments. In each of these examples, movement is suspended if microtubules are disrupted. Transport along microtubules is under the control of proteins called **motor proteins**, which use ATP in moving the larger structures. **Kinesins** carry material away from the MTOC near the nucleus toward the plus end of

FIGURE **2–23** **Dynamic instability of microtubules.**

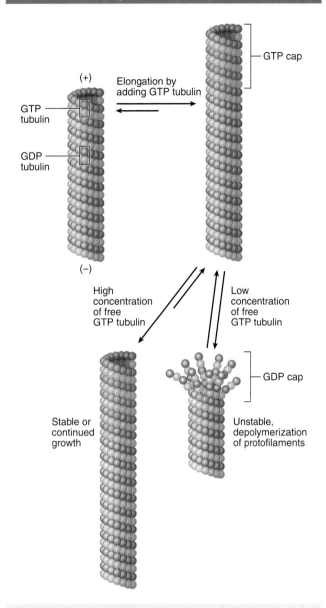

(+)

Elongation by
adding GTP tubulin

GTP cap

GTP
tubulin

GDP
tubulin

(–)

High
concentration
of free
GTP tubulin

Low
concentration
of free
GTP tubulin

GDP cap

Stable or
continued
growth

Unstable,
depolymerization
of protofilaments

At stable tubulin concentrations some microtubules grow
while others shrink, each existing in a condition called dynamic
instability. In cytoplasmic areas where the tubulin concentration is high, tubulin GTP is added at a microtubule's (+) end
faster than the incorporated GTP can be hydrolyzed. The
resulting "GTP cap" stabilizes that end of the microtubule and
promotes further rapid growth. As free tubulin concentrations
decrease, the rate of growth also decreases, thereby allowing GTP hydrolysis to catch up. The resulting "GDP cap" at the
microtubule end is unstable and favors rapid depolymerization
(termed "catastrophe"). This increases the local concentration
of free, monomeric tubulin that "rescues" the microtubule
before it completely disappears and produces another short
period of microtubule elongation.

Dynamic instability allows the growing ends of microtubules to explore the cytoplasm and become stabilized when
they contact stabilizing structures, such as kinetochores on
chromosomes early in mitosis (see Chapter 3).

microtubules (anterograde transport); **cytoplasmic dyneins**
carry material along microtubules in the opposite direction
(retrograde transport), generally toward the nucleus. Important roles for this system include extending the ER from the
nuclear envelope to the plasmalemma and moving vesicles to
and through the Golgi apparatus.

Microfilaments (Actin Filaments)

Microfilaments are composed of **actin** subunits and allow
motility and most contractile activity in cells, using reversible
assembly of the actin filaments and interactions between these
filaments and associated **myosin** family proteins. Actin filaments are thin (5-7 nm diameter), polarized polymers, shorter
and more flexible than microtubules (Figure 2–22). They are
composed of globular G-actin monomers that assemble in
the presence of K^+ and Mg^{2+} into a double-stranded helix of
filamentous F-actin (Table 2–4). G-actin is generally added to
preexisting filaments, but new filaments can be formed from
a pool of G-actin by the action of nucleating proteins, such as
formin. Another nucleating factor, a complex of polypeptides
called Arp2/3, also binds to the side of preexisting actin filaments and induces a new F-actin branch, a process which can
lead to formation of a microfilament network.

Like microtubules actin filaments are highly dynamic.
Monomers are added rapidly at the (+) or barbed end, with
ATP hydrolysis at each addition; at the same time monomers
dissociate at the (–) or pointed end. This leads to migration of
subunits through the polymer, which occurs rapidly in a process called *treadmilling* (Figure 2–25). In cells both the assembly and disassembly of subunits from F-actin are promoted by
other proteins, such as *profilin* and *cofilin*, respectively.

Actin is very abundant in all cells and is concentrated in
networks of actin filaments and as free G-actin subunits near
the cell membrane, a cytoplasmic region often called the **cell
cortex**. Arp2/3 activity produces an important branched
network of microfilaments in this region. Microfilament-
rich cellular extensions, including surface folds or ruffles, are
important for endocytosis and microfilaments of the cortex
underlie endosomal trafficking. Cell movements on firm substrates involve sheet-like protrusions, or lamellipodia, in which
the concentrated actin filaments are continuous with deeper
parallel F-actin bundles called **stress fibers** (Figure 2–13c).

Actin-binding proteins, such as formin and others just
mentioned, change the dynamic physical properties of microfilaments, particularly their lengths and interactions with

FIGURE **2–24** Centrosome.

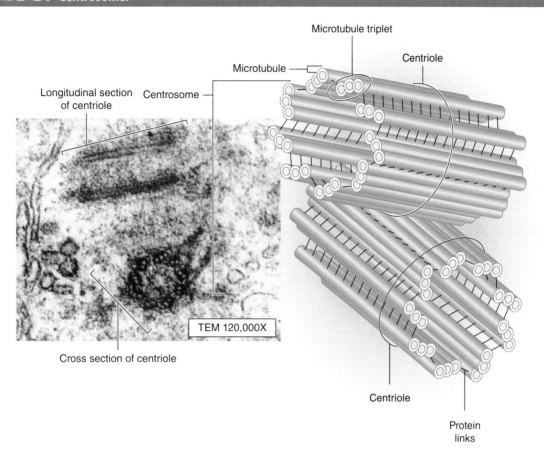

The **centrosome** is the microtubule-organizing center for the mitotic spindle and consists of paired centrioles. The TEM reveals that the two centrioles in a centrosome exist at right angles to each other in a dense matrix of free tubulin subunits and other proteins. Each centriole consists of **nine microtubular triplets**. In a poorly understood process, the centrosome duplicates itself and is divided equally during a cell's interphase, each half having a duplicated centriole pair. At the onset of mitosis, the two daughter centrosomes move to opposite sides of the nucleus and become the two poles of the mitotic spindle of microtubules attaching to chromosomes.

other structures, and this determines the viscosity and other mechanical properties of the local cytoplasm. Cross-linking within networks of F-actin increases cytoplasmic viscosity, while severing (and capping) the filaments tends to decrease viscosity. The lengths and other physical properties of actin filaments are controlled by various other types of actin-binding proteins, including those indicated in Figure 2–26.

Just as the molecular motors kinesin and dynein move structures along microtubules, various **myosin motors** use ATP to transport cargo along F-actin. Movement is usually toward the barbed (+) ends of actin filaments; myosin VI is the only known myosin that moves in the other direction. Interactions between F-actin and myosins form the basis for various cell movements:

■ Transport of organelles, vesicles, and granules in the process of *cytoplasmic streaming*
■ Contractile rings of microfilaments with myosin II constricting to produce two cells by *cytokinesis* during mitosis

■ Membrane-associated molecules of myosin I whose movements along microfilaments produce the cell surface changes during *endocytosis*

Stabilized arrays of actin filaments integrated with arrays of thicker (16-nm) myosin filaments permit very forceful contractions in specialized cells such as those of muscle (see Chapter 10).

Intermediate Filaments

The third class of cytoskeletal components includes filaments intermediate in size between the other two, with a diameter averaging 10 nm (Table 2–4). Unlike microtubules and actin filaments these **intermediate filaments** are stable, confer increased mechanical stability to cell structure, and are made up of different protein subunits in different cell types. More than a dozen proteins, ranging in size from 40 to 230 kDa, serve as subunits of various intermediate filaments and can be localized by immunohistochemistry in various cells.

FIGURE **2–25** Actin filament treadmilling.

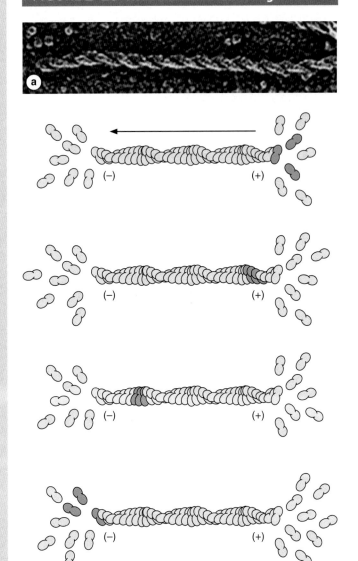

(a) Actin filaments or microfilaments are helical two-stranded polymers assembled from **globular actin subunits**.

(b) Assembly of actin filaments (F-actin) is polarized, with G-actin subunits added to the plus (+) end and removed at the minus (–) end. Even actin filaments of a constant length are highly dynamic structures, balancing G-actin assembly and disassembly at the opposite ends, with a net movement or flow along the polymer known as **treadmilling**.

(Figure 2–25a, used with permission from John Heuser, Washington University School of Medicine, St. Louis, MO.)

As indicated in Table 2–4, nearly all these subunits are coiled, rod-like dimers that form antiparallel tetramers, which self-assemble into large cable-like bundles or protofibrils stabilized by further lateral interactions. Table 2–5 lists six classes of intermediate filament proteins forming rod-like subunits, their sizes and cell distributions, and diseases that result from their disruption.

FIGURE **2–26** Roles of major actin-binding proteins in regulating the organization of microfilaments.

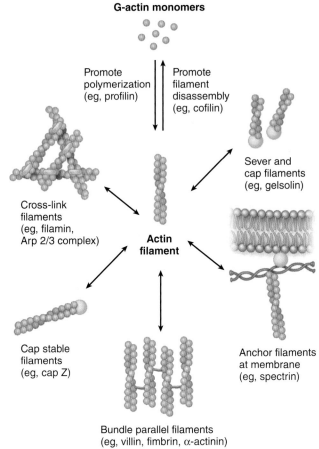

A large number of proteins regulate the assembly of microfilaments and the interactions of these filaments with one another. By altering microfilament length and cross-linking, such proteins greatly influence physical properties of the local cytoplasm.

Intermediate filament proteins with particular biological, histological, or pathological importance include the following:

■ **Keratins** (Gr. *keras*, horn) or **cytokeratins** are a diverse family of acidic and basic isoforms that compose heterodimer subunits of intermediate filaments in all epithelial cells (see Chapter 4). They are encoded by over 30 related genes and produce filaments with different chemical and immunologic properties for various functions. Intermediate filaments of keratins form large bundles (tonofibrils) that attach to certain junctions between epithelial cells (Figure 2–27). In skin epidermal cells, cytokeratins accumulate during differentiation in the process of *keratinization,* producing an outer layer of nonliving cells that reduces dehydration. Keratinization of skin made terrestrial life possible in the course of evolution. Keratinization also provides some protection from minor abrasions and produces various hard

TABLE **2-5**	Major classes and representatives of intermediate filament proteins, their sizes and locations.			
Class	Protein	Size (kDa)	Cell Distribution	Disease Involvement (If Known)
I	Acidic cytokeratin	40-65	Epithelial cells	Certain skin-blistering disorders
II	Basic cytokeratin	51-68	Epithelial cells	Keratoderma; corneal dystrophy
III	Desmin	53	Muscle cells	Myopathies
	Synemin	190	Muscle cells	
	GFAP	50	Astrocytes (less in other glial cells)	Alexander disease
	Peripherin	57	Neurons	
	Vimentin	54	Mesenchymal cells	
IV	NF-L	68	Neurons	
	NF-M	110	Neurons	
	NF-H	130	Neurons	
	α-internexin	55	Embryonic neurons	
V	Lamins	62-72	Nuclei of all cells	Cardiomyopathy; muscular dystrophies; progeria
VI	Nestin	230	Some stem and embryonic cells	

FIGURE **2-27** Intermediate filaments of keratin.

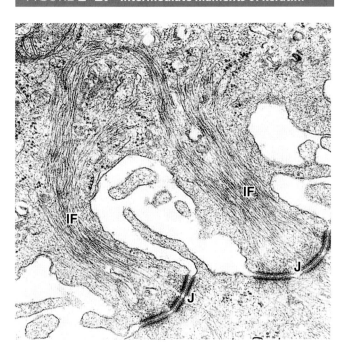

Intermediate filaments (IF) display an average diameter of 8-10 nm, between that of actin filaments and microtubules, and serve to provide mechanical strength or stability to cells. A large and important class of intermediate filaments is composed of **keratin** subunits, which are prominent in epithelial cells. Bundles of keratin filaments called **tonofibrils** associate with certain classes of intercellular junctions **(J)** common in epithelial cells and are easily seen with the TEM, as shown here in two extensions in an epidermal cell bound to a neighboring cell. (60,000X)

protective structures of skin, such as nails (as well as feathers, beaks, horns, and the scales of reptiles).

- **Vimentin** is the most common class III intermediate filament protein and is found in most cells derived from embryonic mesenchyme. Important vimentin-like proteins include **desmin** found in almost all muscle cells and **glial fibrillar acidic protein (GFAP)** found especially in astrocytes, supporting cells of central nervous system tissue. Desmin filaments of a cultured cell are shown immunocytochemically in Figure 1–12a.
- **Neurofilament** proteins of three distinct sizes make heterodimers that form the subunits of the major intermediate filaments of neurons.
- **Lamins** are a family of seven isoforms present in the cell nucleus, where they form a structural framework called the **nuclear lamina** just inside the nuclear envelope (see Chapter 3).

❯❯ MEDICAL APPLICATION

The presence of a specific type of intermediate filament in tumors can often reveal the cellular origin of the tumor, information important for diagnosis and treatment of the cancer. Identification of intermediate filament proteins by means of immunocytochemical methods is a routine procedure. One example is the use of GFAP to identify **astrocytomas**, the most common type of brain tumor.

❯ INCLUSIONS

Cytoplasmic **inclusions** contain accumulated metabolites or other substances, but unlike organelles have little or no metabolic activity themselves. Most inclusions are transitory

FIGURE **2–28** Cellular inclusions.

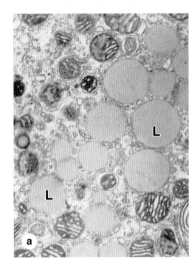
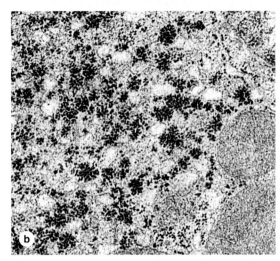
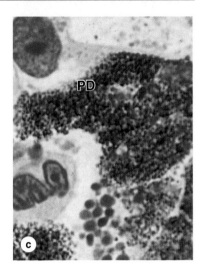

Inclusions are cytoplasmic structures or deposits filled with stored macromolecules and are not present in all cells.

(a) Lipid droplets are abundant in cells of the adrenal cortex and appear with the TEM as small spherical structures with homogenous matrices **(L)**. Mitochondria are also seen here. As aggregates of hydrophobic lipid molecules these inclusions are enclosed by a single monolayer of phospholipids with various peripheral proteins, including enzymes for lipid metabolism. In routine processing of tissue for paraffin sections, fat droplets are generally removed, leaving empty spaces in the cells. Common fat cells have cytoplasm essentially filled with one large lipid droplet. (X19,000)

(b) TEM of a liver cell cytoplasm shows numerous individual or clustered electron-dense particles representing **glycogen granules**, although these granules lack membrane. Glycogen granules

usually form characteristic aggregates such as those shown. Glycogen is a ready source of energy, and such granules are often abundant in cells with high metabolic activity. (X30,000)

(c) Pigment deposits (PD) occur in many cell types and may contain various complex substances, such as **lipofuscin** or melanin. Lipofuscin granules represent an accumulating by-product of lysosomal digestion in long-lived cells, but melanin granules serve to protect cell nuclei from damage to DNA caused by light. Many cells contain pigmented deposits of **hemosiderin granules** containing the protein ferritin, which forms a storage complex for iron. Hemosiderin granules are very electron dense, but with the light microscope they appear brownish and resemble lipofuscin. The liver cells shown have large cytoplasmic regions filled with pigment deposits, which probably represent iron-containing hemosiderin. (X400; Giemsa)

structures not enclosed by membrane. Features of important and commonly seen inclusions are shown Figure 2–28 and include:

- **Lipid droplets**, accumulations of lipid filling adipocytes (fat cells) and present in various other cells,
- **Glycogen granules**, aggregates of the carbohydrate polymer in which glucose is stored, visible as irregular clumps of periodic acid-Schiff (PAS)—positive or electron-dense material in several cell types, notably liver cells, and
- Pigmented deposits of naturally colored material, including **melanin**, dark brown granules which in skin serve to protect cells from ultraviolet radiation; **lipofuscin**, a pale brown granule found in many cells, especially in stable nondividing cells (eg, neurons, cardiac muscle), containing a complex mix of material partly derived from residual bodies after lysosomal digestion; and **hemosiderin**, a dense brown aggregate of denatured ferritin proteins

with many atoms of bound iron, prominent in phagocytic cells of the liver and spleen, where it results from phagocytosis of red blood cells

❯❯ MEDICAL APPLICATION

A condition termed **hemosiderosis**, in which the iron-containing inclusion **hemosiderin** occurs in cells of organs throughout the body, may be seen with increased uptake of dietary iron, impaired iron utilization, or with excessive lysis of red blood cells. Hemosiderosis itself does not damage cell or organ function. However, extreme accumulations of iron in cellular hemosiderin can lead to disorders such as hemochromatosis and iron overload syndrome, in which tissues of the liver and other organs are damaged.

A summary of the major structural and functional features of all cytoplasmic components is presented in Table 2–6.

TABLE **2–6**	Summary of cellular structural components.		
Component	Structure	Major Function	Appearance
Plasma membrane	Phospholipid bilayer containing cholesterol and proteins (integral and peripheral) and some carbohydrates (externally); forms a selectively permeable boundary of the cell	Acts as a physical barrier to enclose cell contents; regulates material movement into and out of the cell; establishes and maintains an electrical charge difference across the plasma membrane; functions in cell communication	
Cilia	Short, numerous membrane extensions supported by microtubules, which occur on exposed membrane surfaces of some cells	Move substances (eg, mucus, and dissolved materials) over the cell surface	
Flagellum	Long, singular membrane extension supported by microtubules; present on sperm cells	Propels sperm	
Microvilli	Numerous thin membrane folds projecting from the free cell surface; supported by microfilaments	Increase membrane surface area for greater absorption	
Nucleus	Large structure enclosed within a double membrane; contains chromatin, nucleolus, and nucleoplasm	Houses the DNA that serves as the genetic material for directing protein synthesis	
Nuclear envelope	Double membrane boundary between cytoplasm and nuclear contents; continuous with rough endoplasmic reticulum	Separates nucleus from cytoplasm	
Nuclear pores	Openings through the nuclear envelope	Allow passage of materials between the cytoplasm and nucleoplasm, including ribonucleic acid (RNA), protein, ions, and small water-soluble molecules	
Nucleolus	Large, prominent structure within the nucleus	Functions in synthesis of ribosomes	
Cytoplasm	Contents of cells between the plasma membrane and nuclear envelope	Responsible for many cellular processes	
Cytosol	Viscous fluid medium with dissolved solutes (eg, ions, proteins, carbohydrates, lipids)	Provides support for organelles; serves as the viscous fluid medium through which diffusion occurs	
Organelles	Membrane-bound and non-membrane-bound structures	Carry out specific metabolic activities of the cell	
Rough endoplasmic reticulum (rough ER)	Extensive interconnected membrane network that varies in shape (eg, cisternae, tubules); ribosomes attached on cytoplasmic surface	Modifies, transports, and stores proteins produced by attached ribosomes; these proteins are secreted, become components of the plasma membrane, or serve as enzymes of lysosomes	
Smooth endoplasmic reticulum (smooth ER)	Extensive interconnected membrane network lacking ribosomes	Synthesizes, transports, and stores lipids (eg, steroids); metabolizes carbohydrates; detoxifies drugs, alcohol, and poisons; forms vesicles and peroxisomes	

(Continued)

TABLE **2–6**	Summary of cellular structural components. (Continued)		
Component	**Structure**	**Major Function**	**Appearance**
Golgi apparatus	Series of several elongated, flattened saclike membranous structures	Modifies, packages, and sorts materials that arrive from the ER in transport vesicles; forms secretory vesicles and lysosomes	
Vesicles	Spherical-shaped membrane-bound sacs; contain various types of materials to be transported through the cell	Transport cellular material	
Lysosomes	Spherical-shaped membrane-bound organelles formed from the Golgi apparatus; contain digestive enzymes	Digest microbes or materials (eg, ingested by the cell, worn-out cellular components, or the entire cell)	
Peroxisomes	Smaller, spherical-shaped membrane-bound organelles formed from the ER or through fission; contain oxidative enzymes	Detoxify specific harmful substances either produced by the cell or taken into the cell; engage in beta oxidation of fatty acids to acetyl CoA	
Mitochondria	Double membrane-bound organelles containing a circular strand of DNA (genes for producing mitochondrial proteins)	Synthesize most ATP during aerobic cellular respiration by digestion of fuel molecules (eg, glucose) in the presence of oxygen	
Ribosomes	Organelles composed of both protein and ribosomal RNA (rRNA) that are organized into both a large and small subunit; may be bound to a membrane or free in cytosol	Engage in protein synthesis: Bound ribosomes produce proteins that are secreted, incorporated into plasma membrane, and within lysosomes; free ribosomes produce proteins used within the cell	Bound ribosomes; Free ribosomes
Cytoskeleton	Organized network of protein filaments and hollow tubules, including microfilaments, intermediate filaments, and microtubules	Maintains intracellular structural support and organization of cells; participates in cell division; facilitates movement	Cytoskeleton; Intermediate filament; Microfilament; Microtubule
Microfilaments	Actin protein monomers organized into two thin, intertwined protein filaments (actin filaments)	Maintain cell shape; support microvilli; separate two cells during cytokinesis (a process of cell division); facilitate change in cell shape; participate in muscle contraction	
Intermediate filaments	Various protein components	Provide structural support; stabilize junctions between cells	
Microtubules	Hollow cylinders composed of tubulin protein	Maintain cell shape and rigidity; organize and move organelles; support cilia and flagella; participate in vesicular transport; separate chromosomes during the process of cell division	
Centrosome	Amorphous region adjacent to nucleus; contains a pair of centrioles	Organizes microtubules; participates in mitotic spindle formation during cell division	Centriole; Centrosome
Proteasomes	Large, barrel-shaped protein complexes located in both the cytosol and nucleus	Degrade and digest damaged or unneeded proteins; ensure quality of exported proteins	
Inclusions	Aggregates of specific types of molecules (eg, melanin protein, glycogen, or lipid)	Serve as temporary storage for these molecules	Variable appearance

The Cytoplasm SUMMARY OF KEY POINTS

- **Cell differentiation** is the process by which cells of an embryo become specialized structurally to augment specific cytoplasmic activities for functions at the level of tissues and organs.
- **Organelles** are metabolically active structures or complexes, with or without membranes, in the cytoplasm of eukaryotic cells.

Plasma Membrane

- The **plasma membrane (cell membrane** or **plasmalemma)** is the lipid bilayer with embedded proteins that surrounds a cell and is seen only with the TEM.
- The **lipid bilayer** forms from amphipathic **phospholipids, stabilized** by **cholesterol**, and contains many **embedded (integral) proteins** and many **peripheral proteins** on its cytoplasmic surface.
- Membrane proteins move laterally within the lipid bilayer, with less movement in areas referred to as **lipid rafts**, which have higher concentrations of cholesterol and saturated fatty acids.
- Integral membrane proteins include **receptors** for external ligands, **channels** for passive or active movement of molecules across the membrane, and **pumps** for active membrane transport.
- **Endocytosis** is cellular uptake of macromolecules or fluid by plasma membrane engulfment or invagination, followed by the "pinching off" of a filled membranous vesicle in the cytoplasm.
- Major types of endocytosis include **phagocytosis** (uptake of particulate material), **pinocytosis** (uptake of dissolved substances), and **receptor-mediated endocytosis** (uptake of specific molecules bound to integral membrane receptor proteins).
- **Exocytosis** is a type of cellular secretion in which cytoplasmic membrane vesicles fuse with the plasma membrane and release their contents to the extracellular space.
- All types of **cell** signaling use membrane receptor proteins that are often linked to enzymes such as kinases or adenylyl cyclase whose activities initiate intracellular signaling pathways.

Ribosomes

- The two **ribosomal subunits**, each a complex of rRNA and many proteins, attach to mRNA and translate that message into protein.
- Multiple ribosomes on the same mRNA make up a **polyribosome (polysome)**, and an abundance of these produces basophilic cytoplasm after H&E staining.

Endoplasmic Reticulum

- The ER is a convoluted network of membrane enclosing continuous spaces called **cisternae** and extending from the nucleus to the plasma membrane.
- **Rough ER** has a **granular, basophilic cytoplasmic surface** due to the presence of polysomes making most membrane proteins, proteins in certain other organelles, or for exocytosis; RER is always well developed in cells actively secreting proteins.
- Proteins to be processed through the RER contain initial **signal peptides** which bind receptors in the ER membrane, localizing them to that organelle.
- After **translocation** across the membrane into the cisterna, the proteins undergo **posttranslational modification and folding** in a process monitored by RER molecular chaperones and enzymes.
- **Smooth ER (SER)** lacks ribosomes, but includes enzymes for **lipid and glycogen metabolism**, for **detoxification reactions**, and for temporary **Ca^{2+} sequestration**.

Golgi Apparatus

- The **Golgi apparatus** is a dynamic organelle consisting of stacked membranous cisternae in which proteins made in RER are **processed** further and **packaged** for secretion or other roles.
- Proteins in **transport vesicles** enter the cis or receiving face of the Golgi, move through medical cisternae of the Golgi network for enzymatic modifications, and are released in other vesicles at the trans face.

- Vesicle movement through the Golgi apparatus is guided by specific **coat proteins** such as COPII and COPI.
- Important protein modifications in the Golgi apparatus include **sulfation** and many **glycosylation** reactions.
- Modified proteins leave the Golgi apparatus after packaging in vesicles with coat proteins that direct movement to lysosomes, the plasma membrane, or secretion by exocytosis.

Lysosomes

- **Primary lysosomes** emerge from the Golgi apparatus containing inactive acid hydrolases specific for degrading a wide variety of cellular macromolecules.
- **Secondary lysosomes** are more heterogeneous, having fused with vesicles produced by endocytosis that contain material to be digested by the hydrolytic enzymes.
- During **autophagy**, lysosomes digest unneeded or nonfunctional organelles after these are surrounded by membrane that then fuses with a lysosome.
- Products of digestion in secondary lysosomes are released to the cytoplasm for reuse; final condensed vesicles containing any indigestible molecules are called **residual bodies**.

Proteasomes

- Proteasomes are small cytoplasmic protein complexes which degrade improperly folded proteins after they are tagged with the polypeptide **ubiquitin**.

Mitochondria

- **Mitochondria** are the major sites of **ATP synthesis** and are abundant in cells or cytoplasmic regions where large amounts of energy are expended.
- Mitochondria are usually **elongated organelles** and form by fission of preexisting mitochondria.
- Mitochondria have two membranes: a **porous outer membrane** encloses the intermembrane space and an **inner membrane with many folds (cristae)** enclosing a gel-like matrix.
- The **mitochondrial matrix** contains enzymes for β-oxidation of fatty acids and the citric acid (Krebs) cycle.
- The inner membrane includes enzyme assemblies of the **electron-transport system and ATP synthase**.
- Mitochondria of stressed cells may release **cytochrome c** from the inner membrane, triggering a regulated series of events culminating in cell death (**apoptosis**).

Peroxisomes

- Peroxisomes are small spherical organelles containing enzymes for various metabolic reactions, notably **for oxidation and detoxification**, and **catalase** that breaks down the H_2O_2 resulting from those reactions.

Cytoskeleton

- The cytoskeleton contains three types of polymers: (1) **microtubules** 25 nm in diameter; (2) actin filaments or **microfilaments** (5-7 nm); and (3) **intermediate filaments** (8-10 nm).
- Microtubules are semirigid tubular structures with walls composed of **polymerized tubulin** heterodimers; their structure is often very dynamic, with steady addition and dissociation of tubulin.
- Microtubules are important in **maintaining cell shape** and as **tracks for transport** of vesicles and organelles by the **motor proteins kinesin** and **dynein**.
- Microfilaments are short, flexible, highly dynamic filaments of **actin subunits**, in which changes in length and interactions with binding proteins regulate cytoplasmic viscosity and movement.
- **Myosins** are **motor proteins** that bind and move along actin filaments, carrying vesicles or producing cytoplasmic movement.

- Movements of cytoplasm produced by actin filaments and myosins are important for endocytosis, cell cleavage after mitosis, and cell locomotion on substrates.
- Intermediate filaments are the **most stable** cytoskeletal component, conferring strong mechanical stability to cells.
- Intermediate filaments are composed of various protein subunits in different cells; they include **vimentin**; nuclear **lamins**; **neurofilament proteins**; and **keratins**, which are especially important in epithelial cells.

Inclusions

- Unlike organelles, inclusions are **not metabolically active** and are primarily **storage sites**, such as lipid droplets, glycogen granules, pigment granules, or residual bodies (also called **lipofuscin**).

The Cytoplasm ASSESS YOUR KNOWLEDGE

1. In transmission EM preparations of cells the cell membrane often appears as a trilaminar structure having two parallel dark-staining components on either side of an unstained middle layer. This central poorly stained region of the membrane is primarily responsible for which of the following functions?

   a. Creation of a barrier to water-soluble molecules
   b. Binding by cellular receptions to specific ligands
   c. Catalyzing membrane-associated activities
   d. Transport of ions
   e. Connections to the cytoskeleton

2. Chaperonins are cytoplasmic proteins most likely to be found in which of the following organelles?

   a. Lysosomes
   b. Golgi complexes
   c. Rough endoplasmic reticulum
   d. Smooth endoplasmic reticulum
   e. Mitochondria

3. Which of the following best defines the term "exocytosis"?

   a. The discharge of ions or small molecules from a cell by protein pumps in the cell membrane
   b. The uptake of material at one domain of a cell's surface and its discharge from the opposite side of the cell
   c. The process by which proteins move from one cytoplasmic compartment to another
   d. The discharge of proteins in cytoplasmic vesicles from a cell following fusion of the vesicles with the plasmalemma
   e. Diffusion of lipid-soluble molecules from a cell across the cell membrane

4. Cytoplasm often stains poorly because its lipid content is removed by the organic solvents used in the clearing step in routine histological preparations. This problem is most likely to occur with cytoplasmic regions rich in which of the following organelles?

   a. Free polysomes
   b. Mitochondria
   c. Lysosomes
   d. Smooth endoplasmic reticulum
   e. Rough endoplasmic reticulum

5. Polarity in microtubules is important in determining which of the following?

   a. The strength of vinblastine binding to microtubules
   b. The velocity of transport along microtubules with myosin motors
   c. The overall dynamic instability of the microtubules
   d. The linkage of microtubules to intermediate filaments
   e. The direction of vesicular transport along microtubules

6. Which of the following proteins is/are most likely to have initially contained a "signal peptide" that bound a "signal recognition particle" during its translation?

   a. An enzyme of the respiratory chain
   b. Lamins
   c. Proteins in secretory granules
   d. F-actin
   e. Proteins in the large ribosomal subunit

7. Vesicles of a Golgi apparatus that are destined to become part of other organelles most likely have which of the following on their membranes?

   a. Channel proteins
   b. Clathrin
   c. COP II
   d. Actin
   e. GTP

8. About 3 years ago, a 39-year-old construction worker became increasingly uncoordinated. His wife describes bouts of depression and apathy beginning about a decade ago. Laboratory tests are normal. MRI and CT reveal striatal and caudate atrophy with "boxcar ventricles." His mini-mental status examination score is 24/30. The cranial nerve examination shows dysarthria, saccadic extraocular eye movements, and a hyperactive gag reflex. There is increased tone in all extremities. Polymerase chain reaction reveals one normal band with 20 CAG (trinucleotide) repeats and the other with 49 CAG repeats. Modulation of respiration and mitochondrial membrane potential, and bioenergetic failure are associated with the abnormal gene in this disease. Which of the following mechanisms used to establish the mitochondrial electrochemical gradient may be altered in this disease?

   a. The action of ATP synthase
   b. Transfer of electrons from NADH to O_2 in the intermembrane space
   c. Pumping of protons into the mitochondrial matrix by respiratory chain activity
   d. Proton-translocating activity in the inner membrane
   e. Transport of ATP out of the matrix compartment by a specific transporter

9. A 56-year-old man has been taking atorvastatin because of a poor lipid profile and a family history of cardiovascular disease. The statin family of drugs enhances endocytosis of low density lipoprotein (LDL) from the blood. Endocytosis of LDL differs from phagocytosis of bacterial cells in which of the following ways?

   a. Use of membrane-enclosed vesicles in the uptake process
   b. Coupling with the lysosomal system
   c. Dependence on acidification
   d. Use of clathrin-coated pits
   e. Use of hydrolases

10. A 14-year-old boy is diagnosed with epidermolysis bullosa simplex (EBS). His skin blisters easily with rubbing or scratching. Blisters occur primarily on his hands and feet and heal without leaving scars. Genetic analysis shows mutations in the *KRT5* and *KRT14* genes, which code keratin 5 and keratin 14. What is the primary function of those proteins?

    a. Generate movement
    b. Provide mechanical stability
    c. Carry out nucleation of microtubules
    d. Stabilize microtubules against disassembly
    e. Transport organelles within the cell

Answers: 1a, 2c, 3d, 4d, 5e, 6c, 7c, 8d, 9d, 10b

3 The Nucleus

Containing the code for all of a cell's enzymes and other proteins, the **nucleus** is the command center of the cell. The nucleus also contains the molecular machinery to replicate the DNA and to synthesize and process all types of RNA. During interphase, pore complexes in the membrane enclosing the nucleus regulate macromolecular transfer between the nuclear and cytoplasmic compartments. Mature RNA molecules pass into the cytoplasm for their roles in protein synthesis, while proteins needed for nuclear activities are imported there from the cytoplasm. Restricting protein synthesis to the cytoplasm helps ensure that newly made RNA molecules do not become involved in translation before processing is complete.

❯ COMPONENTS OF THE NUCLEUS

The nucleus usually appears as a large rounded or oval structure, often near the cell's center (Figure 3-1). Typically the largest structure within a cell, it consists of a **nuclear envelope** containing **chromatin**, the mass of DNA and its associated proteins, with one or more specialized regions of chromatin called **nucleoli**. In specific tissues the size and shape of nuclei normally tend to be uniform.

Nuclear Envelope

The **nuclear envelope** forms a selectively permeable barrier between the nuclear and cytoplasmic compartments. Electron microscopy reveals that the envelope has two concentric membranes separated by a narrow (30-50 nm) **perinuclear space** (Figures 3–2 and 3–3). This space and the outer nuclear membrane are continuous with the extensive cytoplasmic network of the rough endoplasmic reticulum (RER). Closely associated with the inner nuclear membrane is a highly organized meshwork of proteins called the **nuclear lamina** (Figure 3–4),

which stabilizes the nuclear envelope. Major components of this layer are the class of intermediate filament proteins called **lamins** that bind to membrane proteins and associate with chromatin in nondividing cells.

The inner and outer nuclear membranes are bridged at **nuclear pore complexes** (Figures 3–2 through 3–6). Various core proteins of a nuclear pore complex, called **nucleoporins**, display 8-fold symmetry around the lumen. Although ions and small solutes pass through the channels by simple diffusion, the pore complexes regulate movement of macromolecules between the nucleus and cytoplasm. A growing cell has 3000-4000 such channels, each providing passage for up to 1000 macromolecules per second. Individual pores permit molecular transfer in both directions simultaneously. Macromolecules shipped out of the nucleus include ribosomal subunits and other RNAs associated with proteins, while inbound traffic consists of chromatin proteins, ribosomal proteins, transcription factors, and enzymes. Using mechanisms similar to that by which specific proteins are recognized and translocated across the RER membrane, proteins of complexes destined for the cytoplasm have specific nuclear export sequences and proteins to be imported have nuclear localization sequences. Such sequences bind specifically to transport proteins (importins, exportins, etc) that in turn interact with proteins of the pore complexes for transfer across the nuclear envelope. Energy for the transport is derived from guanosine 5′-triphosphate (GTP), with specific GTPases helping provide directionality to the transfer.

Chromatin

Chromatin consists of DNA and all of the associated proteins involved in the organization and function of DNA. In humans each cell's chromatin (except that of eggs and sperm) is divided among 46 **chromosomes** (23 pairs). After DNA replication but before cell division, each chromosome consists of two

FIGURE 3-1 Nuclei of large, active cells.

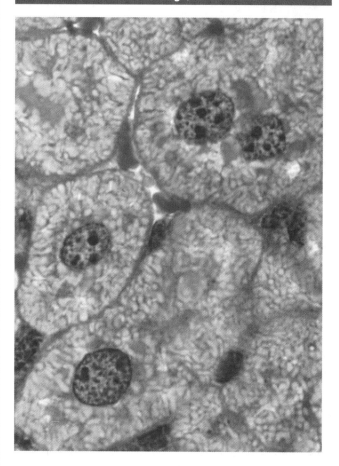

Liver cells have large, central nuclei. One or more highly basophilic nucleoli are visible within each nucleus, indicating intense protein synthesis by these cells. Most of the chromatin is light staining or euchromatic, with small areas of more darkly stained heterochromatin scattered throughout the nucleus and just inside the nuclear envelope. This superficial heterochromatin allows the boundary of the organelle to be seen more easily by light microscopy. One cell here has two nuclei, which is fairly common in the liver. (X500; Pararosaniline–toluidine blue)

identical chromatin units called **chromatids** held together by complexes of cohesin proteins.

DNA of each human cell is approximately 2 m long, with 3.2 billion base pairs (bp), and therefore must be extensively packaged within the nucleus. This occurs initially by the DNA associating with sets of small basic proteins called **histones**. The structural unit of DNA and histones is called the **nucleosome** (Figures 3–7 and 3–8), which has a core of eight histones (two copies each of histones H2A, H2B, H3, and H4), around which is wrapped about 150 bp of DNA. Each nucleosome also has a larger histone (H1) associated with both the wrapped DNA and the surface of the core. In the EM the series of nucleosomes on DNA resembles "beads

on a string", with 50-80 bp of linker DNA separating each bead. Nucleosomes are structurally dynamic; modification and rearrangement of the histones allows temporary unwrapping of the DNA and arrival of enzymes and other proteins required for replication and gene transcription.

DNA wrapped around nucleosomes is coiled further for greater compaction within the nucleus and for general regulation of gene activity. The 10-nm fiber of nucleosomes and DNA undergoes helical folding to yield a fiber with a diameter of 30 nm (Figure 3–8). Beyond the 30-nm fiber chromatin structure is much less well understood, but it does form nearly ubiquitous kilobase-sized and larger loops of coiled DNA, some of which are unstable and may reflect gene activity rather than a distinct hierarchical level of chromatin structure. Many such loops are tethered to a central scaffold containing various proteins, including condensins, which promote compaction of chromatin. Further packaging during the first phase of cell division causes chromosomes to become visible as discrete structures by light microscopy (Figure 3–8).

Microscopically two categories of chromatin can be distinguished in nuclei of most nondividing cells (Figure 3–3). **Euchromatin** is visible as finely dispersed granular material in the electron microscope and as lightly stained basophilic areas in the light microscope. **Heterochromatin** (Gr. *heteros*, other + *chroma*, color) appears as coarse, electron-dense material in the electron microscope and as intensely basophilic clumps in the light microscope.

DNA in the more open structure of euchromatin is rich with genes, although not all of the genes are transcribed in all cells. Heterochromatin is always more compact than euchromatin, shows little or no transcriptional activity, and includes at least two types of genomic material called constitutive and facultative heterochromatin. *Constitutive heterochromatin* is generally similar in all cell types and contains mainly repetitive, gene-poor DNA sequences, including the large chromosomal regions called **centromeres** and **telomeres**, which are located near the middle (most often) and at the ends of chromosomes respectively. *Facultative heterochromatin* contains other regions of DNA with genes where transcription is variably inactivated in different cells by epigenetic mechanisms and can undergo reversible transitions from compact, transcriptionally silent states to more open, transcriptionally active conformations.

The ratio of heterochromatin to euchromatin seen with nuclear staining can provide a rough indicator of a cell's metabolic and biosynthetic activity (Figure 3–3). Euchromatin predominates in active cells such as large neurons, while heterochromatin is more abundant in cells with little synthetic activity such as circulating lymphocytes. Facultative heterochromatin also occurs in the small, dense "sex chromatin" or **Barr body** which is one of the two large X chromosomes present in human females but not males. The Barr body remains tightly coiled, while the other X chromosome is uncoiled, transcriptionally active, and not visible. Cells of males have one X chromosome and one Y chromosome; like the other chromosomes, the single X chromosome remains largely euchromatic.

FIGURE **3–2** **Relationship of nuclear envelope to the rough ER (RER).**

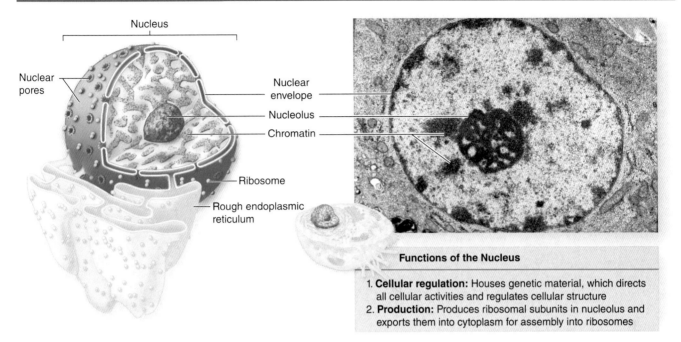

Functions of the Nucleus

1. **Cellular regulation:** Houses genetic material, which directs all cellular activities and regulates cellular structure
2. **Production:** Produces ribosomal subunits in nucleolus and exports them into cytoplasm for assembly into ribosomes

Three-dimensional representation of a cell nucleus shows a single large nucleolus and the distribution of the nuclear pores in the nuclear envelope. The outer membrane of the nuclear envelope is continuous with the RER. (TEM X20,000)

>> **MEDICAL APPLICATION**

Barr bodies or gender chromatin permit gender to be determined microscopically in patients whose external sex organs do not permit that determination, as in hermaphroditism and pseudohermaphroditism. Sex chromatin analysis also helps reveal other anomalies involving the sex chromosomes, such as the presence of XXY chromosomes (Klinefelter syndrome), which causes testicular abnormalities and azoospermia (absence of sperm).

Although much heterochromatin tends to be concentrated near the nuclear lamina, evidence for spatial organization of chromatin is not normally seen. Recent in situ hybridization studies of cultured human fibroblast nuclei, using differently labeled fluorescent probes for sequences on each individual chromosome, have revealed that these structures occupy discrete **chromosomal territories** within dispersed chromatin (Figure 3–9). Such studies show further that chromosomal domains with few genes form a layer beneath the nuclear envelope, while domains with many active genes are located deeper in the nucleus.

The X and Y sex chromosomes contain genes determining whether an individual will develop as a female or a male. In addition to the pair of sex chromosomes, cells contain 22 pairs of autosomes. Each pair contains one chromosome originally derived from the mother and one from the father. The members

of each chromosomal pair are called **homologous** because, although from different parents, they contain forms (alleles) of the same genes. Cells of most tissues (somatic cells) are considered **diploid** because they contain these pairs of chromosomes. Geneticists refer to diploid cells as $2n$, where n is the number of unique chromosomes in a species, 23 in humans. Sperm cells and mature oocytes (germ cells) are **haploid**, with half the diploid number of chromosomes, each pair having been separated during meiosis (described later in the chapter).

Microscopic analysis of chromosomes usually begins with cultured cells arrested in mitotic metaphase by colchicine or other compounds that disrupt microtubules. After processing and staining the cells, the condensed chromosomes of one nucleus are photographed by light microscopy and rearranged digitally to produce a **karyotype** in which stained chromosomes can be analyzed (Figure 3–10).

>> **MEDICAL APPLICATION**

Karyotyping is important for many prenatal diagnoses, in which chromosomal analysis of cultured cells from the fetus or amnion can detect certain genetic anomalies. As with karyotypes of adults, missing or extra chromosomes and chromosomal deletions or translocations are readily seen. New methods of chromosomal staining and molecular techniques such as fluorescence in situ hybridization (FISH) are continuously being developed and used for cytogenetic diagnosis.

FIGURE 3–3 Ultrastructure of a nucleus.

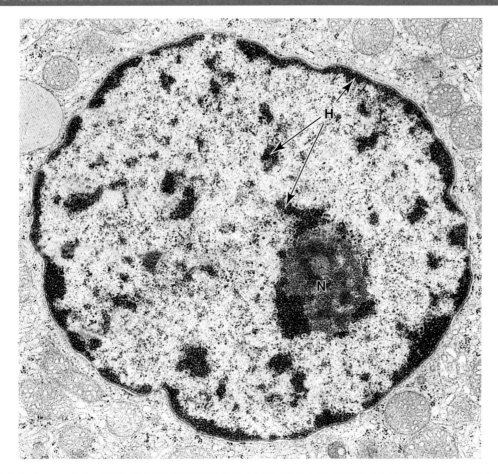

Regions of euchromatin and heterochromatin display variable electron densities with the transmission electron microscope (TEM). An active nucleus typically has much diffuse, light-staining euchromatin and smaller subdomains of electron-dense heterochromatin (**H**), with many of these associated at the periphery associated with the nuclear lamina. The more heterogeneous electron-dense subdomain is the nucleolus (**N**), the site of rRNA synthesis, and ribosomal subunit assembly. (X25,000)

Nucleolus

The **nucleolus** is a generally spherical, highly basophilic subdomain of nuclei in cells actively engaged in protein synthesis (Figures 3–1 through 3–3). The intense basophilia of nucleoli is due not to heterochromatin but to the presence of densely concentrated ribosomal RNA (rRNA) that is transcribed, processed, and assembled into ribosomal subunits. Chromosomal regions with the genes for rRNA organize one or more nucleoli in cells requiring intense ribosome production for protein synthesis during growth or secretion. Ultrastructural analysis of an active nucleolus reveals fibrillar and granular subregions with different staining characteristics that reflect stages of rRNA maturation, as described in Figure 3–11. Molecules of rRNA are processed in the nucleolus and very quickly associate with the ribosomal proteins imported from the cytoplasm via nuclear pores. The newly organized small and large ribosomal subunits are then exported back to the cytoplasm through those same nuclear pores.

> ❯❯ **MEDICAL APPLICATION**
>
> Tissues with either stable or rapidly renewing cell populations can include cells that become transformed to grow at a higher rate and in an uncoordinated manner. Such **neoplastic proliferation** typically follows damage to the DNA of proto-oncogenes and failure of the cells to be eliminated. Neoplastic growth can be either benign (with slow growth and no invasiveness to neighboring organs) or malignant (with rapid growth and great capacity to invade other organs). **Cancer** is the common term for all malignant tumors.

FIGURE **3–4** The nuclear envelope, nuclear lamina, and nuclear pore complexes.

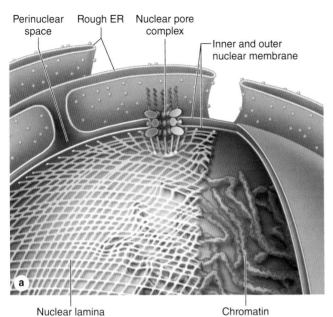

Perinuclear space   Rough ER   Nuclear pore complex

Inner and outer nuclear membrane

Nuclear lamina   Chromatin

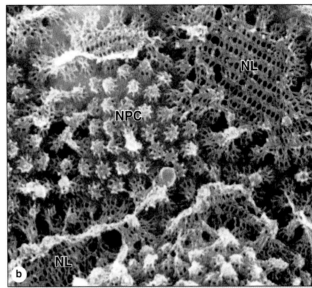

NL

NPC

NL

(a) Bound to the inner membrane of the nuclear envelope is the **nuclear lamina**, a meshwork assembled from lamins (class V intermediate filament proteins). **Nuclear pore complexes** contain more than 30 core proteins (nucleoporins), span both membranes of the nuclear envelope, and regulate the bidirectional transfer of macromolecular complexes between the nucleus and cytoplasm.

(b) Scanning EM of the inner nuclear membrane (nucleoplasmic face) showing portions of the nuclear lamina (**NL**) meshwork with

many embedded nuclear pore complexes (**NPC**). The preparation is from an actively growing amphibian oocyte. Nuclei of these very large cells can be isolated manually, facilitating ultrastructural studies of the nuclear envelope. (X100,000)

(*Used with permission from Dr M.W. Goldberg, Department of Biological and Biomedical Sciences, Durham University, UK.*)

FIGURE **3–5** Nuclear pores.

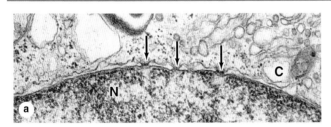

N

C

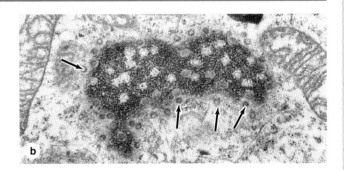

(a) A TEM section through the nuclear envelope between nucleus (**N**) and cytoplasm (**C**) shows its two-membrane structure. The electron-dense nuclear pore complexes bridging the nuclear envelope can also be seen (arrows). Electron-dense heterochromatin is adjacent to the envelope, except at the nuclear pores.

(b) A tangential section through a nuclear envelope shows the

nuclear pore complexes (arrows) and the electron-lucent patches in the peripheral heterochromatin, which represent the areas just inside pores. (X80,000)

FIGURE **3–6** **Cryofracture of nuclear envelope showing nuclear pores.**

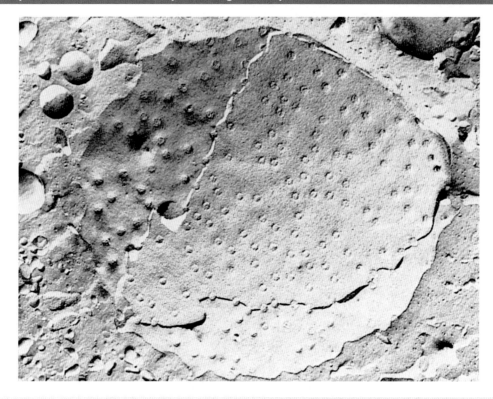

An electron micrograph obtained by freeze-fractured cell shows the two layers of the nuclear envelope and nuclear pores. The fracture plane occurs partly *between* the two nuclear envelope membranes (left) but mostly just *inside* the envelope with the chromatin removed. The size and distribution of the nuclear pore complexes are clearly seen. (X60,000)

❯❯ MEDICAL APPLICATION

Certain mutations in the gene coding for lamin A are associated with a subtype of the disorder progeria, which causes premature aging. In this and other rare "laminopathies," the nuclear envelope is abnormal, but how this is linked to the disorder is unclear. Laminopathies affect some tissues much more than others, although the lamins involved are in all the body's cells.

❯ THE CELL CYCLE

Before differentiation, most cells undergo repeated cycles of macromolecular synthesis (growth) and division (mitosis). The regular sequence of events that produce new cells is termed the **cell cycle**. Improved knowledge about how each phase of the cell cycle is controlled and how the quality of molecular synthesis, particularly DNA replication, is monitored has led to understanding the causes of many types of cancer, in which cells proliferate without those controls.

The cell cycle has four distinct phases: **mitosis** and periods termed G_1 (the time gap between mitosis and the beginning of DNA replication), **S** (the period of DNA synthesis), and

G_2 (the gap between DNA duplication and the next mitosis). The approximate durations of these phases in rapidly dividing human cells are illustrated in Figure 3–12. The G_1 phase, usually the longest and most variable part of the cycle, is a period of active RNA and protein synthesis, including proteins controlling progress through the cell cycle. Also in G_1, the cell volume, reduced by half during mitosis, returns to its previous size. The S phase is characterized by DNA replication, histone synthesis, and the beginning of centrosome duplication. In the relatively short G_2 phase, proteins required for mitosis accumulate. As new postmitotic cells specialize and differentiate, cell cycle activities may be temporarily or permanently suspended, with the cells sometimes referred to as being in the G_0 phase. Some differentiated cells, such as those of the liver, renew cycling under certain conditions; others, including most muscle and nerve cells, are *terminally differentiated*.

Cycling is activated in postmitotic G_0 cells by protein signals from the extracellular environment called **mitogens** or **growth factors** that bind to cell surface receptors and trigger a cascade of kinase signaling in the cells. The cells are then maintained at the *restriction point* at the G_1/S "boundary" until sufficient nutrients and enzymes required for DNA synthesis have accumulated, and when all is ready DNA replication (S phase) begins.

FIGURE **3–7** Components of a nucleosome.

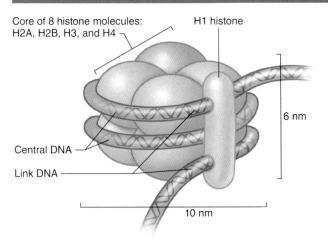

Core of 8 histone molecules: H2A, H2B, H3, and H4

H1 histone

Central DNA

Link DNA

6 nm

10 nm

Nucleosome is a structure that produces the initial organization of free double-stranded DNA into chromatin. Each nucleosome has an octomeric core complex made up of four types of **histones**, two copies each of H2A, H2B, H3, and H4. Around this core is wound DNA approximately 150 base pairs in length. One H1 histone is located outside the DNA on the surface of each nucleosome. DNA associated with nucleosomes in vivo thus resembles a long string of beads. Nucleosomes are very dynamic structures, with H1 loosening and DNA unwrapping at least once every second to allow other proteins, including transcription factors and enzymes, access to the DNA.

As shown in Figure 3–13, entry or progression through other phases of the cycle is also monitored at other specific *checkpoints*, where certain conditions must be met before the cell continues cycling. Overall cycling is regulated by a family of cytoplasmic proteins called **cyclins**. With different cyclins present during different cell cycle phases, each activates one or more specific **cyclin-dependent kinases (CDKs)**. Each activated CDK then phosphorylates specific proteins, including enzymes, transcription factors for specific sets of genes, and cytoskeletal subunits, triggering the activities that characterize the next phase of the cycle. When each successive set of activities is complete, the cyclin controlling that cell cycle phase is ubiquitinated and removed rapidly by proteasomes and a new cyclin that promotes activities for the next phase takes over. In this way diverse cellular activities are coordinated with specific phases of the cell cycle. The major cyclins, CDKs, and important target proteins are summarized in Table 3–1.

❯❯ MEDICAL APPLICATION

Many mitogenic growth factors for research are produced commercially from microorganisms or cells with recombinant DNA, and some have important medical uses. Important examples include analogs of granulocyte colony-stimulating

FIGURE **3–8** From DNA to chromatin.

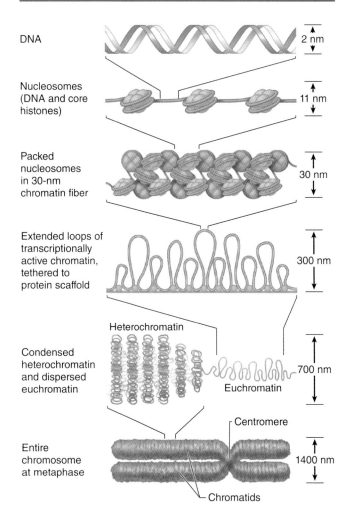

DNA — 2 nm

Nucleosomes (DNA and core histones) — 11 nm

Packed nucleosomes in 30-nm chromatin fiber — 30 nm

Extended loops of transcriptionally active chromatin, tethered to protein scaffold — 300 nm

Heterochromatin

Condensed heterochromatin and dispersed euchromatin — 700 nm

Euchromatin

Centromere

Entire chromosome at metaphase — 1400 nm

Chromatids

Several orders of DNA packing occur in chromatin and during chromatin condensation of mitotic prophase. The top drawing shows the 2-nm DNA double helix, followed by the association of DNA with histones to form 11-nm filaments of **nucleosomes** connected by the DNA ("beads on a string"). Nucleosomes on the DNA then interact in a manner not well understood to form a more compact **30-nm** fiber. DNA in such fibers forms various loops, some of which in **euchromatin** involve gene transcription. Loops remain tethered to and stabilized by interactions with **protein scaffolds** that eventually make up a central framework at the long axis of each chromosome. **Heterochromatin** is not transcribed and remains more highly condensed. The bottom drawing shows a metaphase chromosome, with maximum packing of DNA. The **chromosome** consists of two **chromatids** held together at a constriction called the **centromere**.

factor (G-CSF), which stimulates neutrophil production in immunocompromised patients, and erythropoietin, which can stimulate red blood cell formation in patients with anemia.

FIGURE 3–9   Chromosome territories of a human fibroblast nucleus.

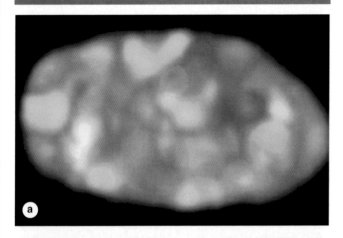

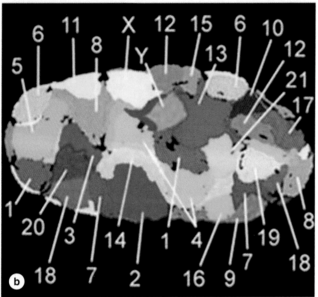

Fluorescence in situ hybridization (FISH) can be used with a combination of labeled probes, each specific for sequences on different chromosomes. A nucleus of a cultured human fibroblast was processed by 24-color FISH, photographed by confocal microscopy in appropriate channels, and the results superimposed to form an RGB (red-green-blue) image **(a)** of the 24 differently labeled chromosome types (1-22, X, and Y). Individual chromosome territories in the image were identified and false-colored after classification by software developed for such analyses **(b)**.

   (*Used with permission from Dr Thomas Cremer, Department of Biology II, Anthropology and Human Genetics, Ludwig Maximilian University, Munich, Germany.*)

Progression through the cell cycle is halted by adverse conditions such as inadequate nutrition (nutrient stress), inappropriate cellular microenvironments, or DNA damage. Nuclear DNA is monitored very closely, and damage here can arrest the cell cycle not only at the **G_1 restriction point** but also during S or at a checkpoint in G_2 (Figure 3–13). G_1 arrest may permit

FIGURE 3–10   Human karyotype.

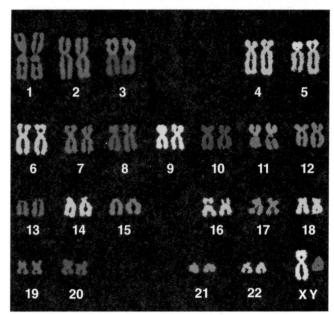

Karyotypes provide light microscopic information regarding the number and morphology of chromosomes in an organism. Such preparations are made by staining and photographing the chromosomes of a cultured cell arrested with colchicine during mitosis, when chromosomes are maximally condensed. From the image individual chromosomes are typically placed together in pairs. With certain stains each chromosome has a particular pattern of banding that facilitates its identification and shows the relationship of the banding pattern to genetic anomalies. Hybridization with fluorescent probes specific for each chromosome (FISH) followed by karyotyping yields an image like that shown here. Note that the 22 pairs of autosomes, as well as the X and Y chromosomes, differ in size, morphology, and location of the centromere.

repair of the damage before the cell enters S phase, so that the damaged DNA does not reproduce gene defects during replication. If the problem encountered at any checkpoint cannot be corrected fairly quickly while cycling is halted, proteins encoded by tumor suppressor genes are activated and that cell's activity is redirected toward cell suicide or apoptosis.

❯❯ MEDICAL APPLICATION

Many genes coding for proteins important in the control of cell proliferation and differentiation are often called **proto-oncogenes**; changes in the structure or expression of these can convert them to **oncogenes** causing uncontrolled cell growth and a potential for cancer. Altered proto-oncogenes are associated with many types of tumors and hematologic cancers. Proto-oncogenes can encode almost any protein involved in the control of mitotic activity, including various specific growth factors, the receptors for growth factors, and various kinases and other proteins involved in intracellular signaling of growth factors.

FIGURE **3–11** Regions within a nucleolus.

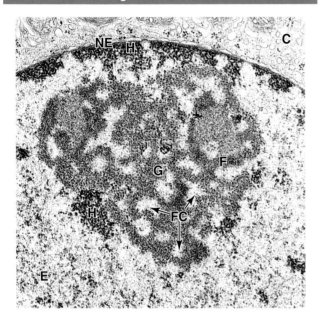

TEM reveals morphologically distinct regions within a nucleolus. Small, light-staining areas are fibrillar centers (**FC**), containing the DNA sequences for the *rRNA* genes (the nucleolar organizers). The darker fibrillar material (**F**) surrounding the fibrillar centers consists of accumulating rRNA transcripts. More granular material (**G**) of the nucleolus contains mainly the large and small ribosomal subunits being assembled from rRNA and ribosomal proteins synthesized in the cytoplasm. Various amounts of heterochromatin (**H**) are also typically found near the nucleolus, scattered in the euchromatin (**E**), and adjacent to the nuclear envelope (**NE**) that separates chromatin from cytoplasm (**C**). (X35,000)

›› MEDICAL APPLICATION

Retinoblastoma is a type of cancer occurring in the eyes, usually in young children. One form of the disease is inherited or familial. Research on the genetic basis of this disease led to the discovery of Rb, a gene coding for a key protein active at the G_1 restriction point that blocks cell cycle progression until a mitogenic stimulus arrives. A kinase activated by a growth factor receptor phosphorylates the Rb protein, causing it to release the E2F transcription factor. This factor then activates genes needed for DNA replication.

DNA changes (mutations) resulting from damage are not always detected and corrected (or eliminated). If such a change occurs in a gene important for cell cycle activities, such as genes for certain growth factors, their receptors or signaling kinases, normal controls on the cell cycle may be affected and growth may occur in a less-regulated manner usually detected by the tumor suppressor proteins such as p53. Failure to detect unregulated cell cycling can lead to additional defects and the cellular changes found in the various types of cancer. In many forms of human cancer, the gene for the key tumor suppressor p53 is itself mutated, thus reducing the ability to eliminate cells with damaged DNA and facilitating proliferation of cells with new genetic defects.

› MITOSIS

The period of cell division, or **mitosis** (Gr. *mitos*, a thread), is the only cell cycle phase that can be routinely distinguished with the light microscope. During mitosis, a parent cell divides and each of the two daughter cells receives a chromosomal set identical to that of the parent cell. The chromosomes replicated during the preceding S phase are distributed to the daughter cells. The long period between mitoses (the G_1, S, and G_2 phases) is also commonly called **interphase**. The events of mitosis are subdivided into four major stages (Figure 3–12). Important details of each mitotic phase are included in Figure 3–14 and summarized here. During the relatively long **prophase**, several changes occur:

- The nucleolus disappears and the replicated chromatin condenses into discrete threadlike chromosomes, each consisting of duplicate sister chromatids joined at the centromere.
- The two centrosomes with their now-duplicated centrioles separate and migrate to opposite poles of the cell and organize the microtubules of the **mitotic spindle**.
- Late in prophase, lamins and inner nuclear membrane are phosphorylated, causing the nuclear lamina and nuclear pore complexes to disassemble and disperse in cytoplasmic membrane vesicles.

During **metaphase**, chromosomes condense further and protein complexes called **kinetochores** (Gr. *kinetos*, moving) at each centromere attach to the mitotic spindle (Figure 3–15). The cell is now more spherical and the chromosomes are moved into alignment at the equatorial plate.

In **anaphase** sister chromatids (now called chromosomes themselves) separate and move toward opposite spindle poles by a combination of microtubule motor proteins and dynamic changes in the lengths of the microtubules as the spindle poles move farther apart.

At **telophase** the following occur:

- The two sets of chromosomes are at the spindle poles and begin reverting to their decondensed state.
- The spindle depolymerizes and the nuclear envelope begins to reassemble around each set of daughter chromosomes.
- A belt-like contractile ring of actin filaments associated with myosins develops in the cortical cytoplasm at the cell's equator. During **cytokinesis** at the end of telophase, constriction of this ring produces a cleavage furrow and progresses until the cytoplasm and its organelles are divided into two daughter cells, each with one nucleus.

FIGURE **3–12** The cell cycle.

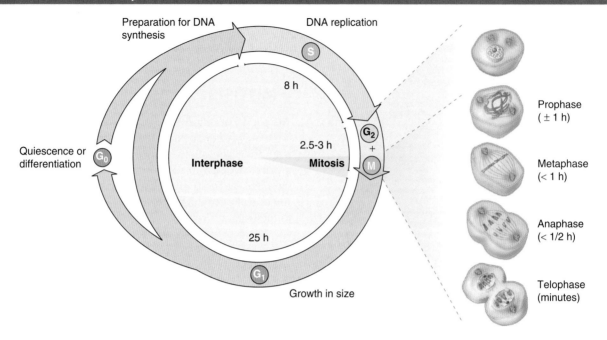

The ability to recognize microscopically cells during both mitosis and DNA replication (by autoradiography after administering radiolabeled thymidine) led to the concept of the cell "cycle" with the phases occurring as shown here. In rapidly dividing cells, **G₁** is a period in which cells accumulate the enzymes and nucleotides required for DNA replication, **S** is the period devoted primarily to DNA replication, **G₂** is a usually short period of preparation for mitosis, and **M** includes all phases of mitosis itself. In rapidly growing human tissues the cell cycle varies from 24 to 36 hours. The length of G₁ depends on many factors and is usually the longest and most variable period; the length of S is largely a function of the genome size. G₂ and mitosis together normally last only 2-3 hours. Differentiating cells in growing tissues may have very long G₁ periods and such cells are often said to be in the **G₀** phase of the cell cycle.

FIGURE **3–13** Controls at cell cycle checkpoints.

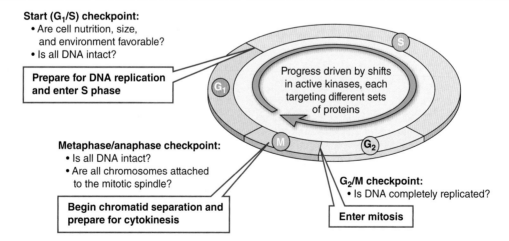

Start (G₁/S) checkpoint:
• Are cell nutrition, size, and environment favorable?
• Is all DNA intact?

Prepare for DNA replication and enter S phase

Progress driven by shifts in active kinases, each targeting different sets of proteins

Metaphase/anaphase checkpoint:
• Is all DNA intact?
• Are all chromosomes attached to the mitotic spindle?

Begin chromatid separation and prepare for cytokinesis

G₂/M checkpoint:
• Is DNA completely replicated?

Enter mitosis

Each phase of the cell cycle has one or more checkpoints where the quality of specific cell activities is checked. Progression to the next phase of the cycle does not occur until all activities of the preceding phase are completed satisfactorily. Three important checkpoints are shown here, including

■ The *start or restriction checkpoint* just before the start of S
■ The *G₂/M checkpoint* that ensures that DNA replication is complete

■ The metaphase *spindle checkpoint* that ensures that all chromosomes will be segregated

Overall progression in the cycle is regulated by proteins called **cyclins** and **cyclin-dependent kinases (CDKs)** that phosphorylate/activate enzymes and other proteins needed for phase-specific functions. Major cyclins, their CDKs, and important protein targets are summarized in Table 3–1.

Table 3–1	Major cyclin and cyclin-dependent kinase complexes regulating the human cell cycle and important target proteins.	
Cycle Phase or Checkpoint	**Active Cyclin-CDK Complex**	**Examples of Target Proteins**
Early G₁	Cyclin D-CDK4 or 6	Phosphorylates Rb protein, releasing E2F, a transcription factor that activates genes for many G₁ activities and for cyclin A
Late G₁/entry of S	Cyclin E-CDK2	Further activation E2F-mediated gene transcription; protein p53; other kinases
Progression through S	Cyclin A-CDK2	DNA polymerase and other proteins for DNA replication
G₂/entry of M	Cyclin A-CDK1	Specific phosphatases and cyclin B
Progression through M	Cyclin B-CDK1	Nuclear lamins; histone H1; chromatin- and centrosome-associated proteins

Most tissues undergo cell turnover with slow cell division and cell death. Nerve tissue and cardiac muscle are exceptions because their differentiated cells cannot undergo mitosis. As discussed in later chapters, a capacity for mitosis within a tissue, either by differentiated cells or by a reserve of stem cells, largely determines the tissue's potential to regeneration. The cell turnover rate is rapid in the epithelium lining the digestive tract and uterus or covering the skin. Mitotic cells are usually difficult to identify conclusively in sections of adult organs but may often be detected in rapidly growing tissues by their condensed chromatin (Figure 3–16).

FIGURE **3–14** Phases of mitosis.

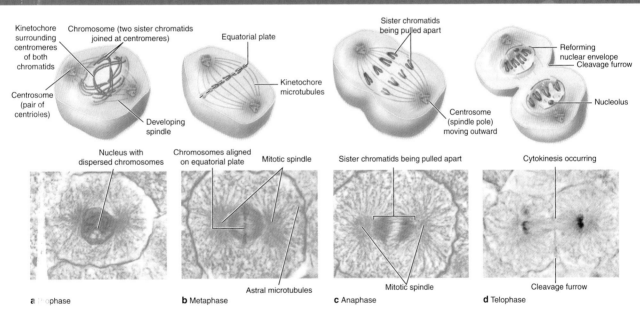

The chromosomal changes that occur during mitosis are easily and commonly studied in the large cells of very early embryos of fish after sectioning, such as the mitotic cells shown here. **(a)** During the relatively long **prophase** the **centrosomes** move to opposite poles, the nuclear envelope disappears by fragmentation, and the chromosomes condense and become visible. Having undergone DNA replication, each chromosome consists of two chromatids joined at their centromere regions by **kinetochore** protein complexes. **(b)** At the short **metaphase** the chromosomes have become aligned at the **equatorial plate** as a result of their attachments to the dynamic microtubules of the **mitotic spindle** organized by the centrosomes. The spindle consists of **kinetochore microtubules**, polar microtubules which interdigitate near the equatorial plate, and shorter astral microtubules anchoring the spindle to the cell membrane. **(c)** During **anaphase** the kinetochores separate and the chromatids (now called chromosomes themselves) are pulled on their microtubules toward each centrosome. **(d)** In **telophase** the cell pinches itself in two by contraction of the F-actin bundle in the cell cortex, after which the chromosomes decondense, transcription resumes, nucleoli reappear, and the nuclear lamina and nuclear envelope reassemble. (All X500; H&E)

FIGURE **3–15** Mitotic spindle and metaphase chromosomes.

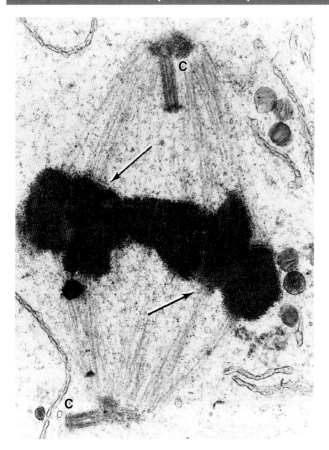

TEM of a sectioned metaphase cell shows several features of the mitotic apparatus, including the very electron-dense chromosomes bound at kinetochores (arrows) to the microtubules of the spindle. The microtubules converge on the centrosomes **(C)**, each containing a pair of centrioles. The flattened membrane vesicles near the mitotic spindle may represent fragments of the nuclear envelope, which begin to reassemble during late telophase. (X19,000)

(*Used with permission from Richard McIntosh, Department of Molecular, Cellular and Developmental Biology, University of Colorado, Boulder.*)

FIGURE **3–16** Mitotic cells in adult tissues.

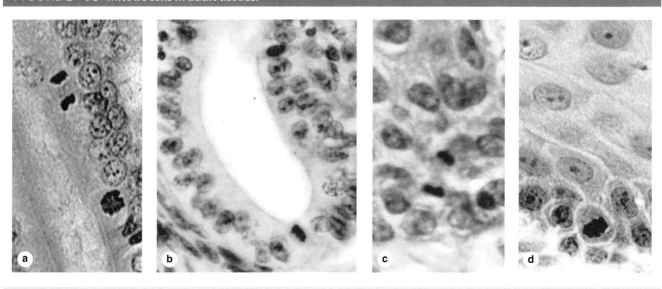

Dividing cells in recognizable stages of mitosis are rarely observed in adult tissues because they are rare and the cells are small, with variable shapes and orientations. However, *mitotic figures*, nuclei with clumped, darkly stained chromatin, can sometimes be identified, as shown here in various rapidly renewing tissues.

(a) In the lining of the small intestine, many mitotic *transit amplifying cells* can be found in the area above the most basal region of the intestinal crypts. Condensed chromosomes of cells in late anaphase and prophase phase can be distinguished here.

(b) Metaphase cell in a gland of proliferating uterine endometrium. **(c)** Telophase cell in the esophagus lining. **(d)** Metaphase in the basal layer of epidermis. (X400; H&E)

❯ STEM CELLS & TISSUE RENEWAL

Throughout an individual's lifetime, many tissues and organs contain a small population of undifferentiated **stem cells** whose cycling serves to renew the differentiated cells of tissues as needed. Many stem cells divide infrequently and the divisions are asymmetric; that is, one daughter cell remains as a stem cell while the other becomes committed to a path that leads to differentiation (Figure 3–17). Stem cells of many tissues are found in specific locations or niches where the microenvironment helps maintain their uniquely undifferentiated properties; they are often rare and inconspicuous by routine histologic methods.

Stem cells are best studied in tissues with *rapidly renewing cell populations*, including blood cells, skin cells, and cells

lining the digestive tract. Most mitotic cells here are not stem cells but the more rapidly dividing progeny of the cells committed to differentiation (Figure 3–17). They are commonly called **progenitor cells** or transit amplifying cells because they are in transit along the path from the stem cell niche to a differentiated state, while still amplifying by mitosis the number of new cells available for the differentiated tissue. Cells formed by progenitor cells may become terminally differentiated, meaning that renewed cycling cannot occur and the specialized cells exist for only a short time.

In tissues with *stable cell populations*, such as most connective tissues, smooth muscle, and the cells lining blood vessels, stem cells are not readily apparent and differentiated cells appear to undergo slow and episodic division to maintain tissue integrity.

❯ MEIOSIS

Meiosis is a specialized process involving *two* unique and closely associated cell divisions that occurs only in the cells that will form sperm and egg cells. Differentiation of these two "germ cells" or **gametes** is discussed fully in Chapters 21 and 22, but the chromosomal aspects of meiosis are described here for comparison with the events of mitosis (Figure 3–18).

Two key features characterize meiosis. (1) Early in the process the homologous chromosomes of each pair (one from the mother, one from the father) come together in an activity termed **synapsis**. During synapsis double-stranded breaks and repairs occur in the DNA, some of which result in reciprocal DNA exchanges called **crossovers** between the aligned homologous chromosomes. This produces new combinations of genes in the chromosomes so that few if any chromosomes in the germ cells are exactly the same as those in the mother and father. (2) The cells produced are **haploid**, having just one chromosome from each pair present in the body's somatic cells. The union of haploid eggs and sperm at fertilization forms a new diploid cell (the zygote) that can develop into a new individual.

As shown in Figure 3–18, the important events of meiosis unfold as follows:

- A cell approaching the first meiotic division has just completed a typical S phase and replicated its DNA; each chromosome contains the two identical DNA molecules called *sister chromatids.*
- During a greatly elongated prophase of the first meiotic division (prophase I), the partially condensed chromatin of homologous chromosomes begins to come together and physically associate along their lengths during synapsis. Because each of the paired chromosomes has two chromatids, geneticists refer to synaptic chromosomes as *tetrads* to emphasize that four

FIGURE **3–17** Stem cells.

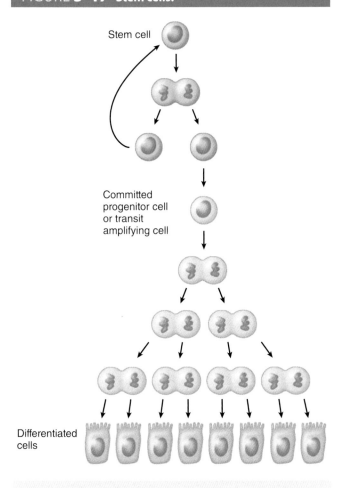

In rapidly growing adult tissues and perhaps in other tissues there are slowly dividing populations of **stem cells**. Stem cells divide asymmetrically, producing one cell that remains as a stem cell and another that becomes committed to a differentiative pathway but divides a few more times at a more rapid rate. Such cells have been termed **progenitor cells**, or "transit amplifying cells," each of which eventually stops dividing and becomes fully differentiated.

FIGURE 3–18 Mitosis and meiosis.

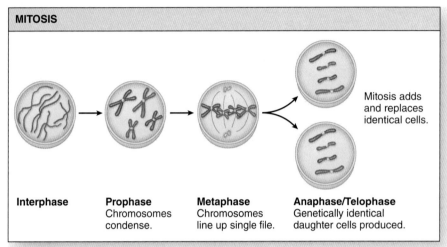

MITOSIS

| Interphase | **Prophase**
Chromosomes
condense. | **Metaphase**
Chromosomes
line up single file. | **Anaphase/Telophase**
Genetically identical
daughter cells produced. |

Mitosis adds and replaces identical cells.

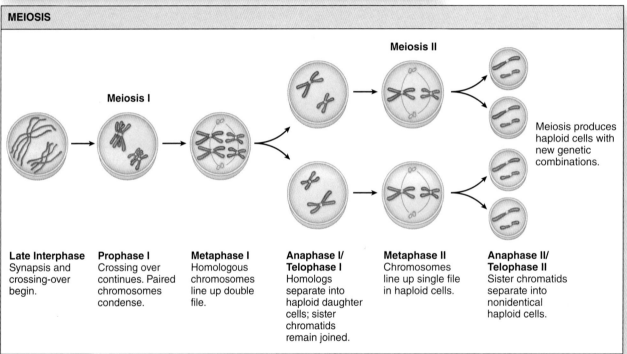

MEIOSIS

Meiosis I

Meiosis II

Meiosis produces haploid cells with new genetic combinations.

| **Late Interphase**
Synapsis and
crossing-over
begin. | **Prophase I**
Crossing over
continues. Paired
chromosomes
condense. | **Metaphase I**
Homologous
chromosomes
line up double
file. | **Anaphase I/
Telophase I**
Homologs
separate into
haploid daughter
cells; sister
chromatids
remain joined. | **Metaphase II**
Chromosomes
line up single file
in haploid cells. | **Anaphase II/
Telophase II**
Sister chromatids
separate into
nonidentical
haploid cells. |

Mitosis and meiosis share many aspects of chromatin condensation and separation but differ in key ways. Mitosis produces *two diploid cells* that are the same genetically. In meiosis, the two homologous maternal and paternal chromosomes physically align in **synapsis** and regions are exchanged during *crossing over* or genetic recombination. This is followed by **two meiotic divisions** with no intervening S phase, producing *four haploid cells*.

copies of each genetic sequence are present. During synapsis recombination or crossing over occurs among the four chromatids, which mixes up the genes inherited from each parent and yields a new and different set of genes to be passed on to the next generation.

In human spermatogenesis prophase I normally lasts for 3 weeks; oocytes arrest in this meiotic phase from the time of their formation in the fetal ovary through the woman's reproductive maturity, that is, for about 12 years to nearly five decades!

- When synapsis and crossing over are completed, the chromosomes become fully condensed and undergo metaphase, anaphase, and telophase as the cell divides. This first meiotic division separates the homologous chromosomes that paired during synapsis; each of the separated chromosomes still contains two chromatids held together at the centromere.
- Each of the two new cells now divides again, much more rapidly and without an intervening S phase. In the second meiotic division the chromatids separate to opposite poles as individual chromosomes. In each new cell a nuclear envelope forms around this new haploid set of chromosomes.

In summary, meiosis and mitosis share many aspects of chromatin condensation and separation (Figure 3–18), but differ in key ways:

- Mitosis is a cell division that produces *two diploid cells*. Meiosis involves two cell divisions and produces *four haploid cells*.
- During meiotic crossing over, new combinations of genes are produced and every haploid cell is genetically unique. Lacking synapsis and the opportunity for DNA recombination, mitosis yields two cells that are the same genetically.

>> **MEDICAL APPLICATION**

In humans chromosome 21 is a very small chromosome and the one most likely to be "overlooked" at the metaphase/anaphase checkpoint. Failure of these homologous chromosomes to separate (nondisjunction) in the first meiotic division also occurs with greater frequency in older oocytes (or sperm progenitor cells). A gamete retaining this chromosome pair forms a viable zygote after fertilization, but the developing **trisomy 21** individual has morphologic and cognitive impairments associated with Down syndrome.

> ## APOPTOSIS

Less evident, but no less important than cell proliferation for body functions, is the process of cell suicide called **apoptosis** (Gr. *apo*, off + *ptosis*, a falling). Apoptosis is a rapid, highly regulated cellular activity that shrinks and eliminates defective and unneeded cells (Figure 3–19). It results in small membrane-enclosed *apoptotic bodies*, which quickly undergo phagocytosis by neighboring cells or cells specialized for debris removal. Apoptotic cells do not rupture and release none of their contents, unlike cells that die as a result of injury and undergo *necrosis*. This difference is highly significant because release of cellular components triggers a local inflammatory reaction and immigration of leukocytes. Such a response is avoided when cells are rapidly eliminated by apoptosis following cell cycle arrest or as part of normal organ development.

FIGURE **3–19** Apoptotic cells.

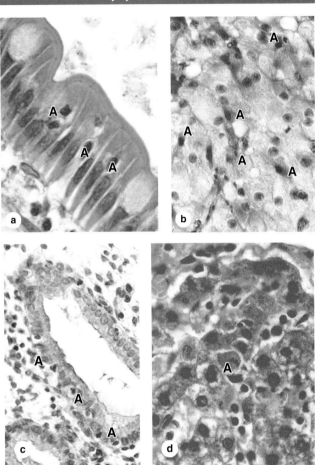

Apoptotic cells in adult tissues are rare because the process is completed very rapidly. Moreover, with their highly condensed chromatin in pyknotic nuclei and rounded shape, cells early in apoptosis may superficially resemble some mitotic cells. Shown here are apoptotic cells (**A**) in the epithelium covering an intestinal villus (**a**), in a corpus luteum beginning to undergo involution (**b**), in the epithelium of a uterine endometrial gland at the onset of menstruation (**c**), and in the liver (**d**). (X400; H&E)

>> **MEDICAL APPLICATION**

Cancer cells often deactivate the genes that control the apoptotic process, thus preventing their elimination in this type of cell death and allowing progression toward a more malignant state. The Bcl-2 family of proteins that controls the onset of apoptosis was first identified by a genetic mutation in a specific B-cell lymphoma, which provided the name for the original protein.

Apoptosis is controlled by cytoplasmic proteins in the **Bcl-2 family**, which regulate the release of death-promoting factors from mitochondria. Activated by either external signals or irreversible internal damage, specific Bcl-2 proteins induce a process with the following features:

- **Loss of mitochondrial function and caspase activation:** Bcl-2 proteins associated with the outer mitochondrial membrane compromise membrane integrity, stopping normal activity and releasing cytochrome c into the cytoplasm where it activates proteolytic enzymes called **caspases**. The initial caspases activate a cascade of other caspases, resulting in protein degradation throughout the cell.
- **Fragmentation of DNA: Endonucleases** are activated, which cleave DNA between nucleosomes into small fragments. (The new ends produced in the fragmented DNA allow apoptotic cells to be stained histochemically using an appropriate enzyme that adds labeled nucleotides at these sites.)
- **Shrinkage of nuclear and cell volumes:** Destruction of the cytoskeleton and chromatin causes the cell to shrink quickly, producing small structures with dense, darkly stained **pyknotic nuclei** that may be identifiable with the light microscope (Figure 3–19).
- **Cell membrane changes:** The plasma membrane of the shrinking cell undergoes dramatic shape changes, such as "blebbing," as membrane proteins are degraded and lipid mobility increases.
- **Formation and phagocytic removal:** Membrane-bound remnants of cytoplasm and nucleus separate as very small **apoptotic bodies** (Figure 3–20). Newly exposed phospholipids on these bodies induce their phagocytosis by neighboring cells or white blood cells.

Nuclei of cells in malignant tumors are often enlarged, abnormally shaped, and extremely dark staining, with abnormal nucleoli, in comparison with nuclei of normal cells. Such changes are useful to pathologists looking for evidence of cancer during microscopic examinations of biopsies.

A few examples of apoptosis emphasize its significance. In the ovary, apoptosis is the mechanism for both the monthly loss of luteal cells and the removal of excess oocytes and their follicles. Apoptosis was first discovered as programmed cell death in embryos, where it is important in shaping developing organs or body regions, such as the free spaces between embryonic fingers and toes. Apoptosis of excess nerve cells plays an important role in the final development of the central nervous system.

Triggered by p53 and other tumor suppressor proteins, apoptosis is the method for eliminating cells whose survival is blocked by lack of nutrients or by damage caused by free radicals or radiation. In all these examples apoptosis occurs very rapidly, in less time than required for mitosis, and the affected cells are removed without a trace.

FIGURE 3–20 Late apoptosis—apoptotic bodies.

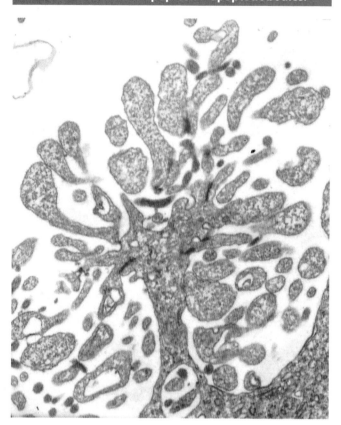

TEM of a cell in late apoptosis shows radical changes in cell shape, with membrane blebbing and the formation of many membrane-bound cytoplasmic regions. These apoptotic bodies may separate from one another but remain enclosed by plasma membrane so that no contents are released into the extracellular space. The membrane changes are recognized by neighboring cells, and macrophages and apoptotic bodies are very rapidly phagocytosed. (X10,000)

The Nucleus SUMMARY OF KEY POINTS

Nuclear Envelope

- Cytoplasm is separated from nucleoplasm by the **nuclear envelope**, a double set of membranes with a narrow perinuclear space; the outer membrane binds ribosomes and is continuous with the RER.
- The nuclear envelope is penetrated by **nuclear pore complexes**, large assemblies of nucleoporins with 8-fold symmetry through which proteins and protein-RNA complexes move in both directions.
- The nuclear envelope is supported internally by a meshwork, the nuclear lamina, composed of intermediate filament subunits called **lamins**.

Chromatin

- Chromatin is the combination of DNA and its associated proteins.
- Chromatin with DNA that is active in transcription stains lightly and is called **euchromatin**; inactive chromatin stains more darkly and is called **heterochromatin**.
- The DNA molecule initially wraps around complexes of basic proteins called **histones** to form **nucleosomes**, producing a structure resembling beads on a string.
- Additional levels of chromatin fiber condensation are less well understood and involve nonhistone proteins, including complexes of condensins.
- The extra X chromosome in cells of female mammals forms facultative heterochromatin and can be seen as the Barr body.

Nucleolus

- The **nucleolus** is a very basophilic or electron-dense area of chromatin localized where **rRNA transcription and ribosomal subunits assembly** occur.
- By TEM, an active nucleolus is seen to have **fibrous and granular parts** where rRNA forms and ribosomal subunits are assembled, respectively.

The Cell Cycle

- The **cell cycle** is the sequence of events that controls cell growth and division.
- The G_1 **phase**, the longest part of the cycle, begins immediately after mitosis and includes all preparations for DNA replication.
- The period of DNA (and histone) synthesis is the **S phase**.
- In a short G_2 **phase** the cell prepares for division during **mitosis (M)**.
- Cell cycling is controlled by the sequential appearance of key cytoplasmic proteins, the **cyclins**, which bind **cyclin-dependent kinases (CDKs)**.
- CDKs phosphorylate and activate the enzymes and transcription factors whose functions characterize each phase of the cell cycle.
- Progress through the cell cycle stages is monitored at checkpoints, including the G_1 **restriction point**; only when each phase's activities are completed are the cyclins changed to trigger those of the next phase.

Mitosis

- Stages of mitotic cell divisions include **prophase**, when chromosomes condense, the nuclear envelope disassembles, and the microtubular spindle forms; **metaphase**, when chromosomes are aligned; **anaphase**, when they begin to separate toward the two centrosomes; and **telophase**, when nuclear envelope re-forms around the separated chromosomes.
- Telophase ends with **cytokinesis** or cell cleavage into two daughter cells by a contractile ring of actin filaments and myosin.

Stem Cells & Tissue Renewal

- **Stem cells** occur in all tissues with rapid cell turnover; they divide slowly in an **asymmetric** manner, with one daughter cell remaining a stem cell and one becoming committed toward differentiation.
- Cells committed to differentiate (**transit amplifying or progenitor cells**) typically divide more rapidly than stem cells before slowing or stopping division to differentiate.

Meiosis

- **Meiosis** is the process by which two successive cell divisions produce cells called **gametes** containing half the number of chromosomes found in somatic cells.
- Prophase of the first meiotic division is a unique, extended period in which homologous chromosomes pair and undergo genetic recombination during the process called **synapsis**.
- Synaptic pairs separate toward two daughter cells at the first meiotic division.
- The second meiotic division occurs with no intervening S phase and separates the sister chromatids into two final cells that are **haploid**.

Apoptosis

- **Apoptosis** is the process by which redundant or defective cells are rapidly eliminated in a manner that does not provoke a local inflammatory reaction in the tissue.
- Apoptosis involves a cascade of events controlled by the **Bcl-2 family of proteins** regulating the release of death-promoting factors from mitochondria.
- Cytochrome c from mitochondria activates cytoplasmic proteases called **caspases**, which degrade proteins of the cytosol, cytoskeleton, and cell membrane.
- **Endonucleases** are activated, which degrade all nuclear DNA.
- Cell and nuclear volumes shrink rapidly, and the cell membrane changes produce extensive blebbing of the cell surface.
- Late in apoptosis, the cell breaks into many small **apoptotic bodies** that undergo **phagocytosis** by neighboring cells.
- Apoptosis occurs rapidly, with little or no release of proteins that would trigger inflammation, unlike the **death of injured cells by necrosis** that typically induces local inflammation.

The Nucleus ASSESS YOUR KNOWLEDGE

1. Which of the following facilitates breakdown of the nuclear envelope during the onset of mitosis?
   a. Disassembly of nucleosomes in the associated constitutive heterochromatin
   b. Increased export of material by the nuclear pore complexes into the perinuclear space
   c. Phosphorylation of lamin subunits by a cyclin-dependent kinase (CDK)
   d. Activities triggered at a restriction point late in G_1
   e. The activity of proteasomes

2. Binding of histone H1 proteins to importins is important for which of the following?
   a. Transport through the nuclear pores complexes
   b. Properly directed vesicular transport through the Golgi apparatus
   c. Transport from the granular part of the nucleolus
   d. Further binding to the "linker DNA" and proper assembly of nucleosomes
   e. Phosphorylation of cyclins

3. Which of the following is a region of chromatin that is well developed in large neurons active in protein synthesis?
   a. Heterochromatin
   b. The nucleolus
   c. The Nissl substance (neuronal RER)
   d. The Barr body
   e. The nucleosome

4. Which of the following is found during meiosis but not mitosis?
   a. Chromatids
   b. Polar microtubules
   c. Metaphase
   d. Synapsis
   e. Cytokinesis

5. Transitions in the cell cycle from one phase to the next are regulated by protein kinases whose activity depends on what other proteins?
   a. Tumor suppressors
   b. Cyclins
   c. Actins
   d. Lamins
   e. Importins

6. Mitotic figures visible in a tissue section from the lining of the small intestine are most likely to belong to which of the following categories?
   a. Terminally differentiated cells
   b. Partially differentiated cells
   c. Blood cells
   d. Stem cells
   e. Progenitor cells

7. Key differences between apoptotic and necrotic cell death include which of the following?
   a. Apoptotic cells do not release factors that induce inflammation.
   b. Necrosis does not trigger inflammation.
   c. Apoptosis does not utilize intracellular proteases.
   d. Apoptosis usually follows lethal physical damage to a cell.
   e. Necrosis is involved in formation of some organs during embryonic development.

8. A 29-year-old woman presents with a 101°F fever, pericardial effusions and Libman-Sachs endocarditis, arthralgia, and facial rash across the malar region ("butterfly rash") that is accentuated by sun exposure. Laboratory tests show creatine 1.7 mg/dL (normal 0.5-1.1 mg/dL), high titers of antinuclear autoantibodies (ANA), Smith antigen, and antinucleosome antibodies in the blood. Which one of the following is most likely to be directly affected by the disruption of nucleosomes in this patient?
   a. Packaging of genetic material in a condensed form
   b. Transcribing DNA
   c. Forming pores for bidirectional nuclear-cytoplasmic transport
   d. Forming the nuclear lamina
   e. Holding together adjacent chromatids

9. A 32-year-old man and his 30-year-old wife are referred for a reproductive endocrinologist infertility (REI) consult after 2 years of "trying to get pregnant." He is diagnosed with oligozoospermia. Ejaculated mature sperm are collected and undergo genetic analysis. Using gene linkage analysis, his REI specialist determines that he has aberrations in spermatogenic meiotic recombination, including both diminished frequency and suboptimal location, resulting in high frequency of aneuploid sperm. In explaining the diagnosis, she explains meiosis and recombination attributing the problem to a specific phase of the meiosis. Which part of meiosis is most closely associated with recombination?
   a. Metaphase I
   b. Anaphase I/Telophase I
   c. Prophase I
   d. Prophase II
   e. Anaphase II/Telophase II

10. A newborn boy is diagnosed with Apert syndrome. He has craniosynostosis, hypoplasia of the middle part of the face with retrusion of the eyes, and syndactyly that includes fusion of the skin, connective tissue, and muscle of the first, middle, and ring fingers with moderate fusion of those digits. There is very limited joint mobility past the first joint. Which one of the following is most likely *decreased* in cells of the interdigital regions of the developing hand of this newborn child?
   a. Random DNA degradation
   b. Inflammation
   c. Cell swelling
   d. Bcl-2
   e. DNA degradation by endonucleases

CHAPTER

4 Epithelial Tissue

Despite its complexity, the organs of the human body are composed of only **four basic tissue types:** epithelial, connective, muscular, and nervous tissues. Each tissue is an assemblage of similarly specialized cells united in performing a specific function. The basic tissues, each containing extracellular matrix (ECM) as well as cells, associate with one another in the variable proportions and morphologies characteristic of each organ. The main features of the basic tissue types are summarized in Table 4–1.

Connective tissue is characterized by cells producing very abundant ECM; muscle tissue is composed of elongated cells specialized for contraction and movement; and nervous tissue is composed of cells with long, fine processes specialized to receive, generate, and transmit nerve impulses. Most organs can be divided into the **parenchyma**, which is composed of the cells responsible for the organ's specialized functions, and the **stroma**, the cells of which have a supporting role in the organ. Except in the brain and spinal cord, the stroma is always connective tissue.

Epithelial tissues are composed of closely aggregated polyhedral cells adhering strongly to one another and to a thin layer of ECM, forming cellular sheets that line the cavities of organs and cover the body surface. Epithelia (Gr. *epi*, upon + *thele*, nipple) line all external and internal surfaces of the body and all substances that enter or leave an organ must cross this type of tissue.

The principal functions of epithelial tissues include the following:

- Covering, lining, and protecting surfaces (eg, epidermis)
- Absorption (eg, the intestinal lining)
- Secretion (eg, parenchymal cells of glands)

Specific cells of certain epithelia may be contractile (myoepithelial cells) or specialized sensory cells, such as those of taste buds or the olfactory epithelium.

TABLE **4–1**	Main characteristics of the four basic types of tissues.		
Tissue	**Cells**	**Extracellular Matrix**	**Main Functions**
Epithelial	Aggregated polyhedral cells	Small amount	Lining of surface or body cavities; glandular secretion
Connective	Several types of fixed and wandering cells	Abundant amount	Support and protection of tissues/organs
Muscle	Elongated contractile cells	Moderate amount	Strong contraction; body movements
Nervous	Elongated cells with extremely fine processes	Very small amount	Transmission of nerve impulses

» CHARACTERISTIC FEATURES OF EPITHELIAL CELLS

The shapes and dimensions of epithelial cells are quite variable, ranging from tall **columnar** to **cuboidal** to low **squamous** cells. The cells' size and morphology are generally dictated by their function. Epithelial cell nuclei vary in shape and may be elliptic (oval), spherical, or flattened, with nuclear shape corresponding roughly to cell shape. Columnar cells generally have elongated nuclei, squamous cells have flattened nuclei, and cuboidal or pyramidal cells have more spherical nuclei (Figure 4–1).

Because the lipid-rich membranes of epithelial cells are frequently indistinguishable by light microscopy, the number and shape of their stained nuclei are important indicators of cell shape and density. The nuclei also allow one to determine the number of cell layers in an epithelium, a primary morphologic criterion for classifying epithelia.

Most epithelia are adjacent to connective tissue containing blood vessels from which the epithelial cells receive nutrients and O_2. Even thick epithelia do not themselves normally contain blood vessels. The connective tissue that underlies the epithelia lining the organs of the digestive, respiratory, and urinary systems is called the **lamina propria**. The area of contact between the two tissues may be increased by small evaginations called **papillae** (L. *papula*, nipple) projecting from the connective tissue into the epithelium. Papillae occur most frequently in epithelial tissues subject to friction, such as the covering of the skin or tongue.

Epithelial cells generally show **polarity**, with organelles and membrane proteins distributed unevenly within the cell. The region of the cell contacting the ECM and connective tissue is called the **basal pole** and the opposite end, usually facing a space, is the **apical pole**, with the two poles differing significantly in both structure and function. Regions of cuboidal or columnar cells that adjoin neighboring cells comprise the cells' **lateral surfaces**; cell membranes here often have numerous folds which increase the area and functional capacity of that surface.

Basement Membranes

The basal surface of all epithelia rests on a thin extracellular, felt-like sheet of macromolecules referred to as the **basement membrane** (Figure 4–1), a semipermeable filter for substances reaching epithelial cells from below. Glycoproteins and other components in this structure can often be stained and visualized with the light microscope (Figure 4–2).

With the transmission electron microscope (TEM) two parts of the basement membrane may be resolved. Nearest the epithelial cells is an electron-dense layer, 20-100 nm thick, consisting of a network of fine fibrils that comprise the **basal lamina** and beneath this layer is a more diffuse and fibrous **reticular lamina** (Figure 4–3a). The terms

FIGURE **4–1** ■ **Epithelia and adjacent connective tissue.**

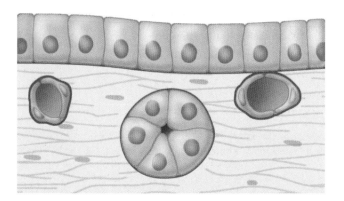

Cuboidal or pyramidal cells of epithelia generally have spherical nuclei, while nuclei of squamous epithelial cells are flattened. An extracellular **basement membrane** (red) always lies at the interface of epithelial cells and connective tissue. Nutrients for epithelial cells must diffuse across the basement membrane. Nerve fibers normally penetrate this structure, but small blood capillaries (being epithelial themselves) normally never enter epithelia.

FIGURE **4–2** **Basement membranes.**

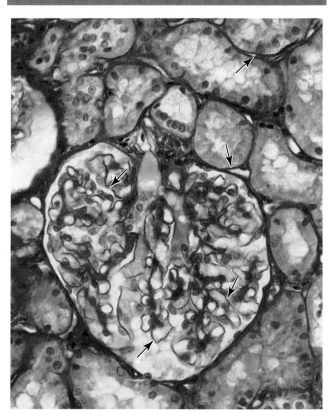

This section of kidney shows the well-stained basement membranes (arrows) of epithelia forming structures within the large, round renal glomerulus and its surrounding tubules. In kidney glomeruli the basement membrane, besides having a supporting function, has a highly developed role as a filter that is key to renal function. (X100; Picrosirius-hematoxylin [PSH])

FIGURE 4–3 Basal and reticular laminae of basement membranes.

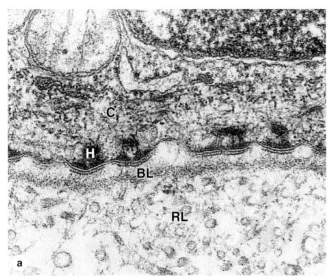

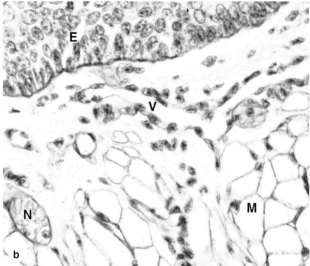

(a) The ultrastructural components of the basement membrane are revealed by TEM. The dense **basal lamina (BL)** may appear with thin clear zones on each side and is anchored to a thicker, more diffuse **reticular lamina (RL)** which contains collagen III reticular fibers. **Hemidesmosomes (H)** bind the basal surface of the epithelial cell (**C**) to the basal lamina. (X54,000)

(b) **Laminin**, a major glycoprotein within basal laminae, is shown here by immunohistochemistry and identifies the basement membranes of the stratified epithelium (**E**) and the simple epithelium lining a small blood vessel (**V**). Laminin also occurs in the **external laminae** surrounding nerves (**N**) and muscle (**M**) fibers, seen here in cross-section. (X200)

"basement membrane" and "basal lamina" are sometimes used interchangeably, but "basal lamina" usually denotes the fine extracellular layer seen ultrastructurally and "basement membrane" the entire structure beneath the epithelial cells visible with the light microscope.

The macromolecules of the basal lamina are secreted from the basal sides of the epithelial cells and form a sheet-like array. ECM components are described more fully in Chapter 5, but those of the basal lamina characteristically include the following:

- **Type IV collagen:** Monomers of type IV collagen self-assemble into a two-dimensional network of evenly spaced subunits.
- **Laminin:** These are large glycoproteins that attach to transmembrane proteins called **integrins** at the cells' basal surface and project through the network of type IV collagen.
- **Nidogen** and **perlecan:** Respectively a short, rod-like protein and a proteoglycan, both of these cross-link laminin to the collagen network and help determine the porosity of the basal lamina and the size of molecules able to filter through it.

Basal laminae often called *external laminae* but with similar composition also exist as sleeves surrounding muscle cells, nerves (Figure 4-3b), and fat-storing cells, where they serve as semipermeable barriers regulating macromolecular exchange between the enclosed cells and connective tissue.

The more diffuse meshwork of the reticular lamina contains **type III collagen** and is bound to the basal lamina by anchoring fibrils of **type VII collagen**, both of which are produced by cells of the connective tissue (Figure 4–3).

Besides acting as filters, functions of basement membranes include helping to provide structural support for epithelial cells and attach epithelia to underlying connective tissue. Basal lamina components help organize integrins and other proteins in the plasma membrane of epithelial cells, maintaining cell polarity and helping to localize endocytosis, signal transduction, and other activities. Basement membrane proteins also mediate many cell-to-cell interactions involving epithelia and mark routes for certain cell migrations along epithelia. Finally, the basement membrane also serves as a scaffold that allows rapid epithelial repair and regeneration.

Intercellular Adhesion & Other Junctions

Several membrane-associated structures provide adhesion and communication between cells. Some are present in other tissues but all are particularly numerous and prominent in epithelia. Epithelial cells adhere strongly to neighboring cells and basal laminae, particularly in epithelia subject to friction or other mechanical forces.

As shown in Figure 4–4 and summarized in Table 4–2, lateral surfaces of epithelial cells have complexes of several specialized intercellular junctions with different functions:

- **Tight** or **occluding junctions** form a seal between adjacent cells.

FIGURE **4–4** **Junctional complexes of epithelial cells.**

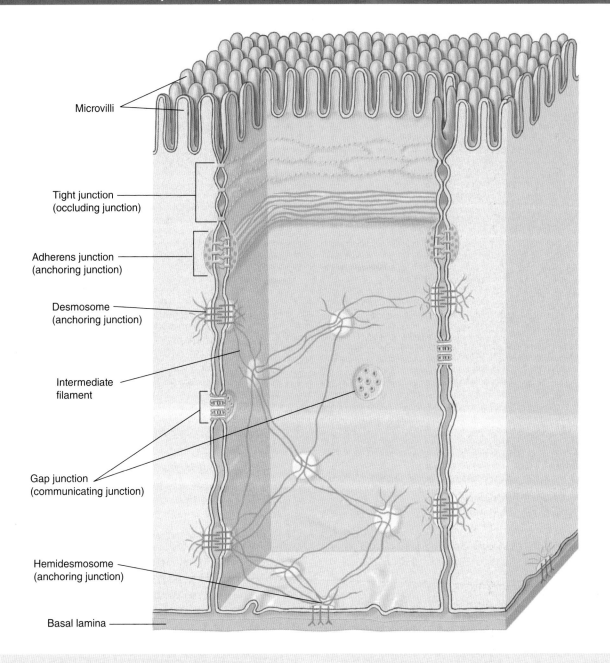

Microvilli

Tight junction
(occluding junction)

Adherens junction
(anchoring junction)

Desmosome
(anchoring junction)

Intermediate
filament

Gap junction
(communicating junction)

Hemidesmosome
(anchoring junction)

Basal lamina

Most cuboidal or columnar epithelial cells have intercellular junctional complexes with the different types of junctions shown schematically here. At the apical end, **tight junctions** (zonulae occludens) and **adherent junctions** (zonulae adherens) are typically close together and each forms a continuous band around the cell. Multiple ridges of the tight junction prevent passive flow of material between the cells but are not very strong; the adhering junctions immediately below them serve to stabilize and strengthen the circular occluding bands and help hold the cells together.

Both **desmosomes** and **gap junctions** are spot-like, not circular, structures between two cells. Bound to intermediate filaments inside the cells, desmosomes form very strong attachment points that supplement the zonulae adherens and play a major role to maintain the integrity of an epithelium. Gap junctions, each a patch of many **connexons** in the adjacent cell membranes, have little strength but serve as intercellular channels for flow of molecules. All of these junctional types are also found in certain other cell types besides epithelia. **Hemidesmosomes** bind epithelial cells to the underlying basal lamina.

TABLE **4–2**	Epithelial cell junctions, their major structural features and functions, and medical significance.				
Junction	Tight Junction (Zonula Occludens)	Adherent Junction (Zonula Adherens)	Desmosome (Macula Adherens)	Hemidesmosome	Gap Junction (Nexus)
Major transmembrane link proteins	Occludins, claudins, ZO proteins	E-cadherin, catenin complexes	Cadherin family proteins (desmogleins, desmocollin)	Integrins	Connexin
Cytoskeletal components	Actin filaments	Actin filaments	Intermediate filaments (keratins)	Intermediate filaments	None
Major functions	Seals adjacent cells to one another, controlling passage of molecules between them; separates apical and basolateral membrane domains	Provides points linking the cytoskeletons of adjacent cells; strengthens and stabilizes nearby tight junctions	Provides points of strong intermediate filament coupling between adjacent cells, strengthening the tissue	Anchors cytoskeleton to the basal lamina	Allows direct transfer of small molecules and ions from one cell to another
Medical significance	Defects in occludins may compromise the fetal blood-brain barrier, leading to severe neurologic disorders	Loss of E-cadherin in epithelial cell tumors (carcinomas) promotes tumor invasion and the shift to malignancy	Autoimmunity against desmoglein I leads to dyshesive skin disorders characterized by reduced cohesion of epidermal cells	Mutations in the integrin-$\beta4$ gene are linked to some types of epidermolysis bullosa, a skin blistering disorder	Mutations in various connexin genes have been linked to certain types of deafness and peripheral neuropathy

- **Adherent** or **anchoring junctions** are sites of strong cell adhesion.
- **Gap junctions** are channels for communication between adjacent cells.

In many epithelia these junctions are present in a definite order at the apical end of the cells. **Tight junctions**, also called zonulae occludens, are the most apical of the junctions. The term "zonula" indicates that the junction forms a band completely encircling each cell. In TEM the adjacent membranes at these junctions appear fused or very tightly apposed (Figures 4–5). The seal between the two cell membranes is due to tight interactions between the transmembrane proteins **claudin** and **occludin**. Tight junctions are clearly seen after cryofracture of epithelia (Figure 4–6), where they appear as a band of branching strands in the membrane around each cell's apical end. The intercellular seal of tight junctions ensures that molecules crossing an epithelium in either direction do so by going *through* the cells (a transcellular path) rather than *between* them (the paracellular pathway). Epithelia with one or very few fused sealing strands (eg, proximal renal tubule) are more permeable to water and solutes than are epithelia with many fused strands (eg, the lining of the urinary bladder).

Epithelial tight junctions also serve a related purpose: these continuous zones within cell membranes serve as fences restricting movements of membrane lipids and proteins at the apical cell surface into the lateral and basal surfaces, and vice versa. The tight junctions thus maintain two distinct membrane domains (apical and basolateral) with different sets of

components, which allows these two sides of the epithelium to display different receptors and other proteins and to function differently. Apical cell membranes of epithelia are part of the luminal compartment of a tissue or organ, while the basolateral domains are part of a basal compartment that also encompasses the underlying connective tissue.

>> MEDICAL APPLICATION

Proteins of **tight junctions** provide the targets for certain common bacteria of medical importance. The enterotoxin secreted by *Clostridium perfringens*, which causes "food poisoning," binds claudin molecules of intestinal cells, prevents insertion of these proteins during maintenance of tight junctions, and causes loss of tissue fluid into the intestinal lumen via the paracellular pathway.

Similarly, *Helicobacter pylori*, which is important in the etiology of **gastric ulcers**, binds the extracellular domains of tight-junction proteins in cells of the stomach and inserts a protein into these cells, which targets ZO-1 and disrupts signaling from the junction.

The second type of junction is the **adherens junction** or zonula adherens (Figures 4–4 and 4–5), which also encircles the epithelial cell, usually immediately below the tight junction. This is an adherent junction, firmly anchoring a cell to its neighbors. Cell adhesion is mediated by **cadherins**, transmembrane glycoproteins of each cell that bind each other in

FIGURE **4–5** Epithelial cell junctional complex.

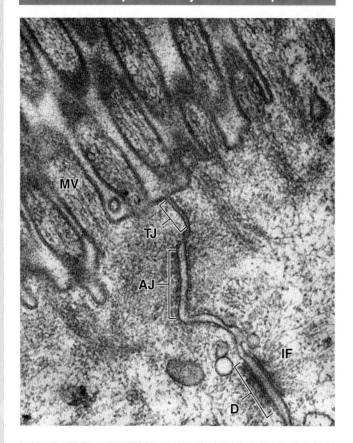

Ultrastructural view of the apical region near microvilli (**MV**) of two epithelial cells, revealing a junctional complex with a tight junction (**TJ**) or zonula occludens, an adherent junction (**AJ**) or zonula adherens, and a desmosome (**D**) associated with intermediate filaments (**IF**). The functions and major protein components of these junction types are summarized in Table 4–2. (X195,000)

FIGURE **4–6** View of a tight junction after cryofracture.

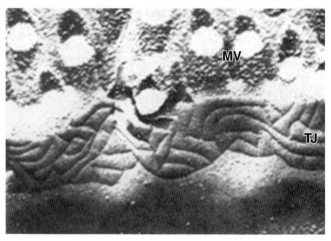

Just below the apical microvilli (**MV**) of this epithelial cell, a cryofracture plane splitting fused cell membranes reveals the fused strands of transmembrane proteins forming the tight junction. (zonula occludens; X100,000)

proteins called desmoplakins in an electron-dense plaque. Desmoplakins in turn bind intermediate filament proteins rather than actins. Epithelial desmosomes attach to cable-like filaments of cytokeratin, sometimes referred to tonofilaments. Such intermediate filaments are very strong and desmosomes provide firm cellular adhesion and strength throughout the epithelium.

❯❯ MEDICAL APPLICATION

Various **blistering (bullous) diseases**, such as pemphigus vulgaris, involving the epidermis or stratified squamous epithelia of the oral mucosa, are due to abnormal desmosome function caused by autoimmune reactions against specific desmogleins that reduce cell-to-cell adhesion. Similar disorders arise with genetic mutations for various junctional proteins.

the presence of Ca^{2+}. At their cytoplasmic ends, cadherins bind **catenins** that link to actin filaments with actin-binding proteins. The actin filaments linked to the adherens junctions form part of the "terminal web," a cytoskeletal feature at the apical pole in many epithelial cells. Together, the tight and adherent junctions encircling the apical ends of epithelial cells function like the plastic bands that hold a six-pack of canned drinks together.

Another anchoring junction is the **desmosome** (Gr., *desmos*, binding and *soma*, body) or macula adherens (L. *macula*, spot). As the name implies this junction resembles a single "spot-weld" and does not form a belt around the cell. Desmosomes are disc-shaped structures at the surface of one cell that are matched with identical structures at an adjacent cell surface (Figures 4–4 and 4–5). Desmosomes contain larger members of the cadherin family called desmogleins and desmocollins. The cytoplasmic ends of these clustered transmembrane proteins bind plakoglobins, catenin-like proteins which link to larger

Gap junctions, shown in Figure 4–7, mediate intercellular communication rather than adhesion or occlusion between cells. Abundant in many epithelia, gap junctions are also functionally important in nearly all mammalian tissues. Cryofracture preparations show that gap junctions consist of aggregated transmembrane protein complexes that form circular patches in the plasma membrane (Figure 4–7b).

The transmembrane gap junction proteins, **connexins**, form hexameric complexes called **connexons**, each of which has a central hydrophilic pore about 1.5 nm in diameter. When two cells attach, connexins in the adjacent cell membranes move laterally and align to produce connexons between the two cells (Figures 4–4 and 4–7a), with each junction having dozens or hundreds of aligned connexon pairs. Gap junctions permit

FIGURE **4–7** Gap junctions.

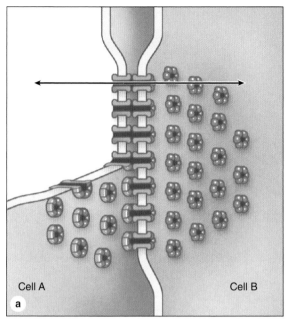
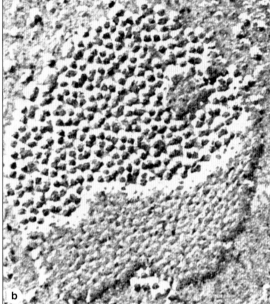

Cell A   Cell B

(a) A diagram of a gap junction shows the structural elements that allow the exchange of nutrients and signal molecules between cells without loss of material into the intercellular space. The communicating channels are formed by pairs of abutting particles **(connexons)**, which are in turn each composed of six protein subunits (connexins) that span the lipid bilayer of each cell membrane. The channel formed by paired connexons (arrow) is about 1.5 nm in diameter, limiting the size of transmitted molecules.

(b) A cryofracture preparation of a gap junction, showing the patch of aggregated transmembrane protein complexes, the connexons. (X150,000) **(c)** A section perpendicular to a gap junction between two cells shows that their cell membranes are very closely apposed, separated only by a 2-nm-wide electron-dense space. Individual connexons are not resolved in sections prepared for TEM. (X150,000)

intercellular exchange of molecules with small (< 1.5 nm) diameters. Some molecules mediating signal transduction, such as cyclic nucleotides and ions, move rapidly through gap junctions, allowing cells in many tissues to act in a coordinated manner rather than as independent units. For example, in heart and visceral muscles gap junctions help produce rhythmic contractions.

On the basal epithelial surface, cells (Figure 4–4) attach to the basal lamina by anchoring junctions called **hemidesmosomes** (Gr. *hemi*, half + *desmos* + *soma*), which can be seen by TEM (Figure 4–3). These adhesive structures resemble a half-desmosome ultrastructurally, but unlike desmosomes the clustered transmembrane proteins that indirectly link to cytokeratin intermediate filaments are integrins rather than cadherins. The integrins of hemidesmosomes bind primarily to laminin molecules in the basal lamina.

Another basal anchoring junction found in cells that are moving during epithelial repair or reorganization is the **focal adhesion**, or focal contact. Although resembling hemidesmosomes superficially, focal adhesions are smaller, more numerous, and consist of integrins linked indirectly to bundled actin filaments, not intermediate filaments. Importantly, integrins of focal adhesions are also linked via paxillin to **focal adhesion kinase**, a signaling protein which upon integrin binding to laminin or other specific ECM proteins initiates a cascade

of intracellular protein phosphorylation affecting cell adhesion, mobility, and gene expression. Focal adhesions are also important in migrating nonepithelial cells such as fibroblasts.

SPECIALIZATIONS OF THE APICAL CELL SURFACE

The apical ends of many columnar and cuboidal epithelial cells have specialized structures projecting from the cells. These function either to increase the apical surface area for better absorption or to move substances along the epithelial surface.

Microvilli

Many cells have cytoplasmic projections best seen with the electron microscope. Such extensions usually reflect the movements and activity of actin filaments and are both temporary and variable in their length, shape, and number. However, in epithelia specialized for absorption the apical cell surfaces are often filled with an array of projecting **microvilli** (L. *villus*, tuft), usually of uniform length. In cells such as those lining the small intestine, densely packed microvilli are visible as a **brush** or **striated border** projecting into the lumen (Figure 4–8). The average microvillus is about 1-μm long and 0.1-μm wide, but

FIGURE **4–8** Microvilli.

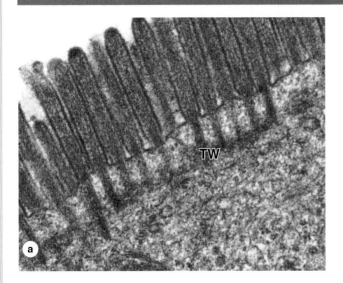

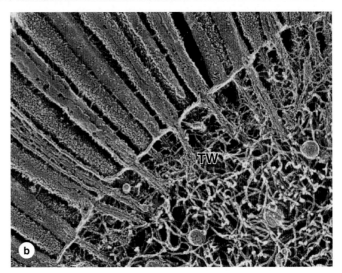

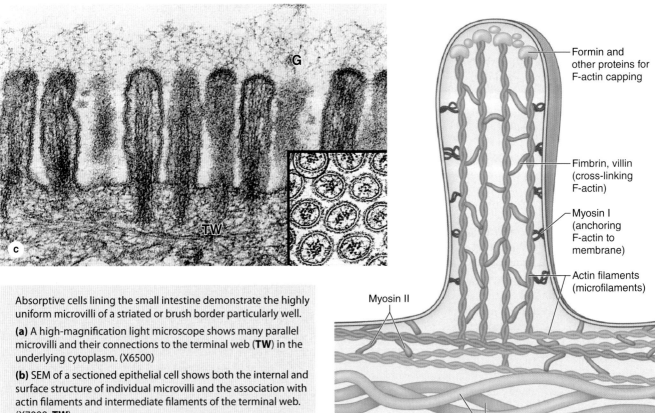

Formin and
other proteins for
F-actin capping

Fimbrin, villin
(cross-linking
F-actin)

Myosin I
(anchoring
F-actin to
membrane)

Actin filaments
(microfilaments)

Myosin II

d   Intermediate filaments (keratins)

Absorptive cells lining the small intestine demonstrate the highly uniform microvilli of a striated or brush border particularly well.

(a) A high-magnification light microscope shows many parallel microvilli and their connections to the terminal web (**TW**) in the underlying cytoplasm. (X6500)

(b) SEM of a sectioned epithelial cell shows both the internal and surface structure of individual microvilli and the association with actin filaments and intermediate filaments of the terminal web. (X7000; **TW**)

(*Figure 4–8b, used with permission from Dr John Heuser, Washington University School of Medicine, St. Louis, MO.*)

(c) TEM of microvilli sectioned longitudinally and transversely (inset) reveals the microfilament arrays that form the core of these projections. The terminal web (**TW**) of the cytoskeleton is also seen. The glycocalyx (**G**) extending from glycoproteins and glyco-lipids of the microvilli plasmalemma contains certain enzymes for late stages of macromolecule digestion. (X15,000)

(d) The diagram shows a few microfilaments in a microvillus, with various actin-binding proteins important for F-actin assembly,

capping, cross-linking, and movement. Like microfilaments in other regions of the cytoskeleton, those of microvilli are highly dynamic, with treadmilling and various myosin-based interactions. Myosin motors import various microvilli components along the actin filaments.

with hundreds or thousands present on the end of each absorptive cell, the total surface area can be increased by 20- or 30-fold. The thick glycocalyx covering microvilli of the intestinal brush border includes membrane-bound proteins and enzymes for digestion of certain macromolecules.

Each microvillus contains bundled actin filaments capped and bound to the surrounding plasma membrane by actin-binding proteins (Figure 4–8d). Although microvilli are relatively stable, the microfilament arrays are dynamic and undergo various myosin-based movements, which help maintain optimal conditions for absorption via numerous channels, receptors, and other proteins in the plasmalemma. The actin filaments insert into the **terminal web** of cortical micro-filaments at the base of the microvilli.

>> **MEDICAL APPLICATION**

Celiac disease, also called **gluten-sensitive enteropathy** or **sprue**, is a disorder of the small intestine in which one of the first pathologic changes is loss of the microvilli brush border of the absorptive cells. This is caused by an immune reaction against the wheat protein gluten during its digestion, which produces diffuse enteritis (intestinal inflammation), changes to the epithelial cells leading to malabsorption, and eventually to pathologic changes in the intestinal wall. The malabsorption problems and structural changes are reversible when gluten is removed from the diet.

Stereocilia

Stereocilia are a much less common type of apical process, best seen on the absorptive epithelial cells lining the male reproductive system (Figure 4–9). Like microvilli, stereocilia increase the cells' surface area, facilitating absorption. More specialized stereocilia with a motion-detecting function are important components of inner ear sensory cells.

Stereocilia resemble microvilli in containing arrays of microfilaments and actin-binding proteins, with similar diameters, and with similar connections to the cell's terminal web. However, stereocilia are typically much longer and less motile than microvilli, and may show branching distally.

Cilia

Cilia are long, highly motile apical structures, larger than microvilli, containing internal arrays of microtubules not microfilaments (Figure 4–10). In addition to cilia on epithelial cells, most (if not all) other cell types have at least one short projection called a *primary cilium*, which is not motile but is enriched with receptors and signal transduction complexes for detection of light, odors, motion, and flow of liquid past the cells.

Motile cilia are abundant on cuboidal or columnar cells of many epithelia. Typical cilia are 5-10 μm long and 0.2 μm in diameter, which is much longer and two times wider than

FIGURE **4–9** Stereocilia.

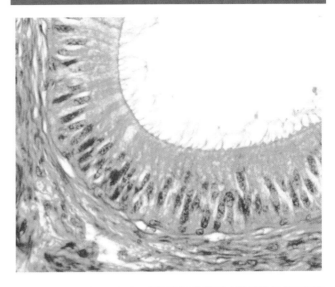

At the apical ends of the tall epithelial cells lining organs such as the epididymis (shown here) are numerous very long stereocilia, which increase the surface area available for absorption. Stereocilia are much longer than microvilli and often have distal branching. (X400; H&E)

a typical microvillus. As shown in Figure 4–11, each cilium has a core structure consisting of nine peripheral microtubule doublets (in which a few tubulin protofilaments are shared) arrayed around two central microtubules. This **9 + 2 assembly** of microtubules is called an **axoneme** (Gr. *axon*, axis + *nema*, thread). As with other microtubules, kinesin and cytoplasmic dynein motors move along the peripheral microtubules for the transport of molecular components into and out of these structures.

Microtubules of axonemes are continuous with those in **basal bodies**, which are apical cytoplasmic structures just below the cell membrane (Figures 4–10 and 4–11). Basal bodies have a structure similar to that of centrioles, with triplets of microtubules and dynamic tubulin protofilaments forming rootlets anchoring the entire structure to the cytoskeleton.

Cilia exhibit rapid beating patterns that move a current of fluid and suspended matter in one direction along the epithelium. Ciliary motion occurs through successive changes in the conformation of the axoneme, in which various accessory proteins make each cilium relatively stiff, but elastic. Complexes with **axonemal dynein** bound to one microtubule in each doublet extend as "arms" toward a microtubule of the next doublet. With energy from ATP dynein-powered sliding of adjacent doublets relative to each other bends the axoneme and a rapid series of these sliding movements produces the beating motion of the cilium. The long flagellum that extends from each fully differentiated sperm cell has an axonemal structure like that of a cilium and moves with a similar mechanism.

FIGURE **4–10** Cilia.

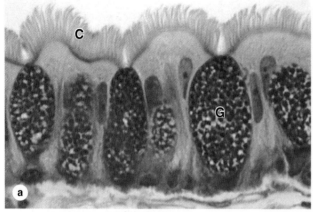

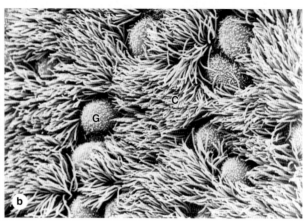

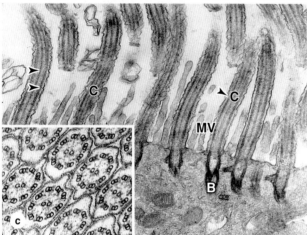

Epithelial cells lining the respiratory tract have many very well-developed cilia.

(a) By light microscopy cilia (**C**) on the columnar cells appear as a wave of long projections, interrupted by nonciliated, mucus-secreting goblet cells (**G**). (X400; Toluidine blue)

(b) SEM of the apical surfaces of this epithelium shows the density of the cilia (**C**) and the scattered goblet cells (**G**). (X600)

(c) TEM of cilia (**C**) sectioned longitudinally reveals central and peripheral microtubules (arrowheads) of the axonemes, with cross sections (inset) clearly showing the 9 + 2 array of the microtubule doublets. At the base of each cilium is a basal body (**B**) anchoring the axoneme to the apical cytoplasm. Much shorter microvilli (**MV**) can be seen between the cilia. (X59,000; Inset: X80,000)

❯❯ MEDICAL APPLICATION

Several mutations have been described in the proteins of the cilia and flagella. They are responsible for the **immotile cilia syndrome** (Kartagener syndrome), whose symptoms are chronic respiratory infections caused by the lack of the cleansing action of cilia in the respiratory tract and immotile spermatozoa, causing male infertility.

❯ TYPES OF EPITHELIA

Epithelia can be divided into two main groups: **covering** (or **lining**) **epithelia** and **secretory (glandular) epithelia**. This is an arbitrary functional division for there are lining epithelia in which all the cells also secrete (eg, the lining of the stomach) or in which glandular cells are distributed among the lining cells (eg, mucous cells in the small intestine or trachea).

Covering or Lining Epithelia

Cells of covering epithelia are organized into one or more layers that cover the surface or line the cavities of an organ. As summarized in Table 4–3, such epithelia are classified according to the number of cell layers and the cell morphology in the outer layer. **Simple epithelia** contain one cell layer and **stratified epithelia** contain two or more layers.

Based on cell shape, simple epithelia are further classified as **squamous** (thin cells), **cuboidal** (cell width and thickness roughly similar) or **columnar** (cells taller than they are wide). Examples of these epithelial types are shown in Figures 4–12 through 4–14.

Most stratified epithelia (Figure 4–15) are classified according to the cell shape of the superficial outer layer(s): **squamous**, **cuboidal**, or **columnar**.

The very thin surface cells of stratified squamous epithelia can be "keratinized" (packed with keratin filaments) or "nonkeratinized" (with relatively sparse keratin). **Stratified squamous keratinized epithelium** is found mainly in the epidermis of skin, where it helps prevent dehydration from the tissue (Figure 4–15a). Its cells form many layers, with the less differentiated cuboidal cells near the basement membrane. These cells have many desmosomes and become more irregular in shape and then flatten as they accumulate keratin in the process of **keratinization** and are moved progressively toward the skin surface, where they become thin, metabolically inactive packets (squames) of keratin lacking nuclei. As discussed

FIGURE **4–11**   Ciliary axoneme.

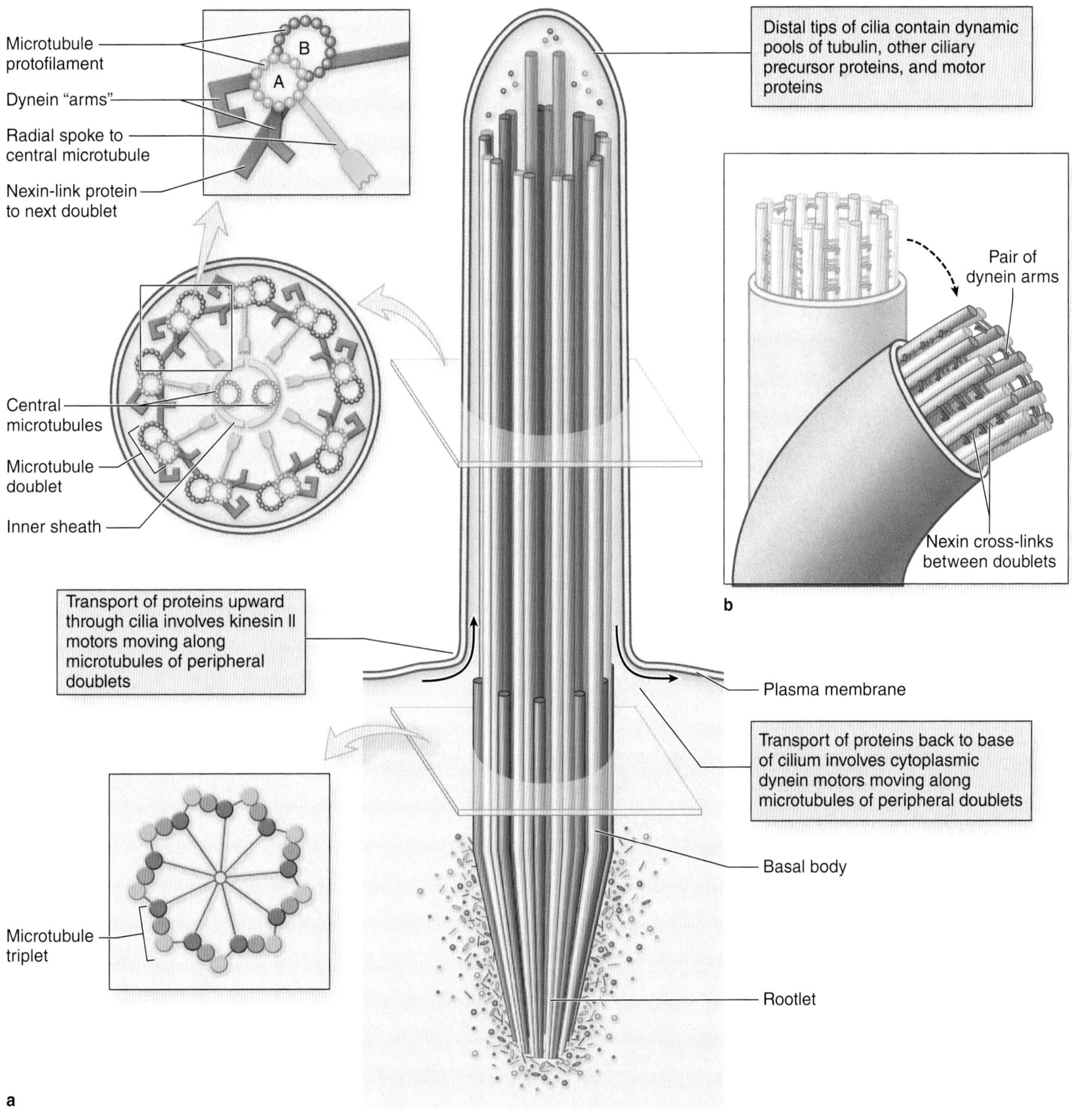

Microtubule protofilament

Dynein "arms"

Radial spoke to central microtubule

Nexin-link protein to next doublet

Central microtubules

Microtubule doublet

Inner sheath

Transport of proteins upward through cilia involves kinesin II motors moving along microtubules of peripheral doublets

Microtubule triplet

Distal tips of cilia contain dynamic pools of tubulin, other ciliary precursor proteins, and motor proteins

Pair of dynein arms

Nexin cross-links between doublets

b

Plasma membrane

Transport of proteins back to base of cilium involves cytoplasmic dynein motors moving along microtubules of peripheral doublets

Basal body

Rootlet

a

(a) A diagram of a cilium with the **axoneme** consisting of **two central microtubules** surrounded by **nine peripheral microtubular doublets** associated with other proteins. In the doublets, microtubule A is complete, consisting of 13 protofilaments, whereas microtubule B shares some of A's protofilament heterodimers. The axoneme is elastic but relatively stiff, with its structure maintained by nexins linking the peripheral doublets and other protein complexes forming a sheath and radial spokes between the doublets and the central microtubules.

The axoneme is continuous with a **basal body** located in the apical cytoplasm. Basal bodies are structurally very similar to centrioles, consisting of nine relatively short **microtubular triplets** linked together in a pinwheel-like arrangement. A dynamic pool of tubulin and other proteins exists distally in cilia, and proteins are transported into and out of the structure by **kinesin** and **cytoplasmic dynein** motors moving along the peripheral doublets of microtubules.

(b) Ciliary movement involves a rapid series of changes in the shape of the axoneme. Along the length of each doublet, a series of paired "arms" with **axonemal dynein** is bound to microtubule A, with each pair extended toward microtubule B of the next doublet. When activated by ATP, the dynein arms briefly bind the neighboring microtubule and the doublets slide past each other slightly. The sliding motion is restricted by nexin cross-links between the doublets, causing the axoneme to bend. A rapid succession of this movement along the axoneme produces ciliary motion.

TABLE **4–3**	Common types of covering epithelia.		
Major Feature	**Cell Form**	**Examples of Distribution**	**Main Function**
Simple (one layer of cells)	Squamous	Lining of vessels (endothelium); Serous lining of cavities: pericardium, pleura, peritoneum (mesothelium)	Facilitates the movement of the viscera (mesothelium), active transport by pinocytosis (mesothelium and endothelium), secretion of biologically active molecules (mesothelium)
	Cuboidal	Covering the ovary, thyroid	Covering, secretion
	Columnar	Lining of intestine, gallbladder	Protection, lubrication, absorption, secretion
Stratified (two or more layers of cells)	Squamous keratinized (dry)	Epidermis	Protection; prevents water loss
	Squamous nonkeratinized (moist)	Mouth, esophagus, larynx, vagina, anal canal	Protection, secretion; prevents water loss
	Cuboidal	Sweat glands, developing ovarian follicles	Protection, secretion
	Transitional	Bladder, ureters, renal calyces	Protection, distensibility
	Columnar	Conjunctiva	Protection
Pseudostratified (layers of cells with nuclei at different levels; not all cells reach surface but all adhere to basal lamina)		Lining of trachea, bronchi, nasal cavity	Protection, secretion; cilia-mediated transport of particles trapped in mucus out of the air passages

FIGURE **4–12** Simple squamous epithelium.

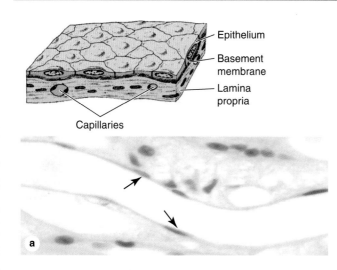

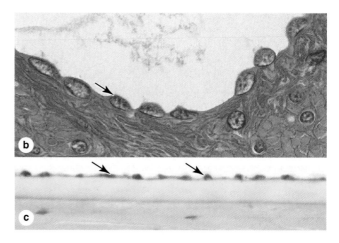

This is a single layer of thin cells, in which the **cell nuclei** (arrows) are the thickest and most visible structures. Simple epithelia are typically specialized as lining of vessels and cavities, where they regulate passage of substances into the underlying tissue. The thin cells often exhibit transcytosis. Examples shown here are those lining the thin renal loops of Henle **(a)**, covering the outer wall of the intestine **(b)**, and lining the inner surface of the cornea **(c)**. (a, c X400; b X600; H&E)

FIGURE **4–13** Simple cuboidal epithelium.

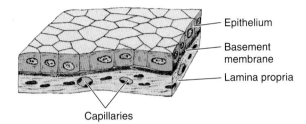

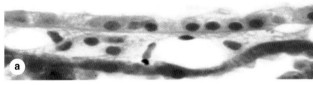

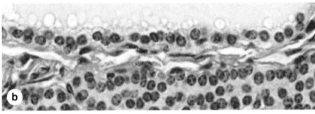

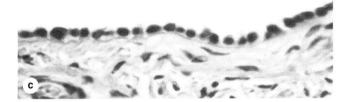

Cells here are roughly as tall as they are wide. Their greater thickness allows cytoplasm to be rich in mitochondria and other organelles for a high level of active transport across the epithelium and other functions. Examples shown here are from a renal collecting tubule **(a)**, a large thyroid follicle **(b)**, and the thick mesothelium covering an ovary **(c)**. (All X400; H&E)

FIGURE **4–14** Simple columnar epithelium.

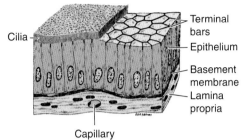

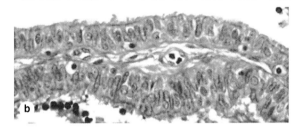

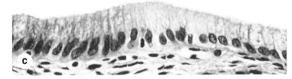

Cells here are always taller than they are wide, with apical cilia or microvilli, and are often specialized for absorption. Complexes of tight and adherent junctions, sometimes called "terminal bars" in light microscopic images, are present at the apical ends of cells. The examples shown here are from a renal collecting duct **(a)**, the oviduct lining, with both secretory and ciliated cells **(b)**, and the lining of the gall bladder **(c)**. (All X400; H&E)

with skin, this surface layer of cells helps protect against water loss across this epithelium. **Stratified squamous nonkeratinized epithelium** (Figure 4–15b) lines moist internal cavities (eg, mouth, esophagus, and vagina) where water loss is not a problem. Here the flattened cells of the surface layer retain their nuclei and most metabolic functions.

Stratified cuboidal and **stratified columnar epithelia** are both relatively rare. Stratified cuboidal epithelium occurs in the excretory ducts of salivary and sweat glands (Figure 4–15d). Stratified columnar epithelium is seen in the conjunctiva lining the eyelids, where it is both protective and mucus secreting.

Unique **transitional epithelium** or **urothelium** lines much of the urinary tract, extending from the kidneys to the proximal part of the urethra, and is characterized by a superficial layer of large, dome-like cells sometimes called umbrella cells (Figure 4–16). As discussed further with the urinary system, these cells are specialized to protect underlying tissues from the hypertonic and potentially cytotoxic effects of urine. Importantly, unique morphological features of the cells allow *distension* of transitional epithelium as the urinary bladder fills.

›› MEDICAL APPLICATION

In individuals with chronic vitamin A deficiency, epithelial tissues of the type found in the bronchi and urinary bladder may gradually be replaced by stratified squamous epithelium.

FIGURE **4–15** Stratified epithelium.

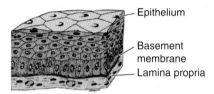

Epithelium
Basement membrane
Lamina propria

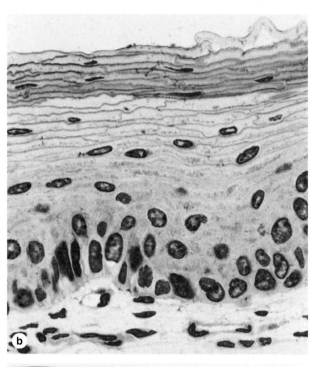

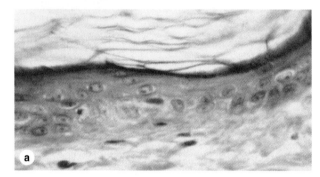

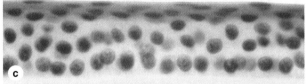

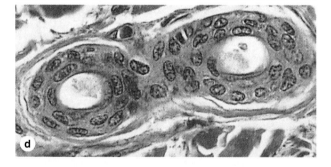

Stratified squamous epithelia usually have protective functions: protection against easy invasion of underlying tissue by microorganisms and protection against water loss. These functions are particularly important in the epidermis **(a)** in which differentiating cells become **keratinized**, that is, filled with keratin and other substances, eventually lose their nuclei and organelles, and form superficial layers flattened squames that impede water loss. Keratinized cells are sloughed off and replaced by new cells from more basal layers, which are discussed fully with the skin in Chapter 18.

Nonkeratinized epithelia occur in many organs, such as the esophageal lining **(b)** or outer covering of the cornea **(c)**. Here cells accumulate much less keratin and retain their nuclei but still provide protection against microorganisms.

Stratified cuboidal or columnar epithelia are fairly rare but occur in excretory ducts of certain glands, such as sweat glands **(d)** where the double layer of cells allows additional functions. All X400; (b) PT, (a, c, and d) H&E.

A final morphological type of epithelium is called **pseudostratified columnar epithelium** (Figure 4–17). Here tall, irregular cells all are attached to the basement membrane but their nuclei are at different levels and not all cells extend to the free surface, giving a stratified appearance. A good example of pseudostratified columnar epithelium is that lining the upper respiratory tract, where the cells are also heavily ciliated.

Secretory Epithelia & Glands

Epithelial cells that function mainly to produce and secrete various macromolecules may occur in epithelia with other major functions or comprise specialized organs called **glands**.

❯❯ MEDICAL APPLICATION

In **chronic bronchitis**, common among habitual smokers, the number of goblet cells in the lining of airways in the lungs often increases greatly. This leads to excessive mucus production in areas where there are too few ciliated cells for its rapid removal and contributes to obstruction of the airways. The ciliated pseudostratified epithelium lining the bronchi of smokers can also be transformed into stratified squamous epithelium by metaplasia.

Secretory cells may synthesize, store, and release proteins (eg, in the pancreas), lipids (eg, adrenal, sebaceous glands), or complexes of carbohydrates and proteins (eg, salivary glands). Epithelia of mammary glands secrete all three substances. The

FIGURE **4–16** Transitional epithelium or urothelium.

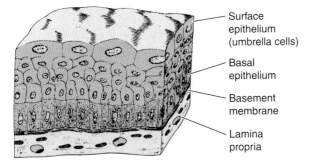

Surface epithelium (umbrella cells)
Basal epithelium
Basement membrane
Lamina propria

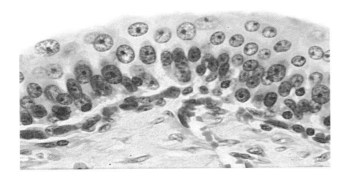

Urothelium is stratified and lines much of the urinary tract. The superficial cells are rounded or dome-shaped, and have specialized membrane features enabling them to withstand the hypertonic effects of urine and protect underlying cells from this toxic solution. Cells of this epithelium are also able to adjust their relationships with one another and undergo a transition in their appearance as the urinary bladder fills and the wall is distended. These unique features of transitional epithelium are discussed more extensively in Chapter 19. (X400; H&E)

FIGURE **4–17** Pseudostratified epithelium.

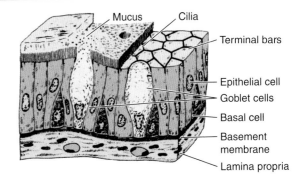

Mucus
Cilia
Terminal bars
Epithelial cell
Goblet cells
Basal cell
Basement membrane
Lamina propria

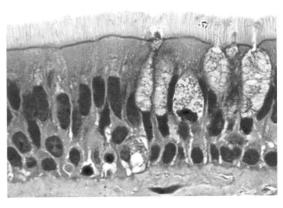

Cells of pseudostratified epithelia appear to be in several layers, but their basal ends all rest on the basement membrane. The pseudostratified columnar epithelium of the upper respiratory tract shown here contains many ciliated cells, as well as other cells with their nuclei at different levels. (X400; H&E)

cells of some glands (eg, sweat glands) have little synthetic activity and secrete mostly water and electrolytes (ions) transferred from the blood.

Scattered secretory cells, sometimes called unicellular glands, are common in simple cuboidal, simple columnar, and pseudostratified epithelia. An important, easily seen example is the **goblet cell** abundant in the lining of the small intestine (Figure 4–18) and respiratory tract (Figure 4–17), which secretes lubricating mucus that aids the function of these organs.

Glands develop from covering epithelia in the fetus by cell proliferation and growth into the underlying connective tissue, followed by further differentiation (Figure 4–19). **Exocrine glands** remain connected with the surface epithelium, the connection forming the tubular ducts lined with epithelium which deliver the secreted material where it is used. **Endocrine glands** lose the connection to their original epithelium and therefore lack ducts. Thin-walled blood vessels (capillaries) adjacent to endocrine cells absorb their

secreted hormone products for transport in blood to target cells throughout the body.

As shown in Figure 4–20, epithelia of exocrine glands are organized as a continuous system of many small **secretory portions** and **ducts** that transport the secretion out of the gland. In both exocrine and endocrine glands the secretory units are supported by a stroma of connective tissue. In larger glands layers of connective tissue also surround the larger ducts, form partitions or *septa* separating the gland into lobules, each containing secretory units connected to a small part of the duct system, and enclose the entire gland as its *capsule* (Figure 4–20).

The structures of their secretory portions and ducts allow exocrine glands to be classified as shown schematically in Table 4–4. Although the three-dimensional morphology is often not prominent in histologic sections, the key points are summarized as follows:

- Glands can be **simple** (ducts not branched) or **compound** (ducts with two or more branches).
- Secretory portions can be **tubular** (either short or long and **coiled**) or **acinar** (rounded and saclike); either

FIGURE 4–18   Goblet cells: unicellular glands.

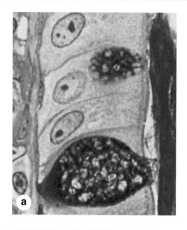
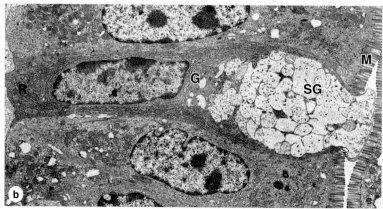

The simple columnar epithelium lining the small intestine shows many isolated goblet cells secreting mucus into the lumen. **(a)** With a stain for the oligosaccharide components of mucin glycoproteins, the cytoplasmic secretory granules of two goblet cells and secreted mucus are stained purple. (X600; PAS-PT) **(b)** As shown ultrastructurally, goblet cells always have basal nuclei surrounded by RER (**R**), a large Golgi complex (**G**), and abundant apical cytoplasm filled with large secretory granules (**SG**). After exocytosis mucin components are hydrated and become mucus. A brush border of microvilli (**M**) is seen on neighboring columnar cells. (X17,000)

type of secretory unit may be **branched**, even if the duct is not branched.

■ **Compound** glands can have branching ducts and can have multiple tubular, acinar, or tubuloacinar secretory portions.

Three basic mechanisms for releasing the product are commonly used by cells specialized for secretion (Figure 4–21), and cells engaged in each type of secretion can be distinguished histologically:

1. **Merocrine secretion:** This is the most common method of protein or glycoprotein secretion and involves typical exocytosis from membrane-bound vesicles or secretory granules.

2. **Holocrine secretion:** Here cells accumulate product continuously as they enlarge and undergo terminal differentiation, culminating in complete cell disruption which releases the product and cell debris into the gland's lumen. This is best seen in the sebaceous glands producing lipid-rich material in skin (Figure 4–22).

3. **Apocrine secretion:** Here product accumulates at the cells' apical ends, portions of which are then extruded to release the product together with small amounts of cytoplasm and cell membrane. Lipid droplets are secreted in the mammary gland in this manner (Figure 4–23).

Exocrine glands with merocrine secretion can be further categorized as either **serous** or **mucous** according to the nature of their secretory products, which give distinct staining properties to the cells. Serous cells synthesize proteins that are mostly not glycosylated, such as digestive enzymes. The cells

> **›› MEDICAL APPLICATION**
>
> The holocrine sebaceous glands are the primary structure involved in the common form of **acne**, acne vulgaris. Excessive holocrine secretion of sebum and keratin triggered by the surge of the steroid hormone testosterone that occurs in both genders at puberty frequently leads to blocked ducts within the gland. Activity of the normal commensal skin bacterium *Propionibacterium acnes* within the blocked duct commonly produces localized inflammation.

have well-developed RER and Golgi complexes and are filled apically with secretory granules in different stages of maturation (Figure 4–24). Serous cells therefore stain intensely with basophilic or acidophilic stains. Acini of the pancreas and parotid salivary glands are composed of serous cells.

Mucous cells, such as goblet cells, also have RER and Golgi complexes and are filled apically with secretory granules, but these contain heavily glycosylated proteins called **mucins**. When mucins are released from the cell, they become hydrated and form a layer of **mucus**. The hydrophilic mucins are usually washed from cells during routine histological preparations, causing the secretory granules to stain poorly with eosin (Figure 4–25). Sufficient oligosaccharides remain in developing mucinogen granules, however, to allow mucous cells to be stained by the PAS method (Figure 4–18a).

Some salivary glands are mixed **seromucous glands**, having both serous acini and mucous tubules with clustered serous cells (see Figure 16–5). The product of such glands is a mixture of digestive enzymes and watery mucus.

FIGURE **4–19** Formation of glands from covering epithelia.

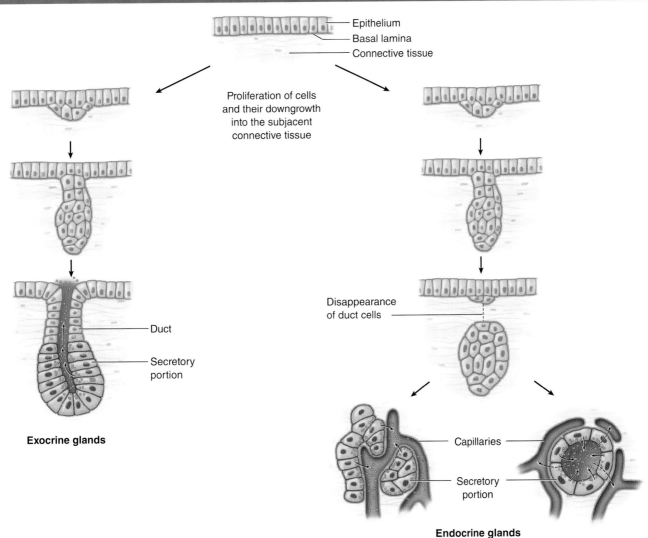

Epithelium
Basal lamina
Connective tissue

Proliferation of cells
and their downgrowth
into the subjacent
connective tissue

Duct

Secretory
portion

Exocrine glands

Disappearance
of duct cells

Capillaries

Secretory
portion

Endocrine glands

During fetal development epithelial cells proliferate and penetrate the underlying connective tissue. These cells may—or may not—maintain a connection with the surface epithelium. The connection is maintained to form a duct in exocrine glands; it is lost as endocrine glands develop. Exocrine glands secrete substances to specific organs via duct systems. Endocrine glands produce hormones and are always rich in capillaries. Hormones are released outside the cells and picked up by these blood vessels for distribution throughout the body, where specific target cells are identified by receptors for the hormones. Endocrine glands can have secretory cells arranged as irregular cords (left) or as rounded follicles (right) with lumens for temporary storage of the secretory product.

In addition to secretory cells, epithelia of many exocrine glands (eg, sweat, lachrymal, salivary, and mammary glands) contain contractile **myoepithelial cells** at the basal ends of the secretory cells (Figure 4–26). Long processes of these cells embrace an acinus as an octopus might embrace a rounded boulder. Bound to the basal lamina by hemidesmosomes and connected to the other epithelial cells by both gap junctions and desmosomes, myoepithelial cells are rich in actin filaments and myosins. Strong contractions in these cells serve to help propel secretory products from acini into the duct system.

Endocrine glands lack myoepithelial cells and are specialized for either protein or steroid **hormone** synthesis, with cytoplasmic staining characteristic of RER or SER, respectively. The proteins are released by exocytosis and the lipophilic steroids by diffusion through the cell membrane for uptake by binding proteins outside the cell. As mentioned previously, endocrine signaling involves hormone transport in the blood to target cells throughout the body, often within other endocrine glands. The receptors may also be on cells very close to the hormone-secreting cell or on the secreting cell itself, signaling which is termed paracrine or autocrine, respectively.

Important but inconspicuous endocrine or paracrine cells also occur singly or in small groups in epithelia of the

FIGURE **4–20**   **General structure of exocrine glands.**

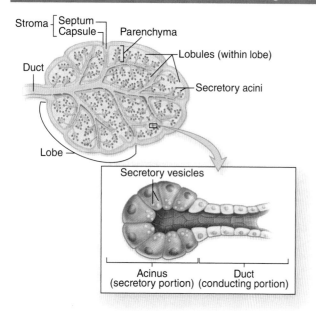

Exocrine glands by definition have ducts that lead to another organ or the body surface. Inside the gland the duct runs through the connective tissue of septa and branches repeatedly, until its smallest branches end in the secretory portions of the gland.

digestive, respiratory, and other organ systems. Hormones are also secreted from some cells specialized for other functions, such as certain cardiac muscle cells or fat cells. The pancreas contains both endocrine and exocrine cells. Liver cells exert both functions in the same cells, secreting bile components into a duct system and releasing other products to the bloodstream.

❯ TRANSPORT ACROSS EPITHELIA

Many cells have the ability to actively transport certain ions against concentration and electrical potential gradients. An important example is the extrusion of Na^+ from cells by the transmembrane protein Na^+/K^+-ATPase, also called the Na^+/K^+ pump, which allows cells to maintain the required low intracellular sodium concentration (5-15 mmol/L vs ~140 mmol/L in extracellular fluid).

Some epithelial cells specialize in the transfer of ions (by ion pumps) and water (via the membrane channels called aquaporins) in either direction across the epithelium, the process known as **transcellular transport** (Figure 4–27). Apical tight junctions prevent paracellular diffusion or backflow *between* the cells.

Epithelia of kidney tubules are key sites for ion and water transport, maintaining the body's overall balance of

salts and water. Cells of the proximal renal tubules are specialized structurally for transcellular transport. The apical surface at the tubule lumen is freely permeable to Na^+, and the basolateral cell membranes have sodium pumps for the active extrusion of Na^+ into the interstitial fluid outside the tubules. Osmotic and electrical balance is maintained by the passive transfer of chloride ions (Cl^-) and water into the cell. The basal membrane of these cells is elaborately folded, with mitochondria located between the folds to supply ATP for Na^+/K^+ pumps (Figure 4–28). Lateral membrane folds interdigitating between the cells further increase the surface area for transport. Regulated transfer of ions and water by various epithelial cells along the renal tubules maintains the ionic balance within the body and allows excretion of excess water and salts in the urine.

All cells can also internalize extracellular molecules and fluid using endocytosis and formation of cytoplasmic, membrane-bound vesicles. This activity is clearly observed in the simple squamous epithelial cells lining blood and lymphatic capillaries (endothelia) or body cavities (mesothelia). These thin cells have few organelles other than the abundant pinocytotic vesicles, which cross the thin cells in both directions and release their contents on the opposite side by exocytosis. This process of **transcytosis** also occurs between the apical and basolateral membranes domains in cells of simple cuboidal and columnar epithelia and is important in many physiologic processes.

❯ RENEWAL OF EPITHELIAL CELLS

Epithelial tissues are relatively labile structures whose cells are renewed continuously by mitotic activity and stem cell populations. The rate of renewal varies widely; it can be fast in tissues such as the intestinal epithelium, which is replaced every week, or slow, as in the large glands. In stratified epithelial tissues, stem cells and mitosis occur only within the basal layer in contact with the basal lamina. In some functionally complex epithelia, stem cells are located only in restricted niches some distance from the transit amplifying cells and differentiating cells. For example, the epithelium lining the small intestine is derived completely from stem cells found in the simple glands between the intestinal villi. In the epidermis, many stem cells are located at a characteristic position along the wall of hair follicles.

❯❯ MEDICAL APPLICATION

Both benign and malignant tumors can arise from most types of epithelial cells. Malignant tumors of epithelial origin are called **carcinomas** (Gr. *karkinos*, cancer + *oma*, tumor). Malignant tumors derived from glandular epithelial tissue are called **adenocarcinomas** (Gr. *adenos*, gland + *karkinos*). Adenocarcinomas are by far the most common tumors in adults after age 45.

TABLE **4–4**	Structural classes of exocrine glands, features of each class, and examples.

SIMPLE Glands (Ducts Do Not Branch)

Class	Simple Tubular	Branched Tubular	Coiled Tubular	Acinar (or Alveolar)	Branched Acinar
Features	Elongated secretory portion; duct usually short or absent	Several long secretory parts joining to drain into 1 duct	Secretory portion is very long and coiled	Rounded, saclike secretory portion	Multiple saclike secretory parts entering the same duct
Examples	Mucous glands of colon; intestinal glands or crypts (of Lieberkühn)	Glands in the uterus and stomach	Sweat glands	Small mucous glands along the urethra	Sebaceous glands of the skin

COMPOUND Glands (Ducts from Several Secretory Units Converge into Larger Ducts)

Class	Tubular	Acinar (Alveolar)	Tubuloacinar
Features	Several *elongated* coiled secretory units and their ducts converge to form larger ducts	Several *saclike* secretory units with small ducts converge at a larger duct	Ducts of both tubular and acinar secretory units converge at larger ducts
Examples	Submucosal mucous glands (of Brunner) in the duodenum	Exocrine pancreas	Salivary glands

Epithelia are normally capable of rapid repair and replacement of apoptotic or damaged cells. In some large glands, most notably the liver, mitotic activity is normally rare but is actively renewed following major damage to the organ. When a portion of liver tissue is removed surgically or lost by the acute effects of toxic substances, cells of undamaged regions quickly begin active proliferation and a mass of liver tissue with normal function is regenerated.

›› MEDICAL APPLICATION

Some epithelial cells are prone to abnormal growth or dysplasia, which can progress to precancerous growth called **neoplasia**. Early neoplastic growth is often reversible and does not always result in cancer.

Under certain abnormal conditions, one type of epithelial tissue may undergo transformation into another type in another reversible process called **metaplasia**. In heavy cigarette smokers, the ciliated pseudostratified epithelium lining the bronchi can be transformed into stratified squamous epithelium.

FIGURE 4–21 Mechanisms of exocrine gland secretion.

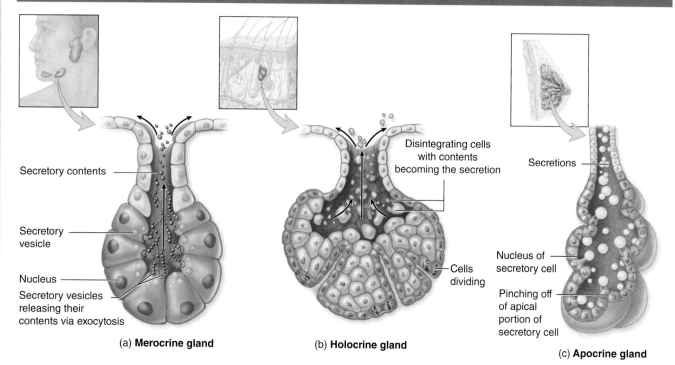

Secretory contents

Secretory vesicle

Nucleus

Secretory vesicles releasing their contents via exocytosis

(a) **Merocrine gland**

Disintegrating cells with contents becoming the secretion

Cells dividing

(b) **Holocrine gland**

Secretions

Nucleus of secretory cell

Pinching off of apical portion of secretory cell

(c) **Apocrine gland**

Three basic types of secretion are used by cells of exocrine glands, depending on what substance is being secreted.

(a) Merocrine secretion releases products, usually containing proteins, by means of exocytosis at the apical end of the secretory cells. Most exocrine glands are merocrine.

(b) Holocrine secretion is produced by the disintegration of the secretory cells themselves as they complete their terminal differentiation, which involves becoming filled with product. Sebaceous glands of hair follicles are the best examples of holocrine glands.

(c) Apocrine secretion involves loss of membrane-enclosed apical cytoplasm, usually containing one or more lipid droplets. Apocrine secretion, along with merocrine secretion, is seen in mammary glands.

Epithelial Tissue SUMMARY OF KEY POINTS

- An **epithelium** is a tissue in which cells are bound tightly together structurally and functionally to form a sheetlike or tubular structure with little extracellular material between the cells.
- Cells in epithelia each have an **apical side** facing the sheet's free surface and a **basal side** facing a basement membrane and underlying connective tissue.
- Epithelia are often specialized for absorption or **transcytosis**, pinocytosis of material at the apical side and exocytosis at the basolateral side (or vice versa).
- Cells of most epithelia exhibit **continuous renewal**, with the locations of stem cells and rates of cell turnover variable in various specialized epithelia.

Basement Membrane

- The **basement membrane** of all epithelia is a thin extracellular layer of specialized proteins, usually having two parts: a basal lamina and a more fibrous reticular lamina.
- The **basal lamina** is a thin meshwork of type IV collagen and laminin produced by the epithelial cells.
- The **reticular lamina** contains type III collagen and anchoring fibrils of VII collagen, all secreted by cells of the immediately adjacent connective tissue.

- Together, these components **attach** epithelia to connective tissue, regulate (**filter**) substances passing from connective tissue into epithelia, provide a guide or **scaffold** during tissue regeneration after injury, and **compartmentalize** epithelial cells from other tissues.

Intercellular Junctions

- Intercellular junctions are well developed in epithelia and consist of three major types, with different functions.
- **Tight or occluding junctions** are formed by interacting transmembrane proteins such as **claudin** and **occludin**; linear arrangements of these linked proteins surround the apical ends of the cells and **prevent paracellular passage** of substances (between the cells.)
- **Adherent or anchoring junctions**, formed by interacting proteins of the **cadherin** family, are points of strong **attachment** holding together cells of the epithelium.
- Adherent junctions may form **zonula adherens** that encircle epithelial cells just below their tight junctions or scattered, spot-like attachment sites called **desmosomes** or **maculae adherens**, both of which are attached to cytoplasmic **keratins**.

FIGURE **4–22** Holocrine secretion in a sebaceous gland.

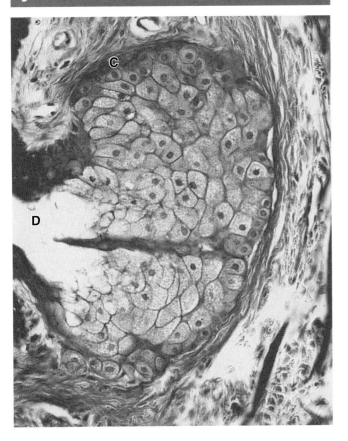

In holocrine secretion, best seen in the sebaceous gland adjacent to hair follicles, entire cells fill with a lipid-rich product as they differentiate. Mature (terminally differentiated) cells separate and completely disintegrate, releasing the lipid that serves to protect and lubricate adjacent skin and hair. Sebaceous glands lack myoepithelial cells; cell proliferation inside a dense, inelastic connective tissue capsule continuously forces product into the duct. (X200; H&E)

FIGURE **4–23** Apocrine secretion in the mammary gland.

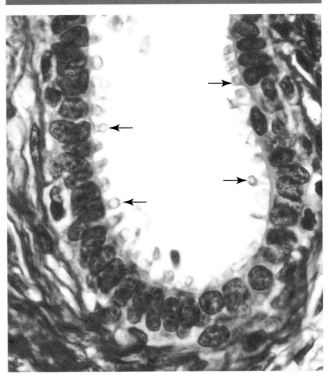

The secretory portions of a mammary gland demonstrate apocrine secretion, characterized by extrusion of the secretion product along with a bit of apical cytoplasm (**arrows**). The released portion of cell contains lipid droplet(s). Merocrine secretion also occurs from the same and other cells of the gland. (X400; PSH)

- **Hemidesmosomes** composed of transmembrane **integrins** attach cells to proteins of the basal lamina.
- **Gap or communicating junctions** are points of cell contact where both plasma membranes have numerous hexameric complexes of transmembrane **connexons**, each forming a channel allowing passage of small molecules from one cell to the other.

Apical Structures of Epithelial Cells

- **Microvilli** are small membrane projections with cores of **actin filaments** that generally function to increase epithelial cells' apical surface area for **absorption**.
- **Stereocilia** are long microvilli with specialized mechanosensory function in cells of the inner ear and for absorption in tissues of the male reproductive tract.
- **Cilia** are larger projecting structures with a well-organized core of **microtubules** (in a 9 + 2 arrangement called the **axoneme**) in which restricted, dynein-based sliding of microtubules causes ciliary movement that propel material along an epithelial surface.

Morphological Types of Epithelia

- An epithelium in which the basement membrane has one cell layer is **simple**; the cells of different simple epithelia range widely in height, from very thin or **squamous**, to roughly **cuboidal**, to very tall or **columnar**.
- Epithelia with two or more layers of cells are **stratified** and almost all such epithelia are stratified squamous, in which the outer cell layers are thin and flattened.
- Cells of stratified squamous epithelia move gradually from the basal to the surface layers, changing shape and becoming filled with **keratin** intermediate filaments.
- Stratified squamous epithelia such as the epidermis cover the body surface, **protecting** underlying tissues from excess water loss (dehydration) and microbial invasion.
- **Pseudostratified epithelia** are thick and appear to have several cell layers; all cells attach to the basal lamina but not all extend to the free epithelial surface.

FIGURE 4–24 Serous cells.

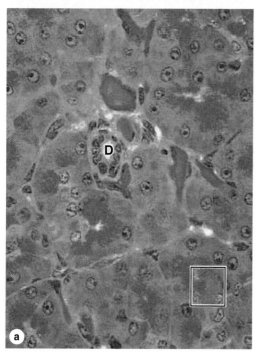
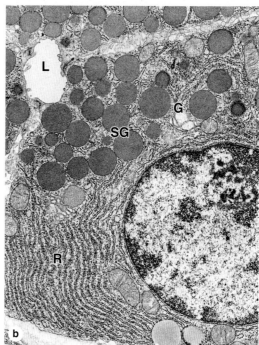

The small serous acini of the exocrine pancreas each have 5-10 cells facing a very small central lumen. Each acinar cell is roughly pyramidal, with its apex at the lumen. **(a)** As seen by light microscopy, the apical ends are very eosinophilic due to the abundant secretory granules present there. The cells' basal ends contain the nuclei and an abundance of RER, making this area basophilic. A small duct **(D)** is seen, but lumens of acini are too small to be readily visible. The enclosed area is comparable to that shown in part **b**. (X300; H&E) **(b)** A portion of one acinar cell is shown ultrastructurally, indicating the abundant RER **(R)**, a Golgi complex **(G)**, apical secretory granules **(SG)** and the small acinar lumen **(L)**. (X13,000)

- **Transitional epithelium or urothelium**, found only in the lining of the urinary system, is stratified, with large rounded surface cells protective against urine.

Epithelial Secretion/Glands

- The major function in many epithelial cells is synthesis and secretion of specialized products; organs composed primarily of such epithelia are called **glands**.
- **Exocrine glands** have epithelial ducts carrying secretions to specific sites; the ducts of **simple glands** are unbranched and those of **compound glands** are branched.
- The secretory portions of exocrine glands may form round, saclike **acini** (also called **alveoli**) or elongated **tubules**; both types of secretory units may themselves branch.

- **Endocrine glands** lack ducts; secreted substances are hormones carried throughout the body by the interstitial fluid and blood, with specificity produced by the hormone receptors of target cells.
- Glands have three basic secretory mechanisms: **merocrine**, which uses exocytosis; **holocrine**, in which terminally differentiated cells filled with lipid product are released; and **apocrine**, in which apical, product-filled areas of cells are extruded.
- Exocrine glands producing mucus, or similar individual cells called **goblet cells**, are called **mucous glands**; oligosaccharide components of mucus stain poorly with routine dyes but stain well with PAS stain.
- Exocrine glands producing largely enzymes (proteins) are called **serous glands** and stain darkly with H&E due to the cells' content of RER and secretory granules.

FIGURE **4–25** Mucous cells.

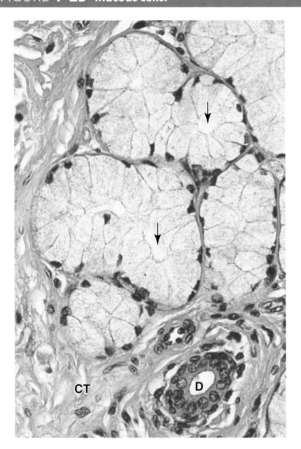

Mucous cells of salivary glands are typically larger than serous cells, with flattened basal nuclei. Most of the cytoplasm is filled with secretory granules containing mucinogen like that of goblet cells. The RER and Golgi complexes of mucous cells produce heavily glycosylated glycoproteins with water-binding properties. The lumens (arrows) of mucous tubules are larger than those of serous acini. Much connective tissue surrounds the mucous tubules and ducts (**D**). (X200; PT)

FIGURE **4–26** Myoepithelial cells.

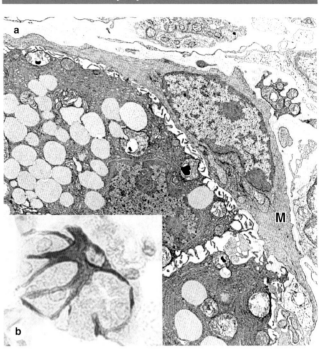

(a) The TEM shows two salivary gland cells containing secretory granules, with an associated myoepithelial cell (**M**). (X20,000) **(b)** A myoepithelial cell immunostained brown with antibodies against actin shows its association with cells of an acinus stained by H&E. Contraction of the myoepithelial cell compresses the acinus and aids in the expulsion of secretory products into the duct. (X200)

Epithelial Tissue   ASSESS YOUR KNOWLEDGE

1. Functions of the basement membrane include which of the following?
   a. Contractility
   b. Molecular filtering
   c. Active ion transport
   d. Excitability
   e. Modification of secreted proteins

2. Using immunohistochemistry a population of cells is shown to be positive for the protein connexin. From this we can infer that the cells are connected by what type of junction?
   a. Tight (occluding) junctions
   b. Zonula adherens
   c. Gap junctions
   d. Hemidesmosomes
   e. Desmosomes (macula adherens)

3. An individual genetically unable to synthesize normal occludin is likely to have epithelia with defective regulation in which of the following?
   a. Material crossing the epithelium between the cells (paracellular movement)
   b. Communication between the cells
   c. Attachment to the basement membrane
   d. Strong attachment to neighboring cells
   e. Movement of membrane proteins in the apical domains of cells

4. An intermediate filament protein found in cytoplasm of most epithelial cells is which of the following?
   a. Actin
   b. Vimentin
   c. Laminin
   d. Myosin
   e. Keratin

FIGURE **4–27** Ion and water absorption and secretion.

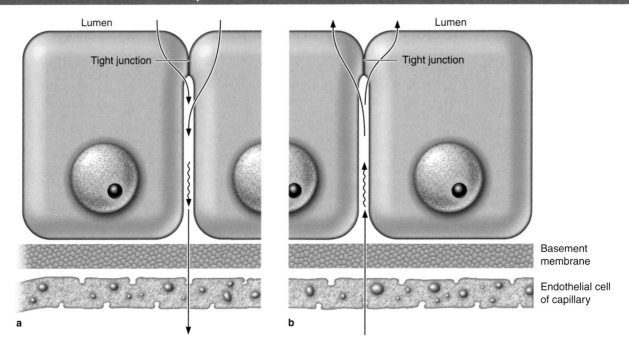

Ion and water transport across epithelia can occur in either direction, depending on the organ involved. **(a) Absorption** is the process of transport from an organ or duct's lumen to capillaries near the epithelial basement membrane and involves movement from the apical to the basolateral cell membrane domains. Absorption occurs for example in the epithelium of the gallbladder and intestine where it serves to concentrate bile or obtain water and ions from digested material.

(b) Secretion involves transport in the other direction from the capillaries into a lumen, as in many glands and the choroid plexus. Secretion by epithelial cells removes water from the neighboring interstitial fluid or plasma and releases it as part of the specialized aqueous fluids in such organs.

No matter whether an epithelium is involved in absorption or secretion, apical occluding junctions are necessary to maintain tight separation of the apical and basolateral compartments of either side of the epithelium.

5. Which of the following cellular features is used in naming types of epithelia?
   a. Shape of cells in the basal layer
   b. Number of cell layers
   c. Presence of a basal lamina
   d. Size of the nuclei
   e. Nature of the cell junctions that are present

6. The release of lipid droplets from cells is which type of secretion?
   a. Merocrine
   b. Serous
   c. Apocrine
   d. Mucous
   e. Holocrine

7. Exocrine glands in which the acini all produce a secretion of heavily glycosylated, hydrophilic proteins are an example of which type of gland?
   a. Serous gland
   b. Mixed gland
   c. Mucous gland
   d. Tubuloacinar gland
   e. Simple gland

8. With a 5-year history of chronic respiratory infections, a 23-year-old, non-smoking man is referred to an otolaryngologist. A bronchial biopsy indicates altered structures in the epithelial cells. Which of the following, if altered to reduce function, is most likely involved in this patient's condition?
   a. Hemidesmosomes
   b. Cilia
   c. Basolateral cell membrane folds
   d. Microvilli
   e. Tight junctions

FIGURE **4–28** Features of absorptive cells.

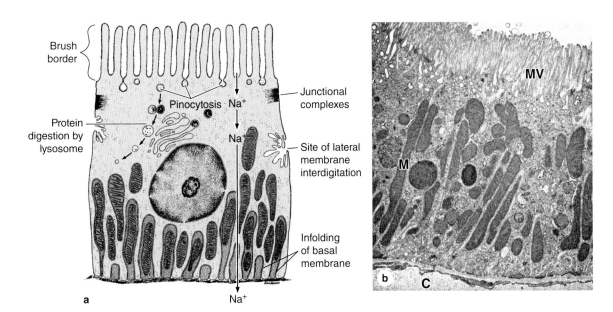

A diagram and TEM photo showing the major ultrastructural features of a typical epithelial cell highly specialized for absorption, cells of proximal convoluted tubule of the kidney. The apical cell surface has a brush border consisting of uniform microvilli (**MV**) which increase the area of that surface to facilitate all types of membrane transport. Vesicles formed during pinocytosis may fuse with lysosomes as shown in (**a**) or mediate transcytosis by secreting their contents at the basolateral cell membrane. The basal cell surface is also enlarged, here by invaginations of the cell membrane which are associated with mitochondria (**M**) providing ATP for active transport. Basolateral membrane infoldings from neighboring cells (the more heavily stippled structures) also with mitochondria interdigitate with those of this cell. Various ions entering through the apical membranes of renal epithelial cells undergo active transport out of the cells across the basolateral membrane. Immediately below the basal lamina shown in (**b**) is a capillary (**C**) that removes water and other substances absorbed across the epithelium. Junctional complexes between individual cells separate the apical and basolateral compartments on either side of the epithelium. Epithelial cells also show lateral membrane interdigitations with neighboring cells. (X9600)

9. An 11-month-old girl is referred to a pediatric gastroenterology clinic due to a history of generalized weakness, slow growth, and refractory diarrhea. For the past month she has been hospitalized regularly to receive parenteral nutrition. Examination of the epithelium lining her small intestine confirms that the failure to absorb nutrients is most likely due to a significant decrease in which of the following?

    a. Microvilli
    b. Gap junctions
    c. Cilia
    d. Cell layers
    e. Basement membrane thickness

10. A 42-year-old woman of Mediterranean descent presents with multiple oral blisters and a few cutaneous blisters on her back and buttocks. The superficial bullae are fragile, some have unroofed to form ulcerated lesions, and there is a positive Nikolsky sign. Blood tests reveal antibodies to a subfamily of cadherins and immunohistochemical staining of a biopsy from the oral mucosa shows distribution of the antigen throughout the epithelium. In what structures is the defect that is causing this patient's condition?

    a. Desmosomes
    b. Tight junctions
    c. Hemidesmosomes
    d. Gap junctions
    e. Reticular lamina

Answers: 1b, 2c, 3a, 4e, 5b, 6c, 7c, 8b, 9a, 10a

5 Connective Tissue

Connective tissue provides a matrix that supports and physically connects other tissues and cells together to form the organs of the body. The interstitial fluid of connective tissue gives metabolic support to cells as the medium for diffusion of nutrients and waste products.

Unlike the other tissue types (epithelium, muscle, and nerve), which consist mainly of cells, the major constituent of connective tissue is the **extracellular matrix (ECM)**. Extracellular matrices consist of different combinations of **protein fibers** (collagen and elastic fibers) and **ground substance**. Ground substance is a complex of anionic, hydrophilic proteoglycans, glycosaminoglycans (GAGs), and multiadhesive glycoproteins (laminin, fibronectin, and others). As described briefly in Chapter 4 with the basal lamina, such glycoproteins help stabilize the ECM by binding to other matrix components and to integrins in cell membranes. Water within this ground substance allows the exchange of nutrients and metabolic wastes between cells and the blood supply.

The variety of connective tissue types in the body reflects differences in composition and amount of the cells, fibers, and ground substance which together are responsible for the remarkable structural, functional, and pathologic diversity of connective tissue.

All connective tissues originate from embryonic **mesenchyme**, a tissue developing mainly from the middle layer of the embryo, the mesoderm. Mesenchyme consists largely of viscous ground substance with few collagen fibers (Figure 5–1). **Mesenchymal cells** are undifferentiated and have large nuclei, with prominent nucleoli and fine chromatin. They are often said to be "spindle-shaped," with their scant cytoplasm extended as two or more thin cytoplasmic processes. Mesodermal cells

migrate from their site of origin in the embryo, surrounding and penetrating developing organs. In addition to producing all types of connective tissue proper and the specialized connective tissues bone and cartilage, the embryonic mesenchyme includes stem cells for other tissues such as blood, the vascular endothelium, and muscle. This chapter describes the features of soft, supportive connective tissue proper.

> ## ❯❯ MEDICAL APPLICATION
>
> Some cells in mesenchyme are **multipotent stem cells** potentially useful in **regenerative medicine** after grafting to replace damaged tissue in certain patients. Mesenchyme-like cells remain present in some adult connective tissues, including that of tooth pulp and some adipose tissue, and are being investigated as possible sources of stem cells for therapeutic repair and organ regeneration.

❯ CELLS OF CONNECTIVE TISSUE

Fibroblasts are the key cells in connective tissue proper (Figure 5–2 and Table 5–1). Fibroblasts originate locally from mesenchymal cells and are permanent residents of connective tissue. Other cells found here, such as **macrophages**, **plasma cells**, and **mast cells**, originate from hematopoietic stem cells in bone marrow, circulate in the blood, and then move into connective tissue where they function. These and other white blood cells (leukocytes) are transient cells of most connective tissues, where they perform various functions for a short period as needed and then die by apoptosis.

FIGURE **5–1** Embryonic mesenchyme.

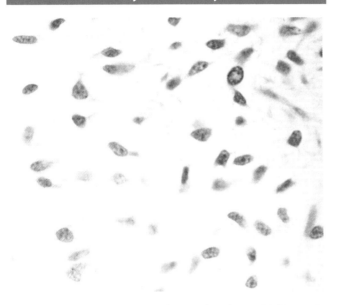

Mesenchyme consists of a population of undifferentiated cells, generally elongated but with many shapes, having large euchromatic nuclei and prominent nucleoli that indicate high levels of synthetic activity. These cells are called **mesenchymal cells**. Mesenchymal cells are surrounded by an ECM that they produced and that consists largely of a simple ground substance rich in hyaluronan (hyaluronic acid), but with very little collagen. (X200; Mallory trichrome)

Fibroblasts are targets of many families of proteins called **growth factors** that influence cell growth and differentiation. In adults, connective tissue fibroblasts rarely undergo division. However, stimulated by locally released growth factors, cell cycling and mitotic activity resume when the tissue requires additional fibroblasts, for example, to repair a damaged organ. Fibroblasts involved in wound healing, sometimes called **myofibroblasts**, have a well-developed contractile function and are enriched with a form of actin also found in smooth muscle cells.

❯❯ MEDICAL APPLICATION

The regenerative capacity of connective tissue is clearly observed in organs damaged by ischemia, inflammation, or traumatic injury. Spaces left after such injuries, especially in tissues whose cells divide poorly or not at all (eg, cardiac muscle), are filled by connective tissue, forming dense irregular **scar tissue**. The healing of surgical incisions and other wounds depends on the reparative capacity of connective tissue, particularly on activity and growth of fibroblasts.

In some rapidly closing wounds, a cell called the myofibroblast, with features of both fibroblasts and smooth muscle cells, is also observed. These cells have most of the morphologic characteristics of fibroblasts but contain increased amounts of actin microfilaments and myosin and behave much like smooth muscle cells. Their activity is important for the phase of tissue repair called **wound contraction**.

Fibroblasts

Fibroblasts (Figure 5–3), the most common cells in connective tissue proper, produce and maintain most of the tissue's extracellular components. Fibroblasts synthesize and secrete collagen (the most abundant protein of the body) and elastin, which both form large fibers, as well as the GAGs, proteoglycans, and multiadhesive glycoproteins that comprise the ground substance. As described later, most of the secreted ECM components undergo further modification outside the cell before assembling as a matrix.

Distinct levels of fibroblast activity can be observed histologically (Figure 5–3b). Cells with intense synthetic activity are morphologically different from the quiescent fibroblasts that are scattered within the matrix they have already synthesized. Some histologists reserve the term "fibroblast" to denote the active cell and "fibrocyte" to denote the quiescent cell. The active fibroblast has more abundant and irregularly branched cytoplasm, containing much rough endoplasmic reticulum (RER) and a well-developed Golgi apparatus, with a large, ovoid, euchromatic nucleus and a prominent nucleolus. The quiescent cell is smaller than the active fibroblast, is usually spindle-shaped with fewer processes, much less RER, and a darker, more heterochromatic nucleus.

Adipocytes

Adipocytes (L. *adeps*, fat + Gr. *kytos*, cell), or fat cells, are found in the connective tissue of many organs. These large, mesenchymally derived cells are specialized for cytoplasmic storage of lipid as neutral fats, or less commonly for the production of heat. Tissue with a large population of adipocytes, called adipose connective tissue, serves to cushion and insulate the skin and other organs. Adipocytes have major metabolic significance with considerable medical importance and are described and discussed separately in Chapter 6.

Macrophages & the Mononuclear Phagocyte System

Macrophages have highly developed phagocytic ability and specialize in turnover of protein fibers and removal of dead cells, tissue debris, or other particulate material, being especially abundant at sites of inflammation. Size and shape vary considerably, corresponding to their state of functional activity. A typical macrophage measures between 10 and 30 μm in diameter and has an eccentrically located, oval or kidney-shaped nucleus. Macrophages are present in the connective tissue of most organs and are sometimes referred to by pathologists as "histiocytes."

FIGURE 5–2 Cellular and extracellular components of connective tissue.

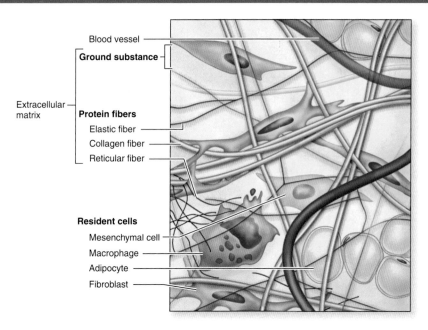

Blood vessel

Ground substance

Extracellular matrix

Protein fibers
Elastic fiber
Collagen fiber
Reticular fiber

Resident cells
Mesenchymal cell
Macrophage
Adipocyte
Fibroblast

Connective tissue is composed of **fibroblasts** and other cells and an **extracellular matrix (ECM)** of various protein fibers, all of which are surrounded by watery **ground substance**. In all types of connective tissue the extracellular volume exceeds that of the cells.

TABLE 5–1	Functions of cells in connective tissue proper.
Cell Type	**Major Product or Activity**
Fibroblasts (fibrocytes)	Extracellular fibers and ground substance
Plasma cells	Antibodies
Lymphocytes (several types)	Various immune/defense functions
Eosinophilic leukocytes	Modulate allergic/vasoactive reactions and defense against parasites
Neutrophilic leukocytes	Phagocytosis of bacteria
Macrophages	Phagocytosis of ECM components and debris; antigen processing and presentation to immune cells; secretion of growth factors, cytokines, and other agents
Mast cells and basophilic leukocytes	Pharmacologically active molecules (eg, histamine)
Adipocytes	Storage of neutral fats

❯❯ MEDICAL APPLICATION

Besides their function in turnover of ECM fibers, macrophages are key components of an organism's innate immune defense system, removing cell debris, neoplastic cells, bacteria, and other invaders. Macrophages are also important antigen-presenting cells required for the activation and specification of lymphocytes.

When macrophages are stimulated (by injection of foreign substances or by infection), they change their morphologic characteristics and properties, becoming **activated macrophages**. In addition to showing an increase in their capacity for phagocytosis and intracellular digestion, activated macrophages exhibit enhanced metabolic and lysosomal enzyme activity. Macrophages are also secretory cells producing an array of substances, including various enzymes for ECM breakdown and various growth factors or cytokines that help regulate immune cells and reparative functions.

When adequately stimulated, macrophages may increase in size and fuse to form **multinuclear giant cells**, usually found only in pathologic conditions.

FIGURE **5–3** Fibroblasts.

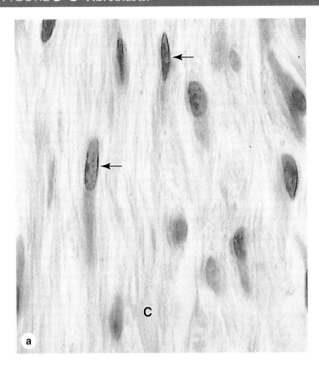

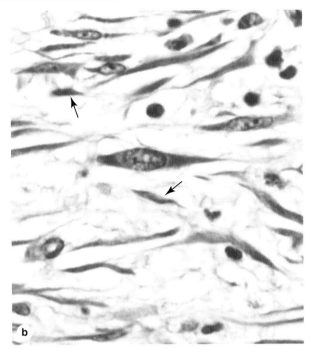

(a) Fibroblasts typically have large active nuclei and eosinophilic cytoplasm that tapers off in both directions along the axis of the nucleus, a morphology often referred to as "spindle-shaped." Nuclei (**arrows**) are clearly seen, but the eosinophilic cytoplasmic processes resemble the collagen bundles (**C**) that fill the ECM and are difficult to distinguish in H&E-stained sections.

(b) Both active and quiescent fibroblasts may sometimes be distinguished, as in this section of dermis. Active fibroblasts have large, euchromatic nuclei and basophilic cytoplasm, while inactive fibroblasts (or fibrocytes) are smaller with more heterochromatic nuclei (**arrows**). The round, very basophilic round cells are in leukocytes. (Both X400; H&E)

In the TEM, macrophages are shown to have a characteristic irregular surface with pleats, protrusions, and indentations, features related to their active pinocytotic and phagocytic activities (Figure 5–4). They generally have well-developed Golgi complexes and many lysosomes.

Macrophages derive from bone marrow precursor cells called **monocytes** that circulate in the blood. These cells cross the epithelial wall of small venules to enter connective tissue, where they differentiate, mature, and acquire the morphologic features of phagocytic cells. Therefore, monocytes and macrophages are the same cell at different stages of maturation. Macrophages play a very important role in the early stages of repair and inflammation after tissue damage. Under such conditions these cells accumulate in connective tissue by local proliferation of macrophages and recruitment of more monocytes from the blood. Macrophages are distributed throughout the body and are normally present in the stroma of most organs. Along with other monocyte-derived cells, they comprise a family of cells called the **mononuclear phagocyte system** (Table 5–2). All of these macrophage-like cells are derived from monocytes, but have different names in various organs, for example, Kupffer cells in the liver, microglial cells in the central nervous system,

Langerhans cells in the skin, and osteoclasts in bone. All are long-living cells and may survive in the tissues for months. In addition to debris removal, these cells are highly important for the uptake, processing, and presentation of antigens for lymphocyte activation, a function discussed later with the immune system. The transformation from monocytes to macrophages in connective tissue involves increases in cell size, increased protein synthesis, and increases in the number of Golgi complexes and lysosomes.

Mast Cells

Mast cells are oval or irregularly shaped cells of connective tissue, between 7 and 20 μm in diameter, filled with basophilic secretory granules which often obscure the central nucleus (Figure 5–5). These granules are electron-dense and of variable size, ranging from 0.3 to 2.0 μm in diameter. Because of the high content of acidic radicals in their sulfated GAGs, mast cell granules display **metachromasia**, which means that they can change the color of some basic dyes (eg, toluidine blue) from blue to purple or red. The granules are poorly preserved by common fixatives, so that mast cells may be difficult to identify in routinely prepared slides.

FIGURE **5–4** **Macrophage ultrastructure.**

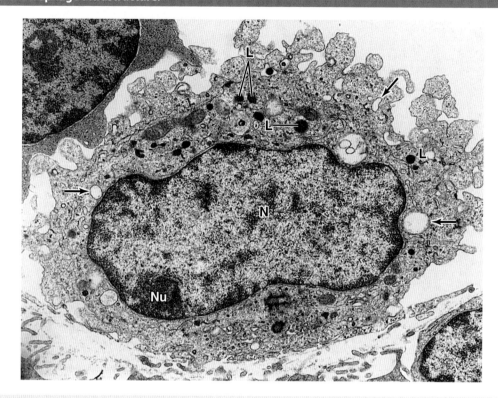

Characteristic features of macrophages seen in this TEM of one such cell are the prominent nucleus (**N**) and the nucleolus (**Nu**) and the numerous secondary lysosomes (**L**). The arrows indicate phagocytic vacuoles near the protrusions and indentations of the cell surface. (X10,000)

Mast cells function in the localized release of many bioactive substances important in the local inflammatory response, innate immunity, and tissue repair. A partial list of molecules released from these cells' secretory granules includes the following:

■ **Heparin**, a sulfated GAG that acts locally as an anticoagulant

■ **Histamine**, which promotes increased vascular permeability and smooth muscle contraction
■ **Serine proteases**, which activate various mediators of inflammation
■ **Eosinophil** and **neutrophil chemotactic factors**, which attract those leukocytes

TABLE **5–2**   Distribution and main functions of the cells of the mononuclear phagocyte system.

Cell Type	Major Location	Main Function
Monocyte	Blood	Precursor of macrophages
Macrophage	Connective tissue, lymphoid organs, lungs, bone marrow, pleural and peritoneal cavities	Production of cytokines, chemotactic factors, and several other molecules that participate in inflammation (defense), antigen processing, and presentation
Kupffer cell	Liver (perisinusoidal)	Same as macrophages
Microglial cell	Central nervous system	Same as macrophages
Langerhans cell	Epidermis of skin	Antigen processing and presentation
Dendritic cell	Lymph nodes, spleen	Antigen processing and presentation
Osteoclast (from fusion of several macrophages)	Bone	Localized digestion of bone matrix
Multinuclear giant cell (several fused macrophages)	In connective tissue under various pathological conditions	Segregation and digestion of foreign bodies

FIGURE **5–5** Mast cells.

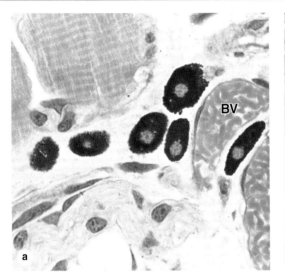

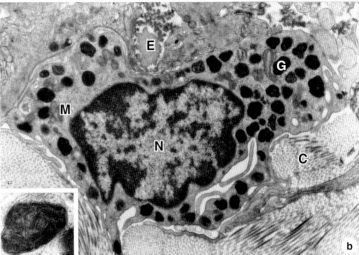

Mast cells are components of loose connective tissues, often located near small blood vessels (**BV**). **(a)** They are typically oval shaped, with cytoplasm filled with strongly basophilic granules. (X400; PT)

(b) Ultrastructurally mast cells show little else around the nucleus (**N**) besides these cytoplasmic granules (**G**), except for occasional

mitochondria (**M**). The granule staining in the TEM is heterogeneous and variable in mast cells from different tissues; at higher magnifications some granules may show a characteristic scroll-like substructure (inset) that contains preformed mediators such as histamine and proteoglycans. The ECM near this mast cell includes elastic fibers (**E**) and bundles of collagen fibers (**C**).

- ▪ **Cytokines**, polypeptides directing activities of leukocytes and other cells of the immune system
- ▪ **Phospholipid** precursors, which are converted to prostaglandins, leukotrienes, and other important lipid mediators of the inflammatory response.

Occurring in connective tissue of many organs, mast cells are especially numerous near small blood vessels in skin and mesenteries (*perivascular* mast cells) and in the tissue that lines digestive and respiratory tracts (*mucosal* mast cells); the granule content of the two populations differs somewhat. These major locations suggest that mast cells place themselves strategically to function as sentinels detecting invasion by microorganisms.

Release of certain chemical mediators stored in mast cells promotes the allergic reactions known as **immediate hypersensitivity reactions** because they occur within a few minutes after the appearance of an antigen in an individual previously sensitized to that antigen. There are many examples of immediate hypersensitivity reaction; a dramatic one is anaphylactic shock, a potentially fatal condition. Anaphylaxis consists of the following sequential events (Figure 5–6). The first exposure to an antigen (allergen), such as bee venom, causes antibody-producing cells to produce an immunoglobulin of the IgE class which binds avidly to receptors on the surface of mast cells. Upon a second exposure to

the antigen it reacts with the IgE on the mast cells, triggering rapid release of histamine, leukotrienes, chemokines, and heparin from the mast cell granules which can produce the sudden onset of the allergic reaction. Degranulation of mast cells also occurs as a result of the action of the complement molecules that participate in the immunologic reactions described in Chapter 14.

Like macrophages, mast cells originate from progenitor cells in the bone marrow, which circulate in the blood, cross the wall of small vessels called venules, and enter connective tissues, where they differentiate. Although mast cells are in many respects similar to basophilic leukocytes, they appear to have a different lineage at least in humans.

Plasma Cells

Plasma cells are lymphocyte–derived, antibody-producing cells. These relatively large, ovoid cells have basophilic cytoplasm rich in RER and a large Golgi apparatus near the nucleus that may appear pale in routine histologic preparations (Figure 5–7).

The nucleus of the plasma cell is generally spherical but eccentrically placed. Many of these nuclei contain compact, peripheral regions of heterochromatin alternating with lighter areas of euchromatin. At least a few plasma cells are present in most connective tissues. Their average lifespan is only 10-20 days.

FIGURE 5–6 Mast cell secretion.

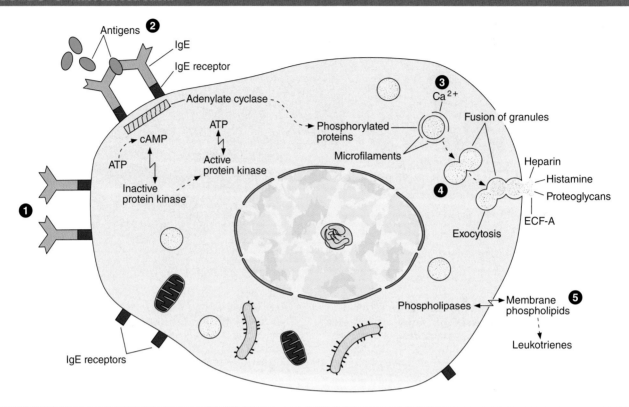

Mast cell secretion is triggered by reexposure to certain antigens and allergens. Molecules of IgE antibody produced in an initial response to an allergen such as pollen or bee venom are bound to surface receptors for IgE (**1**), of which 300,000 are present per mast cell.

When a second exposure to the allergen occurs, IgE molecules bind this antigen and a few IgE receptors very rapidly become cross-linked (**2**). This activates adenylate cyclase, leading to phosphorylation of specific proteins (**3**), entry of Ca^{2+} and rapid

exocytosis of some granules (**4**). In addition, phospholipases act on specific membrane phospholipids, leading to production and release of leukotrienes (**5**).

The components released from granules, as well as the leukotrienes, are immediately active in the local microenvironment and promote a variety of controlled local reactions that together normally comprise part of the inflammatory process called the **immediate hypersensitivity reaction**. "ECF-A" is the eosinophil chemotactic factor of anaphylaxis.

›› MEDICAL APPLICATION

Plasma cells are derived from B lymphocytes and are responsible for the synthesis of immunoglobulin antibodies. Each antibody is specific for the one antigen that stimulated the clone of B cells and reacts only with that antigen or molecules resembling it (see Chapter 14). The results of the antibody-antigen reaction are variable, but they usually neutralize harmful effects caused by antigens. An antigen that is a toxin (eg, tetanus, diphtheria) may lose its capacity to do harm when it is bound by a specific antibody. Bound antigen-antibody complexes are quickly removed from tissues by phagocytosis.

Leukocytes

Other white blood cells, or **leukocytes**, besides macrophages and plasma cells normally comprise a population of wandering cells in connective tissue. Derived from circulating blood cells, they leave blood by migrating between the endothelial cells of venules to enter connective tissue. This process increases greatly during inflammation, which is a vascular and cellular defensive response to injury or foreign substances, including pathogenic bacteria or irritating chemical substances.

Inflammation begins with the local release of chemical mediators from various cells, the ECM, and blood plasma proteins. These substances act on local blood vessels, mast cells, macrophages, and other cells to induce events characteristic of

FIGURE 5–7 Plasma cells.

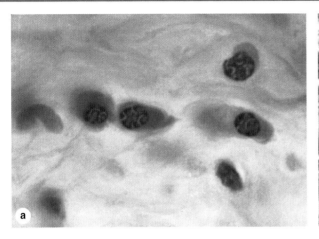

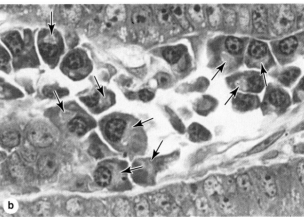

Antibody-secreting plasma cells are present in variable numbers in the connective tissue of many organs.

(a) Plasma cells are large, ovoid cells, with basophilic cytoplasm. The round nuclei frequently show peripheral clumps of heterochromatin, giving the structure a "clock-face" appearance. (X640; H&E)

(b) Plasma are often more abundant in infected tissues, as in the inflamed lamina propria shown here. A large pale Golgi apparatus (arrows) at a juxtanuclear site in each cell is actively involved in the terminal glycosylation of the antibodies (glycoproteins). Plasma cells leave their sites of origin in lymphoid tissues, move to connective tissue, and produce antibodies that mediate immunity. (X400 PT)

inflammation, for example, increased blood flow and vascular permeability, entry and migration of leukocytes, and activation of macrophages for phagocytosis.

Most leukocytes function in connective tissue only for a few hours or days and then undergo apoptosis. However, as discussed with the immune system, some lymphocytes and phagocytic antigen-presenting cells normally leave the interstitial fluid of connective tissue, enter blood or lymph, and move to selected lymphoid organs.

❯❯ MEDICAL APPLICATION

Increased vascular permeability is caused by the action of vasoactive substances such as histamine released from mast cells during **inflammation**. Classically, the major signs of inflamed tissues include "redness and swelling with heat and pain" (*rubor et tumor cum calore et dolore*). Increased blood flow and vascular permeability produce local tissue **swelling (edema)**, with increased redness and warmth. **Pain** is due mainly to the action of the chemical mediators on local sensory nerve endings. All these activities help protect and repair the inflamed tissue. **Chemotaxis** (Gr. *chemeia*, alchemy + *taxis*, orderly arrangement), the phenomenon by which specific cell types are attracted by specific molecules, draws much larger numbers of leukocytes into inflamed tissues.

❯ FIBERS

The fibrous components of connective tissue are elongated structures formed from proteins that polymerize after secretion from fibroblasts (Figure 5–2). The three main types of fibers include **collagen**, **reticular**, and **elastic fibers**. Collagen and reticular fibers are both formed by proteins of the collagen family, and elastic fibers are composed mainly of the protein **elastin**. These fibers are distributed unequally among the different types of connective tissue, with the predominant fiber type conferring most specific tissue properties.

Collagen

The **collagens** constitute a family of proteins selected during evolution for their ability to form various extracellular fibers, sheets, and networks, all of which extremely strong and resistant to normal shearing and tearing forces. Collagen is a key element of all connective tissues, as well as epithelial basement membranes and the external laminae of muscle and nerve cells.

Collagen is the most abundant protein in the human body, representing 30% of its dry weight. A major product of fibroblasts, collagens are also secreted by several other cell types and are distinguishable by their molecular compositions, morphologic characteristics, distribution, functions, and pathologies. A family of 28 collagens exists in vertebrates, numbered in the order they were identified, and the most important are listed in Table 5–3. They can be categorized according to the structures formed by their interacting α-chains subunits:

- **Fibrillar collagens**, notably **collagen types I, II, and III**, have polypeptide subunits that aggregate to form large fibrils clearly visible in the electron or light microscope (Figure 5–8). Collagen type I, the most abundant and widely distributed collagen, forms large, eosinophilic bundles usually called **collagen fibers**. These often densely fill the connective tissue, forming structures such as tendons, organ capsules, and dermis.

- **Network or sheet-forming collagens** such as **type IV collagen** have subunits produced by epithelial cells and are major structural proteins of external laminae and all epithelial basal laminae.

- **Linking/anchoring collagens** are short collagens that link fibrillar collagens to one another (forming larger fibers) and to other components of the ECM. **Type VII collagen** binds type IV collagen and anchors the basal lamina to the underlying reticular lamina in basement membranes (see Figure 4–3).

TABLE **5–3**	Collagen types.				
Type	**α-Chain Composition**	**Structure**	**Optical Microscopy**	**Major Location**	**Main Function**
Fibril-Forming Collagens					
I	$[\alpha1\,(I)]_2[\alpha2\,(I)]$	300-nm molecule, 67-nm banded fibrils	Thick, highly picrosirius birefringent, fibers	Skin, tendon, bone, dentin	Resistance to tension
II	$[\alpha1\,(II)]_3$	300-nm molecule, 67-nm banded fibrils	Loose aggregates of fibrils, birefringent	Cartilage, vitreous body	Resistance to pressure
III	$[\alpha1\,(III)]_3$	67-nm banded fibrils	Thin, weakly birefringent, argyrophilic (silver-binding) fibers	Skin, muscle, blood vessels, frequently together with type I	Structural maintenance in expansible organs
V	$[\alpha1\,(V)]_3$	390-nm molecule, N-terminal globular domain	Frequently forms fiber together with type I	Fetal tissues, skin, bone, placenta, most interstitial tissues	Participates in type I collagen function
XI	$[\alpha1\,(XI)]\,[\alpha2\,(XI)]\,[\alpha3\,(XI)]$	300-nm molecule	Small fibers	Cartilage	Participates in type II collagen function
Network-Forming Collagens					
IV	$[\alpha1\,(IV)]_2\,[\alpha2\,(IV)]$	2-dimensional cross-linked network	Detected by immunocytochemistry	All basal and external laminae	Support of epithelial cells; filtration
X	$[\alpha1(X)]_3$	Hexagonal lattices	Detected by immunocytochemistry	Hypertrophic cartilage involved in endochondral bone formation	Increases density of the matrix
Linking/Anchoring Collagens					
VII	$[\alpha1\,(VII)]_3$	450 nm, globular domain at each end	Detected by immunocytochemistry	Epithelial basement membranes	Anchors basal laminae to underlying reticular lamina
IX	$[\alpha1\,(IX)]\,[\alpha2\,(IX)]\,[\alpha3\,(IX)]$	200-nm molecule	Detected by immunocytochemistry	Cartilage, vitreous body	Binds various proteoglycans; associated with type II collagen
XII	$[\alpha1\,(XII)]_3$	Large N-terminal domain	Detected by immunocytochemistry	Placenta, skin, tendons	Interacts with type I collagen
XIV	$[\alpha1\,(XIV)]_3$	Large N-terminal domain; cross-shaped molecule	Detected by immunocytochemistry	Placenta, bone	Binds type I collagen fibrils, with types V and XII, strengthening fiber formation

FIGURE **5–8** Type I collagen.

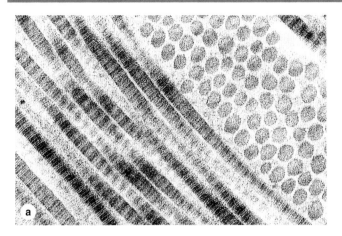

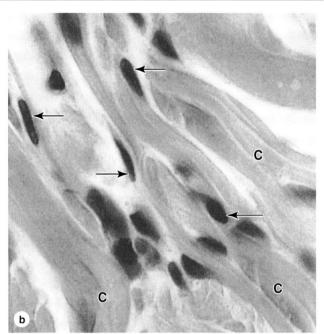

Subunits of type I collagen, the most abundant collagen, assemble to form extremely strong fibrils, which are then bundled together further by other collagens into much larger structures called **collagen fibers.**

(a) TEM shows fibrils cut longitudinally and transversely. In longitudinal sections fibrils display alternating dark and light bands; in cross section the cut ends of individual collagen molecules appear as dots. Ground substance completely surrounds the fibrils. (X100,000)

(b) The large bundles of type I collagen fibrils (**C**) appear as acidophilic collagen fibers in connective tissues, where they

may fill the extracellular space. Subunits for these fibers were secreted by the fibroblasts (**arrows**) associated with them. (X400; H&E)

Collagen synthesis occurs in many cell types but is a specialty of fibroblasts. The initial **procollagen α chains** are polypeptides made in the RER. Several different α chains of variable lengths and sequences can be synthesized from the related collagen genes. In the ER three α chains are selected, aligned, and stabilized by disulfide bonds at their carboxyl terminals, and folded as a **triple helix**, another defining feature of collagens. The triple helix

undergoes exocytosis and is cleaved to a rodlike **procollagen molecule** (Figure 5–9) that is the basic subunit from which the fibers or sheets are assembled. These subunits may be homotrimeric, with all three chains identical, or heterotrimeric, with two or all three chains having different sequences. Different combinations of procollagen α chains produce the various types of collagen with different structures and functional properties.

FIGURE **5–9** The collagen subunit.

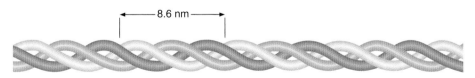

In the most abundant form of collagen, type I, each procollagen molecule or subunit has two α1- and one α2-peptide chains, each with a molecular mass of approximately 100 kDa, intertwined in a right-handed helix and held together by hydrogen bonds and

hydrophobic interactions. The length of each molecule (sometimes called tropocollagen) is 300 nm, and its width is 1.5 nm. Each complete turn of the helix spans a distance of 8.6 nm.

An unusually large number of posttranslational processing steps are required to prepare collagen for its final assembly in the ECM. These steps have been studied most thoroughly for type I collagen, which accounts for 90% of all the body's collagen. The most important parts of this process are summarized in Figure 5–10 and described briefly here:

1. The procollagen α chains are produced on polyribosomes of the RER and translocated into the cisternae. These typically have long central domains rich in proline and lysine; in type I collagen every third amino acid is glycine.

2. **Hydroxylase** enzymes in the ER cisternae add hydroxyl (-OH) groups to some prolines and lysines in reactions that require O_2, Fe^{2+}, and ascorbic acid (vitamin C) as cofactors.

3. Glycosylation of some hydroxylysine residues also occurs, to different degrees in various collagen types.

4. Both the amino- and carboxyl-terminal sequences of α chains have globular structures that lack the gly-X-Y repeats. In the RER the C-terminal regions of three selected α chains (α1, α2) are stabilized by cysteine disulfide bonds, which align the three polypeptides and facilitates their central domains folding as the triple helix. With its globular terminal sequences intact, the trimeric procollagen molecule is transported through the Golgi apparatus, packaged in vesicles and secreted.

5. Outside the cell, specific proteases called **procollagen peptidases** remove the terminal globular peptides, converting the procollagen molecules to collagen molecules. These now self-assemble (an entropy-driven process) into polymeric collagen fibrils, usually in specialized niches near the cell surface.

6. Certain proteoglycans and other collagens (eg, types V and XII) associate with the new collagen fibrils, stabilize these assemblies, and promote the formation of larger fibers from the fibrils.

7. Fibrillar structure is reinforced and disassembly is prevented by the formation of covalent cross-links between the collagen molecules, a process catalyzed by **lysyl oxidase**.

The other fibrillar and sheetlike collagens are formed in processes similar to that described for collagen type I and stabilized by linking or anchoring collagens. Because there are so many steps in collagen biosynthesis, there are many points at which the process can be interrupted or changed by defective enzymes or by disease processes (Table 5–4).

Type I collagen fibrils have diameters ranging from 20 to 90 nm and can be several micrometers in length. Adjacent rod-like collagen subunits of the fibrils are staggered by 67 nm, with small gaps (lacunar regions) between their ends (Figure 5–11). This structure produces a characteristic feature of type I collagen visible by EM: transverse striations with a regular periodicity (Figure 5–11). Type I collagen fibrils assemble further to form large, extremely strong collagen fibers that may be further bundled by linking collagens and proteoglycans. Collagen type II (present in cartilage) occurs as fibrils but does not form fibers or bundles. Sheet-forming collagen type IV subunits assemble as a lattice-like network in epithelial basal laminae.

When they fill the ECM (eg, in tendons or the sclera of the eye), bundles of collagen appear white. The highly regular orientation of subunits makes collagen fibers birefringent with polarizing microscopy (see Figure 1–7). In routine light microscopy collagen fibers are acidophilic, staining pink with eosin, blue with Mallory trichrome stain, and red with Sirius red. Because collagen bundles are long and tortuous, their length and diameter are better studied in spread preparations rather than sections, as shown in Figure 1–7a. Mesentery is frequently used for this purpose; when spread on a slide, this structure is sufficiently thin to let the light pass through; it can be stained and examined directly under the microscope.

Collagen turnover and renewal in normal connective tissue is generally a very slow but ongoing process. In some organs, such as tendons and ligaments, the collagen is very stable, whereas in others, as in the periodontal ligament surrounding teeth, the collagen turnover rate is high. To be renewed, the collagen must first be degraded. Degradation is initiated by specific enzymes called **collagenases**, which are members of an enzyme class called **matrix metalloproteinases (MMPs)**, which clip collagen fibrils or sheets in such a way that they are then susceptible to further degradation by nonspecific proteases. Various MMPs are secreted by macrophages and play an important role in remodeling the ECM during tissue repair.

Reticular Fibers

Found in delicate connective tissue of many organs, notably in the immune system, **reticular fibers** consist mainly of collagen type III, which forms an extensive network (reticulum) of

FIGURE **5–10** Collagen synthesis.

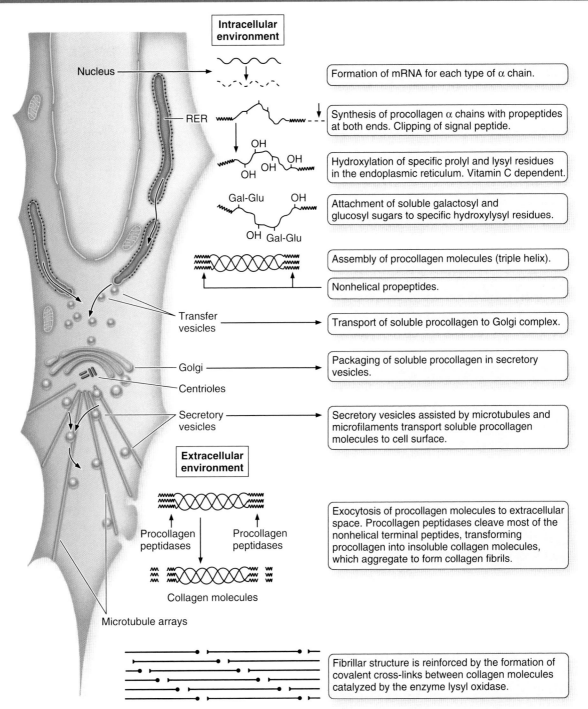

Intracellular environment

Nucleus → Formation of mRNA for each type of α chain.

RER → Synthesis of procollagen α chains with propeptides at both ends. Clipping of signal peptide.

OH OH OH OH → Hydroxylation of specific prolyl and lysyl residues in the endoplasmic reticulum. Vitamin C dependent.

Gal-Glu OH OH Gal-Glu → Attachment of soluble galactosyl and glucosyl sugars to specific hydroxylysyl residues.

Assembly of procollagen molecules (triple helix).

Nonhelical propeptides.

Transfer vesicles → Transport of soluble procollagen to Golgi complex.

Golgi → Packaging of soluble procollagen in secretory vesicles.

Centrioles

Secretory vesicles → Secretory vesicles assisted by microtubules and microfilaments transport soluble procollagen molecules to cell surface.

Extracellular environment

Procollagen peptidases   Procollagen peptidases → Exocytosis of procollagen molecules to extracellular space. Procollagen peptidases cleave most of the nonhelical terminal peptides, transforming procollagen into insoluble collagen molecules, which aggregate to form collagen fibrils.

Collagen molecules

Microtubule arrays

Fibrillar structure is reinforced by the formation of covalent cross-links between collagen molecules catalyzed by the enzyme lysyl oxidase.

Hydroxylation and glycosylation of procollagen α chains and their assembly into triple helices occur in the RER, and further assembly into fibrils occurs in the ECM after secretion of procollagen. Because there are many slightly different genes for procollagen α chains and collagen production depends on several posttranslational events involving several other enzymes, many diseases involving defective collagen synthesis have been described.

TABLE **5–4**	Examples of clinical disorders resulting from defects in collagen synthesis.	
Disorder	**Defect**	**Symptoms**
Ehlers-Danlos type IV	Faulty transcription or translation of collagen type III	Aortic and/or intestinal rupture
Ehlers-Danlos type VI	Faulty lysine hydroxylation	Increased skin elasticity, rupture of eyeball
Ehlers-Danlos type VII	Decrease in procollagen peptidase activity	Increased articular mobility, frequent luxation
Scurvy	Lack of vitamin C, a required cofactor for prolyl hydroxylase	Ulceration of gums, hemorrhages
Osteogenesis imperfecta	Change of 1 nucleotide in genes for collagen type I	Spontaneous fractures, cardiac insufficiency

FIGURE **5–11** Assembly of type I collagen.

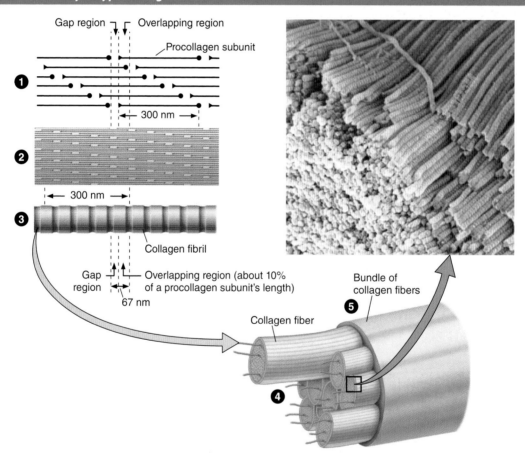

Shown here are the relationships among type I collagen molecules, fibrils, fibers, and bundles.

1. Rodlike triple-helix collagen molecules, each 300-nm long, self-assemble in a highly organized, lengthwise arrangement of overlapping regions.
2. The regular, overlapping arrangement of subunits continues as large collagen fibrils are assembled.
3. This structure causes fibrils to have characteristic cross striations with alternating dark and light bands when observed in the EM.
4. Fibrils assemble further and are linked together in larger collagen fibers visible by light microscopy.
5. Type I fibers often form into still larger aggregates bundled and linked together by other collagens.

The photo shows an SEM view of type I collagen fibrils closely aggregated as part of a collagen fiber. Striations are visible on the surface of the fibrils.

FIGURE **5–12** Reticular fibers.

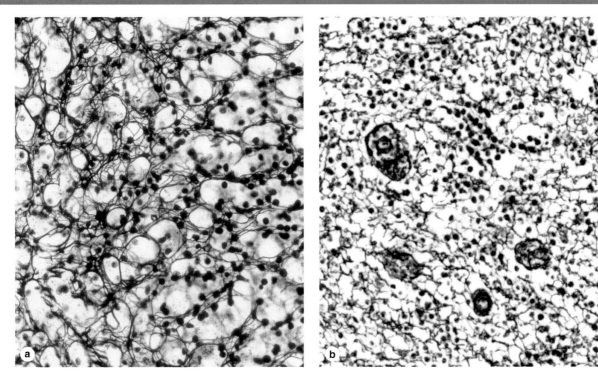

In these silver-stained sections of adrenal cortex **(a)** and lymph node **(b)**, networks of delicate, black **reticular fibers** are prominent. These fibers serve as a supportive stroma in most lymphoid and hematopoietic organs and many endocrine glands. The fibers consist of type III collagen that is heavily glycosylated, producing the black argyrophilia. Cell nuclei are also dark, but cytoplasm is unstained. (X100) Fibroblasts specialized for reticular fiber production in hematopoietic and lymphoid organs are often called reticular cells.

thin (diameter 0.5-2 μm) fibers for the support of many different cells. Reticular fibers are seldom visible in hematoxylin and eosin (H&E) preparations but are characteristically stained black after impregnation with silver salts (Figure 5–12) and are thus termed **argyrophilic** (Gr. *argyros*, silver). Reticular fibers are also periodic acid-Schiff (PAS) positive, which, like argyrophilia, is due to the high content of sugar chains bound to type III collagen α chains. Reticular fibers contain up to 10% carbohydrate as opposed to 1% in most other collagen fibers.

Reticular fibers produced by fibroblasts occur in the reticular lamina of basement membranes and typically also surround adipocytes, smooth muscle and nerve fibers, and small blood vessels. Delicate reticular networks serve as the supportive stroma for the parenchymal secretory cells and rich microvasculature of the liver and endocrine glands. Abundant reticular fibers also characterize the stroma of hemopoietic tissue (bone marrow), the spleen, and lymph nodes where they support rapidly changing populations of proliferating cells and phagocytic cells.

Elastic Fibers

Elastic fibers are also thinner than the type I collagen fibers and form sparse networks interspersed with collagen bundles in many organs, particularly those subject to regular stretching or bending. As the name implies, elastic fibers have rubberlike properties that allow tissue containing these fibers, such as the stroma of the lungs, to be stretched or distended and return to their original shape. In the wall of large blood vessels, especially arteries, elastin also occurs as fenestrated sheets called **elastic lamellae**. Elastic fibers and lamellae are not strongly acidophilic and stain poorly with H&E; they are stained more darkly than collagen with other stains such as orcein and aldehyde fuchsin (Figure 5–13).

Elastic fibers (and lamellae) are a composite of **fibrillin** (350 kDa), which forms a network of **microfibrils**, embedded in a larger mass of cross-linked **elastin** (60 kDa). Both proteins are secreted from fibroblasts (and smooth muscle cells in vascular walls) and give rise to elastic fibers in a stepwise manner are shown in Figure 5–14. Initially, microfibrils with diameters of 10 nm form from fibrillin and various glycoproteins. The microfibrils act as scaffolding upon which elastin is then deposited. Elastin accumulates around the microfibrils, eventually making up most of the elastic fiber, and is responsible for the rubberlike property.

The elastic properties of these fibers and lamellae result from the structure of the elastin subunits and the unique

FIGURE **5–13** Elastic fibers.

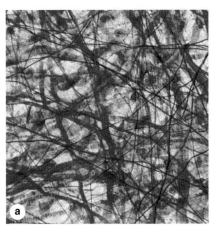
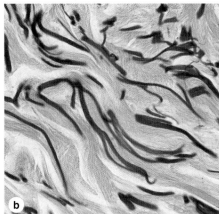
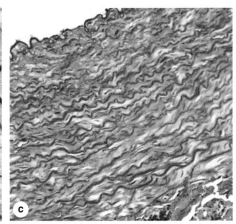

Elastic fibers or lamellae (sheets) add resiliency to connective tissue. Such fibers may be difficult to discern in H&E-stained tissue, but elastin has a distinct, darker-staining appearance with other staining procedures.

(a) The length, diameter, distribution, and density of dark **elastic fibers** are easily seen in this spread preparation of nonstretched connective tissue in a mesentery. (X200; Hematoxylin and orcein)

(b) In sectioned tissue at higher magnification, **elastic fibers** can be seen among the acidophilic collagen bundles of dermis. (X400; Aldehyde fuchsin)

(c) Elastic lamellae in the wall of the aorta are more darkly stained, incomplete sheets of elastin between the layers of eosinophilic smooth muscle. (X80; H&E)

FIGURE **5–14** Formation of elastic fibers.

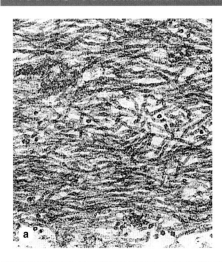
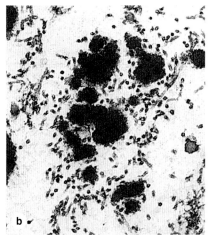
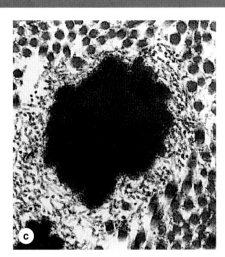

Stages in the formation of elastic fibers can be seen by TEM.

(a) Initially, a developing fiber consists of many 10-nm-diameter **microfibrils** composed of **fibrillin** subunits secreted by fibroblasts and smooth muscle cells.

(b) Elastin is deposited on the scaffold of microfibrils, forming growing, amorphous composite structures. The elastin molecules are also secreted by the fibroblasts and quickly become covalently cross-linked into larger assemblies.

(c) Elastin accumulates and ultimately occupies most of the electron-dense center of the single elastic fiber shown here. Fibrillin microfibrils typically remain visible at the fiber surface. Collagen fibrils, seen in cross section, are also present surrounding the elastic fiber. (All X50,000)

FIGURE 5–15 Molecular basis of elastic fiber elasticity.

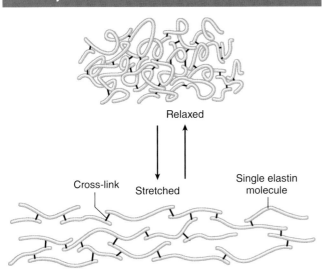

Relaxed

Cross-link   Stretched   Single elastin molecule

The diagram shows a small piece of an elastic fiber, in two conformations. Elastin polypeptides, the major components of elastic fibers, have multiple random-coil domains that straighten or stretch under force, and then relax. Most of the cross-links between elastin subunits consist of the covalent, cyclic structure **desmosine**, each of which involves four converted lysines in two elastin molecules. This unusual type of protein cross-link holds the aggregate together with little steric hindrance to elastin movements. These properties give the entire network its elastic quality.

cross-links holding them together. Elastin molecules have many lysine-rich regions interspersed with hydrophobic domains rich in lysine and proline which are thought to form extensible, random-coil conformations (like natural rubber). During deposition on the fibrillin microfibrils, lysyl oxidase converts the lysines' amino groups to aldehydes and four oxidized lysines on neighboring elastin molecules then condense covalently as a **desmosine** ring, cross-linking the polypeptides. Bound firmly by many desmosine rings, but maintaining the rubberlike properties of their hydrophobic domains, elastic fibers stretch reversibly when force is applied (Figure 5–15). Elastin resists digestion by most proteases, but it is hydrolyzed by pancreatic **elastase**.

❯❯ MEDICAL APPLICATION

Fibrillins comprise a family of proteins involved in making the scaffolding necessary for the deposition of elastin. Mutations in the fibrillin genes result in **Marfan syndrome**, a disease characterized by a lack of resistance in tissues rich in elastic fibers. Because the walls of large arteries are rich in elastic components and because the blood pressure is high in the aorta, patients with this disease often experience aortic swellings called **aneurysms**, which are life-threatening conditions.

❯ GROUND SUBSTANCE

The **ground substance** of the ECM is a highly hydrated (with much bound water), transparent, complex mixture of three major kinds of macromolecules: **glycosaminoglycans (GAGs)**, **proteoglycans**, and **multiadhesive glycoproteins**. Filling the space between cells and fibers in connective tissue, ground substance allows diffusion of small molecules and, because it is viscous, acts as both a lubricant and a barrier to the penetration of invaders. Physical properties of ground substance also profoundly influence various cellular activities. When adequately fixed for histologic analysis, its components aggregate as fine, poorly resolved material that appears in TEM preparations as electron-dense filaments or granules (Figure 5–16a).

GAGs (also called mucopolysaccharides) are long polymers of repeating disaccharide units, usually a hexosamine and uronic acid. The hexosamine can be glucosamine or galactosamine, and the uronic acid can be glucuronate or iduronate. The largest and most ubiquitous GAG is **hyaluronan** (also called hyaluronate or hyaluronic acid). With a molecular weight from 100s to 1000s of kDa, hyaluronan is a very long polymer of the disaccharide glucosamine-glucuronate. Uniquely among GAGs, hyaluronan is synthesized directly into the ECM by an enzyme complex, **hyaluronan synthase**, located in the cell membrane of many cells. Hyaluronan forms a viscous, pericellular network which binds a considerable amount of water, giving it an important role in allowing molecular diffusion through connective tissue and in lubricating various organs and joints.

All other GAGs are much smaller (10-40 kDa), sulfated, bound to proteins (as parts of proteoglycans), and are synthesized in Golgi complexes. The four major GAGs found in proteoglycans are **dermatan sulfate**, **chondroitin sulfates**, **keratan sulfate**, and **heparan sulfate**, all of which have different disaccharide units modified further with carboxyl and sulfate groups and different tissue distributions (Table 5–5). Their high negative charge forces GAGs to an extended conformation and causes them to sequester cations as well as water. These features provide GAGs with space-filling, cushioning, and lubricant functions.

Proteoglycans consist of a core protein to which are covalently attached various numbers and combinations of the sulfated GAGs. Like glycoproteins, they are synthesized on RER, mature in the Golgi apparatus, where the GAG side-chains are added, and secreted from cells by exocytosis. Unlike glycoproteins, proteoglycans have attached GAGs which often comprise a greater mass than the polypeptide core. As shown in Figure 5–16b, after secretion proteoglycans become bound to the hyaluronan by link proteins and their GAG side-chains associate further with collagen fibers and other ECM components.

Proteoglycans are distinguished by their diversity, which is generated in part by enzymatic differences in the Golgi complexes. A region of ECM may contain several different core proteins, each with one or many sulfated GAGs of different lengths and composition. As mentioned with epithelia, **perlecan** is the key proteoglycan in all basal laminae. One of the

FIGURE 5–16 Ground substance of the extracellular matrix (ECM).

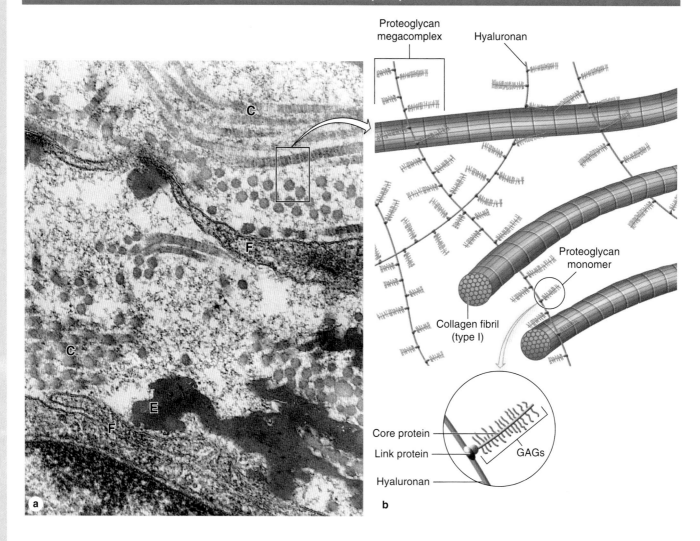

(a) TEM of connective tissue ECM reveals **ground substance** as areas containing only fine granular material among the collagen (**C**) fibers, elastic (**E**) fibers and fibroblast processes (**F**). X100,000.

(b) As shown here schematically, connective tissue ground substance contains a vast complex of **proteoglycans** linked to very long **hyaluronan** molecules. Each proteoglycan monomer has a **core protein** with a few or many side chains of the sulfated glycosaminoglycans (GAGs) listed in Table 5–5. Synthesized in the RER and Golgi apparatus like glycoproteins, proteoglycan monomers are distinguished by often being more heavily glycosylated and by the addition and sulfation of GAGs, which vary significantly among proteoglycans in their number, length, and the degree to which the sugar polymers are modified. The large proteoglycan **aggrecan** (25,000 kDa) typically has about 50 chains of keratan sulfate chains and twice that number of chondroitin sulfate.

best studied proteoglycans, **aggrecan,** is very large (250 kDa), having a core protein heavily bound with chondroitin and keratan sulfate chains. A link protein joins aggrecan to hyaluronan (Figure 5–16b). Abundant in cartilage, aggrecan-hyaluronan complexes fill the space between collagen fibers and cells and contribute greatly to the physical properties of this tissue. Other proteoglycans include decorin, with very few GAG side chains that binds the surface of type I collagen fibrils, and syndecan, with an integral membrane core protein providing an additional attachment of ECM to cell membranes.

Embryonic mesenchyme (Figure 5–1) is very rich in hyaluronan and water, producing the characteristic wide spacing of cells and a matrix ideal for cell migrations and growth. In both developing and mature connective tissues, core proteins and GAGs (especially heparan sulfate) of many proteoglycans bind and sequester various growth factors and other signaling proteins. Degradation of such proteoglycans during the early phase of tissue repair releases these stored growth factors which then help stimulate new cell growth and ECM synthesis.

TABLE **5–5**	Composition and distribution of glycosaminoglycans in connective tissue and their interactions with collagen fibers.			
	Repeating Disaccharides			
Glycosaminoglycan	**Hexuronic Acid**	**Hexosamine**	**Distribution**	**Electrostatic Interaction with Collagen**
Hyaluronic acid	D-glucuronic acid	D-glucosamine	Umbilical cord, synovial fluid, vitreous humor, cartilage	
Chondroitin 4-sulfate	D-glucuronic acid	D-galactosamine	Cartilage, bone, cornea, skin, notochord, aorta	High levels of interaction, mainly with collagen type II
Chondroitin 6-sulfate	D-glucuronic acid	D-galactosamine	Cartilage, umbilical cord, skin, aorta (media)	High levels of interaction, mainly with collagen type II
Dermatan sulfate	L-iduronic acid or D-glucuronic acid	D-galactosamine	Skin, tendon, aorta (adventitia)	Low levels of interaction, mainly with collagen type I
Heparan sulfate	D-glucuronic acid or L-iduronic acid	D-galactosamine	Aorta, lung, liver, basal laminae	Intermediate levels of interaction, mainly with collagen types III and IV
Keratan sulfate	D-galactose	D-glucosamine	Cartilage, nucleus pulposus, annulus fibrosus	None

›› MEDICAL APPLICATION

The degradation of proteoglycans is carried out by several cell types and depends in part on the presence of several lysosomal enzymes. Several disorders have been described, including a deficiency in certain lysosomal enzymes that degrade specific GAGs, with the subsequent accumulation of these macromolecules in tissues. The lack of specific hydrolases in the lysosomes has been found to be the cause of several disorders, including the **Hurler**, **Hunter**, **Sanfilippo**, and **Morquio syndromes**.

Because of their high viscosity, hyaluronan and proteoglycans tend to form a barrier against bacterial penetration of tissues. Bacteria that produce hyaluronidase, an enzyme that hydrolyzes hyaluronan and disassembles proteoglycans complexes, reduce the viscosity of the connective tissue ground substance and have greater invasive power.

Making up the third major class of ground substance macromolecules, **multiadhesive glycoproteins** all have multiple binding sites for cell surface integrins and for other matrix macromolecules. The adhesive glycoproteins are large molecules with branched oligosaccharide chains and allow adhesion of cells to their substrate. An example is the large (200-400 kDa), trimeric glycoprotein **laminin** with binding sites for integrins, type IV collagen, and specific proteoglycans, providing adhesion for epithelial and other cells. As described in the previous chapter, all basal and external laminae are rich in laminin, which is essential for the assembly and maintenance of these structures.

Another glycoprotein, **fibronectin** (L. *fibra*, fiber + *nexus*, interconnection), is a 235-270 kDa dimer synthesized largely by fibroblasts, with binding sites for collagens and certain GAGs, and forms insoluble fibrillar networks throughout connective tissue (Figure 5–17). The fibronectin substrate provides specific binding sites for integrins and is important both for cell adhesion and cellular migration through the ECM.

As briefly described in Chapter 2 integrins are integral membrane proteins that act as matrix receptors for specific sequences on laminin, fibronectin, some collagens, and certain other ECM proteins. Integrins bind their ECM ligands with relatively low affinity, allowing cells to explore their environment without losing attachment to it or becoming glued to it. All are heterodimers with two transmembrane polypeptides: the α and β chains. Great diversity in the subsets of integrin α and β chains which cells express allows cells to have different specific ECM ligands.

Integrin-microfilament complexes are clustered in fibroblasts and other mesenchymal cells to form structures called **focal adhesions** that can be seen by TEM or immunocytochemistry. As mentioned in Chapter 4 this type of adhesive junction is typically present at the ends of actin filaments bundled by α-actinin as cytoplasmic stress fibers and focal adhesion kinases provide a mechanism by which pulling forces or other physical properties of the ECM can change various cellular activities.

Water in the ground substance of connective tissue is referred to as **interstitial fluid** and has an ion composition similar to that of blood plasma. Interstitial fluid also contains plasma proteins of low molecular weight that pass through

FIGURE **5–17** Fibronectin localization.

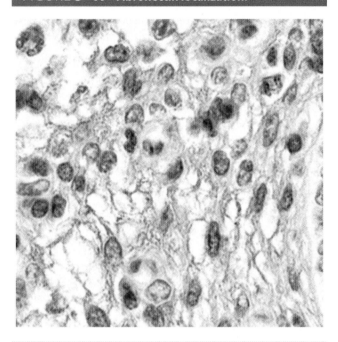

Like laminin of basement membranes, **fibronectin** is a multiadhesive glycoprotein, with binding sites for ECM components and for integrins at cell surfaces, and has important roles in cell migration and the maintenance of tissue structure. As shown here by immunohistochemistry, fibronectin forms a fine network throughout the ECM of connective tissue. (X400)

the thin walls of the smallest blood vessels, the capillaries. Although only a small proportion of connective tissue proteins are plasma proteins, it is estimated that as much as one-third of the body's plasma proteins are normally found in the interstitial fluid of connective tissue because of its large volume and wide distribution.

>> **MEDICAL APPLICATION**

Edema is the excessive accumulation of interstitial fluid in connective tissue. This water comes from the blood, passing through the capillary walls that become more permeable during inflammation and normally produces at least slight swelling.

Capillaries in connective tissue also bring the various nutrients required by cells and carry away their metabolic waste products to the detoxifying and excretory organs, the liver and kidneys. Interstitial fluid is the solvent for these substances.

As shown in Figure 5–18, two main forces act on the water in capillaries:

- The **hydrostatic pressure** of the blood caused by the pumping action of the heart, which forces water out across the capillary wall
- The colloid **osmotic pressure** produced by plasma proteins such as albumin, which draws water back into the capillaries

The colloid osmotic pressure exerted by the blood proteins—which are unable to pass through the capillary walls—tends to pull back into the capillary the water forced out by hydrostatic pressure (Figure 5–18). (Because the ions and low-molecular-weight compounds that pass easily through the capillary walls have similar concentrations inside and outside these blood vessels, the osmotic pressures they exert are approximately equal on either side of the capillaries and cancel each other.)

The quantity of water drawn back into capillaries is often less than that which was forced out. This excess fluid does not normally accumulate in connective tissue but drains continuously into lymphatic capillaries that eventually return it to the blood. Discussed later with the lymphoid system, lymphatic capillaries originate in connective tissue as delicate endothelial tubes (Figure 5–18).

❯ TYPES OF CONNECTIVE TISSUE

Different combinations and densities of the cells, fibers, and other ECM components produce graded variations in histological structure within connective tissue. Descriptive names or classifications used for the various types of connective tissue typically denote either a structural characteristic or a major component. Table 5–6 gives a classification commonly used for the main types of connective tissue. Adipose tissue, an important specialized connective tissue, and two other supporting tissues, cartilage and bone, are covered in Chapters 6, 7, and 8.

Connective Tissue Proper

Connective tissue proper is broadly classified as "loose" or "dense," terms which refer to the amount of collagen present (Figure 5–19). **Loose connective tissue** is common, forming a layer beneath the epithelial lining of many organs and filling the spaces between fibers of muscle and nerve (Figure 5–19).

Also called **areolar tissue**, the loose connective tissue typically contains cells, fibers, and ground substance in roughly equal parts. The most numerous cells are fibroblasts, but the other types of connective tissue cells are also normally found, along with nerves and small blood vessels. Collagen fibers predominate, but elastic and reticular fibers are also present. With at least a moderate amount of ground substance,

FIGURE 5–18 Movement of fluid in connective tissue.

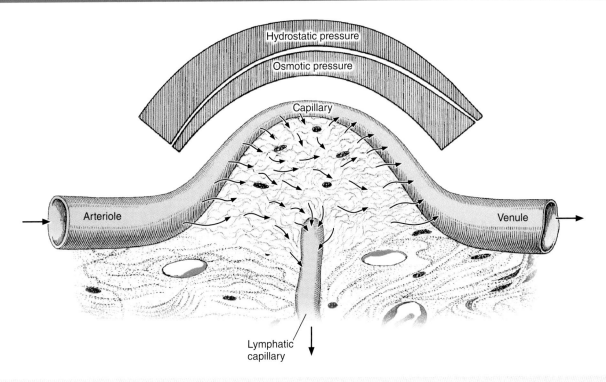

Water normally passes through capillary walls into the ECM of surrounding connective tissues primarily at the arterial end of a capillary, because the **hydrostatic pressure** is greater than the colloid **osmotic pressure**. However, hydrostatic pressure decreases toward the venous end of the capillary, as indicated at the top of the figure. The fall in hydrostatic pressure parallels a rise in osmotic pressure of the capillary blood because the plasma protein concentration increases as water is pushed out across the capillary wall.

As a result of the increased protein concentration and decreased hydrostatic pressure, osmotic pressure at the venous end is greater than hydrostatic pressure and water is drawn back into the capillary. In this way plasma and interstitial fluid constantly mix, nutrients in blood circulate to cells in connective tissue, and cellular wastes are removed.

Not all water that leaves capillaries by hydrostatic pressure is drawn back in by osmotic pressure. This excess tissue fluid is normally drained by the lymphatic capillaries, open-ended vessels that arise in connective tissue and enter the one-way lymphatic system that eventually delivers the fluid (now called **lymph**) back to veins.

loose connective tissue has a delicate consistency; it is flexible and not very resistant to stress.

Dense connective tissue has similar components as loose connective tissue, but with fewer cells, mostly fibroblasts, and a clear predominance of bundled type I collagen fibers over ground substance. The abundance of collagen here protects organs and strengthens them structurally. In **dense irregular connective tissue** bundles of collagen fibers appear randomly interwoven, with no definite orientation. The tough three-dimensional collagen network provides resistance to stress from all directions. Examples of dense irregular connective tissue include the deep dermis layer of skin and capsules surrounding most organs. Dense irregular and loose connective tissues are often closely associated, with the two types grading into each other and making distinctions between them somewhat arbitrary (Figure 5–19).

Dense regular connective tissue consists mostly of type I collagen bundles and fibroblasts aligned in parallel for great resistance to prolonged or repeated stresses from the same direction (Figure 5–20).

The best examples of dense regular connective tissue are the very strong and flexible **tendons** (Figure 5–20), cords connecting muscles to bones; **aponeuroses**, which are sheet-like tendons; and **ligaments**, bands or sheets that hold together components of the skeletal system. Consisting almost entirely of densely packed parallel collagen fibers separated by very little ground substance and having very few blood vessels, these inextensible structures are white in

TABLE **5–6**	Classification of connective or supporting tissues.		
	General Organization	**Major Functions**	**Examples**
Connective Tissue Proper			
Loose (areolar) connective tissue	Much ground substance; many cells and little collagen, randomly distributed	Supports microvasculature, nerves, and immune defense cells	Lamina propria beneath epithelial lining of digestive tract
Dense irregular connective tissue	Little ground substance; few cells (mostly fibroblasts); much collagen in randomly arranged fibers	Protects and supports organs; resists tearing	Dermis of skin, organ capsules, submucosa layer of digestive tract
Dense regular connective tissue	Almost completely filled with parallel bundles of collagen; few fibroblasts, aligned with collagen	Provide strong connections within musculoskeletal system; strong resistance to force	Ligaments, tendons, aponeuroses, corneal stroma
Embryonic Connective Tissues			
Mesenchyme	Sparse, undifferentiated cells, uniformly distributed in matrix with sparse collagen fibers	Contains stem/progenitor cells for all adult connective tissue cells	Mesodermal layer of early embryo
Mucoid (mucous) connective tissue	Random fibroblasts and collagen fibers in viscous matrix	Supports and cushions large blood vessels	Matrix of the fetal umbilical cord
Specialized Connective Tissues			
Reticular connective tissue (see Chapter 14)	Delicate network of reticulin/collagen III with attached fibroblasts (reticular cells)	Supports blood-forming cells, many secretory cells, and lymphocytes in most lymphoid organs	Bone marrow, liver, pancreas, adrenal glands, all lymphoid organs except the thymus
Adipose Tissue (Chapter 6)			
Cartilage (Chapter 7)			
Bone (Chapter 8)			
Blood (Chapter 12)			

the fresh state. Fibrocytes with elongated nuclei lie parallel to the collagen fibers of dense regular connective tissue, with cytoplasmic folds enveloping portions of the collagen bundles (Figure 5–20b). Cytoplasm in these "**tendinocytes**" is difficult to distinguish in H&E-stained preparations because it is very sparse and has acidophilia like that of the collagen. In aponeuroses the parallel bundles of collagen exist as multiple layers alternating at 90° angles to one another. Some ligaments, such as the elastic ligaments along the vertebral column, contain besides collagen many parallel bundles of elastic fibers.

On their outer surface tendons and ligaments have a layer of dense irregular connective tissue that is continuous with the outermost layers of the adjacent muscles and bones. Collagen bundles vary in size in different tendons and ligaments, but all regular connective tissue structures are poorly vascularized and repair of damage in this tissue is usually slow. Ligaments and tendons will be discussed again in Chapters 8 and 10, respectively, with bone, joints, and muscle.

> **》》 MEDICAL APPLICATION**
>
> Overuse of tendon-muscle units can result in **tendonitis**, characterized by inflammation of the tendons and their attachments to muscle. Common locations are the elbow, the Achilles tendon of the heel, and the shoulder rotator cuff. The swelling and pain produced by the localized inflammation restricts the affected area's normal range of motion and can be relieved by injections of anti-inflammatory agents such as cortisone. Fibroblasts eventually repair damaged collagen bundles of the area.

Reticular Tissue

Reticular tissue is characterized by abundant fibers of type III collagen (Figure 5–12) forming a delicate network that supports various types of cells. This collagen is also known as

FIGURE **5–19** Loose connective tissue and dense irregular connective tissue.

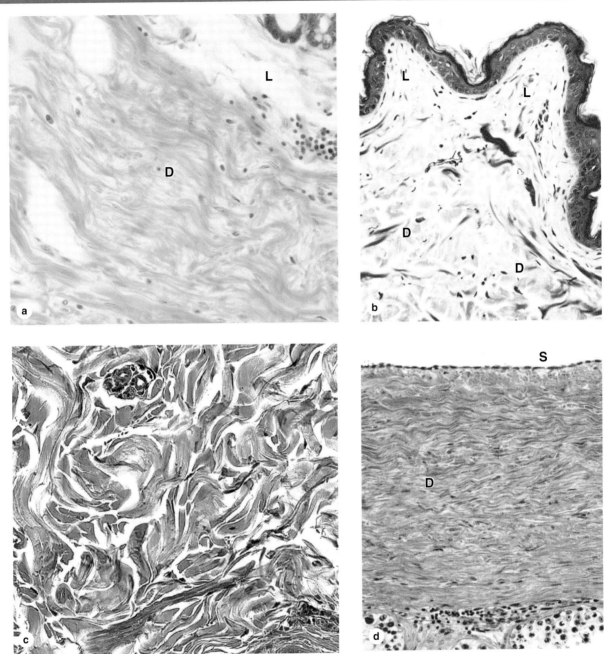

Examples of these connective tissue types shown here indicate the close association that often occurs between these two types.

(a) Loose connective tissue (**L**) of a gland contains faintly stained ground substance with fine fibers of collagen and frequently forms a thin layer near epithelia, while dense irregular connective tissue (**D**) forms a thicker layer and is invariably much richer in larger bundles of collagen. Scattered leukocytes can be seen in both connective tissues, along with the large irregular spaces of lymphatic vessels (left). (X100; H&E)

(b) Trichrome staining of a section from skin demonstrates the blue staining of collagen with this method and its relative density in loose (**L**) and dense irregular (**D**) connective tissue. (X100; Mallory trichrome)

(c) Another example of dense irregular connective tissue, showing the randomly arranged large collagen bundles. The arrangement of collagen strengthens the tissues and resists tearing from all directions. (X150; H&E)

(d) Dense irregular connective tissue (**D**) forms a thick, protective capsule around many organs such as the testis shown here. Here the capsule is covered by a simple squamous epithelium of serous mesothelial cells (**S**), which produce a hyaluronate-rich lubricant around such organs. (X150; H&E)

FIGURE **5–20** Dense regular connective tissue.

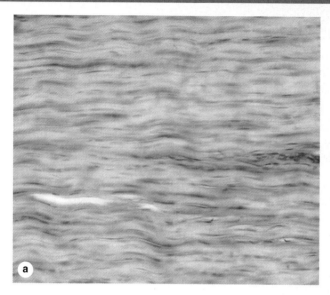

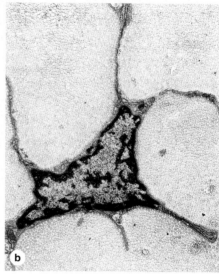

(a) Micrograph shows a longitudinal section of dense regular connective tissue in a tendon. Long, parallel bundles of collagen fibers fill the spaces between the elongated nuclei of fibrocytes. (X100; H&E stain)

(b) The electron micrograph shows one fibrocyte in a cross section of tendon, revealing that the sparse cytoplasm of the fibrocytes is divided into numerous thin cytoplasmic processes extending among adjacent collagen fibers. (X25,000)

FIGURE **5–21** Reticular tissue.

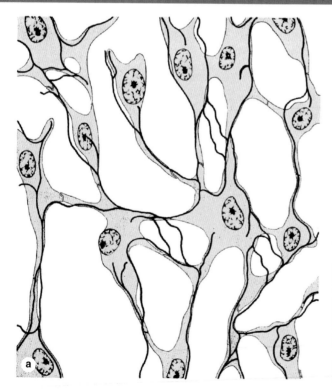

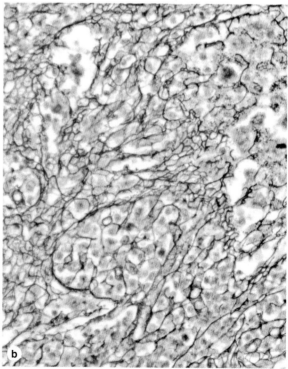

(a) The diagram shows only the fibers and attached reticular cells (free, transient cells are not represented). Reticular fibers of type III collagen (also referred to as "reticulin") are produced and enveloped by the reticular cells, forming an elaborate network through which interstitial fluid or lymph and wandering cells from blood pass continuously.

(b) The micrograph shows a silver-stained section of lymph node in which reticular fibers are seen as irregular black lines. Reticular cells are also heavily stained and dark. Most of the smaller, more lightly stained cells are lymphocytes passing through the lymph node. (X200; Silver)

FIGURE **5–22** Mucoid tissue.

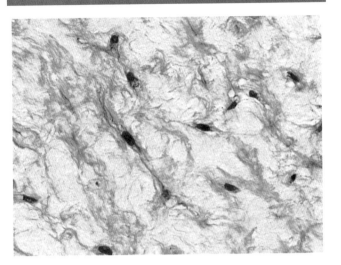

A section of umbilical cord shows large fibroblasts surrounded by a large amount of very loose ECM containing mainly ground substances very rich in hyaluronan, with wisps of collagen. Histologically mucoid (or mucous) connective tissue resembles embryonic mesenchyme in many respects and is rarely found in adult organs. (X200; H&E)

reticulin and is produced by modified fibroblasts often called **reticular cells** that remain associated with and partially cover the fibers (Figure 5–21). The loose disposition of glycosylated reticular fibers provides a framework with specialized microenvironments for cells in hemopoietic tissue and some lymphoid organs (bone marrow, lymph nodes, and spleen). The resulting cell-lined system creates a meshwork for the passage of leukocytes and lymph. Macrophages and dendritic cells (also in the mononuclear phagocyte family) are also dispersed within these reticular tissues to monitor cells formed there or passing through and to remove debris.

Mucoid Tissue

Mucoid (or mucous) connective tissue is the principal component of the fetal umbilical cord, where it is referred to as **Wharton's jelly**. With abundant ground substance composed chiefly of hyaluronan, mucoid tissue is gelatinous, with sparse collagen fibers and scattered fibroblasts (Figure 5–22). Included among the fibroblastic cells are many mesenchymal stem cells, which are being studied for their potential in regenerative medicine. Mucoid connective tissue is similar to the tissue found in the vitreous chambers of eyes and pulp cavities of young teeth.

Connective Tissue SUMMARY OF KEY POINTS

- Connective tissue is specialized to physically **support** and **connect** other tissues and maintain the water required for metabolite diffusion to and from cells.
- Connective tissues all consist primarily of **extracellular** material rather than cells.
- Within most organs connective tissue proper forms the supportive **stroma**, which supports the organ's unique functional components or **parenchyma**.
- The **extracellular matrix (ECM)** of connective tissue proper usually consists of both large protein **fibers** and nonfibrous areas of unstained **ground substance** rich in various GAGs and water.
- All adult connective tissues are derived from an embryonic form of connective tissue called **mesenchyme**, which contains uniformly undifferentiated cells scattered in a gel-like matrix.

Cells of Connective Tissue

- **Fibroblasts** (fibrocytes), the major cells of connective tissue proper, are elongated, irregularly shaped cells with oval nuclei that synthesize and secrete most components of the ECM.
- **Adipocytes** (fat cells) are very large cells specialized for storage of triglycerides; they predominate in a specialized form of connective tissue called **adipose tissue**.
- **Macrophages** are short-lived cells that differentiate in connective tissue from precursor cells called **monocytes** circulating in the blood; they function in ECM turnover, phagocytosis of dead cells and debris, and antigen presentation to lymphocytes.
- **Mast cells** also originate from blood cell precursors and are filled with granules for the release of various vasoactive agents and other substances during inflammatory and allergic reactions.
- **Plasma cells** are short-lived cells that differentiate from B lymphocytes and are specialized for the abundant secretion of specific antibodies (immunoglobulins).

- Besides macrophages and plasma cells, other **leukocytes** normally wander through all types of connective tissue proper, providing surveillance against bacterial invaders and stimulating tissue repair.

Fibers of Connective Tissue

- The most important and abundant fibers of connective tissue are composed of the protein **collagen**, of which there are some 20 related types.
- Synthesis of collagen by fibroblasts and certain other cells involves posttranslational modifications in the RER, notably **hydroxylation** of the numerous prolines and lysines, and formation of helical trimeric subunits of **procollagen**.
- Upon exocytosis, the nonhelical ends of the procollagen subunits are removed, forming trimeric **collagen molecules** that aggregate and become covalently bound together in large **collagen fibrils**.
- The highly regular assembly of collagens in the fibrils produces a characteristic pattern of **crossbanding** visible ultrastructurally along the fibrils of some collagen types.
- Fibrils of type I collagen are bundled together by other forms of nonfibrillar, linking collagens to produce large **collagen bundles**.
- Collagen fibrils are degraded by collagenase enzymes classified as **matrix metalloproteinases (MMPs)**, produced primarily by macrophages.
- Type III collagen produces a network of delicate **reticular fibers**, which stain very dark with silver stains and are abundant in immune and lymphoid tissues.
- **Elastic fibers**, or sheets called **elastic lamellae**, are composed of the proteins **elastin** and **fibrillin**, which exist in a stretchable conformation that provides elastic properties to connective tissues rich in this material.

Ground Substance

- **Ground substance** is the watery, largely unstained extracellular material that is more abundant than fibers in some types of connective tissue proper.
- Ground substance is rich in **hydrated glycosaminoglycans (GAGs)**, **proteoglycans**, and **multiadhesive glycoproteins**.
- The major types of GAGs are **hyaluronan** (hyaluronic acid), which is a very long polymer of the disaccharide glucosamine-glucuronate, and various shorter chains of **sulfated GAGs** composed of other disaccharide polymers.
- Sulfated GAGs such as **chondroitin sulfate** and **keratan sulfate** have various sizes and compositions, but they are all bound to the core proteins of **proteoglycans** and are produced in the Golgi apparatus before secretion.
- Proteoglycans attach to polymers of HA via **linker proteins** to form huge complexes in ground substance that bind water and other substances, including certain polypeptide growth factors that help regulate fibroblast proliferation.
- **Multiadhesive glycoproteins** such as fibronectin and laminin have binding sites for collagens and for integrin proteins in cell membranes, thus allowing temporary attachments between cells and the ECM required for cell migration and positioning.

Types of Connective Tissue

- **Connective tissue proper** is usually classified as loose or dense according to the amount of collagen and ground substance present.
- **Loose connective tissue** (or **areolar tissue**) has relatively more ground substance than collagen, and it typically surrounds small blood vessels and occupies areas adjacent to other types of epithelia.
- **Dense irregular connective tissue** is filled primarily with randomly distributed bundles of type I collagen, with some elastic fibers, providing resistance to tearing from all directions as well as some elasticity.
- **Dense regular connective tissue**, prominent in tendons and ligaments, features bundles of essentially parallel type I collagen, providing great strength (but little stretch) in binding together components of the musculoskeletal system.
- **Reticular tissue** consists of delicate networks of type III collagen and is most abundant in certain lymphoid organs where the fibers form attachment sites for lymphocytes and other immune cells.
- **Mucoid tissue** is a gel-like connective tissue with few cells found most abundantly around blood vessels in the umbilical cord.

Connective Tissue ASSESS YOUR KNOWLEDGE

1. Which of the following connective tissue components is located in the ECM but not in the ground substance?
   a. Collagen bundles
   b. Fibronectin
   c. GAGs
   d. Hyaluronan
   e. Proteoglycans

2. What cells numerous in loose connective tissue are filled with secretory granules and stain with metachromasia?
   a. Macrophages
   b. Mast cells
   c. Fibrocytes
   d. Active fibroblasts
   e. Leukocytes

3. What is the first step of collagen production that occurs after the protein undergoes exocytosis?
   a. Cross-linking of collagen fibrils with a short linking collagen
   b. Removal of the terminal nonhelical domains by peptidases
   c. Hydroxylation of lysine and proline
   d. Assembly of subunits to form a larger structure
   e. Disulfide bond formation

4. What is an important part of the role played by macrophages during maintenance and renewal of strong extracellular fibers in connective tissue?
   a. Storage for a major energy source needed for ECM maintenance
   b. Production of specific collagen subunits
   c. A sentinel function against invaders entering the ECM
   d. Secretion of matrix metalloproteinases
   e. Presentation of antigens important for assembly of collagen bundles

5. Sulfated GAGs are important constituents of what extracellular structures?
   a. Hyaluronan
   b. Elastic fibers
   c. Type I collagen
   d. Proteoglycans
   e. Multiadhesive glycoproteins

6. Which of the following contains binding sites for integrins and is an important part of the ECM in both loose connective tissue and dense irregular connective tissue?
   a. Aggrecan
   b. Fibronectin
   c. Perlecan
   d. Fibrillin
   e. Most types of collagen

7. Dense regular connective tissue typically involves which of the following features?
   a. Contains mostly synthetically active fibroblasts
   b. Contains much ground substance
   c. Contains a similar cell population as areolar connective tissue
   d. Predominant tissue type in the stroma of most organs
   e. Predominantly located in tendons and ligaments

8. Research scientists at a small biotech firm are investigating new methods of controlling the growth and metastasis of malignant cells in patients diagnosed with breast cancer. They have developed a novel peptide-based drug, potentially deliverable therapeutically, that disrupts the tumor cells' ability to adhere to the ECM, which in turn triggers apoptosis. Which of the following is a most likely target of such drugs?
   a. Cadherins
   b. Adhesins
   c. Integrins
   d. Glycolipids of the cell membrane
   e. Fibrillin

9. A 36-year-old man is referred by his family physician to the pulmonary clinic. He complains of shortness of breath following physical activity and decreased capacity for exercise. He says that strenuous exercise including yard work is impossible without sitting down and resting every few minutes. After taking several deep breaths during the physical examination, he begins to wheeze. He is not a smoker and works in an office not exposed to dust, fumes, or other irritants. He appears slightly jaundiced. Serum alpha-1-antitrypsin (AAT) concentration analysis is below normal and is followed up with AAT phenotype and DNA testing which indicates one copy of S and one of Z mutations with 40% abnormal AAT production. Urinalysis shows elevated levels of desmosine and isodesmosine. These excreted compounds normally contribute to efficient lung function by which of the following mechanisms?

   a. Post-translational modification of fibrillin
   b. Cross-linking elastin
   c. Activating elastase
   d. Activating AAT
   e. Binding type IV collagen to elastin

10. A 33-year-old homeless woman has been living in an abandoned building eating dried meat and bread from the dumpster behind a delicatessen. She smokes cigarettes "bummed" from others. She presents at a free clinic with bleeding under the skin, particularly around hair follicles, and bruises on her arms and legs. She is irritable, clinically depressed, and fatigued with general muscle weakness. Her gums are bleeding, swollen, purple, and spongy, with several loose teeth. She has an infected toe, which may be broken. She is afebrile, a glucose finger-stick is normal, and the urine dipstick shows no sugar, protein, or ketones. You suspect a vitamin deficiency. What might be the underlying mechanism for this patient's symptoms?

   a. Decreased degradation of collagen
   b. Stimulation of prolyl hydroxylase
   c. Formation of unstable collagen helices
   d. Excessive callus formation in healing fractures
   e. Organ fibrosis

Answers: 1a, 2b, 3b, 4d, 5d, 6b, 7e, 8c, 9b, 10c

6 Adipose Tissue

Connective tissue in which fat-storing cells or **adipocytes** predominate is called **adipose tissue**. These large cells are typically found isolated or in small groups within loose or dense irregular connective tissue but occur in large aggregates in adipose tissue or "fat" in many organs and body regions. Adipose tissue normally represents 15%-20% of the body weight in men, somewhat more in women. Besides serving as storage depots for neutral fats, chiefly triglycerides (long-chain fatty acyl esters of glycerol), adipocytes function as key regulators of the body's overall energy metabolism. With a growing epidemic of obesity and its associated health problems, including diabetes and heart disease, adipocytes and adipose tissue now constitute a major area of medical research.

Two properties of triglyceride lipids explain their selection as the preferred form of nutrient storage. Insoluble in water, lipids can be concentrated with no adverse osmotic effects on cells. Also, the caloric density of triglycerides (9.3 kcal/g) is twice that of proteins or carbohydrates, including glycogen, making these simple lipids the most efficient means of storing calories. Adipocytes specialize in concentrating triglycerides as lipid droplet(s), with other cells normally accumulating relatively little lipid.

Adipocytes are active cells metabolically, responding to both nervous and hormonal stimuli. They release hormones and various other important substances and adipose tissue is now recognized as an endocrine organ at the center of nutritional homeostasis. With its unique physical properties, tissue rich in fat conducts heat poorly and provides thermal insulation for the body. Adipose tissue also fills spaces between other tissues, helping to keep some organs in place. Subcutaneous layers of adipose tissue help shape the body surface, and cushion regions subject to repeated mechanical stress such as the palms, heels, and toe pads.

There are two major types of adipose tissue with different locations, structures, colors, and functions. **White adipose tissue**, the more common type specialized for fat storage, consists of cells each containing one large cytoplasmic droplet of whitish-yellow fat. **Brown adipose tissue** contains cells with multiple lipid droplets interspersed among abundant mitochondria, which helps give this tissue a darker appearance. Brown adipocytes release heat and function to warm the blood. Both types of adipose tissue have a rich blood supply and the adipocytes, unlike other cells of connective tissue proper, are individually surrounded by a thin external lamina containing type IV collagen.

› WHITE ADIPOSE TISSUE

Specialized for relatively long-term energy storage, adipocytes of white adipose tissue are spherical when isolated but are polyhedral when closely packed in situ. When completely developed, a white adipocyte is very large, between 50 and 150 µm in diameter, and contains a single huge droplet of lipid filling almost the entire cell. With the single large droplets of triglycerides, white adipocytes are also called **unilocular** (Figure 6–1). Because lipid is removed from cells by xylene or other solvents used in routine histological techniques, unilocular adipocytes are often empty in standard light microscopy. The cells are sometimes said to have a signet-ring appearance, with the lipid droplet displacing and flattening the nucleus against the cell membrane (Figure 6–1d). This membrane and the thin rim of cytoplasm that remains after dissolution of the stored lipid may shrink, collapse, or rupture, distorting cell and tissue structure.

›› MEDICAL APPLICATION

Unilocular adipocytes can generate benign tumors called **lipomas** that are relatively common, although malignant adipose tumors (**liposarcomas**) occur infrequently. Fetal lipomas of brown fat are sometimes called **hibernomas**.

FIGURE **6–1**　White adipose tissue.

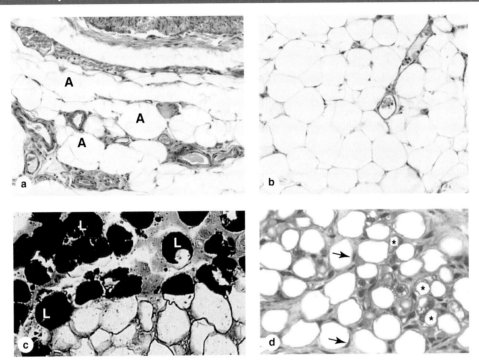

White or unilocular adipose tissue is commonly seen in sections of many human organs.

(a) Large white adipocytes (**A**) are seen in the connective tissue associated with small blood vessels. The fat cells are empty because lipid was dissolved away in slide preparation. Nuclei at the cell membranes are visible in some of the fat cells. (X100; H&E)

(b) Large (empty) adipocytes predominate in this typical white adipose tissue, which shows only a small portion of microvasculature. In a single histologic section, nuclei of most very large adipocytes are not included. (X100; H&E)

(c) Tissue was fixed here with osmium tetroxide, which preserves lipid (**L**) and stains it black. Many adipocytes in this slide retain at least part of their large lipid droplets. (X440; Osmium tetroxide)

(d) In this specimen from a young mammal the smaller adipocytes marked with asterisks are not unilocular, having many lipid droplets of various sizes. Such cells in white fat represent those in which differentiation is incomplete as well as a small subpopulation of beige cells with brown fat-forming potential. The eccentric nuclei of the unilocular cells are indicated by arrowheads. (X200; PT)

Most cytoplasmic organelles in a white adipocyte are near the peripheral nucleus, including mitochondria, a small Golgi apparatus, a few cisternae of RER, and free polyribosomes. The thin, submembranous layer of cytoplasm surrounding the lipid droplet contains cisternae of smooth ER (SER) and pinocytotic vesicles. TEM studies reveal that most adipocytes, especially immature cells, contain minute lipid droplets in addition to the large droplet. The lipid droplet-cytoplasm interface is reinforced only by intermediate filaments of vimentin.

As shown in Figure 6–1 white fat is subdivided into incomplete lobules by partitions of connective tissue containing a vascular bed and nerve network. Fibroblasts, macrophages, and other cells typically comprise about half the total cell number in white adipose tissue. Reticular fibers form a fine interwoven network that supports individual fat cells and binds them together, and the microvasculature between adipocytes may not always be apparent in tissue sections.

The distribution of white adipose tissue changes significantly through childhood and adult life and is partly regulated by sex hormones controlling adipose deposition in the breasts and thighs. The color of freshly dissected white adipose tissue depends on the diet, varying from white to yellow with the amount of carotenoids dissolved in the lipid.

Storage & Mobilization of Lipids

White adipocytes can store triglycerides derived from three sources:

- Dietary fats brought to the cells via the circulation as **chylomicrons**,
- Lipids synthesized in the liver and transported in blood with **very-low-density lipoproteins (VLDLs)**,
- Free fatty acids and glycerol synthesized by the adipocytes.

Chylomicrons (Gr. *chylos*, juice + *micros*, small) are particles of variable size, up to 1200 nm in diameter, formed from ingested lipids in epithelial cells lining the small intestine and transported in the blood and lymph. They consist of a core containing mainly triglycerides, surrounded by a stabilizing monolayer of phospholipids, cholesterol, and several apolipoproteins.

VLDLs are smaller complexes (30-80 nm, providing a greater surface-to-volume ratio), of similar lipid and protein composition to chylomicrons, but are synthesized from lipids in liver cells. Levels of circulating lipoproteins are routinely measured in clinical tests for blood lipids, after fasting to allow depletion of chylomicrons. Varying levels of apoproteins and triglycerides in the complexes allow their categorization according to density, from VLDL to high-density lipoprotein (HDL).

In adipose tissue both chylomicrons and VLDLs are hydrolyzed at the luminal surfaces of blood capillaries by **lipoprotein lipase**, an enzyme synthesized by the adipocytes and transferred to the capillary cell membrane (Figure 6–2). Free fatty acids then enter the adipocytes by both active transport and diffusion. Within the adipocytes, the fatty acids combine with glycerol phosphate, supplied by glucose metabolism, to again form triglycerides, which are then deposited in the growing lipid droplet. **Insulin** stimulates glucose uptake by adipocytes and accelerates its conversion into triglycerides, and the production of lipoprotein lipase.

When adipocytes are stimulated by nerves or various hormones, stored lipids are mobilized and cells release fatty acids and glycerol. **Norepinephrine** released in the adrenal gland and by postganglionic sympathetic nerves in adipose tissue activates a **hormone-sensitive lipase** that breaks down triglycerides at the surface of the stored lipid droplets (Figure 6–2). This lipase activity is also stimulated by growth hormone (GH) from the pituitary gland. The free fatty acids diffuse across the membranes of the adipocyte and the capillary endothelium, and bind the protein albumin in blood for transport throughout the body. The more water-soluble glycerol remains free in blood and is taken up by the liver. Insulin inhibits the hormone-sensitive lipase, reducing fatty acid release, and also stimulates enzymes for lipid synthesis. Besides insulin and GH, other peptide hormones also cooperate in regulating lipid synthesis and mobilization in adipocytes.

Hormonal activity of white adipocytes themselves includes production of the 16-kDa polypeptide hormone **leptin** (Gr. *leptos*, thin), a "satiety factor" with target cells in the hypothalamus, other brain regions, and peripheral organs which helps regulate the appetite under normal conditions and participates in regulating the formation of new adipose tissue.

FIGURE **6–2** **Lipid storage and mobilization from adipocytes.**

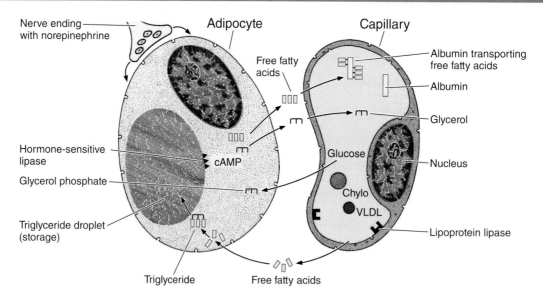

Triglycerides are transported by blood and lymph from the intestine and liver in lipoprotein complexes known as **chylomicrons (Chylo)** and **VLDLs**. In the capillary endothelial cells of adipose tissue, these complexes are partly broken down by **lipoprotein lipase**, releasing free fatty acids and glycerol. The free fatty acids diffuse from the capillary into the adipocyte, where they are reesterified to glycerol phosphate, forming triglycerides that are stored in the lipid droplet until needed.

Norepinephrine from nerve endings stimulates the cyclic AMP (cAMP) system, which activates **hormone-sensitive lipase** to hydrolyze the stored triglycerides to free fatty acids and glycerol. These substances diffuse into the capillary, where the fatty acids bind albumin for transport throughout the body for use as an energy source.

Although white adipose tissue associated with different organs appears histologically similar, differences in gene expression have been noted between visceral deposits (in the abdomen) and subcutaneous deposits of white fat. Such differences may be important in the medical risks of obesity; it is well established that increased visceral adipose tissue raises the risk of diabetes and cardiovascular disease whereas increased subcutaneous fat does not. The release of visceral fat products directly to the portal circulation and liver may also influence the medical importance of this form of obesity.

In response to body needs, lipids are mobilized rather uniformly from white adipocytes in all parts of the body, although adipose tissue in the palms, soles, and fat pads behind the eyes resists even long periods of starvation. During starvation adipocytes can lose nearly all their fat and become polyhedral or spindle-shaped cells with only very small lipid droplets.

Histogenesis of White Adipose Tissue

Like other connective tissue, skeletal and muscle cells, adipocytes develop from mesenchymal stem cells. Adipose development first produces **preadipocytes**, which look rather like larger fibroblasts with cytoplasmic lipid droplets (Figure 6–3). Initially the droplets of white adipocytes are isolated from one another but soon fuse to form the single large droplet (Figure 6–1).

As shown in Figure 6–3 white adipocytes develop together with a smaller population of cells termed beige adipocytes, which remain within white adipose tissue and have histological and metabolic features generally intermediate between white and brown adipocytes. With adaptation to cold temperatures beige adipocytes change reversibly, forming many more small lipid droplets, adopting a gene expression profile more like that of brown fat, and begin to release heat (see below).

FIGURE 6–3 Development of white and brown fat cells.

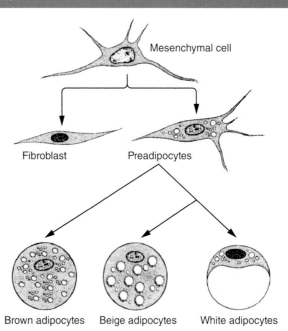

Mesenchymal cell

Fibroblast   Preadipocytes

Brown adipocytes   Beige adipocytes   White adipocytes

Mesenchymal stem cells differentiate as progenitor cells for all types of connective tissue, including **preadipocytes**. These are initially of at least two types. Preadipocytes developing within the lateral mesoderm of the embryo produce large number of **white adipocytes** (forming white adipose tissue) and a smaller number of so-called beige adipocytes with cytological features and gene expression patterns of both white and brown adipocytes. White adipocytes are unilocular, with one large lipid droplet occupying most of the cytoplasm. The white adipocyte is usually much larger than that shown here in relation to the other cell types.

  Brown adipocytes differentiate from another population of preadipocytes located in paraxial embryonic mesoderm and remain multilocular (having many small lipid droplets) with numerous mitochondria (not shown here). Mitochondrial metabolism of lipid in brown adipocytes releases heat rather than ATP. Cells functioning as brown adipocytes can also develop from beige adipocytes during adaptation to cold temperatures.

Humans are born with stores of white adipose tissue, which begin to accumulate by the 14th week of gestation. Both visceral and subcutaneous fat is well-developed before birth. Proliferation of progenitor cells diminishes by late gestation, and adipose tissue increases mainly by the filling of existing adipocytes until around age 10, followed by a period of new fat cell differentiation which lasts through adolescence. New adipocyte formation occurs around small blood vessels, where undifferentiated mesenchymal cells are most abundant.

Excessive adipose tissue accumulation, or obesity, occurs when nutritional intake exceeds energy expenditure, an increasingly common condition in modern, sedentary lifestyles. Although adipocytes can differentiate from mesenchymal stem cells throughout life, adult-onset obesity mainly involves increasing the size of existing adipocytes (hypertrophy). Childhood obesity, in contrast, often involves increases in both adipocyte size and numbers due to the differentiation of more preadipocytes from mesenchymal cells (hyperplasia). Weight loss after dietary changes is due to reductions in adipocyte volume, but not their overall number.

❯ BROWN ADIPOSE TISSUE

Brown adipose tissue constitutes 2%-5% of the newborn body weight, located mainly in the back, neck, and shoulders, but it is greatly reduced during childhood and adolescence. In adults it is found only in scattered areas, especially around the kidneys, adrenal glands, aorta, and mediastinum. The color of **brown fat** is due to both the very abundant mitochondria (containing cytochrome pigment) scattered among the lipid droplets of the fat cells and the large number of blood capillaries in this tissue. Brown adipocytes contain many small lipid inclusions and are therefore called **multilocular** (Figure 6–3). The small lipid droplets, abundant mitochondria, and rich vasculature all help mediate this tissue's principal function of **heat production** and warming the blood.

Cells of brown fat are polygonal and generally smaller than white adipocytes; their smaller lipid droplets allow the nucleus to be more centrally located (Figure 6–4). Brown adipocytes are often closely packed around large capillaries and the tissue is subdivided by connective tissue partitions into

---

FIGURE **6–4** Brown adipose tissue.

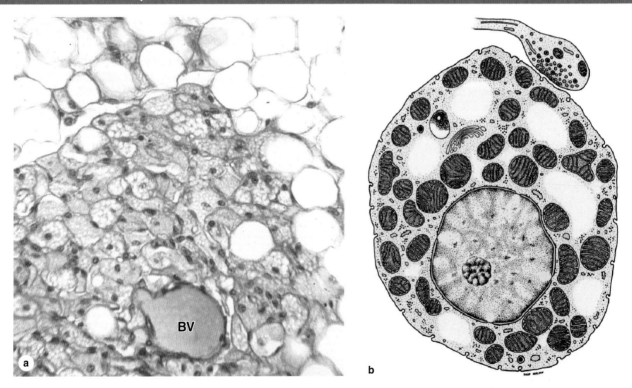

(a) Brown adipose tissue is shown here around a small blood vessel (**BV**) and adjacent white adipose tissue at the top of the photo. Brown adipocytes are slightly smaller and characteristically contain many small lipid droplets and central spherical nuclei. If the lipid has been dissolved from the cells, as shown here, the many mitochondria among the lipid spaces are retained and can be easily discerned. (X200; PT)

(b) A diagram of a single multilocular adipocyte showing the central nucleus, numerous small lipid droplets (yellow), and many mitochondria. Also shown is a sympathetic nerve ending that releases norepinephrine to stimulate mitochondrial production of heat.

lobules that are better delineated than the lobules of white adipose tissue. Cells of this tissue receive direct sympathetic innervation, which regulates their metabolic activity.

Function of Brown Adipocytes

The main function of these multilocular adipose cells is to produce heat by nonshivering thermogenesis. The physiology of brown fat is best understood from studies of the tissue in hibernating species. In animals ending their hibernation period, and in newborn humans, nerve impulses liberate norepinephrine into brown adipose tissue. As in white fat, this neurotransmitter activates the hormone-sensitive lipase of adipocytes, promoting hydrolysis of triglycerides to fatty acids and glycerol. However, unlike the process in white fat, liberated fatty acids of multilocular adipocytes are not released but are quickly metabolized, with a consequent increase in O_2 consumption and heat production. This raises the temperature within the tissue and warms the locally circulating blood, which then distributes the heat throughout the body.

Heat production in brown adipocytes is greater than that of other cells because their inner mitochondrial membranes have greatly upregulated levels of the transmembrane protein uncoupling protein-1 (UCP1) or thermogenin. In the presence of free fatty acids, UCP1 permits the flow of protons from the intermembranous space to the matrix without passing through ATP synthetase complexes. Instead of producing ATP, the energy associated with this proton flow dissipates as heat.

Histogenesis of Brown Adipose Tissue

Brown adipose tissue also develops from mesenchyme, but involves preadipocytes in a different embryonic location (paraxial) from those producing white adipose tissue. Brown adipocytes also emerge earlier than white fat during fetal development. In humans the amount of brown fat is maximal relative to body weight at birth, when thermogenesis is most needed and partially disappears by involution and apoptosis during childhood. In adults the amount and activity of brown fat are higher in lean individuals.

The number of brown adipocytes increases during cold adaptation, usually appearing as clusters of multilocular cells in white adipose tissue. As indicated earlier this increase involves the reversible shift of beige cells to functional brown adipocytes, but may also include proliferation and differentiation of new adipocytes from preexisting progenitor cells. Besides stimulating thermogenic activity, autonomic nerves also promote brown adipocyte differentiation and prevent apoptosis in mature brown fat cells.

Adipose Tissue SUMMARY OF KEY POINTS

- The defining cells of adipose tissue (fat), adipocytes, are very large cells derived from mesenchyme and specialized for energy storage in lipid droplet(s) with triglycerides.
- Adipocytes store lipids from three sources: from dietary fats packaged as chylomicrons in the intestine; from triglycerides produced in the liver and circulating as very-low-density lipoproteins (VLDLs); and from fatty acids synthesized locally.
- Lipids are mobilized from adipocytes by hormone-sensitive lipase activated by norepinephrine released from the adrenal gland and various peptide hormones.
- Cells of adipose tissue are supported by reticular fibers, with connective tissue septa dividing the tissue into lobules of various sizes.
- There are two types of adipose tissue: white fat and brown fat.

White Adipose Tissue

- White adipose tissue is found in many organs throughout the body, typically forming about 20% of the body weight in adults.
- Adipocytes of white fat are typically very large cells, ranging in diameter from 50 to 150 μm.

- These cells each contain primarily one large lipid droplet (they are unilocular), causing the nucleus and remaining cytoplasm to be pushed against the plasmalemma.
- Fatty acids are released from white adipocytes by lipase activity when nutrients are needed and carried throughout the body on plasma proteins such as albumin.
- Leptin is a polypeptide hormone with target cells in the hypothalamus that is released from white adipocytes and helps regulate eating behavior.

Brown Adipose Tissue

- Brown fat comprises up to 5% of the newborn body weight but smaller amounts in adults.
- Adipocytes of this tissue are typically smaller than those of white fat and contain primarily many small lipid droplets (they are multilocular) in cytoplasm containing many mitochondria and a central nucleus.
- Fatty acids released in adipocytes of brown fat are metabolized in mitochondria of these cells for thermogenesis rather than ATP synthesis, using uncoupling protein-1.

Adipose Tissue ASSESS YOUR KNOWLEDGE

1. White adipocytes are derived developmentally from what precursor cells?

   a. Monocytes
   b. Fibroblasts
   c. Mesenchymal cells
   d. Brown adipocytes
   e. Mast cells

2. What are the relatively large particles formed in the intestinal epithelial cells and rich in ingested lipids?

   a. Fatty acids
   b. Chylomicrons
   c. Glycerols
   d. Very-low-density lipoproteins
   e. Adipocytes

3. What substance, released from the adrenal gland and some autonomic neurons, increases lipolytic activity in white adipocytes?

   a. Leptin
   b. Insulin
   c. Norepinephrine
   d. Glycogen
   e. Triglyceride

4. What is the most important form of lipid storage in both white and brown adipocytes?

   a. Free fatty acids
   b. Cholesterol
   c. Chylomicrons
   d. Glycerol
   e. Triglycerides

5. Important target cells of leptin are found in which organ?

   a. Small intestine
   b. White adipose tissue
   c. Large intestine
   d. Hypothalamus
   e. Brown adipose tissue

6. The hormone-sensitive lipase in the cells of adipose tissue acts primarily on what substrate?

   a. Glucose
   b. Free fatty acids
   c. Glycerol
   d. Triglycerides
   e. Very-low-density lipoproteins

7. Applied to adipocytes, the term "multilocular" refers to which of the following?

   a. The large number of small cytoplasmic lipid droplets
   b. The proliferation of the cells in an obese individual
   c. The large number of mitochondria in the cells
   d. The high density of nerves supplying the tissue
   e. The type of mesenchymal cells also present

8. Fully differentiated white adipocytes are large cells, typically having diameters of approximately what size?

   a. 5 μm
   b. 10 μm
   c. 100 μm
   d. 500 μm
   e. 1000 μm

9. Ten days after birth a full-term newborn boy develops firm, erythematous nodules and plaques over his trunk, arms, buttocks, thighs, and cheeks. His mother's pregnancy was complicated by placenta previa and his airway was cleared of aspirated meconium immediately after birth. A biopsy of subcutaneous tissue shows necrosis within the brown adipose tissue. What metabolic activity is liable to be affected in this patient?

   a. Export of fatty acids from fat
   b. Thermal insulation
   c. Oxidation of fatty acids for thermogenesis
   d. Activation of the adenylate cyclase system
   e. Initiation of shivering

10. A 44-year-old African-American woman visits her family physician for a physical examination at the urging of her husband. She has no current complaints and is taking no medications. She is allergic to erythromycin. She works as a software developer and lives with her 52-year-old husband and 12-year-old daughter. She is a nonsmoker and drinks an occasional glass of wine when she and her husband go out to dinner. She is involved in no regular exercise. Her mother is 66 and suffers from type II diabetes, hyperlipidemia, and hypertension and had a myocardial infarction last year. The patient's father died of a stroke last year at the age of 72. On examination, the patient's blood pressure is 155/100 mm Hg, pulse 84, weight 215 lb (increased from 180 lb 3 years ago), and height 5 ft. 7 in. In this patient, during the period of weight gain which one of the following responses would be most likely in her white fat?

   a. Increased synthesis of growth hormone
   b. Decreased synthesis of leptin
   c. Decreased release of leptin to the blood
   d. Conversion of beige adipocytes to unilocular white adipocytes
   e. Increased release of norepinephrine from nerve terminals near white adipocytes

CHAPTER

7 Cartilage

Cartilage is a tough, durable form of supporting connective tissue, characterized by an **extracellular matrix** (ECM) with high concentrations of GAGs and proteoglycans, interacting with collagen and elastic fibers. Structural features of its matrix make cartilage ideal for a variety of mechanical and protective roles within the adult skeleton and elsewhere (Figure 7–1).

Cartilage ECM has a firm consistency that allows the tissue to bear mechanical stresses without permanent distortion. In the respiratory tract, ears, and nose, cartilage forms the framework supporting softer tissues. Because of its resiliency and smooth, lubricated surface, cartilage provides cushioning and sliding regions within skeletal joints and facilitates bone movements. As described in Chapter 8, cartilage also guides development and growth of long bones, both before and after birth.

Cartilage consists of cells called **chondrocytes** (Gr. *chondros*, cartilage + *kytos*, cell) embedded in the ECM which unlike connective tissue proper contains no other cell types. Chondrocytes synthesize and maintain all ECM components and are located in matrix cavities called **lacunae**.

The physical properties of cartilage depend on electrostatic bonds between **type II collagen** fibrils, **hyaluronan**, and the sulfated GAGs on densely packed **proteoglycans**. Its semi-rigid consistency is attributable to water bound to the negatively charged hyaluronan and GAG chains extending from proteoglycan core proteins, which in turn are enclosed within a dense meshwork of thin type II collagen fibrils. The high content of bound water allows cartilage to serve as a shock absorber, an important functional role.

All types of cartilage lack vascular supplies and chondrocytes receive nutrients by diffusion from capillaries in surrounding connective tissue (the perichondrium). In some skeletal elements, large blood vessels do traverse cartilage to supply other tissues, but these vessels release few nutrients to the chondrocytes. As might be expected of cells in an avascular tissue, chondrocytes exhibit low metabolic activity. Cartilage also lacks nerves.

The **perichondrium** (Figure 7–2) is a sheath of dense connective tissue that surrounds cartilage in most places, forming an interface between the cartilage and the tissues supported by the cartilage. The perichondrium harbors the blood supply serving the cartilage and a small neural component. Articular cartilage, which covers the ends of bones in movable joints and which erodes in the course of arthritic degeneration, lacks perichondrium and is sustained by the diffusion of oxygen and nutrients from the synovial fluid.

As shown in Figure 7–1, variations in the composition of the matrix characterize three main types of cartilage: hyaline cartilage, elastic cartilage, and fibrocartilage. Important features of these are summarized in Table 7–1.

>> **MEDICAL APPLICATION**

Many genetic conditions in humans or mice that cause defective cartilage, joint deformities, or short limbs are due to recessive mutations in genes for collagen type II, the aggrecan core protein, the sulfate transporter, and other proteins required for normal chondrocyte function.

HYALINE CARTILAGE

Hyaline (Gr. *hyalos*, glass) cartilage, the most common of the three types, is homogeneous and semitransparent in the fresh state. In adults hyaline cartilage is located in the articular surfaces of movable joints, in the walls of larger respiratory passages (nose, larynx, trachea, bronchi), in the ventral ends of ribs, where they articulate with the sternum, and in the epiphyseal plates of long bones, where it makes possible longitudinal bone growth (Figure 7–1). In the embryo, hyaline cartilage forms the temporary skeleton that is gradually replaced by bone.

FIGURE **7–1** Distribution of cartilage in adults.

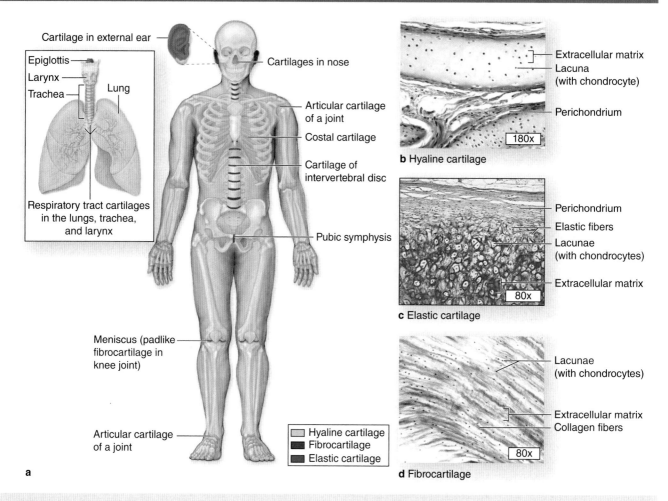

a

b Hyaline cartilage

c Elastic cartilage

d Fibrocartilage

(a) There are three types of adult cartilage distributed in many areas of the skeleton, particularly in joints and where pliable support is useful, as in the ribs, ears, and nose. Cartilage support of other tissues throughout the respiratory tract is also prominent.

The photomicrographs show the main features of (b) hyaline cartilage, (c) elastic cartilage, and (d) fibrocartilage. Dense connective tissue of perichondrium is shown here with hyaline and elastic cartilage.

TABLE **7–1** Important features of the major cartilage types.

	Hyaline Cartilage	Elastic Cartilage	Fibrocartilage
Main features of the extracellular matrix	Homogeneous, with type II collagen and aggrecan	Type II collagen, aggrecan, and darker elastic fibers	Type II collagen and large areas of dense connective tissue with type I collagen
Major cells	Chondrocytes, chondroblasts	Chondrocytes, chondroblasts	Chondrocytes, fibroblasts
Typical arrangement of chondrocytes	Isolated or in small isogenous groups	Usually in small isogenous groups	Isolated or in isogenous groups arranged axially
Presence of perichondrium	Yes (except at epiphyses and articular cartilage)	Yes	No
Main locations or examples	Many components of upper respiratory tract; articular ends and epiphyseal plates of long bones; fetal skeleton	External ear, external acoustic meatus, auditory tube; epiglottis and certain other laryngeal cartilages	Intervertebral discs, pubic symphysis, meniscus, and certain other joints; insertions of tendons
Main functions	Provides smooth, low-friction surfaces in joints; structural support for respiratory tract	Provides flexible shape and support of soft tissues	Provides cushioning, tensile strength, and resistance to tearing and compression

FIGURE **7–2** The structure of cartilage matrix and cells.

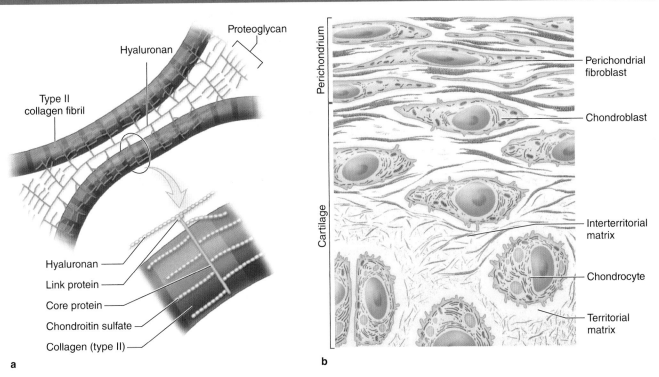

(a) A schematic representation of the most abundant molecules in cartilage matrix shows the interaction between type II collagen fibrils and proteoglycans linked to hyaluronan. Link proteins noncovalently bind the protein core of proteoglycans to the linear hyaluronan molecules. The chondroitin sulfate side chains of the proteoglycan electrostatically bind to the collagen fibrils, forming a cross-linked matrix. The circled area is shown larger in the lower part of the figure. Physical properties of these matrix components produce a highly hydrated, pliable material with great strength. Approximately 75% of the wet weight of hyaline cartilage is water.

(b) A diagram of the transitional area between the perichondrium and the cartilage matrix. Fibroblast-like progenitor cells in the perichondrium give rise to larger chondroblasts, which divide and differentiate as chondrocytes. These functional cells produce matrix components and exist in lacunae surrounded by the matrix. The ECM immediately around each lacuna, called the **territorial matrix**, contains mostly proteoglycans and sparse collagen; that more distant from lacunae, the **interterritorial matrix**, is richer in collagen and may be less basophilic.

❯❯ MEDICAL APPLICATION

Osteoarthritis, a chronic condition that commonly occurs during aging, involves the gradual loss or changed physical properties of the hyaline cartilage that lines the articular ends of bones in joints. Joints that are weight-bearing (knees, hips) or heavily used (wrist, fingers) are most prone to cartilage degeneration. Fragments released by wear-and-tear to the articular cartilage trigger secretion of matrix metalloproteinases and other factors from macrophages in adjacent tissues, which exacerbate damage and cause pain and inflammation within the joint.

Matrix

The dry weight of hyaline cartilage is nearly 40% collagen embedded in a firm, hydrated gel of proteoglycans and structural glycoproteins. In routine histology preparations, the proteoglycans make the matrix generally basophilic and the thin collagen fibrils are barely discernible. Most of the collagen in hyaline cartilage is **type II**, although small amounts of minor collagens are also present.

Aggrecan (250 kDa), with approximately 150 GAG side chains of chondroitin sulfate and keratan sulfate, is the most abundant proteoglycan of hyaline cartilage. Hundreds of these proteoglycans are bound noncovalently by link proteins to long polymers of hyaluronan, as shown schematically in Figure 7–2a and discussed in Chapter 5. These proteoglycan complexes bind further to the surface of type II collagen fibrils (Figure 7–2a). Water bound to GAGs in the proteoglycans constitutes up to 60%-80% of the weight of fresh hyaline cartilage.

Another important component of cartilage matrix is the structural multiadhesive glycoprotein **chondronectin**. Like fibronectin in other connective tissues, chondronectin binds specifically to GAGs, collagen, and integrins, mediating the adherence of chondrocytes to the ECM.

Staining variations within the matrix reflect local differences in its molecular composition. Immediately surrounding each chondrocyte, the ECM is relatively richer in GAGs than collagen, often causing these areas of **territorial matrix** to stain differently from the intervening areas of interterritorial matrix (Figures 7–2b and 7–3).

Chondrocytes

Cells occupy relatively little of the hyaline cartilage mass. At the periphery of the cartilage, young chondrocytes or **chondroblasts** have an elliptic shape, with the long axes parallel to the surface (Figure 7–3). Deeper in the cartilage, they are round and may appear in groups of up to eight cells that originate from mitotic divisions of a single chondroblast and are called **isogenous aggregates**. As the chondrocytes become more active in secreting collagens and other ECM components, the aggregated cells are pushed apart and occupy separate lacunae.

Cartilage cells and matrix may shrink slightly during routine histologic preparation, resulting in both the irregular shape of the chondrocytes and their retraction from the matrix. In living tissue chondrocytes fill their lacunae completely.

Because cartilage matrix is avascular, chondrocytes respire under low-oxygen tension. Hyaline cartilage cells metabolize glucose mainly by anaerobic glycolysis. Nutrients from the blood diffuse to all the chondrocytes from the cartilage surface, with movements of water and solutes in the cartilage matrix promoted by intermittent tissue compression and decompression during body movements. The limits of such diffusion define the maximum thickness of hyaline cartilage, which usually exists as small, thin plates.

≫ MEDICAL APPLICATION

In contrast to other forms of cartilage and most other tissues, hyaline cartilage is susceptible to partial or isolated regions of **calcification** during aging, especially in the costal cartilage adjacent to the ribs. Calcification of the hyaline matrix, accompanied by degenerative changes in the chondrocytes, is a common part of the aging process and in many respects resembles endochondral ossification by which bone is formed.

FIGURE 7–3 Hyaline cartilage.

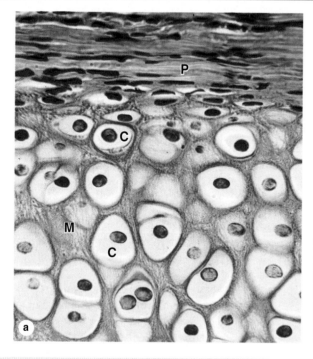
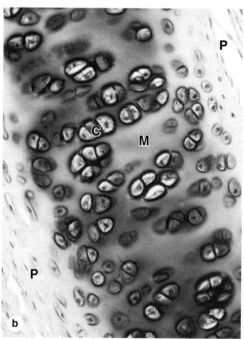

(a) The upper part of the photo shows the perichondrium (**P**), an example of dense connective tissue consisting largely of type I collagen. Among the fibroblastic cells of the perichondrium are indistinguishable mesenchymal stem cells. There is a gradual transition and differentiation of cells from the perichondrium to the cartilage, with some elongated fibroblast-like cells becoming larger and more rounded as chondroblasts and chondrocytes (**C**). These are located within lacunae surrounded by the matrix (**M**) which these cells secreted. (X200; H&E)

(b) The thin region of hyaline cartilage shown here has perichondrium (**P**) on both sides and shows larger lacunae containing isogenous groups of chondrocytes (**C**) within the matrix (**M**). Such groups of two, four, or more cells are produced by mitosis; the cells will separate into individual lacunae as they begin to secrete matrix. Territorial matrix immediately around the chondrocytes is more basophilic than interterritorial matrix farther from the cells. (X160; H&E)

Chondrocyte synthesis of sulfated GAGs and secretion of proteoglycans are accelerated by many hormones and growth factors. A major regulator of hyaline cartilage growth is the pituitary-derived protein called growth hormone or **somato-tropin**. This hormone acts indirectly, promoting the endocrine release from the liver of insulin-like growth factors, or somato-medins, which directly stimulate the cells of hyaline cartilage.

Perichondrium

Except in the articular cartilage of joints, all hyaline car-tilage is covered by a layer of dense connective tissue, the **perichondrium**, which is essential for the growth and main-tenance of cartilage (Figures 7–2b and 7–3). The outer region of the perichondrium consists largely of collagen type I fibers and fibroblasts, but an inner layer adjoining the cartilage matrix also contains mesenchymal stem cells which provide a source for new chondroblasts that divide and differentiate into chondrocytes.

❯ ELASTIC CARTILAGE

Elastic cartilage is essentially similar to hyaline cartilage except that it contains an abundant network of elastic fibers in addition to a meshwork of collagen type II fibrils (Figures 7–4 and 7–1c), which give fresh elastic cartilage a yellowish color. With appropriate staining the elastic fibers usually appear as dark bundles distributed unevenly through the matrix.

More flexible than hyaline cartilage, elastic cartilage is found in the auricle of the ear, the walls of the external auditory

FIGURE **7–4** Elastic cartilage.

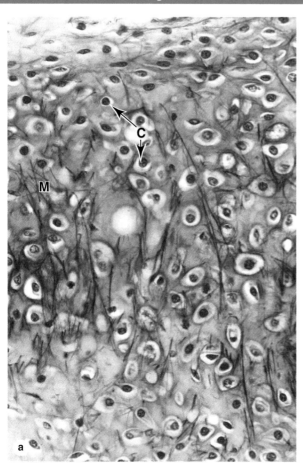
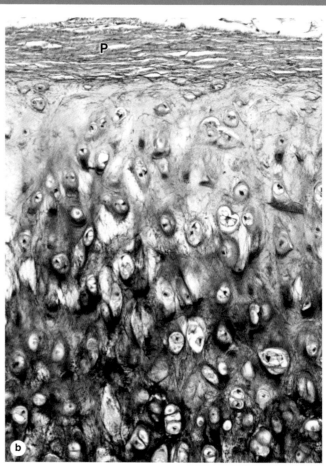

The chondrocytes (**C**) and overall organization of elastic cartilage are similar to those of hyaline cartilage, but the matrix (**M**) also contains elastic fibers that can be seen as darker components with proper staining. The abundant elastic fibers provide greater flexibility to this type of cartilage. The section in part **b** includes perichondrium (**P**) that is also similar to that of hyaline cartilage. (**a**) X160; Hematoxylin and orcein. (**b**) X180; Weigert resorcin and van Gieson.

canals, the auditory (Eustachian) tubes, the epiglottis, and the upper respiratory tract. Elastic cartilage in these locations includes a perichondrium similar to that of most hyaline cartilage. Throughout elastic cartilage the cells resemble those of hyaline cartilage both physiologically and structurally.

› FIBROCARTILAGE

Fibrocartilage takes various forms in different structures but is essentially a mingling of hyaline cartilage and dense connective tissue (Figures 7–5 and 7–1d). It is found in intervertebral discs, in attachments of certain ligaments, and in the pubic symphysis—all places where it serves as very tough, yet cushioning support tissue for bone.

Chondrocytes of fibrocartilage occur singly and often in aligned isogenous aggregates, producing type II collagen and other ECM components, although the matrix around these chondrocytes is typically sparse. Areas with chondrocytes and hyaline matrix are separated by other regions with fibroblasts and dense bundles of type I collagen which confer extra tensile strength to this tissue (Figure 7–5). The relative scarcity of proteoglycans overall makes fibrocartilage matrix more acidophilic than that of hyaline or elastic cartilage. There is no distinct surrounding perichondrium in fibrocartilage.

Intervertebral discs of the spinal column are composed primarily of fibrocartilage and act as lubricated cushions and shock absorbers preventing damage to adjacent vertebrae from abrasive forces or impacts. Held in place by ligaments, intervertebral discs are discussed further with joints in Chapter 8.

Important features of the three major types of cartilage are summarized in Table 7–1.

› CARTILAGE FORMATION, GROWTH, & REPAIR

All cartilage forms from embryonic mesenchyme in the process of **chondrogenesis** (Figure 7–6). The first indication of cell differentiation is the rounding up of the mesenchymal cells, which retract their extensions, multiply rapidly, and become more densely packed together. In general the terms "chondroblasts" and "chondrocytes" respectively refer to the cartilage cells during and after the period of rapid proliferation. At both stages the cells have basophilic cytoplasm rich in RER for collagen synthesis (Figure 7–7). Production of the ECM encloses the cells in their lacunae and then gradually separates chondroblasts from one another. During embryonic development, the cartilage differentiation takes place primarily from the center outward; therefore the more central cells have the characteristics of chondrocytes, whereas the peripheral cells are typical chondroblasts. The superficial mesenchyme develops as the perichondrium.

Once formed, the cartilage tissue enlarges both by **interstitial growth**, involving mitotic division of preexisting chondrocytes, and by **appositional growth**, which involves chondroblast differentiation from progenitor cells in the

FIGURE **7–5** Fibrocartilage.

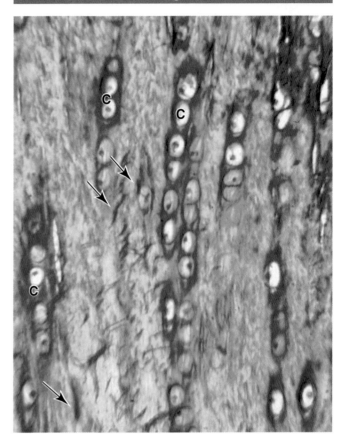

Fibrocartilage varies histologically in different structures, but is always essentially a mixture of hyaline cartilage and dense connective tissue.

In a small region of intervertebral disc, the axially arranged aggregates of chondrocytes (**C**) are seen to be surrounded by small amounts of matrix and separated by larger regions with dense collagen and scattered fibroblasts with elongated nuclei (arrows). (X250; Picrosirius-hematoxylin)

perichondrium (Figure 7–2b). In both cases, the synthesis of matrix contributes greatly to the growth of the cartilage. Appositional growth of cartilage is more important during postnatal development, although as described in Chapter 8, interstitial growth in cartilaginous regions within long bones is important in increasing the length of these structures. In articular cartilage, cells and matrix near the articulating surface are gradually worn away and must be replaced from within, because there is no perichondrium to add cells by appositional growth.

Except in young children, damaged cartilage undergoes slow and often incomplete **repair**, primarily dependent on cells in the perichondrium which invade the injured area and produce new cartilage. In damaged areas the perichondrium produces a scar of dense connective tissue instead of forming new cartilage. The poor capacity of cartilage for repair or regeneration is due in part to its avascularity and low metabolic rate.

FIGURE **7–6** Chondrogenesis.

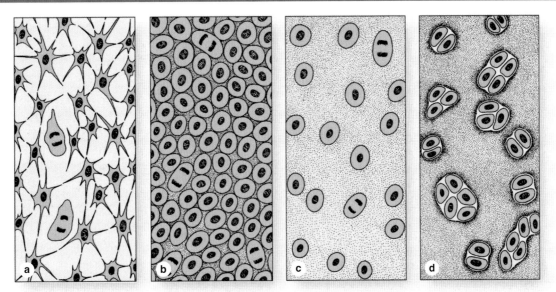

The major stages of embryonic cartilage formation, or chondrogenesis, are shown here.

(a) Mesenchyme is the precursor for all types of cartilage. **(b)** Mitosis and initial cell differentiation produces a tissue with condensations of rounded cells called **chondroblasts**. **(c)** Chondroblasts are then separated from one another again by their production of the various matrix components, which collectively swell

with water and form the very extensive ECM. **(d)** Multiplication of chondroblasts within the matrix gives rise to isogenous cell aggregates surrounded by a condensation of territorial matrix. In mature cartilage, this interstitial mitotic activity ceases and all chondrocytes typically become more widely separated by their production of matrix.

FIGURE **7–7** Chondrocytes in growing cartilage.

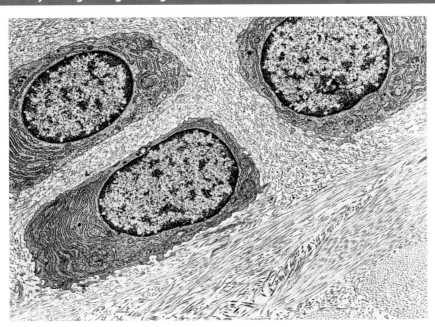

This TEM of **fibrocartilage** shows **chondrocytes** with abundant RER actively secreting the collagen-rich matrix. Bundles of **collagen fibrils**, sectioned in several orientations, are very prominent around the chondrocytes of fibrocartilage. Collagen types I and II

are both present in fibrocartilage. Chondrocytes in growing hyaline and elastic cartilage have more prominent Golgi complexes and synthesize abundant proteoglycans in addition to collagens. (X3750)

Cartilage SUMMARY OF KEY POINTS

- Cartilage is a **tough**, **resilient** type of connective tissue that structurally supports certain soft tissues, notably in the respiratory tract, and provides cushioned, low-friction surfaces in joints.
- Cells of cartilage, **chondrocytes**, make up a small percentage of the tissue's mass, which is mainly a flexible mass of **extracellular matrix (ECM)**.
- Chondrocytes are embedded within **lacunae** surrounded by the ECM.
- Cartilage ECM typically includes **collagen** as well as abundant **proteoglycans**, notably **aggrecan**, which bind a large amount of **water**.
- Cartilage always **lacks blood vessels**, lymphatics, and nerves, but it is usually surrounded by a dense connective tissue **perichondrium** that is vascularized.
- There are three major forms of cartilage: (1) **hyaline cartilage**, (2) **elastic cartilage**, and (3) **fibrocartilage**.

Hyaline Cartilage

- The ECM of hyaline cartilage is **homogenous and glassy**, rich in fibrils of type II collagen and aggrecan complexes with bound water.
- The ECM has less collagen and more proteoglycan immediately around the lacunae, producing slight staining differences in this **territorial matrix**.
- Chondrocytes occur **singly** or in small, mitotically derived **isogenous groups**.
- **Perichondrium** is usually present, but not at the hyaline cartilage of articular surfaces or the epiphyses of growing long bones.

Elastic Cartilage

- Elastic cartilage generally resembles hyaline cartilage in its chondrocytes and major ECM components, but its matrix includes **abundant elastic fibers**, visible with special stains, which increase the tissue's **flexibility**.
- Elastic cartilage provides flexible support for the external ear as well as certain structures of the middle ear and larynx; it is always surrounded by **perichondrium**.

Fibrocartilage

- Fibrocartilage contains varying **combinations of hyaline cartilage** in small amounts of **dense connective tissue**.
- Histologically it consists of small **chondrocytes** in a hyaline matrix, usually layered with larger areas of bundled **type I collagen** with scattered **fibroblasts**.
- Fibrocartilage provides very **tough**, **strong support** at tendon insertions and in **intervertebral discs** and certain other joints.

Cartilage Formation, Growth, & Repair

- All forms of cartilage form from embryonic **mesenchyme**.
- Cartilaginous structures grow by mitosis of existing chondroblasts in lacunae (**interstitial growth**) or formation of new chondroblasts peripherally from progenitor cells in the perichondrium (**appositional growth**).
- Repair or replacement of injured cartilage is very slow and ineffective, due in part to the tissue's **avascularity** and **low metabolic rate**.

Cartilage ASSESS YOUR KNOWLEDGE

1. The molecular basis for the shock absorbing properties of cartilage involves which of the following?
   a. Electrostatic interaction of proteoglycans with type IV collagen
   b. Ability of glycosaminoglycans to bind anions
   c. Noncovalent binding of glycosaminoglycans to protein cores
   d. Sialic acid residues in the glycoproteins
   e. Hydration of glycosaminoglycans

2. What distinguishes cartilage from most other connective tissues?
   a. Its extracellular matrix is rich in collagen.
   b. Its predominant cell type is a mesenchymal derivative.
   c. Its predominant cell type secretes both fibers and proteoglycans.
   d. It lacks blood vessels.
   e. It functions in mechanical support.

3. Which feature is typical of elastic cartilage?
   a. Primary skeletal tissue in the fetus
   b. No identifiable perichondrium
   c. Found in intervertebral discs
   d. Most widely distributed cartilage type in the body
   e. Collagen is mainly type II

4. Which area in cartilage is relatively collagen-poor and proteoglycan-rich?
   a. Fibrocartilage
   b. Territorial matrix
   c. Epiphyseal plate
   d. Interterritorial matrix
   e. Perichondrium

5. What is the source of the mesenchymal progenitor cells activated for the repair of hyaline cartilage of accident-damaged costal cartilages?
   a. Perichondrium
   b. Adjacent loose connective tissue
   c. Bone of the adjacent rib(s) and sternum
   d. Chondrocytes of the injured cartilage
   e. Stem cells circulating with blood

6. How does articular cartilage differ from most other hyaline cartilage?
   a. It undergoes mainly appositional growth.
   b. It contains isogenous groups of chondrocytes.
   c. It lacks a perichondrium.
   d. Its matrix contains aggrecan.
   e. It is derived from embryonic mesenchyme.

7. Which step occurs first in chondrogenesis?
   a. Appositional growth
   b. Conversion of chondroblasts to chondrocytes
   c. Formation of mesenchymal condensations
   d. Interstitial growth
   e. Secretion of collagen-rich and proteoglycan-rich matrix

8. Osteoarthritis is characterized by the progressive erosion of articular cartilage. The matrix metalloproteinases involved in this erosion primarily act on which matrix component?
    a. Aggrecan
    b. Link proteins
    c. Network-forming collagen
    d. Fibril-forming collagen
    e. Chondronectin

9. A 28-year-old woman visits the family medicine clinic complaining of loss of the sense of smell, nosebleeds, problems with swallowing, and hoarseness. She admits to "casual, social use" of cocaine on a regular basis since her sophomore year of college. A complete examination of her nose with a speculum and otoscope shows severe rhinitis (inflammation). There is also perforation and collapse of the nasal cartilage resulting in a "saddle nose" deformity. Erosions in the enamel of her front teeth are noted. The breakdown of the nasal cartilage releases collagen fibers primarily of which type?
    a. Type I
    b. Type II
    c. Type III
    d. Type IV
    e. Type VII

10. A 66-year-old man who suffered from severe osteoarthritis is referred to an orthopedic surgeon for replacement of his right knee. He had been actively involved in both high school and intercollegiate football and had continued running until about the age of 45 as a form of relaxation and exercise. With the patient's permission the removed joint is used by investigators performing a proteomic analysis of different joint tissues. The meniscus was found to contain almost exclusively type I collagen and aggrecan was undetectable. What is the most likely explanation for this result?
    a. The meniscus normally consists of dense regular connective tissue, which contains primarily type I collagen.
    b. The meniscus normally consists of fibrocartilage, which contains only type I collagen.
    c. The meniscus had undergone repeated rounds of repair due to wear-and-tear during which its hyaline cartilage component was replaced by dense connective tissue.
    d. Osteoarthritic injury in the knee resulted in the chondrocytes of the meniscus switching from expression of genes for type II collagen to type I collagen.
    e. Elastic cartilage is normally replaced by fibrocartilage during aging and this process can be accelerated by exercise.

8 Bone

As the main constituent of the adult skeleton, bone tissue (Figure 8–1) provides solid support for the body, protects vital organs such as those in the cranial and thoracic cavities, and encloses internal (medullary) cavities containing bone marrow where blood cells are formed. Bone (or osseous) tissue also serves as a reservoir of calcium, phosphate, and other ions that can be released or stored in a controlled fashion to maintain constant concentrations in body fluids.

In addition, bones form a system of levers that multiply the forces generated during skeletal muscle contraction and transform them into bodily movements. This mineralized tissue therefore confers mechanical and metabolic functions to the skeleton.

Bone is a specialized connective tissue composed of calcified extracellular material, the **bone matrix**, and following three major cell types (Figure 8–2):

- **Osteocytes** (Gr. *osteon*, bone + *kytos*, cell), which are found in cavities (**lacunae**) between bone matrix layers (**lamellae**), with cytoplasmic processes in small **canaliculi** (L. *canalis*, canal) that extend into the matrix (Figure 8–1b)
- **Osteoblasts** (*osteon* + Gr. *blastos*, germ), growing cells which synthesize and secrete the organic components of the matrix
- **Osteoclasts** (*osteon* + Gr. *klastos*, broken), which are giant, multinucleated cells involved in removing calcified bone matrix and remodeling bone tissue

Because metabolites are unable to diffuse through the calcified matrix of bone, the exchanges between osteocytes and blood capillaries depend on communication through the very thin, cylindrical spaces of the canaliculi.

All bones are lined on their internal and external surfaces by layers of connective tissue containing osteogenic cells—**endosteum** on the internal surface surrounding the marrow cavity and **periosteum** on the external surface.

Because of its hardness, bone cannot be sectioned routinely. Bone matrix is usually softened by immersion in a decalcifying solution before paraffin embedding, or embedded in plastic after fixation and sectioned with a specialized microtome.

❯ BONE CELLS

Osteoblasts

Originating from mesenchymal stem cells, **osteoblasts** produce the organic components of bone matrix, including type I collagen fibers, proteoglycans, and matricellular glycoproteins such as osteonectin. Deposition of the inorganic components of bone also depends on osteoblast activity. Active osteoblasts are located exclusively at the surfaces of bone matrix, to which they are bound by integrins, typically forming a single layer of cuboidal cells joined by adherent and gap junctions (Figure 8–3). When their synthetic activity is completed, some osteoblasts differentiate as osteocytes entrapped in matrix-bound lacunae, some flatten and cover the matrix surface as **bone lining cells**, and the majority undergo apoptosis.

During the processes of matrix synthesis and calcification, osteoblasts are polarized cells with ultrastructural features denoting active protein synthesis and secretion. Matrix components are secreted at the cell surface in contact with existing bone matrix, producing a layer of unique collagen-rich material

FIGURE **8-1** Components of bone.

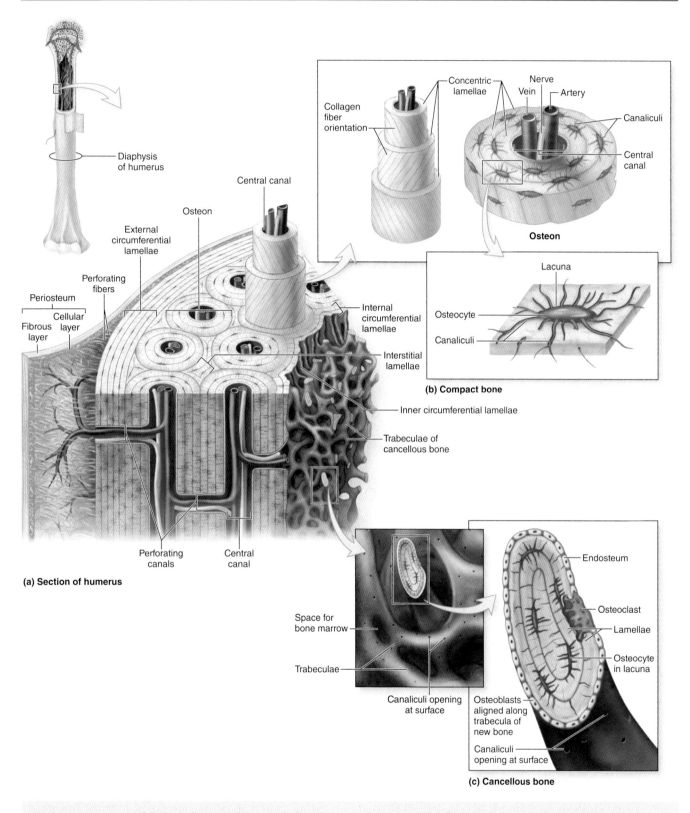

Diaphysis of humerus

Central canal

Osteon

External circumferential lamellae

Perforating fibers

Periosteum

Cellular layer

Fibrous layer

Perforating canals

Central canal

(a) Section of humerus

Collagen fiber orientation

Concentric lamellae

Nerve

Vein — Artery

Canaliculi

Central canal

Osteon

Internal circumferential lamellae

Interstitial lamellae

Lacuna

Osteocyte

Canaliculi

(b) Compact bone

Inner circumferential lamellae

Trabeculae of cancellous bone

Space for bone marrow

Trabeculae

Canaliculi opening at surface

Endosteum

Osteoclast

Lamellae

Osteocyte in lacuna

Osteoblasts aligned along trabecula of new bone

Canaliculi opening at surface

(c) Cancellous bone

A schematic overview of the basic features of bone, including the three key cell types: **osteocytes**, **osteoblasts**, and **osteoclasts**; their usual locations; and the typical **lamellar organization** of bone. Osteoblasts secrete the matrix that then hardens by calcification, trapping the differentiating cells now called **osteocytes** in individual **lacunae**. Osteocytes maintain the calcified matrix and receive nutrients from microvasculature in the central canals of the osteons via very small channels called **canaliculi** that interconnect the lacunae. Osteoclasts are monocyte-derived cells in bone required for bone remodeling.

The **periosteum** consists of dense connective tissue, with a primarily fibrous layer covering a more cellular layer. Bone is vascularized by small vessels that penetrate the matrix from the periosteum. **Endosteum** covers all **trabeculae** around the marrow cavities.

FIGURE **8–2**   Bone tissue.

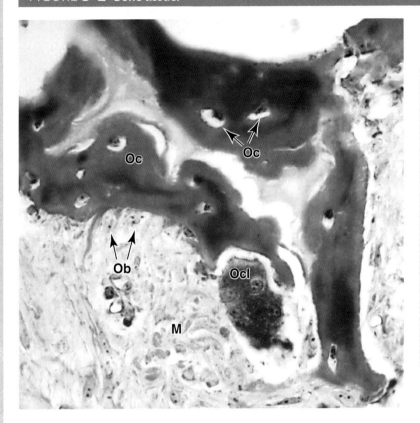

Newly formed bone tissue decalcified for sectioning and stained with trichrome in which the collagen-rich ECM appears bright blue. The tissue is a combination of mesenchymal regions (**M**) containing capillaries, fibroblasts, and osteoprogenitor stem cells and regions of normally calcified matrix with varying amounts of collagen and the three major cell types found in all bone tissue.

Bone-forming osteoblasts (**Ob**) differentiate from osteoprogenitor cells in the periosteum and endosteum, and cover the surfaces of existing bone matrix. Osteoblasts secrete **osteoid** rich in collagen type I, but also containing proteoglycans and other molecules. As osteoid undergoes calcification and hardens, it entraps some osteoblasts which then differentiate further as osteocytes (**Oc**) occupying lacunae surrounded by bony matrix. The much less numerous large, multinuclear osteoclasts (**Ocl**), produced by the fusion of blood monocytes, reside on bony surfaces and erode the matrix during bone remodeling. (400X; Mallory trichrome)

FIGURE **8–3**   Osteoblasts, osteocytes, and osteoclasts.

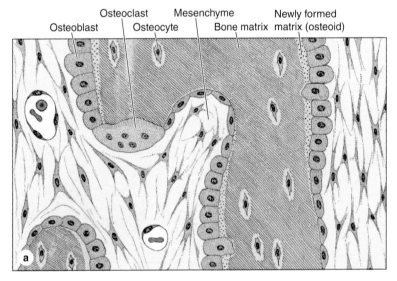

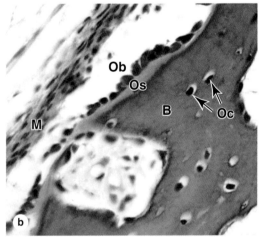

(a) Diagram showing the relationship of osteoblasts to the newly formed matrix called "osteoid," bone matrix, and osteocytes. Osteoblasts and most of the larger osteoclasts are part of the endosteum covering the bony trabeculae.

(b) The photomicrograph of developing bone shows the location and morphologic differences between active osteoblasts (**Ob**) and osteocytes (**Oc**). Rounded osteoblasts, derived from progenitor cells in the adjacent mesenchyme (**M**), cover a thin layer of lightly stained osteoid (**Os**) on the surface of the more heavily stained bony matrix (**B**). Most osteoblasts that are no longer actively secreting osteoid will undergo apoptosis; others differentiate either as flattened bone lining cells on the trabeculae of bony matrix or as osteocytes located within lacunae surrounded by bony matrix. (X300; H&E)

called **osteoid** between the osteoblast layer and the preexisting bone surface (Figure 8–3). This process of bone appositional growth is completed by subsequent deposition of calcium salts into the newly formed matrix.

The process of matrix mineralization is not completely understood, but basic aspects of the process are shown in Figure 8–4. Prominent among the noncollagen proteins secreted by osteoblasts is the vitamin K–dependent polypeptide **osteocalcin**, which together with various glycoproteins binds Ca^{2+} ions and concentrates this mineral locally. Osteoblasts also release membrane-enclosed **matrix vesicles** rich in alkaline phosphatase and other enzymes whose activity raises the local concentration of PO_4^{3-} ions. In the microenvironment with high concentrations of both these ions, matrix vesicles serve as foci for the formation of hydroxyapatite

$[Ca_{10}(PO_4)_6(OH)_2]$ crystals, the first visible step in calcification. These crystals grow rapidly by accretion of more mineral and eventually produce a confluent mass of calcified material embedding the collagen fibers and proteoglycans (Figure 8–4).

>> **MEDICAL APPLICATION**

Cancer originating directly from bone cells (a primary bone tumor) is fairly uncommon (0.5% of all cancer deaths), although a cancer called **osteosarcoma** can arise in osteoprogenitor cells. The skeleton is often the site of secondary, **metastatic tumors**, however, arising when cancer cells move into bones via small blood or lymphatic vessels from malignancies in other organs, most commonly the breast, lung, prostate gland, kidney, or thyroid gland.

FIGURE **8–4** **Mineralization in bone matrix.**

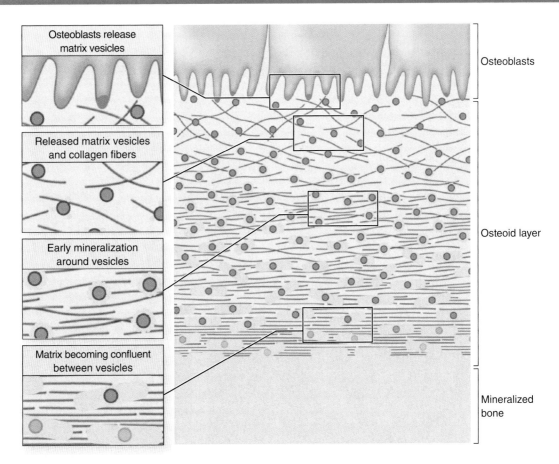

From their ends adjacent to the bone matrix, osteoblasts secrete type I collagen, several glycoproteins, and proteoglycans. Some of these factors, notably **osteocalcin** and certain glycoproteins, bind Ca^{2+} with high affinity, raising the local concentration of these ions. Osteoblasts also release very small membrane-enclosed **matrix vesicles** containing alkaline phosphatase and other enzymes. These enzymes hydrolyze PO_4^- ions from various matrix macromolecules, creating a high concentration of these ions locally. The high Ca^{2+}

and PO_4^- ion concentrations cause calcified nanocrystals to form in and around the matrix vesicles. The crystals grow and mineralize further with formation of small growing masses of calcium hydroxyapatite $[Ca_{10}(PO_4)_6(OH)_2]$, which surround the collagen fibers and all other macromolecules. Eventually the masses of hydroxyapatite merge as a confluent solid bony matrix as calcification of the matrix is completed.

Osteocytes

As mentioned some osteoblasts become surrounded by the material they secrete and then differentiate as **osteocytes** enclosed singly within the **lacunae** spaced throughout the mineralized matrix. During the transition from osteoblasts to osteocytes, the cells extend many long dendritic processes, which also become surrounded by calcifying matrix. The processes thus come to occupy the many canaliculi, 250-300 nm in diameter, radiating from each lacuna (Figures 8–5 and 8–1b).

Diffusion of metabolites between osteocytes and blood vessels occurs through the small amount of interstitial fluid in the canaliculi between the bone matrix and the osteocytes and their processes. Osteocytes also communicate with one another and ultimately with nearby osteoblasts and bone lining cells via gap junctions at the ends of their processes.

Normally the most abundant cells in bone, the almond-shaped osteocytes exhibit significantly less RER, smaller Golgi complexes, and more condensed nuclear chromatin than osteoblasts (Figure 8–5a). Osteocytes maintain the calcified matrix, and their death is followed by rapid matrix resorption. While sharing most matrix-related activities with osteoblasts, osteocytes also express many different proteins, including factors with paracrine and endocrine effects that help regulate bone remodeling. The extensive lacunar-canalicular network of these cells and their communication with all other bone cells allow osteocytes to serve as sensitive detectors of stress- or fatigue-induced microdamage in bone and to trigger remedial activity in osteoblasts and osteoclasts.

> ### ›› MEDICAL APPLICATION
>
> The network of dendritic processes extending from osteocytes has been called a "mechanostat," monitoring areas within bones where loading has been increased or decreased, and signaling cells to adjust ion levels and maintain the adjacent bone matrix accordingly. **Lack of exercise** (or the weightlessness experienced by astronauts) leads to **decreased bone density**, due in part to the lack of mechanical stimulation of these cells.

FIGURE 8–5   Osteocytes in lacunae.

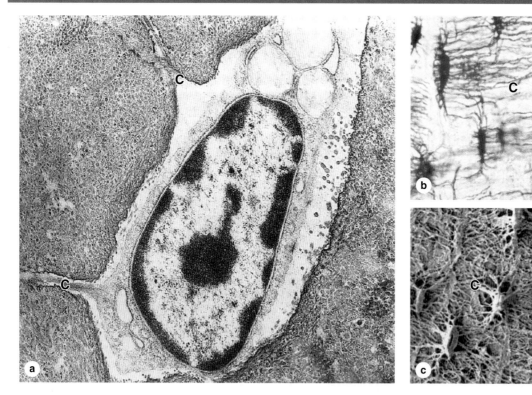
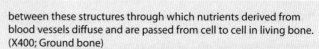

(a) TEM showing an osteocyte in a lacuna and two dendritic processes in canaliculi (**C**) surrounded by bony matrix. Many such processes are extended from each cell as osteoid is being secreted; this material then undergoes calcification around the processes, giving rise to canaliculi. (X30,000)

(b) Photomicrograph of bone, not decalcified or sectioned, but ground very thin to demonstrate lacunae and canaliculi. The lacunae and canaliculi (**C**) appear dark and show the communication between these structures through which nutrients derived from blood vessels diffuse and are passed from cell to cell in living bone. (X400; Ground bone)

(c) SEM of non-decalcified, sectioned, and acid-etched bone showing lacunae and canaliculi (**C**). (X400)

(*Figure 8-5c, used with permission from Dr Matt Allen, Indiana University School of Medicine, Indianapolis.*)

Osteoclasts

Osteoclasts are very large, motile cells with multiple nuclei (Figure 8–6) which are essential for matrix resorption during bone growth and remodeling. The large size and multinucleated condition of osteoclasts are due to their origin from the fusion of bone marrow–derived monocytes. Osteoclast development requires two polypeptides produced by osteoblasts: macrophage-colony–stimulating factor (M-CSF; discussed with hemopoiesis, Chapter 13) and the receptor activator of nuclear factor-κB ligand (RANKL). In areas of bone undergoing resorption, osteoclasts on the bone surface lie within enzymatically etched depressions or cavities in the matrix known as **resorption lacunae** (or **Howship lacunae**).

In an active osteoclast the membrane domain that contacts the bone forms a circular **sealing zone** which binds the cell tightly to the bone matrix and surrounds an area with many surface projections, called the **ruffled border**. This circumferential sealing zone allows the formation of a specialized microenvironment between the osteoclast and the matrix in which bone resorption occurs (Figure 8–6b).

Into this subcellular pocket the osteoclast pumps protons to acidify and promote dissolution of the adjacent hydroxyapatite, and releases matrix metalloproteinases and other hydrolytic enzymes from lysosome-related secretory vesicles for the localized digestion of matrix proteins. Osteoclast activity is controlled by local signaling factors from other bone cells. Osteoblasts activated by parathyroid hormone produce M-CSF, RANKL, and other factors that regulate the formation and activity of osteoclasts.

❯❯ MEDICAL APPLICATION

In the genetic disease **osteopetrosis**, which is characterized by dense, heavy bones ("marble bones"), the osteoclasts lack ruffled borders and bone resorption is defective. This disorder results in overgrowth and thickening of bones, often with obliteration of the marrow cavities, depressing blood cell formation and causing anemia and the loss of white blood cells. The defective osteoclasts in most patients with osteopetrosis have mutations in genes for the cells' proton-ATPase pumps or chloride channels.

❯ BONE MATRIX

About 50% of the dry weight of bone matrix is inorganic materials. Calcium hydroxyapatite is most abundant, but bicarbonate, citrate, magnesium, potassium, and sodium ions are also found. Significant quantities of noncrystalline calcium phosphate are also present. The surface of hydroxyapatite crystals are hydrated, facilitating the exchange of ions between the mineral and body fluids.

The organic matter embedded in the calcified matrix is 90% type I collagen, but also includes mostly small proteoglycans and multiadhesive glycoproteins such as **osteonectin**. Calcium-binding proteins, notably osteocalcin, and the phosphatases released from cells in matrix vesicles promote calcification of the matrix. Other tissues rich in type I collagen lack osteocalcin and matrix vesicles and therefore do not normally become calcified.

The association of minerals with collagen fibers during calcification provides the hardness and resistance required for bone function. If a bone is decalcified by a histologist, its shape is preserved but it becomes soft and pliable like other connective tissues. Because of its high collagen content, decalcified bone matrix is usually acidophilic.

❯ PERIOSTEUM & ENDOSTEUM

External and internal surfaces of all bones are covered by connective tissue of the periosteum and endosteum respectively (Figures 8–1a and 8–1c). The **periosteum** is organized much like the perichondrium of cartilage, with an outer fibrous layer of dense connective tissue, containing mostly bundled type I collagen, but also fibroblasts and blood vessels. Bundles of periosteal collagen, called **perforating** (or **Sharpey**) **fibers**, penetrate the bone matrix and bind the periosteum to the bone. Periosteal blood vessels branch and penetrate the bone, carrying metabolites to and from bone cells.

The periosteum's inner layer is more cellular and includes osteoblasts, bone lining cells, and mesenchymal stem cells referred to as **osteoprogenitor cells**. With the potential to proliferate extensively and produce many new osteoblasts, osteoprogenitor cells play a prominent role in bone growth and repair.

Internally the very thin **endosteum** covers small **trabeculae** of bony matrix that project into the marrow cavities (Figure 8–1). The endosteum also contains osteoprogenitor cells, osteoblasts, and bone lining cells, but within a sparse, delicate matrix of collagen fibers.

❯❯ MEDICAL APPLICATION

Osteoporosis, frequently found in immobilized patients and in postmenopausal women, is an imbalance in skeletal turnover so that bone resorption exceeds bone formation. This leads to calcium loss from bones and reduced **bone mineral density** (BMD). Individuals at risk for osteoporosis are routinely tested for BMD by **dual-energy x-ray absorptiometry** (DEXA scans).

❯ TYPES OF BONE

Gross observation of a bone in cross section (Figure 8–7) shows a dense area near the surface corresponding to **compact (cortical) bone**, which represents 80% of the total bone mass, and deeper areas with numerous interconnecting cavities, called **cancellous (trabecular) bone**, constituting about 20% of total bone mass. Histological features and important locations of the major types of bone are summarized in Table 8–1.

FIGURE **8–6** Osteoclasts and their activity.

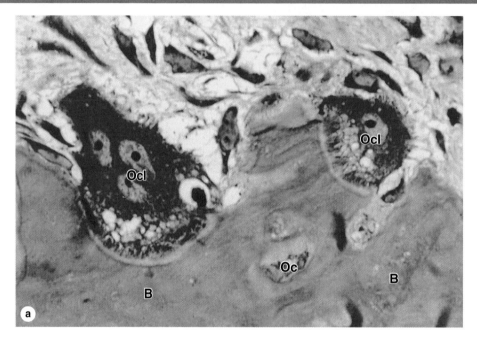

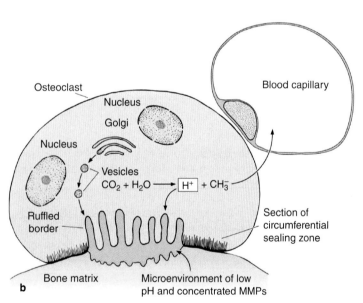

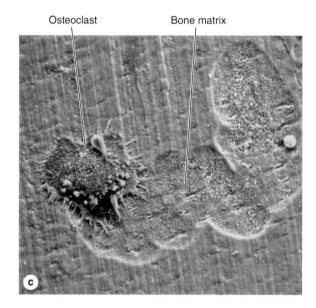

Osteoclast   Bone matrix

Osteoclasts are large multinucleated cells which are derived by the fusion in bone of several blood-derived monocytes. **(a)** Photo of bone showing two osteoclasts (**Ocl**) digesting and resorbing bone matrix (**B**) in relatively large resorption cavities (or Howship lacunae) on the matrix surface. An osteocyte (**Oc**) in its smaller lacuna is also shown. (X400; H&E)

(b) Diagram showing an osteoclast's circumferential **sealing zone** where integrins tightly bind the cell to the bone matrix. The sealing zone surrounds a **ruffled border** of microvilli and other cytoplasmic projections close to this matrix. The sealed space between the cell and the matrix is acidified to ~pH 4.5 by proton pumps in the ruffled part of the cell membrane and receives secreted matrix

metalloproteases and other hydrolytic enzymes. Acidification of the sealed space promotes dissolution of hydroxyapatite from bone and stimulates activity of the protein hydrolases, producing localized matrix resorption. The breakdown products of collagen fibers and other polypeptides are endocytosed by the osteoclast and further degraded in lysosomes, while Ca^{2+} and other ions are released directly and taken up by the blood.

(c) SEM showing an active osteoclast cultured on a flat substrate of bone. A trench is formed on the bone surface by the slowly migrating osteoclast. (X5000)

(Figure 8–6c, used with permission from Alan Boyde, Centre for Oral Growth and Development, University of London.)

FIGURE **8–7** Compact and cancellous bone.

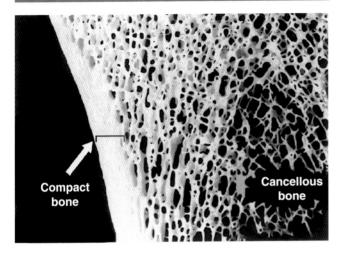

Compact
bone

Cancellous
bone

Macroscopic photo of a thick section of bone showing the cortical **compact bone** and the lattice of trabeculae in **cancellous bone** at the bone's interior. The small trabeculae that make up highly porous cancellous bone serve as supportive struts, collectively providing considerable strength, without greatly increasing the bone's weight. The compact bone is normally covered externally with periosteum and all trabecular surfaces of the cancellous bone are covered with endosteum. (X10)

In long bones, the bulbous ends—called **epiphyses** (Gr. *epiphysis*, an excrescence)—are composed of cancellous bone covered by a thin layer of compact cortical bone. The cylindrical part—the **diaphysis** (Gr. *diaphysis*, a growing between)—is almost totally dense compact bone, with a thin region of cancellous bone on the inner surface around the central **marrow cavity** (Figure 8–1). Short bones such as those of the wrist and ankle usually have cores of cancellous bone surrounded completely by compact bone. The flat bones that form the calvaria (skullcap) have two layers of compact bone called **plates**, separated by a thicker layer of cancellous bone called the **diploë**.

At the microscopic level both compact and cancellous bone typically show two types of organization: mature **lamellar bone**, with matrix existing as discrete sheets, and **woven bone**, newly formed with randomly arranged components.

Lamellar Bone

Most bone in adults, compact or cancellous, is organized as **lamellar bone**, characterized by multiple layers or **lamellae** of calcified matrix, each 3-7 μm thick. The lamellae are organized as parallel sheets or concentrically around a central canal. In each lamella, type I collagen fibers are aligned, with the pitch of the fibers' orientation shifted orthogonally (by about 90 degrees) in successive lamellae (Figure 8–1a). This highly ordered organization of collagen within lamellar bone causes birefringence with polarizing light microscopy; the alternating bright and dark layers are due to the changing orientation of collagen fibers in the lamellae (Figure 8–8). Like the orientation of wood fibers in plywood the highly ordered, alternating organization of collagen fibers in lamellae adds greatly to the strength of lamellar bone.

An **osteon** (or **Haversian system**) refers to the complex of concentric lamellae, typically 100-250 μm in diameter, surrounding a central canal that contains small blood vessels, nerves, and endosteum (Figures 8–1 and 8–9). Between successive lamellae are lacunae, each with one osteocyte, all interconnected by the canaliculi containing the cells' dendritic processes (Figure 8–9). Processes of adjacent cells are in contact via gap junctions, and all cells of an osteon receive nutrients and oxygen from vessels in the central canal (Figure 8–1). The outer boundary of each osteon is a layer called the cement line which includes many more noncollagen proteins in addition to mineral and collagen.

Each osteon is a long, sometimes bifurcated, cylinder generally parallel to the long axis of the diaphysis. Each has 5-20 concentric lamellae around the central canal which communicates with the marrow cavity and the periosteum. Canals also communicate with one another through transverse **perforating canals** (or **Volkmann canals**) which have few, if any, concentric lamellae (Figures 8–1 and 8–10). All central osteonic

Type of Bone	Histological Features	Major Locations	Synonyms
Woven bone, newly calcified	Irregular and random arrangement of cells and collagen; lightly calcified	Developing and growing bones; hard callus of bone fractures	Immature bone; primary bone; bundle bone
Lamellar bone, remodeled from woven bone	Parallel bundles of collagen in thin layers (lamellae), with regularly spaced cells between; heavily calcified	All normal regions of adult bone	Mature bone; secondary bone
Compact bone, ~80% of all lamellar bone	Parallel lamellae or densely packed osteons, with interstitial lamellae	Thick, outer region (beneath periosteum) of bones	Cortical bone
Cancellous bone, ~20% of all lamellar bone	Interconnected thin spicules or trabeculae covered by endosteum	Inner region of bones, adjacent to marrow cavities	Spongy bone; trabecular bone; medullary bone

Table 8–1   Summary of bone types and their organization.

FIGURE 8–8 Lamellar bone.

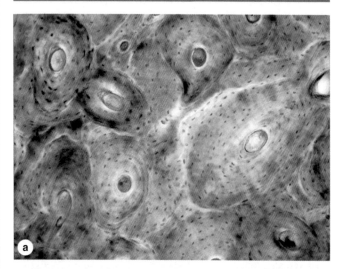

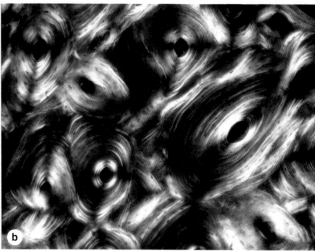

Two photographs of the same area of an unstained section of compact bone, showing osteons with concentric lamellae around central canals. Lamellae are seen only faintly by bright-field microscopy **(a)**, but they appear as alternating bright and dark bands under the polarizing light microscope **(b)**. Bright bands are due to birefringence from the highly ordered collagen fibers in a lamella. Alternating bright and dark bands indicate that fibers in successive lamellae have different orientations, an organization that makes lamellar bone very strong. (Both X100)
(*Used with permission from Dr Matt Allen, Indiana University School of Medicine, Indianapolis.*)

FIGURE 8–9 An osteon.

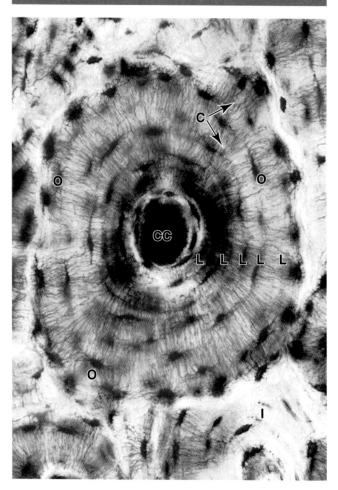

Osteons (Haversian systems) constitute most of the compact bone. Shown here is an osteon with four to five concentric lamellae **(L)** surrounding the central canal **(CC)**. Osteocytes **(O)** in lacunae are in communication with each other and with the central canal and periphery of the osteon via through hundreds of dendritic processes located within fine canaliculi **(C)**. Also shown are the partial, interstitial lamellae **(I)** of an osteon that was eroded when the intact osteon was formed. (Ground bone; X500)

canals and perforating canals form when matrix is laid down around areas with preexisting blood vessels.

Scattered among the intact osteons are numerous irregularly shaped groups of parallel lamellae called **interstitial lamellae**. These structures are lamellae remaining from osteons partially destroyed by osteoclasts during growth and remodeling of bone (Figure 8–10).

Compact bone (eg, in the diaphysis of long bones) also includes parallel lamellae organized as multiple **external**

circumferential lamellae immediately beneath the periosteum and fewer **inner circumferential lamellae** around the marrow cavity (Figure 8–1a). The lamellae of these outer and innermost areas of compact bone enclose and strengthen the middle region containing vascularized osteons.

Bone remodeling occurs continuously throughout life. In compact bone, remodeling resorbs parts of old osteons and produces new ones. As shown in Figure 8–11 osteoclasts remove old bone and form small, tunnel-like cavities. Such tunnels are quickly invaded by osteoprogenitor cells from the endosteum or periosteum and sprouting loops of capillaries. Osteoblasts develop, line the wall of the tunnels, and begin to secrete osteoid in a cyclic manner, forming a new osteon with concentric

FIGURE **8–10** Lamellar bone: Perforating canals and interstitial lamellae.

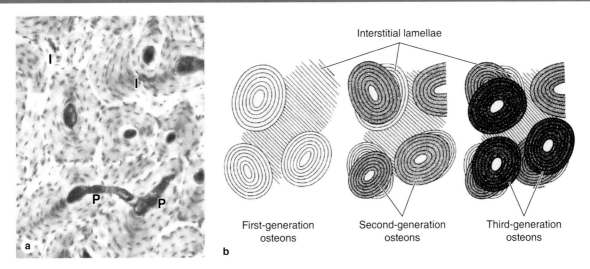

Interstitial lamellae

First-generation
osteons

Second-generation
osteons

Third-generation
osteons

a   b

(a) Transverse perforating (Volkmann) canals (**P**) connecting adjacent osteons are shown in this micrograph of compact lamellar bone. Such canals "perforate" lamellae and provide another source of microvasculature for the central canals of osteons. Among the intact osteons are also found remnants of eroded osteons, seen as irregular interstitial lamellae (**I**). (Ground bone; X100)

(b) Diagram showing the remodeling of compact lamellar bone with three generations of osteons and their successive contributions to the formation of interstitial lamellae. The shading indicates that successive generations of osteons have different degrees of mineralization, with the most newly formed being the least mineralized. Remodeling is a continuous process that involves the coordinated activity of osteoblasts and osteoclasts, and is responsible for adaptation of bone to changes in stress, especially during the body's growth.

FIGURE **8–11** Development of an osteon.

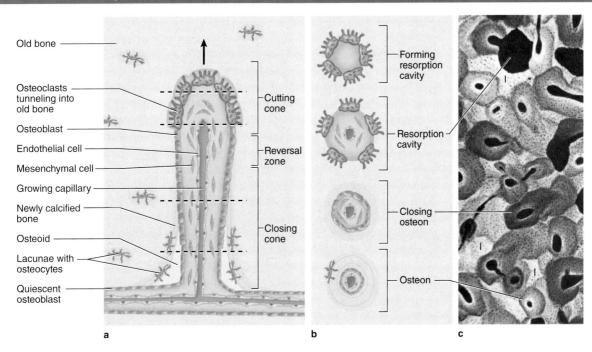

Old bone

Osteoclasts
tunneling into
old bone

Osteoblast

Endothelial cell

Mesenchymal cell

Growing capillary

Newly calcified
bone

Osteoid

Lacunae with
osteocytes

Quiescent
osteoblast

Cutting
cone

Reversal
zone

Closing
cone

Forming
resorption
cavity

Resorption
cavity

Closing
osteon

Osteon

a   b   c

During remodeling of compact bone, osteoclasts act as a cutting cone that tunnels into existing bone matrix. Behind the osteoclasts, a population of osteoblast progenitors enters the newly formed tunnel and lines its walls. The osteoblasts secrete osteoid in a cyclic manner, producing layers of new matrix (lamellae) and trapping some cells (future osteocytes) in lacunae. The tunnel becomes constricted with multiple concentric layers of new matrix, and its lumen finally exists as only a narrow central canal with small blood vessels. The dashed lines in **(a)** indicate the levels of the structures shown in cross section **(b)**. An x-ray image **(c)** shows the different degrees of mineralization in osteons and in interstitial lamellae (**I**).

lamellae of bone and trapped osteocytes (Figures 8–11). In healthy adults 5%-10% of the bone turns over annually.

Woven Bone

Woven bone is nonlamellar and characterized by random disposition of type I collagen fibers and is the first bone tissue to appear in embryonic development and in fracture repair. Woven bone is usually temporary and is replaced in adults by lamellar bone, except in a very few places in the body, for example, near the sutures of the calvaria and in the insertions of some tendons.

In addition to the irregular, interwoven array of collagen fibers, woven bone typically has a lower mineral content (it is more easily penetrated by x-rays) and a higher proportion of osteocytes than mature lamellar bone. These features reflect the facts that immature woven bone forms more quickly but has less strength than lamellar bone.

› OSTEOGENESIS

Bone development or **osteogenesis** occurs by one of two processes:

- **Intramembranous ossification**, in which osteoblasts differentiate directly from mesenchyme and begin secreting osteoid

FIGURE **8–12** **Tetracycline localization of new bone matrix.**

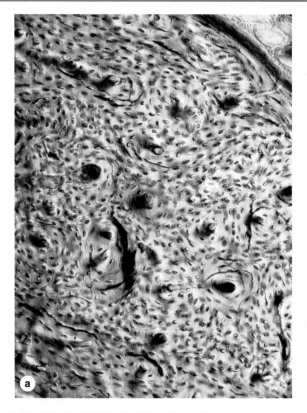
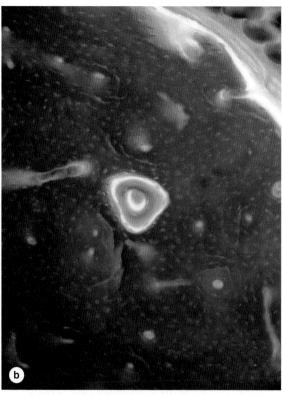

Newly formed bone can be labeled with tetracycline, which forms fluorescent complexes with calcium at ossification sites and provides an in vivo tracer by which newly formed bone can be localized. A group of osteons in bone after tetracycline incorporation in vivo seen with bright-field **(a)** and fluorescent microscopy

(b) reveals active ossification in one osteon (center) and in the external circumferential lamellae (upper right).

(Used with permission from Dr Matt Allen, Indiana University School of Medicine, Indianapolis.)

- **Endochondral ossification**, in which a preexisting matrix of *hyaline cartilage* is eroded and invaded by osteoblasts, which then begin osteoid production.

The names refer to the mechanisms by which the bone forms initially; in both processes woven bone is produced first and is soon replaced by stronger lamellar bone. During growth of all bones, areas of woven bone, areas of bone resorption, and areas of lamellar bone all exist contiguous to one another.

≫ MEDICAL APPLICATION

Osteogenesis imperfecta, or "brittle bone disease," refers to a group of related congenital disorders in which the osteoblasts produce deficient amounts of type I collagen or defective type I collagen due to genetic mutations. Such defects lead to a spectrum of disorders, all characterized by significant fragility of the bones. The fragility reflects the deficit in normal collagen, which normally reinforces and adds a degree of resiliency to the mineralized bone matrix.

Intramembranous Ossification

Intramembranous ossification, by which most flat bones begin to form, takes place within condensed sheets ("membranes") of embryonic mesenchymal tissue. Most bones of the skull and jaws, as well as the scapula and clavicle, are formed embryonically by intramembranous ossification.

Within the condensed mesenchyme bone formation begins in **ossification centers**, areas in which osteoprogenitor cells arise, proliferate, and form incomplete layers of osteoblasts around a network of developing capillaries. Osteoid secreted by the osteoblasts calcifies as described earlier, forming small irregular areas of woven bone with osteocytes in lacunae and canaliculi (Figure 8–13). Continued matrix secretion and calcification enlarges these areas and leads to the fusion of neighboring ossification centers. The anatomical bone forms gradually as woven bone matrix is replaced by compact bone that encloses a region of cancellous bone with marrow and larger blood vessels. Mesenchymal regions that do not undergo ossification give rise to the endosteum and the periosteum of the new bone.

In cranial flat bones, lamellar bone formation predominates over bone resorption at both the internal and external surfaces. Internal and external plates of compact bone arise, while the central portion (diploë) maintains its cancellous nature. The fontanelles or "soft spots" on the heads of newborn infants are areas of the skull in which the membranous tissue is not yet ossified.

Endochondral Ossification

Endochondral (Gr. *endon*, within + *chondros*, cartilage) ossification takes place within hyaline cartilage shaped as a small version, or model, of the bone to be formed. This type of ossification forms most bones of the body and is especially well studied in developing long bones, where it consists of the sequence of events shown in Figure 8–14.

FIGURE 8–13   Intramembranous ossification.

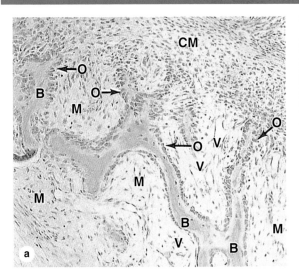
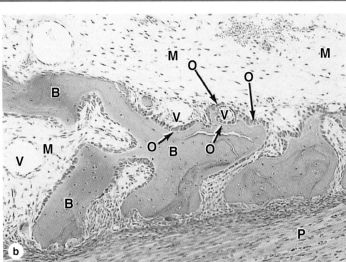

A section of fetal pig mandible developing by intramembranous ossification. **(a)** Areas of typical mesenchyme (**M**) and condensed mesenchyme (**CM**) are adjacent to layers of new osteoblasts (**O**). Some osteoblasts have secreted matrices of bone (**B**), the surfaces of which remain covered by osteoblasts. Between these thin regions of new woven bone are areas with small blood vessels (**V**). (X40; H&E)

(b) At higher magnification another section shows these same structures, but also includes the developing periosteum (**P**) adjacent to the masses of woven bone that will soon merge to form a continuous plate of bone. The larger mesenchyme-filled region at the top is part of the developing marrow cavity. Osteocytes in lacunae can be seen within the bony matrix. (X100; H&E)

FIGURE 8–14 Osteogenesis of long bones by endochondral ossification.

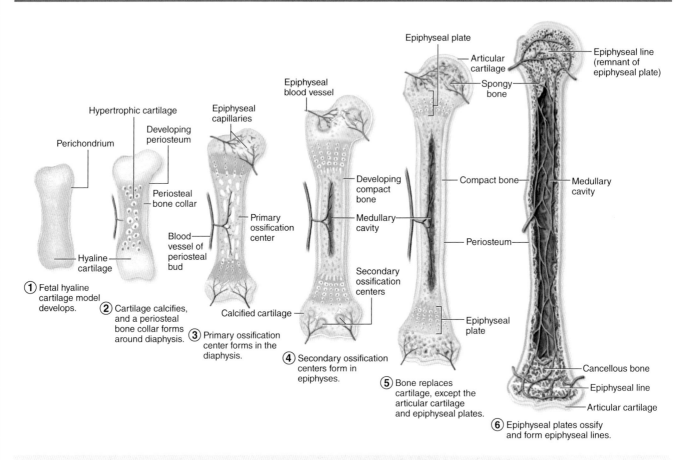

This process, by which most bones form initially, begins with embryonic models of the skeletal elements made of hyaline cartilage (**1**). Late in the first trimester, a bone collar develops beneath the perichondrium around the middle of the cartilage model, causing chondrocyte hypertrophy in the underlying cartilage (**2**).

This is followed by invasion of that cartilage by capillaries and osteoprogenitor cells from what is now the periosteum to produce a **primary ossification center** in the diaphysis (**3**). Here osteoid is deposited by the new osteoblasts,

undergoes calcification into woven bone, and is then remodeled as compact bone.

(**4**) Around the time of birth **secondary ossification centers** begin to develop by a similar process in the bone's epiphyses. During childhood the primary and secondary ossification centers gradually come to be separated only by the **epiphyseal plate** (**5**) which provides for continued bone elongation. The two ossification centers do not merge until the epiphyseal plate disappears (**6**) when full stature is achieved.

In this process ossification first occurs within a **bone collar** produced by osteoblasts that differentiate within the perichondrium (transitioning to periosteum) around the cartilage model diaphysis. The collar impedes diffusion of oxygen and nutrients into the underlying cartilage, causing local chondrocytes to swell up (hypertrophy), compress the surrounding matrix, and initiate its calcification by releasing osteocalcin and alkaline phosphatase. The hypertrophic chondrocytes eventually die, creating empty spaces within the calcified matrix. One or more blood vessels from the perichondrium (now the periosteum) penetrate the bone collar, bringing osteoprogenitor cells to the porous central region. Along with the vasculature newly formed osteoblasts move into all available spaces and produce woven bone. The remnants of calcified cartilage at this stage are basophilic and the new bone is more acidophilic (Figure 8–15).

This process in the diaphysis forms the **primary ossification center** (Figure 8–14), beginning in many embryonic bones as early as the first trimester. **Secondary ossification centers** appear later at the epiphyses of the cartilage model and develop in a similar manner. During their expansion and remodeling both the primary and secondary ossification centers produce cavities that are gradually filled with bone marrow and trabeculae of cancellous bone.

With the primary and secondary ossification centers, two regions of cartilage remain:

- **Articular cartilage** within the joints between long bones (Figure 8–14), which normally persists through adult life
- The specially organized **epiphyseal cartilage** (also called the **epiphyseal plate** or growth plate), which connects each epiphysis to the diaphysis and allows longitudinal bone growth (Figure 8–14).

FIGURE **8–15** Cells and matrices of a primary ossification center.

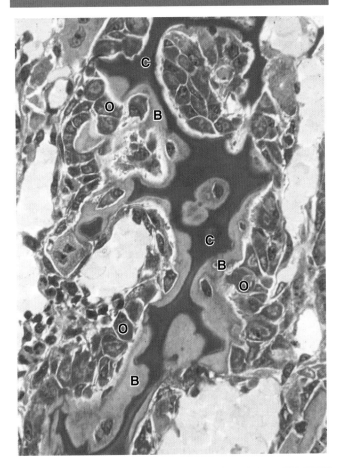

A small region of a primary ossification center showing key features of endochondral ossification. Compressed remnants of calcified cartilage matrix (**C**) are basophilic and devoid of chondrocytes. This material becomes enclosed by more lightly stained osteoid and woven bone (**B**) which contains osteocytes in lacunae. The new bone is produced by active osteoblasts (**O**) arranged as a layer on the remnants of old cartilage. (X200; Pararosaniline–toluidine blue)

2. In the **proliferative zone**, the cartilage cells divide repeatedly, enlarge and secrete more type II collagen and proteoglycans, and become organized into columns parallel to the long axis of the bone.

3. The **zone of hypertrophy** contains swollen, terminally differentiated chondrocytes which compress the matrix into aligned spicules and stiffen it by secretion of type X collagen. Unique to the hypertrophic chondrocytes in developing (or fractured) bone, type X collagen limits diffusion in the matrix and with growth factors promotes vascularization from the adjacent primary ossification center.

4. In the **zone of calcified cartilage** chondrocytes about to undergo apoptosis release matrix vesicles and osteocalcin to begin matrix calcification by the formation of hydroxyapatite crystals.

5. In the **zone of ossification** bone tissue first appears. Capillaries and osteoprogenitor cells invade the now vacant chondrocytic lacunae, many of which merge to form the initial marrow cavity. Osteoblasts settle in a layer over the spicules of calcified cartilage matrix and secrete osteoid which becomes woven bone (Figures 8–16 and 8–17). This woven bone is then remodeled as lamellar bone.

In summary, longitudinal growth of a bone occurs by cell proliferation in the epiphyseal plate cartilage. At the same time, chondrocytes in the diaphysis side of the plate undergo hypertrophy, their matrix becomes calcified, and the cells die. Osteoblasts lay down a layer of new bone on the calcified cartilage matrix. Because the rates of these two opposing events (proliferation and destruction) are approximately equal, the epiphyseal plate does not change thickness, but is instead displaced away from the center of the diaphysis as the length of the bone increases.

Growth in the circumference of long bones does not involve endochondral ossification but occurs through the activity of osteoblasts developing from osteoprogenitor cells in the periosteum by a process of **appositional growth** which begins with formation of the bone collar on the cartilaginous diaphysis. As shown in Figure 8–18 the increasing bone circumference is accompanied by enlargement of the central marrow cavity by the activity of osteoclasts in the endosteum.

❭❭ MEDICAL APPLICATION

Calcium deficiency in children can lead to **rickets**, a disease in which the bone matrix does not calcify normally and the epiphyseal plate can become distorted by the normal strains of body weight and muscular activity. Ossification processes are consequently impeded, which causes bones to grow more slowly and often become deformed. The deficiency can be due either to insufficient calcium in the diet or a failure to produce the steroid prohormone vitamin D, which is important for the absorption of Ca^{2+} by cells of the small intestine.

In adults calcium deficiency can give rise to **osteomalacia** (osteon + Gr. *malakia*, softness), characterized by deficient calcification of recently formed bone and partial decalcification of already calcified matrix.

The epiphyseal cartilage is responsible for the growth in length of the bone and disappears upon completion of bone development at adulthood. Elimination of these epiphyseal plates ("epiphyseal closure") occurs at various times with different bones and by about age 20 is complete in all bones, making further growth in bone length no longer possible. In forensics or through x-ray examination of the growing skeleton, it is possible to determine the "bone age" of a young person, by noting which epiphyses have completed closure.

An epiphyseal growth plate shows distinct regions of cellular activity and is often discussed in terms of overlapping but histologically distinct zones (Figures 8–16 and 8–17), starting with the cartilage farthest from the ossification center in the diaphysis:

1. The **zone of reserve** (or **resting**) **cartilage** is composed of typical hyaline cartilage.

FIGURE 8–16   Epiphyseal growth plate: Locations and zones of activity.

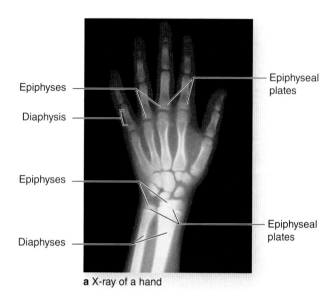

Epiphyses

Diaphysis

Epiphyses

Diaphyses

Epiphyseal plates

Epiphyseal plates

a X-ray of a hand

Zone of reserve cartilage

Zone of proliferation

Zone of hypertrophy

Zone of calcified cartilage

Zone of ossification

b Epiphyseal plate

The large and growing primary ossification center in long bone diaphyses and the secondary ossification centers in epiphyses are separated in each developing bone by a plate of cartilage called the **epiphyseal plate**.

(a) Epiphyseal plates can be identified in an x-ray of a child's hand as marrow regions of lower density between the denser ossification centers. Cells in epiphyseal growth plates are responsible for continued elongation of bones until the body's full size is reached. Developmental activities in the epiphyseal growth plate occur in overlapping zones with distinct histological appearances.

(b) From the epiphysis to the diaphysis, five general zones have cells specialized for the following: (**1**) a reserve of normal hyaline cartilage, (**2**) cartilage with proliferating chondroblasts aligned as axial aggregates in lacunae, (**3**) cartilage in which the aligned cells are hypertrophic and the matrix condensed, (**4**) an area in which the chondrocytes have disappeared and the matrix is undergoing calcification, and (**5**) an ossification zone in which blood vessels and osteoblasts invade the lacunae of the old cartilage, producing marrow cavities and osteoid for new bone. (X100; H&E)

❯ BONE REMODELING & REPAIR

Bone growth involves both the continuous resorption of bone tissue formed earlier and the simultaneous laying down of new bone at a rate exceeding that of bone removal. The sum of osteoblast and osteoclast activities in a growing bone constitutes osteogenesis or the process of bone modeling, which maintains each bone's general shape while increasing its mass. The rate of **bone turnover** is very active in young children, where it can be 200 times faster than that of adults. In adults the skeleton is also renewed continuously in a process of **bone remodeling** which involves the coordinated, localized cellular activities for bone resorption and bone formation shown in the diagram of Figure 8–11.

The constant remodeling of bone ensures that, despite its hardness, this tissue remains plastic and capable of adapting its internal structure in the face of changing stresses. A well-known example of bone plasticity is the ability of the positions of teeth in the jawbone to be modified by the lateral pressures produced by orthodontic appliances. Bone forms on the side where traction is applied and is resorbed on the opposite side where pressure is exerted. In this way, teeth are moved within the jaw while the bone is being remodeled.

Because it contains osteoprogenitor stem cells in the periosteum, endosteum, and marrow and is very well vascularized, bone normally has an excellent capacity for repair. **Bone repair** after a fracture or other damage uses cells, signaling molecules, and processes already active in bone remodeling. Surgically created gaps in bone can be filled with new bone, especially when periosteum is left in place. The major phases that occur typically during bone fracture repair include initial formation of fibrocartilage and its replacement with a temporary **callus** of woven bone, as shown in Figure 8–19.

❯❯ MEDICAL APPLICATION

Bone fractures are repaired by a developmental process involving fibrocartilage formation and osteogenic activity of the major bone cells (Figure 8–19). Bone fractures disrupt blood vessels, causing bone cells near the break to die. The damaged blood vessels produce a localized hemorrhage or hematoma. Clotted blood is removed along with tissue debris by macrophages and the matrix of damaged, cell-free bone is resorbed by osteoclasts.

FIGURE **8–17** Details of the epiphyseal growth plate.

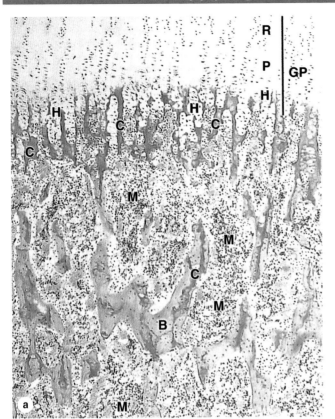

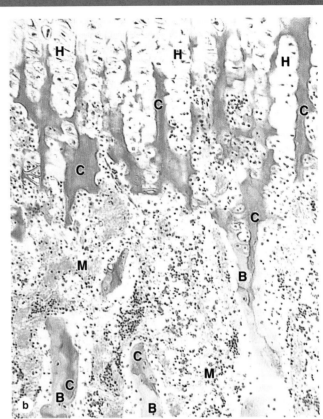

(a) At the top of the micrograph the growth plate (**GP**) shows its zones of hyaline cartilage with chondrocytes at rest (**R**), proliferating (**P**), and hypertrophying (**H**). As the chondrocytes swell they release alkaline phosphatase and type X collagen, which initiates hydroxyapatite formation and strengthens the adjacent calcifying spicules (**C**) of old cartilage matrix. The tunnel-like lacunae in which the chondrocytes have undergone apoptosis are invaded from the diaphysis by capillaries that begin to convert these spaces into marrow (**M**) cavities. Endosteum with osteoblasts also moves in from the diaphyseal primary ossification center, covering the spicules of calcified cartilage and laying down layers of osteoid to form a matrix of woven bone (**B**). (X40; H&E)

(b) Higher magnification shows more detail of the cells and matrix spicules in the zones undergoing hypertrophy (**H**) and ossification. Staining properties of the matrix clearly change as it is compressed and begins to calcify (**C**), and when osteoid and bone (**B**) are laid down. The large spaces between the ossifying matrix spicules become the marrow cavity (**M**), in which pooled masses of eosinophilic red blood cells and aggregates of basophilic white blood cell precursors can be distinguished. Still difficult to see at this magnification is the thin endosteum between the calcifying matrices and the marrow. (X100; H&E)

The periosteum and the endosteum at the fracture site respond with intense proliferation and produce a soft callus of fibrocartilage-like tissue that surrounds the fracture and covers the extremities of the fractured bone.

The fibrocartilaginous callus is gradually replaced in a process that resembles a combination of endochondral and intramembranous ossification. This produces a hard callus of woven bone around the fractured ends of bone.

Stresses imposed on the bone during repair and during the patient's gradual return to activity serve to remodel the bone callus. The immature, woven bone of the callus is gradually resorbed and replaced by lamellar bone, remodeling and restoring the original bone structure.

❯ METABOLIC ROLE OF BONE

Calcium ions are required for the activity of many enzymes and many proteins mediating cell adhesion, cytoskeletal movements, exocytosis, membrane permeability, and other cellular functions. The skeleton serves as the calcium reservoir, containing 99% of the body's total calcium in hydroxyapatite crystals. The concentration of calcium in the blood (9-10 mg/dL) and tissues is generally quite stable because of a continuous interchange between blood calcium and bone calcium.

The principal mechanism for raising blood calcium levels is the mobilization of ions from hydroxyapatite to interstitial fluid, primarily in cancellous bone. Ca^{2+} mobilization is regulated mainly by paracrine interactions among bone cells, many of which are not well understood, but two

FIGURE 8–18 Appositional bone growth

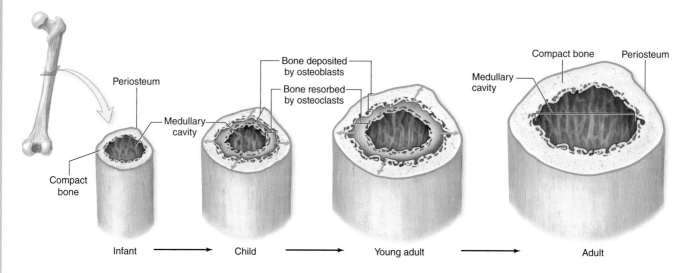

Bones increase in diameter as new bone tissue is added beneath the periosteum in a process of appositional growth. Also called radial bone growth, such growth in long bones begins with formation of the bone collar early in endochondral ossification. During radial bone growth formation of new bone at the periosteal surface occurs concurrently with bone removal at the endosteal surface around the large medullary, enlarging this marrow-filled region and not greatly increasing the bone's weight.

FIGURE 8–19 Main features of bone fracture repair.

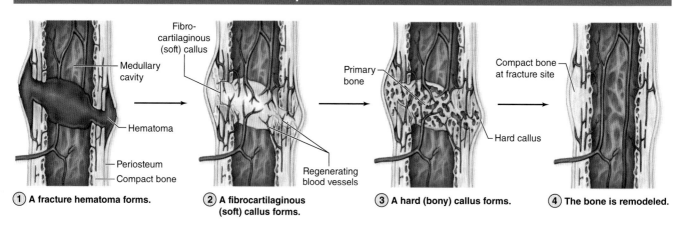

① A fracture hematoma forms.   ② A fibrocartilaginous (soft) callus forms.   ③ A hard (bony) callus forms.   ④ The bone is remodeled.

Repair of a fractured bone occurs through several stages but utilizes the cells and mechanisms already in place for bone growth and remodeling. (1) Blood vessels torn within the fracture release blood that clots to produce a large fracture hematoma. (2) This is gradually removed by macrophages and replaced by a soft fibrocartilage-like mass called procallus tissue. If torn by the break the periosteum reestablishes its continuity over this tissue. (3) The procallus is invaded by regenerating blood vessels and proliferating osteoblasts. In the next few weeks the fibrocartilage is gradually replaced by woven bone which forms a hard callus throughout the original area of fracture. (4) The woven bone is then remodeled as compact and cancellous bone in continuity with the adjacent uninjured areas and fully functional vasculature is reestablished.

polypeptide hormones also target bone cells to influence calcium homeostasis:

- **Parathyroid hormone (PTH)** from the parathyroid glands raises low blood calcium levels by stimulating osteoclasts and osteocytes to resorb bone matrix and release Ca^{2+}. The PTH effect on osteoclasts is indirect; PTH receptors occur on *osteoblasts*, which respond by secreting RANKL and other paracrine factors that stimulate osteoclast formation and activity.
- **Calcitonin**, produced within the thyroid gland, can reduce elevated blood calcium levels by opposing the effects of PTH in bone. This hormone directly targets osteoclasts to slow matrix resorption and bone turnover.

❯❯ MEDICAL APPLICATION

In addition to PTH and calcitonin, several other hormones act on bone. The anterior lobe of the pituitary synthesizes growth hormone (GH or somatotropin), which stimulates the liver to produce insulin-like growth factor-1 (IGF-1 or somatomedin). IGF has an overall growth-promoting effect, especially on the epiphyseal cartilage. Consequently, lack of growth hormone during the growing years causes **pituitary dwarfism**; an excess of growth hormone causes excessive growth of the long bones, resulting in **gigantism**. Adult bones cannot increase in length even with excess IGF because they lack epiphyseal cartilage, but they do increase in width by periosteal growth. In adults, an increase in GH causes **acromegaly**, a disease in which the bones—mainly the long ones—become very thick.

❯❯ MEDICAL APPLICATION

In **rheumatoid arthritis** chronic inflammation of the synovial membrane causes thickening of this connective tissue and stimulates the macrophages to release collagenases and other hydrolytic enzymes. Such enzymes eventually cause destruction of the articular cartilage, allowing direct contact of the bones projecting into the joint.

❯ JOINTS

Joints are regions where adjacent bones are capped and held together firmly by other connective tissues. The type of joint determines the degree of movement between the bones. Joints classified as **synarthroses** (Gr. *syn*, together + *arthrosis*, articulation) allow very limited or no movement and are subdivided into fibrous and cartilaginous joints, depending on the type of tissue joining the bones. Major subtypes of synarthroses include the following:

- **Synostoses** involve bones linked to other bones and allow essentially no movement. In older adults synostoses unite the skull bones, which in children and young adults are held together by **sutures**, or thin layers of dense connective tissue with osteogenic cells.

- **Syndesmoses** join bones by dense connective tissue only. Examples include the interosseous ligament of the inferior tibiofibular joint and the posterior region of the sacroiliac joints.
- **Symphyses** have a thick pad of fibrocartilage between the thin articular cartilage covering the ends of the bones. All symphyses, such as the intervertebral discs and pubic symphysis, occur in the midline of the body.

Intervertebral discs (Figure 8–20) are large symphyses between the articular surfaces of successive bony

FIGURE **8–20 Intervertebral disc.**

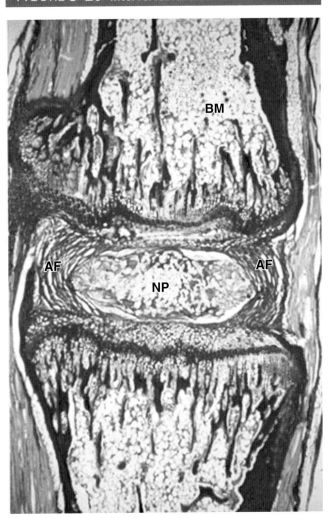

Section of a rat tail showing an intervertebral disc and the two adjacent vertebrae with bone marrow (**BM**) cavities. The disc consists of concentric layers of fibrocartilage, comprising the annulus fibrosus (**AF**), which surrounds the nucleus pulposus (**NP**). The nucleus pulposus contains scattered residual cells of the embryonic notochord embedded in abundant gel-like matrix. The intervertebral discs function primarily as shock absorbers within the spinal column and allow greater mobility within the spinal column. (X40; PSH)

vertebral bodies. Held in place by ligaments these discoid components of the intervertebral joints cushion the bones and facilitate limited movements of the vertebral column. Each disc has an outer portion, the **annulus fibrosus**, consisting of concentric fibrocartilage laminae in which collagen bundles are arranged orthogonally in adjacent layers. The multiple lamellae of fibrocartilage produce a disc with unusual toughness able to withstand pressures and torsion generated within the vertebral column.

Situated in the center of the annulus fibrosus, a gel-like body called the **nucleus pulposus** allows each disc to function as a shock absorber (Figure 8–20). The nucleus pulposus consists of a viscous fluid matrix rich in hyaluronan and type II collagen fibers, but also contains scattered, vacuolated cells derived from the embryonic notochord, the only cells of that structure to persist postnatally. The nucleus pulposus is large in children, but these structures gradually become smaller with age and are partially replaced by fibrocartilage.

>> **MEDICAL APPLICATION**

Within an intervertebral disc, collagen loss or other degenerative changes in the annulus fibrosus are often accompanied by displacement of the nucleus pulposus, a condition variously called a **slipped** or **herniated disc**. This occurs most frequently on the posterior region of the intervertebral disc where there are fewer collagen bundles. The affected disc frequently dislocates or shifts slightly from its normal position. If it moves toward nerve plexuses, it can compress the nerves and result in severe pain and other neurologic disturbances. The pain accompanying a slipped disc may be perceived in areas innervated by the compressed nerve fibers—usually the lower lumbar region.

Joints classified as **diarthroses** permit free bone movement. Diarthroses (Figure 8–21) such as the elbow and knee generally unite long bones and allow great mobility. In a diarthrosis ligaments and a capsule of dense connective tissue

FIGURE 8–21 Diarthroses or synovial joints.

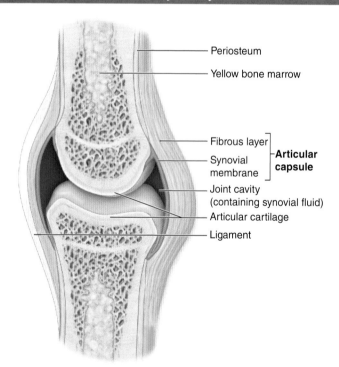

a Typical synovial joint

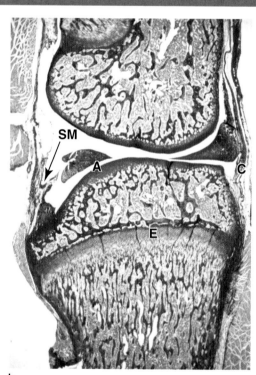

b

Diarthroses are joints that allow free movement of the attached bones, such as knuckles, knees, and elbows. **(a)** Diagram showing major components of a diarthrosis, including the **articular capsule** continuous with a ligament inserting into the periosteum of both bones; the **joint cavity** containing synovial fluid lubricant; and the ends of epiphyses covered by **articular cartilage**. The **synovial membrane** lines the capsule and produces the synovial fluid.

(b) Longitudinal section through a diarthrosis with the growing bones of a mouse knee, showing the position near the boundaries of the capsule (**C**) of the epiphyseal growth plate (**E**) where endochondral ossification occurs. Also shown are the articular cartilage (**A**) and the folds of synovial membrane (**SM**), which extend prominently into the joint cavity from connective tissue of the capsule for production of synovial fluid. (X10; PSH stain)

maintain proper alignment of the bones. The capsule encloses a sealed **joint cavity** containing a clear, viscous liquid called **synovial fluid**. The joint cavity is lined, not by epithelium, but by a specialized connective tissue called the **synovial membrane** which extends folds and villi into the joint cavity and produces the lubricant synovial fluid.

In different diarthrotic joints the synovial membrane may have prominent regions with dense connective tissue or fat. The superficial regions of this tissue however are usually well vascularized, with many porous (fenestrated) capillaries. Besides having cells typical of connective tissue proper and a changing population of leukocytes, this area of a synovial membrane is characterized by two specialized cells with distinctly different functions and origins (Figure 8–22):

- **Macrophage-like synovial cells**, also called **type A cells**, are derived from blood monocytes and remove wear-and-tear debris from the synovial fluid. These

modified macrophages, which represent approximately 25% of the cells lining the synovium, are important in regulating inflammatory events within diarthrotic joints.

- **Fibroblastic synovial cells**, or **type B cells**, produce abundant hyaluronan and smaller amounts of proteoglycans. Much of this material is transported by water from the capillaries into the joint cavity to form the synovial fluid, which lubricates the joint, reducing friction on all internal surfaces, and supplies nutrients and oxygen to the articular cartilage.

The collagen fibers of the hyaline articular cartilage are disposed as arches with their tops near the exposed surface which, unlike most hyaline cartilage, is not covered by perichondrium (Figure 8–23). This arrangement of collagen helps distribute more evenly the forces generated by pressure on joints. The resilient articular cartilage efficiently absorbs the intermittent mechanical pressures to which many joints are subjected.

FIGURE **8–22** Synovial membrane.

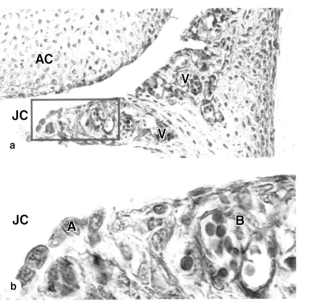

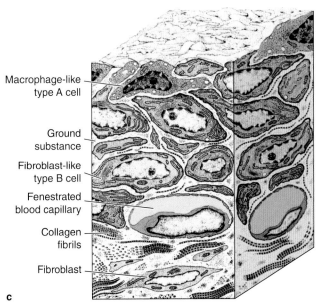

The synovial membrane is a specialized connective tissue that lines capsules of synovial joints and contacts the synovial fluid lubricant, which it is primarily responsible for maintaining.

(a) The synovial membrane projects folds into the joint cavity (**JC**) and these contain many small blood vessels (**V**). The joint cavity surrounds the articular cartilage (**AC**). (X100; Mallory trichrome)

(b) Higher magnification of the fold showing a high density of capillaries and two specialized types of cells called **synoviocytes**. Contacting the synovial fluid at the tissue surface are many rounded **macrophage-like synovial cells (type A)** derived from blood monocytes. These cells bind, engulf, and remove tissue debris from synovial fluid. These cells often form a layer at the tissue surface

(A) and can superficially resemble an epithelium, but there is no basal lamina and the cells are not joined together by cell junctions. **Fibroblast-like (type B) synovial cells** (**B**) are mesenchymally derived and specialized for synthesis of hyaluronan that enters the synovial fluid, replenishing it. (X400)

(c) Schematic representation of synovial membrane histology. Among the macrophage-like and fibroblast-like synovial cells are collagen fibers and other typical components of connective tissue. Surface cells have no basement membrane or junctional complexes denoting an epithelium, despite the superficial resemblance. Blood capillaries are fenestrated, which facilitates exchange of substances between blood and synovial fluid.

FIGURE **8–23** Articular cartilage.

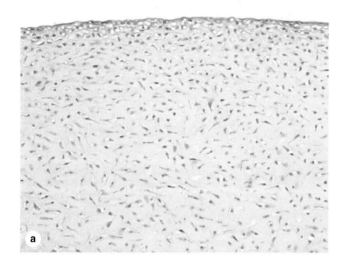

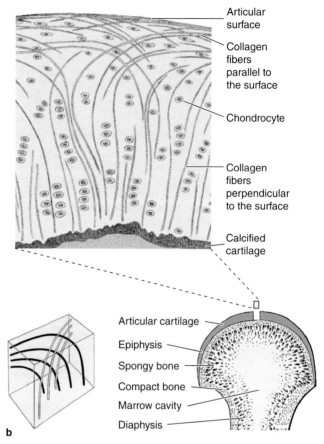

Labels on diagram (b), top: Articular surface; Collagen fibers parallel to the surface; Chondrocyte; Collagen fibers perpendicular to the surface; Calcified cartilage.

Labels on diagram (b), lower right: Articular cartilage; Epiphysis; Spongy bone; Compact bone; Marrow cavity; Diaphysis.

(a) Articular surfaces of a diarthrosis are made of hyaline cartilage that lacks the usual perichondrium covering (X40; H&E).
(b) The top diagram here shows a small region of articular cartilage in which type II collagen fibers run perpendicular to the tissue surface and then bend gradually in a broad arch. The lower left diagram shows a 3D view of arched collagen fibers in articular cartilage. Proteoglycan aggregates bound to hyaluronan fill the space among the collagen fibers and form a hydrated megacomplex that acts as a biomechanical spring. When pressure is applied a small amount of water is forced out of the cartilage matrix into the synovial fluid. When pressure is released water is attracted back into the interstices of the matrix. Such movements of water occur constantly with normal use of the joint and are essential for nutrition of the articular cartilage and for facilitating the interchange of O_2, CO_2, and metabolites between synovial fluid and chondrocytes.

Bone SUMMARY OF KEY POINTS

- Bone is a type of connective tissue with a **calcified** extracellular matrix (ECM), specialized to **support** the body, **protect** many internal organs, and act as the body's **Ca^{2+} reservoir**.

Major Cells & Matrix Components of Bone

- **Osteoblasts** differentiate from (stem) osteoprogenitor cells and secrete components of the initial matrix, called **osteoid**, that allow matrix mineralization to occur.
- Important components of osteoid include type I collagen, the protein **osteocalcin**, which binds Ca^{2+} and **matrix vesicles** with enzymes generating PO_4^-.
- High concentrations of Ca^{2+} and PO_4^- ions cause formation of **hydroxyapatite** crystals, whose growth gradually calcifies the entire matrix.
- **Osteocytes** differentiate further from osteoblasts when they become enclosed within matrix **lacunae** and act to maintain the matrix and detect mechanical stresses on bone.

- Osteocytes maintain communication with adjacent cells via a network of long **dendritic processes** that extend through the matrix via narrow **canaliculi** radiating from each lacuna.
- **Osteoclasts** are very large cells, formed by fusion of several blood monocytes, which locally erode bone matrix during osteogenesis and bone remodeling.

Periosteum & Endosteum

- **Periosteum** is a layer of dense connective tissue on the outer surface of bone, bound to bone matrix by bundles of type I collagen called **perforating** (or **Sharpey**) **fibers**.
- Regions of periosteum adjacent to bone are rich in **osteoprogenitor cells** and **osteoblasts** that mediate much bone growth and remodeling.
- The **endosteum** is a thin layer of active and inactive osteoblasts, which lines all the internal surfaces within bone; osteoblasts here are also required for bone growth.

Types & Organization of Bone (Table 8-1)

- Dense bone immediately beneath the periosteum is called **compact bone**; deep to the compact bone are small bony trabeculae or spicules of **cancellous** (or spongy) **bone**.
- In long bones of the limbs these two types of mature bone tissue occur in both the knobby, bulbous ends, called **epiphyses**, and in the intervening shaft or **diaphysis**.
- Immature bone, called **woven bone**, is formed during osteogenesis or repair and has a calcified matrix with randomly arranged collagen fibers.
- By the action of osteoclasts and osteoblasts, woven bone undergoes rapid turnover and is remodeled into **lamellar bone** with new matrix deposited in distinct layers with parallel collagen bundles; both compact and cancellous bone is lamellar bone.
- Most lamellar bone consists of lamellae organized concentrically around small **central canals** containing blood vessels and nerves; this organization is called an **osteon or Haversian system**.
- Within each osteon osteocytic lacunae occur between the lamellae, with **canaliculi** radiating through the lamellae, which allow all cells to communicate with the central canal.

Osteogenesis

- Bones of the skull and jaws form initially by **intramembranous ossification**, with osteoblasts differentiating directly from progenitor cells in condensed **"membranes" of mesenchyme**.
- All other bones form by **endochondral ossification**, in which osteoprogenitor cells surround and then invade hyaline **cartilage models** of the skeletal elements in the embryo.
- **Primary ossification centers** in diaphyses of fetal long bones form when chondrocytes die after enclosure of the cartilage within a collar of woven bone, creating an initial cavity that is entered by periosteal osteoblasts and vasculature.
- Later, **secondary ossification centers** develop similarly within the epiphyses, with cartilage of the **epiphyseal growth plate** between the primary and secondary ossification sites.
- The growth plates are the key to **bone elongation** during childhood and are organized as an interrelated series of developing zones.
- Most distally is a **"resting"** or **reserve zone** of typical hyaline cartilage.
- In an adjacent **zone of proliferation**, chondrocytes undergo mitosis and appear stacked within elongated lacunae.
- The most mature chondrocytes in these lacunae swell up, compress the matrix, and undergo apoptosis in a **zone of hypertrophy** closer to the large primary ossification center.
- Spaces created in the matrix by these events characterize the **zone of cartilage calcification** when they are invaded by osteoblasts, osteoclasts, and vasculature from the primary center.
- In the **zone of ossification** woven bone is laid down initially by osteoblasts and remodeled into lamellae bone.

- **Appositional bone growth** increases the circumference of a bone by osteoblast activity at the periosteum and is accompanied by enlargement of the medullary marrow cavity.

Bone Growth, Remodeling, & Repair

- **Growth** of bones occurs throughout life, with cells and matrix turning over continuously through activities of osteoblasts and osteoclasts.
- Lamellae and osteons are temporary structures and are replaced and rebuilt continuously in a process of **bone remodeling** by which bones change size and shape according to changes in mechanical stress.
- **Bone repair** after fracture or other injury involves the activation of periosteal fibroblasts to produce an initial **soft callus of fibrocartilage-like tissue**.
- The soft callus is gradually replaced by a **hard callus of woven bone** that is soon remodeled to produce stronger lamellar bone.

Metabolic Role of Bone

- Ca^{2+}, a key ion for all cells, is **stored** in bone when dietary calcium is adequate and **mobilized** from bone when dietary calcium is deficient.
- Maintenance of proper **blood calcium levels** involves activity of all three major bone cells and is largely regulated by subtle paracrine interaction among these and other cells.
- Hormones affecting calcium deposition and removal from bone include **parathyroid hormone (PTH)**, which indirectly stimulates osteoclasts to elevate levels of calcium in blood, and **calcitonin**, which can inhibit osteoclast activity, lowering blood calcium levels.

Joints

- **Joints** are places where bones meet, or articulate, allowing at least the potential for bending or movement in that portion of the skeleton.
- Joints with very limited or no movement are classified collectively as **synarthroses** and freely mobile joints are called **diarthroses**.
- **Intervertebral discs** are synarthroses in the vertebral column which cushion adjacent vertebrae.
- Each intervertebral disc consists of a thick outer layer of fibrocartilage forming a tough **annulus fibrosus** and a shock-absorbing inner, gel-like core, the **nucleus pulposus**.
- Diarthroses have a **joint cavity** filled with lubricant **synovial fluid**, enclosed within a tough, fibrous **articular capsule**; ends of the bones involved are covered with hyaline **articular cartilage**.
- Specialized connective tissue of the **synovial membrane** lines the capsule, with folds extended into some areas of the joint cavity.
- **Macrophage-like synovial cells** of the synovial membrane remove wear-and-tear debris from synovial fluid.
- **Fibroblast-like synovial cells** of the synovial membrane synthesize hyaluronan which moves into the synovial fluid with water from local capillaries to lubricate and nourish the articular cartilage.

Bone ASSESS YOUR KNOWLEDGE

1. Which component of bone impedes the distribution of nutrients and oxygen to osteocytes?
   a. Extracellular matrix
   b. Canaliculi
   c. Periosteum
   d. Cell processes
   e. Haversian canals

2. Which if the following most accurately describes compact bone?
   a. Predominant bone type in the epiphyses of adult long bones
   b. Also known as cancellous bone
   c. Characterized by the presence of osteons
   d. Lines the medullary (marrow) cavity
   e. Forms the diploë in cranial bones

3. In healthy bone canaliculi are likely to contain which one of the following?
   a. Capillaries
   b. Nerve axons
   c. Osteocytic processes
   d. Osteoid
   e. Osteoclasts in resorption lacunae

4. Which of the following most accurately describes the endosteum?
   a. Composed of two layers: osteogenic and fibrous
   b. Continuous with the joint capsule
   c. Attached to the bone surface by collagen bundles called Sharpey fibers
   d. Lines the medullary cavity
   e. Contains mature osteocytes

5. In the diaphysis of a typical long bone which of the following structures is in closest proximity to the trabeculae of cancellous bone?
   a. Interstitial lamellae
   b. Osteons
   c. Sharpey fibers
   d. Outer circumferential lamellae
   e. Inner circumferential lamellae

6. Which "zone" of endochondral ossification in the growing femur of an adolescent is the farthest from that bone's secondary ossification center?
   a. Zone of hypertrophy
   b. Zone of reserve cartilage
   c. Zone of calcified cartilage
   d. Zone of ossification
   e. Zone of proliferation

7. The major lubricant for diarthrotic joints is synthesized by cells located in which joint structure?
   a. Nucleus pulposus
   b. Synovial membrane
   c. Articular cartilage
   d. Annulus fibrosus
   e. Fibrous capsule

8. A 25-year-old man presents with persistent joint pain and a history of recurrent fractures of each humerus. His hematocrit and complete blood count (CBC) are normal, but blood calcium levels are high. Hormone levels are all within normal ranges except parathyroid hormone (PTH) which exceeds normal by 3-fold. Which if the following could be prescribed to offset the effects of the elevated PTH?
   a. Vitamin D
   b. Vitamin C
   c. Recombinant RANK ligand
   d. Somatotrophin (growth hormone)
   e. Calcitonin

9. A 42-year-old woman, who has been a type I diabetic for 30 years, falls when she trips over the vacuum cleaner hose. She tried to break her fall by placing her hand out to save herself and in the process her wrist was forced backward, breaking her radius near the wrist. Which of the following is produced in the first step in healing this bone injury?
   a. Osteoid
   b. Hematoma
   c. Bony callus
   d. Fibrocartilage
   e. Compact bone

10. A 46-year-old woman presents with pain in the left leg that worsens on weight bearing. An x-ray shows demineralization, and a bone biopsy decalcified with EDTA shows reduction in bone quantity. The patient had undergone menopause at age 45 without estrogen replacement. She reports long-standing diarrhea. In addition, laboratory tests show low levels of vitamin D, calcium and phosphorus, and elevated alkaline phosphatase. A second bone biopsy was taken but not decalcified, which showed extensive deposition of uncalcified osteoid on all the bone surfaces. On the basis of these data, the best diagnosis would be which of the following?
   a. Osteoporosis
   b. Scurvy
   c. Osteomalacia
   d. Rickets
   e. Hypoparathyroidism

CHAPTER

9 Nerve Tissue & the Nervous System

The human nervous system, by far the most complex system in the body, is formed by a network of many billion nerve cells (**neurons**), all assisted by many more supporting cells called **glial cells**. Each neuron has hundreds of interconnections with other neurons, forming a very complex system for processing information and generating responses.

Nerve tissue is distributed throughout the body as an integrated communications network. Anatomically, the general organization of the nervous system (Figure 9–1) has two major divisions:

- **Central nervous system (CNS)**, consisting of the brain and spinal cord
- **Peripheral nervous system (PNS)**, composed of the cranial, spinal, and peripheral nerves conducting impulses to and from the CNS (sensory and motor nerves, respectively) and **ganglia** that are small aggregates of nerve cells outside the CNS.

Cells in both central and peripheral nerve tissue are of two kinds: **neurons**, which typically have numerous long processes, and various **glial cells** (Gr. *glia*, glue), which have short processes, support and protect neurons, and participate in many neural activities, neural nutrition, and defense of cells in the CNS.

Neurons respond to environmental changes (**stimuli**) by altering the ionic gradient that exists across their plasma membranes. All cells maintain such a gradient, also called an electrical potential, but cells that can rapidly change this potential in response to stimuli (eg, neurons, muscle cells, some gland cells) are said to be **excitable** or irritable. Neurons react promptly to stimuli with a reversal of the ionic gradient (**membrane depolarization**) that generally spreads from the place that received the stimulus and is propagated across the neuron's entire plasma membrane. This propagation, called the **action potential**, the **depolarization wave**, or the **nerve impulse**, is capable of traveling long distances along neuronal processes, transmitting such signals to other neurons, muscles, and glands.

By collecting, analyzing, and integrating information in such signals, the nervous system continuously stabilizes the intrinsic conditions of the body (eg, blood pressure, O_2 and CO_2 content, pH, blood glucose levels, and hormone levels) within normal ranges and maintains behavioral patterns (eg, feeding, reproduction, defense, interaction with other living creatures).

DEVELOPMENT OF NERVE TISSUE

The nervous system develops from the outermost of the three early embryonic layers, the ectoderm, beginning in the third week of development (Figure 9–2). With signals from the underlying axial structure, the notochord, ectoderm on the mid-dorsal side of the embryo thickens to form the epithelial **neural plate**. The sides of this plate fold upward and grow toward each other medially, and within a few days fuse to form

FIGURE **9–1** The general organization of the nervous system.

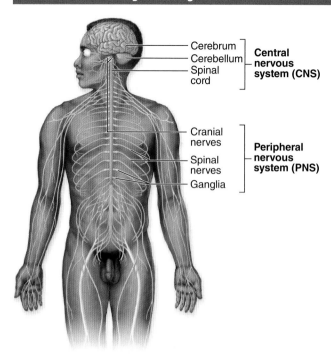

Cerebrum
Cerebellum — **Central nervous system (CNS)**
Spinal cord

Cranial nerves — **Peripheral nervous system (PNS)**
Spinal nerves
Ganglia

Anatomically the nervous system is divided into the **CNS** and **PNS**, which have the major components shown in the diagram.

Functionally the nervous system consists of:

1. **Sensory division (afferent)**

    A. **Somatic** – sensory input perceived consciously (eg, from eyes ears, skin, musculoskeletal structures)

    B. **Visceral** – sensory input *not* perceived consciously (eg, from internal organs and cardiovascular structures)

2. **Motor division (efferent)**

    A. **Somatic** – motor output controlled consciously or voluntarily (eg, by skeletal muscle effectors)

    B. **Autonomic** – motor output *not* controlled consciously (eg, by heart or gland effectors)

The autonomic motor nerves, comprising what is often called the **autonomic nervous system (ANS)**, all have pathways involving two neurons: a **preganglionic neuron** with the cell body in the CNS and a **postganglionic neuron** with the cell body in a ganglion. The ANS has two divisions: (1) The **parasympathetic division**, with its ganglia within or near the effector organs, maintains normal body homeostasis. (2) The **sympathetic division** has its ganglia close to the CNS and controls the body's responses during emergencies and excitement. ANS components located in the wall of the digestive tract are sometimes referred to as the enteric nervous system.

FIGURE **9–2** Neurulation in the early embryo.

Neural groove

Neural folds
Notochord

Neural crest

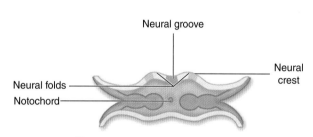

(1) Neural folds and neural groove form from the neural plate.

Neural groove
Ectoderm

Neural crest cells

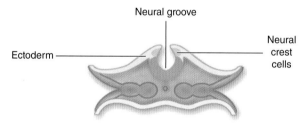

(3) As neural folds prepare to fuse and form the neural tube and dorsal epidermis, neural crest cells loosen and become mesenchymal.

Neural groove

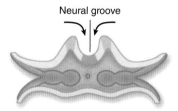

(2) Neural folds elevate and approach one another.

Neural tube

Neural crest cells

Developing epidermis

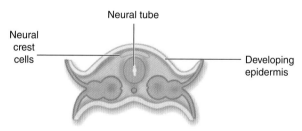

(4) The mass of neural crest cells initially lies atop the newly formed neural tube.

Stages in the process of **neurulation**, by which cells of the CNS and PNS are produced, are shown in diagrammatic cross sections of a 3- and 4-week human embryo with the extraembryonic membranes removed. Under an inductive influence from the medial notochord, the overlying layer of ectodermal cells thickens as a bending **neural plate,** with a medial neural groove and lateral neural folds **(1)**. All other ectoderm will become epidermis. The plate bends further, making the **neural folds** and **groove** more prominent **(2)**. The neural folds rise and fuse at the midline **(3)**, converting the groove into the **neural tube (4)**, which is large at the cranial end of the embryo and much narrower caudally. The neural tube will give rise to the entire CNS.

As the neural tube detaches from the now overlying ectoderm, many cells separate from it and produce a mass of mesenchymal cells called the **neural crest**. Located initially above the neural tube, neural crest cells immediately begin migrating laterally. Cell derived from the neural crest will form all components of the PNS and also contribute to certain non-neural tissues.

the **neural tube**. Cells of this tube give rise to the entire CNS, including neurons and most glial cells.

As the folds fuse and the neural tube separates from the now overlying surface ectoderm that will form epidermis, a large population of developmentally important cells, the **neural crest**, separates from the neuroepithelium and becomes mesenchymal. Neural crest cells migrate extensively and differentiate as all the cells of the PNS, as well as a number of other non-neuronal cell types.

⟩ NEURONS

The functional unit in both the CNS and PNS is the **neuron**. Some neuronal components have special names, such as "neurolemma" for the cell membrane. Most neurons have three main parts (Figure 9–3):

- The **cell body** (also called the **perikaryon** or **soma**) which contains the nucleus and most of the cell's organelles and serves as the synthetic or trophic center for the entire neuron.
- The **dendrites**, which are the numerous elongated processes extending from the perikaryon and specialized to receive stimuli from other neurons at unique sites called **synapses**.
- The **axon** (Gr. *axon*, axis), which is a single long process ending at synapses specialized to generate and conduct nerve impulses to other cells (nerve, muscle, and gland cells). Axons may also receive information from other neurons, information that mainly modifies the transmission of action potentials to those neurons.

Neurons and their processes are extremely variable in size and shape. Cell bodies can be very large, measuring up to 150 μm in diameter. Other neurons, such as the cerebellar granule cells, are among the body's smallest cells.

Neurons can be classified according to the number of processes extending from the cell body (Figure 9–4):

- **Multipolar neurons**, each with one axon and two or more dendrites, are the most common.
- **Bipolar neurons**, with one dendrite and one axon, comprise the sensory neurons of the retina, the olfactory epithelium, and the inner ear.
- **Unipolar** or **pseudounipolar neurons**, which include all other sensory neurons, each have a single process that bifurcates close to the perikaryon, with the longer branch extending to a peripheral ending and the other toward the CNS.
- **Anaxonic neurons**, with many dendrites but no true axon, do not produce action potentials, but regulate electrical changes of adjacent CNS neurons.

Because the fine processes emerging from cell bodies are seldom seen in sections of nervous tissue, it is difficult to classify neurons structurally by microscopic inspection.

Nervous components can also be subdivided functionally (Figure 9–1). **Sensory neurons** are **afferent**, receiving stimuli from receptors throughout the body. **Motor neurons** are **efferent**, sending impulses to effector organs such as muscle fibers and glands. **Somatic** motor nerves are under voluntary control and typically innervate skeletal muscle; **autonomic** motor nerves control the involuntary or unconscious activities of glands, cardiac muscle, and most smooth muscle.

Interneurons establish relationships among other neurons, forming complex functional networks or **circuits** in the CNS. Interneurons are either multipolar or anaxonic and comprise 99% of all neurons in adults.

In the CNS most neuronal perikarya occur in the **gray matter**, with their axons concentrated in the **white matter**. These terms refer to the general appearance of unstained CNS tissue caused in part by the different densities of nerve cell bodies. In the PNS cell bodies are found in ganglia and in some sensory regions, such as the olfactory mucosa, and axons are bundled in **nerves**.

⟩⟩ MEDICAL APPLICATION

Parkinson disease is a slowly progressing disorder affecting muscular activity characterized by tremors, reduced activity of the facial muscles, loss of balance, and postural stiffness. It is caused by gradual loss by apoptosis of dopamine-producing neurons whose cell bodies lie within the nuclei of the CNS substantia nigra. Parkinson disease is treated with L-dopa (L-3,4-dihydroxyphenylalanine), a precursor of dopamine which augments the declining production of this neurotransmitter.

Cell Body (Perikaryon or Soma)

The neuronal **cell body** contains the nucleus and surrounding cytoplasm, exclusive of the cell processes (Figure 9–3). It acts as a trophic center, producing most cytoplasm for the processes. Most cell bodies are in contact with a great number of nerve endings conveying excitatory or inhibitory stimuli generated in other neurons. A typical neuron has an unusually large, euchromatic nucleus with a prominent nucleolus, indicating intense synthetic activity.

Cytoplasm of perikarya often contains numerous free polyribosomes and highly developed RER, indicating active production of both cytoskeletal proteins and proteins for transport and secretion. Histologically these regions with concentrated RER and other polysomes are basophilic and are distinguished as **chromatophilic substance** (or **Nissl substance, Nissl bodies**) (Figure 9–3). The amount of this material varies with the type and functional state of the neuron and is particularly abundant in large nerve cells such as motor neurons (Figure 9–3b). The Golgi apparatus is located only in the cell body, but mitochondria can be found throughout the cell and are usually abundant in the axon terminals.

In both perikarya and processes microtubules, actin filaments, and intermediate filaments are abundant, with the

FIGURE **9–3** Structures of a typical neuron.

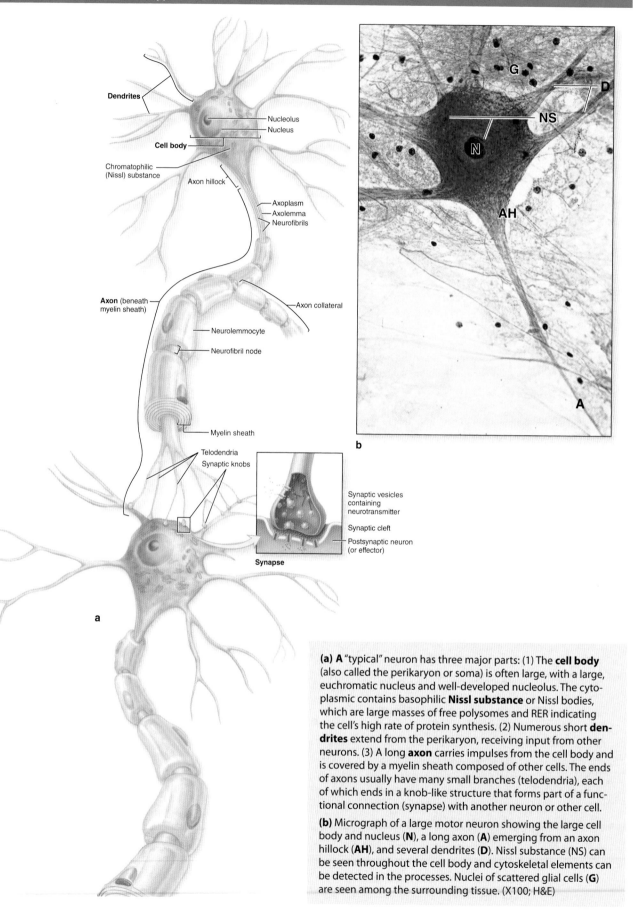

Dendrites

Nucleolus
Nucleus

Cell body

Chromatophilic
(Nissl) substance

Axon hillock

Axoplasm
Axolemma
Neurofibrils

Axon (beneath
myelin sheath)

Axon collateral

Neurolemmocyte

Neurofibril node

Myelin sheath

Telodendria
Synaptic knobs

Synaptic vesicles
containing
neurotransmitter

Synaptic cleft

Postsynaptic neuron
(or effector)

Synapse

a

b

G

D

NS

N

AH

A

(a) A "typical" neuron has three major parts: (1) The **cell body** (also called the perikaryon or soma) is often large, with a large, euchromatic nucleus and well-developed nucleolus. The cyto-plasmic contains basophilic **Nissl substance** or Nissl bodies, which are large masses of free polysomes and RER indicating the cell's high rate of protein synthesis. (2) Numerous short **den-drites** extend from the perikaryon, receiving input from other neurons. (3) A long **axon** carries impulses from the cell body and is covered by a myelin sheath composed of other cells. The ends of axons usually have many small branches (telodendria), each of which ends in a knob-like structure that forms part of a func-tional connection (synapse) with another neuron or other cell.

(b) Micrograph of a large motor neuron showing the large cell body and nucleus (**N**), a long axon (**A**) emerging from an axon hillock (**AH**), and several dendrites (**D**). Nissl substance (**NS**) can be seen throughout the cell body and cytoskeletal elements can be detected in the processes. Nuclei of scattered glial cells (**G**) are seen among the surrounding tissue. (X100; H&E)

FIGURE **9–4** Structural classes of neurons.

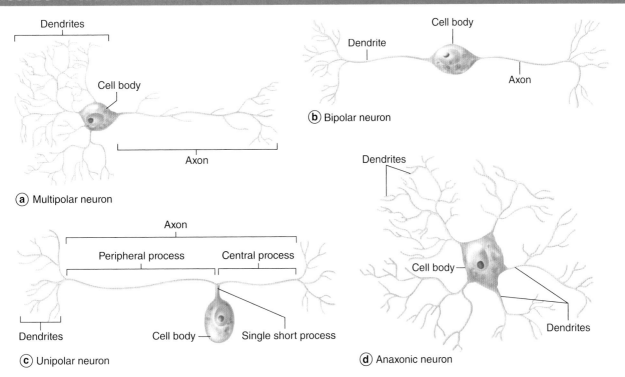

Shown are the four main types of neurons, with short descriptions. **(a)** Most neurons, including all motor neurons and CNS interneurons, are **multipolar**. **(b) Bipolar neurons** include sensory neurons of the retina, olfactory mucosa, and inner ear. **(c)** All other sensory neurons are **unipolar** or **pseudounipolar**. **(d) Anaxonic** neurons of the CNS lack true axons and do not produce action potentials, but regulate local electrical changes of adjacent neurons.

latter formed by unique protein subunits and called **neurofilaments** in this cell type. Cross-linked with certain fixatives and impregnated with silver stains, neurofilaments are also referred to as neurofibrils by light microscopists. Some nerve cell bodies also contain inclusions of pigmented material, such as lipofuscin, consisting of residual bodies left from lysosomal digestion.

Dendrites

Dendrites (Gr. *dendron*, tree) are typically short, small processes emerging and branching off the soma (Figure 9–3). Usually covered with many synapses, dendrites are the principal signal reception and processing sites on neurons. The large number and extensive arborization of dendrites allow a single neuron to receive and integrate signals from many other nerve cells. For example, up to 200,000 axonal endings can make functional contact with the dendrites of a single large Purkinje cell of the cerebellum.

Unlike axons, which maintain a nearly constant diameter, dendrites become much thinner as they branch, with cytoskeletal elements predominating in these distal regions. In the CNS most synapses on dendrites occur on **dendritic spines**, which are dynamic membrane protrusions along the small dendritic branches, visualized with silver staining (Figure 9–5)

and studied by confocal or electron microscopy. Dendritic spines serve as the initial processing sites for synaptic signals and occur in vast numbers, estimated to be on the order of 10^{14} for cells of the human cerebral cortex. Dendritic spine morphology depends on actin filaments and changes continuously as synaptic connections on neurons are modified. Changes in dendritic spines are of key importance in the constant changes of the **neural plasticity** that occurs during embryonic brain development and underlies adaptation, learning, and memory postnatally.

Axons

Most neurons have only one **axon**, typically longer than its dendrites. Axonal processes vary in length and diameter according to the type of neuron. Axons of the motor neurons that innervate the foot muscles have lengths of nearly a meter; large cell bodies are required to maintain these axons, which contain most of such neurons' cytoplasm. The plasma membrane of the axon is often called the **axolemma** and its contents are known as **axoplasm**.

Axons originate from a pyramid-shaped region of the perikaryon called the **axon hillock** (Figure 9–3), just beyond which the axolemma has concentrated ion channels which

FIGURE **9–5** **Dendrites and dendritic spines.**

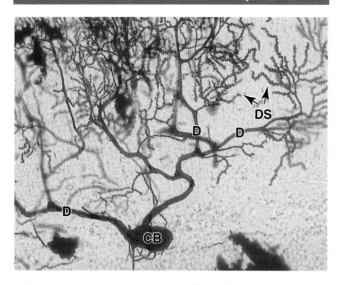

The large Purkinje neuron in this silver-impregnated section of cerebellum has many dendrites (**D**) emerging from its cell body (**CB**) and forming branches. The small dendritic branches each have many tiny projecting dendritic spines (**DS**) spaced closely along their length, each of which is a site of a synapse with another neuron. Dendritic spines are highly dynamic, the number of synapses changing constantly. (X650; Silver stain)

generate the action potential. At this initial segment of the axon the various excitatory and inhibitory stimuli impinging on the neuron are algebraically summed, resulting in the decision to propagate—or not to propagate—a nerve impulse.

Axons generally branch less profusely than dendrites, but do undergo **terminal arborization** (Figure 9–3). Axons of interneurons and some motor neurons also have major branches called **collaterals** that end at smaller branches with synapses influencing the activity of many other neurons. Each small axonal branch ends with a dilation called a **terminal bouton** (Fr. *bouton*, button) that contacts another neuron or non-nerve cell at a synapse to initiate an impulse in that cell.

Axoplasm contains mitochondria, microtubules, neurofilaments, and transport vesicles, but very few polyribosomes or cisternae of RER, features which emphasize the dependence of axoplasm on the perikaryon. If an axon is severed from its cell body its distal part quickly degenerates and undergoes phagocytosis.

Lively bidirectional transport of molecules large and small occurs within axons. Organelles and macromolecules synthesized in the cell body move by **anterograde transport** along axonal microtubules via **kinesin** from the perikaryon to the synaptic terminals. **Retrograde transport** in the opposite direction along microtubules via **dynein** carries certain other macromolecules, such as material taken up by endocytosis (including viruses and toxins), from the periphery to the cell body. Retrograde transport can be used to study the pathways of neurons: if peroxidase or another marker is

injected into regions with axon terminals, its later distribution throughout the neurons serving such regions can be determined histochemically.

Anterograde and retrograde transports both occur fairly rapidly, at rates of 50-400 mm/d. A much slower anterograde stream, moving only a few millimeters per day, involves movement of the axonal cytoskeleton itself. This slow axonal transport corresponds roughly to the rate of axon growth.

Nerve Impulses

A **nerve impulse**, or **action potential**, travels along an axon like a spark moves along an explosive's fuse. It is an electrochemical process initiated at the axon hillock when other impulses received at the cell body or dendrites meet a certain threshold. The action potential is propagated along the axon as a wave of membrane depolarization produced by **voltage-gated Na⁺ and K⁺ channels** in the axolemma that allow diffusion of these ions into and out of the axoplasm. The extracellular compartment around all regions of the neuron is a very thin zone immediately outside the cell that is formed by enclosing glial cells which also regulate its ionic contents.

In unstimulated neurons ATP-dependent Na-K pumps and other membrane proteins maintain an axoplasmic Na⁺ concentration only one-tenth of that outside the cell and a K⁺ level many times greater than the extracellular concentration. This produces a potential electrical difference across the axolemma of about −65 mV, with the inside negative to the outside. This difference is the axon's **resting potential**.

When the threshold for triggering an impulse is met, channels at the axon's initial segment open and allow a very rapid influx of extracellular Na⁺ that makes the axoplasm positive in relation to the extracellular environment and shifts (depolarizes) the resting potential from negative to positive, to +30 mV. Immediately after the membrane depolarization, the voltage-gated Na⁺ channels close and those for K⁺ open, which rapidly returns the membrane to its resting potential. This cycle of events occurs in less than 1 millisecond.

Depolarization stimulates adjacent portions of the axolemma to depolarize and return immediately to the resting potential, which causes a nerve impulse, or wave of depolarization, to move rapidly along the axon. After a refractory period also measured in milliseconds, the neuron is ready to repeat this process and generate another action potential. Impulses arriving at the synaptic nerve endings promote the discharge of stored neurotransmitter that stimulates or inhibits action potentials in another neuron or a non-neural cell.

Synaptic Communication

Synapses (Gr. *synapsis*, union) are sites where nerve impulses are transmitted from one neuron to another, or from neurons and other effector cells. The structure of a synapse (Figure 9–6) ensures that transmission is unidirectional. Synapses convert an electrical signal (nerve impulse) from the **presynaptic cell** into a chemical signal that affects the **postsynaptic cell**. Most synapses act by releasing **neurotransmitters**, which are usually small molecules that bind specific receptor proteins to either open or close ion channels or initiate second-messenger cascades. A synapse (Figure 9–6a) has the following components:

- The **presynaptic axon terminal (terminal bouton)** contains mitochondria and numerous **synaptic vesicles** from which neurotransmitter is released by exocytosis.

- The **postsynaptic cell membrane** contains receptors for the neurotransmitter, and ion channels or other mechanisms to initiate a new impulse.
- A 20- to 30-nm-wide intercellular space called the **synaptic cleft** separates these presynaptic and postsynaptic membranes.

At the presynaptic region the nerve impulse briefly opens calcium channels, promoting a Ca^{2+} influx that triggers neurotransmitter release by exocytosis or similar mechanisms. Immediately the released neurotransmitter molecules diffuse across the synaptic cleft and bind receptors at the postsynaptic region. This produces either an excitatory or an inhibitory effect at the postsynaptic membrane, as follows:

- Neurotransmitters from **excitatory synapses** cause postsynaptic Na^+ channels to open, and the resulting Na^+

FIGURE **9–6** Major components of a synapse.

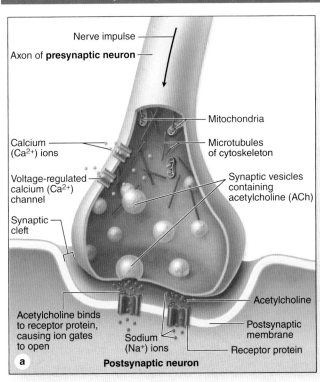

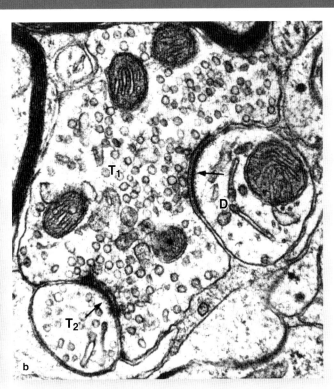

(a) Diagram showing a synapse releasing neurotransmitters by exocytosis from the terminal bouton. Presynaptic terminals always contain a large number of **synaptic vesicles** containing neurotransmitters, numerous **mitochondria**, and smooth ER as a source of new membrane. Some neurotransmitters are synthesized in the cell body and then transported in vesicles to the presynaptic terminal. Upon arrival of a nerve impulse, voltage-regulated Ca^{2+} channels permit Ca^{2+} entry, which triggers neurotransmitter release into the synaptic cleft. Excess membrane accumulating at the presynaptic region as a result of exocytosis is recycled by clathrin-mediated endocytosis, which is not depicted here.

(b) The TEM shows a large presynaptic terminal (**T₁**) filled with synaptic vesicles and asymmetric electron-dense regions around 20- to 30-nm-wide synaptic clefts (**arrows**). The postsynaptic membrane contains the neurotransmitter receptors and mechanisms to initiate an impulse at the postsynaptic neuron. The postsynaptic membrane on the right is part of a dendrite (**D**), associated with fewer vesicles of any kind, showing this to be an axodendritic synapse. On the left is another presynaptic terminal (**T₂**), suggesting an axoaxonic synapse with a role in modulating activity of the other terminal. (X35,000)

influx initiates a depolarization wave in the postsynaptic neuron or effector cell as just described.

- At **inhibitory synapses** neurotransmitters open Cl⁻ or other anion channels, causing influx of anions and **hyperpolarization** of the postsynaptic cell, making its membrane potential more negative and more resistant to depolarization.

Interplay between excitatory and inhibitory effects on postsynaptic cells allows synapses to process neuronal input and fine-tune the reaction of the effector cell. Impulses passing from presynaptic neurons to postsynaptic cells are usually modified at the synapse by similar connections there with other neurons (Figure 9–6b). The response in postsynaptic neurons is determined by the summation of activity at hundreds of synapses on that cell. Three common morphological types of synapses occur between neurons of the CNS and are shown in Figure 9–7.

The chemical transmitter used at neuromuscular junctions and some synapses of the CNS is **acetylcholine**. Within the CNS other major categories of neurotransmitters include:

- Certain **amino acids** (often modified), such as glutamate and γ-aminobutyrate (GABA)
- **Monoamines**, such as serotonin (5-hydroxytryptamine or 5-HT) and **catecholamines**, such as dopamine, all of which are synthesized from amino acids
- Small **polypeptides**, such as endorphins and substance P.

Important actions of these and other common neurotransmitters are summarized in Table 9–1. Different receptors and second messenger systems often occur for the same transmitter, greatly multiplying the possible effects of these molecules. After their release transmitters are removed quickly by enzymatic breakdown, by glial activity, or by endocytotic recycling involving presynaptic membrane receptors.

❯❯ MEDICAL APPLICATION

Levels of neurotransmitters in the synaptic cleft and available for binding postsynaptic receptors are normally regulated by several local mechanisms. **Selective serotonin reuptake inhibitors (SSRIs)**, a widely used class of drugs for treatment of depression and anxiety disorders, were designed to augment levels of this neurotransmitter at the postsynaptic membrane of serotonergic CNS synapses by specifically inhibiting its reuptake at the presynaptic membrane.

❯ GLIAL CELLS & NEURONAL ACTIVITY

Glial cells support neuronal survival and activities, and are ten times more abundant than neurons in the mammalian brain. Like neurons most glial cells develop from progenitor cells of the embryonic neural plate. In the CNS glial cells surround both the neuronal cell bodies, which are often larger than the glial cells, and the processes of axons and dendrites occupying the spaces between neurons. Except around the

larger blood vessels, the CNS has only a very small amount of connective tissue and collagen. Glial cells substitute for cells of connective tissue in some respects, supporting neurons and creating immediately around those cells microenvironments that are optimal for neuronal activity. The fibrous intercellular network of CNS tissue superficially resembles collagen by light microscopy, but is actually the network of fine cellular processes emerging from neurons and glial cells. Such processes are collectively called the **neuropil** (Figure 9–8).

There are six major kinds of glial cells, as shown schematically in Figure 9–9, four in the CNS, two in the PNS. Their main functions, locations, and origins are summarized in Table 9–2.

Oligodendrocytes

Oligodendrocytes (Gr. *oligos*, small, few + *dendron*, tree + *kytos*, cell) extend many processes, each of which becomes sheet-like and wraps repeatedly around a portion of a nearby CNS axon (Figure 9–9a). During this wrapping most cytoplasm gradually moves out of the growing extension, leaving multiple compacted layers of cell membrane collectively termed **myelin**. An axon's full length is covered by the action of many oligodendrocytes. The resulting **myelin sheath** electrically insulates the axon and facilitates rapid transmission of nerve impulses. Found only in the CNS oligodendrocytes are the predominant glial cells in white matter, which is white because of the lipid concentrated in the wrapped membrane sheaths. The processes and sheaths are not visible by routine light microscope staining, in which oligodendrocytes usually appear as small cells with rounded, condensed nuclei and unstained cytoplasm (Figure 9–8a).

Astrocytes

Also unique to the CNS **astrocytes** (Gr. *astro-*, star + *kytos*) have a large number of long radiating, branching processes (Figures 9–9a and 9–10). Proximal regions of the astrocytic processes are reinforced with bundles of intermediate filaments made of **glial fibrillary acid protein (GFAP)**, which serves as a unique marker for this glial cell. Distally the processes lack GFAP, are not readily seen by microscopy, and form a vast network of delicate terminals contacting synapses and other structures. Terminal processes of a single astrocyte typically occupy a large volume and associate with over a million synaptic sites.

Astrocytes originate from progenitor cells in the embryonic neural tube and are by far the most numerous glial cells of the brain, as well as the most diverse structurally and functionally. **Fibrous astrocytes**, with long delicate processes, are abundant in white matter; those with many shorter processes are called **protoplasmic astrocytes** and predominate in the gray matter. The highly variable and dynamic processes mediate most of these cells' many functions.

❯❯ MEDICAL APPLICATION

Most brain tumors are **astrocytomas** derived from fibrous astrocytes. These are distinguished pathologically by their expression of GFAP.

FIGURE **9–7** Types of synapses.

Axosomatic synapse

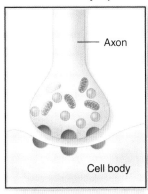

Axon

Cell body

Axodendritic synapse

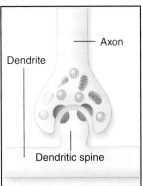

Axon

Dendrite

Dendritic spine

Axoaxonic synapse

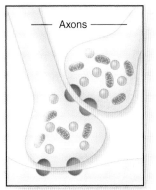

Axons

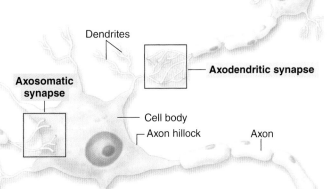

Dendrites

Axodendritic synapse

Axosomatic synapse

Cell body

Axon hillock

Axon

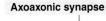

Axoaxonic synapse

Terminal arborizations

The diagrams show three common morphologic types of synapses. Branched axon terminals usually associate with and transmit a nerve impulse to another neuron's cell body (or soma) or a dendritic spine. These types of connections are termed an **axosomatic synapse** and an **axodendritic synapse**, respectively. Less frequently, an axon terminal forms a synapse with an axon terminal of another neuron; such an **axoaxonic synapse** functions to modulate synaptic activity in the other two types.

All three morphologic types of synapses have the features of all true synapses: a presynaptic axon terminal that releases a transmitter; a postsynaptic cell membrane with receptors for the transmitter; and an intervening synaptic cleft.

Synaptic structure usually cannot be resolved by light microscopy, although components such as dendritic spines may be shown with special techniques (Figure 9–5).

TABLE **9–1**	Common neurotransmitters and their actions.
Neurotransmitter	**Description/Action**

ACETYLCHOLINE (ACh)

	Chemical structure significantly different from that of other neurotransmitters; active in CNS and in both somatic and autonomic parts of PNS; binds to ACh receptors (cholinergic receptors) in PNS to open ion channels in postsynaptic membrane and stimulate muscle contraction

AMINO ACIDS

	Molecules with both carboxyl (—COOH) and amine (—NH$_2$) groups and various R groups; act as important transmitters in the CNS
Glutamate	Excites activity in neurons to promote cognitive function in the brain (learning and memory); most common neurotransmitter in the brain; opens Na$^+$ channels
Gamma-aminobutyric acid (GABA)	Synthesized from glutamate; primary inhibitory neurotransmitter in the brain; also influences muscle tone; opens or closes various ion channels
Glycine	Inhibits activity between neurons in the CNS, including retina; opens Cl$^-$ channels

MONOAMINES

	Molecules synthesized from an amino acid by removing the carboxyl group and retaining the single amine group; also called biogenic amines
Serotonin or 5-hydroxytryptamine (5-HT)	Has various functions in the brain related to sleep, appetite, cognition (learning, memory), and mood; modulates actions of other neurotransmitters
Catecholamines	A distinct group of monoamines
Dopamine	Produces inhibitory activity in the brain; important roles in cognition (learning, memory), motivation, behavior, and mood; opens K$^+$ channels, closes Ca^{2+} channels
Norepinephrine (noradrenaline)	Neurotransmitter of PNS (sympathetic division of autonomic nervous system) and specific CNS regions
Epinephrine (adrenaline)	Has various effects in the CNS, especially the spinal cord, thalamus, and hypothalamus

NEUROPEPTIDES

	Small polypeptides act as signals to assist in and modulate communication among neurons in the CNS
Enkephalin	Helps regulate response to noxious and potentially harmful stimuli
Neuropeptide Y	Involved in memory regulation and energy balance (increased food intake and decreased physical activity)
Somatostatin	Inhibits activities of neurons in specific brain areas
Substance P	Assists with pain information transmission into the brain
Cholecystokinin (CCK)	Stimulates neurons in the brain to help mediate satiation (fullness) and repress hunger
Beta-endorphin	Prevents release of pain signals from neurons and fosters a feeling of well-being
Neurotensin	Helps control and moderate the effects of dopamine

OTHERS

Adenosine	Also part of a nucleotide, inhibits activities in certain CNS neurons
Nitric oxide	Involved in learning and memory; relaxes muscle in the digestive tract; important for relaxation of smooth muscle in blood vessels (vasodilation)

FIGURE **9–8** Neurons, neuropil, and the common glial cells of the CNS.

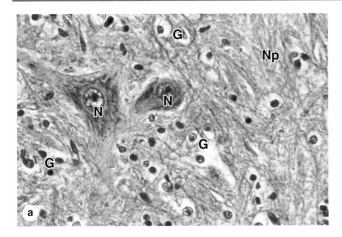

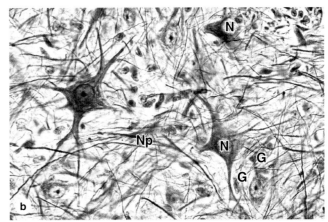

(a) Most neuronal cell bodies (**N**) in the CNS are larger than the much more numerous glial cells (**G**) that surround them. The various types of glial cells and their relationships with neurons are difficult to distinguish by most routine light microscopic methods. However, oligodendrocytes have condensed, rounded nuclei and unstained cytoplasm due to very abundant Golgi complexes, which stain poorly and are very likely represented by the cells with

those properties seen here. The other glial cells seen here similar in overall size, but with very little cytoplasm and more elongated or oval nuclei, are mostly astrocytes. Routine H&E staining does not allow neuropil to stand out well. (X200; H&E)

(b) With the use of gold staining for neurofibrils, neuropil (**Np**) is more apparent. (X200; Gold chloride and hematoxylin)

Functions attributed to astrocytes of various CNS regions include the following:

- Extending processes that associate with or cover synapses, affecting the formation, function, and plasticity of these structures
- Regulating the extracellular ionic concentrations around neurons, with particular importance in buffering extracellular K+ levels
- Guiding and physically supporting movements and locations of differentiating neurons during CNS development

- Extending fibrous processes with expanded **perivascular feet** that cover capillary endothelial cells and modulate blood flow and help move nutrients, wastes, and other metabolites between neurons and capillaries (Figure 9–9a)
- Forming a barrier layer of expanded protoplasmic processes, called the **glial limiting membrane**, which lines the meninges at the external CNS surface
- Filling tissue defects after CNS injury by proliferation to form an **astrocytic scar**.

TABLE **9–2** | Origin, location and principal functions of neuroglial cells.

Glial Cell Type	Origin	Location	Main Functions
Oligodendrocyte	Neural tube	CNS	Myelin production, electrical insulation
Astrocyte	Neural tube	CNS	Structural and metabolic support of neurons, especially at synapses; repair processes
Ependymal cell	Neural tube	Line ventricles and central canal of CNS	Aid production and movement of CSF
Microglia	Bone marrow (monocytes)	CNS	Defense and immune-related activities
Schwann cell	Neural crest	Peripheral nerves	Myelin production, electrical insulation
Satellite cells (of ganglia)	Neural crest	Peripheral ganglia	Structural and metabolic support for neuronal cell bodies

FIGURE **9–9** Glial cells of the CNS and PNS.

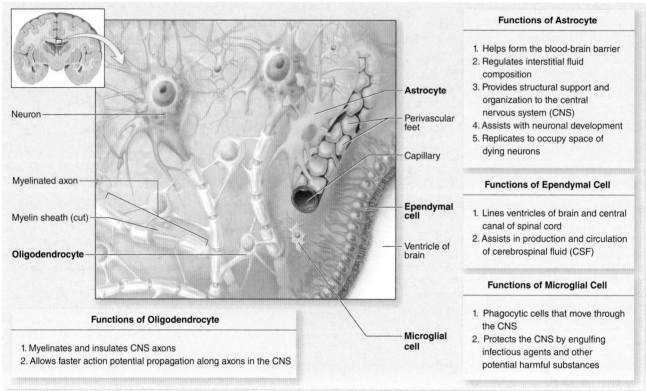

Functions of Astrocyte

1. Helps form the blood-brain barrier
2. Regulates interstitial fluid composition
3. Provides structural support and organization to the central nervous system (CNS)
4. Assists with neuronal development
5. Replicates to occupy space of dying neurons

Functions of Ependymal Cell

1. Lines ventricles of brain and central canal of spinal cord
2. Assists in production and circulation of cerebrospinal fluid (CSF)

Functions of Microglial Cell

1. Phagocytic cells that move through the CNS
2. Protects the CNS by engulfing infectious agents and other potential harmful substances

Functions of Oligodendrocyte

1. Myelinates and insulates CNS axons
2. Allows faster action potential propagation along axons in the CNS

(a)

Functions of Satellite Cell

1. Electrically insulates PNS cell bodies.
2. Regulates nutrient and waste exchange for cell bodies in ganglia

Functions of Neurolemmocyte

1. Surround and insulate PNS axons and myelinate those having large diameters
2. Allows for faster action potential propagation along an axon in the PNS

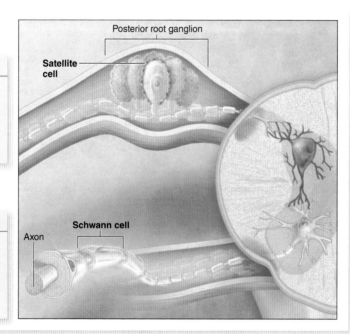

(b)

(a) There are four major kinds of glial cells in the CNS: **oligodendrocytes, astrocytes, ependymal cells, and microglial cells**. The interrelationships and major functions of these cells are shown diagrammatically here.

(b) Two glial cells occur in the PNS: **Schwann cells** (sometimes called neurolemmocytes), which surround peripheral nerve fibers, and **satellite cells**, which surround the nerve cell bodies and are thus found only in ganglia. Major functions of these cells are indicated.

FIGURE 9–10 Astrocytes.

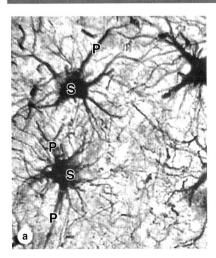
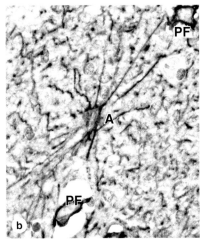
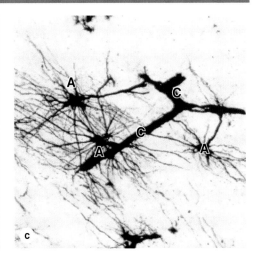

(a) Astrocytes are the most abundant glial cells of the CNS and are characterized by numerous cytoplasmic processes (**P**) radiating from the glial cell body or soma (**S**). Astrocytic processes are not seen with routine light microscope staining but are easily seen after gold staining. Morphology of the processes allows astrocytes to be classified as fibrous (relatively few and straight processes) or protoplasmic (numerous branching processes), but functional differences between these types are not clear. (X500; Gold chloride)

(b) All astrocytic processes contain intermediate filaments of GFAP, and antibodies against this protein provide a simple method to stain these cells, as seen here in a fibrous astrocyte (**A**) and its

processes. The small pieces of other GFAP-positive processes in the neuropil around this cell give an idea of the density of this glial cell and its processes in the CNS. Astrocytes form part of the blood-brain barrier (BBB) and help regulate entry of molecules and ions from blood into CNS tissue. Capillaries at the extreme upper right and lower left corners are enclosed by GFAP-positive perivascular feet (**PF**) at the ends of numerous astrocytic processes. (X500; Anti-GFAP immunoperoxidase and hematoxylin counterstain)

(c) A length of capillary (**C**) is shown here completely covered by silver-stained terminal processes extending from astrocytes (**A**). (X400; Rio Hortega silver)

Finally, astrocytes communicate directly with one another via gap junctions, forming a very large cellular network for the coordinated regulation of their various activities in different brain regions.

Alzheimer disease, a common type of dementia in the elderly, affects both neuronal perikarya and synapses within the cerebrum. Functional defects are due to **neurofibrillary tangles,** which are accumulations of tau protein associated with microtubules of the neuronal perikaryon and axon hillock regions, and **neuritic plaques,** which are dense aggregates of β-amyloid protein that form around the outside of these neuronal regions.

Ependymal Cells

Ependymal cells are columnar or cuboidal cells that line the fluid-filled ventricles of the brain and the central canal of the spinal cord (Figures 9–9a and 9–11). In some CNS locations, the apical ends of ependymal cells have cilia, which facilitate the movement of cerebrospinal fluid (CSF), and long microvilli, which are likely involved in absorption.

Ependymal cells are joined apically by apical junctional complexes similar to those of epithelial cells. However, unlike a true epithelium there is no basal lamina. Instead, the basal

ends of ependymal cells are elongated and extend branching processes into the adjacent neuropil.

Microglia

Less numerous than oligodendrocytes or astrocytes but nearly as common as neurons in some CNS regions, **microglia** are small cells with actively mobile processes evenly distributed throughout gray and white matter (Figures 9–9a and 9–12). Unlike other glial cells microglia migrate, with their processes scanning the neuropil and removing damaged or effete synapses or other fibrous components. Microglial cells also constitute the major mechanism of immune defense in the CNS, removing any microbial invaders and secreting a number of immunoregulatory cytokines. Microglia do not originate from neural progenitor cells like other glia, but from circulating blood monocytes, belonging to the same family as macrophages and other antigen-presenting cells.

Nuclei of microglial cells can often be recognized in routine hematoxylin and eosin (H&E) preparations by their small, dense, slightly elongated structure, which contrasts with the larger, spherical, more lightly stained nuclei of other glial cells. Immunohistochemistry using antibodies against cell surface antigens of immune cells demonstrates microglial processes. When activated by damage or microorganisms microglia retract their processes, proliferate, and assume the morphologic characteristics and functions of antigen-presenting cells (see Chapter 14).

FIGURE **9–11** Ependymal cells.

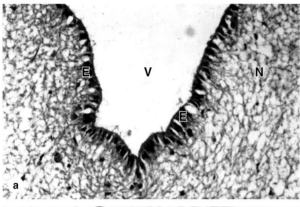

a

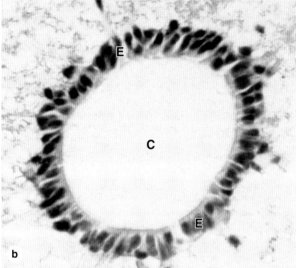

b

Ependymal cells are epithelial-like cells that form a single layer lining the fluid-filled ventricles and central canal of the CNS.
(a) Lining the ventricles of the cerebrum, columnar ependymal cells (**E**) extend cilia and microvilli from the apical surfaces into the ventricle (**V**). These modifications help circulate the CSF and monitor its contents. Ependymal cells have junctional complexes at their apical ends like those of epithelial cells but lack a basal lamina. The cells' basal ends are tapered, extending processes that branch and penetrate some distance into the adjacent neuropil (**N**). Other areas of ependyma are responsible for production of CSF. (X100; H&E)

(b) Ependymal cells (**E**) lining the central canal (**C**) of the spinal cord help move CSF in that CNS region. (X200; H&E)

>> MEDICAL APPLICATION

In **multiple sclerosis** (MS) the myelin sheaths surrounding axons are damaged by an autoimmune mechanism that interferes with the activity of the affected neurons and produces various neurologic problems. T lymphocytes and microglia, which phagocytose and degrade myelin debris, play major roles in progression of this disease. In MS, destructive actions of these cells exceed the capacity of oligodendrocytes to produce myelin and repair the myelin sheaths.

FIGURE **9–12** Microglial cells.

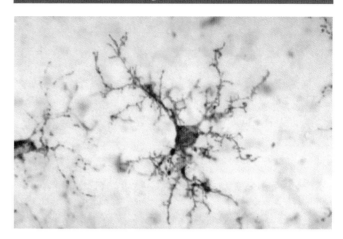

Microglia are monocyte-derived, antigen-presenting cells of the CNS, less numerous than astrocytes but nearly as common as neurons and evenly distributed in both gray and white matter. By immunohistochemistry, here using a monoclonal antibody against human leukocyte antigens (HLA) of immune-related cells, the short branching processes of microglia can be seen. Routine staining demonstrates only the small dark nuclei of the cells. Unlike other glia of the CNS, microglia are not interconnected; they are motile cells, constantly used in immune surveillance of CNS tissues. When activated by products of cell damage or by invading microorganisms, the cells retract their processes, begin phagocytosing the damage- or danger-related material, and behave as antigen-presenting cells. (X500; Antibody against HLA-DR and peroxidase)

(Used with permission from Wolfgang Streit, Department of Neuroscience, University of Florida College of Medicine, Gainesville.)

Schwann Cells

Schwann cells (named for 19th century German histologist Theodor Schwann), sometimes called neurolemmocytes, are found only in the PNS and differentiate from precursors in the neural crest. Schwann cells are the counterparts to oligodendrocytes of the CNS, having trophic interactions with axons and most importantly forming their **myelin sheaths**. However unlike an oligodendrocyte, a Schwann cell forms myelin around a portion of only one axon. Figure 9–9b shows a series of Schwann cells sheathing the full length of an axon, a process described more fully with peripheral nerves.

Satellite Cells of Ganglia

Also derived from the embryonic neural crest, small **satellite cells** form a thin, intimate glial layer around each large neuronal cell body in the ganglia of the PNS (Figures 9–9b and 9–13). Satellite cells exert a trophic or supportive effect on these neurons, insulating, nourishing, and regulating their microenvironments.

FIGURE **9–13** Satellite cells around neurons of ganglia in the PNS.

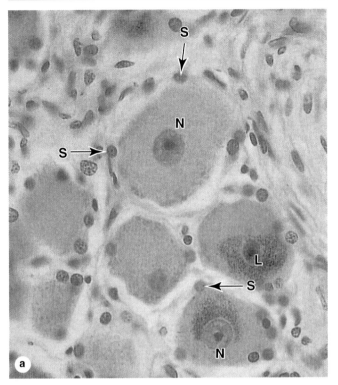

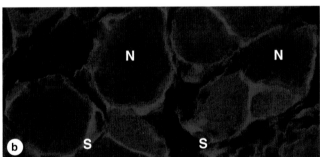

Satellite cells are very closely associated with neuronal cell bodies in sensory and autonomic ganglia of the PNS and support these cells in various ways.

(a) Nuclei of the many satellite cells (**S**) surrounding the perikarya of neurons (**N**) in an autonomic ganglion can be seen by light microscopy, but their cytoplasmic extensions are too thin to see with H&E staining. These long-lived neurons commonly accumulate brown lipofuscin (**L**). (X560; H&E)

(b) Immunofluorescent staining of satellite cells (**S**) reveals the cytoplasmic sheets extending from these cells and surrounding the neuronal cell bodies (**N**). The layer of satellite cells around each soma is continuous with the myelin sheath around the axon. Like the effect of Schwann cells on axons, satellite glial cells insulate, nourish, and regulate the microenvironment of the neuronal cell bodies. (X600; Rhodamine red-labeled antibody against glutamine synthetase)

(*Used with permission from Menachem Hanani, Laboratory of Experimental Surgery, Hadassah University Hospital, Jerusalem, Israel.*)

CENTRAL NERVOUS SYSTEM

The major structures comprising the CNS are the **cerebrum**, **cerebellum**, and **spinal cord** (Figure 9–1). The CNS is completely covered by connective tissue layers, the meninges, but CNS tissue contains very little collagen or similar material, making it relatively soft and easily damaged by injuries affecting the protective skull or vertebral bones. Most CNS neurons and their functional organization are more appropriately covered in neuroscience rather than histology courses, but certain important cells and basic topics will be introduced here.

Many structural features of CNS tissues can be seen in unstained, freshly dissected specimens. Many regions show organized areas of **white matter** and **gray matter**, differences caused by the differential distribution of lipid-rich myelin. The main components of white matter are myelinated axons (Figure 9–14), often grouped together as **tracts**, and the myelin-producing oligodendrocytes. Astrocytes and microglia are also present, but very few neuronal cell bodies. Gray matter contains abundant neuronal cell bodies, dendrites, astrocytes, and microglial cells, and is where most synapses occur. Gray matter makes up the thick cortex or surface layer of both the cerebrum and the cerebellum; most white matter is found in deeper regions. Deep within the brain are localized, variously shaped darker areas called the **cerebral nuclei**, each containing large numbers of aggregated neuronal cell bodies.

In the folded **cerebral cortex** neuroscientists recognize six layers of neurons with different sizes and shapes. The most conspicuous of these cells are the efferent **pyramidal neurons** (Figure 9–15). Neurons of the cerebral cortex function in the integration of sensory information and the initiation of voluntary motor responses.

The sharply folded **cerebellar cortex** coordinates muscular activity throughout the body and is organized with three layers (Figure 9–16):

- A thick outer **molecular layer** has much neuropil and scattered neuronal cell bodies.
- A thin middle layer consists only of very large neurons called **Purkinje cells** (named for the 19th century Czech histologist Jan Purkinje). These are conspicuous even in H&E-stained sections, and their dendrites extend throughout the molecular layer as a branching basket of nerve fibers (Figures 9–16c and d).
- A thick inner **granular layer** contains various very small, densely packed neurons (including granule cells, with diameters of only 4-5 μm) and little neuropil.

In cross sections of the **spinal cord** the white matter is peripheral and the gray matter forms a deeper, H-shaped mass (Figure 9–17). The two anterior projections of this gray matter, the **anterior horns**, contain cell bodies of very large motor neurons whose axons make up the ventral roots of spinal nerves. The two **posterior horns** contain interneurons which receive sensory fibers from neurons in the spinal (dorsal root) ganglia. Near the middle of the cord the gray matter surrounds a small **central canal**, which develops from the lumen of the neural tube, is continuous with the ventricles of the brain, is lined by ependymal cells, and contains CSF.

FIGURE **9–14** White versus gray matter.

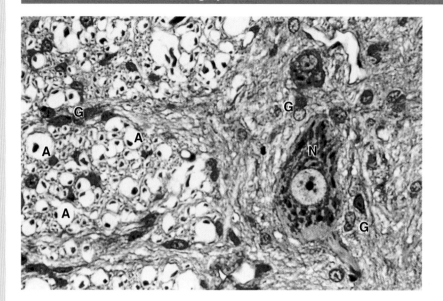

A cross section of H&E-stained spinal cord shows the transition between white matter (left region) and gray matter (right). The gray matter has many glial cells (**G**), neuronal cell bodies (**N**), and neuropil; white matter also contains glia (**G**) but consists mainly of axons (**A**) whose myelin sheaths were lost during preparation, leaving the round empty spaces shown. Each such space surrounds a dark-stained spot that is a small section of the axon. (X400)

FIGURE **9–15** Cerebral cortex.

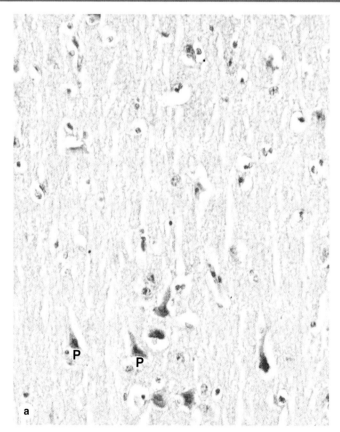

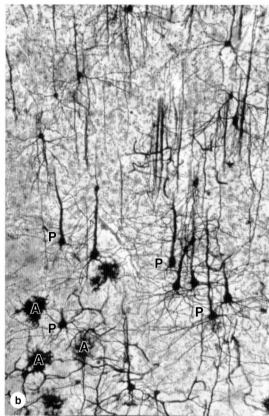

(a) Important neurons of the cerebrum are the pyramidal neurons (**P**), which are arranged vertically and interspersed with numerous smaller glial cells, mostly astrocytes, in the eosinophilic neuropil. (X200; H&E)

(b) From the apical ends of pyramidal neurons (**P**), long dendrites extend in the direction of the cortical surface, which can be best seen in thick silver-stained sections in which only a few other protoplasmic astrocytes (**A**) cells are seen. (X200; Silver)

FIGURE **9–16** Cerebellum.

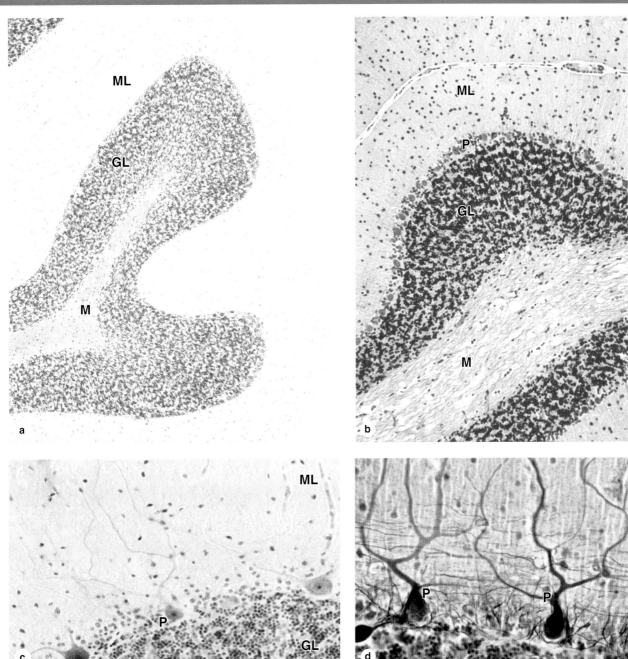

(a) The cerebellar cortex is convoluted with many distinctive small folds, each supported at its center by tracts of white matter in the cerebellar medulla (**M**). Each fold has distinct molecular layers (**ML**) and granular layers (**GL**). (X6; Cresyl violet)

(b) Higher magnification shows that the granular layer (**GL**) immediately surrounding the medulla (**M**) is densely packed with several different types of very small rounded neuronal cell bodies. The outer molecular layer (**ML**) consists of neuropil with fewer, much more scattered small neurons. At the interface of these two regions a layer of large Purkinje neuron (**P**) perikarya can be seen. (X20; H&E)

(c) A single intervening layer contains the very large cell bodies of unique Purkinje neurons (**P**), whose axons pass through the granular layer (**GL**) to join tracts in the medulla and whose multiple branching dendrites ramify throughout the molecular layer (**ML**). Dendrites are not seen well with H&E staining. (X40; H&E)

(d) With appropriate silver staining dendrites from each large Purkinje cell (**P**) are shown to have hundreds of small branches, each covered with hundreds of dendritic spines. Axons from the small neurons of the granular layer are unmyelinated and run together into the molecular layer where they form synapses with the dendritic spines of Purkinje cells. (X40; Silver)

FIGURE **9–17** Spinal cord.

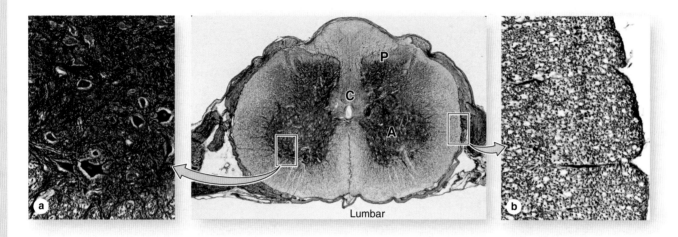

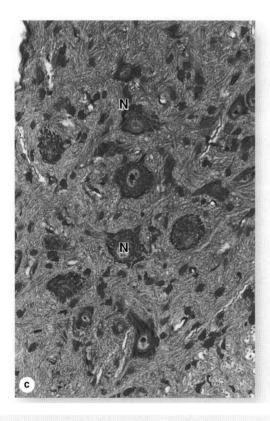

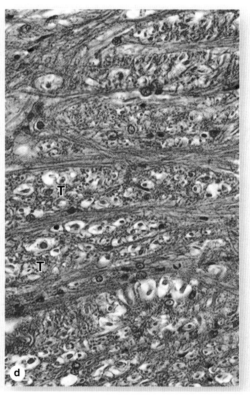

The spinal cord varies slightly in diameter along its length but in cross section always shows bilateral symmetry around the small, CSF-filled central canal (**C**). Unlike the cerebrum and cerebellum, in the spinal cord the gray matter is internal, forming a roughly H-shaped structure that consists of two posterior (**P**) horns (sensory) and two anterior (**A**) (motor) horns, all joined by the gray commissure around the central canal.

(a) The gray matter contains abundant astrocytes and large neuronal cell bodies, especially those of motor neurons in the ventral horns.

(b) The white matter surrounds the gray matter and contains primarily oligodendrocytes and tracts of myelinated axons

running along the length of the cord. (Center X5, a, b X100; All silver-stained)

(c) With H&E staining the large motor neurons (**N**) of the ventral horns show large nuclei, prominent nucleoli, and cytoplasm rich in Nissl substance, all of which indicate extensive protein synthesis to maintain the axons of these cells that extend great distances.

(d) In the white commissure ventral to the central canal, tracts (**T**) run lengthwise along the cord, seen here in cross section with empty myelin sheaths surrounding axons, as well as small tracts running from one side of the cord to the other. (Both X200; H&E)

Meninges

The skull and the vertebral column protect the CNS, but between the bone and nervous tissue are membranes of connective tissue called the **meninges**. Three meningeal layers are distinguished: the dura, arachnoid, and pia maters (Figures 9–18 and 9–19).

Dura Mater

The thick external **dura mater** (L. *dura mater*, tough mother) consists of dense irregular connective tissue organized as an outer periosteal layer continuous with the periosteum of the skull, and an inner meningeal layer. These two layers are usually fused, but along the superior sagittal surface and other specific areas around the brain they separate to form the blood-filled **dural venous sinuses** (Figure 9–19). Around the spinal cord the dura mater is separated from the periosteum of the vertebrae by the **epidural space**, which contains a plexus of thin-walled veins and loose connective tissue (Figure 9–18). The dura mater may be separated from the arachnoid by formation of a thin subdural space.

Arachnoid

The **arachnoid** (Gr. *arachnoeides*, spider web-like) has two components: (1) a sheet of connective tissue in contact with the dura mater and (2) a system of loosely arranged trabeculae composed of collagen and fibroblasts, continuous with the underlying pia mater layer. Surrounding these trabeculae is a large, sponge-like cavity, the **subarachnoid space**, filled with CSF. This fluid-filled space helps cushion and protect the CNS from minor trauma. The subarachnoid space communicates with the ventricles of the brain where the CSF is produced.

The connective tissue of the arachnoid is said to be avascular because it lacks nutritive capillaries, but larger blood vessels run through it (Figures 9–18 and 9–19). Because the arachnoid has fewer trabeculae in the spinal cord, it can be more clearly distinguished from the pia mater in that area. The arachnoid and the pia mater are intimately associated and are often considered a single membrane called the pia-arachnoid.

In some areas, the arachnoid penetrates the dura mater and protrudes into blood-filled dural venous sinuses located there (Figure 9–19). These CSF-filled protrusions, which are

FIGURE **9–18** **Spinal cord and meninges.**

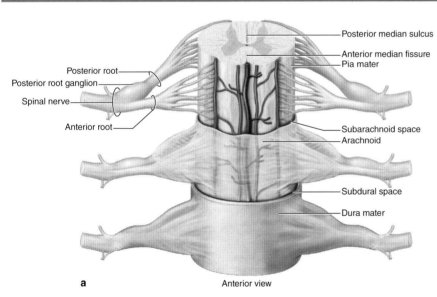
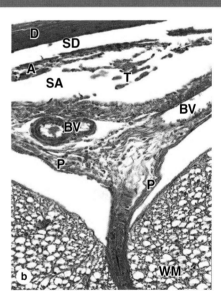

a   Anterior view

(a) A diagram of the spinal cord indicates the relationship of the three meningeal layers of connective tissue: the innermost **pia mater**, the **arachnoid**, and the **dura mater**. Also depicted are the blood vessels coursing through the subarachnoid space and the nerve rootlets that fuse to form the posterior and anterior roots of the spinal nerves. The posterior root ganglia contain the cell bodies of sensory nerve fibers and are located in intervertebral foramina.

(b) Section of an area near the anterior median fissure showing the tough dura mater (**D**). Surrounding the dura, the epidural space (not shown) contains cushioning adipose tissue and vascular plexuses. The subdural space (**SD**) is an artifact created by separation of the dura from underlying tissue. The middle meningeal layer

is the thicker weblike arachnoid mater (**A**) containing the large subarachnoid space (**SA**) and connective tissue trabeculae (**T**). The subarachnoid space is filled with CSF and the arachnoid acts as a shock-absorbing pad between the CNS and bone. Fairly large blood vessels (**BV**) course through the arachnoid. The innermost pia mater (**P**) is thin and is not clearly separate from the arachnoid; together, they are sometimes referred to as the pia-arachnoid or the leptomeninges. The space between the pia and the white matter (**WM**) of the spinal cord here is an artifact created during dissection; normally the pia is very closely applied to a layer of astrocytic processes at the surface of the CNS tissue. (X100; H&E)

FIGURE **9–19** Meninges around the brain.

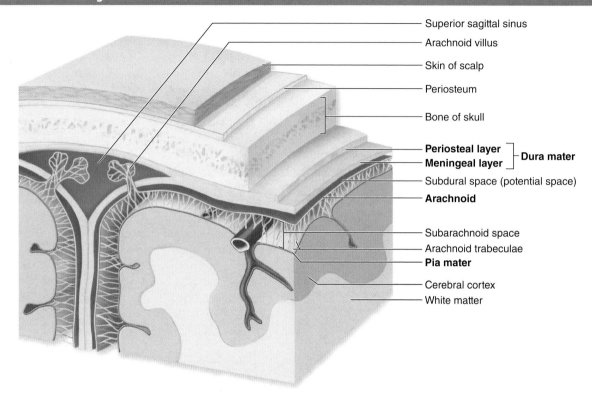

- Superior sagittal sinus
- Arachnoid villus
- Skin of scalp
- Periosteum
- Bone of skull
- **Periosteal layer** ⎤
- **Meningeal layer** ⎦ ⎱ **Dura mater**
- Subdural space (potential space)
- **Arachnoid**
- Subarachnoid space
- Arachnoid trabeculae
- **Pia mater**
- Cerebral cortex
- White matter

The **dura, arachnoid, and pia maters** also surround the brain and as shown here the relationships among the cranial meninges are similar to those of the spinal cord. The diagram includes **arachnoid villi**, which are outpocketings of arachnoid away from the brain, which penetrate the dura mater and enter blood-filled **venous sinuses** located within that layer. The arachnoid villi function in releasing excess CSF into the blood. Blood vessels from the arachnoid branch into smaller arteries and veins that enter brain tissue carrying oxygen and nutrients. These small vessels are initially covered with pia mater, but as capillaries they are covered only by the perivascular feet of astrocytes.

covered by the vascular endothelial cells lining the sinuses, are called **arachnoid villi** and function as sites for absorption of CSF into the blood of the venous sinuses.

Pia Mater

The innermost **pia mater** (L. *pia mater*, tender mother) consists of flattened, mesenchymally derived cells closely applied to the entire surface of the CNS tissue. The pia does not directly contact nerve cells or fibers, being separated from the neural elements by the very thin superficial layer of astrocytic processes (the glial limiting membrane, or glia limitans), which adheres firmly to the pia mater. Together, the pia mater and the layer of astrocytic end feet form a physical barrier separating CNS tissue from CSF in the subarachnoid space (Figure 9–19).

Blood vessels penetrate CNS tissue through long **perivascular spaces** covered by pia mater, although the pia disappears when the blood vessels branch to form the small capillaries. However, these capillaries remain completely covered by the perivascular layer of astrocytic processes (Figures 9–9a and 9–10c).

Blood-Brain Barrier

The **blood-brain barrier** (BBB) is a functional barrier that allows much tighter control than that in most tissues over the passage of substances moving from blood into the CNS tissue. The main structural component of the BBB is the **capillary endothelium**, in which the cells are tightly sealed together with well-developed occluding junctions, with little or no transcytosis activity, and surrounded by the basement membrane. The **limiting layer of perivascular astrocytic feet** that envelops the basement membrane of capillaries in most CNS regions (Figure 9–10c) contributes to the BBB and further regulates passage of molecules and ions from blood to brain.

The BBB protects neurons and glia from bacterial toxins, infectious agents, and other exogenous substances, and helps maintain the stable composition and constant balance of ions in the interstitial fluid required for normal neuronal function. The BBB is not present in regions of the hypothalamus where plasma components are monitored, in the posterior pituitary which releases hormones, or in the choroid plexus where CSF is produced.

Choroid Plexus

The **choroid plexus** consists of highly vascular tissue, elaborately folded and projecting into the large ventricles of the brain (Figure 9–20a). It is found in the roofs of the third and fourth ventricles and in parts of the two lateral ventricular walls, all regions in which the ependymal lining directly contacts the pia mater.

Each villus of the choroid plexus contains a thin layer of well-vascularized pia mater covered by cuboidal ependymal cells (Figure 9–20b). The function of the choroid plexus is to remove water from blood and release it as the **CSF**. CSF is clear, contains Na⁺, K⁺, and Cl⁻ ions but very little protein, and its only cells are normally very sparse lymphocytes. It is produced continuously and it completely fills the ventricles, the central canal of the spinal cord, the subarachnoid and perivascular spaces. It provides the ions required for CNS neuronal activity and in the arachnoid serves to help absorb mechanical shocks. Arachnoid villi (Figure 9–19) provide the main pathway for absorption of CSF back into the venous circulation. There are very few lymphatic vessels in CNS tissue.

> ## ❯❯ MEDICAL APPLICATION
>
> A decrease in the absorption of CSF or a blockage of outflow from the ventricles during fetal or postnatal development results in the condition known as **hydrocephalus** (Gr. *hydro*, water + *kephale*, head), which promotes a progressive enlargement of the head followed by mental impairment.

FIGURE 9–20   Choroid plexus.

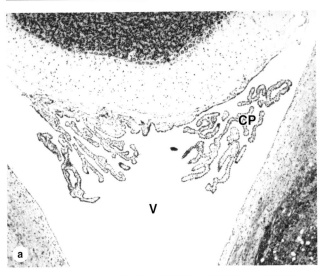

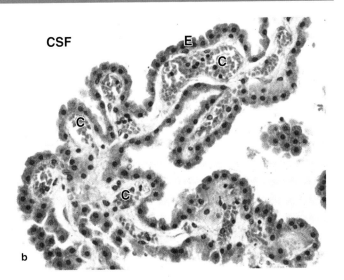

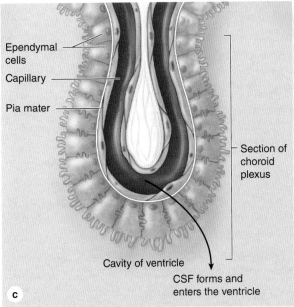

Ependymal cells
Capillary
Pia mater
Section of choroid plexus
Cavity of ventricle
CSF forms and enters the ventricle

The choroid plexus consists of ependyma and vascularized pia mater and projects many thin folds from certain walls of the ventricles.

(a) Section of the bilateral choroid plexus (**CP**) projecting into the fourth ventricle (**V**) near the cerebellum. (X12; Kluver-Barrera stain)

(b) At higher magnification each fold of choroid plexus is seen to be well-vascularized with large capillaries (**C**) and covered by a continuous layer of cuboidal ependymal cells (**E**). (X150)

(c) The choroid plexus is specialized for transport of water and ions across the capillary endothelium and ependymal layer and the elaboration of these as CSF.

› PERIPHERAL NERVOUS SYSTEM

The main components of the peripheral nervous system (PNS) are the **nerves**, **ganglia**, and **nerve endings**. Nerves are bundles of nerve fibers (axons) surrounded by Schwann cells and layers of connective tissue.

Nerve Fibers

Nerve fibers are analogous to tracts in the CNS, containing axons enclosed within sheaths of glial cells specialized to facilitate axonal function. In peripheral nerve fibers, axons are sheathed by **Schwann cells**, or neurolemmocytes (Figure 9–9b). The sheath may or may not form myelin around the axons, depending on their diameter.

Myelinated Fibers

As axons of large diameter grow in the PNS, they are engulfed along their length by a series of differentiating neurolemmocytes and become **myelinated nerve fibers**. The plasma membrane of each covering Schwann cell fuses with itself at an area termed the mesaxon and a wide, flattened process of the cell continues to extend itself, moving circumferentially around the axon many times (Figure 9–21). The multiple layers of Schwann cell membrane unite as a thick **myelin sheath**. Composed mainly of lipid bilayers and membrane proteins, myelin is a large lipoprotein complex that, like cell membranes, is partly removed by standard histologic procedures (Figures 9–14 and 9–17d). Unlike oligodendrocytes of the CNS, a Schwann cell forms myelin around only a portion of one axon.

With high-magnification TEM, the myelin sheath appears as a thick electron-dense axonal covering in which the concentric membrane layers may be visible (Figure 9–22). The prominent electron-dense layers visible ultrastructurally in the sheath, the **major dense lines**, represent the fused, protein-rich cytoplasmic surfaces of the Schwann cell membrane. Along the myelin sheath, these surfaces periodically separate slightly to allow transient movement of cytoplasm for membrane maintenance; at these **myelin clefts** (or Schmidt-Lanterman clefts) the major dense lines temporarily disappear (Figure 9–23).

FIGURE **9–21** Myelination of large-diameter PNS axons.

① Schwann cell becomes aligned along the axon and extends a wide cytoplasmic process to encircle it.

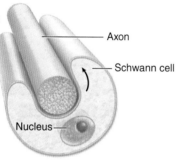

Axon

Schwann cell

Nucleus

② The growing process completely encloses the axon but continues its spiral extension.

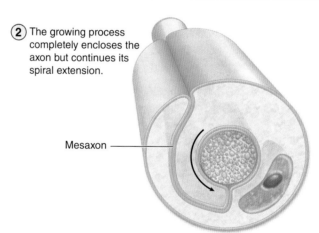

Mesaxon

③ The spiral wrappings become compacted layers of cell membrane (myelin) as cytoplasm leaves the growing process.

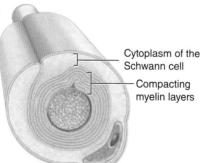

Cytoplasm of the Schwann cell

Compacting myelin layers

④ The mature Schwann cell myelin sheath has up to 100 lamellae, with most cytoplasm in the outermost layer with the cell body.

Myelin sheath

Schwann cell nucleus

A Schwann cell (neurolemmocyte) engulfs one portion along the length of a large-diameter axon. The Schwann cell membrane fuses around the axon and one thin extension of the Schwann cell elongates greatly and wraps itself repeatedly around the axon to form multiple, compacted layers. The Schwann cell membrane wrappings constitute the myelin sheath, with the Schwann cell body always on its outer surface. The myelin layers are very rich in lipid, and provide insulation and facilitate formation of action potentials along the axolemma.

FIGURE **9–22** Ultrastructure of myelinated and unmyelinated fibers.

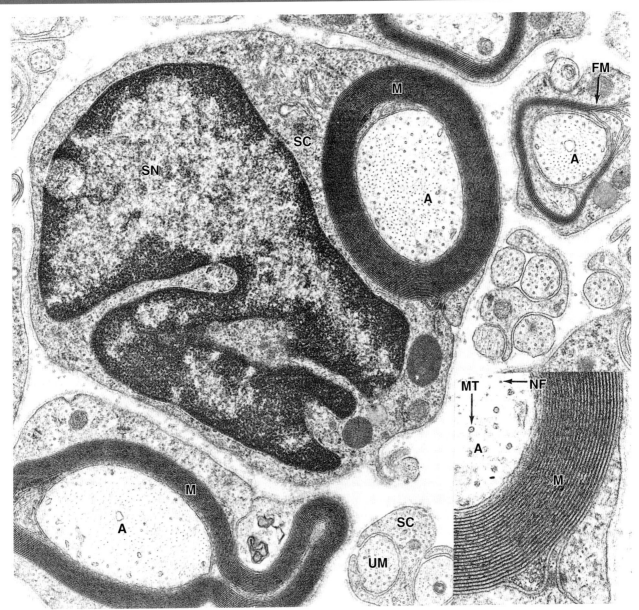

Cross section of PNS fibers in the TEM reveals differences between myelinated and unmyelinated axons. Large axons (**A**) are wrapped in a thick myelin sheath (**M**) of multiple layers of Schwann cell membrane.

The inset shows a portion of myelin at higher magnification in which the major dense lines of individual membrane layers can be distinguished, as well as the neurofilaments (**NF**) and microtubules (**MT**) in the axoplasm (**A**). At the center of the photo is a Schwann cell showing its active nucleus (**SN**) and Golgi-rich cytoplasm (**SC**). At the right is an axon around which myelin is still forming (**FM**).

Unmyelinated axons (**UM**) are much smaller in diameter, and many such fibers may be engulfed by a single Schwann cell (**SC**). The glial cell does not form myelin wrappings around such small axons but simply encloses them. Whether it forms myelin or not, each Schwann cell is surrounded, as shown, by an external lamina containing type IV collagen and laminin like the basal laminae of epithelial cells. (X28,000, inset X70,000)

(*Used with permission from Dr Mary Bartlett Bunge, The Miami Project to Cure Paralysis, University of Miami Miller School of Medicine, Miami, FL.*)

FIGURE **9–23** **Myelin maintenance and nodes of Ranvier.**

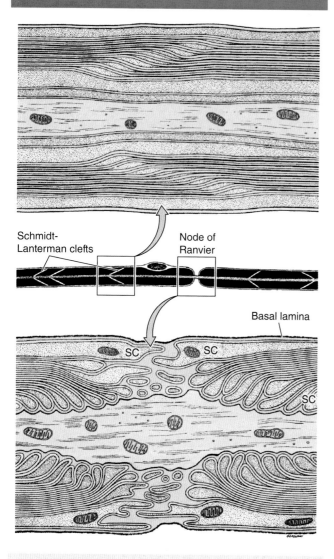

Schmidt-Lanterman clefts

Node of Ranvier

Basal lamina

SC

SC

SC

The middle diagram shows schematically a myelinated peripheral nerve fiber as seen under the light microscope. The axon is enveloped by the myelin sheath, which, in addition to membrane, contains some Schwann cell cytoplasm in spaces called **Schmidt-Lanterman or myelin clefts** between the major dense lines of membranes.

The upper diagram shows one set of such clefts ultrastructurally. The clefts contain Schwann cell cytoplasm that was not displaced to the cell body during myelin formation. This cytoplasm moves slowly along the myelin sheath, opening temporary spaces (the clefts) that allow renewal of some membrane components as needed for maintenance of the sheath.

The lower diagram depicts the ultrastructure of a single node of Ranvier or nodal gap. Interdigitating processes extending from the outer layers of the Schwann cells (SC) partly cover and contact the axolemma at the nodal gap. This contact acts as a partial barrier to the movement of materials in and out of the periaxonal space between the axolemma and the Schwann sheath. The basal or external lamina around Schwann cells is continuous over the nodal gap. The axolemma at nodal gaps has abundant voltage-gated Na+ channels important for impulse conductance in these axons.

Faintly seen ultrastructurally in the light staining layers are the intraperiod lines that represent the apposed outer bilayers of the Schwann cell membrane.

Membranes of Schwann cells have a higher proportion of lipids than do other cell membranes, and the myelin sheath serves to insulate axons and maintain a constant ionic microenvironment most suitable for action potentials. Between adjacent Schwann cells on an axon the myelin sheath shows small **nodes of Ranvier** (or **nodal gaps**, Figures 9–9b, 9–23 and 9–24), where the axon is only partially covered by interdigitating Schwann cell processes. At these nodes the axolemma is exposed to ions in the interstitial fluid and has a much higher concentration of voltage-gated Na+ channels, which renew the action potential and produce **saltatory conduction** (L. *saltare*, to jump) of nerve impulses, their rapid movement from node to node. The length of axon ensheathed by one Schwann cell, the **internodal segment**, varies directly with axonal diameter and ranges from 300 to 1500 μm.

Unmyelinated Fibers

Unlike the CNS where many short axons are not myelinated at all but run free among the other neuronal and glial processes, the smallest-diameter axons of peripheral nerves are still enveloped within simple folds of Schwann cells (Figure 9–25). In these **unmyelinated fibers** the glial cell does not form the multiple wrapping of a myelin sheath (Figure 9–21). In unmyelinated fibers, each Schwann cell can enclose portions of many axons with small diameters. Without the thick myelin sheath, nodes of Ranvier are not seen along unmyelinated nerve fibers. Moreover, these small-diameter axons have evenly distributed voltage-gated ion channels; their impulse conduction is not saltatory and is much slower than that of myelinated axons.

Nerve Organization

In the PNS nerve fibers are grouped into bundles to form **nerves**. Except for very thin nerves containing only unmyelinated fibers, nerves have a whitish, glistening appearance because of their myelin and collagen content.

Axons and Schwann cells are enclosed within layers of connective tissue (Figures 9–24, 9–26, and 9–27). Immediately around the external lamina of the Schwann cells is a thin layer called the **endoneurium**, consisting of reticular fibers, scattered fibroblasts, and capillaries. Groups of axons with Schwann cells and endoneurium are bundled together as **fascicles** by a sleeve of **perineurium**, containing flat fibrocytes with their edges sealed together by tight junctions. From two to six layers of these unique connective tissue cells regulate diffusion into the fascicle and make up the **blood-nerve barrier** that helps maintain the fibers' microenvironment. Externally, peripheral nerves have a dense, irregular fibrous coat called the **epineurium**, which extends deeply to fill the space between fascicles.

Very small nerves consist of one fascicle (Figure 9–28). Small nerves can be found in sections of many organs and often show a winding disposition in connective tissue.

FIGURE **9–24** Node of Ranvier and endoneurium.

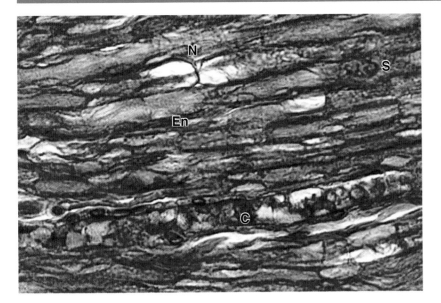

A longitudinally oriented nerve shows one node of Ranvier (**N**) with the axon visible. Collagen of the sparse endoneurium (**En**), blue in this trichrome stain, surrounds the Schwann cells and a capillary (**C**). At least one Schwann cell nucleus (**S**) is also clearly seen. (X400; Mallory trichome)

Peripheral nerves establish communication between centers in the CNS and the sense organs and effectors (muscles, glands, etc). They generally contain both afferent and efferent fibers. **Afferent** fibers carry information from internal body regions and the environment to the CNS. **Efferent** fibers carry impulses from the CNS to effector organs commanded by these centers. Nerves possessing only sensory fibers are called **sensory nerves**; those composed only of fibers carrying impulses to the effectors are called **motor nerves**. Most nerves have both sensory and motor fibers and are called **mixed nerves**, usually also with both myelinated and unmyelinated axons.

FIGURE **9–25** Unmyelinated nerves.

Unmyelinated axons

① Schwann cell starts to envelop multiple axons.

② The unmyelinated axons are enveloped by the Schwann cell, but there are *no* myelin sheath wraps around each axon.

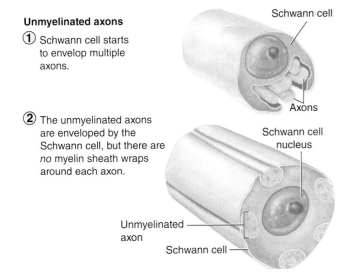

During development, portions of several small-diameter axons are engulfed by one Schwann cell. Subsequently the axons are separated and each typically becomes enclosed within its own fold of Schwann cell surface. No myelin is formed by wrapping. Small-diameter axons utilize action potentials whose formation and maintenance do not depend on the insulation provided by the myelin sheath required by large-diameter axons.

Ganglia

Ganglia are typically ovoid structures containing neuronal cell bodies and their surrounding glial satellite cells supported by delicate connective tissue and surrounded by a denser capsule. Because they serve as relay stations to transmit nerve impulses, at least one nerve enters and another exits from each ganglion. The direction of the nerve impulse determines whether the ganglion will be a **sensory** or an **autonomic** ganglion.

Sensory Ganglia

Sensory ganglia receive afferent impulses that go to the CNS. Sensory ganglia are associated with both cranial nerves (cranial ganglia) and the dorsal roots of the spinal nerves (spinal ganglia). The large neuronal cell bodies of ganglia (Figure 9–29) are associated with thin, sheet-like extensions of small glial **satellite cells** (Figures 9–9b and 9–13). Sensory ganglia are supported by a distinct connective tissue capsule and an internal framework continuous with the connective tissue layers of the nerves. The neurons of these ganglia are pseudounipolar and relay information from the ganglion's nerve endings to the gray matter of the spinal cord via synapses with local neurons.

FIGURE 9–26 Peripheral nerve connective tissue: Epi-, peri-, and endoneurium.

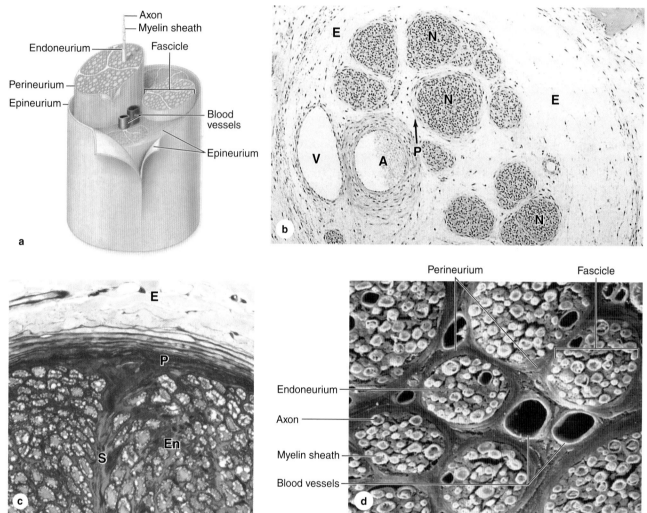

(a) The diagram shows the relationship among these three connective tissue layers in large peripheral nerves. The epineurium (E) consists of a dense superficial region and a looser deep region that contains the larger blood vessels.

(b) The micrograph shows a small vein (V) and artery (A) in the deep epineurium (E). Nerve fibers (N) are bundled in fascicles. Each fascicle is surrounded by the perineurium (P), consisting of a few layers of unusual squamous fibroblastic cells that are all joined at the peripheries by tight junctions. The resulting blood-nerve barrier helps regulate the microenvironment inside the fascicle. Axons and Schwann cells are in turn surrounded by a thin layer of endoneurium. (X140; H&E)

(c) As shown here and in the diagram, septa (S) of connective tissue often extend from the perineurium into larger fascicles. The endoneurium (En) and lamellar nature of the perineurium (P) are also shown at this magnification, along with some adjacent epineurium (E). (X200; PT)

(d) SEM of transverse sections of a large peripheral nerve showing several fascicles, each surrounded by perineurium and packed with endoneurium around the individual myelin sheaths. Each fascicle contains at least one capillary. Endothelial cells of these capillaries are tightly joined as part of the blood-nerve barrier and regulate the kinds of plasma substances released to the endoneurium. Larger blood vessels course through the deep epineurium that fills the space around the perineurium and fascicles. (X450)

Autonomic Ganglia

Autonomic (Gr. *autos,* self + *nomos,* law) nerves effect the activity of smooth muscle, the secretion of some glands, heart rate, and many other involuntary activities by which the body maintains a constant internal environment (**homeostasis**).

Autonomic ganglia are small bulbous dilations in autonomic nerves, usually with multipolar neurons. Some are located within certain organs, especially in the walls of the digestive tract, where they constitute the **intramural ganglia**. The capsules of these ganglia may be poorly defined among

FIGURE **9–27** Peripheral nerve ultrastructure.

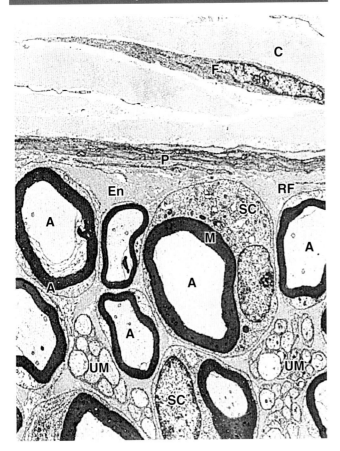

This low-magnification TEM shows a fibroblast (**F**) surrounded by collagen (**C**) in the epineurium (**E**) and three layers of flattened cells in the perineurium (**P**) which form another part of the blood-nerve barrier. Inside the perineurium the endoneurium (**En**) is rich in reticulin fibers (**RF**) that surround all Schwann cells. Nuclei of two Schwann cells (**SC**) of myelinated axons (**A**) are visible as well as many unmyelinated axons (**UM**) within Schwann cells. (X1200)

the local connective tissue. A layer of satellite cells also envelops the neurons of autonomic ganglia (Figure 9–29), although these may also be inconspicuous in intramural ganglia.

Autonomic nerves use two-neuron circuits. The first neuron of the chain, with the **preganglionic fiber**, is located in the CNS. Its axon forms a synapse with **postganglionic fibers** of the second multipolar neuron in the chain located in a peripheral ganglion system. The chemical mediator present in the synaptic vesicles of all preganglionic axons is acetylcholine.

As indicated earlier autonomic nerves make up the **autonomic nervous system**. This has two parts: the **sympathetic** and the **parasympathetic divisions**. Neuronal cell bodies of preganglionic sympathetic nerves are located in the thoracic and lumbar segments of the spinal cord and those of the parasympathetic division are in the medulla and midbrain and in

the sacral portion of the spinal cord. Sympathetic second neurons are located in small ganglia along the vertebral column, while second neurons of the parasympathetic series are found in very small ganglia always located near or within the effector organs, for example in the walls of the stomach and intestines. Parasympathetic ganglia may lack distinct capsules altogether, perikarya and associated satellite cells simply forming a loosely organized plexus within the surrounding connective tissue.

▶ NEURAL PLASTICITY & REGENERATION

Despite its general stability, the nervous system exhibits neuronal differentiation and formation of new synapses even in adults. Embryonic development of the nervous system produces an excess of differentiating neurons, and the cells that do not establish correct synapses with other neurons are eliminated by apoptosis. In adult mammals after an injury, the neuronal circuits may be reorganized by the growth of neuronal processes, forming new synapses to replace ones lost by injury. Thus, new communications are established with some degree of functional recovery. This **neural plasticity** and reformation of processes are controlled by several growth factors produced by both neurons and glial cells in a family of proteins called **neurotrophins**.

Neuronal stem cells are present in the adult CNS, located in part among the cells of the ependyma, which can supply new neurons, astrocytes, and oligodendrocytes. Fully differentiated, interconnected CNS neurons cannot temporarily disengage these connections and divide to replace cells lost by injury or disease; the potential of neural stem cells to allow tissue regeneration and functional recovery within the CNS components is a subject of intense investigation. Astrocytes do proliferate at injured sites and these growing cells can interfere with successful axonal regeneration in structures such as spinal cord tracts.

In the histologically much simpler peripheral nerves, injured axons have a much greater potential for regeneration and return of function. If the cell bodies are intact, damaged, or severed PNS axons can regenerate as shown in the sequence of diagrams in Figure 9–30. Distal portions of axons, isolated from their source of new proteins and organelles, degenerate; the surrounding Schwann cells dedifferentiate, shed the myelin sheaths, and proliferate within the surrounding layers of connective tissue. Cellular debris including shed myelin is removed by blood-derived macrophages, which also secrete neurotrophins to promote anabolic events of axon regeneration.

The onset of regeneration is signaled by changes in the perikaryon that characterize the process of **chromatolysis**: the cell body swells slightly, Nissl substance is initially diminished, and the nucleus migrates to a peripheral position within the perikaryon. The proximal segment of the axon close to the wound degenerates for a short distance, but begins to grow again distally as new Nissl substance appears and debris is removed. The new Schwann cells align to serve as guides for the regrowing axons and produce polypeptide factors that

FIGURE **9–28** Small nerves.

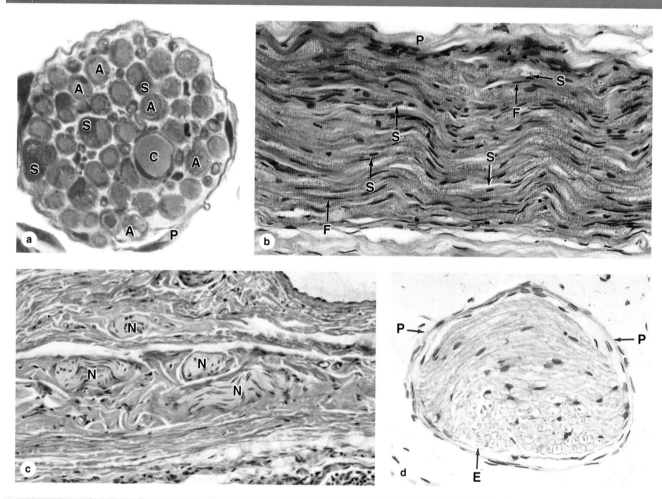

Small nerves can be seen in sections from most organs.

(a) In cross section an isolated, resin-embedded nerve is seen to have a thin perineurium (**P**), one capillary (**C**), and many large axons (**A**) associated with Schwann cells (**S**). A few nuclei of fibroblasts can be seen in the endoneurium between the myelinated fibers. (X400; PT)

(b) In longitudinal sections the flattened nuclei of endoneurial fibroblasts (**F**) and more oval nuclei of Schwann cells (**S**) can be distinguished. Nerve fibers are held rather loosely in the endoneurium and in low-magnification longitudinal section are seen to be wavy rather than straight. This indicates a slackness of fibers within the

nerve, which allows nerves to stretch slightly during body movements with no potentially damaging tension on the fibers. (X200; H&E)

(c) In sections of mesentery and other tissues, a highly wavy or tortuous disposition of a single small nerve (**N**) will be seen as multiple oblique or transverse pieces as the nerve enters and leaves the area in the section. (X200; H&E)

(d) Often, a section of small nerve will have some fibers cut transversely and others cut obliquely within the same fascicle, again suggesting the relatively unrestrained nature of the fibers within the endoneurium (**E**) and perineurium (**P**). (X300; H&E)

promote axonal outgrowth. Motor axons reestablish synaptic connections with muscles and function is restored.

❯❯ MEDICAL APPLICATION

Regeneration of peripheral nerves is functionally efficient only when the fibers and the columns of Schwann cells are directed properly. In a mixed nerve, if regenerating sensory

fibers grow into columns formerly occupied by motor fibers connected to motor end plates, the function of the muscle will not be reestablished. When there is an extensive gap between the distal and proximal segments of cut or injured peripheral nerves or when the distal segment disappears altogether (as in the case of amputation of a limb), the newly growing axons may form a swelling, or **neuroma**, that can be the source of spontaneous pain.

FIGURE **9–29** Ganglia.

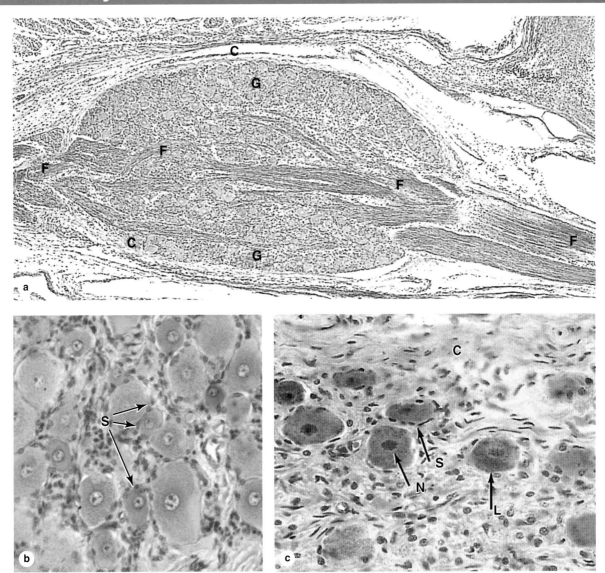

(a) A sensory ganglion (**G**) has a distinct connective tissue capsule (**C**) and internal framework continuous with the epineurium and other components of peripheral nerves, except that no perineurium is present and that there is no blood-nerve barrier function. Fascicles of nerve fibers (**F**) enter and leave these ganglia. (X56; Kluver-Barrera stain)

(b) Higher magnification shows the small, rounded nuclei of glia cells called satellite cells (**S**) that produce thin, sheet-like cytoplasmic extensions that completely envelop each large neuronal perikaryon. (X400; H&E)

(c) Sympathetic ganglia are smaller than most sensory ganglia but similar in having large neuronal cell bodies (**N**), some containing lipofuscin (**L**). Sheets from satellite cells (**S**) enclose each neuronal cell body with morphology slightly different from that of sensory ganglia. Autonomic ganglia generally have less well-developed connective tissue capsules (**C**) than sensory ganglia. (X400; H&E)

FIGURE **9–30** Regeneration in peripheral nerves.

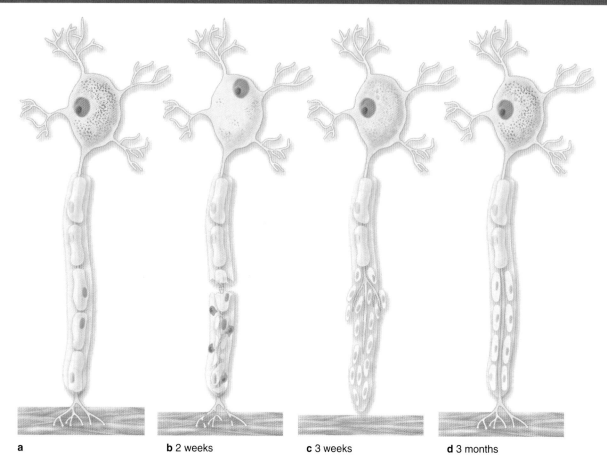

a   b 2 weeks   c 3 weeks   d 3 months

In an injured or cut peripheral nerve, proximal axon segments can regenerate from their cut ends after a delay. The main changes that take place in an injured nerve fiber are shown here.

(a) Normal nerve fiber, with its perikaryon, extensive RER (Nissl substance), and effector cell (muscle).

(b) When the axon is injured, the RER is greatly reduced initially and the nerve fiber distal to the injury degenerates along with its myelin sheath. Debris is phagocytosed by macrophages (shown in purple).

(c) In the following weeks after injury, muscle fiber shows denervation atrophy, but Schwann cells proliferate to form a compact cord penetrated by the regrowing axon. The axon grows at the rate of 0.5-3 mm/d.

(d) After some months, the nerve fiber regeneration is successful and functional connections with the muscle fiber are restored.

Nervous System   SUMMARY OF KEY POINTS

Development of Nerve Tissue

- Nervous tissue develops in the early embryo when the dorsal ectoderm **neural plate** folds lengthwise to form the **neural tube**, the precursor of the CNS, and releases **neural crest cells**, precursors for much of the PNS.

Neurons

- There are many kinds of **neurons**, but all consist of a **cell body (perikaryon)** containing the nucleus, a long cytoplasmic extension called the **axon**, and one or more shorter processes called **dendrites**.
- Neurons use the common cell property of **excitability** to produce and move an **action potential (nerve impulse)** along the axon to excite another neuron or other effector cell.

- Such nerve communication is transmitted to another neuron or effector cell via a **synapse**, where **neurotransmitter** is released at the **presynaptic membrane** and binds receptors on the **postsynaptic cell**, initiating a new action potential there.

Glial Cells

- **Glial cells (glia)**, required to support neurons in many ways, consist of six major types:
  - **Oligodendrocytes** wrap processes around portions of axons in the CNS, forming **myelin sheaths** that insulate the axons and facilitate nerve impulses.
  - **Astrocytes**, the most numerous cell of the CNS, all produce hundreds of processes to cover and provide regulated microenvironments for neuronal perikarya, synapses, and capillaries.

- **Ependymal cells** are epithelial-like cells, lacking basement membranes, which line the fluid-filled cerebral ventricles and central canal of the spinal cord.
- **Microglia** differs from all other glial cells in originating from blood monocytes, not from neural tissue precursors; they mediate immune defense activity within the CNS.
- **Schwann cells (neurolemmocytes)** enclose all axons in nerves of the PNS, producing **myelin sheaths** around large-diameter axons, whose impulse conductivity is augmented at the **nodes of Ranvier** between successive Schwann cells.
- **Satellite cells** are located within PNS **ganglia**, aggregated sensory or autonomic neuronal cell bodies, where they enclose each perikaryon and regulate its microenvironment.

Central Nervous System

- Within the brain and spinal cord, regions rich in neuronal perikarya and astrocytes comprise the **gray matter** and regions containing tracts of myelinated axons comprise **white matter**.
- Hundreds of different neurons make up the CNS; large, unique **Purkinje neurons** characterize the cortex of the cerebellum, and layers of small **pyramidal neurons** form the cerebral cortex.
- The CNS is completely enclosed by three connective tissue layers called **meninges**: (1) the tough external **dura mater**; (2) the middle **arachnoid layer**; and (3) the delicate **pia mater** that directly contacts neural tissue.
- The **arachnoid layer** contains much **CSF**, which helps **cushion** the CNS within its bony enclosure.
- The **choroid plexus** consists of elaborate folds of vascularized **pia mater** covered by **ependyma** that project from walls of the cerebral ventricles; there water is removed from capillaries and transferred into the ventricles as **cerebrospinal fluid (CSF)**.
- In most CNS regions, neurons are also protected by the blood-brain barrier, consisting of the **perivascular feet of astrocytic processes** and the nonfenestrated capillary endothelial cells' **tight junctions**.

Peripheral Nervous System

- **Peripheral nerves** consist of axons from motor neurons (in the spinal cord), sensory neurons, and autonomic neurons (in ganglia); all the axons are enclosed within a series of **Schwann cells**, but only large (myelinated) axons have myelin sheaths and nodes of Ranvier.
- **Endoneurium** is a thin connective tissue layer immediately surrounding Schwann cells in peripheral nerves, containing a few nonfenestrated capillaries and much reticulin.
- Groups of axons (with Schwann cells and endoneurium) are surrounded by **perineurium**, consisting of layered, squamous fibroblastic cells joined by **tight junctions** to make a **blood-nerve barrier**.
- In large peripheral nerves, groups of axons are subdivided as **fascicles**, each of which is surrounded by perineurium.
- Surrounding the perineurium is a thick, outermost layer of dense irregular connective tissue, the **epineurium**.
- **Ganglia**, which can be either sensory or autonomic, contain neuronal cell bodies and their **satellite cells** and are surrounded by connective tissue continuous with that of nerves.

Neural Plasticity & Regeneration

- Certain regions of the CNS, such as near the ependyma, retain rare **neural stem and progenitor cells** that allow some replacement of neurons throughout life; **neural plasticity** involving formation and remodeling of synaptic connections is also prevalent throughout life.
- The complexity and distances of the neuronal and glial interconnections with the CNS make regeneration and restoration of function within this tissue after major injury very difficult.
- The more simply organized peripheral nerves have better capacity for **axonal regeneration**, a process involving reactivation of the perikaryon, Schwann cells, and macrophages.

Nervous System ASSESS YOUR KNOWLEDGE

1. Which of the following is characteristic of the chromatophilic material called Nissl substance in neural tissue?
   a. Found throughout neurons
   b. Site of mRNA translation for proteins of the axolemma
   c. Most abundant in unipolar neurons
   d. Becomes more abundant as an individual gets older
   e. An example of intermediate filament proteins

2. Which of the following events occurs immediately after an action potential reaches a synapse at an axon terminal?
   a. Vesicle fusion with the presynaptic terminal membrane
   b. Calcium ion influx at the presynaptic terminal
   c. Neurotransmitter binding to receptors on the postsynaptic membrane
   d. Neurotransmitter release into the synaptic cleft
   e. Binding of the neurotransmitter at the presynaptic terminal

3. A report from a hospital pathology laboratory indicates that a microscope slide with a small specimen of neural tissue contains "numerous GFAP-positive" cells. What is the most likely source of this specimen?
   a. A region of white matter
   b. A sensory ganglion
   c. An autonomic ganglion
   d. A region of gray matter
   e. Pia mater

4. In the choroid plexus water from capillaries is transported directly into the cerebrospinal fluid by what structure(s)?
   a. Ependyma
   b. Astrocytes
   c. Cells of the arachnoid mater
   d. Lining of the central canal
   e. Microglial cells

5. What term applies to collections of neuronal cell bodies (somata) in the central nervous system?
   a. Ganglia
   b. Neuroglia
   c. Nodes
   d. White matter
   e. Nuclei

6. Which structure contains trabeculae around which cerebrospinal fluid (CSF) flows?
   a. Arachnoid mater
   b. Ependyma
   c. Dura mater
   d. Pia mater
   e. Gray matter

7. Which of the following is a characteristic of the connective tissue layer that surrounds individual fascicles in large peripheral nerves?
   a. A delicate region of connective tissue in contact with Schwann cells
   b. Called the dura mater
   c. Important as part of the blood-nerve barrier in the nerve
   d. Rich in myelin
   e. The thickest sheath of connective tissue in the nerve

8. A 35-year-old woman presents with weakness and spasticity in the lower left extremity, visual impairment and throbbing in the left eye, difficulties with balance, fatigue, and malaise. There is an increase in cerebrospinal fluid (CSF) protein, elevated gamma globulin, and moderate pleocytosis. MRI confirms areas of demyelination in the anterior corpus callosum. Imaging identifies plaques which are hyperintense on T2-weighted and fluid attenuated inversion recovery (FLAIR) images, and hypointense on T1-weighted scans. Which of the following cells are specifically targeted in her condition?
   a. Microglia
   b. Oligodendrocytes
   c. Astrocytes
   d. Schwann cells
   e. Multipolar neurons

9. A 22-year-old man receives a severe, traumatic compression injury to his radial nerve during a motorcycle crash. He shows an advancing Tinel sign. Which one of the following characterizes regeneration of axons after this nerve injury?
   a. It occurs in the absence of motor nerve action potentials.
   b. It occurs at a rate of about 100 mm/d.
   c. It occurs in the segment distal to the site of axon damage.
   d. It occurs by a process that involves Schwann cell proliferation.
   e. It occurs in conjunction with degeneration and phagocytosis of the endoneurium.

10. A 2-year-old boy presents with hearing impairment, poliosis (a white shock of hair), complete heterochromia and sectoral heterochromia, hypertelorism, a low hairline with eyebrows that touch in the middle, white pigmentation of the skin, and suspected neurologic deficits. He is diagnosed with Waardenburg syndrome with a mutation in the *PAX-3* gene that affects neural crest differentiation. Which of the following structures would most likely also be affected in this patient?
   a. Purkinje cells
   b. Pyramidal neurons
   c. Ventral horns of the spinal cord
   d. Astrocytes
   e. Neurons and satellite cells of the spinal ganglion

CHAPTER

10 Muscle Tissue

Muscle tissue, the fourth basic tissue type with epithelia, connective tissues, and nervous tissue, is composed of cells that optimize the universal cell property of **contractility**. As in all cells, actin microfilaments and associated proteins generate the forces necessary for the muscle contraction, which drives movement within organ systems, of blood, and of the body as a whole. Essentially all muscle cells are of mesodermal origin and differentiate by a gradual process of cell lengthening with abundant synthesis of the myofibrillar proteins actin and myosin.

Three types of muscle tissue can be distinguished on the basis of morphologic and functional characteristics (Figure 10–1), with the structure of each adapted to its physiologic role.

- **Skeletal muscle** contains bundles of very long, multinucleated cells with cross-striations. Their contraction is quick, forceful, and usually under voluntary control.
- **Cardiac muscle** also has cross-striations and is composed of elongated, often branched cells bound to one another at structures called **intercalated discs** that are unique to cardiac muscle. Contraction is involuntary, vigorous, and rhythmic.
- **Smooth muscle** consists of collections of fusiform cells that lack striations and have slow, involuntary contractions.

In all types of muscle, contraction is caused by the sliding interaction of thick myosin filaments along thin actin filaments. The forces necessary for sliding are generated by other proteins affecting the weak interactions in the bridges between actin and myosin.

As with neurons, muscle specialists refer to certain muscle cell organelles with special names. The cytoplasm of muscle cells is often called **sarcoplasm** (Gr. *sarkos*, flesh + *plasma*,

thing formed), the smooth ER is the **sarcoplasmic reticulum**, and the muscle cell membrane and its external lamina are the **sarcolemma** (*sarkos* + Gr. *lemma*, husk).

>> **MEDICAL APPLICATION**

The variation in diameter of muscle fibers depends on factors such as the specific muscle, age, gender, nutritional status, and physical training of the individual. Exercise enlarges the skeletal musculature by stimulating formation of new myofibrils and growth in the diameter of individual muscle fibers. This process, characterized by increased cell volume, is called **hypertrophy** (Gr. *hyper*, above + *trophe*, nourishment). Tissue growth by an increase in the number of cells is termed hyperplasia (hyper + Gr. *plasis*, molding), which takes place very readily in smooth muscle, whose cells have not lost the capacity to divide by mitosis.

SKELETAL MUSCLE

Skeletal (or **striated**) **muscle** consists of **muscle fibers**, which are long, cylindrical multinucleated cells with diameters of 10-100 μm. During embryonic muscle development, mesenchymal myoblasts (L. *myo*, muscle) fuse, forming myotubes with many nuclei. Myotubes then further differentiate to form striated muscle fibers (Figure 10–2). Elongated nuclei are found peripherally just under the sarcolemma, a characteristic nuclear location unique to skeletal muscle fibers/cells. A small population of reserve progenitor cells called muscle **satellite cells** remains adjacent to most fibers of differentiated skeletal muscle.

FIGURE **10–1** Three types of muscle.

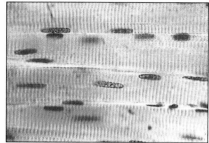
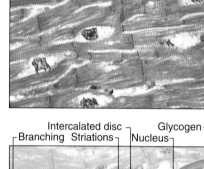
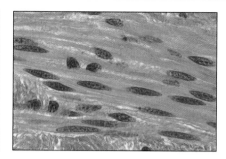
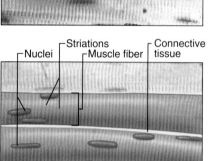
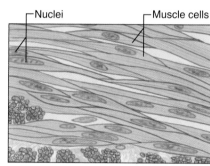

(a) Skeletal muscle

(b) Cardiac muscle

(c) Smooth muscle

Light micrographs of each type, accompanied by labeled drawings. **(a) Skeletal muscle** is composed of large, elongated, multinucleated fibers that show strong, quick, voluntary contractions. **(b) Cardiac muscle** is composed of irregular branched cells bound together longitudinally by intercalated discs and shows strong, involuntary contractions. **(c) Smooth muscle** is composed of grouped, fusiform cells with weak, involuntary contractions. The density of intercellular packing seen reflects the small amount of extracellular connective tissue present. ([a, b]: X200; [c]: X300; All H&E)

Organization of a Skeletal Muscle

Thin layers of connective tissue surround and organize the contractile fibers in all three types of muscle, and these layers are seen particularly well in skeletal muscle (Figures 10–3 and 10–4). The concentric organization given by these supportive layers resembles that in large peripheral nerves:

- The **epimysium**, an external sheath of dense irregular connective tissue, surrounds the entire muscle. Septa of this tissue extend inward, carrying the larger nerves, blood vessels, and lymphatics of the muscle.
- The **perimysium** is a thin connective tissue layer that immediately surrounds each bundle of muscle fibers termed a **fascicle** (Figure 10–3). Each fascicle of muscle fibers makes up a functional unit in which the fibers work together. Nerves, blood vessels, and lymphatics penetrate the perimysium to supply each fascicle.
- Within fascicles a very thin, delicate layer of reticular fibers and scattered fibroblasts, the **endomysium**, surrounds the external lamina of individual muscle fibers. In addition to nerve fibers, capillaries form a rich network in the endomysium bringing O_2 to the muscle fibers (Figure 10–5).

Collagens in these connective tissue layers of muscle serve to transmit the mechanical forces generated by the contracting muscle cells/fibers; individual muscle fibers seldom extend from one end of a muscle to the other.

All three layers, plus the dense irregular connective tissue of the deep fascia which overlies the epimysium, are continuous with the tough connective tissue of a tendon at **myotendinous junctions** which join the muscle to bone, skin, or another muscle (Figures 10–3 and 10–4c). Ultrastructural studies show that in these transitional regions, collagen fibers from the tendon insert themselves among muscle fibers and associate directly with complex infoldings of sarcolemma.

Organization Within Muscle Fibers

Longitudinally sectioned skeletal muscle fibers show striations of alternating light and dark bands (Figure 10–6a). The sarcoplasm is highly organized, containing primarily long cylindrical filament bundles called **myofibrils** that run parallel to the long axis of the fiber (Figure 10–6b). The dark bands on the myofibrils are called **A bands** (*a*nisotropic or birefringent in polarized light microscopy); the light bands are called **I bands** (*i*sotropic, do not alter polarized light). In the TEM (Figure 10–6c), each I

FIGURE **10–2** Development of skeletal muscle.

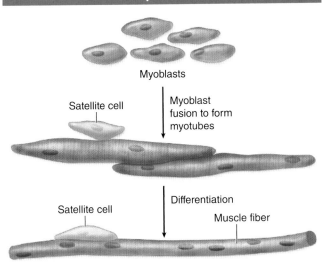

Myoblasts

Satellite cell

Myoblast fusion to form myotubes

Satellite cell

Differentiation

Muscle fiber

Skeletal muscle begins to differentiate when mesenchymal cells, called **myoblasts**, align and fuse together to make longer, multinucleated tubes called **myotubes**. Myotubes synthesize the proteins to make up myofilaments and gradually begin to show cross-striations by light microscopy. Myotubes continue differentiating to form functional myofilaments, and the nuclei are displaced against the sarcolemma.

Part of the myoblast population does not fuse and differentiate but remains as a group of mesenchymal cells called muscle **satellite cells** located on the external surface of muscle fibers inside the developing external lamina. Satellite cells proliferate and produce new muscle fibers following muscle injury.

FIGURE **10–3** Organization of skeletal muscle.

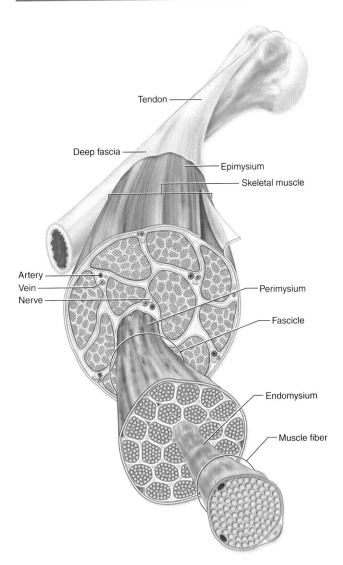

Tendon

Deep fascia

Epimysium

Skeletal muscle

Artery
Vein
Nerve

Perimysium

Fascicle

Endomysium

Muscle fiber

An entire skeletal muscle is enclosed within a thick layer of dense connective tissue called the **epimysium** that is continuous with fascia and the tendon binding muscle to bone. Large muscles contain several **fascicles** of muscle tissue, each wrapped in a thin but dense connective tissue layer called the **perimysium**. Within fascicles individual muscle fibers (elongated multinuclear cells) are surrounded by a delicate connective tissue layer, the **endomysium**.

band is seen to be bisected by a dark transverse line, the **Z disc** (Ger. *zwischen*, between). The repetitive functional subunit of the contractile apparatus, the **sarcomere**, extends from Z disc to Z disc (Figure 10–6c) and is about 2.5-μm long in resting muscle.

Mitochondria and sarcoplasmic reticulum are found between the myofibrils, which typically have diameters of 1-2 μm. Myofibrils consist of an end-to-end repetitive arrangement of sarcomeres (Figure 10–7); the lateral registration of sarcomeres in adjacent myofibrils causes the entire muscle fiber to exhibit a characteristic pattern of transverse striations.

The A and I banding pattern in sarcomeres is due mainly to the regular arrangement of thick and thin **myofilaments**, composed of **myosin** and **F-actin**, respectively, organized within each myofibril in a symmetric pattern containing thousands of each filament type (Figure 10–7).

The thick myosin filaments are 1.6-μm long and 15-nm wide; they occupy the A band at the middle region of the sarcomere. **Myosin** is a large complex (~500 kDa) with two identical heavy chains and two pairs of light chains. Myosin heavy chains are thin, rodlike motor proteins (150-nm long and 2-3 nm thick) twisted together as myosin tails (Figure 10–7). Globular projections containing the four myosin light chains form a head at one end of each heavy chain. The myosin heads bind both actin, forming transient crossbridges between the

thick and thin filaments, and ATP, catalyzing energy release (**actomyosin ATPase activity**). Several hundred myosin molecules are arranged within each thick filament with overlapping rodlike portions and the globular heads directed toward either end (Figure 10–7a).

The thin, helical actin filaments are each 1.0-μm long and 8-nm wide and run between the thick filaments. Each G-actin monomer contains a binding site for myosin (Figure 10–7b).

FIGURE **10–4** Skeletal muscle.

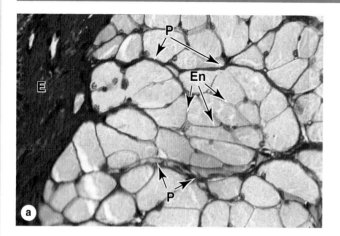

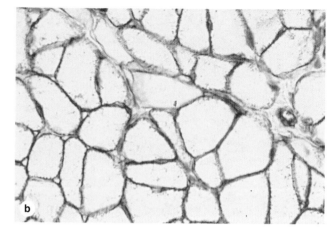

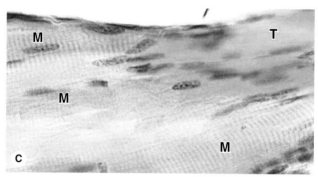

(a) A cross section of striated muscle demonstrating all three layers of connective tissue and cell nuclei. The endomysium (**En**) surrounds individual muscle, and perimysium (**P**) encloses a group of muscle fibers comprising a fascicle. A thick epimysium (**E**) surrounds the entire muscle. All three of these tissues contain collagen types I and III (reticulin). (X200; H&E)

(b) An adjacent section immunohistochemically stained for laminin, which specifically stains the external laminae of the muscle fibers, surrounded by endomysium. (X400; Immunoperoxidase)

(c) Longitudinal section of a **myotendinous junction**. Tendons develop together with skeletal muscles and join muscles to the periosteum of bones. The dense collagen fibers of a tendon (**T**) are continuous with those in the three connective tissue layers around muscle fibers (**M**), forming a strong unit that allows muscle contraction to move other structures. (X400; H&E)

FIGURE **10–5** Capillaries of skeletal muscle.

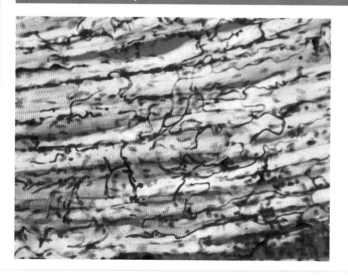

The blood vessels were injected with a dark plastic polymer before the muscle was collected and sectioned longitudinally. A rich network of capillaries in endomysium surrounding muscle fibers is revealed by this method. (X200; Giemsa with polarized light)

FIGURE **10–6** **Striated skeletal muscle in longitudinal section.**

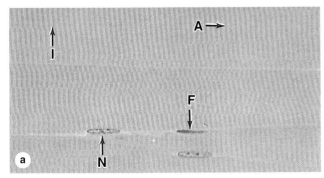

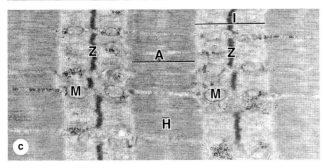

Longitudinal sections reveal the striations characteristic of skeletal muscle.

(a) Parts of three muscle fibers are separated by very thin endomysium that includes one fibroblast nucleus (**F**). Muscle nuclei (**N**) are found against the sarcolemma. Along each fiber thousands of dark-staining **A** bands alternate with lighter **I** bands. (X200; H&E)

(b) At higher magnification, each fiber can be seen to have three or four myofibrils, here with their striations slightly out of alignment with one another. Myofibrils are cylindrical bundles of thick and thin myofilaments that fill most of each muscle fiber. (X500; Giemsa)

(c) TEM showing one contractile unit (**sarcomere**) in the long series that comprises a myofibril. In its middle is an electron-dense **A** band bisected by a narrow, less dense region called the **H** zone. On each side of the A band are the lighter-stained **I** bands, each bisected by a dense **Z disc** which marks one end of the sarcomere. Mitochondria (**M**), glycogen granules, and small cisternae of SER occur around the Z disc. (X24,000)

(Figure 10–6c, used with permission from Mikel H. Snow, Department of Cell and Neurobiology, Keck School of Medicine at the University of Southern California, Los Angeles.)

The thin filaments have two tightly associated regulatory proteins (Figure 10–7b):

- **Tropomyosin**, a 40-nm-long coil of two polypeptide chains located in the groove between the two twisted actin strands
- **Troponin**, a complex of three subunits: TnT, which attaches to tropomyosin; TnC, which binds Ca^{2+}; and TnI, which regulates the actin-myosin interaction

Troponin complexes attach at specific sites regularly spaced along each tropomyosin molecule.

The organization of important myofibril components is shown in Figure 10–8. I bands consist of the portions of the thin filaments that do not overlap the thick filaments in the A bands, which is why I bands stain more lightly than A bands. Actin filaments are anchored perpendicularly on the Z disc by the actin-binding protein **α-actinin** and exhibit opposite polarity on each side of this disc (Figure 10–8c). An important accessory protein in I bands is **titin** (3700 kDa), the largest protein in the body, with scaffolding and elastic properties, which supports the thick myofilaments and connects them to the Z disc (Figure 10–8c). Another large accessory protein, nebulin, binds each thin myofilament laterally, helps anchor them to α-actinin, and specifies the length of the actin polymers during myogenesis.

The A bands contain both the thick filaments and the overlapping portions of thin filaments. Close observation of the A band shows the presence of a lighter zone in its center, the H zone, corresponding to a region with only the rod-like portions of the myosin molecule and no thin filaments (Figure 10–8c). Bisecting the H zone is the M line (Ger. *Mitte*, middle; Figure 10–8d), containing a myosin-binding protein **myomesin** that holds the thick filaments in place, and **creatine kinase**. This enzyme catalyzes transfer of phosphate groups from phosphocreatine, a storage form of high-energy phosphate groups, to ADP, helping to supply ATP for muscle contraction.

Despite the many proteins present in sarcomeres, myosin and actin together represent over half of the total protein in striated muscle. The overlapping arrangement of thin and thick filaments within sarcomeres produces in TEM cross sections hexagonal patterns of structures that were important in determining the functions of the filaments and other proteins in the myofibril (Figures 10–8b and 10–8e).

Sarcoplasmic Reticulum & Transverse Tubule System

In skeletal muscle fibers the membranous smooth ER, called here **sarcoplasmic reticulum**, contains pumps and other proteins for Ca^{2+} sequestration and surrounds the myofibrils (Figure 10–9). Calcium release from cisternae of the sarcoplasmic reticulum through voltage-gated Ca^{2+} channels is triggered by membrane depolarization produced by a motor nerve.

To trigger Ca^{2+} release from sarcoplasmic reticulum throughout the muscle fiber simultaneously and produce

FIGURE **10–7** **Molecules composing thin and thick filaments.**

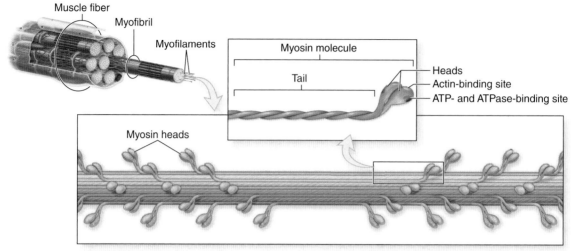

Muscle fiber

Myofibril

Myofilaments

Myosin molecule

Tail

Heads
Actin-binding site
ATP- and ATPase-binding site

Myosin heads

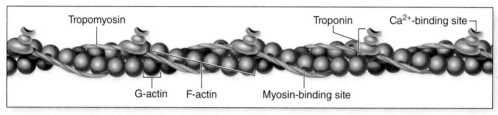

a Thick filament

Tropomyosin

Troponin

Ca²⁺-binding site

G-actin   F-actin   Myosin-binding site

b Thin filament

Myofilaments, which include both thick and thin filaments, consist of contractile protein arrays bundled within myofibrils. **(a)** A thick myofilament contains 200-500 molecules of **myosin**. **(b)** A thin filament contains **F-actin**, **tropomyosin**, and **troponin**.

uniform contraction of all myofibrils, the sarcolemma has tubular infoldings called **transverse** or **T-tubules** (Figures 10–9 and 10–10). These long fingerlike invaginations of the cell membrane penetrate deeply into the sarcoplasm and encircle each myofibril near the aligned A- and I-band boundaries of sarcomeres.

Adjacent to each T-tubule are expanded **terminal cisternae** of sarcoplasmic reticulum. In longitudinal TEM sections, this complex of a T-tubule with two terminal cisternae is called a **triad** (Figures 10–9 and 10–10). The triad complex allows depolarization of the sarcolemma in a T-tubule to affect the sarcoplasmic reticulum and trigger release of Ca²⁺ ions into cytoplasm around the thick and thin filaments, which initiates contraction of sarcomeres.

Mechanism of Contraction

Figure 10–11 summarizes the key molecular events of muscle contraction. During this process neither the thick nor thin filaments change their length. Contraction occurs as the overlapping thin and thick filaments of each sarcomere slide past one another.

Contraction is induced when an action potential arrives at a synapse, the **neuromuscular junction (NMJ)**, and is transmitted along the T-tubules to terminal cisternae of the sarcoplasmic reticulum to trigger Ca²⁺ release. In a resting muscle, the myosin heads cannot bind actin because the binding sites are blocked by the troponin-tropomyosin complex on the F-actin filaments. Calcium ions released upon neural stimulation bind troponin, changing its shape and moving tropomyosin on the F-actin to expose the myosin-binding active sites and allow crossbridges to form. Binding actin produces a conformational change or pivot in the myosins, which pulls the thin filaments farther into the A band, toward the Z disc (Figure 10–11).

Energy for the myosin head pivot which pulls actin is provided by hydrolysis of ATP bound to the myosin heads, after which myosin binds another ATP and detaches from actin. In the continued presence of Ca²⁺ and ATP, these attach-pivot-detach events occur in a repeating cycle, each lasting about

FIGURE **10–8** Structure of a myofibril: A series of sarcomeres.

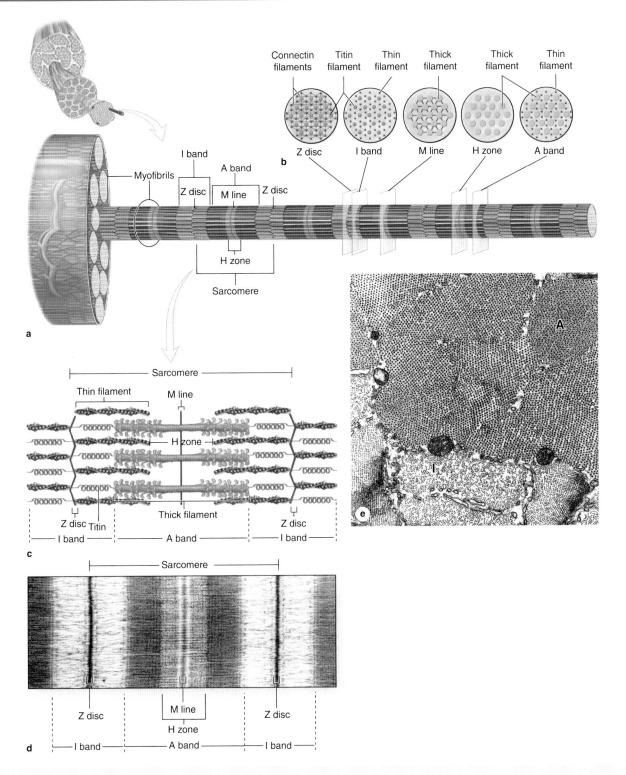

(a) The diagram shows that each muscle fiber contains several parallel bundles called **myofibrils**.

(b) Each myofibril consists of a long series of sarcomeres, separated by Z discs and containing thick and thin filaments that overlap in certain regions.

(c) Thin filaments are actin filaments with one end bound to α-**actinin** in the Z disc. Thick filaments are bundles of myosin, which span the entire A band and are bound to proteins of the M line and to the Z disc across the I bands by a very large protein called **titin**, which has springlike domains.

(d) The molecular organization of the sarcomeres produces staining differences that cause the dark- and light-staining bands seen by light microscopy and TEM. (X28,000)

(e) With the TEM an oblique section of myofibrils includes both **A** and **I** bands and shows hexagonal patterns that indicate the relationships between thin and thick myofilaments and other proteins, as shown in part **b** of this figure. Thin and thick filaments are arranged so that each myosin bundle contacts six actin filaments. Large mitochondria in cross section and SER cisternae are seen between the myofibrils. (X45,000)

FIGURE **10–9**   **Organization of a skeletal muscle fiber**

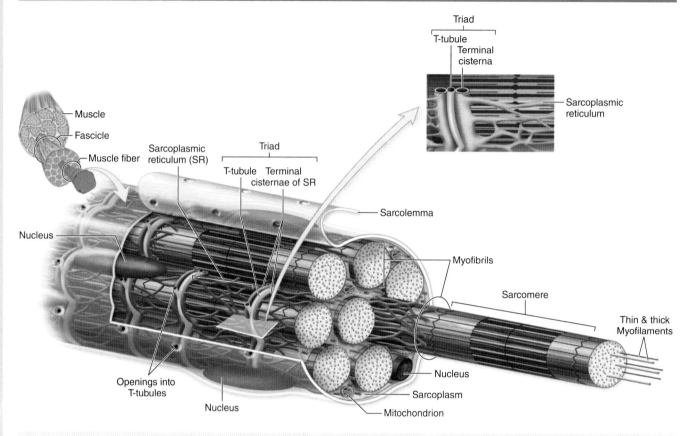

Skeletal muscle fibers are composed mainly of myofibrils. Each myofibril extends the length of the fiber and is surrounded by parts of the sarcoplasmic reticulum. The sarcolemma has deep invaginations called T-tubules, each of which becomes associated with two terminal cisternae of the sarcoplasmic reticulum. A T-tubule and its two associated terminal cisterna comprise a "triad" of small spaces along the surface of the myofibrils.

50 milliseconds, which rapidly shorten the sarcomere and contract the muscle (Figures 10–11 and 10–12). A single muscle contraction results from hundreds of these cycles.

When the neural impulse stops and levels of free Ca²⁺ ions diminish, tropomyosin again covers the myosin-binding sites on actin and the filaments passively slide back and sarcomeres return to their relaxed length (Figure 10–11). In the absence of ATP, the actin-myosin crossbridges become stable, which accounts for the rigidity of skeletal muscles (**rigor mortis**) that occurs as mitochondrial activity stops after death.

Innervation

Myelinated motor nerves branch out within the perimysium, where each nerve gives rise to several unmyelinated terminal twigs that pass through endomysium and form synapses with individual muscle fibers. Schwann cells enclose the small axon branches and cover their points of contact with the muscle cells (Figure 10–13); the external lamina of the Schwann cell fuses with that of the sarcolemma. Each axonal branch forms a dilated termination situated within a trough on the muscle cell surface, which are part of the synapses termed the neuromuscular junctions, or **motor end plates (MEP)** (Figure 10–13). As in all synapses the axon terminal contains mitochondria and numerous synaptic vesicles; here the vesicles contain the neurotransmitter **acetylcholine**. Between the axon and the muscle is the **synaptic cleft**. Adjacent to the synaptic cleft, the sarcolemma is thrown into numerous deep **junctional folds**, which provide for greater postsynaptic surface area and more transmembrane acetylcholine receptors.

When a **nerve action potential** reaches the MEP, acetylcholine is liberated from the axon terminal, diffuses across the cleft, and binds to its receptors in the folded sarcolemma. The **acetylcholine receptor** contains a nonselective cation channel that opens upon neurotransmitter binding, allowing influx

FIGURE **10–10** Transverse tubule system and triads.

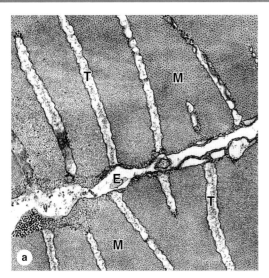 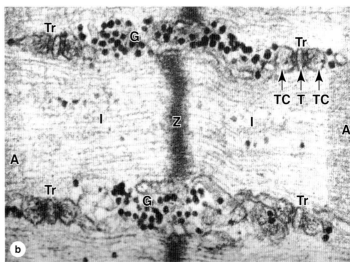

Transverse tubules are invaginations of the sarcolemma that penetrate deeply into the muscle fiber around all myofibrils.

(a) TEM cross section of fish muscle shows portions of two fibers and the endomysium (**E**) between them. Several transverse or T-tubules (**T**) are shown, perpendicular to the fiber surface, penetrating between myofibrils (**M**). (X50,000)

(b) Higher-magnification TEM of skeletal muscle in longitudinal section shows four membranous triads (**Tr**) cut transversely near the **A**-band–**I**-band junctions. Each triad consists of a central

transverse tubule (**T**) and two adjacent terminal cisterns (**TC**) extending from the sarcoplasmic reticulum. Centrally located is the **Z** disc. Besides elements of the triad, sarcoplasm surrounding the myofibril also contains dense glycogen granules (**G**).

Components of the triad are responsible for the cyclic release of Ca^{2+} from the cisternae and its sequestration again that occurs during muscle contraction and relaxation. The association between SR cisternae and T-tubules is shown diagrammatically in Figure 10–11. (X90,000)

of cations, depolarizing the sarcolemma, and producing the **muscle action potential**. Acetylcholine quickly dissociates from its receptors, and free neurotransmitter is removed from the synaptic cleft by the extracellular enzyme **acetylcholinesterase**, preventing prolonged contact of the transmitter with its receptors.

As discussed with Figure 10–11, the muscle action potential moves along the sarcolemma and along T-tubules that penetrate deeply into sarcoplasm. At triads the depolarization signal triggers the release of Ca^{2+} from terminal cisterns of the sarcoplasmic reticulum, initiating the contraction cycle.

An axon from a single motor neuron can form MEPs with one or many muscle fibers. Innervation of single muscle fibers by single motor neurons provides precise control of muscle activity and occurs, for example, in the extraocular muscles for eye movements. Larger muscles with coarser movements have motor axons that typically branch profusely and innervate 100 or more muscle fibers. In this case the single axon and all the muscle fibers in contact with its branches make up a **motor unit**. Individual striated muscle fibers do not show graded contraction—they contract either all the way or not at all. To vary the force of contraction, the fibers within a muscle fascicle do not all contract at the same time. With large

muscles composed of many motor units, the firing of a single motor axon will generate tension proportional to the number of muscle fibers it innervates. Thus, the number of motor units and their variable size control the intensity and precision of a muscle contraction.

Key features of skeletal muscle cells, connective tissue, contraction, and innervation are summarized in Table 10–1.

❯❯ MEDICAL APPLICATION

Myasthenia gravis is an autoimmune disorder that involves circulating antibodies against proteins of acetylcholine receptors. Antibody binding to the antigenic sites interferes with acetylcholine activation of their receptors, leading to intermittent periods of skeletal muscle weakness. As the body attempts to correct the condition, junctional folds of sarcolemma with affected receptors are internalized, digested by lysosomes, and replaced by newly formed receptors. These receptors, however, are again made unresponsive to acetylcholine by similar antibodies, and the disease follows a progressive course. The extraocular muscles of the eyes are commonly the first affected.

FIGURE **10–11** Events of muscle contraction.

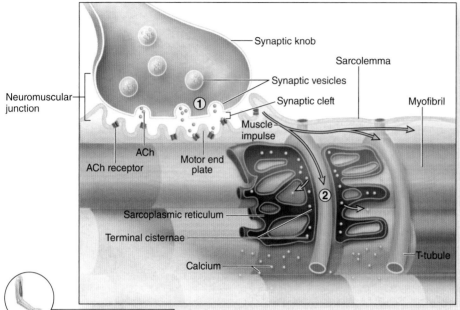

① A nerve impulse triggers release of ACh from the synaptic knob into the synaptic cleft. ACh binds to ACh receptors in the motor end plate of the neuromuscular junction, initiating a muscle impulse in the sarcolemma of the muscle fiber.

② As the muscle impulse spreads quickly from the sarcolemma along T tubules, calcium ions are released from terminal cisternae into the sarcoplasm.

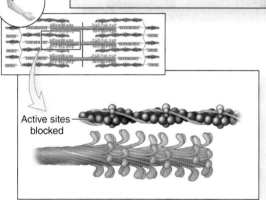

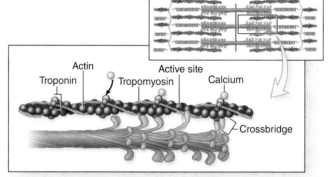

⑤ When the impulse stops, calcium ions are actively transported into the sarcoplasmic reticulum, tropomyosin re-covers active sites, and filaments passively slide back to their relaxed state.

③ Calcium ions bind to troponin. Troponin changes shape, moving tropomyosin on the actin to expose active sites on actin molecules of thin filaments. Myosin heads of thick filaments attach to exposed active sites to form crossbridges.

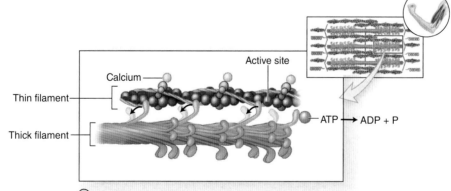

④ Myosin heads pivot, moving thin filaments toward the sarcomere center. ATP binds myosin heads and is broken down into ADP and P. Myosin heads detach from thin filaments and return to their prepivot position. The repeating cycle of *attach-pivot-detach-return* slides thick and thin filaments past one another. The sarcomere shortens and the muscle contracts. The cycle continues as long as calcium ions remain bound to troponin to keep active sites exposed.

FIGURE **10–12** Sliding filaments and sarcomere shortening in contraction.

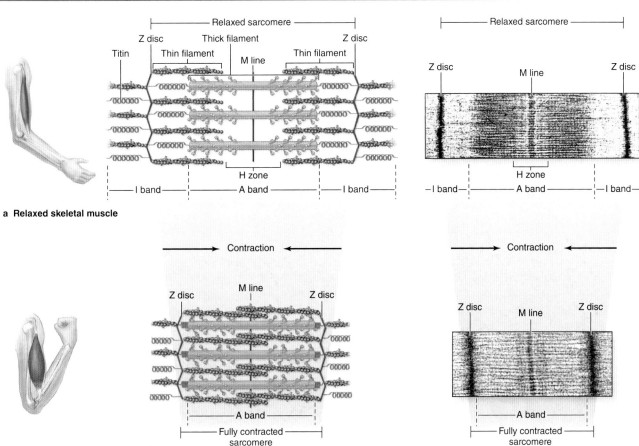

a Relaxed skeletal muscle

b Fully contracted skeletal muscle

Diagrams and TEM micrographs show sarcomere shortening during skeletal muscle contraction. **(a)** In the relaxed state the sarcomere, I band, and H zone are at their expanded length. The springlike action of titin molecules, which span the I band, helps pull thin and thick filaments past one another in relaxed muscle. **(b)** During muscle contraction, the Z discs at the sarcomere boundaries are drawn closer together as they move toward the ends of thick filaments in the A band. Titin molecules are compressed during contraction.

Muscle Spindles & Tendon Organs

Striated muscles and myotendinous junctions contain sensory receptors acting as proprioceptors (L. *proprius*, one's own + *capio*, to take), providing the central nervous system (CNS) with data from the musculoskeletal system. Among the muscle fascicles are stretch detectors known as **muscle spindles**, approximately 2-mm long and 0.1-mm wide (Figure 10–14a). A muscle spindle is encapsulated by modified perimysium, with concentric layers of flattened cells, containing interstitial fluid and a few thin muscle fibers filled with nuclei and called **intrafusal fibers** (Figure 10–14). Several sensory nerve axons penetrate each muscle spindle and wrap around individual intrafusal fibers. Changes in length (distension) of the surrounding (extrafusal) muscle fibers caused by body movements are detected by the muscle spindles and the sensory nerves relay this information to the spinal cord. Different types of sensory and intrafusal fibers mediate reflexes of varying complexity to help maintain posture and to regulate the activity of opposing muscle groups involved in motor activities such as walking.

A similar role is played by **Golgi tendon organs**, much smaller encapsulated structures that enclose sensory axons penetrating among the collagen bundles at the myotendinous junction (Figure 10–14a). Tendon organs detect changes in tension within tendons produced by muscle contraction and act to inhibit motor nerve activity if tension becomes excessive. Because both of these proprioceptors detect increases in tension, they help regulate the amount of effort required to perform movements that call for variable amounts of muscular force.

FIGURE **10–13** The neuromuscular junction (NMJ).

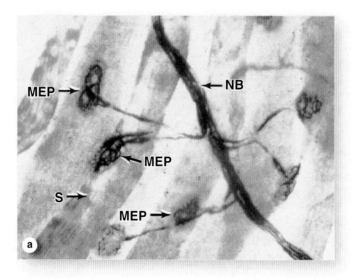

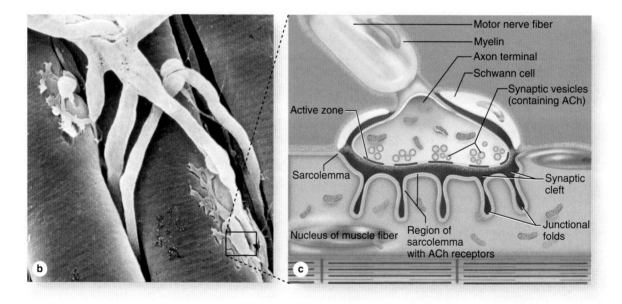

Before it terminates in a skeletal muscle, each motor axon bundled in the nerve forms many branches, each of which forms a synapse with a muscle fiber.

(a) Silver staining can reveal the nerve bundle (**NB**), the terminal axonal twigs, and the motor end plates (**MEP**, also called neuromuscular junctions or NMJ) on striated muscle fibers (**S**). (X1200)

(b) An SEM shows the branching ends of a motor axon, each covered by an extension of the last Schwann cell and expanded

terminally as an MEP embedded in a groove in the external lamina of the muscle fiber.

(c) Diagram of enclosed portion of the SEM indicating key features of a typical MEP: synaptic vesicles of acetylcholine (**ACh**), a synaptic cleft, and a postsynaptic membrane. This membrane, the sarcolemma, is highly folded to increase the number of ACh receptors at the MEP. Receptor binding initiates muscle fiber depolarization, which is carried to the deeper myofibrils by the T-tubules.

TABLE **10–1**	Important comparisons of the three types of muscle.		
	Skeletal Muscle	**Cardiac Muscle**	**Smooth Muscle**
Fibers	Single multinucleated cells	Aligned cells in branching arrangement	Single small, closely packed fusiform cells
Cell/fiber shape and size	Cylindrical, 10-100 μm diameter, many cm long	Cylindrical, 10-20 μm diameter, 50-100 μm long	Fusiform, diameter 0.2-10 μm, length 50-200 μm
Striations	Present	Present	Absent
Location of nuclei	Peripheral, adjacent to sarcolemma	Central	Central, at widest part of cell
T tubules	Center of triads at A-I junctions	In dyads at Z discs	Absent; caveolae may be functionally similar
Sarcoplasmic reticulum (SR)	Well-developed, with two terminal cisterns per sarcomere in triads with T tubule	Less well-developed, one small terminal cistern per sarcomere in dyad with T tubule	Irregular smooth ER without distinctive organization
Special structural features	Very well-organized sarcomeres, SR, and transverse tubule system	Intercalated discs joining cell, with many adherent and gap junctions	Gap junctions, caveolae, dense bodies
Control of contraction	Troponin C binds Ca^{2+}, moving tropomyosin and exposing actin for myosin binding	Similar to that of skeletal muscle	Actin-myosin binding occurs with myosin phosphorylation by MLCK triggered when calmodulin binds Ca^{2+}
Connective tissue organization	Endomysium, perimysium, and epimysium	Endomysium; subendocardial and subpericardial CT layers	Endomysium and less-organized CT sheaths
Major locations	Skeletal muscles, tongue, diaphragm, eyes, and upper esophagus	Heart	Blood vessels, digestive and respiratory tracts, uterus, bladder, and other organs
Key function	Voluntary movements	Automatic (involuntary) pumping of blood	Involuntary movements
Efferent innervation	Motor	Autonomic	Autonomic
Contractions	All-or-none, triggered at motor end plates	All-or-none, intrinsic (beginning at nodes of conducting fibers)	Partial, slow, often spontaneous, wavelike and rhythmic
Cell response to increased load	Hypertrophy (increase in fiber size)	Hypertrophy	Hypertrophy and hyperplasia (increase in cell/fiber number)
Capacity for regeneration	Limited, involving satellite cells mainly	Very poor	Good, involving mitotic activity of muscle cells

>> MEDICAL APPLICATION

Dystrophin is a large actin-binding protein located just inside the sarcolemma of skeletal muscle fibers which is involved in the functional organization of myofibrils. Research on **Duchenne muscular dystrophy** revealed that mutations of the dystrophin gene can lead to defective linkages between the cytoskeleton and the extracellular matrix (ECM). Muscle contractions can disrupt these weak linkages, causing the atrophy of muscle fibers typical of this disease.

Skeletal Muscle Fiber Types

Skeletal muscles such as those that move the eyes and eyelids need to contract rapidly, while others such as those for bodily posture must maintain tension for longer periods while resisting fatigue. These metabolic differences are possible because of varied expression in muscle fibers of contractile or regulatory protein isoforms and other factors affecting oxygen delivery and use. Different types of fibers can be identified on the basis of (1) their maximal rate of contraction (fast or slow fibers) and (2) their major pathway for ATP synthesis (oxidative

FIGURE 10–14   Sensory receptors associated with skeletal muscle.

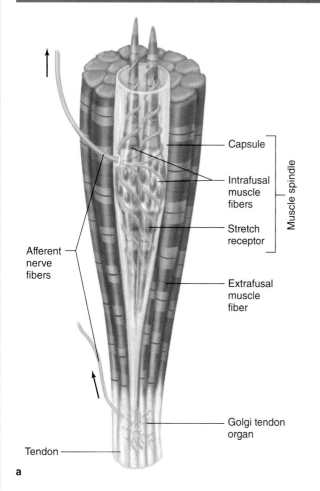

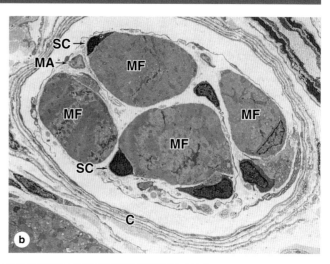

(a) The diagram shows both a **muscle spindle** and a **tendon organ**. Muscle spindles have **afferent sensory** and efferent motor nerve fibers associated with the **intrafusal fibers**, which are modified muscle fibers. The size of the spindle is exaggerated relative to the extrafusal fibers to show better the nuclei packed in the intrafusal fibers. Both types of sensory receptors provide the CNS with information concerning degrees of stretch and tension within the musculoskeletal system.

(b) A TEM cross section near the end of a muscle spindle shows the capsule (**C**), lightly myelinated axons (**MA**) of a sensory nerve, and the intrafusal muscle fibers (**MF**). These thin fibers differ from the ordinary skeletal muscle fibers in having very few myofibrils. Their many nuclei can either be closely aligned (nuclear chain fibers) or piled in a central dilation (nuclear bag fibers). Muscle satellite cells (**SC**) are also present within the external lamina of the intrafusal fibers. (X3600)

phosphorylation or glycolysis). Fast versus slow rates of fiber contraction are due largely to myosin isoforms with different maximal rates of ATP hydrolysis.

Histochemical staining is used to identify fibers with differing amounts of "fast" and "slow" ATPases (Figure 10–15). Other histological features reflecting metabolic differences among muscle fibers include the density of surrounding capillaries, the number of mitochondria, and levels of glycogen and **myoglobin**, a globular sarcoplasmic protein similar to hemoglobin which contains iron atoms and allows for O_2 storage.

Each of these features exists as a continuum in skeletal muscle fibers, but fiber diversity is divided into three major types:

- **Slow oxidative** muscle fibers are adapted for slow contractions over long periods without fatigue, having many mitochondria, many surrounding capillaries, and much myoglobin, all features that make fresh tissue rich in these fibers dark or red in color.

- **Fast glycolytic** fibers are specialized for rapid, short-term contraction, having few mitochondria or capillaries and depending largely on anaerobic metabolism of glucose derived from stored glycogen, features that make such fibers appear white. Rapid contractions lead to rapid fatigue as lactic acid produced by glycolysis accumulates.

- **Fast oxidative-glycolytic** fibers have physiological and histological features intermediate between those of the other two types.

Table 10–2 summarizes these and other characteristics of the three skeletal muscle fiber types. The metabolic type of each fiber is determined by the rate of impulse conduction along its motor nerve supply, so that all fibers of a motor unit are similar. Most skeletal muscles receive motor input from multiple nerves and contain a mixture of fiber types (Figure 10–15). Determining the fiber types in needle biopsies of skeletal muscle helps in the diagnosis of specific

FIGURE **10–15** Skeletal muscle fiber types.

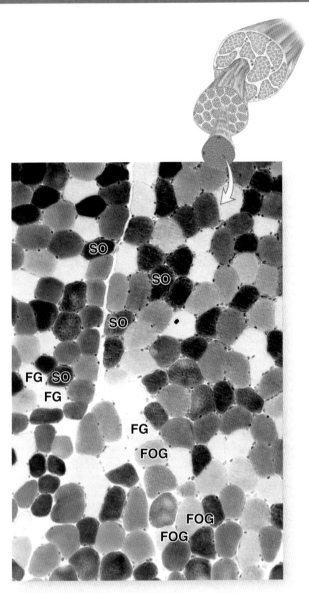

Cross section of a skeletal muscle stained histochemically for myosin ATPase at acidic pH, which reveals activity of the "slow" ATPase and shows the distribution of the three main fiber types. Slow oxidative (**SO**) or type I fibers have high levels of acidic ATPase activity and stain the darkest. Fast glycolytic (**FG**) or type IIb fibers stain the lightest. Fast oxidative-glycolytic (**FOG**) or type IIa fibers are intermediate between the other two types (X40). ATPase histochemistry of unfixed, cryostat section, pH 4.2.

myopathies (*myo* + Gr. *pathos*, suffering), motor neuron diseases, and other causes of muscle atrophy. Different fiber types also exist in cardiac muscle at various locations within the heart and in smooth muscle of different organs.

▶ CARDIAC MUSCLE

During embryonic development mesenchymal cells around the primitive heart tube align into chainlike arrays. Rather than fusing into multinucleated cells/fibers as in developing skeletal muscle fibers, **cardiac muscle** cells form complex junctions between interdigitating processes (Figure 10–16). Cells within one fiber often branch and join with cells in adjacent fibers. Consequently, the heart consists of tightly knit bundles of cells, interwoven in spiraling layers that provide for a characteristic wave of contraction that resembles wringing out of the heart ventricles.

Mature cardiac muscle cells are 15-30 μm in diameter and 85-120 μm long, with a striated banding pattern comparable to that of skeletal muscle. Unlike skeletal muscle, however, each cardiac muscle cell usually has only one nucleus and is centrally located. Surrounding the muscle cells is a delicate sheath of endomysium with a rich capillary network. A thicker perimysium separates bundles and layers of muscle fibers and in specific areas (described in Chapter 11) forms larger masses of fibrous connective tissue comprising the "cardiac skeleton."

A unique characteristic of cardiac muscle is the presence of transverse lines that cross the fibers at irregular intervals where the myocardial cells join. These **intercalated discs** represent the interfaces between adjacent cells and consist of many junctional complexes (Figures 10–16). Transverse regions of these irregular, steplike discs are composed of many **desmosomes** and **fascia adherens** junctions, which together provide strong intercellular adhesion during the cells' constant contractile activity. The less abundant, longitudinally oriented regions of each intercalated disc run parallel to the myofibrils and are filled with **gap junctions** that provide ionic continuity between the cells. These regions serve as "electrical synapses," promoting rapid impulse conduction through many cardiac muscle cells simultaneously and contraction of many adjacent cells as a unit.

The structure and function of the contractile apparatus in cardiac muscle cells are essentially the same as in skeletal muscle (Figure 10–17). Mitochondria occupy up to 40% of the cell volume, higher than in slow oxidative skeletal muscle fibers. Fatty acids, the major fuel of the heart, are stored as triglycerides in small lipid droplets. Glycogen granules as well as perinuclear lipofuscin pigment granules may also be present.

Muscle of the heart ventricles is much thicker than that of the atria, reflecting its role in pumping blood through the cardiovascular system. T-tubules in ventricular muscle fibers are well-developed, with large lumens and penetrate the sarcoplasm in the vicinity of the myofibrils' Z discs. In atrial muscle T-tubules are much smaller or entirely absent. Sarcoplasmic reticulum is less well-organized in cardiac compared to skeletal muscle fibers. The junctions between its terminal cisterns and T-tubules typically involve only one structure of each type, forming profiles called **dyads** rather than triads in TEM

TABLE **10–2** Major characteristics of skeletal muscle fiber types.			
	Slow, Oxidative Fibers (Type I)	Fast, Oxidative-Glycolytic Fibers (Type IIa)	Fast, Glycolytic Fibers (Type IIb)
Mitochondria	Numerous	Numerous	Sparse
Capillaries	Numerous	Numerous	Sparse
Fiber diameter	Small	Intermediate	Large
Size of motor unit	Small	Intermediate	Large
Myoglobin content	High (red fibers)	High (red fibers)	Low (white fibers)
Glycogen content	Low	Intermediate	High
Major source of ATP	Oxidative phosphorylation	Oxidative phosphorylation	Anaerobic glycolysis
Glycolytic enzyme activity	Low	Intermediate	High
Rate of fatigue	Slow	Intermediate	Fast
Myosin-ATPase activity	Low	High	High
Speed of contraction	Slow	Fast	Fast
Typical major locations	Postural muscles of back	Major muscles of legs	Extraocular muscles

sections. Components of this cardiac muscle transverse tubule system have the same basic functions as their counterparts in skeletal muscle fibers.

Cardiac muscle fiber contraction is intrinsic and spontaneous, as evidenced by the continued contraction of the cells in tissue culture. Impulses for the rhythmic contraction (or heartbeat) are initiated, regulated, and coordinated locally by nodes of unique myocardial fibers specialized for impulse generation and conduction, which are discussed in Chapter 11. As with skeletal muscle fibers, contraction of individual myocardial fibers is all-or-none. The rate of contraction is modified by autonomic innervation at the nodes of conducting cells, with the sympathetic nerve supply accelerating and the parasympathetic supply decreasing the frequency of the impulses.

Secretory granules about 0.2-0.3 μm in diameter are found near atrial muscle nuclei and are associated with small Golgi complexes (Figure 10–17b). These granules release the peptide hormone atrial natriuretic factor (ANF) that acts on target cells in the kidney to affect Na^+ excretion and water balance. The contractile cells of the heart's atria thus also serve an endocrine function.

Key features of cardiac muscle cells, with comparisons to those of skeletal muscle, are summarized in Table 10–1.

>> MEDICAL APPLICATION

The most common injury sustained by cardiac muscle is that due to **ischemia**, or tissue damage due to lack of oxygen when coronary arteries are occluded by heart disease. Lacking muscle satellite cells, adult mammalian cardiac muscle has little potential to regenerate after injury. However, certain fish and amphibians, as well as newborn mice, do form new muscle when the heart is partially removed, despite the lack of satellite cells. Research on the possibility of mammalian **heart muscle regeneration** builds on work with the animal models, focusing primarily on the potential of mesenchymal stem cells to form new, site-specific muscle.

>> SMOOTH MUSCLE

Smooth muscle is specialized for slow, steady contraction under the influence of autonomic nerves and various hormones. This type of muscle is a major component of blood vessels and of the digestive, respiratory, urinary, and reproductive tracts and their associated organs. Fibers of smooth muscle (also called **visceral muscle**) are elongated, tapering, and unstriated cells, each of which is enclosed by an external lamina and a network of type I and type III collagen fibers comprising the endomysium (Figure 10–18).

Smooth muscle cells range in length from 20 μm in small blood vessels to 500 μm in the pregnant uterus. At each cell's central, broadest part, where its diameter is 5-10 μm, is a single elongated nucleus. The cells stain uniformly along their lengths, and close packing is achieved with the narrow ends of each cell adjacent to the broad parts of neighboring cells. With this arrangement cross sections of smooth muscle show a range of cell diameters, with only the largest profiles containing a nucleus (Figures 10–18 and 10–19a). All cells are linked by numerous gap junctions. The borders of the cell become scalloped when smooth muscle contracts and the nucleus becomes distorted (Figure 10–20). Concentrated near the nucleus are mitochondria,

FIGURE **10–16** Cardiac muscle.

(a)

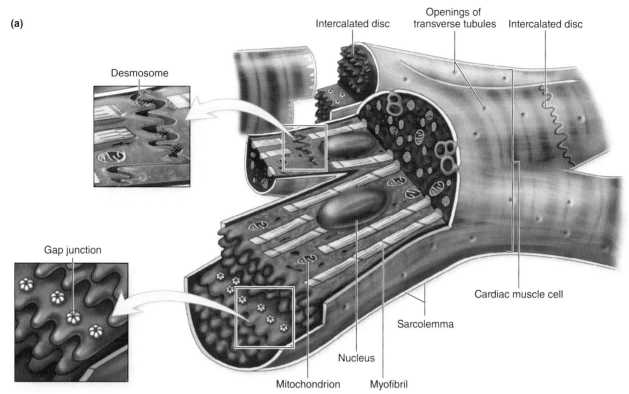

Desmosome

Gap junction

Intercalated disc

Openings of transverse tubules

Intercalated disc

Cardiac muscle cell

Sarcolemma

Nucleus

Mitochondrion

Myofibril

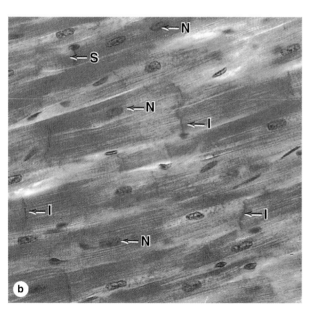

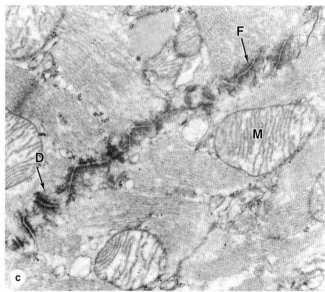

(a) The diagram of cardiac muscle cells indicates their characteristic features. The fibers consist of separate cells in a series joined at interdigitating regions called the **intercalated discs**, which cross an entire fiber between two cells. The transverse regions of the steplike intercalated disc have abundant **desmosomes** and other adherent junctions for firm adhesion, while longitudinal regions of the discs are filled with **gap junctions**.

Cardiac muscle cells have central nuclei and myofibrils that are usually sparser and less well-organized than those of skeletal muscle. Also, the cells are often branched, allowing the muscle fibers to interweave in a more complicated arrangement within fascicles that produces an efficient contraction mechanism for emptying the heart.

(b) Light microscopy of cardiac muscle in longitudinal section show nuclei (**N**) in the center of the muscle fibers and widely

spaced intercalated discs (**I**) that cross the fibers. These irregular intercalated discs should not be confused with the repetitive, much more closely spaced striations (**S**), which are similar to those of skeletal muscle but less well-organized. Nuclei of fibroblasts in endomysium are also present. (X200; H&E)

(c) TEM showing an electron-dense intercalated disc with a step-like structure along the short interdigitating processes of adjacent cardiac muscle cells. As shown here transverse disc regions have many desmosomes (**D**) and adherent junctions called **fascia adherentes (F)** which join the cells firmly. Other regions of the disc have abundant gap junctions which join the cells physiologically. The sarcoplasm has numerous mitochondria (**M**) and myofibrillar structures similar to those of skeletal muscle but slightly less organized. (X31,000)

FIGURE 10–17 Cardiac muscle ultrastructure.

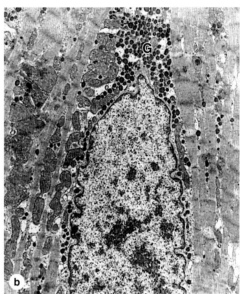

(a) TEM of cardiac muscle shows abundant mitochondria (**M**) and rather sparse sarcoplasmic reticulum (**SR**) in the areas between myofibrils. T-tubules are less well-organized and are usually associated with one expanded terminal cistern of SR, forming dyads (**D**) rather than the triads of skeletal muscle. Functionally, these structures are similar in these two muscle types. (X30,000)

(b) Muscle cells from the heart atrium show the presence of membrane-bound granules (**G**), mainly aggregated at the nuclear poles. These granules are most abundant in muscle cells of the right atrium (~600 per cell), but smaller quantities are also found in the left atrium and the ventricles. The atrial granules contain the precursor of a polypeptide hormone, **atrial natriuretic factor (ANF)**. ANF targets cells of the kidneys to bring about sodium and water loss (natriuresis and diuresis). This hormone thus opposes the actions of aldosterone and antidiuretic hormone, whose effects on kidneys result in sodium and water conservation. (X10,000)

(Figure 10–17b, used with permission from Dr J. C. Nogueira, Department of Morphology, Federal University of Minas Gerais, Belo Horizonte, Brazil.)

polyribosomes, RER, and vesicles of a Golgi apparatus. The short plasmalemma invaginations resembling **caveolae** are often numerous at the surface of smooth muscle cells.

The fibers have rudimentary sarcoplasmic reticulum, but lack T-tubules; their function is unnecessary in these smaller, tapering cells with many gap junctions. Caveolae of smooth muscle cells contain the major ion channels that control Ca^{2+} release from sarcoplasmic cisternae at myofibrils which initiates contraction. The characteristic contractile activity of smooth muscle is generated by myofibrillar arrays of actin and myosin organized somewhat differently from those of striated muscle. In smooth muscle cells bundles of thin and thick myofilaments crisscross the sarcoplasm obliquely. The myosin filaments have a less regular arrangement among the thin filaments and fewer crossbridges than in striated muscle. Moreover smooth muscle actin filaments are not associated with troponin and tropomyosin, using instead **calmodulin** and Ca^{2+}-sensitive **myosin light-chain kinase (MLCK)** to produce contraction. The contraction mechanism, however, is basically similar to that in striated muscle.

As shown in Figure 10–20 the actin myofilaments insert into anchoring cytoplasmic and plasmalemma-associated **dense bodies** which contain α-actinin and are functionally similar to the Z discs of striated and cardiac muscle. Smooth muscle cells also have an elaborate array of 10-nm intermediate filaments, composed of desmin, which also attach to the dense bodies. The submembranous dense bodies include cadherins of desmosomes linking adjacent smooth muscle cells. Dense bodies in smooth muscle cells thus serve as points for transmitting the contractile force not only within the cells, but also between adjacent cells (Figure 10–20). The endomysium and other connective tissue layers help combine the force generated by the smooth muscle fibers into a concerted action, for example peristalsis in the intestine.

Smooth muscle is not under voluntary motor control and its fibers typically lack well-defined neuromuscular junctions. Contraction is most commonly stimulated by autonomic nerves, but in the gastrointestinal tract smooth muscle is also controlled by various paracrine secretions and in the uterus by oxytocin from the pituitary gland.

Axons of autonomic nerves passing through smooth muscle have periodic swellings or varicosities that lie in close contact with muscle fibers. Synaptic vesicles in the varicosities release a neurotransmitter, usually acetylcholine or

FIGURE **10–18** Smooth muscle.

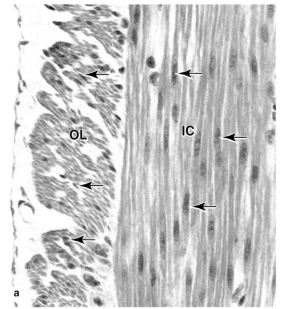

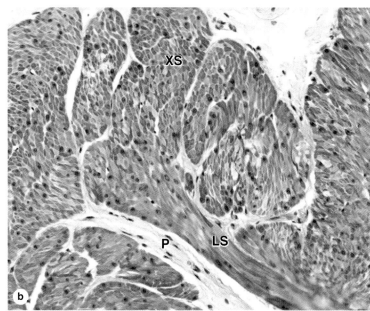

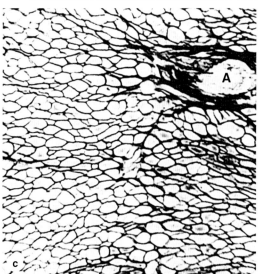

Cells or fibers of smooth muscle are long, tapering structures with elongated nuclei centrally located at the cell's widest part.

(a) In most of the digestive tract and certain similar structures smooth muscle is organized into two layers which contract in a coordinated manner to produce a wave that moves the tract's contents in a process termed **peristalsis**. In smooth muscle of the small intestine wall cut in cross section, cells of the inner circular (**IC**) layer are cut lengthwise and cells of the outer longitudinal layer (**OL**) are cut transversely. Only some nuclei (arrows) of the latter cells are in the plane of section, so that many cells appear to be devoid of nuclei. (X140; H&E)

(b) Section of smooth muscle in bladder shows interwoven bundles of muscle fibers in cross section (**XS**) and longitudinal section (**LS**) with the same fascicle. There is much collagen in the branching perimysium (**P**), but the endomysium can barely be seen by routine staining. (X140; Mallory trichrome)

(c) Section stained only for reticulin reveals the thin endomysium around each fiber, with more reticulin in the connective tissue of small arteries (**A**). Reticulin fibers associated with the basal laminae of smooth muscle cells help hold the cells together as a functional unit during the slow, rhythmic contractions of this tissue. (X200; Silver)

norepinephrine, which diffuses and binds receptors in the sarcolemmae of numerous muscle cells. There is little or no specialized structure to such junctions. As in cardiac muscle, stimulation is propagated to more distant fibers via gap junctions which allow all the smooth muscle cells to contract synchronously or in a coordinated manner.

In addition to contractile activity, smooth muscle cells also supplement fibroblast activity, synthesizing collagen, elastin, and proteoglycans, with a major influence on the extracellular matrix (ECM) in tissues where these contractile cells are abundant. Active synthesis of ECM by the small cells/fibers of smooth muscle may reflect less specialization for strong contractions than in skeletal and cardiac muscle and is similar to this synthetic function in other contractile cells, such as myofibroblasts and pericytes.

Key histologic and functional features of smooth muscle, with comparisons to those of skeletal and cardiac muscle, are summarized in Table 10–1.

❯❯ MEDICAL APPLICATION

Benign tumors called **leiomyomas** commonly develop from smooth muscle fibers but are seldom problematic. They most frequently occur in the wall of the uterus, where they are more commonly called **fibroids** and where they can become sufficiently large to produce painful pressure and unexpected bleeding.

FIGURE **10–19** Smooth muscle ultrastructure.

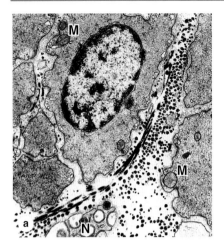

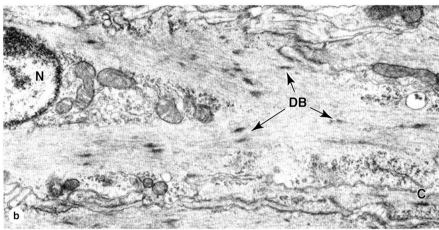

(a) TEM of a transverse section of smooth muscle showing several cells sectioned at various points along their lengths, yielding profiles of various diameters with only the largest containing a nucleus. Thick and thin filaments are not organized into myofibril bundles, and there are few mitochondria (**M**). There is evidence of a sparse external lamina around each cell, and reticular fibers are abundant in the ECM. A small unmyelinated nerve (**N**) is also seen between the cells. (X6650)

(b) Longitudinal section showing several dense bodies (**DB**) in the cytoplasm and at the cell membrane. Thin filaments and intermediate filaments both attach to the dense bodies. In the cytoplasm near the nucleus (**N**) are mitochondria, glycogen granules, and Golgi complexes. In the lower right corner of the photo the cell membrane shows invaginations called caveolae (**C**) that may regulate release of Ca²⁺ from sarcoplasmic reticulum. (X9000)

FIGURE **10–20** Smooth muscle contraction.

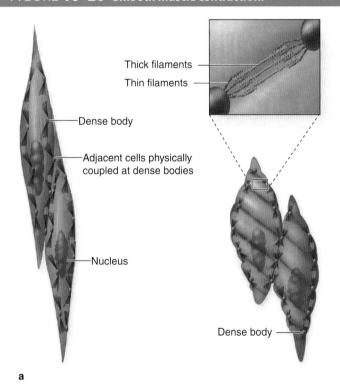

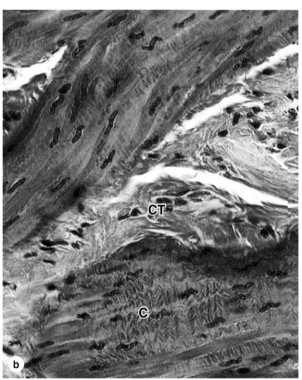

Most molecules that allow contraction are similar in the three types of muscle, but the filaments of smooth muscle are arranged differently and appear less organized.

(a) The diagram shows that thin filaments attach to **dense bodies** located at the cell membrane and deep in the cytoplasm. Dense bodies contain α-actinin for thin filament attachment. Dense bodies at the membrane are also attachment sites for intermediate filaments and for adhesive junctions between cells. This arrangement of both

the cytoskeleton and contractile apparatus allows the multicellular tissue to contract as a unit, providing better efficiency and force.

(b) Micrograph showing a contracted (**C**) region of smooth muscle, with contraction decreasing the cell length and deforming the nuclei. The long nuclei of individual fibers assume a cork-screw shape when the fibers contract, reflecting the reduced cell length at contraction. Connective tissue (**CT**) of the perimysium outside the muscle fascicle is stained blue. (X240; Mallory trichrome)

REGENERATION OF MUSCLE TISSUE

The three types of adult muscle have different potentials for regeneration after injury which are also summarized in Table 10–1.

In skeletal muscle, although the multinucleated cells cannot undergo mitosis, the tissue can still display limited regeneration. The source of regenerating cells is the sparse population of mesenchymal **satellite cells** lying inside the external lamina of each muscle fiber. Satellite cells are inactive, reserve myoblasts that persist after muscle differentiation. After injury the normally quiescent satellite cells become activated, proliferating and fusing to form new skeletal muscle fibers. Similar activity of satellite cells has been implicated in muscle growth after extensive exercise, a process in which they fuse with existing fibers to increase muscle mass beyond that which occurs by cell hypertrophy. Following major traumatic injuries, scarring and excessive connective tissue growth interferes with skeletal muscle regeneration.

Cardiac muscle lacks satellite cells and shows very little regenerative capacity beyond early childhood. Defects or damage (eg, infarcts) to heart muscle are generally replaced by proliferating fibroblasts and growth of connective tissue, forming only myocardial scars.

Smooth muscle, composed of simpler, smaller, mononucleated cells, is capable of a more active regenerative response. After injury, viable smooth muscle cells undergo mitosis and replace the damaged tissue. As discussed in Chapter 11, contractile pericytes from the walls of small blood vessels participate in the repair of vascular smooth muscle.

Muscle Tissue SUMMARY OF KEY POINTS

- There are three major types of muscle: (1) **skeletal** or striated muscle, (2) **cardiac** muscle, and (3) **smooth** or visceral muscle.
- **Skeletal muscle cells** are very **long, multinucleated fibers, cylindrically shaped** and with **diameters up to 100 μm**.
- The **sarcolemma** of each fiber is surrounded by an external lamina and thin connective tissue, **endomysium**, containing capillaries.

Organization of Skeletal Muscle Fibers

- Groups of fibers called **fascicles** are surrounded by **perimysium**; all fascicles are enclosed within a dense connective tissue **epimysium**.
- Internally each muscle fiber is filled with myofibrils, composed of thousands of **thick myosin filaments** and **thin actin filaments**, highly organized into contractile units called **sarcomeres**.
- Within sarcomeres thick and thin filaments **interdigitate**; globular myosin heads project from the thick filaments toward the F-actin filaments, which are associated with **tropomyosin** and **troponin**.
- Sarcomeres are separated by **Z discs** that bisect the **light-staining I bands** that contain mainly the thin filaments attached to **α-actinin** in the Z disc.
- Between the two I bands of a sarcomere is the **dark-staining A band** with the thick myosin filaments; alternating light and dark bands appear as microscopic **striations** along the fibers.

Sarcoplasmic Reticulum & Transverse Tubule System

- In the sarcoplasm between parallel myofibrils are **mitochondria** and cisternae of smooth ER, called the **sarcoplasmic reticulum (SR)** specialized for Ca^{2+} **sequestration and release**.
- At each sarcomere, two **terminal cisterns of SR** contact a deep invagination of the sarcolemma called a **transverse** or **T-tubule**, forming a **triad** that triggers Ca^{2+} release when the sarcolemma is depolarized.

Mechanism of Contraction

- Ca^{2+} **binding to troponin** causes **tropomyosin to change shape** and allow the **myosin heads to bind the actin** subunits, forming **crossbridges** between thick and thin filaments.
- The myosin heads then pivot with **ATP hydrolysis**, which pulls the thin filaments along the thick filaments.
- With Ca^{2+} and ATP present, a **contraction cycle** ensues in which **myosin heads repeatedly attach, pivot, detach, and return**, causing the **filaments to slide past one another**, shortening the sarcomere.

- When the membrane **depolarization ends, Ca^{2+} is again sequestered**, ending contraction and allowing the sarcomeres to lengthen again as the **muscle relaxes**.
- Synapses of motor axons with skeletal muscle are called **motor end plates (MEPs)**, **neuromuscular junctions (NMJs)**, or myoneural junctions; the neurotransmitter is **acetylcholine**.
- A motor axon may form many terminal branches, each ending on an MEP of a muscle fiber; all fibers innervated by branches of that axon comprise a **motor unit**.

Muscle Spindles & Tendon Organs

- These are both **sensory proprioceptors** in which sensory **axons wrap around intrafusal fibers** in small specialized fascicles or around **myotendinous collagen bundles**, respectively.

Muscle Fiber Types

- Skeletal muscles contain fibers that can be physiologically classified as the three main types: (1) **slow, oxidative (type I)**; (2) **fast, intermediate oxidative-glycolytic (type IIa)**; and (3) **fast, glycolytic (type IIb)**.

Cardiac Muscle

- **Cardiac muscle fibers** are also **striated**, but they consist of **individual cylindrical cells**, each containing **one (or two) central nuclei** and linked by adherent and gap junctions at prominent **intercalated discs**.
- **Sarcomeres** of cardiac muscle are organized and function similarly to those of skeletal muscle.
- Contraction of cardiac muscle is **intrinsic** at nodes of **impulse-generating pacemaker muscle fibers**; autonomic nerves regulate the rate of contraction.

Smooth Muscle

- **Smooth muscle fibers** are individual **small, fusiform (tapering) cells**, linked by numerous gap junctions.
- **Thin and thick filaments** in smooth muscle fibers do not form sarcomeres, and **no striations** are present.
- Thin actin filaments attach to α-actinin located in **dense bodies** that are located throughout the sarcoplasm and near the sarcolemma; contraction causes cells to shorten individually.
- Sarcoplasmic reticulum is less well-organized in smooth muscle fibers, and there is **no transverse tubule system**.

■ Troponin is lacking in smooth muscle; proteins controlling the sliding filaments here include **myosin light-chain kinase (MLCK)** and the Ca^{2+}-binding protein **calmodulin**.

Regeneration of Muscle Tissue

■ Repair and regeneration can occur in **skeletal muscle** because of a population of **reserve muscle satellite cells** that can proliferate, fuse, and form new muscle fibers.

■ Cardiac muscle **lacks satellite cells** and has little capacity for regeneration.

■ Regeneration is rapid in smooth muscle because the cells/fibers are small and relatively less differentiated, which allow **renewed mitotic activity** after injury.

Muscle Tissue ASSESS YOUR KNOWLEDGE

1. The basal lamina of a muscle fiber is part of which structure?
   a. Perimysium
   b. Epimysium
   c. Fascia
   d. Endomysium
   e. Sarcoplasmic reticulum

2. With the transmission electron microscope skeletal muscle fibers can be seen to contain structures called triads. What do the two lateral components of a triad represent?
   a. Attachment sites for thick myofilaments
   b. Sites for calcium sequestration and release
   c. Sites for impulse conduction into the fiber
   d. Sites for ATP production
   e. Sites for synthesis of proteins to be secreted outside the cell

3. Which characteristic is unique to cardiac muscle?
   a. Contain centrally located nuclei
   b. Striated
   c. Often branched
   d. Multinucleated
   e. Lack T-tubules

4. In smooth muscle calcium released by the smooth ER initiates contraction by binding to what protein?
   a. Actin
   b. Calmodulin
   c. Desmin
   d. Myosin light chain kinase
   e. Tropomyosin

5. Which feature typifies T-tubules?
   a. Evaginations of the sarcoplasmic reticulum
   b. Sequester calcium during muscle relaxation, releasing it during contraction
   c. Carry depolarization to the muscle fiber interior
   d. Overlie the A-I junction in cardiac muscle cells
   e. Rich supply of acetylcholine receptors

6. Which characteristic is unique to smooth muscle?
   a. T-tubules lie across Z lines
   b. Each thick filament is surrounded by six thin filaments
   c. Thin filaments attach to dense bodies
   d. Cells are multinucleated
   e. Cells have centrally located nuclei

7. In one type of muscle, numerous gap junctions, desmosomes, and adherens junctions are specifically localized in which structures?
   a. Myofilaments
   b. Dense bodies
   c. Sarcomeres
   d. Neuromuscular spindles
   e. Intercalated discs

8. A 66-year-old man who lives alone has a severe myocardial infarction and dies during the night. The medical examiner's office is called the following morning and describes the man's body as being in *rigor mortis*. This state of *rigor mortis* is due to which one of the following?
   a. Inhibition of Ca^{2+} leakage from the extracellular fluid and sarcoplasmic reticulum
   b. Enhanced retrieval of Ca^{2+} by the sarcoplasmic reticulum
   c. Failure to disengage tropomyosin and troponin from the myosin active sites
   d. Absence of ATP preventing detachment of the myosin heads from actin
   e. Increased lactic acid production

9. A 5-year-old boy sustains a small tear in his gastrocnemius muscle when he is involved in a bicycle accident. Regeneration of the muscle will occur through which of the following mechanisms?
   a. Dedifferentiation of muscle cells into myoblasts
   b. Differentiation of muscle satellite cells
   c. Fusion of damaged myofibers to form new myotubes
   d. Hyperplasia of existing muscle fibers
   e. Differentiation of fibroblasts to form myoblasts

10. A healthy 32-year-old man lifts weights regularly as part of his workout. In one of his biceps muscle fibers at rest, the length of the I band is 1.0 μm and the A band is 1.5 μm. Contraction of that muscle fiber results in a 10% shortening of the length of the sarcomere. What is the length of the A band after the shortening produced by muscle contraction?
    a. 1.50 μm
    b. 1.35 μm
    c. 1.00 μm
    d. 1.90 μm
    e. 0.45 μm

11 The Circulatory System

The circulatory system pumps and directs blood cells and substances carried in blood to all tissues of the body. It includes both the blood and lymphatic vascular systems, and in an adult the total length of its vessels is estimated at between 100,000 and 150,000 kilometers. The **blood vascular system**, or **cardiovascular system** (Figure 11–1), consists of the following structures:

- The **heart** propels blood through the system.
- **Arteries**, a series of vessels efferent from the heart that become smaller as they branch into the various organs, carry blood to the tissues.
- **Capillaries**, the smallest vessels, are the sites of O_2, CO_2, nutrient, and waste product exchange between blood and tissues. Together with the smallest arterial and venous branches carrying blood to and from them, capillaries in almost every organ form a complex network of thin, anastomosing tubules called the **microvasculature** or microvascular bed.
- **Veins** result from the convergence of venules into a system of larger channels that continue enlarging as they approach the heart, toward which they carry the blood to be pumped again.

As shown in Figure 11–1, two major divisions of arteries, microvasculature, and veins make up the **pulmonary circulation**, where blood is oxygenated in the lungs, and the **systemic circulation**, where blood brings nutrients and removes wastes in tissues throughout the body.

The **lymphatic vascular system**, introduced with the discussion of interstitial fluid in Chapter 5, begins with the **lymphatic capillaries**, which are thin-walled, closed-ended tubules carrying lymph, that merge to form vessels of steadily increasing size. The largest lymph vessels connect with the blood vascular system and empty into the large veins near the heart. This returns fluid from tissue spaces all over the body to the blood.

The internal surface of all components of the blood and lymphatic systems is lined by a simple squamous epithelium called **endothelium**. As the interface between blood and the organs, cardiovascular endothelial cells have crucial physiologic and medical importance. Not only must endothelial cells maintain a selectively permeable, antithrombogenic (inhibitory to clot formation) barrier, they also determine when and where white blood cells leave the circulation for the interstitial space of tissues and secrete a variety of paracrine factors for vessel dilation, constriction, and growth of adjacent cells.

❯ HEART

Cardiac muscle in the four chambers of the **heart** wall contracts rhythmically, pumping the blood through the circulatory system (Figure 11–2). The right and left **ventricles** propel blood to the pulmonary and systemic circulation, respectively; right and left **atria** receive blood from the body and the pulmonary veins, respectively. The walls of all four heart chambers consist of three major layers: the internal endocardium; the middle myocardium; and the external epicardium.

- The **endocardium** consists of a thin inner layer of endothelium and supporting connective tissue, a middle myoelastic layer of smooth muscle fibers and connective tissue, and a deep layer of connective tissue called the **subendocardial layer** that merges with the myocardium. Branches of the heart's impulse-conducting system, consisting of modified cardiac muscle fibers, are also located in the subendocardial layer (Figure 11–3).
- The thickest layer, the **myocardium**, consists mainly of cardiac muscle with its fibers arranged spirally around each heart chamber. Because strong force is required to

FIGURE 11–1 Diagram of the cardiovascular system.

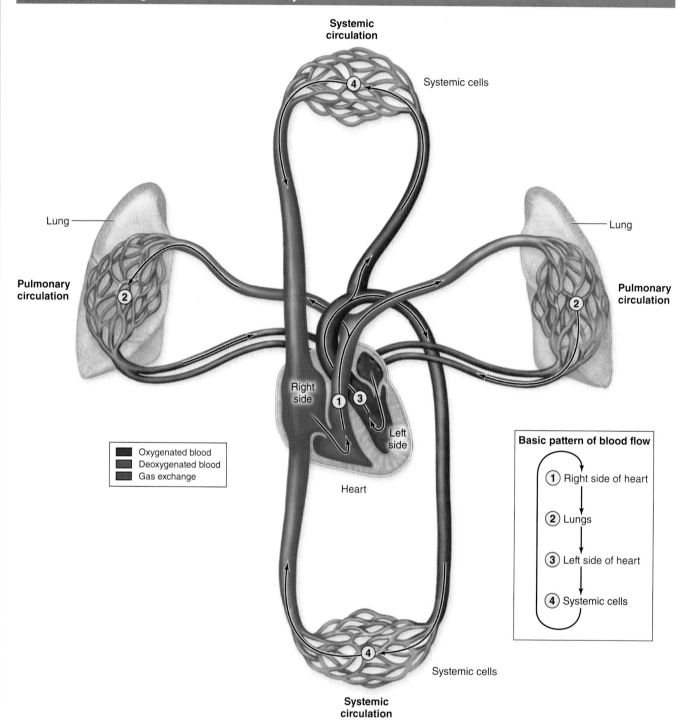

Systemic circulation

Systemic cells

Lung

Pulmonary circulation

Lung

Pulmonary circulation

Right side

Left side

Heart

Oxygenated blood
Deoxygenated blood
Gas exchange

Systemic cells

Systemic circulation

Basic pattern of blood flow

1 Right side of heart

2 Lungs

3 Left side of heart

4 Systemic cells

The system consisting of the **heart, arteries, veins**, and **micro-vascular beds** is organized as the **pulmonary circulation** and the **systemic circulation**. In the pulmonary circulation the right side of the heart pumps blood through pulmonary vessels, through the lungs for oxygenation, and back to the left side of the heart. The larger systemic circulation pumps blood from the left side of the heart through vessels supplying either the head and arms or the lower body, and back to the right side of the heart.

When the body is at rest, approximately 70% of the blood moves through the systemic circulation, about 18% through the pulmonary circulation, and 12% through the heart.

FIGURE **11–2** **Overview of the heart.**

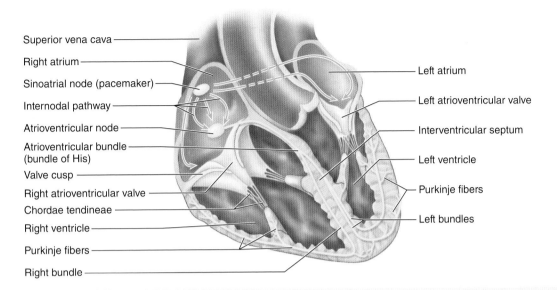

As seen in the diagram, the human heart has two **atria** and two **ventricles**. The myocardium of the ventricular walls is thicker than that of the atria. The **valves** are basically flaps of connective tissue anchored in the heart's dense connective tissue, or **cardiac skeleton**, concentrated in the regions shown in white. This fibrous tissue includes the chordae tendineae, cords that extend from the cusps of both atrioventricular valves and attach to papillary muscles, preventing the valves from turning inside-out during ventricular contraction. Valves and cords are covered by the nonthrombogenic endothelium.

Shown in yellow are parts of the cardiac **conducting system**, which initiates the electrical impulse for contraction (heartbeat) and spreads it through the ventricular myocardium. Both the **sinoatrial (SA) node (pacemaker)**, in the right atrial wall, and the **atrioventricular (AV) node**, in the floor of the right atrium, consist of myocardial tissue that is difficult to distinguish histologically from surrounding cardiac muscle. The AV node is continuous with a specialized bundle of cardiac muscle fibers, the **AV bundle** (of His), which gives rise to right and left bundle branches that run along the interventricular septum to the apex of the heart. At the apex the bundle branches subdivide further as **conducting (Purkinje) fibers** that extend into myocardium of the ventricles.

pump blood through the systemic and pulmonary circulations, the myocardium is much thicker in the walls of the ventricles, particularly the left, than in the atrial walls (Figure 11–3).

▪ The **epicardium** is a simple squamous **mesothelium** supported by a layer of loose connective tissue containing blood vessels and nerves (Figure 11–4). The epicardium corresponds to the **visceral layer of the pericardium**, the membrane surrounding the heart. Where the large vessels enter and leave the heart, the epicardium is reflected back as the **parietal layer** lining the pericardium. During heart movements, underlying structures are cushioned by deposits of adipose tissue in the epicardium and friction within the pericardium is prevented by lubricant fluid produced by both layers of serous mesothelial cells.

Within these major layers the heart contains other structures important for its overall function of moving blood. Dense fibrous connective tissue of the **cardiac skeleton** forms part of the interventricular and interatrial septa, surrounds all valves of the heart, and extends into the valve cusps and the chordae tendineae to which they are attached

(Figures 11–2 and 11–5). These regions of dense irregular connective tissue perform the following functions:

▪ Anchoring and supporting the heart valves
▪ Providing firm points of insertion for cardiac muscle
▪ Helping coordinate the heartbeat by acting as electrical insulation between atria and ventricles

Within the subendocardial layer and adjacent myocardium, modified cardiac muscle cells make up the impulse **conducting system of the heart**, which generates and propagates waves of depolarization that spread through the myocardium to stimulate rhythmic contractions. This system (Figure 11–2) consists of two nodes of specialized myocardial tissue in the right atrium: the **sinoatrial (SA) node** (or **pacemaker**) and the **atrioventricular (AV) node**, followed by the **AV bundle** (of His) and the **subendocardial conducting network**.

Located in the right atrial wall near the superior vena cava, the SA node is a 6- to 7-mm³ mass of cardiac muscle cells with smaller size, fewer myofibrils, and fewer typical intercalated disks than the neighboring muscle fibers. Impulses initiated by these cells move along the myocardial fibers of both atria, stimulating their contraction. When the impulses reach

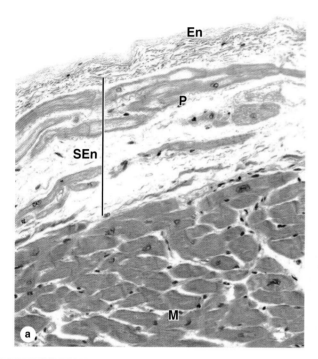

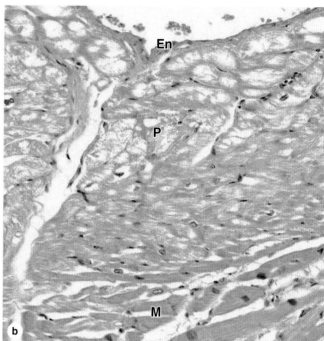

The lining layer of the heart, the **endocardium**, includes the endothelium and its supportive subendothelial connective tissue; a middle myoelastic layer of smooth muscle cells and connective tissue; and a deeper connective tissue layer of variable thickness called the **subendocardial layer**.

(a) Located below the endothelium (**En**) and myoelastic layer, the subendocardial layer (**SEn**) in the ventricles contains the conducting (Purkinje) fibers (**P**) of the heart's impulse conducting

network. These fibers are modified cardiac muscle cells joined by intercalated disks but specialized for impulse conduction rather than contraction. With glycogen filling much of the cytoplasm and displacing myofibrils to the periphery, Purkinje fibers typically are paler staining than contractile cardiac muscle fibers (**M**).

(b) In the atria Purkinje fibers (**P**) are often closer to the endothelium (**En**) and intermingle with the contractile fibers within the myocardium (**M**). (Both X200; H&E)

the slightly smaller AV node, located in the floor of the right atrium near the AV valve and composed of cells similar to those of the SA node, they stimulate depolarization of those cells. Conducting muscle fibers from the AV node form the AV bundle, pass through an opening in the cardiac skeleton into the interventricular septum, and bifurcate into the wall of each ventricle as the left and right bundle branches.

At the apex of the heart, these bundles subdivide further into a subendocardial conducting network of myofibers, usually called **Purkinje fibers**. These are pale-staining fibers, larger than the adjacent contractile muscle fibers, with sparse, peripheral myofibrils and much glycogen (Figure 11–3). Purkinje fibers mingle distally with contractile fibers of both ventricles and trigger waves of contraction through both ventricles simultaneously.

Both parasympathetic and sympathetic neural components innervate the heart. Ganglionic nerve cells and nerve fibers are present in the regions close to the SA and AV nodes, where they affect heart rate and rhythm, such as during physical exercise and emotional stress. Stimulation of the parasympathetic division (vagus nerve) slows the heartbeat, whereas stimulation of the sympathetic nerve accelerates activity of the

pacemaker. Between fibers of the myocardium are afferent free nerve endings that register pain, such as the discomfort called angina pectoris that occurs when partially occluded coronary arteries cause local oxygen deprivation.

❯❯ MEDICAL APPLICATION

Abnormalities in the structure of heart valves can be produced by developmental defects, scarring after certain infections, or cardiovascular problems such as hypertension. Such abnormal valves may not close tightly, allowing slight regurgitation and backflow of blood. This produces an abnormal heart sound referred to as a **heart murmur**. If the valve defect is severe, the heart will have to work harder to circulate the normal amount of blood, eventually enlarging to accommodate the increased workload. Defective heart valves often may be repaired surgically or replaced by an artificial valve or one from a large animal donor. Because such valve replacements lack a complete endothelial covering, the patients require exogenous anticoagulant agents to prevent thrombus formation at these sites.

FIGURE **11–4** Epicardium or visceral pericardium.

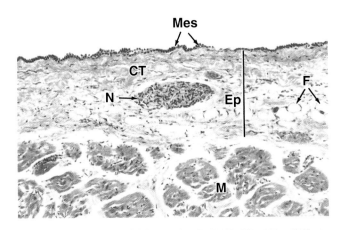

The external tunic of the heart, the epicardium, is the site of the coronary vessels and contains considerable adipose tissue. This section of atrium shows part of the myocardium (**M**) and epicardium (**Ep**). The epicardium consists of loose connective tissue (**CT**) containing autonomic nerves (**N**) and variable amounts of fat (**F**). The epicardium is the visceral layer of the pericardium and is covered by the simple mesothelium (**Mes**) that also lines the pericardial space. The mesothelial cells secrete a lubricant fluid that prevents friction as the beating heart contacts the parietal pericardium on the other side of the pericardial cavity. (X100; H&E)

FIGURE **11–5** Valve leaflet and cardiac skeleton.

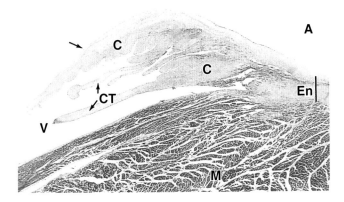

The fibrous **cardiac skeleton** consists of dense irregular connective tissue, primarily in the endocardium (**En**), which anchors the valves and surrounds the two atrioventricular canals, maintaining their proper shape. The micrograph shows a section through a cusp of an **atrioventricular valve** (arrow) and attached chordae tendineae (**CT**). These structures are largely dense connective tissue (**C**) covered with a thin layer of endothelium. The collagen-rich connective tissue of the valves is stained pale blue here and is continuous with the fibrous ring of connective tissue at the base of the valves, which fills the endocardium between the atrium (**A**) and ventricle (**V**). The thick ventricular myocardium (**M**) is also shown. (X20; Masson trichrome)

❯ TISSUES OF THE VASCULAR WALL

Walls of all blood vessels except capillaries contain smooth muscle and connective tissue in addition to the endothelial lining. The amount and arrangement of these tissues in vessels are influenced by **mechanical factors**, primarily blood pressure, and **metabolic factors** reflecting the local needs of tissues.

The **endothelium** is a specialized epithelium that acts as a semipermeable barrier between two major internal compartments: the blood and the interstitial tissue fluid. Vascular endothelial cells are squamous, polygonal, and elongated with the long axis in the direction of blood flow. Endothelium with its basal lamina is highly differentiated to mediate and actively monitor the bidirectional exchange of molecules by simple and active diffusion, receptor-mediated endocytosis, transcytosis, and other mechanisms discussed in Chapter 4.

Besides their key role in metabolite exchanges between blood and tissues, endothelial cells have several other functions:

■ The endothelium presents a **nonthrombogenic surface** on which blood will not clot and actively secretes agents that control local clot formation (such as heparin, tissue plasminogen activator, and von Willebrand factor).

■ The cells regulate local **vascular tone and blood flow** by secreting various factors that stimulate smooth muscle contraction (such as endothelin 1 and angiotensin-converting enzyme [ACE]) or relaxation (including nitric oxide [NO] and prostacyclin).

■ Endothelium has several roles in **inflammation and local immune responses**. In venules endothelial cells induce specific white blood cells to stop and undergo transendothelial migration at sites of injury or infection. Under those conditions **P-selectin** is expressed rapidly on the luminal surface when unique elongated granules, called **Weibel-Palade bodies**, fuse with the cell membrane. As described further in Chapter 12, adhesion to selectins is the first step in the activation of white blood cells specifically where they are needed. Endothelial cells also secrete various factors called **interleukins** that affect the activity of local white blood cells during inflammation.

■ Under various conditions endothelial cells secrete various **growth factors**, including proteins promoting proliferation of specific white blood cell lineages and cells that make up the vascular wall.

Growth factors such as vascular endothelial growth factor (VEGF) stimulate formation of the vascular system from embryonic mesenchyme (**vasculogenesis**), help maintain the vasculature in adults, and promote capillary sprouting and outgrowth from small existing vessels (**angiogenesis**) during normal growth, during tissue repair and regeneration, and in tumors and other pathological conditions. In both processes other growth factors, called **angiopoietins**, stimulate endothelial cells to recruit smooth muscle cells and fibroblasts to form the other tissues of the vascular wall.

The normal vascular endothelium is antithrombogenic, allowing adhesion of no blood cells or platelets and preventing blood clot formation. When endothelial cells of the microvasculature are damaged by tissue injury, collagen is exposed in the subendothelial tissues and induces the aggregation of blood platelets. These platelets release factors that initiate a cascade of events that produce fibrin from circulating plasma fibrinogen. An intravascular clot, or **thrombus** (plural, thrombi), with a fibrin framework quickly forms to stop blood loss from the damaged vessels.

From large thrombi, solid masses called **emboli** (singular, embolus) may detach and be carried by the blood to obstruct distant vessels. In both cases vascular flow may be blocked, producing a potentially life-threatening condition. Thus, the integrity of the endothelial layer preventing contact between platelets and the subendothelial connective tissue is an important antithrombogenic mechanism.

Individuals in the initial stages of medical conditions involving thrombus formation, such as myocardial infarct, stroke, or pulmonary embolism, are treated intravenously with tissue plasminogen activator, commonly abbreviated as tPA. This is a serine protease that breaks down fibrin and quickly dissolves the clot.

Smooth muscle fibers occur in the walls of all vessels larger than capillaries and are arranged helically in layers. In arterioles and small arteries, the smooth muscle cells are connected by many more gap junctions and permit vasoconstriction and vasodilation which are of key importance in regulating the overall blood pressure.

Connective tissue components are present in vascular walls in variable amounts and proportions based on local functional requirements. Collagen fibers are found in the subendothelial layer, between the smooth muscle layers, and in the outer covering. Elastic fibers provide the resiliency required for the vascular wall to expand under pressure. Elastin is a major component in large arteries where it forms parallel lamellae, regularly distributed between the muscle layers. Variations in the amount and composition of ground substance components such as proteoglycans and hyaluronate also contribute to the physical and metabolic properties of the wall in different vessels, especially affecting their permeability.

The walls of all blood vessels larger than the microvasculature have many components in common and similar organization. Branching of the vessels helps produce reductions in their size which are accompanied by gradual changes in the composition of the vascular wall. Transitions such as those from "small arteries" to "arterioles" are not clear-cut. However, all of these larger vessels have walls with three concentric layers, or tunics (L. *tunica*, coat), as shown in the diagram of Figure 11–6 and in the micrographs of Figures 11–7 through 11–9.

- The innermost tunica **intima** consists of the endothelium and a thin subendothelial layer of loose connective tissue sometimes containing smooth muscle fibers (Figure 11–7). In arteries the intima includes a thin layer, the **internal elastic lamina**, composed of elastin, with holes allowing better diffusion of substances from blood deeper into the wall.

- The tunica **media**, the middle layer, consists chiefly of concentric layers of helically arranged smooth muscle cells (Figures 11–6 and 11–7). Interposed among the muscle fibers are variable amounts of elastic fibers and elastic lamellae, reticular fibers, and proteoglycans, all of which are produced by the smooth muscle cells. In arteries the media may also have an external elastic lamina separating it from the outermost tunic.

- The outer **adventitia**, or tunica externa, is connective tissue consisting principally of type I collagen and elastic fibers (Figures 11–7 and 11–8). The adventitia is continuous with and bound to the stroma of the organ through which the blood vessel runs.

Just as the heart wall is supplied with its own coronary vasculature for nutrients and O_2, large vessels usually have **vasa vasorum** ("vessels of the vessel"): arterioles, capillaries, and venules in the adventitia and outer part of the media (Figure 11–8). The vasa vasorum are required to provide metabolites to cells in those tunics in larger vessels because the wall is too thick to be nourished solely by diffusion from the blood in the lumen. Luminal blood alone does provide the needs of cells in the intima. Because they carry deoxygenated blood, large veins commonly have more vasa vasorum than arteries.

The adventitia of larger vessels also contains a network of unmyelinated autonomic nerve fibers, the vasomotor nerves (Figure 11–8), which release the vasoconstrictor norepinephrine. The density of this innervation is greater in arteries than in veins.

》 VASCULATURE

Large blood vessels and those of the microvasculature branch frequently and undergo gradual transitions into structures with different histologic features and functions. For didactic purposes vessels can be classified arbitrarily as the types discussed here and listed in Table 11–1.

Elastic Arteries

Elastic arteries are the aorta, the pulmonary artery, and their largest branches; these large vessels are also called **conducting arteries** because their major role is to carry blood to smaller arteries. As shown in Figure 11–7a, the most prominent feature of elastic arteries is the thick tunica media in which elastic lamellae alternate with layers of smooth muscle fibers. The adult aorta has about 50 elastic lamellae (more if the individual is hypertensive).

The tunica intima is well developed, with many smooth muscle cells in the subendothelial connective tissue, and often shows folds in cross section as a result of the loss of blood

FIGURE **11–6** Walls of arteries and veins.

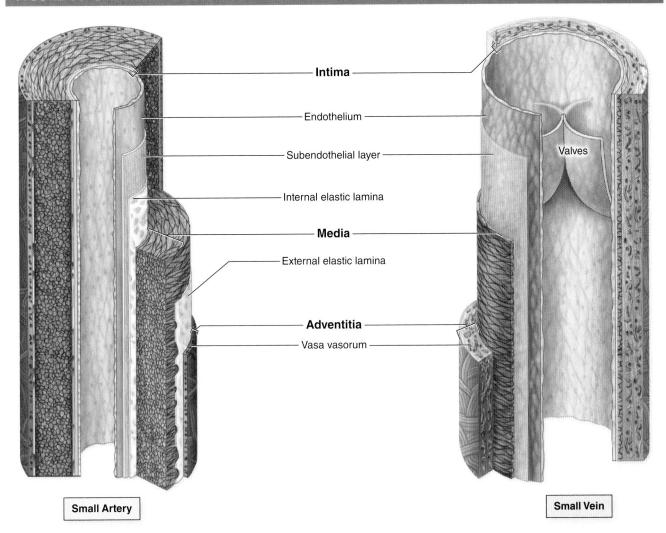

Intima

Endothelium

Subendothelial layer

Internal elastic lamina

Media

External elastic lamina

Adventitia

Vasa vasorum

Valves

Small Artery

Small Vein

Walls of both arteries and veins have three tunics called the **intima**, **media**, and the **adventitia** (or externa), which correspond roughly to the heart's endocardium, myocardium, and epicardium. An artery has a thicker media and relatively narrow lumen. A vein has a larger lumen and its adventitia is the thickest layer. The intima of veins is often folded to form **valves**. Capillaries have only an endothelium, with no subendothelial layer or other tunics.

pressure and contraction of the vessel at death (Figure 11–8). Between the intima and the media is the internal elastic lamina, which is more well-defined than the elastic laminae of the media (Figures 11–7a and 11–9). The adventitia is much thinner than the media.

The numerous elastic laminae of these arteries contribute to their important function of making the blood flow more uniform. During ventricular contraction (systole) blood is moved through the arteries forcefully and the elastin is stretched, distending the wall within the limit set by the wall's collagen. When the ventricles relax (diastole) ventricular pressure drops to a low level, but the elastin rebounds passively, helping to maintain arterial pressure. The aortic and pulmonary valves prevent backflow of blood into the heart, so the

rebound continues the blood flow away from the heart. Arterial blood pressure and blood velocity decrease and become less variable as the distance from the heart increases.

❯❯ MEDICAL APPLICATION

Atherosclerosis (Gr. *athero*, gruel or porridge, and *scleros*, hardening) is a disease of elastic arteries and large muscular arteries that may play a role in nearly half of all deaths in developed parts of the world. It is initiated by damaged or dysfunctional endothelial cells oxidizing low-density lipoproteins (LDLs) in the tunica intima, which induces adhesion and intima entry of monocytes/macrophages to remove the

FIGURE **11–7** Tunics of the vascular wall.

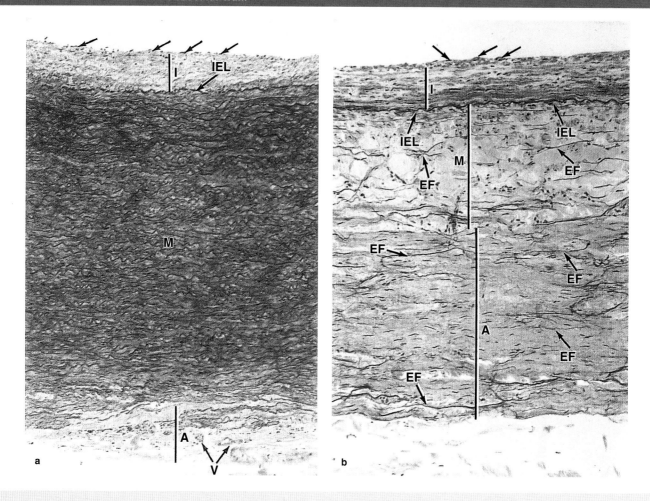

Comparison of the three major layers or tunics in the largest artery and vein. **(a)** Aorta, **(b)** vena cava. Simple squamous endothelial cells (arrows) line the intima (**I**) that also has subendothelial connective tissue and in arteries is separated from the media by an internal elastic lamina (**IEL**), a structure absent in all but the largest veins. The media (**M**) contains many elastic lamellae and elastic fibers (**EF**) alternating with layers of smooth muscle. The media is much thicker in large arteries than veins, with relatively more elastin. Elastic fibers are also present in the outer tunica adventitia (**A**), which is relatively thicker in large veins. Vasa vasorum (**V**) are seen in the adventitia of the aorta. The connective tissue of the adventitia always merges with the less dense connective tissue around it. (Both X122; Elastic stain)

modified LDL. Lipid-filled macrophages (called **foam cells**) accumulate and, along with the free LDL, produce a pathologic sign of early atherosclerosis called **fatty streaks**. During disease progression these develop into **fibro-fatty plaques**, or **atheromas**, consisting of a gruel-like mix of smooth muscle cells, collagen fibers, and lymphocytes with necrotic regions of lipid, debris, and foam cells. Predisposing factors include dyslipidemia (> 3:1 ratios of LDL to HDL [high-density lipoprotein]), hyperglycemia of diabetes, hypertension, and the presence of toxins introduced by smoking.

In elastic arteries atheromas produce localized destruction within the wall, weakening it and causing arterial bulges or **aneurysms** that can rupture. In muscular arteries such as the coronary arteries, atheromas can occlude blood flow to downstream vessels, leading to ischemic heart disease.

Arterial Sensory Structures

Carotid sinuses are slight dilations of the bilateral internal carotid arteries where they branch from the (elastic) common carotid arteries; they act as important **baroreceptors** monitoring arterial blood pressure. At these sinuses the tunica media is thinner, allowing greater distension when blood pressure rises, and the adventitia contains many sensory nerve endings from cranial nerve IX, the glossopharyngeal nerve. The brain's vasomotor centers process these afferent impulses and adjust vasoconstriction, maintaining normal blood pressure. Functionally similar baroreceptors present in the aortic arch transmit signals pertaining to blood pressure via cranial nerve X, the vagus nerve.

Histologically more complex **chemoreceptors** which monitor blood CO_2 and O_2 levels, as well as its pH, are found in the **carotid bodies** and **aortic bodies**, located in the walls

FIGURE **11–8** Vasa vasorum.

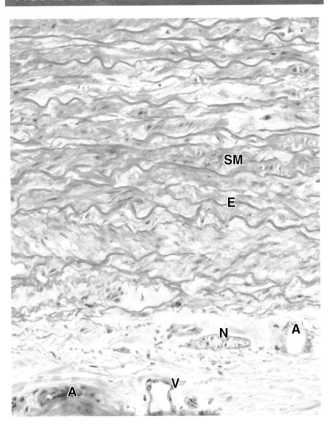

The adventitia of the larger arteries contains a supply of micro-vasculature to bring O_2 and nutrients to local cells that are too far from the lumen to be nourished by blood there. These arteri-oles (**A**), capillaries, and venules (**V**) constitute the vasa vasorum (vessels of vessels). The adventitia of large arteries is also sup-plied more sparsely with small sympathetic nerves (**N**) for con-trol of vasoconstriction. Above the adventitia in this section can be seen muscle fibers (**SM**) and elastic lamellae (**E**) in the media. (X100; H&E)

FIGURE **11–9** Elastic artery.

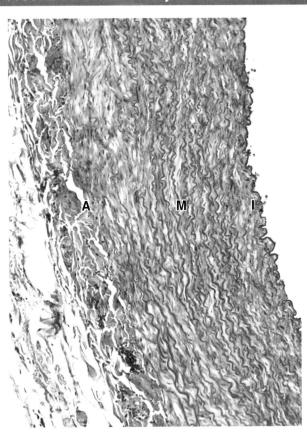

The largest arteries contain considerable elastic material and expand with blood when the heart contracts. A transverse section through part of a large elastic artery shows a thick media (**M**) consisting largely of many well-developed elastic lamellae. Strong pressure of blood pulsating into such arteries during systole expands the arterial wall, reducing the pressure and allowing strong blood flow to continue during diastole. The intima (**I**) of the empty aorta is typically folded, and the dense irregular connective tissue of the adventitia (**A**) is thinner than the media. (X200; PT)

of the carotid sinuses and aortic arch, respectively. These struc-tures are parts of the autonomic nervous system called **para-ganglia** with rich capillary networks. The capillaries are closely surrounded by large, neural crest-derived **glomus cells** filled with dense-core vesicles containing dopamine, acetylcholine, and other neurotransmitters, which are supported by smaller satellite cells (Figure 11–10). Ion channels in the glomus cell membranes respond to stimuli in the arterial blood, primar-ily hypoxia (low O_2), hypercapnia (excess CO_2), or acidosis, by activating release of neurotransmitters. Sensory fibers branch-ing from the glossopharyngeal nerve form synapses with the glomus cells and signal brain centers to initiate cardiovascular and respiratory adjustments that correct the condition.

Muscular Arteries

The muscular arteries, also called distributing arteries, dis-tribute blood to the organs and help regulate blood pressure

by contracting or relaxing the smooth muscle in the media. The intima has a thin subendothelial layer and a prominent internal elastic lamina (Figure 11–11). The media may con-tain up to 40 layers of large smooth muscle cells interspersed with a variable number of elastic lamellae (depending on the size of the vessel). An external elastic lamina is present only in the larger muscular arteries. The adventitial connective tissue contains lymphatic capillaries, vasa vasorum, and nerves, all of which may penetrate to the outer part of the media.

Arterioles

Muscular arteries branch repeatedly into smaller and smaller arteries, until reaching a size with three or four layers of medial smooth muscle. The smallest arteries branch as **arte-rioles**, which have only one or two smooth muscle layers;

TABLE 11–1	Size ranges, major features, and important roles of major blood vessel types.				
Type of Artery	Outer Diameter (Approx. Range)	Intima	Media	Adventitia	Roles in Circulatory System
Elastic arteries	> 10 mm	Endothelium; connective tissue with smooth muscle	Many elastic lamellae alternating with smooth muscle	Connective tissue, thinner than media, with vasa vasorum	Conduct blood from heart and with elastic recoil help move blood forward under steady pressure
Muscular arteries	10-1 mm	Endothelium; connective tissue with smooth muscle, internal elastic lamina prominent	Many smooth muscle layers, with much less elastic material	Connective tissue, thinner than media; vasa vasorum maybe present	Distribute blood to all organs and maintain steady blood pressure and flow with vasodilation and constriction
Small arteries	1-0.1 mm	Endothelium; connective tissue less smooth muscle	3-10 layers of smooth muscle	Connective tissue, thinner than media; no vasa vasorum	Distribute blood to arterioles, adjusting flow with vasodilation and constriction
Arterioles	100-10 μm	Endothelium; no connective tissue or smooth muscle	1-3 layers of smooth muscle	Very thin connective tissue layer	Resist and control blood flow to capillaries; major determinant of systemic blood pressure
Capillaries	10-4 μm	Endothelium only	A few pericytes only	None	Exchange metabolites by diffusion to and from cells
Venules (postcapillary, collecting, and muscular)	10-100 μm	Endothelium; no valves	Pericytes and scattered smooth muscle cells	None	Drain capillary beds; site of leukocyte exit from vasculature
Small veins	0.1-1 mm	Endothelium; connective tissue with scattered smooth muscle fibers	Thin, 2-3 loose layers of smooth muscle cells	Connective tissue, thicker than media	Collect blood from venules
Medium veins	1-10 mm	Endothelium; connective tissue, with valves	3-5 more distinct layers of smooth muscle	Thicker than media; longitudinal smooth muscle may be present	Carry blood to larger veins, with no backflow
Large veins	> 10 mm	Endothelium; connective tissue, smooth muscle cells; prominent valves	> 5 layers of smooth muscle, with much collagen	Thickest layer, with bundled longitudinal smooth muscle	Return blood to heart

these indicate the beginning of an organ's **microvasculature** (Figures 11–12 and 11–13) where exchanges between blood and tissue fluid occur. Arterioles are generally less than 0.1 mm in diameter, with lumens approximately as wide as the wall is thick (Figure 11–14). The subendothelial layer is very thin, elastic laminae are absent, and the media consists of the circularly arranged smooth muscle cells. In both small arteries and arterioles the adventitia is very thin and inconspicuous.

Arterioles almost always branch to form anastomosing networks of capillaries that surround the parenchymal cells of the organ. At the ends of arterioles the smooth muscle fibers act as sphincters and produce periodic blood flow into capillaries (Figure 11–13). Muscle tone normally keeps arterioles partially closed, resisting blood flow, which makes these vessels the major determinants of systemic blood pressure.

FIGURE **11–10** Cells and capillaries in a glomus body.

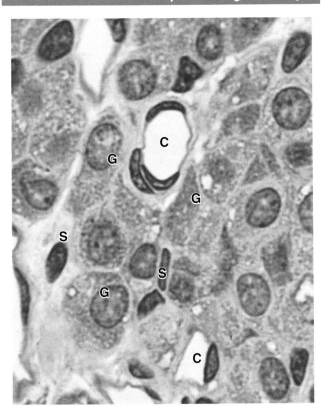

Specialized regions in the walls of certain elastic arteries contain tissues acting as chemoreceptors that provide information to the brain regarding blood chemistry. The glomus bodies are two small (0.5-5 mm diameter) ganglion-like structures found near the common carotid arteries. They contain many large capillaries (**C**) intermingled with clusters of large glomus cells (**G**) filled with vesicles of various neurotransmitters. Supportive satellite cells (**S**) with elongated nuclei ensheath each glomus cell. Glomus cells form synaptic connections with sensory fibers. Significant changes in the blood CO_2, O_2, or H^+ concentrations are detected by the chemoreceptive glomus cells, which then release a neurotransmitter that activates the sensory nerve to relay this information to the brain. (X400; PT)

❯❯ MEDICAL APPLICATION

Blood pressure depends on cardiac output and the total peripheral resistance to blood flow, which is mostly due to the resistance of arterioles. **Hypertension** or elevated blood pressure may occur secondarily to renal or endocrine problems, but is more commonly essential hypertension, due to a wide variety of mechanisms that increase arteriolar constriction.

In certain tissues and organs arterioles deviate from this simple path to accommodate various specialized functions (Figure 11–15). For example, thermoregulation by the skin involves arterioles that can bypass capillary networks and connect directly to venules. The media and adventitia are thicker

FIGURE **11–11** Muscular artery.

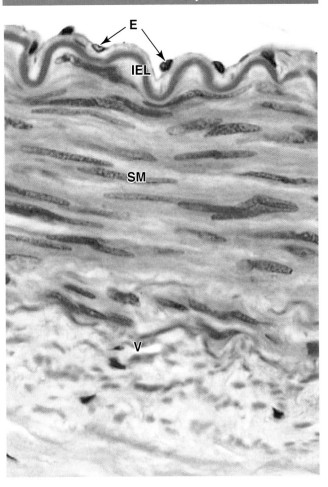

With distance from the heart, arteries gradually have relatively less elastin and more smooth muscle in their walls. Most arteries, large enough to have names, are of the muscular type. A transverse section through a muscular (medium-caliber) artery shows a slightly folded intima with only sparse connective tissue between the endothelial cells (**E**) and internal elastic lamina (**IEL**). Multiple layers of smooth muscle (**SM**) in the media are thicker than the elastic lamellae and fibers with which they intersperse. Vasa vasorum (**V**) are seen in the adventitia. (X100; H&E)

in these **arteriovenous shunts** (or arteriovenous anastomoses) and richly innervated by sympathetic and parasympathetic nerve fibers. The autonomic fibers control the degree of vasoconstriction at the shunts, regulating blood flow through the capillary beds. High capillary blood flow in the skin allows more heat dissipation from the body, while reduced capillary blood flow conserves heat—important functions when the environmental temperature is hot or cold, respectively.

Another important alternative microvascular pathway is a venous **portal system** (Figure 11–15), in which blood flows through two successive capillary beds separated by a **portal vein**. This arrangement allows for hormones or nutrients picked

FIGURE **11–12** Microvasculature.

FIGURE **11–12** Microvasculature.

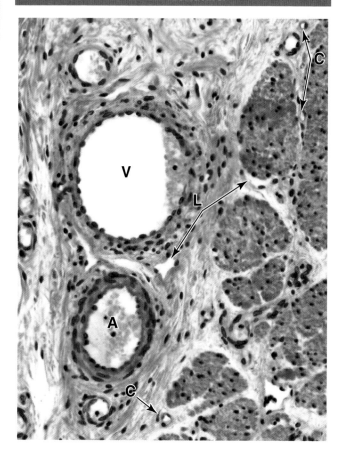

Arterioles (**A**), capillaries (**C**), and venules (**V**) comprise the micro-vasculature where, in almost every organ, molecular exchange takes place between blood and the interstitial fluid of the sur-rounding tissues. Lacking media and adventitia tunics and with diameters of only 4-10 μm, capillaries (**C**) in paraffin sections can be recognized by nuclei adjacent to small lumens or by highly eosinophilic red blood cells in the lumen. As described in Figure 5–20, not all interstitial fluid formed at capillary beds is drained into venules; the excess is called **lymph** and collects in thin-walled, irregularly shaped lymphatic vessels (**L**), such as those seen in connective tissue and smooth muscle here. (200X; H&E)

FIGURE **11–13** Microvascular bed structure and perfusion.

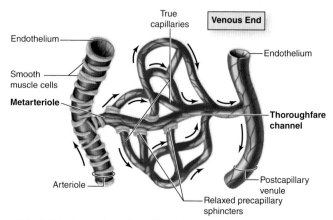

(a) Sphincters relaxed; capillary bed well perfused

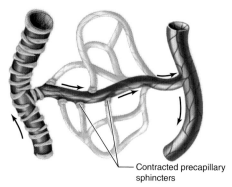

(b) Sphincters contracted; blood bypasses capillary bed

Arterioles supplying a capillary bed typically form smaller branches called **metarterioles** in which the **smooth muscle cells** are dispersed as bands that act as **precapillary sphinc-ters**. The distal portion of the metarteriole, sometimes called a **thoroughfare channel**, lacks smooth muscle cells and merges with the **postcapillary venule**. Branching from the metarteriole and thoroughfare channel are the smallest vessels, **true capil-laries**, which lack smooth muscle cells (although pericytes may be present). The precapillary sphincters regulate blood flow into the true capillaries.

Part **a** shows a well-perfused capillary bed with all the sphincters relaxed and open; part **b** shows a capillary bed with the blood shunted away by contracted sphincters. At any given moment, most sphincters are at least partially closed and blood enters the capillary bed in a pulsatile manner for maximally efficient exchange of nutrients, wastes, O_2, and CO_2 across the endothelium. Except in the pulmonary circulation (Figure 11–1), blood enters the microvasculature well oxygenated and leaves poorly oxygenated.

up by the blood in the first capillary network to be delivered most efficiently to cells around the second capillary bed before the blood is returned to the heart for general distribution. The best examples are the hepatic portal system of the liver and the hypothalamic-hypophyseal portal system in the anterior pitu-itary gland, both of which have major physiologic importance.

Capillary Beds

Capillaries permit and regulate metabolic exchange between blood and surrounding tissues. These smallest blood vessels always function in networks called **capil-lary beds**, whose size and overall shape conforms to that of the structure supplied. The density of the capillary bed is related to the metabolic activity of the tissues. Tissues with high metabolic rates, such as the kidney, liver, and cardiac and skeletal muscle, have abundant capillaries; the opposite is true of tissues with low metabolic rates, such as smooth muscle and dense connective tissue.

Capillary beds are supplied preferentially by one or more terminal arteriole branches called **metarterioles**, which

FIGURE 11–14 Arterioles.

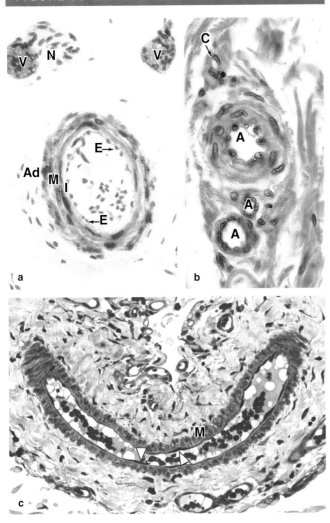

(a) Arterioles are microvessels with an intima (**I**) consisting only of endothelium (**E**), in which the cells may have rounded nuclei. They have media (**M**) tunics with only one or two layers of smooth muscle, and usually thin, inconspicuous adventitia (**Ad**). (X350; Masson trichrome)

(b) Three arterioles (**A**) of various sizes and a capillary (**C**) are shown here. (X400; H&E)

(c) A large mesenteric arteriole cut obliquely and longitudinally clearly shows the endothelial cells (arrow heads) and one or two layers of smooth muscle cells (**M**) cut transversely. Adventitia merges imperceptibly with neighboring connective tissue. (X300; PT)

FIGURE 11–15 Comparison of the simple microvascular pathway with arteriovenous shunts and portal systems.

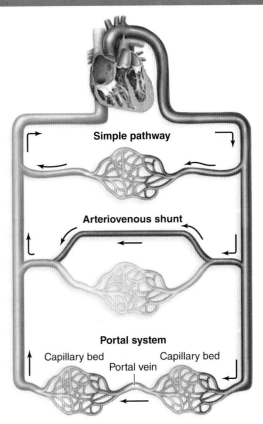

Most capillary beds are supplied by arterioles and drain into venules, but alternative pathways are found in certain organs. In skin blood flow can be varied according to external conditions by **arteriovenous (AV) shunts**, or anastomoses, commonly coiled, which directly connect the arterial and venous systems and temporarily bypass capillaries.

In **venous portal systems** one capillary bed drains into a vein that then branches again into another capillary bed. This arrangement allows molecules entering the blood in the first set of capillaries to be delivered quickly and at high concentrations to surrounding tissues at the second capillary bed, which is important in the anterior pituitary gland and liver.

Not shown are **arterial portal systems** (afferent arteriole → capillaries → efferent arteriole) which occur in the kidney.

are continuous with **thoroughfare channels** connected with the **postcapillary venules** (Figure 11–13). Capillaries branch from the metarterioles, which are encircled by scattered smooth muscle cells, and converge into the thoroughfare channels, which lack muscle. The metarteriole muscle cells act as **precapillary sphincters** that control blood flow into the capillaries. These sphincters contract and relax cyclically, with 5-10 cycles per minute, causing blood to pass through

capillaries in a pulsatile manner. When the sphincters are closed blood flows directly from the metarterioles and thoroughfare channels into postcapillary venules.

Capillaries are composed of the simple layer of **endothelial cells** rolled up as a tube surrounded by basement membrane (Figure 11–16). The average diameter of capillaries varies from 4 to 10 μm, which allows transit of blood cells only one at a time, and their individual length is usually not more than 50 μm. These minute vessels make up over 90% of the body's vasculature, with a total length of more than 100,000 km

FIGURE **11–16** Capillary with pericytes.

FIGURE **11–16** Capillary with pericytes.

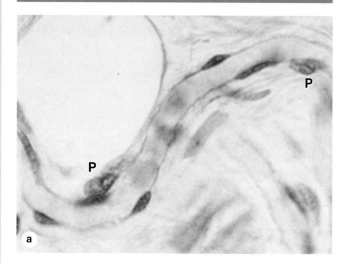

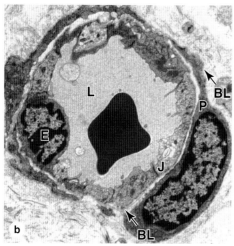

Capillaries consist only of an endothelium rolled as a tube, across which molecular exchange occurs between blood and tissue fluid. **(a)** Capillaries are normally associated with perivascular contractile cells called **pericytes (P)** that have a variety of functions. The more flattened nuclei belong to endothelial cells. (X400; H&E of a spread mesentery preparation)

(b) TEM of a capillary cut transversely, showing the nucleus of one thin capillary endothelial cell (**E**). Endothelial cells form the capillary lumen (**L**), are covered by a basal lamina (**BL**), and are bound tightly together with junctional complexes (**J**). One pericyte (**P**) is shown, surrounded by its own basal lamina (**BL**) and with cytoplasmic extensions which surround the endothelial cells. (X13,000)

In addition to the endothelial properties mentioned earlier in this chapter, capillary cells have many features specialized for molecular transfer by mechanisms ranging from simple diffusion to transcytosis. The average thickness of the cells is only 0.25 μm and their nuclei are often distinctively curved to accommodate the very small tubular structure (Figure 11–10). The cytoplasm contains mitochondria and most other organelles, as well as a large population of membranous vesicles typically. Along with the basal lamina, junctional complexes between the cells maintain the tubular structure, with variable numbers of tight junctions having an important role in capillary permeability.

Major structural variations in capillaries occur in organs with various functions that permit very different levels of metabolic exchange. Capillaries are generally grouped into three histologic types, depending on the continuity of the endothelial cells and their basement membrane (Figures 11–17 through 11–20).

- **Continuous capillaries** (Figure 11–17a) have many tight, well-developed occluding junctions between slightly overlapping endothelial cells, which provide for continuity along the endothelium and well-regulated metabolic exchange across the cells. This is the most common type of capillary and is found in muscle, connective tissue, lungs, exocrine glands, and nervous tissue. Ultrastructural studies show numerous vesicles indicating transcytosis of macromolecules in both directions across the endothelial cell cytoplasm.

- **Fenestrated capillaries** (Figure 11–17b) have a sievelike structure that allows more extensive molecular exchange across the endothelium. The endothelial cells are penetrated by numerous small circular openings or fenestrations (L. *fenestra*, perforation), approximately 80 nm in diameter. Some fenestrations are covered by very thin diaphragms of proteoglycans (Figure 11–19); others may represent membrane invaginations during transcytosis that temporarily involve both sides of the very thin cells. The basement membrane however is continuous and covers the fenestrations. Fenestrated capillaries are found in organs with rapid interchange of substances between tissues and the blood, such as the kidneys, intestine, choroid plexus, and endocrine glands.

- **Discontinuous capillaries**, commonly called **sinusoids** (Figure 11–17c), permit maximal exchange of macromolecules as well as allow easier movement of cells between tissues and blood. The endothelium here has large perforations without diaphragms and irregular intercellular clefts, forming a discontinuous layer with spaces between and through the cells. Unlike other capillaries sinusoids also have highly discontinuous basement membranes and much larger diameters, often 30-40 μm, which slows blood flow. Sinusoidal capillaries of this type are found in the liver, spleen, some endocrine organs, and bone marrow (Figure 11–20).

and a total surface area of approximately 5000 m². Because of the cyclical opening and closing of the sphincters, most capillaries are essentially empty at any given time, with only about 5% (~300 mL in an adult) of the total blood volume moving through these structures. Their thin walls, extensive surface area, and slow, pulsatile blood flow optimize capillaries for the exchange of water and solutes between blood and tissues.

FIGURE **11–17** Types of capillaries.

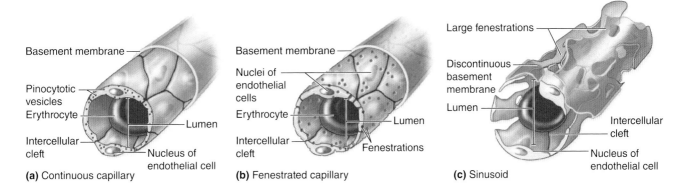

(a) Continuous capillary

Basement membrane

Pinocytotic vesicles
Erythrocyte
Intercellular cleft
Lumen
Nucleus of endothelial cell

(b) Fenestrated capillary

Basement membrane
Nuclei of endothelial cells
Erythrocyte
Intercellular cleft
Lumen
Fenestrations

(c) Sinusoid

Large fenestrations
Discontinuous basement membrane
Lumen
Intercellular cleft
Nucleus of endothelial cell

The vessels between arterioles and venules can be any of three types. **(a) Continuous capillaries**, the most common type, have tight, occluding junctions sealing the intercellular clefts between all the endothelial cells to produce minimal fluid leakage. All molecules exchanged across the endothelium must cross the cells by diffusion or transcytosis.

(b) Fenestrated capillaries also have tight junctions, but perforations (fenestrations) through the endothelial cells allow greater exchange across the endothelium. The basement membrane is continuous in both these capillary types. Fenestrated capillaries are found in organs where molecular exchange with the blood is important, such as endocrine organs, intestinal walls, and choroid plexus.

(c) Sinusoids, or discontinuous capillaries, usually have a wider diameter than the other types and have discontinuities between the endothelial cells, large fenestrations through the cells, and a partial, discontinuous basement membrane. Sinusoids are found in organs where exchange of macromolecules and cells occurs readily between tissue and blood, such as in bone marrow, liver, and spleen.

At various locations along continuous capillaries and postcapillary venules are mesenchymal cells called **pericytes** (Gr. *peri*, around + *kytos*, cell), with long cytoplasmic processes partly surrounding the endothelial layer. Pericytes secrete many ECM components and form their own basal lamina, which fuses with the basement membrane of the endothelial cells (Figure 11–16). Well-developed cytoskeletal networks of myosin, actin, and tropomyosin indicate that pericytes also dilate or constrict capillaries, helping to regulate blood flow in some organs. Within the CNS pericytes are important for maintaining the endothelial blood-brain barrier. After injuries pericytes proliferate and differentiate to form smooth muscle and other cells in new vessels as the microvasculature is reestablished. In many organs the pericyte population also includes mesenchymal stem cells important for regeneration of other tissues.

›› MEDICAL APPLICATION

The **hyperglycemia** or excessive blood sugar that occurs with diabetes commonly leads to **diabetic microangiopathy**, a diffuse thickening of capillary basal laminae and concomitant decrease in metabolic exchange at these vessels, particularly in the kidneys, retina, skeletal muscle, and skin.

Venules

The transition from capillaries to venules occurs gradually. Postcapillary venules (Figure 11–21a) are similar to capillaries with pericytes but larger, ranging in diameter from 15 to 20 μm. As described with blood in Chapter 12, postcapillary venules are the primary site at which white blood cells adhere to endothelium and leave the circulation at sites of infection or tissue damage.

Postcapillary venules converge into larger **collecting venules** that have more distinct contractile cells. With increasing size venules become surrounded by a recognizable tunica media with two or three smooth muscle layers and are called **muscular venules**. A characteristic feature of all venules is the large diameter of the lumen compared to the overall thinness of the wall (Figure 11–21).

Veins

Veins carry blood back to the heart from microvasculature all over the body. Blood entering veins is under very low pressure and moves toward the heart by contraction of the smooth muscle fibers in the media and by external compressions from surrounding skeletal muscles and other organs. Most veins are classified as **small** or **medium veins** (Figure 11–22), with diameters of 10 mm or less (Table 11–1). These veins are usually located close and parallel to corresponding muscular arteries. The tunica intima is usually thin, the media has small bundles of smooth muscle cells mixed with a network of reticular fibers and delicate elastic fibers, and the collagenous adventitial layer is thick and well developed.

The big venous trunks, paired with elastic arteries close to the heart, are the **large veins** (Figure 11–7b). These have well-developed intimal layers, but relatively thin media with alternating smooth muscle and connective tissue. The tunica adventitia is thicker than the media in large veins and frequently contains longitudinal bundles of smooth muscle.

FIGURE 11–18   Continuous capillary.

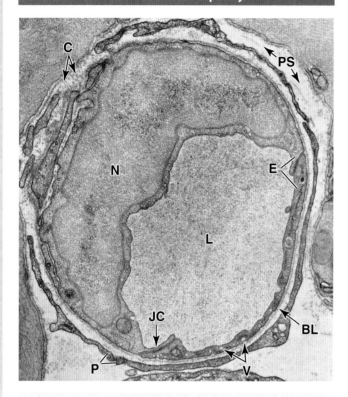

Continuous capillaries exert the tightest control over what mole-cules leave and enter the capillary lumen (**L**). The TEM shows a con-tinuous capillary in transverse section. An endothelial cell nucleus (**N**) is prominent, and tight or occluding junctions are abundant in the junctional complexes (**JC**) at overlapping folds between the endothelial cells (**E**). Numerous transcytotic vesicles (**V**) are evident. All material that crosses continuous capillary endothelium must pass *through* the cells, usually by diffusion or transcytosis.

Around the capillary are a basal lamina (**BL**) and thin cyto-plasmic extensions from pericytes (**P**). Collagen fibers (**C**) and other extracellular material are present in the perivascular space (**PS**). (X10,000)

Both the media and adventitia contain elastic fibers, and an internal elastic lamina like those of arteries may be present.

An important feature of large and medium veins are **valves**, which consist of thin, paired folds of the tunica intima projecting across the lumen, rich in elastic fibers and covered on both sides by endothelium (Figures 11–22 and 11–23). The valves, which are especially numerous in veins of the legs, help keep the flow of venous blood directed toward the heart.

〉〉 MEDICAL APPLICATION

Junctions between endothelial cells of postcapillary venules are the loosest of the microvasculature. This facilitates tran-sendothelial migration of leukocytes at these locations during **inflammation**, as well as a characteristic loss of fluid here during the inflammatory response, leading to tissue **edema**.

FIGURE 11–19   Fenestrated capillary.

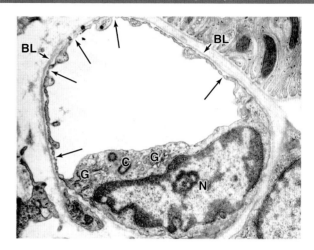

Fenestrated capillaries are specialized for uptake of molecules such as hormones in endocrine glands or for outflow of molecules such as in the kidney's filtration system. TEM of a transversely sec-tioned fenestrated capillary in the peritubular region of the kidney shows many typical fenestrae closed by diaphragms (arrows), with a continuous basal lamina surrounding the endothelial cell (**BL**). In this cell the Golgi apparatus (**G**), nucleus (**N**), and centrioles (**C**) can also be seen. Fenestrated capillaries allow a freer exchange of molecules than continuous capillaries and are found in the intesti-nal wall, kidneys, and endocrine glands. (X10,000)

(*Used with permission from Dr Johannes Rhodin, Department of Cell Biology, New York University School of Medicine.*)

FIGURE 11–20   Sinusoidal capillary.

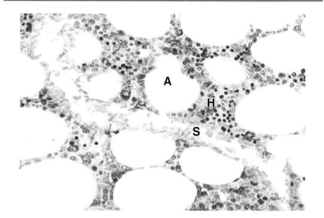

Sinusoidal capillaries or sinusoids generally have much greater diameters than most capillaries and are specialized not only for maximal molecular exchange between blood and surround-ing tissue but also for easy movement of blood cells across the endothelium. The sinusoid (**S**) shown here is in bone marrow and is surrounded by tissue containing adipocytes (**A**) and masses of hematopoietic cells (**H**). The endothelial cells are very thin and cell nuclei are more difficult to find than in smaller cap-illaries. Ultrastructurally sinusoidal capillaries are seen to have large fenestrations through the cells and large discontinuities between the cells and through the basal lamina. (X200; H&E)

FIGURE **11–21** Venules.

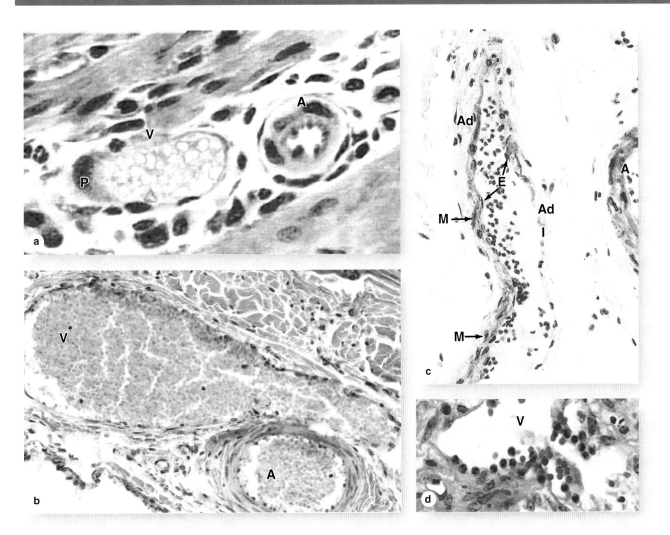

A series of increasingly larger and more organized venules lie between capillaries and veins.

(a) Compared to arterioles (**A**), postcapillary venules (**V**) have large lumens and an intima of simple endothelial cells, with occasional pericytes (**P**). (X400; Toluidine blue [TB])

(b) Larger collecting venules (**V**) have much greater diameters than arterioles (**A**), but the wall is still very thin, consisting of an endothelium with more numerous pericytes or smooth muscle cells. (X200; H&E)

(c) The muscular venule cut lengthwise here has a better defined tunica media, with as many as three layers of smooth muscle (**M**) in

some areas, a very thin intima (**I**) of endothelial cells (**E**), and a more distinct adventitia (**Ad**). Part of an arteriole (**A**) shows a thicker wall than the venule. (X200; Masson trichrome)

As discussed with white blood cells in Chapter 12, postcapillary venules are important as the site in the vasculature where these cells leave the circulation to become functional in the interstitial space of surrounding tissues when such tissues are inflamed or infected.

(d) Postcapillary venule (**V**) from an infected small intestine shows several leukocytes adhering to and migrating across the intima. (X200; H&E)

❯ LYMPHATIC VASCULAR SYSTEM

In addition to the blood vasculature, the body has a system of very thin-walled channels, the **lymphatic capillaries**, which collect excess interstitial fluid from the tissue spaces as **lymph** and return it to the blood. Like the interstitial fluid, lymph is usually rich in lightly staining proteins but does not normally

contain red blood cells, although lymphocytes and other white blood cells may normally be present (Figure 11–24a). With exceptions such as the bone marrow and most of the CNS, most tissues with blood microvasculature also contain lymphatic capillaries (or lymphatics)

Lymphatic capillaries originate locally as tubes of very thin endothelial cells which lack tight junctions and rest on a

FIGURE **11–22** Veins.

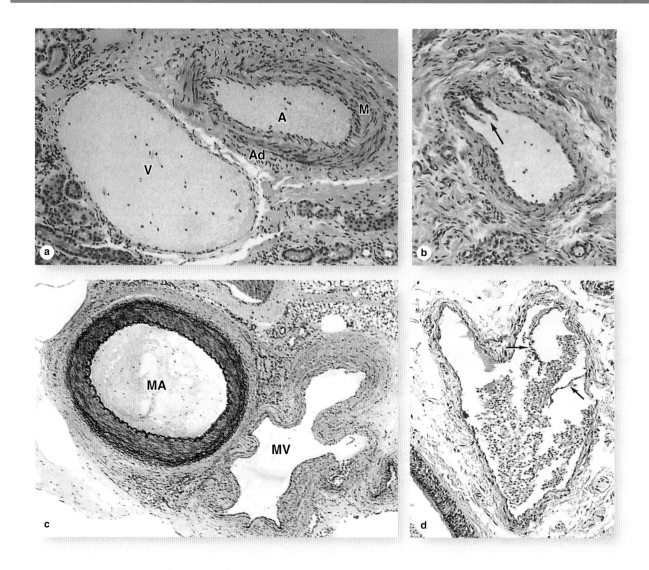

Veins usually travel as companions to arteries and are classified as small, medium, or large based on size and development of the tunics.

(a) Micrograph of small vein (**V**) shows a relatively large lumen compared to the small muscular artery (**A**) with its thick media (**M**) and adventitia (**Ad**). The wall of a small vein is very thin, containing only two or three layers of smooth muscle. (X200; H&E)

(b) Micrograph showing valve in an oblique section of a small vein (arrow). Valves are thin folds of intima projecting well into the

lumen, which act to prevent backflow of blood. (X200; Aldehyde fuchsin & van Gieson)

(c) Micrograph of a medium vein (**MV**) shows a thicker wall but still less prominent than that of the accompanying muscular artery (**MA**). Both the media and adventitia are better developed, but the wall is often folded around the relatively large lumen (X100; Aldehyde fuchsin & van Gieson).

(d) Micrograph of a medium vein contains blood and shows valve folds (arrows). (X200; Masson trichrome)

discontinuous basal lamina. Fine anchoring filaments of collagen extend from the basal lamina to the surrounding connective tissue, preventing collapse of the vessels. Interstitial fluid enters lymphatic capillaries by flowing between endothelial cells and by transcytosis. Specific domains of adjacent endothelial cells also lack hemidesmosome connections to the basal lamina and extend into the lumen to form leaflets of valves

facilitating fluid entry and preventing most backflow of lymph (Figure 11–24b).

Lymphatic capillaries converge into larger **lymphatic vessels** with thin walls and increasing amounts of connective tissue and smooth muscle which never form clearly distinct outer tunics (Figure 11–25). Like veins lymphatic vessels have valves comprised of complete intimal folds. Interposed

FIGURE **11–23** Wall of large vein with valve.

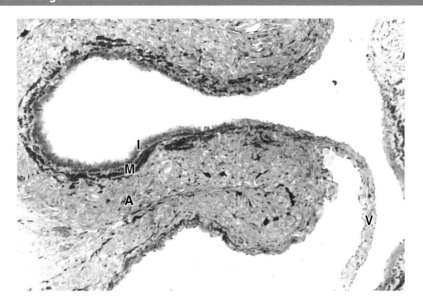

Large veins have a muscular media layer (**M**) that is very thin compared to the surrounding adventitia (**A**) of dense irregular connective tissue. The wall is often folded as shown here, with the intima (**I**) projecting into the lumen as a valve (**V**) composed of the subendothelial connective tissue with endothelium on both sides. (X100; PT)

FIGURE **11–24** Lymphatic capillary.

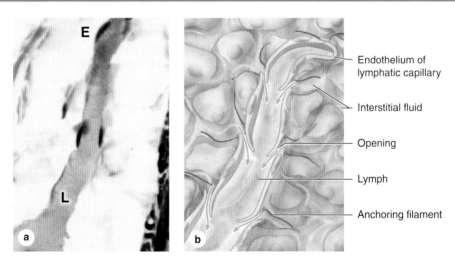

Endothelium of lymphatic capillary

Interstitial fluid

Opening

Lymph

Anchoring filament

Lymphatic capillaries drain interstitial fluid produced when the plasma forced from the microvasculature by hydrostatic pressure does not all return to blood by the action of osmotic pressure. **(a)** Micrograph shows a lymphatic capillary filled with this fluid called lymph (**L**). Lymphatics are blind-ended vessels with a wall of very thin endothelial cells (**E**) and are quite variable in diameter (10-50 μm). Lymph is rich in proteins and other material and often stains somewhat better than the surrounding ground substance, as seen here. (X200; Mallory trichrome)

(b) Diagram indicating more details about lymphatics, including the **openings between the endothelial cells**. The openings are held in place by **anchoring filaments** containing elastin and are covered by extensions of the endothelial cells. **Interstitial fluid** enters primarily via these openings, and the endothelial folds prevent backflow of lymph into tissue spaces. Lymphatic endothelial cells are typically larger than those of blood capillaries.

FIGURE **11–25** Lymphatic vessels and valve.

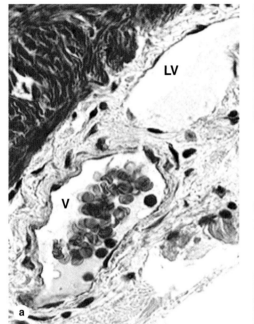

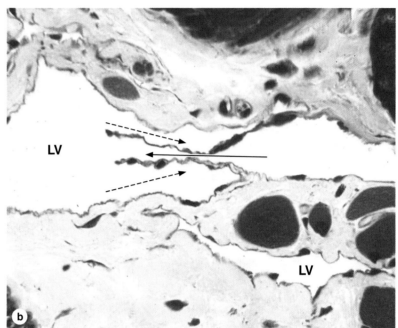

Lymphatic vessels are formed by the merger of lymphatic capillaries, but their walls remain extremely thin. **(a)** Cross section shows a lymphatic vessel (**LV**) near a venule (**V**), whose wall is thick by comparison. Lymphatic vessels normally do not contain red blood cells, which provides another characteristic distinguishing them from venules. (X200; Mallory trichrome)

(b) Lymphatic vessel (**LV**) in muscle cut longitudinally shows a valve, the structure responsible for the unidirectional flow of lymph. The solid arrow shows the direction of the lymph flow, and the dotted arrows show how the valves prevent lymph backflow. The lower small lymphatic vessel is a lymphatic capillary with a wall consisting only of endothelium. (X200; PT)

in the path of these larger lymphatic vessels are **lymph nodes**, where lymph is processed by cells of the immune system (discussed in Chapter 14). In histological sections lymphatic vessels are often dilated with lymph. As in veins, lymphatic circulation is aided by external forces (eg, contraction of surrounding skeletal muscle) with the valves keeping lymph flow unidirectional.

Lymphatic vessels ultimately converge as two large trunks: the **thoracic duct** and the **right lymphatic duct**, which empty lymph back into the blood. The thoracic duct connects with the blood circulatory system near the junction of the left internal jugular vein with the left subclavian vein, whereas the right lymphatic duct enters near the confluence of the right subclavian vein and the right internal jugular vein. The structure of these largest lymphatic vessels is similar to that of small veins. The adventitia is relatively underdeveloped, but contains vasa vasorum and a neural network.

Besides gathering interstitial fluid as lymph and returning it to the blood, the lymphatic vascular system is a major distributor of lymphocytes, antibodies, and other immune components which are carried through many organs to and from lymph nodes and other lymphoid tissues.

❯❯ MEDICAL APPLICATION

Lymphatics and larger lymphatic vessels are clinically important because (among other reasons) they facilitate the spread of pathogens, parasites, and malignant cells in the body. Surgical removal of lymph nodes, standard procedure to determine the occurrence of cancer metastasis, can disrupt the lymphatic drainage and produce swelling or **lymphedema**, in tissues of the affected region.

The Circulatory System SUMMARY OF KEY POINTS

Heart

- The heart has three major layers: (1) the inner **endocardium** of endothelium and subendothelial connective tissue; (2) the **myocardium** of cardiac muscle; and (3) the **epicardium**, connective tissue with many adipocytes and covered by mesothelium.
- The **cardiac conducting system** stimulates rhythmic contractions and consists of modified cardiac muscle fibers forming the **sinoatrial (SA) and atrioventricular (AV) nodes**, the **atrioventricular bundle (of His)**, left and right bundle branches, and **Purkinje fibers**.
- **Purkinje fibers**, located just beneath the endocardium of both ventricles, are distinguished from contractile fibers by their greater diameter, abundant glycogen, and more sparse bundles of myofibrils.
- Masses of dense irregular connective tissue make up the **cardiac skeleton**, which surrounds the bases of all heart valves, separates the atria from the ventricles, and provides insertions for cardiac muscle.

Vasculature

- Macroscopically visible blood vessels have three major layers or **tunics**: (1) The **intima** includes the endothelium, connective tissue, and an internal elastic lamina in larger vessels; (2) the **media** contains alternating layers of smooth muscle and collagen or elastic lamellae; and (3) the **adventitia** (or externa) contains connective tissue, small vessels (**vasa vasorum**), and nerves.
- Through the vasculature, **endothelial cells** are not simply heart and vessel liners; they actively produce factors that prevent blood clotting, factors that cause adjacent smooth muscle cells to contract or relax, and factors that initiate inflammation at sites of damage or infection.
- Arteries are grouped by size and wall composition: (1) **large elastic arteries**, with fenestrated elastic laminae in the thick tunica media; (2) **muscular, medium-sized arteries**; and (3) **small arteries**, with fewer than 10 layers of smooth muscle in the media.
- A **microvasculature** too small for surgical manipulation permeates most organs and consists of (1) **arterioles**, with one to three smooth muscle layers; (2) **capillaries**, consisting only of an intima endothelial layer; and (3) **venules**, with large lumens and thin walls, which drain capillaries.

- Terminal arterioles branch into **metarterioles**, in which smooth muscle **sphincters** contract to resist blood flow and relax cyclically to allow pulsatile flow of blood into an anastomosing **capillary bed**, where **metabolic exchange** with surrounding cells occurs.
- Capillaries are classified as three structural and functional types, with features that allow different degrees of molecular or even cellular exchange: (1) **continuous capillaries** with many tight junctions so that all exchange must occur through the cells; (2) **fenestrated capillaries** with small pores or fenestrations through the cells; and (3) **discontinuous capillaries**, or **sinusoids**, with larger lumens, large spaces between the endothelial cells, and a discontinuous basal lamina.
- Capillary beds generally drain into **venules**, the last segment of the microvasculature; **postcapillary venules** are the sites at which white blood cells enter damaged or infected tissues.
- The endothelium of continuous capillaries and postcapillary venules is frequently surrounded by thin cells called **pericytes**, whose contractions facilitate blood flow and which can give rise to smooth muscle and connective tissue during microvascular remodeling or repair.
- Two alternative microvascular pathways include **arteriovenous anastomoses**, or **AV shunts**, in which arterioles can bypass a capillary bed, and **venous portal systems**, in which venules draining a capillary bed quickly branch again to form another capillary bed.
- **Small, medium, and large veins**, all with lumen diameters exceeding the thickness of the wall, carry blood back to the heart, with intimal **valves** preventing backflow, and have increasingly well-developed tunics.

Lymphatic Vessels

- Interstitial fluid that is not pulled into venules by colloidal osmotic pressure drains as **lymph** into blind vessels called **lymphatics**, or **lymphatic capillaries**, which have very thin endothelial cell walls with spaces between the cells.
- Lymphatics converge into larger, thin-walled **lymphatic vessels** in which lymph is propelled by movements of surrounding muscles and organs, with intimal **valves** keeping the flow unidirectional.
- The largest lymphatic vessels, the **thoracic duct** and **right lymphatic duct**, both with walls having tunics like those of veins, return lymph to the circulatory system by joining veins near the heart.

The Circulatory System ASSESS YOUR KNOWLEDGE

1. Vasa vasorum serve a function analogous to that of which of the following?
   a. Valves
   b. Basal lamina
   c. Coronary arteries
   d. Endothelial diaphragms
   e. Arterioles

2. What tissue is directly associated with and extends into the heart valves?
   a. Myocardium
   b. Epicardium
   c. Atrioventricular bundle of His
   d. Cardiac skeleton
   e. Pericardium

3. Which of the following is true for ventricles?
   a. Located at the base of the heart
   b. Myocardial cells contains abundant granules
   c. Receive blood directly from the venae cavae and pulmonary veins
   d. Walls contain Purkinje fibers of the right and left branches from the atrioventricular bundle
   e. Contain more elastic fibers than the atria

4. Individuals with Marfan syndrome have mutations in the fibrillin gene and commonly experience aortic aneurysms. What portion of the arterial wall is most likely to be affected by the malformed fibrillin?
   a. Endothelium
   b. Tunica intima
   c. Tunica media
   d. Tunica adventitia
   e. Vasa vasorum

5. Which description is true of continuous capillaries?
   a. Unusually wide lumens
   b. Most common in both brain and muscle
   c. Abundant fenestrations
   d. Lack a complete basement membrane
   e. Phagocytic cells often seen inserted in the intercellular clefts

6. Which of the following is true of pericytes?
   a. Are associated with the basal lamina of capillary endothelial cells
   b. Have similar histological features as contractile cells of the myocardium
   c. Form a layer of cells joined by gap junctions
   d. Are terminally differentiated
   e. Capable of forming multinucleated muscle fibers

7. During light microscopic examination of a tissue, you note a vessel that has no smooth muscle but a large amount of connective tissue at its periphery. Which of the following vessels are you examining?
   a. Arteriole
   b. Venule
   c. Elastic artery
   d. Capillary
   e. Large vein

8. A 43-year-old woman notices a lump in her left breast which upon pathological examination of a needle biopsy is diagnosed as stage 3 adenocarcinoma of the mammary gland. She elects to have a single mastectomy and the surgeon also removes several axillary lymph nodes to be examined to determine the tumor's state of metastasis. The patient recovers well from the surgery, but at a 6-month follow-up visit at the clinic her upper left arm is seen to be swollen and the surgeon prescribes a bandage wrap for "lymph edema." This condition likely resulted from which of the following?
   a. Angiogenesis from arterial branches that brought blood to the left breast
   b. Growth of cancer cells and blockage of lymphatic drainage from the left arm
   c. Surgical disruption of the left arm's lymphatic drainage by removal of lymph nodes
   d. Surgical damage to the thoracic duct during lymph node removal
   e. Hypertrophy of the vessels in the upper arm to accommodate blood otherwise flowing to the left breast

9. A 66-year-old man diagnosed with type II diabetes 10 years earlier presents with an aching pain in the muscles of his lower extremities. He says the pain is relieved by rest and worsened by physical activity. His lower limbs appear cold, pale, discolored, and he has a sore on the skin of his left heel. He has a weak tibial pulse on both sides and poor skin filling from dermal capillaries. The problems with blood distribution in this patient's leg are most likely associated with what vascular structures?
   a. Veins and venules
   b. Arterioles
   c. Branches of the aorta
   d. Lymphatic vessels
   e. Ventricles

10. A 62-year-old African American man presents with exercise-induced angina. His serum cholesterol is 277 mg/dL (normal < 200), LDL is 157 (normal < 100), HDL is 43 (normal > 35), and triglycerides 170 (normal < 150). His body mass index (BMI) is 34 and his coronary risk ratio is 6.84 (normal < 5). Cardiac catheterization reveals an occlusion of the left anterior descending and the origin of the right coronary artery. This disease process initially involved which one of the following?
   a. Smooth muscle cell proliferation
   b. Formation of an intimal plaque
   c. Intimal thickening through addition of collagen and elastin
   d. Adventitial proliferation of fibroblasts
   e. Injury to endothelial cells

12 Blood

Blood is a specialized connective tissue consisting of cells and fluid extracellular material called **plasma**. Propelled mainly by rhythmic contractions of the heart, about 5 L of blood in an average adult moves unidirectionally within the closed circulatory system. The so-called **formed elements** circulating in the plasma are **erythrocytes** (red blood cells), **leukocytes** (white blood cells), and **platelets**.

When blood leaves the circulatory system, either in a test tube or in the extracellular matrix (ECM) surrounding blood vessels, plasma proteins react with one another to produce a clot, which includes formed elements and a pale yellow liquid called **serum**. Serum contains growth factors and other proteins released from platelets during clot formation, which confer biological properties very different from those of plasma.

Collected blood in which clotting is prevented by the addition of anticoagulants (eg, heparin or citrate) can be separated by centrifugation into layers that reflect its heterogeneity (Figure 12–1). Erythrocytes comprise the sedimented material and their volume, normally about 44% of the total blood volume in healthy adults, is called the **hematocrit**.

The straw-colored, translucent, slightly viscous supernatant comprising 55% at the top half of the centrifugation tube is the plasma. A thin gray-white layer called the **buffy coat** between the plasma and the hematocrit, about 1% of the volume, consists of leukocytes and platelets, both less dense than erythrocytes.

Blood is a distributing vehicle, transporting O_2, CO_2, metabolites, hormones, and other substances to cells throughout the body. Most O_2 is bound to hemoglobin in erythrocytes and is much more abundant in arterial than venous blood (Figure 12–2), while CO_2 is carried in solution as CO_2 or HCO_3^-, in addition to being hemoglobin-bound. Nutrients are distributed from their sites of synthesis or absorption in the gut, while metabolic residues are collected from cells throughout the body and removed from the blood by the excretory organs. Hormone distribution in blood permits the exchange of chemical messages between distant organs regulating normal organ function. Blood also participates in heat distribution, the regulation of body temperature, and the maintenance of acid-base and osmotic balance.

Leukocytes have diverse functions and are one of the body's chief defenses against infection. These cells are generally spherical and inactive while suspended in circulating blood, but, when called to sites of infection or inflammation, they cross the wall of venules, become motile and migrate into the tissues, and display their defensive capabilities.

COMPOSITION OF PLASMA

Plasma is an aqueous solution, pH 7.4, containing substances of low or high molecular weight that make up 7% of its volume. As summarized in Table 12–1, the dissolved components are mostly plasma proteins, but they also include nutrients, respiratory gases, nitrogenous waste products, hormones, and inorganic ions collectively called **electrolytes**. Through the capillary walls, the low-molecular-weight components of plasma are in equilibrium with the interstitial fluid of the tissues. The composition of plasma is usually an indicator of the mean composition of the extracellular fluids in tissues.

The major plasma proteins include the following:

- **Albumin,** the most abundant plasma protein, is made in the liver and serves primarily to maintain the osmotic pressure of the blood.
- **Globulins (α- and β-globulins),** made by liver and other cells, include transferrin and other transport

FIGURE 12–1 Composition of whole blood.

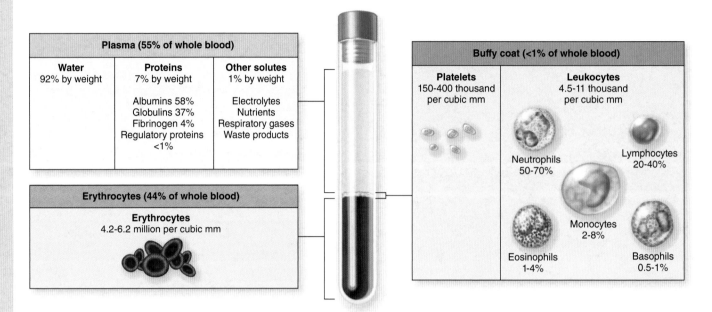

A tube of blood after centrifugation (center) has nearly half of its volume represented by erythrocytes in the bottom half of the tube, a volume called the **hematocrit**. Between the sedimented erythrocytes and the supernatant light-colored plasma is a thin layer of leukocytes and platelets called the **buffy coat**. The concentration ranges of erythrocytes, platelets, and leukocytes in normal blood are included here, along with the **differential count** or percent range for each type of leukocyte represented in the buffy coat. A cubic millimeter of blood is equivalent to a microliter (µL). (Complete blood count [CBC] values in this chapter are those used by the US National Board of Medical Examiners.)

FIGURE 12–2 Blood O_2 content in each type of blood vessel.

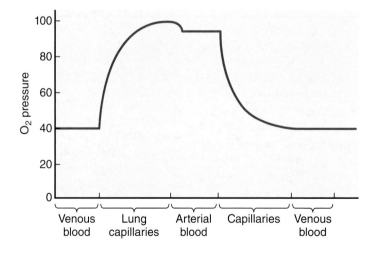

The amount of O_2 in blood (the O_2 pressure) is highest in arteries and lung capillaries and decreases in tissue capillaries, where exchange of O_2 and CO_2 occurs between blood and tissues.

TABLE 12–1	The composition of blood plasma.
Plasma Component (Percentage of Plasma)	**Functions**
Water (~92% of plasma)	Is the solvent in which formed elements are suspended and proteins and solutes are dissolved
Plasma proteins (~7% of plasma)	All proteins serve to buffer against pH changes
Albumin (~58% of plasma proteins)	Exerts osmotic force to retain fluid within the microvasculature
	Contributes to blood's viscosity
	Binds and transports some fatty acids, electrolytes, hormones and drugs
Globulins (~37% of plasma proteins)	α-Globulins transport lipids and some metal ions
	β-Globulins transport iron ions and lipids in bloodstream
	γ-Globulins are antibodies with various immune functions
Fibrinogen (~4% of plasma proteins)	Participates in blood coagulation (clotting); precursor of fibrin
Regulatory proteins (>1% of plasma proteins)	Consists of enzymes, proenzymes, hormones, and the complement system
Other Solutes (~1% of Blood Plasma)	
Electrolytes (eg, sodium, potassium, calcium, chloride, iron, bicarbonate, and hydrogen)	Help establish and maintain membrane potentials, maintain pH balance, and regulate osmosis (control of the percentages of water and salt in the blood)
Nutrients (eg, amino acids, glucose, cholesterol, vitamins, fatty acids)	Energy source; precursor for synthesizing other molecules
Respiratory gases (eg, oxygen: > 2% dissolved in plasma, 98% bound to hemoglobin within erythrocytes; and carbon dioxide: ~7% dissolved in plasma, ~27% bound to hemoglobin within erythrocytes, ~66% converted to HCO_3^-)	Oxygen is needed for aerobic cellular respiration; carbon dioxide is a waste product produced by cells during this process
Wastes (breakdown products of metabolism) (eg, lactic acid, creatinine, urea, bilirubin, ammonia)	Waste products serve no function in the blood plasma; they are merely being transported to the liver and kidneys where they can be removed from the blood

factors; fibronectin; prothrombin and other coagulation factors; lipoproteins and other proteins entering blood from tissues.

- **Immunoglobulins** (**antibodies** or **γ-globulins**) secreted by plasma cells in many locations.
- **Fibrinogen**, the largest plasma protein (340 kD), also made in the liver, which, during clotting, polymerizes as insoluble, cross-linked fibers of fibrin that block blood loss from small vessels.
- **Complement proteins**, which comprise a defensive system important in inflammation and destruction of microorganisms.

❯ BLOOD CELLS

Blood cells can be studied histologically in smears prepared by spreading a drop of blood in a thin layer on a microscope slide (Figure 12–3). In such films the cells are clearly visible and distinct from one another, facilitating observation of their nuclei and cytoplasmic characteristics. Blood smears are routinely stained with mixtures of acidic (eosin) and basic (methylene blue) dyes. These mixtures may also contain dyes called azures that are more useful in staining cytoplasmic granules containing charged proteins and proteoglycans. Azurophilic granules produce metachromasia in stained leukocytes like that seen with mast cells in connective tissue. Some of these special stains, such as Giemsa and Wright stain, are named after hematologists who introduced their own modifications into the original mixtures.

Erythrocytes

Erythrocytes (red blood cells or RBCs) are terminally differentiated structures lacking nuclei and completely filled with the O_2-carrying protein hemoglobin. RBCs are the only blood cells whose function does not require them to leave the vasculature.

❯❯ MEDICAL APPLICATION

Anemia is the condition of having a concentration of erythrocytes below the normal range. With fewer RBCs per milliliter of blood, tissues are unable to receive adequate O_2. Symptoms of anemia include lethargy, shortness of breath, fatigue, skin pallor, and heart palpitations. Anemia may result from insufficient red cell production, due, for example, to iron deficiency, or from blood loss with a stomach ulcer or excessive menses.

An increased concentration of erythrocytes in blood (**erythrocytosis**, or **polycythemia**) may be a physiologic adaptation found, for example, in individuals who live at high altitudes, where O_2 tension is low. Elevated hematocrit increases blood viscosity, putting strain on the heart, and, if severe, can impair circulation through the capillaries.

FIGURE **12–3** **Preparing a blood smear.**

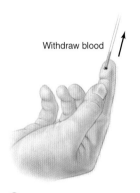

Withdraw blood

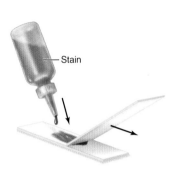

Stain

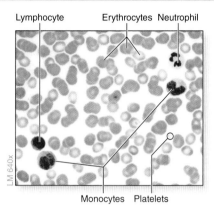

Lymphocyte   Erythrocytes Neutrophil

Monocytes   Platelets

① Prick finger and collect a small amount of blood using a micropipette.

② Place a drop of blood on a slide.

③a Using a second slide, pull the drop of blood across the first slide's surface, leaving a thin layer of blood on the slide.

③b After the blood dries, apply a stain briefly and rinse. Place a coverslip on top.

④ When viewed under the microscope, blood smear reveals the components of the formed elements.

Human erythrocytes suspended in an isotonic medium are flexible biconcave discs (Figure 12–4). They are approximately 7.5 μm in diameter, 2.6-μm thick at the rim, but only 0.75-μm thick in the center. Because of their uniform dimensions and their presence in most tissue sections, RBCs can often be used by histologists as an internal standard to estimate the size of other nearby cells or structures.

The biconcave shape provides a large surface-to-volume ratio and facilitates gas exchange. The normal concentration of erythrocytes in blood is approximately 3.9-5.5 million per microliter (μL, or mm³) in women and 4.1-6.0 million/μL in men.

Erythrocytes are normally quite flexible, which permits them to bend and adapt to the small diameters and irregular

FIGURE **12–4** **Normal human erythrocytes.**

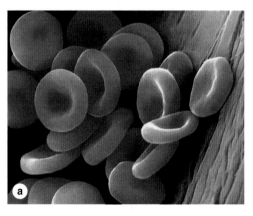

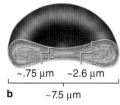

Sectional view

~.75 μm   ~2.6 μm

b   ~7.5 μm

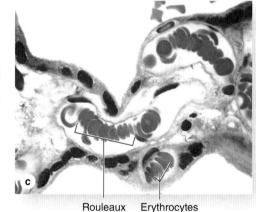

Rouleaux   Erythrocytes

(a) Colorized SEM micrograph of normal **erythrocytes** with each side concave. (X1800)

(b) Diagram of an erythrocyte giving the cell's dimensions. The biconcave shape gives the cells a very high surface-to-volume ratio and places most hemoglobin within a short distance from the cell surface, both qualities that provide maximally efficient O_2 transport.

Erythrocytes are also quite flexible and can easily bend to pass through small capillaries.

(c) In small vessels red blood cells also often stack up in loose aggregates called **rouleaux**. The standard size of RBCs allows one to estimate that the vessel seen is approximately 15 mm in diameter. (X250; H&E)

turns of capillaries. Observations in vivo show that at the angles of capillary bifurcations, erythrocytes with normal adult hemoglobin frequently assume a cuplike shape. In larger blood vessels RBCs may adhere to one another loosely in stacks called rouleaux (Figure 12–4c).

The erythrocyte plasmalemma, because of its ready availability, is the best-known membrane of any cell. It consists of about 40% lipid, 10% carbohydrate, and 50% protein. Most of the latter are integral membrane proteins (see Chapter 2), including ion channels, the anion transporter called **band 3 protein**, and **glycophorin A**. The glycosylated extracellular domains of the latter proteins include antigenic sites that form the basis for the ABO blood typing system. Several peripheral proteins are associated with the inner surface of the membrane, including **spectrin**, dimers of which form a lattice bound to underlying actin filaments, and **ankyrin**, which anchors the spectrin lattice to the glycophorins and band 3 proteins. This submembranous meshwork stabilizes the membrane, maintains the cell shape, and provides the cell elasticity required for passage through capillaries.

Erythrocyte cytoplasm lacks all organelles but is densely filled with **hemoglobin**, the tetrameric O_2-carrying protein that accounts for the cells' uniform acidophilia. When combined with O_2 or CO_2, hemoglobin forms oxyhemoglobin or carbaminohemoglobin, respectively. The reversibility of these combinations is the basis for the protein's gas-transporting capacity.

Erythrocytes undergo terminal differentiation (discussed in Chapter 13) which includes loss of the nucleus and organelles shortly before the cells are released by bone marrow into the circulation. Lacking mitochondria, erythrocytes rely on anaerobic glycolysis for their minimal energy needs. Lacking nuclei, they cannot replace defective proteins.

Human erythrocytes normally survive in the circulation for about 120 days. By this time defects in the membrane's cytoskeletal lattice or ion transport systems begin to produce swelling or other shape abnormalities, as well as changes in the cells' surface oligosaccharide complexes. Senescent or worn-out RBCs displaying such changes are recognized and removed from circulation, mainly by macrophages of the spleen, liver, and bone marrow.

Leukocytes

Leukocytes (white blood cells or WBCs) leave the blood and migrate to the tissues where they become functional and perform various activities related to immunity. Leukocytes are divided into two major groups, **granulocytes** and **agranulocytes**, based on the density of their cytoplasmic granules (Table 12–2). All are rather spherical while suspended in blood plasma, but they become amoeboid and motile after leaving the blood vessels and invading the tissues. Their estimated sizes mentioned here refer to observations in blood smears in which the cells are spread and appear slightly larger than they are in the circulation.

Granulocytes possess two major types of abundant cytoplasmic granules: lysosomes (often called **azurophilic granules** in blood cells) and **specific granules** that bind neutral, basic, or acidic stains and have specific functions.

Granulocytes also have **polymorphic nuclei** with two or more distinct (almost separated) lobes and include the **neutrophils, eosinophils**, and **basophils** (Figure 12–1 and Table 12–2). All granulocytes are also terminally differentiated cells with a life span of only a few days. Their Golgi complexes and rough ER are poorly developed, and with few mitochondria they depend largely on glycolysis for their energy needs. Most granulocytes undergo apoptosis in the connective tissue and billions of neutrophils alone die each day in adults. The resulting cellular debris is removed by macrophages and, like all apoptotic cell death, does not itself elicit an inflammatory response.

Agranulocytes lack specific granules, but do contain some azurophilic granules (lysosomes). The nucleus is spherical or indented but not lobulated. This group includes the **lymphocytes** and **monocytes** (Figure 12–1 and Table 12–2). The differential count (percentage of all leukocytes) for each type of leukocyte is also presented in Table 12–2.

All leukocytes are key players in the constant defense against invading microorganisms and in the repair of injured tissues, specifically leaving the microvasculature in injured or infected tissues. At such sites factors termed **cytokines** are released from various sources and these trigger loosening of intercellular junctions in the endothelial cells of local postcapillary venules (Figure 12–6). Simultaneously the cell adhesion protein **P-selectin** appears on the endothelial cells' luminal surfaces following exocytosis from cytoplasmic Weibel-Palade bodies. The surfaces of neutrophils and other leukocytes display glycosylated ligands for P-selectin, and their interactions cause cells flowing through the affected venules to slow down, like rolling tennis balls arriving at a patch of velcro. Other cytokines stimulate the now slowly rolling leukocytes to express integrins and other adhesion factors that produce firm attachment to the endothelium (see Figure 11–21d). In a process called **diapedesis** (Gr. *dia*, through + *pedesis*, to leap), the leukocytes send extensions through the openings between the endothelial cells, migrate out of the venules into the surrounding tissue space, and head directly for the site of injury or invasion. The attraction of neutrophils to bacteria involves chemical mediators in a process of **chemotaxis**, which causes leukocytes to rapidly accumulate where their defensive actions are specifically needed.

The number of leukocytes in the blood varies according to age, sex, and physiologic conditions. Healthy adults have 4500-11,000 leukocytes per microliter of blood.

Neutrophils (Polymorphonuclear Leukocytes)

Mature neutrophils constitute 50%-70% of circulating leukocytes, a figure that includes slightly immature forms released

TABLE 12–2   Leukocytes: Numbers, structural features, and major functions.

Eosinophil

Neutrophil

Basophil

Granulocytes

Agranulocytes

Lymphocyte

Monocyte

Type	Nucleus	Specific Granules[a]	Differential Count[b] (%)	Life Span	Major Functions
Granulocytes					
Neutrophils	3-5 lobes	Faint/light pink	50-70	1-4 d	Kill and phagocytose bacteria
Eosinophils	Bilobed	Red/dark pink	1-4	1-2 wk	Kill helminthic and other parasites; modulate local inflammation
Basophils	Bilobed or S-shaped	Dark blue/purple	0.5-1	Several months	Modulate inflammation, release histamine during allergy
Agranulocytes					
Lymphocytes	Rather spherical	(none)	20-40	Hours to many years	Effector and regulatory cells for adaptive immunity
Monocytes	Indented or C-shaped	(none)	2-8	Hours to years	Precursors of macrophages and other mononuclear phagocytic cells

[a]Color with routine blood smear stains. There are typically 4500-11,000 total leukocytes/μL of blood in adults, higher in infants and young children.

[b]The percentage ranges given for each type of leukocyte are those used by the US National Board of Medical Examiners. The value for neutrophils includes 3%-5% circulating, immature band forms.

All micrographs X1600.

FIGURE **12–5** Sickle cell erythrocyte.

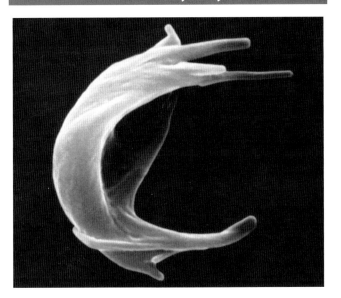

A single nucleotide substitute in the hemoglobin gene produces a version of the protein that polymerizes to form rigid aggregates, leading to greatly misshapen cells with reduced flexibility. In individuals homozygous for the mutated *HbS* gene, this can lead to greater blood viscosity, and poor microvascular circulation, both features of sickle cell disease. (X6500)

to the circulation. Neutrophils are 12-15 μm in diameter in blood smears, with nuclei having two to five lobes linked by thin nuclear extensions (Table 12–2; Figure 12–7). In females, the inactive X chromosome may appear as a drumstick-like appendage on one of the lobes of the nucleus (Figure 12–7c) although this characteristic is not always seen. Neutrophils are inactive and spherical while circulating but become amoeboid and highly active during diapedesis and upon adhering to ECM substrates such as collagen.

Neutrophils are usually the first leukocytes to arrive at sites of infection where they actively pursue bacterial cells using chemotaxis and remove the invaders or their debris by phagocytosis.

The cytoplasmic granules of neutrophils provide the cells' functional activities and are of two main types (Figure 12–8). **Azurophilic primary granules** or lysosomes are large, dense vesicles with a major role in both killing and degrading engulfed microorganisms. They contain proteases and antibacterial proteins, including the following:

- **Myeloperoxidase** (MPO), which generates hypochlorite and other agents toxic to bacteria
- **Lysozyme**, which degrades components of bacterial cell walls
- **Defensins**, small cysteine-rich proteins that bind and disrupt the cell membranes of many types of bacteria and other microorganisms.

Several kinds of **neutrophil defects**, often genetic in origin, can affect function of these cells, for example, by decreasing adhesion to the wall of venules, by causing the absence of specific granules, or with deficits in certain factors of the azurophilic granules. Individuals with such disorders typically experience more frequent and more persistent bacterial infections, although macrophages and other leukocytes may substitute for certain neutrophil functions.

Specific secondary granules are smaller and less dense, stain faintly pink, and have diverse functions, including secretion of various ECM-degrading enzymes such as collagenases, delivery of additional bactericidal proteins to phagolysosomes, and insertion of new cell membrane components.

Activated neutrophils at infected or injured sites also have important roles in the inflammatory response that begins the process of restoring the normal tissue microenvironment. They release many polypeptide **chemokines** that attract other leukocytes and **cytokines** that direct activities of these and local cells of the tissue. Important **lipid mediators** of inflammation are also released from neutrophils.

Neutrophils contain glycogen, which is broken down into glucose to yield energy via the glycolytic pathway. The citric acid cycle is less important, as might be expected in view of the paucity of mitochondria in these cells. The ability of neutrophils to survive in an anaerobic environment is highly advantageous, because they can kill bacteria and help clean up debris in poorly oxygenated regions, for example, damaged or necrotic tissue lacking normal microvasculature.

Neutrophils are short-lived cells with a half-life of 6-8 hours in blood and a life span of 1-4 days in connective tissues before dying by apoptosis.

Neutrophils look for bacteria to engulf by pseudopodia and internalize them in vacuoles called **phagosomes**. Immediately thereafter, specific granules fuse with and discharge their contents into the phagosomes which are then acidified by proton pumps. Azurophilic granules then discharge their enzymes into this acidified vesicle, killing and digesting the engulfed microorganisms.

During phagocytosis, a burst of O_2 consumption leads to the formation of superoxide anions (O_2^-) and hydrogen peroxide (H_2O_2). O_2^- is a short-lived, highly reactive free radical that, together with MPO and halide ions, forms a powerful microbial killing system inside the neutrophils. Besides the activity of lysozyme cleaving cell wall peptidoglycans to kill certain bacteria, the protein lactoferrin avidly binds iron, a crucial element in bacterial nutrition whose lack of availability then causes bacteria to die. A combination of these mechanisms will kill most microorganisms, which are then digested by lysosomal enzymes. Apoptotic neutrophils, bacteria,

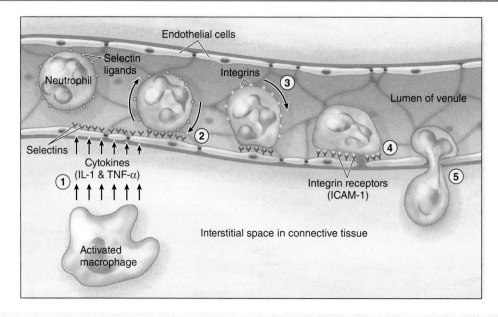

Locations in connective tissue with injuries or infection require the rapid immigration of various leukocytes to initiate cellular events for tissue repair and removal of the invading microorganisms. The cytokines and cell binding proteins target various leukocytes and are best known for neutrophils. The major initial events of neutrophil migration during inflammation are summarized here:

1. Local macrophages activated by bacteria or tissue damage release **proinflammatory cytokines** such as interleukin-1 (IL-1) or tumor necrosis factor-α (TNF-α) that signal endothelial cells of nearby postcapillary venules to rapidly insert glycoprotein selectins on the luminal cell surfaces.

2. Passing neutrophils with appropriate cell surface glycoproteins bind the **selectins**, which causes such cells to adhere loosely to the endothelium and "roll" slowly along its surface.

3. Exposure to these and other cytokines causes expression of new **integrins** on the rolling leukocytes and expression of the integrin ligand **ICAM-1** (intercellular adhesion molecule-1) on the endothelial cells. Junctional complexes between the endothelial cells are selectively downregulated, loosening these cells.

4. Integrins and their ligands provide firm endothelial adhesion of neutrophils to the endothelium, allowing the leukocytes to receive further stimulation from the local cytokines.

5. Neutrophils become motile, probe the endothelium with pseudopodia, and, being attracted by other local injury-related factors called chemokines, finally migrate by **diapedesis** between the loosened cells of the venule. Rapid transendothelial migration of neutrophils is facilitated by the cells' elongated and segmented nuclei. All leukocytes first become functional in the ECM after emerging from the circulation by this process.

semidigested material, and tissue-fluid form a viscous, usually yellow collection of fluid called **pus**.

Several neutrophil hereditary dysfunctions have been described. In one of them, actin does not polymerize normally, reducing neutrophil motility. With a NADPH oxidase deficiency, there is a failure to produce H_2O_2 and hypochlorite, reducing the cells' microbial killing power. Children with such dysfunctions can experience more persistent bacterial infections.

Eosinophils

Eosinophils are far less numerous than neutrophils, constituting only 1%-4% of leukocytes. In blood smears, this cell is about the same size as a neutrophil or slightly larger, but with a characteristic bilobed nucleus (Table 12–2; Figure 12–9).

The main identifying characteristic is the abundance of large, acidophilic specific granules typically staining pink or red.

Ultrastructurally the eosinophilic specific granules are seen to be oval in shape, with flattened crystalloid cores (Figure 12–9c) containing **major basic proteins (MBP)**, an arginine-rich factor that accounts for the granule's acidophilia and constitutes up to 50% of the total granule protein. MBPs, along with eosinophilic peroxidase, other enzymes and toxins, act to kill parasitic worms or helminths. Eosinophils also modulate inflammatory responses by releasing chemokines, cytokines, and lipid mediators, with an important role in the inflammatory response triggered by allergies. The number of circulating eosinophils increases during helminthic infections and allergic reactions. These leukocytes also remove antigen-antibody complexes from interstitial fluid by phagocytosis.

Eosinophils are particularly abundant in connective tissue of the intestinal lining and at sites of chronic inflammation, such as lung tissues of asthma patients.

FIGURE **12–7** Neutrophils.

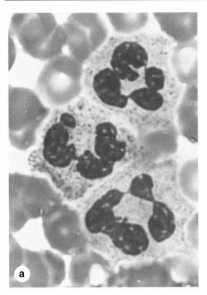
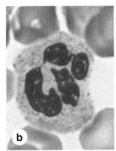
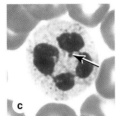

(a) In blood smears neutrophils can be identified by their multi-lobulated nuclei, with lobules held together by very thin strands. With this feature, the cells are often called **polymorphonuclear leukocytes**, PMNs, or just **polymorphs**. The cells are dynamic and the nuclear shape changes frequently. (X1500; Giemsa)

(b) Neutrophils typically have diameters ranging from 12 to 15 μm, approximately twice that of the surrounding erythrocytes. The cytoplasmic granules are relatively sparse and have heterogeneous staining properties, although generally pale and not obscuring the nucleus. (X1500; Giemsa)

(c) Micrograph showing a neutrophil from a female in which the condensed X chromosome appears as a drumstick appendage to a nuclear lobe (arrow). (X1500; Wright)

›› MEDICAL APPLICATION

An increase in the number of eosinophils in blood (**eosinophilia**) is associated with allergic reactions and helminthic infections. In patients with such conditions, eosinophils are found in the connective tissues underlying epithelia of the bronchi, gastrointestinal tract, uterus, and vagina, and surrounding any parasitic worms present. In addition, these cells produce substances that modulate inflammation by inactivating the leukotrienes and histamine produced by other cells. Corticosteroids (hormones from the adrenal cortex) produce a rapid decrease in the number of blood eosinophils, probably by interfering with their release from the bone marrow into the bloodstream.

Basophils

Basophils are also 12-15 μm in diameter but make up less than 1% of circulating leukocytes and are therefore difficult to find in normal blood smears. The nucleus is divided into two irregular lobes, but the large specific granules overlying the nucleus usually obscure its shape.

The specific granules (0.5 μm in diameter) typically stain purple with the basic dye of blood smear stains and are fewer, larger, and more irregularly shaped than the granules of other granulocytes (Table 12–2; Figure 12–10). The strong basophilia of the granules is due to the presence of **heparin** and other sulfated GAGs. Basophilic specific granules also contain much **histamine** and various other mediators of inflammation, including platelet activating factor, eosinophil chemotactic factor, and the enzyme phospholipase A that catalyzes an initial step in producing lipid-derived proinflammatory factors called **leukotrienes**.

By migrating into connective tissues, basophils appear to supplement the functions of mast cells, which are described in Chapter 5. Both basophils and mast cells have metachromatic granules containing heparin and histamine, have surface **receptors for immunoglobulin E (IgE)**, and secrete their granular components in response to certain antigens and allergens.

›› MEDICAL APPLICATION

In some individuals a second exposure to a strong allergen, such as that delivered in a bee sting, may produce an intense, adverse systemic response. Basophils and mast cells may rapidly degranulate, producing vasodilation in many organs, a sudden drop in blood pressure, and other effects comprising a potentially lethal condition called **anaphylaxis** or **anaphylactic shock**.

Basophils and mast cells also are central to immediate or **type 1 hypersensitivity**. In some individuals substances such as certain pollen proteins or specific proteins in food are allergenic, that is, elicit production of specific IgE antibodies, which then bind to receptors on mast cells and immigrating basophils. Upon subsequent exposure, the allergen combines with the receptor-bound IgE molecules, causing them to cross-link and aggregate on the cell surfaces and triggering rapid exocytosis of the cytoplasmic granules. Release of the inflammatory mediators in this manner can result in **bronchial asthma**, cutaneous **hives**, **rhinitis**, **conjunctivitis**, or **allergic gastroenteritis**.

Lymphocytes

By far the most numerous type of agranulocyte in normal blood smears, lymphocytes constitute a family of leukocytes with spherical nuclei (Table 12–2; Figure 12–11). Lymphocytes are typically the smallest leukocytes and constitute approximately a third of these cells. Although they are morphologically similar, mature lymphocytes can be subdivided into functional groups by distinctive surface molecules (called "cluster of differentiation" or **CD markers**) that can be distinguished using antibodies with immunocytochemistry or flow cytometry. Major classes include **B lymphocytes**,

FIGURE **12–8** Neutrophil ultrastructure.

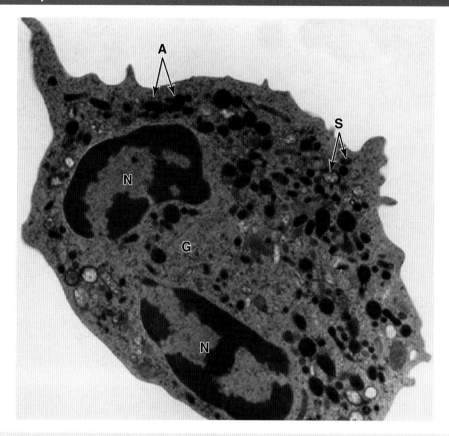

A TEM of a sectioned human neutrophil reveals the two types of cytoplasmic granules: the small, pale, more variably stained specific granules (**S**) and the larger, electron-dense azurophilic granules (**A**).

Specific granules undergo exocytosis during and after diapedesis, releasing many factors with various activities, including enzymes to digest ECM components and bactericidal factors.

Azurophilic granules are modified lysosomes with components to kill engulfed bacteria.

The nucleus (**N**) is lobulated and the central Golgi apparatus (**G**) is small. Rough ER and mitochondria are not abundant, because this cell utilizes glycolysis and is in the terminal stage of its differentiation. (X25,000)

helper and cytotoxic **T lymphocytes** (CD4$^+$ and CD8$^+$, respectively), and **natural killer (NK) cells**. These and other types of lymphocytes have diverse roles in immune defenses against invading microorganisms and certain parasites or abnormal cells. T lymphocytes, unlike B cells and all other circulating leukocytes, differentiate outside the bone marrow in the thymus. Functions and formation of lymphocytes are discussed with the immune system in Chapter 14.

Although generally small, circulating lymphocytes have a wider range of sizes than most leukocytes. Small, newly released lymphocytes have diameters similar to those of RBCs; medium and large lymphocytes are 9-18 μm in diameter, with the latter representing activated lymphocytes or NK cells. The small lymphocytes are characterized by spherical nuclei with highly condensed chromatin and only a thin surrounding rim of scant cytoplasm, making them easily distinguishable from granulocytes. Larger lymphocytes have larger, slightly

indented nuclei and more cytoplasm that is slightly basophilic, with a few azurophilic granules, mitochondria, free polysomes, and other organelles (Figure 12–11d).

Lymphocytes vary in life span according to their specific functions; some live only a few days and others survive in the circulating blood or other tissues for many years.

❯❯ MEDICAL APPLICATION

Given their central roles in immunity, lymphocytes are obviously important in many diseases. **Lymphomas** are a group of disorders involving neoplastic proliferation of lymphocytes or the failure of these cells to undergo apoptosis. Although often slow-growing, all lymphomas are considered malignant because they can very easily become widely spread throughout the body.

FIGURE **12–9** Eosinophils.

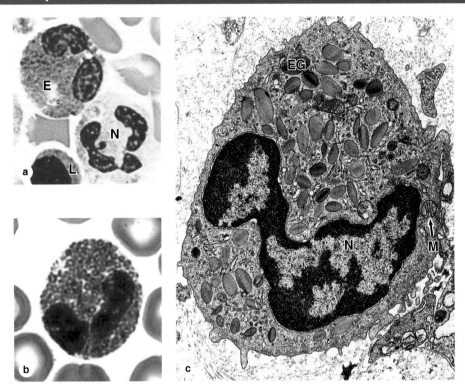

Eosinophils are about the same size as neutrophils but have bilobed nuclei and more abundant coarse cytoplasmic granules. The cytoplasm is often filled with brightly eosinophilic specific granules, but it also includes some azurophilic granules. **(a)** Micrograph shows an eosinophil (**E**) next to a neutrophil (**N**) and a small lymphocyte (**L**). (X1500; Wright)

(b) Even with granules filling the cytoplasm, the two nuclear lobes of eosinophils are usually clear. (X1500; Giemsa)

(c) Ultrastructurally a sectioned eosinophil clearly shows the unique specific eosinophilic granules (**EG**), as oval structures with disc-shaped electron-dense, crystalline cores. These granules, along with a few lysosomes and mitochondria (**M**), fill the cytoplasm around the bilobed nucleus (**N**). (X20,000)

Monocytes

Monocytes are agranulocytes that are precursor cells of macrophages, osteoclasts, microglia, and other cells of the **mononuclear phagocyte system** in connective tissue (see Chapter 5). All monocyte-derived cells are antigen-presenting cells and have important roles in immune defense of tissues. Circulating monocytes have diameters of 12-15 μm, but macrophages are often somewhat larger. The monocyte nucleus is large and usually distinctly indented or C-shaped (Figure 12–12). The chromatin is less condensed than in lymphocytes and typically stains lighter than that of large lymphocytes.

The cytoplasm of the monocyte is basophilic and contains many small lysosomal azurophilic granules, some of which are at the limit of the light microscope's resolution. These granules are distributed through the cytoplasm, giving it a bluish-gray color in stained smears. Mitochondria and small areas of rough ER are present, along with a Golgi apparatus involved in the formation of lysosomes (Figure 12–12e).

›› MEDICAL APPLICATION

Extravasation or the accumulation of immigrating monocytes occurs in the early phase of inflammation following tissue injury. **Acute inflammation** is usually short-lived as macrophages undergo apoptosis or leave the site, but chronic inflammation usually involves the continued recruitment of monocytes. The resulting continuous presence of macrophages can lead to excessive tissue damage that is typical of chronic inflammation.

Platelets

Blood platelets (or **thrombocytes**) are very small non-nucleated, membrane-bound cell fragments only 2-4 μm in diameter (Figure 12–13a). As described in Chapter 13, platelets originate by separation from the ends of cytoplasmic processes extending from giant polyploid bone marrow cells called

FIGURE **12–10** Basophils.

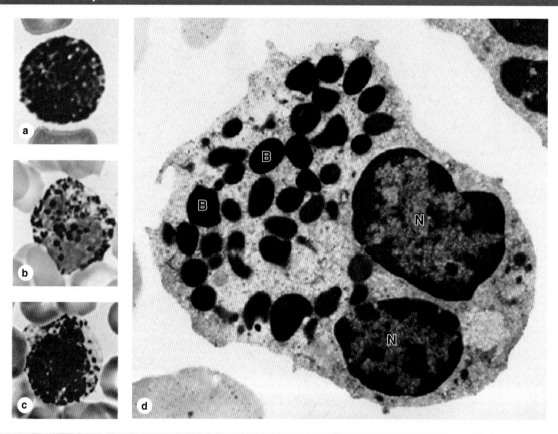

(a-c) Basophils are also approximately the same size as neutrophils and eosinophils, but they have large, strongly basophilic specific granules that usually obstruct the appearance of the nucleus which usually has two large irregular lobes. (a and b: X1500, Wright; c: X1500, Giemsa)

(d) A TEM of a sectioned basophil reveals the single bilobed nucleus (**N**) and the large, electron-dense specific basophilic granules (**B**). Basophils exert many activities modulating the immune response and inflammation and have many functional similarities with mast cells, which are normal, longer-term residents of connective tissue. (X25,000)

megakaryocytes. Platelets promote blood clotting and help repair minor tears or leaks in the walls of small blood vessels, preventing loss of blood from the microvasculature. Normal platelet counts range from 150,000 to 400,000/μL (mm³) of blood. Circulating platelets have a life span of about 10 days.

In stained blood smears, platelets often appear in clumps. Each individual platelet is generally discoid, with a very lightly stained peripheral zone, the **hyalomere**, and a darker-staining central zone rich in granules, called the **granulomere**. A sparse glycocalyx surrounding the platelet plasmalemma is involved in adhesion and activation during blood coagulation.

Ultrastructural analysis (Figure 12–13b) reveals a peripheral **marginal bundle** of microtubules and microfilaments, which helps to maintain the platelet's shape. Also in the hyalomere are two systems of membrane channels. An **open canalicular system** of vesicles is connected to invaginations of the plasma membrane, which may facilitate platelets' uptake of factors from plasma. A much less prominent set of irregular tubular vesicles comprising the dense tubular system is derived from the ER and stores Ca^{2+} ions. Together, these two membranous systems facilitate the extremely rapid exocytosis

of proteins from platelets (degranulation) upon adhesion to collagen or other substrates outside the vascular endothelium.

Besides specific granules, the central granulomere has a sparse population of mitochondria and glycogen particles (Figure 12–13b). Electron-dense **delta granules (δG)**, 250-300 nm in diameter, contain ADP, ATP, and serotonin (5-hydroxytryptamine) taken up from plasma. **Alpha granules (αG)** are larger (300-500 nm in diameter) and contain platelet-derived growth factor (PDGF), platelet factor 4, and several other platelet-specific proteins. Most of the stained granules seen in platelets with the light microscope are alpha granules.

The role of platelets in controlling blood loss (hemorrhage) and in wound healing can be summarized as follows:

- **Primary aggregation:** Disruptions in the microvascular endothelium, which are very common, allow the platelet glycocalyx to adhere to collagen in the vascular basal lamina or wall. Thus, a **platelet plug** is formed as a first step to stop bleeding (Figure 12–14).
- **Secondary aggregation:** Platelets in the plug release a specific adhesive glycoprotein and ADP, which induce

FIGURE **12–11** Lymphocytes.

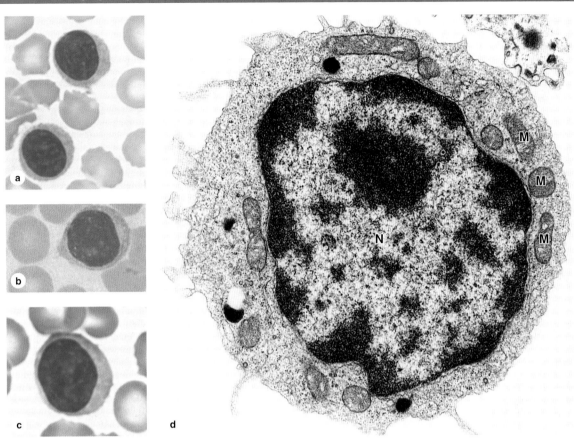

Lymphocytes are agranulocytes and lack the specific granules characteristic of granulocytes. Lymphocytes circulating in blood generally range in size from 6 to 15 μm in diameter and are sometimes classified arbitrarily as small, medium, and large.

(a) The most numerous small lymphocytes shown here are slightly larger than the neighboring erythrocytes and have only a thin rim of cytoplasm surrounding the spherical nucleus. (X1500; Giemsa)

(b) Medium lymphocytes are distinctly larger than erythrocytes. (X1500; Wright)

(c) Large lymphocytes, much larger than erythrocytes, may represent activated cells that have returned to the circulation. (X1500; Giemsa)

(d) Ultrastructurally a medium-sized lymphocytes is seen to be mostly filled with a euchromatic nucleus (**N**) surrounded by cytoplasm containing mitochondria (**M**), free polysomes, and a few dark lysosomes (azurophilic granules). (X22,000)

further platelet aggregation and increase the size of the platelet plug.

- **Blood coagulation:** During platelet aggregation, **fibrinogen** from plasma, **von Willebrand factor** and other proteins released from the damaged endothelium, and **platelet factor 4** from platelet granules promote the sequential interaction (cascade) of plasma proteins, giving rise to a **fibrin** polymer that forms a three-dimensional network of fibers trapping red blood cells and more platelets to form a **blood clot**, or **thrombus** (Figure 12–14). Platelet factor 4 is a chemokine for monocytes, neutrophils, and fibroblasts and proliferation of the fibroblasts is stimulated by PDGF.
- **Clot retraction:** The clot that initially bulges into the blood vessel lumen contracts slightly due to the activity of platelet-derived actin and myosin.
- **Clot removal:** Protected by the clot, the endothelium and surrounding tunic are restored by new tissue, and

the clot is then removed, mainly dissolved by the proteolytic enzyme **plasmin**, which is formed continuously through the local action of **plasminogen activators** from the endothelium on **plasminogen** from plasma.

❯❯ MEDICAL APPLICATION

Aspirin and other nonsteroidal anti-inflammatory agents have **an** inhibitory effect on platelet function and blood coagulation because they block the local prostaglandin synthesis that is needed for platelet aggregation, contraction, and exocytosis at sites of injury. **Bleeding disorders** result from abnormally slow blood clotting. One such disease directly related to a defect in the platelets is a rare autosomal recessive **glycoprotein Ib deficiency**, involving a factor on the platelet surface needed to bind subendothelial collagen and begin the cascade of events leading to clot formation.

FIGURE **12–12** Monocytes.

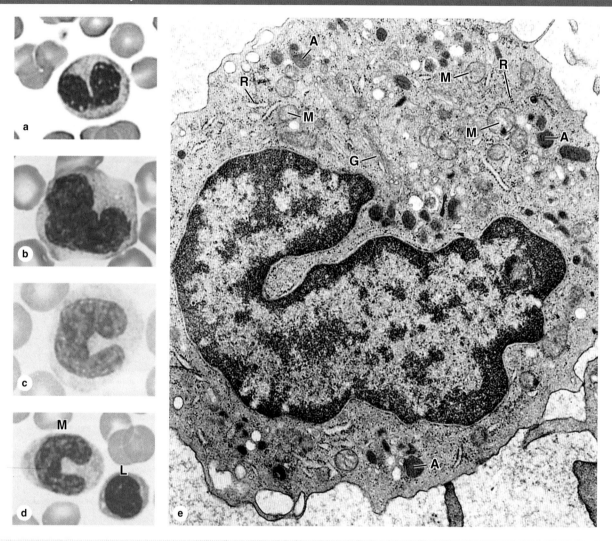

Monocytes are large agranulocytes with diameters from 12 to 20 μm that circulate as precursors to macrophages and other cells of the mononuclear phagocyte system.

(a-d) Micrographs of monocytes showing their distinctive nuclei which are indented, kidney-shaped, or C-shaped. (a: X1500, Giemsa; b-d: X1500, Wright)

(e) Ultrastructurally the cytoplasm of a monocyte shows a Golgi apparatus (**G**), mitochondria (**M**), and lysosomes or azurophilic granules (**A**). Rough ER is poorly developed and there are some free polysomes (**R**). (X22,000)

(Figure 12-12e, used with permission from D.F. Bainton and M.G. Farquhar, Department of Pathology, University of California at San Francisco, CA.)

Blood SUMMARY OF KEY POINTS

- The liquid portion of circulating blood is **plasma**, while the cells and platelets comprise the **formed elements**; upon clotting, some proteins are removed from plasma and others are released from platelets, forming a new liquid termed **serum**.
- Important protein components of plasma include **albumin**, diverse α- and β-globulins, proteins of the **complement** system, and **fibrinogen**, all of which are secreted within the liver, as well as the **immunoglobulins**.
- **Red blood cells** or **erythrocytes**, which make up the **hematocrit** portion (~45%) of a blood sample, are **enucleated**, **biconcave discs** 7.5 μm in diameter, filled with **hemoglobin** for the uptake,

transport, and release of O_2, and with a normal life span of about **120 days**.
- **White blood cells** or **leukocytes** are broadly grouped as **granulocytes** (neutrophils, eosinophils, basophils) or **agranulocytes** (lymphocytes, monocytes).
- All leukocytes become active outside the circulation, specifically leaving the microvasculature in a process involving cytokines, selective adhesion, changes in the endothelium, and **transendothelial migration** or **diapedesis**.
- All granulocytes have specialized lysosomes called **azurophilic granules** and smaller **specific granules** with proteins for various cell-specific functions.

FIGURE **12–13** Platelets.

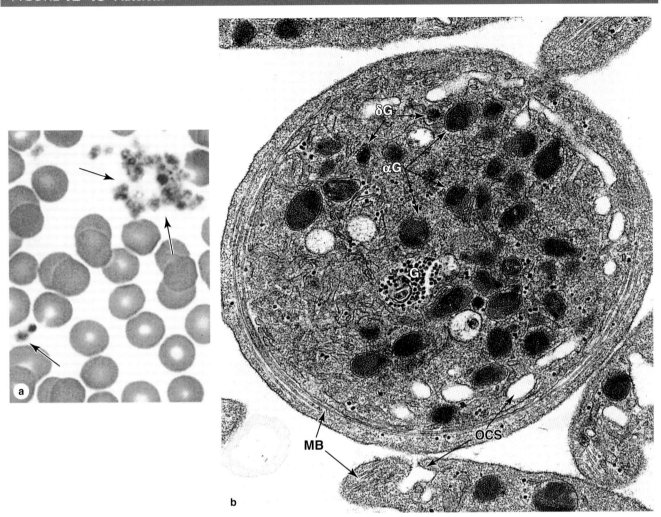

Platelets are cell fragments 2-4 μm in diameter derived from megakaryocytes of bone marrow. Their primary function is to rapidly release the content of their granules upon contact with collagen (or other materials outside of the endothelium) to begin the process of clot formation and reduce blood loss from the vasculature.

(a) In a blood smear, platelets (arrows) are often found as aggregates. Individually they show a lightly stained **hyalomere** region surrounding a more darkly stained central **granulomere** containing membrane-enclosed granules. (X1500; Wright)

(b) Ultrastructurally a platelet shows a system of microtubules and actin filaments near the periphery, called the **marginal bundle (MB)**, which is formed as the platelet pinches off from megakaryocyte (Chapter 13), and helps maintain its shape. An open canalicular system (**OCS**) of invaginating membrane vesicles continuous with the plasmalemma facilitates rapid degranulation upon activation and Ca^{2+} release. The central granulomere region contains small dense delta granules (**δG**), larger and more numerous alpha granules (**αG**), and glycogen (**G**). (X40,000)

(Figure 12-13b, used with permission from Dr M. J. G. Harrison, Middlesex Hospital and University College London, UK.)

- **Neutrophils**, the most abundant type of leukocyte, have **polymorphic**, **multilobed nuclei**, and faint pink cytoplasmic granules that contain many factors for highly efficient phagolysosomal killing and removal of bacteria.
- **Eosinophils** have bilobed nuclei and eosinophilic specific granules containing factors for destruction of helminthic parasites and for modulating inflammation.
- **Basophils**, the rarest type of circulating leukocyte, have irregular bilobed nuclei and resemble mast cells with strongly basophilic specific granules containing factors important in allergies and chronic inflammatory conditions, including **histamine**, **heparin**, **chemokines**, and various **hydrolases**.

- **Lymphocytes**, agranulocytes with many functions as T- and B-cell subtypes in the immune system, range widely in size, depending on their activation state, and have roughly spherical nuclei with little cytoplasm and few organelles.
- **Monocytes** are larger agranulocytes with distinctly indented or C-shaped nuclei that circulate as precursors of macrophages and other cells of the **mononuclear phagocyte system**.
- **Platelets** are small (2-4 μm) cell fragments derived from **megakaryocytes** in bone marrow, with a **marginal bundle** of actin filaments, **alpha granules** and **delta granules**, and an **open canalicular system** of membranous vesicles; rapid degranulation on contact with collagen triggers blood clotting.

FIGURE 12–14 Platelet aggregation, degranulation, and fibrin clot formation.

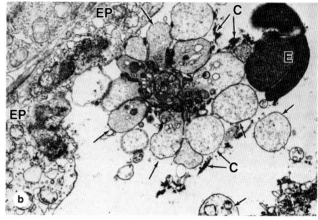

Minor trauma to vessels of the microvasculature is a routine occurrence in active individuals and quickly results in a fibrin clot, shown here by SEM **(a)**. Upon contact with collagen in the vascular basement membrane, platelets **(P)** aggregate, swell, and release factors that trigger formation of a fibrin meshwork **(F)** that traps erythrocytes **(E)** and more degranulating platelets. Platelets in various states of degranulation are shown. Such a clot grows until blood loss from the vasculature stops. After repair of the vessel wall, fibrin clots are removed by proteolysis due primarily to locally generated plasmin, a nonspecific protease. (X4100)

(b) Platelets aggregate at the onset of clot formation. This TEM section shows platelets in a platelet plug adhering to collagen **(C)**. Upon adhering to collagen, platelets are activated and their granules undergo exocytosis into the open canalicular system, which facilitates extremely rapid release of factors involved in blood coagulation. When their contents are completely released, the swollen degranulated platelets **(arrows)** remain as part of the aggregate until the clot is removed. Several other key proteins for blood coagulation are released locally from adjacent endothelial cell processes **(EP)** and from the plasma. Part of an erythrocyte **(E)** is seen at the right. (X7500)

Blood ASSESS YOUR KNOWLEDGE

1. Which biochemical component of the erythrocyte cell surface is primarily responsible for determining blood type (eg, the A-B-O system).
   a. Fatty acid
   b. Carbohydrate
   c. Nucleic acid
   d. Protein
   e. Cholesterol

2. What cell in circulating blood is the precursor to microglia and most antigen-presenting cells?
   a. Eosinophil
   b. Basophil
   c. Lymphocyte
   d. Monocyte
   e. Mast cell

3. What is the approximate life span of a circulating erythrocyte?
   a. 8 days
   b. 20 days
   c. 5 weeks
   d. 4 months
   e. 1 year

4. Which cell type has cytoplasmic granules that contain heparin and histamine?
   a. Eosinophils
   b. Basophils
   c. Lymphocytes
   d. Monocytes
   e. Neutrophils

5. A differential cell count of a blood smear from a patient with a parasitic infection is likely to reveal an increase in the circulating numbers of which cell type?
   a. Neutrophils
   b. Lymphocytes
   c. Monocytes
   d. Basophils
   e. Eosinophils

6. Which of the following blood cells differentiate outside of the bone marrow?
   a. Neutrophils
   b. Basophils
   c. Eosinophils
   d. T lymphocytes
   e. Megakaryocytes

7. Examination of a normal peripheral blood smear reveals a cell more than twice the diameter of an erythrocyte with a kidney-shaped nucleus. There cells are < 10% of the total leukocytes. Which of the following cell types is being described?

   a. Monocyte
   b. Basophil
   c. Eosinophil
   d. Neutrophil
   e. Lymphocyte

8. A 43-year-old anatomy professor is working in her garden, pruning rose bushes without gloves, when a thorn deeply penetrates her forefinger. The next day the area has become infected. She removes the tip of the thorn, but there is still pus remaining at the wound site. Which of the following cells function in the formation of pus?

   a. Cells with spherical nuclei and scant cytoplasm
   b. Biconcave cells with no nuclei
   c. Cells with bilobed nuclei and many acidophilic cytoplasmic granules
   d. Very small, cell-like elements with no nuclei but many granules
   e. Cells with polymorphic, multiply lobed nuclei

9. A 35-year-old woman's physician orders laboratory blood tests. Her fresh blood is drawn and centrifuged in the presence of heparin as an anticoagulant to obtain a hematocrit. From top to bottom, the fractions resulting from centrifugation are which of the following?

   a. Serum, packed erythrocytes, and leukocytes
   b. Leukocytes, erythrocytes, and serum proteins
   c. Plasma, buffy coat, and packed erythrocytes
   d. Fibrinogen, platelets, buffy coat, and erythrocytes
   e. Albumin, plasma lipoproteins, and erythrocytes

10. A hematologist diagnoses a 34-year-old woman with idiopathic thrombobocytic purpura (ITP). Which of the following symptoms/ characteristics would one expect in this patient?

    a. Normal blood count
    b. Hypercoagulation
    c. Decreased clotting time
    d. Abnormal bruising
    e. Light menstrual periods

Answers: 1b, 2d, 3d, 4b, 5e, 6d, 7a, 8e, 9c, 10d

13 Hemopoiesis

Mature blood cells have a relatively short life span and must be continuously replaced with new cells from precursors developing during **hemopoiesis** (Gr. *haima*, blood + *poiesis*, a making). In the early embryo these blood cells arise in the **yolk sac** mesoderm. In the second trimester, hemopoiesis (also called **hematopoiesis**) occurs primarily in the developing **liver**, with the spleen playing a minor role (Figure 13–1). Skeletal elements begin to ossify and **bone marrow** develops in their medullary cavities, so that in the third trimester marrow of specific bones becomes the major hemopoietic organ.

Throughout childhood and adult life, erythrocytes, granulocytes, monocytes, and platelets continue to form from stem cells located in bone marrow. The origin and maturation of these cells are termed, respectively, **erythropoiesis** (Gr. *erythros*, red + *poiesis*), **granulopoiesis, monocytopoiesis**, and **thrombocytopoiesis**. As described in Chapter 14 on the immune system, **lymphopoiesis** or lymphocyte development occurs in the marrow and in the lymphoid organs to which precursor cells migrate from marrow.

This chapter describes the stem and progenitor cells of hemopoiesis, the histology of bone marrow, the major stages of red and white blood cell differentiation, and platelet formation.

❯ STEM CELLS, GROWTH FACTORS, & DIFFERENTIATION

As discussed in Chapter 3, stem cells are **pluripotent** cells capable of asymmetric division and self-renewal. Some of their daughter cells form specific, irreversibly committed progenitor cells, and other daughter cells remain as a small pool of slowly dividing stem cells.

Hemopoietic stem cells can be isolated by using fluorescence-labeled antibodies to mark specific cell surface antigens and passing the cell population through a fluorescence-activated cell-sorting (FACS) instrument. Stem cells are studied using experimental techniques that permit analysis of hemopoiesis in vivo and in vitro.

In vivo techniques include injecting the bone marrow of normal donor mice into irradiated mice whose hematopoietic cells have been destroyed. In these animals, only the transplanted bone marrow cells produce hematopoietic colonies in the bone marrow and spleen, simplifying studies of this process. This work led to the clinical use of bone marrow transplants to treat potentially lethal hemopoietic disorders.

In vitro techniques using semisolid tissue culture media containing substances produced by marrow stromal cells are used to identify and study the cytokines promoting hemopoietic cell growth and differentiation.

Hemopoietic Stem Cells

All blood cells arise from a single type of pluripotent **hemopoietic stem cell** in the bone marrow that can give rise to all the blood cell types (Figure 13–2). These pluripotent stem cells are rare, proliferate slowly and give rise to two major lineages of progenitor cells with restricted potentials (committed to produce specific blood cells): one for **lymphoid cells** (lymphocytes) and another for **myeloid cells** (Gr. *myelos*, marrow) that develop in bone marrow. Myeloid cells include granulocytes, monocytes, erythrocytes, and megakaryocytes. As described in Chapter 14 on the immune system, the lymphoid progenitor cells migrate from the bone marrow to the thymus or the lymph nodes, spleen, and other lymphoid structures, where they proliferate and differentiate.

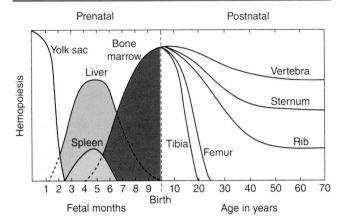

FIGURE **13–1** Shifting locations of hemopoiesis during development and aging.

Hemopoiesis, or blood cell formation, first occurs in a mesodermal cell population of the embryonic yolk sac, and shifts during the second trimester mainly to the developing liver, before becoming concentrated in newly formed bones during the last 2 months of gestation. Hemopoietic bone marrow occurs in many locations through puberty, but then becomes increasingly restricted to components of the axial skeleton.

Progenitor & Precursor Cells

The progenitor cells for blood cells are often called **colony-forming units (CFUs)**, because they give rise to colonies of only one cell type when cultured in vitro or injected into a spleen. As shown in Figure 13–2, there are four major types of progenitor cells/CFUs:

- Erythroid lineage of erythrocytes
- Thrombocytic lineage of megakaryocytes for platelet formation
- Granulocyte-monocyte lineage of all three granulocytes and monocytes
- Lymphoid lineage of B lymphocytes, T lymphocytes, and natural killer cells

Each progenitor cell lineage produces precursor cells (or blasts) that gradually assume the morphologic characteristics of the mature, functional cell types they will become (Figure 13–2). In contrast, stem and progenitor cells cannot be morphologically distinguished and simply resemble large lymphocytes. While stem cells divide at a rate only sufficient to maintain their relatively small population, progenitor and precursor cells divide more rapidly, producing large numbers of differentiated, mature cells (3×10^9 erythrocytes and 0.85×10^9 granulocytes/kg/d in human bone marrow). The changing potential and activities of cells during hemopoiesis are shown graphically in Figure 13–3.

Hemopoiesis depends on a microenvironment, or niche, with specific endocrine, paracrine, and juxtacrine factors. These requirements are provided largely by the local cells and extracellular matrix (ECM) of the hemopoietic organs, which together create the niches in which stem cells are maintained and progenitor cells develop.

Hemopoietic growth factors, often called **colony-stimulating factors (CSF)** or cytokines, are glycoproteins that stimulate proliferation of progenitor and precursor cells and promote cell differentiation and maturation within specific lineages. Cloning of the genes for several important hematopoietic growth factors has significantly advanced study of blood formation and permitted the production of clinically useful factors for patients with hemopoietic disorders. The major activities, target cells, and sources of several well-characterized cytokines promoting hemopoiesis are presented in Table 13–1.

> >> **MEDICAL APPLICATION**
>
> Hemopoietic growth factors are important products of biotechnology companies. They are used clinically to increase marrow cellularity and blood cell counts in patients with conditions such as severe anemia or during chemo- or radiotherapy, which lower white blood cell counts (leukopenia). Such cytokines may also increase the efficiency of marrow transplants by enhancing cell proliferation, enhance host defenses in patients with infectious and immunodeficient diseases, and improve treatment of some parasitic diseases.

❯ BONE MARROW

Under normal conditions, the production of blood cells by the bone marrow is adjusted to the body's needs, increasing its activity several-fold in a very short time. Bone marrow is found in the medullary canals of long bones and in the small cavities of cancellous bone, with two types based on their appearance at gross examination: blood-forming **red bone marrow**, whose color is produced by an abundance of blood and hemopoietic cells, and **yellow bone marrow**, which is filled with adipocytes that exclude most hemopoietic cells. In the newborn all bone marrow is red and active in blood cell production, but as the child grows, most of the marrow changes gradually to the yellow variety. Under certain conditions, such as severe bleeding or hypoxia, yellow marrow reverts to red.

Red bone marrow (Figure 13–4) contains a reticular connective tissue **stroma** (Gr. *stroma*, bed), **hemopoietic cords** or **islands** of cells, and **sinusoidal capillaries**. The stroma is a meshwork of specialized fibroblastic cells called **stromal cells** (also called **reticular** or **adventitial cells**) and a delicate web of reticular fibers supporting the hemopoietic cells and macrophages. The matrix of bone marrow also contains collagen type I, proteoglycans, fibronectin, and laminin, the latter glycoproteins interacting with integrins to bind cells to the matrix. Red marrow is also a site where older, defective erythrocytes undergo phagocytosis by macrophages, which then reprocess heme-bound iron for delivery to the differentiating erythrocytes.

FIGURE 13–2 Origin and differentiative stages of blood cells.

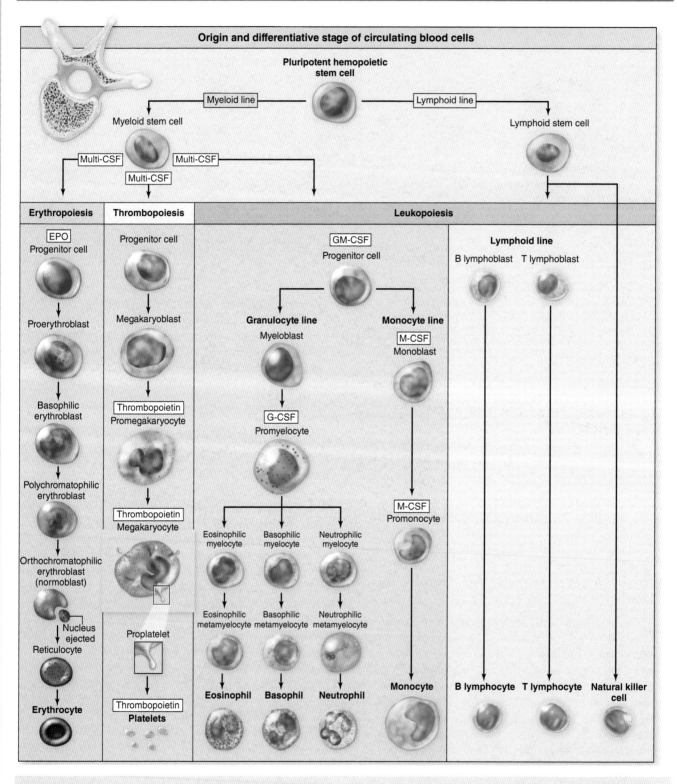

The rare pluripotent hemopoietic stem cells divide slowly, maintain their own population, and give rise to two major cell lineages of progenitor cells: the myeloid and lymphoid stem cells. The myeloid lineage includes precursor cells (blasts) for erythropoiesis, thrombopoiesis, granulopoiesis, and monocytopoiesis, all in the bone marrow.

The lymphoid lineage forms B and T lymphocytes and related cells called natural killer cells, with the later differentiative stages occurring in lymphoid organs. Erythropoietin (EPO), colony stimulating factors (CSF), cytokines and growth factors promote growth and differentiation throughout these developmental processes.

FIGURE **13–3** Major changes in developing hemopoietic cells.

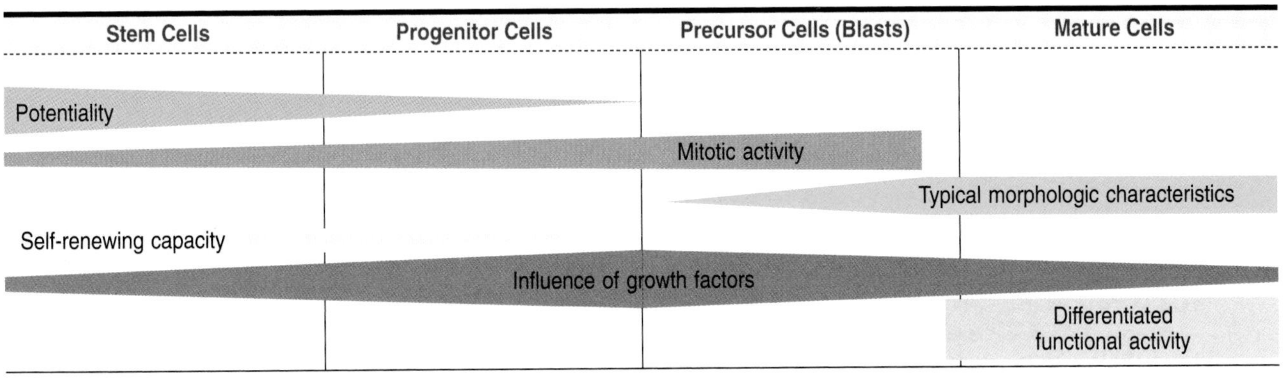

As blood cells in each lineage develop the stem cells' pluri-potentiality and capacity for self-renewal become restricted. Progenitor and precursor cells undergo more rapid mitotic activity than their stem cells but then terminally differentiate with characteristic morphological features that underlie specific functional properties. Within each lineage specific protein and glycoprotein growth factors and cytokines promote the growth and development.

TABLE 13-1 Major hemopoietic cytokines (growth factors or colony-stimulating factors).

Cytokine	Major Activities and Target Cells[a]	Important Sources
Stem cell factor (SCF)	Mitogen for all hemopoietic progenitor cells	Stromal cells of bone marrow
Erythropoietin (EPO)	Mitogen for all erythroid progenitor and precursor cells, also promoting their differentiation	Peritubular endothelial cells of the kidney; hepatocytes
Thrombopoietin (TPO)	Mitogen for megakaryoblasts and their progenitor cells	Kidney and liver
Granulocyte-macrophage colony-stimulating factor (GM-CSF)	Mitogen for all myeloid progenitor cells	Endothelial cells of bone marrow and T lymphocytes
Granulocyte colony-stimulating factor (G-CSF or filgrastim)	Mitogen for neutrophil precursor cells	Endothelial cells of bone marrow and macrophages
Monocyte colony-stimulating factor (M-CSF)	Mitogen for monocyte precursor cells	Endothelial cells of marrow and macrophages
Interleukin-1 (IL-1)	Regulates activities and cytokine secretion of many leukocytes and other cells	Macrophages and T helper cells
Interleukin-2 (IL-2)	Mitogen for activated T and B cells; promotes differentiation of NK cells	T helper cells
Interleukin-3 (IL-3)	Mitogen for all granulocyte and megakaryocyte progenitor cells	T helper cells
Interleukin-4 (IL-4)	Promotes development of basophils and mast cells and B-lymphocyte activation	T helper cells
Interleukin-5 (IL-5) or eosinophil differentiation factor (EDF)	Promotes development and activation of eosinophils	T helper cells
Interleukin-6 (IL-6)	Mitogen for many leukocytes; promotes activation of B cells and regulatory T cells	Macrophages, neutrophils, local endothelial cells
Interleukin-7 (IL-7)	Major mitogen for all lymphoid stem cells	Stromal cells of bone marrow

[a]Most of the cytokines listed here target all the cells of specific lineages, Including the progenitor cells and the precursor cells that are committed and maturing but still dividing. Many promote both mitosis and differentiation in target cells.

FIGURE **13–4**   Red bone marrow (active in hemopoiesis).

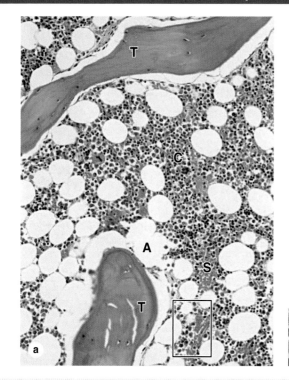

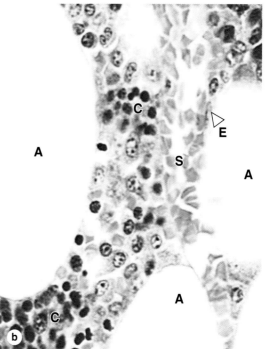

Red bone marrow contains adipocytes but is primarily active in hemopoiesis, with several cell lineages usually present. It can be examined histologically in sections of bones or in biopsies, but its cells can also be studied in smears. Marrow consists of capillary sinusoids running through a stroma of specialized, fibroblastic stromal cells and an ECM meshwork with reticular fibers. Stromal cells produce the ECM; both stromal and bone cells secrete various CSFs, creating the microenvironment for hemopoietic stem cell maintenance, proliferation, and differentiation.	**(a)** Sections of red bone marrow include trabeculae (**T**) of cancellous bone, adipocytes (**A**), and blood-filled sinusoids (**S**) between hemopoietic cords (**C**) or islands of developing blood cells. (X140; H&E) **(b)** At higher magnification the flattened nuclei of sinusoidal endothelial cells (**E**) can be distinguished, as well as the variety of densely packed hemopoietic cells in the cords (**C**) between the sinusoids (**S**) and adipocytes (**A**). Most stromal cells and specific cells of the hemopoietic lineages are difficult to identify with certainty in routinely stained sections of marrow. (X400; H&E)

The hematopoietic niche in marrow includes the stroma, osteoblasts, and megakaryocytes. Between the hematopoietic cords run the sinusoids, which have discontinuous endothelium, through which newly differentiated blood cells and platelets enter the circulation (Figure 13–5).

❯❯ MEDICAL APPLICATION

Red bone marrow also contains stem cells that can produce other tissues in addition to blood cells. These pluripotent cells may make it possible to generate specialized cells that are not rejected by the body because they are produced from stem cells from the marrow of the same patient. The procedure is to collect bone marrow stem cells, cultivate them in appropriate medium for their differentiation to the cell type needed for transplant, and then use the resulting cells to replace defective cells. These studies in **regenerative medicine** are at early stages, but results with animal models are promising.

❯ MATURATION OF ERYTHROCYTES

A mature cell is one that has differentiated to the stage at which it can carry out its specific functions. Erythrocyte maturation is an example of terminal cell differentiation involving hemoglobin synthesis and formation of a small, enucleated, biconcave corpuscle. Several major changes take place during **erythropoiesis** (Figures 13–6 and 13–7). Cell and nuclear volumes decrease, while the nucleoli diminish in size and disappear. Chromatin density increases until the nucleus presents a pyknotic appearance and is finally extruded from the cell. There is a gradual decrease in the number of polyribosomes (basophilia), with a simultaneous increase in the amount of hemoglobin (a highly eosinophilic protein). Mitochondria and other organelles gradually disappear.

Erythropoiesis requires approximately a week and involves three to five cell divisions between the progenitor cell stage and the release of functional cells into the circulation. The glycoprotein **erythropoietin**, a growth factor produced by cells

FIGURE **13–5** Sinusoidal endothelium in active marrow.

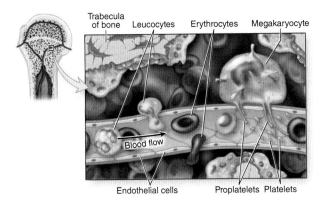

The diagram shows that mature, newly formed erythrocytes, leukocytes, and platelets in marrow enter the circulation by passing through the discontinuous sinusoidal endothelium. All leukocytes cross the wall of the sinusoid by their own activity, but the non-motile erythrocytes cannot migrate through the wall actively and enter the circulation pushed by a pressure gradient across the wall. Megakaryocytes form thin processes (proplatelets) that also pass through such apertures and liberate platelets at their tips.

in the kidneys, stimulates production of mRNA for the protein components of hemoglobin and is essential for erythrocyte production.

The distinct erythroid progenitor cell (Figure 13–6) is the **proerythroblast**, a large cell with loose, lacy chromatin, nucleoli, and basophilic cytoplasm. The next stage is represented by the early **basophilic erythroblast**, slightly smaller with cytoplasmic basophilia and a more condensed nucleus. The basophilia is caused by the large number of free polysomes synthesizing hemoglobin. During the next stage cell volume is reduced, polysomes decrease, and some cytoplasmic areas begin to be filled with hemoglobin, producing regions of both basophilia and acidophilia in the cell and the name **polychromatophilic erythroblast**. Cell and nuclear volumes continue to condense and basophilia is gradually lost, producing cells with uniformly acidophilic cytoplasm—the **orthochromatophilic erythroblasts** (also called normoblasts). Late in this stage the cell nucleus is ejected and undergoes phagocytosis by macrophages. The cell still retains a few polyribosomes which, when treated with the dye brilliant cresyl blue, form a faintly stained network and the cells are termed **reticulocytes** (Figure 13-7b). These cells enter the circulation (where they may constitute 1% of the red blood cells), quickly lose all polyribosomes, and mature as erythrocytes.

FIGURE **13–6** Summary of erythrocyte maturation.

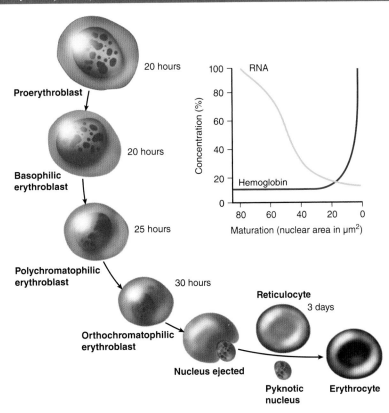

The color change in the cytoplasm shows the continuous decrease in basophilia and the increase in hemoglobin concentration from proerythroblast to erythrocyte. There is also a gradual decrease in nuclear volume and an increase in chromatin condensation, followed by extrusion of a pyknotic nucleus. The times indicate the average duration of each cell type. In the graph, 100% represents the highest recorded concentrations of hemoglobin and RNA.

FIGURE **13–7** Erythropoiesis: Major erythrocyte precursors.

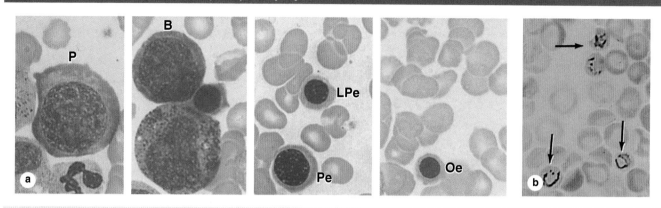

(a) Micrographs showing a very large and scarce proerythroblast (**P**), a slightly smaller basophilic erythroblast (**B**) with very basophilic cytoplasm, typical and late polychromatophilic erythroblasts (**Pe** and **LPe**) with both basophilic and acidophilic cytoplasmic regions, and a small orthochromatophilic erythroblast (**Oe**) with

cytoplasm nearly like that of the mature erythrocytes in the field. (All X1400; Wright)

(b) Micrograph containing reticulocytes (arrows) that have not yet completely lost the polyribosomes used to synthesize globin, as demonstrated by a stain for RNA. (X1400; Brilliant cresyl blue)

›› MATURATION OF GRANULOCYTES

Granulopoiesis involves cytoplasmic changes dominated by synthesis of proteins for the **azurophilic granules** and **specific granules**. These proteins are produced in the rough ER and the prominent Golgi apparatus in two successive stages (Figure 13–8). Formed first are the azurophilic granules, which contain lysosomal hydrolases, stain with

basic dyes, and are generally similar in all three types of granulocytes. Golgi activity then changes to package proteins for the specific granules, whose contents differ in each of the three types of granulocytes and endow each type with certain different properties (see Chapter 12). In sections of bone marrow cords of granulopoietic cells can be distinguished from erythropoietic cords by their granule-filled cytoplasm (Figure 13–9).

FIGURE **13–8** Granulopoiesis: Formation of granules.

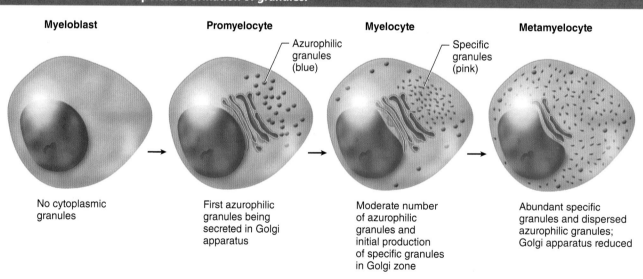

Illustrated is the sequence of cytoplasmic events in the maturation of granulocytes from **myeloblasts**. Modified lysosomes or azurophilic granules form first at the **promyelocyte** stage and are shown in blue; the specific granules of the particular cell type form at

the **myelocyte** stage and are shown in pink. All granules are fully dispersed at the **metamyelocyte** stage, when indentation of the nucleus begins.

FIGURE **13–9** Developing erythrocytes and granulocytes in marrow.

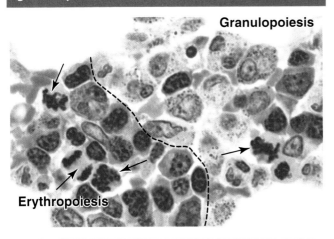

Precursor cells of different hemopoietic lineages develop side by side with some intermingling as various cell islands or cords in the bone marrow. This plastic section of red bone marrow shows mitotic figures (arrows) and fairly distinct regions of **erythropoiesis** and **granulopoiesis**. Most immature granulocytes are in the myelocyte stage: their cytoplasm contains large, dark-stained azurophilic granules and small, less darkly stained specific granules. The large white areas shown peripherally are sites of fat cells. (X400; Giemsa)

FIGURE **13–10** Granulopoiesis: Major granulocyte precursors.

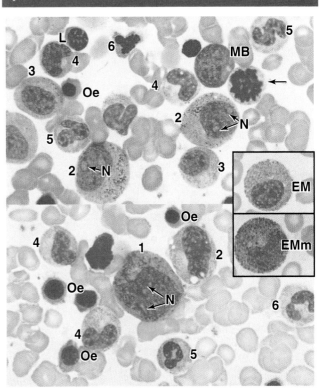

Two micrographs from smears of bone marrow show the major cells of the neutrophilic granulocyte lineage. Typical precursor cells shown are labeled as follows: myeloblast (**MB**); promyelocyte (**1**); myelocytes (**2**); late myelocyte (**3**); metamyelocytes (**4**); band cells (**5**); nearly mature segmented neutrophils (**6**). Some of the early stages show faint nucleoli (**N**). Inset: Eosinophilic myelocytes (**EM**) and metamyelocytes (**EMm**) with their specific granules having distinctly different staining. These and cells of the basophilic lineage are similar to developing neutrophils, except for their specific staining granules and lack of the stab cell form. Also seen among the erythrocytes of these marrow smears are some orthochromatophilic erythroblasts (**Oe**), a small lymphocyte (**L**), and a cell in mitosis (arrow). (All X1400; Wright)

The **myeloblast** is the most immature recognizable cell in the myeloid series (Figures 13–2 and 13–10). Typically these have finely dispersed chromatin, and faint nucleoli. In the next stage, the **promyelocyte** is characterized by basophilic cytoplasm and azurophilic granules containing lysosomal enzymes and myeloperoxidase. Different promyelocytes activate different sets of genes, resulting in lineages for the three types of granulocytes (Figure 13–2). The first visible sign of this differentiation appears in the **myelocyte** stage (Figure 13–11), in which specific granules gradually increase in number and eventually occupy most of the cytoplasm at the **metamyelocyte** stage. These neutrophilic, basophilic, and eosinophilic metamyelocytes mature with further condensation of their nuclei. Before its complete maturation the neutrophilic granulocyte passes through an intermediate stage, the **band cell** (Figure 13–10), in which the nucleus is elongated but not yet polymorphic.

❯❯ MEDICAL APPLICATION

The appearance of large numbers of immature neutrophils (band cells) in the blood, sometimes called a "shift to the left," is clinically significant, usually indicating a bacterial infection.

The vast majority of granulocytes are neutrophils and the total time required for a myeloblast to produce mature,

circulating neutrophils ranges from 10 to 14 days. Five mitotic divisions normally occur during the myeloblast, promyelocyte, and neutrophilic myelocyte stages. As diagrammed in Figure 13–12, developing and mature neutrophils exist in four functionally and anatomically defined compartments: (1) the granulopoietic compartment in active marrow; (2) storage as mature cells in marrow until release; (3) the circulating population; and (4) a population undergoing margination, a process in which neutrophils adhere loosely and accumulate transiently along the endothelial surface in venules and small veins. Margination of neutrophils in some organs can persist for several hours and is not always followed by the cells' emigration from the microvasculature.

FIGURE **13–11** Neutrophilic myelocyte.

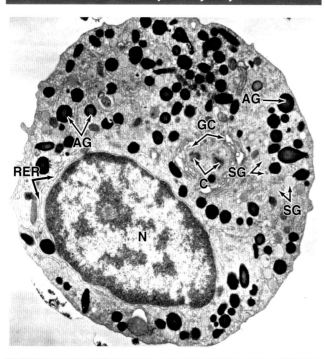

At the myelocyte stage lysosomes (azurophilic granules) have formed and production of specific secretory granules is under way. This micrograph shows ultrastructurally a peroxidase-stained section of a neutrophilic myelocyte with cytoplasm containing both large, peroxidase-positive azurophilic granules (**AG**) and smaller specific granules (**SG**), which do not stain for peroxidase. The peroxidase reaction product is present only in mature azurophilic granules and is not seen in the rough ER (**RER**) or Golgi cisternae (**GC**), which are located around the centriole (**C**) near the nucleus (**N**). (X15,000)

(*Used with permission from Dr Dorothy F. Bainton, Department of Pathology, University of California at San Francisco.*)

At sites of injury or infection, neutrophils and other granulocytes enter the connective tissues by migrating through intercellular junctions between endothelial cells of postcapillary venules in diapedesis. Inflamed connective tissues thus form a fifth terminal compartment for neutrophils, where the cells reside for a few days and then die by apoptosis, regardless of whether they have performed their major function of bacterial phagocytosis.

❯❯ MEDICAL APPLICATION

Changes in the number of neutrophils in the blood must be evaluated by taking all their compartments into consideration. Thus, **neutrophilia**, an increase in the number of circulating neutrophils, does not necessarily imply an increase in granulopoiesis. Intense muscular activity or the

FIGURE **13–12** Compartments of neutrophils in the body.

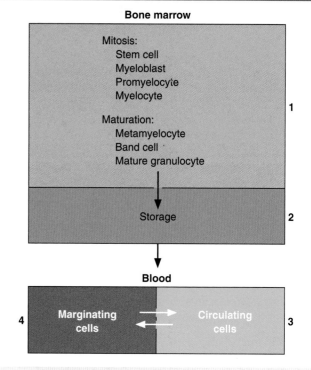

Neutrophils exist in at least four anatomically and functionally distinct compartments, whose sizes reflect the number of cells:

(1) A **granulopoietic** compartment in bone marrow with developing progenitor cells.

(2) A **storage** (reserve) compartment, also in red marrow, acts as a buffer system, capable of releasing large numbers of mature neutrophils as needed. Trillions of neutrophils typically move from marrow to the bloodstream every day.

(3) A **circulating** compartment throughout the blood.

(4) A **marginating** compartment, in which cells temporarily do not circulate but rather accumulate temporarily at the surface of the endothelium in venules and small veins.

The marginating and circulating compartments are actually of about equal size, and there is a constant interchange of cells between them, with the half-life of cells in these two compartments less than 10 hours. The granulopoietic and storage compartments together include cells in approximately the first 14 days of their existence and are about 10 times larger than the circulating and marginating compartments.

administration of epinephrine can cause neutrophils in the marginating compartment to move into the circulating compartment, producing neutrophilia even though granulopoiesis has not increased. However, glucocorticoids (adrenal hormones) such as cortisone increase the mitotic activity of neutrophil precursors and this also increases the blood count of neutrophils.

Transitory neutrophilia may also result from liberation of greater numbers of neutrophils from the medullary storage compartment and is typically followed by a recovery period during which no neutrophils are released.

The neutrophilia that occurs during bacterial infections is due to an increase in production of neutrophils and a shorter duration of these cells in the medullary storage compartment. In such cases, immature forms such as band or stab cells, neutrophilic metamyelocytes, and even myelocytes may appear in the bloodstream. The neutrophilia occurring during infection is typically of much longer duration than that occurring as a result of intense muscular activity.

❯ MATURATION OF AGRANULOCYTES

The precursor cells of monocytes and lymphocytes do not show specific cytoplasmic granules or nuclear lobulation, both of which facilitate the distinction of cells in the granulopoietic series. Monocytes and lymphocytes in smear preparations are discriminated mainly on the basis of size and nuclear shape.

Monocytes

The **monoblast** is a committed progenitor cell that is virtually identical to the myeloblast morphologically. Further differentiation leads to the **promonocyte**, a large cell (up to 18 μm in diameter) with basophilic cytoplasm and a large, slightly indented nucleus (Figures 13–2 and 12–12). The chromatin is lacy and nucleoli are evident. Promonocytes divide twice as they develop into **monocytes**. Differentiating monocytes contain extensive RER and large Golgi complexes forming lysosomes, which are observed as fine azurophilic granules at maturity. Monocytes circulate in blood for several hours and enter tissues where they mature as **macrophages** (or other phagocytic cells) and function for up to several months.

Lymphocytes

As discussed with the immune system (see Chapter 14), circulating lymphocytes originate mainly in the thymus and the peripheral lymphoid organs (eg, spleen, lymph nodes, tonsils, etc). However, lymphocyte progenitor cells originate in the bone marrow. Some of these lymphocytes migrate to the thymus, where they acquire the properties of T lymphocytes. Subsequently, T lymphocytes populate specific regions of peripheral lymphoid organs. Other bone marrow lymphocytes differentiate into B lymphocytes in the bone marrow and then migrate to peripheral lymphoid organs, where they inhabit and multiply within their own niches.

The first identifiable progenitor of lymphoid cells is the **lymphoblast**, a large cell capable of dividing two or three times to form **lymphocytes** (Figures 13–2 and 12–11). As lymphocytes develop their nuclei become smaller, nucleoli disappear, and cell size decreases. In the bone marrow and in the thymus, these cells synthesize the specific cell surface proteins that characterize B or T lymphocytes, respectively. Mature and functionally active B and T cells are generally larger than newly formed lymphocytes. Subsets of lymphocytes acquire distinctive cell surface and other proteins during differentiation that can be detected by immunocytochemical techniques and used to sort the specific lymphocytic types.

❯❯ MEDICAL APPLICATION

Abnormal proliferation of stem cells in bone marrow can produce a range of myeloproliferative disorders. **Leukemias** are malignant clones of leukocyte precursors. They can occur in both lymphoid tissue (**lymphoblastic leukemias**) and bone marrow (**myelogenous leukemias**). In these diseases, there is usually a release of large numbers of immature cells into the blood and an overall shift in hemopoiesis, with a lack of some cell types and excessive production of others. The patient is usually anemic and prone to infection.

Diagnosis of leukemias and other bone marrow disturbances involves **bone marrow aspiration**. A needle is introduced through the compact bone, typically at the iliac crest, and a sample of marrow is withdrawn. Immunocytochemistry with labeled monoclonal antibodies specific to membrane proteins of precursor blood cells contributes to a more precise diagnosis of the leukemia.

❯ ORIGIN OF PLATELETS

The membrane-enclosed cell fragments called **platelets** or thrombocytes originate in the red bone marrow by dissociating from mature **megakaryocytes** (Gr. *megas*, big + *karyon*, nucleus, + *kytos*), which in turn differentiate from **megakaryoblasts** in a process driven by **thrombopoietin**. The megakaryoblast is 25-50 μm in diameter and has a large ovoid or kidney-shaped nucleus (Figure 13–13), often with several small nucleoli. Before differentiating, these cells undergo endomitosis, with repeated rounds of DNA replication not separated by cell divisions, resulting in a nucleus that is highly polyploid (from 8N to 64N). The cytoplasm of this cell is homogeneous and highly basophilic.

Megakaryocytes are giant cells, up to 150 μm in diameter, and the polyploid nuclei are large and irregularly lobulated with coarse chromatin. Their cytoplasm contains numerous mitochondria, a well-developed RER, and an extensive Golgi apparatus from which arise the conspicuous specific granules of platelets (see Chapter 12). They are widely scattered in marrow, typically near sinusoidal capillaries.

To form platelets, megakaryocytes extend several long (>100 μm), wide (2-4 μm) branching processes called **proplatelets**. These cellular extensions penetrate the sinusoidal endothelium and are exposed in the circulating blood of the sinusoids (Figure 13–5). Internally proplatelets have a framework of actin filaments and loosely bundled, mixed polarity microtubules along which membrane vesicles and

FIGURE **13–13** Megakaryoblast and megakaryocytes.

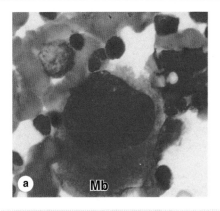
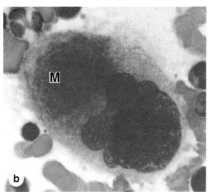
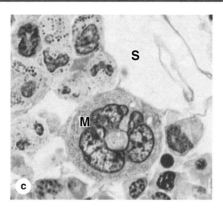

(a) Megakaryoblasts (**Mb**) are very large, fairly rare cells in bone marrow, with very basophilic cytoplasm. (X1400; Wright)

(b) Megakaryoblasts undergo endomitosis (DNA replication without intervening cell divisions), becoming polyploid as they differentiate into megakaryocytes (**M**). These cells are even larger but with cytoplasm that is less intensely basophilic. (X1400; Wright)

(c) Micrograph of sectioned bone marrow in which a megakaryocyte (**M**) is shown near sinusoids (**S**). (X400; Giemsa) Megakaryocytes produce all the characteristic components of platelets (membrane vesicles, specific granules, marginal microtubule bundles, etc) and in a complex process extend many long, branching pseudopodia-like projections called **proplatelets**, from the ends of which platelets are pinched off almost fully formed.

specific granules are transported. A loop of microtubules forms a teardrop-shaped enlargement at the distal end of the proplatelet, and cytoplasm within these loops is pinched off to form platelets with their characteristic marginal bundles of microtubules and actin filaments surrounding cytoplasmic granules and vesicles of the open canalicular system (see Figure 12–13b).

During proplatelet growth microtubules polymerize in both directions. Proplatelet elongation depends on both this polymerization and dynein-based sliding of microtubules past one another. Mature megakaryocytes have numerous invaginations of plasma membrane ramifying throughout the cytoplasm, called **demarcation membranes** (Figure 13–14), which were formerly considered "fracture lines" or "perforations" for the release of platelets but are now thought to represent a membrane reservoir that facilitates the continuous rapid elongation of proplatelets. Each megakaryocyte produces a few thousand platelets, after which the remainder of the cell shows apoptotic changes and is removed by macrophages.

>> MEDICAL APPLICATION

Some bleeding disorders result from **thrombocytopenia**, a reduction in the number of circulating platelets. One cause of thrombocytopenia is **ineffective megakaryopoiesis** resulting from deficiencies of folic acid or vitamin B_{12}. In different types of **thrombocytopenic purpura** (L. purple, the color of small spots or petechiae in the skin from poorly inhibited bleeding), platelet function is compromised, usually by autoimmune reactions.

FIGURE **13–14** Megakaryocyte ultrastructure.

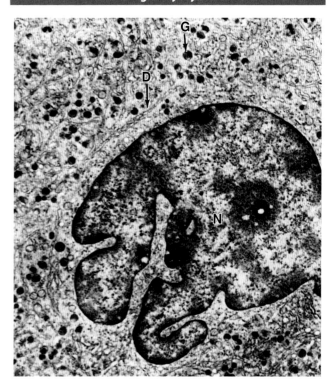

This TEM of a megakaryocyte shows the lobulated nucleus (**N**), numerous cytoplasmic granules (**G**), and an extensive system of demarcation membranes (**D**) through the cytoplasm. The system of demarcation membranes is considered to serve as a reservoir to facilitate rapid elongation of the numerous proplatelets extending from the megakaryocyte surface. (X10,000)

Hemopoiesis SUMMARY OF KEY POINTS

- **Pluripotent stem cells** for blood cell formation, or **hemopoiesis**, occur in the bone marrow of children and adults.
- **Progenitor cells**, committed to forming each type of mature blood cell, proliferate and differentiate within microenvironmental niches of **stromal cells**, other cells, and ECM with specific growth factors.
- These progenitor cells are also known as **colony-forming units (CFUs)** and the growth factors are also called **colony-stimulating factors (CSFs)** or cytokines.
- **Red bone marrow** is active in hemopoiesis; **yellow bone marrow** consists mostly of adipose tissue.
- **Erythropoietic islands** or cords within marrow contain the red blood cell lineage: **proerythroblasts, erythroblasts** with succeeding developmental stages called **basophilic, polychromatophilic,** and **orthochromatophilic** that reflect the cytoplasmic transition from RNA-rich to hemoglobin-filled.
- At the last stage of erythropoiesis cell nuclei are extruded, producing **reticulocytes** that still contain some polyribosomes but are released into the circulation.

- **Granulopoiesis** includes **myeloblasts**, which have large nuclei and relatively little cytoplasm; **promyelocytes**, in which lysosomal **azurophilic granules** are produced; **myelocytes**, in which **specific granules** for one of the three types of granulocytes are formed; and **metamyelocytes**, in which the characteristic changes in nuclear morphology occur.
- Immature neutrophilic metamyelocytes called **band (stab) cells** are released prematurely when the compartment of circulating neutrophils is deleted during bacterial infections.
- **Monoblasts** produce **monocytes** in red marrow, but **lymphoblasts** give rise to **lymphocytes** primarily in the lymphoid tissues in processes involving acquired immunity.
- **Megakaryocytes**, large polyploid cells of red bone marrow, produce **platelets**, or **thrombocytes**, by releasing them from the ends of cytoplasmic processes called **proplatelets**.
- All these **formed elements** of blood enter the circulation by crossing the **discontinuous endothelium** of sinusoids in the red marrow.

Hemopoiesis ASSESS YOUR KNOWLEDGE

1. In which of the following cells involved in erythropoiesis does hemoglobin synthesis begin?
   a. Orthochromatic erythroblast
   b. Polychromatophilic erythroblast
   c. Reticulocyte
   d. Basophilic erythroblast
   e. Proerythroblast

2. Which of the following can be used to describe megakaryocytes?
   a. Multinucleated
   b. Formed by fusion of haploid cells
   c. Precursors to bone marrow macrophages
   d. A minor but normal formed element found in the circulation
   e. Possess dynamic cell projections from which one type of formed element is released

3. Which cytoplasmic components are the main constituents of the dark precipitate that forms in reticulocytes upon staining with the dye cresyl blue?
   a. Golgi complexes
   b. Hemoglobin
   c. Nucleoli
   d. Nuclear fragments
   e. Polyribosomes

4. Which process occurs during granulopoiesis but not during erythropoiesis?
   a. Cells lose their capacity for mitosis
   b. Euchromatin content increases
   c. Nucleus becomes increasingly lobulated
   d. Overall cell diameter decreases
   e. Overall nuclear diameter decreases

5. What fate often awaits granulocytes that have entered the marginating compartment?
   a. Undergo mitosis
   b. Crossing the wall of a venule to enter connective tissue
   c. Cannot reenter the circulation
   d. Differentiate into functional macrophages
   e. Begin to release platelets

6. What is the earliest stage at which specific granulocyte types can be distinguished from one another?
   a. Myelocyte
   b. Band form
   c. Reticulocyte
   d. Metamyelocyte
   e. Promyelocyte

7. Which cell type is capable of further mitosis after leaving the hemopoietic organ in which it is formed?
   a. Basophil
   b. Eosinophil
   c. Reticulocyte
   d. Lymphocyte
   e. Neutrophil

8. Shortly after her birth a baby is diagnosed with a mutation in the erythropoietin receptor gene which leads to familial erythrocytosis (familial polycythemia). During the seventh to ninth months of fetal development, the primary effect on her red blood cell production was in which of the following?
   a. Liver
   b. Yolk sac
   c. Spleen
   d. Thymus
   e. Bone marrow

9. A 54-year-old man presents with recurrent breathlessness and chronic fatigue. After routine tests followed by a bone marrow biopsy he is diagnosed with lymphocytic leukemia. Chemotherapy is administered to remove the cancerous cells, which also destroys the precursor cells of erythrocytes. To reestablish the erythrocytic lineage, which of the following cells should be transplanted?

a. Reticulocytes
b. Orthochromatophilic erythroblasts
c. Megakaryoblasts
d. Basophilic erythroblasts
e. Metamyelocytes

10. A smear of blood from a 70-year-old leukemia patient reveals a larger than normal population of cells that have large, round nuclei with 1 or 2 nucleoli. The cytoplasm of these cells shows azurophilic granules. Which of the following forms of leukemia would you suspect?

a. Promyelocytic leukemia
b. Basophilic leukemia
c. Lymphoblastic leukemia
d. Stem cell leukemia
e. Eosinophilic leukemia

CHAPTER

14 The Immune System & Lymphoid Organs

The **immune system** provides defense or **immunity** against infectious agents ranging from viruses to multicellular parasites. Histologically this system consists of a large, diverse population of leukocytes located within every tissue of the body and **lymphoid organs** interconnected only by the blood and lymphatic circulation. Immunity obviously has tremendous medical importance, one part of which focuses on autoimmune diseases in which immune cells begin to function abnormally and attack molecular components of the body's own organs.

Immunologists recognize two partially overlapping lines of defense against invaders and/or other abnormal, potentially harmful cells: **innate immunity** and **adaptive immunity**. The first of these is nonspecific, involves a wide variety of effector mechanisms, and is evolutionarily older than the second type. Among the cells mediating innate immunity are most of the granulocytes and other leukocytes described in Chapters 12 and 13. Conversely, adaptive immunity aims at specific microbial invaders, is mediated by lymphocytes and **antigen-presenting cells (APCs)** discussed in this chapter, and produces memory cells that permit a similar, very rapid response if that specific microbe appears again.

The lymphocytes and APCs for adaptive immunity are distributed throughout the body in the blood, lymph, and epithelial and connective tissues. Lymphocytes are formed initially in **primary lymphoid organs** (the thymus and bone marrow), but most lymphocyte activation and proliferation occur in **secondary lymphoid organs** (the lymph nodes, the spleen, and diffuse lymphoid tissue found in the mucosa

of the digestive system, including the tonsils, Peyer patches, and appendix). The immune cells located diffusely in the digestive, respiratory, or urogenital mucosae comprise what is collectively known as **mucosa-associated lymphoid tissue (MALT)**. Proliferating B lymphocytes in the secondary structures of MALT are arranged in small spherical **lymphoid nodules**. The wide distribution of immune system cells and the constant traffic of lymphocytes through the blood, lymph, connective tissues, and secondary lymphoid structures provide the body with an elaborate and efficient system of surveillance and defense (Figure 14–1).

›› INNATE & ADAPTIVE IMMUNITY

The system of defenses termed **innate immunity** involves immediate, nonspecific actions, including **physical barriers** such as the skin and mucous membranes of the gastrointestinal, respiratory, and urogenital tracts that prevent infections or penetration of the host body. Bacteria, fungi, and parasites that manage to penetrate these barriers are quickly removed by **neutrophils** and other leukocytes in the adjacent connective tissue. **Toll-like receptors (TLRs)** on leukocytes allow the recognition and binding of surface components of such invaders. Other leukocytes orchestrate the defenses at sites of penetration. **Natural killer (NK) cells** destroy various unhealthy host cells, including those infected with virus or bacteria, as well as certain potentially tumorigenic cells.

Leukocytes and specific cells of the tissue barriers also produce a wide variety of antimicrobial chemicals that

FIGURE **14–1** **The lymphoid organs and main paths of lymphatic vessels.**

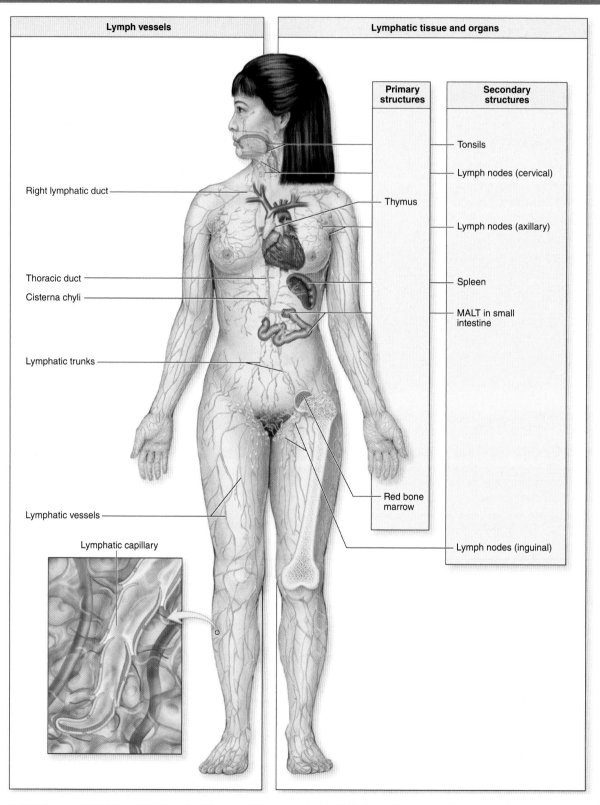

The lymphatic system is composed of lymphatic vessels that transport interstitial fluid (as lymph) back to the blood circulation, and the lymphoid organs that house lymphocytes and other cells of the body's immune defense system. Primary lymphoid organs are the bone marrow and thymus, where B and T lymphocytes are formed, respectively. The secondary lymphoid organs include the lymph nodes, mucosa-associated lymphoid tissue (MALT), and spleen.

also form a major part of innate immunity, including the following:

- **Hydrochloric acid (HCl)** and **organic acids** in specific regions lower the pH locally to either kill entering microorganisms directly or inhibit their growth.
- **Defensins**, short cationic polypeptides produced by neutrophils and various epithelial cells that kill bacteria by disrupting the cell walls.
- **Lysozyme**, an enzyme made by neutrophils and cells of epithelial barriers, which hydrolyzes bacterial cell wall components, killing those cells.
- **Complement**, a system of proteins in blood plasma, mucus, and macrophages that react with bacterial surface components to aid removal of bacteria.
- **Interferons**, paracrine factors from leukocytes and virus-infected cells that signal NK cells to kill such cells and adjacent cells to resist viral infection.

›› MEDICAL APPLICATION

Some pathogenic bacteria, such as *Haemophilus influenzae* and *Streptococcus pneumoniae*, **avoid phagocytosis** by granulocytes and macrophages of **innate immunity** by covering their cell walls with a "capsule" of polysaccharide. The capsule inhibits recognition and binding to the phagocytes' receptors. Eventually such bacteria can be removed by antibody-based mechanisms, including phagocytosis after **opsonization**, but in the interim of several days the cells proliferate undisturbed and establish a more dangerous infection. Elderly or immunocompromised patients, with reduced **adaptive immunity**, are particularly susceptible to infections with such bacteria.

Adaptive immunity, acquired gradually by exposure to microorganisms, is more specific, slower to respond, and an evolutionarily more recent development than innate immunity. The adaptive immune response involves **B** and **T lymphocytes**, whose origins are described in this chapter, which become activated against specific invaders by being presented with specific molecules from those cells by **APCs**, which are usually derived from monocytes. Unlike innate immunity, adaptive immune responses are aimed at specific microbial invaders and involve production of **memory lymphocytes** so that a similar response can be mounted very rapidly if that invader ever appears again.

› CYTOKINES

Within lymphoid organs and during inflammation at sites of infection or tissue injury cells in the immune system communicate with each other primarily via **cytokines** to coordinate

TABLE 14–1	Examples of cytokines, grouped by their main function.
Cytokineᵃ	**Main Functions**
GM-CSF, M-CSF	Growth and differentiation factors for leukocyte progenitor cells in bone marrow
TNF-α, TGF-β, IL-1	Stimulation of inflammation and fever
IL-12	Stimulation of growth in T lymphocytes and NK cells
IL-2, IL-4	Growth factors for T helper cells and B lymphocytes
IL-5	Eosinophil proliferation, differentiation, and activation
Interferon-γ, IL-4	Activation of macrophages
IL-10	Inhibition of macrophages and specific adaptive immune responses
Interferon-α, interferon-β	Antiviral activity
IL-8	Chemokine for neutrophils and T lymphocytes

ᵃGM-CSF, granulocyte-macrophage colony-stimulating factor; IL, Interleukin; M-CSF, macrophage colony-stimulating factor; TGF, transforming growth factor; TNF, tumor necrosis factor.

defensive measures. Involved in both innate and adaptive immunity, cytokines are a diverse group of peptides and glycoproteins, usually with low molecular masses (between 8 and 80 kDa) and a paracrine mode of action. They coordinate cell activities in the innate and adaptive immune responses. Examples of several important cytokines are given in Table 14–1. Major responses induced in target cells by such factors are the following:

- Directed cell movements, or **chemotaxis**, toward and cell accumulation at sites of inflammation, for example, during diapedesis. Cytokines producing this effect are also called **chemokines**.
- Increased mitotic activity in certain leukocytes, both locally and in the bone marrow.
- Stimulation or suppression of lymphocyte activities in adaptive immunity. A group of cytokines with such effects were named **interleukins** because they were thought to be produced by and to target only leukocytes.
- Stimulated phagocytosis or directed cell killing by innate immune cells.

Most cytokines have multiple target cells in which they exert several effects. Some are produced by and target cells besides immune cells, including endothelial cells, certain autonomic neurons, and cells of the endocrine system. The broad range of cytokine actions greatly extends the physiologic effects of infections and other stressors.

❯ ANTIGENS & ANTIBODIES

A molecule that is recognized by cells of the adaptive immune system is called an **antigen** and typically elicits a response from these cells. Antigens may consist of soluble molecules (such as proteins or polysaccharides) or molecules that are still components of intact cells (bacteria, protozoa, or tumor cells). Immune cells recognize and react to small molecular domains of the antigen known as antigenic determinants or **epitopes**. The immune response to antigens may be cellular (in which lymphocytes are primarily in charge of eliminating the antigen), humoral (in which antibodies are primarily responsible for the response), or both.

An **antibody** is a glycoprotein of the **immunoglobulin** family that interacts specifically with an antigenic determinant. Antibodies are secreted by plasma cells that arise by terminal differentiation of clonally proliferating B lymphocytes whose receptors recognize and bind specific epitopes. Antibodies either accumulate in the blood plasma and interstitial fluid of tissues or are transported across epithelia into the secretion of glands such as mucous, salivary, and mammary glands. Other antibodies are membrane proteins on the surface of B lymphocytes or other leukocytes. In all these situations each antibody combines with the epitope that it specifically recognizes.

Immunoglobulins of all antibody molecules have a common design, consisting of two identical light chains and two identical heavy chains bound by disulfide bonds (Figure 14–2). The isolated carboxyl-terminal portion of the heavy-chain molecules is called the constant **Fc region**. The Fc regions of some immunoglobulins are recognized by cell surface receptors on basophils and mast cells, localizing these antibodies to the surface of these cells. The first 110 amino acids near the amino-terminal ends of the light and heavy chains vary widely among different antibody molecules, and this region is called the **variable region**. The variable portions of one heavy and one light chain make up an antibody's **antigen-binding site**. DNA sequences coding for these regions undergo recombination and rearrangement after B lymphocytes are activated against a specific antigen and the progeny of those cells all produce antibodies that specifically bind that antigen. Each antibody has two antigen-binding sites, both for the same antigen.

Classes of Antibodies

Immunoglobulins of humans fall into five major classes, listed in Table 14–2 with their structural features, abundance in plasma, major locations, and functions. The classes are called **immunoglobulin G (IgG), IgA, IgM, IgE**, and **IgD**, and key aspects for each include the following:

- **IgG** is the most abundant class representing 75%-85% of the immunoglobulin in blood. Production increases during immune responses following infections, etc. Unlike the other classes of antibodies, IgG is highly soluble, stable (half-life > 3 weeks), and crosses the placental barrier into the fetal circulation. This confers **passive immunity** against certain infections until the newborn's own adaptive immune system is acquired.

- **IgA** is present in almost all exocrine secretions as a dimeric form in which the heavy chains of two monomers are united by a polypeptide called the **J chain**. IgA is produced by plasma cells in mucosae of the digestive, respiratory, and reproductive tracts. Another protein bound to this immunoglobulin, the **secretory component**, is released by the epithelial cells as IgA undergoes transcytosis. The resulting structure is relatively resistant to proteolysis and reacts with microorganisms in milk, saliva, tears, and mucus coating the mucosae in which it is made.

- **IgM** constitutes 5%-10% of blood immunoglobulin and usually exits in a pentameric form united by a J chain. IgM is mainly produced in an initial response to an antigen. IgM bound to antigen is the most effective antibody class in activating the complement system.

- **IgE**, usually a monomer, is much less abundant in the circulation and exists bound at its Fc region to receptors on the surface of mast cells and basophils. When this IgE encounters the antigen that elicited its production, the antigen-antibody complex triggers the liberation of several biologically active substances, such as histamine, heparin, and leukotrienes. This characterizes an **allergic reaction**, which is thus mediated by the binding of cell-bound IgE with the antigens (**allergens**) that stimulated the IgE to be synthesized initially (see Mast Cells in Chapter 5).

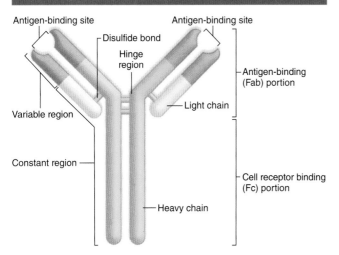

FIGURE **14–2** **Basic structure of an immunoglobulin (antibody).**

Antigen-binding site

Antigen-binding site

Disulfide bond

Hinge region

Antigen-binding (Fab) portion

Light chain

Variable region

Constant region

Cell receptor binding (Fc) portion

Heavy chain

Two light chains and two heavy chains form an **antibody** molecule ("monomer"). The chains are linked by disulfide bonds. The **variable portions** (Fab) near the amino end of the light and heavy chains bind the antigen. The **constant region** (or Fc) of the molecule may bind to surface receptors of several cell types.

TABLE 14–2	Important features of the antibody classes in humans.				
	IgG	**IgM**	**IgA**	**IgD**	**IgE**
Structure			secretory component		
	Monomer	Pentamer	Dimer with J chain and secretory component	Monomer	Monomer
Antibody percentage in the plasma	75% -85%	5%-10%	10%-15%	0.001%	0.002%
Presence in sites other than blood, connective tissue, and lymphoid organs	Fetal circulation in pregnant women	B lymphocyte surface (as a monomer)	Secretions (saliva, milk, tears, etc)	Surface of B lymphocytes	Bound to the surface of mast cells and basophils
Known functions	Activates phagocytosis, neutralizes antigens	First antibody produced in initial immune response; activates complement	Protects mucosae	Antigen receptor triggering initial B cell activation	Destroys parasitic worms and participates in allergies

- **IgD**, the least abundant immunoglobulin in plasma, is also the least understood class of antibody. Monomers of IgD are bound to the surface of B lymphocytes where they (along with IgM monomers) act as antigen receptors in triggering B-cell activation.

Actions of Antibodies

As shown in Figure 14–3a, antigen-binding sites of IgG and IgA antibodies are able to bind specifically and **neutralize** certain viral particles and bacterial toxins, **agglutinate** many bacterial cells, and **precipitate** most soluble antigens. In addition, the Fc portions of these and other antibodies also bind receptors for this sequence and thereby optimize three important actions of innate immunity (Figure 14–3b):

- **Complement activation:** Antigen-antibody complexes containing IgG or IgM bind polypeptides of the **complement system**, a group of around 20 plasma proteins produced mainly in the liver, and activate them through a cascade of enzymatic reactions. After activation, specific complement components bind and rupture membranes of invading cells, clump antigen-bearing bacteria or cells, and elicit arrival of relevant leukocytes.
- **Opsonization:** This refers to the ability of receptors on macrophages, neutrophils, and eosinophils to recognize and bind the Fc portions of antibodies attached to surface antigens of microorganisms. Opsonization greatly increases the efficiency of **phagocytosis** by these leukocytes at sites of infection.
- **NK cells activation:** Antibodies bound to antigens on virus-infected cells of the body are recognized by the primitive lymphocytes called **NK cells**, which are then

activated to kill the infected cell by releasing **perforin** and various **granzymes**. These two proteins together enter the infected cell via other receptors and cause apoptosis.

❯ ANTIGEN PRESENTATION

Antigens recognized by lymphocytes are often bound to specialized integral membrane protein complexes on cell surfaces. These abundant antigen-presenting proteins are parts of the **major histocompatibility complex (MHC)** that includes the two key types called MHC class I and class II. As the name implies, these proteins were first recognized by their roles in the immune rejection of grafted tissue or organs. Proteins of both classes, which on human cells are often called human leukocyte antigens (HLAs), are encoded by genes in large chromosomal loci having very high degrees of allelic variation between different individuals. T lymphocytes are specialized to recognize both classes of MHC proteins and the antigens they present. If the MHCs on cells of a tissue graft are not similar to those that T lymphocytes encountered during their development, the grafted cells will induce a strong immune reaction by T cells of the recipient. To these lymphocytes, the unfamiliar MHC epitopes on the graft's cells are recognized as markers of potentially tumorigenic, infected, or otherwise abnormal ("non-self") cells that they must eliminate.

Like all integral membrane protein complexes, MHC molecules are made in the rough ER and Golgi apparatus. Before leaving the ER, **MHC class I** proteins bind a wide variety of proteasome-derived peptide fragments representing the range of all proteins synthesized in that cell. All nucleated cells

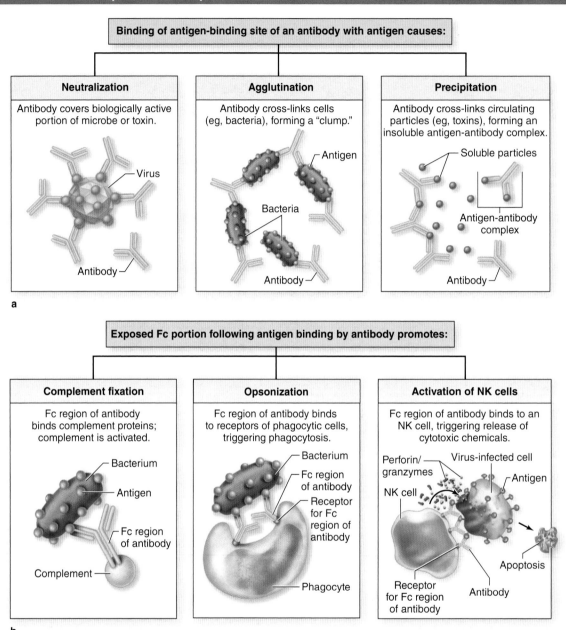

FIGURE **14–3** Various specific and nonspecific functions of antibodies.

Shown here are important mechanisms by which the most common antibodies act in immunity. **(a)** Specific binding of antigens can **neutralize** or **precipitate** antigens, or cause microorganisms bearing the antigens to clump (**agglutinate**) for easier removal.

(b) Complement proteins and surface receptors on many leukocytes bind the Fc portions of antibodies attached to cell-surface antigens, producing **active complement**, more efficient phagocytosis (**opsonization**), and **NK-cell activation**.

produce and expose on their surfaces MHC class I molecules presenting such "self-antigens," which T cells recognize as a signal to ignore those cells. By this same mechanism, some virally infected cells or cells with proteins altered by gene mutation also have MHC class I proteins displaying peptides that T cells do *not* recognize as "self," helping lead to the elimination of such cells.

MHC class II proteins are synthesized and transported to the cell surface similarly but only in cells of the mononuclear phagocyte system and certain other cells under some conditions. Before joining the plasmalemma, the Golgi-derived vesicles with the MHC class II complexes first fuse with endolysosomal vesicles containing antigens ingested by receptor-mediated endocytosis, pinocytosis, or phagocytosis. This allows the class II proteins to bind fragments of whatever proteins the cells had ingested, including those from dead, infected, or abnormal cells and atypical proteins of all kinds. At the surface of these cells, the class II complexes display the peptides from these potentially pathogenic cells, signaling T lymphocytes and activating their responses against sources of these antigens.

CELLS OF ADAPTIVE IMMUNITY

Described in Chapter 12 with blood, lymphocytes and the monocyte-derived cells specialized for antigen presentation to lymphocytes are the major players in adaptive immune responses.

>> MEDICAL APPLICATION

Tissue grafts and organ transplants are classified as **autografts** when the donor and the host are the same individual, such as a burn patient for whom skin is moved from an undamaged to the damaged body region; **isografts** are those involving identical twins. Neither of these graft types is immunologically rejected. **Homografts** (or **allografts**), which involve two related or unrelated individuals, consist of cells with MHC class I molecules and contain dendritic cells with MHC class II molecules, all presenting peptides that the host's T cells recognize as "foreign," leading to immune rejection of the graft.

Development of **immunosuppressive drugs** such as the **cyclosporins** that inhibit the activation of cytotoxic T cells has allowed the more widespread use of allografts or even **xenografts** taken from an animal donor if allografts are in short supply. Such immunosuppression can however lead to other immune-related problems, such as certain opportunistic infections or cancers.

Antigen-Presenting Cells

Most specialized antigen-presenting cells (APCs) are part of the mononuclear phagocyte system, including all types of macrophages and specialized **dendritic cells** in lymphoid organs. Features common to all APCs are an active endocytotic system and expression of MHC class II molecules for presenting peptides of exogenous antigens. Besides dendritic cells (not to be confused with cells of nervous tissue) and all monocyte-derived cells, "professional" APCs also include thymic epithelial cells (discussed below with Thymus).

During inflammation transient expression of MHC class II is induced by interferon-γ in certain local cells that can be considered "nonprofessional" APCs, including fibroblasts and vascular endothelial cells.

Lymphocytes

Lymphocytes both regulate and carry out adaptive immunity. In adults stem cells for all lymphocytes are located in the red

	TABLE 14–3	**Approximate percentages of B and T cells in lymphoid organs.**

Lymphoid Organ	T Lymphocytes (%)	B Lymphocytes (%)
Thymus	100	0
Bone marrow	10	90
Spleen	45	55
Lymph nodes	60	40
Blood	70	30

bone marrow, but cells of the major lymphoid lineages mature and become functional in two different central or **primary lymphoid organs**. Cells destined to become B lymphocytes remain and differentiate further in the **bone marrow**. Progenitors of T lymphocytes move via the circulation into the developing **thymus**. After maturation in these primary structures, B and T cells circulate to the peripheral **secondary lymphoid organs**, which include the **MALT**, the **lymph nodes**, and the **spleen** (Figure 14–1). Lymphocytes do not stay long in the lymphoid organs; they continuously recirculate through the body in connective tissues, blood, and lymph. Because of the constant mobility of lymphocytes and APCs, the cellular locations and microscopic details of lymphoid organs differ from one day to the next. However, the relative percentages of T and B lymphocytes in these compartments are relatively steady (Table 14–3).

Lymphoid tissue is usually reticular connective tissue filled with large numbers of lymphocytes. It can be either diffuse within areas of loose connective tissue or surrounded by capsules, forming discrete (secondary) lymphoid organs. Because lymphocytes have prominent basophilic nuclei and very little cytoplasm, lymphoid tissue packed with such cells usually stains dark blue in hematoxylin and eosin (H&E)–stained sections. In all secondary lymphoid tissue the lymphocytes are supported by a rich reticulin fiber network of type III collagen (Figure 14–4a). The fibers are produced by fibroblastic **reticular cells**, which extend numerous processes along and around the fibers (Figure 14–4b). Besides lymphocytes and reticular cells, lymphoid tissue typically contains various APCs and plasma cells.

Although most lymphocytes are morphologically indistinguishable in either the light or electron microscope, various surface proteins ("cluster of differentiation" or **CD markers**) allow them to be distinguished as B cells and subcategories of T cells by immunocytochemical methods. Key features of B and T lymphocytes also include the surface receptors involved in activating their different responses to antigens (Figure 14–5). Receptors of B cells are immunoglobulins that bind antigens directly; those on T cells react only with antigen on MHC molecules and this requires the additional cell surface proteins CD4 or CD8.

FIGURE **14–4** Reticular fibers and cells of lymphoid tissue.

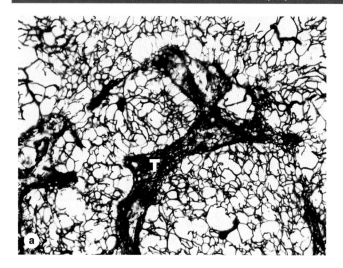

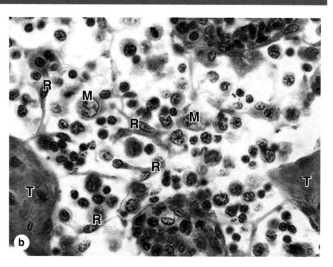

(a) A three-dimensional framework of reticular fibers (collagen type III) supports the cells of most lymphoid tissues and organs (except the thymus). Areas with larger spaces between the fibers offer more mobility to cells than areas in which the fiber meshwork is denser, such as in trabeculae (**T**) where fewer lymphocytes are aggregated and cells are generally more stationary. (X140; Silver impregnation)

(b) Cells of typical lymphoid tissue include the fibroblast-like reticular cells (**R**) that produce and maintain the trabeculae (**T**) and reticulin framework. Many cells are loosely attached to the reticulin fibers, including macrophages (**M**) and many lymphocytes. (X240; H&E)

(*Used with permission from Paulo A. Abrahamsohn, Institute of Biomedical Sciences, University of São Paulo, Brazil.*)

FIGURE **14–5** Specific receptors on T and B lymphocytes.

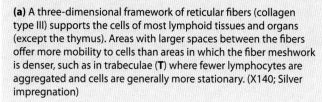

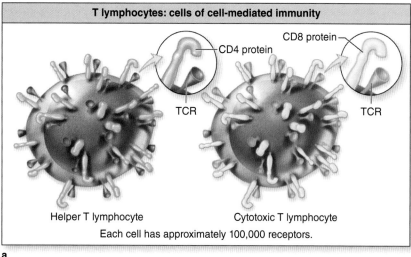

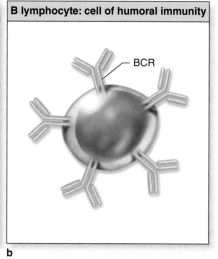

(a) All T lymphocytes have cell surface protein receptors (TCRs) with variable regions that recognize specific antigens. Cell activation requires **costimulation** by the **TCR** and either **CD4 or CD8**, which characterize helper and cytotoxic T cells, respectively.

(b) B-cell receptors (BCRs) are **immunoglobulin** molecules projecting from the plasmalemma.

Lymphocytes in the marrow and thymus of a newborn infant not yet exposed to antigens are immunocompetent but naive and unable to recognize antigens. After circulating to the various secondary lymphoid structures, lymphocytes are exposed to antigens on APCs and become activated, proliferating to produce a clone of lymphocytes all able to recognize that antigen.

T Lymphocytes

T cells are long-lived lymphocytes and constitute nearly 75% of the circulating lymphocytes. They recognize antigenic epitopes via surface protein complexes termed **T-cell receptors (TCRs)**. Most TCRs include two glycoproteins called the α and β **chains**, each with variable regions produced similarly to those of immunoglobulins. Because TCRs only recognize antigenic peptides when presented as part of MHC molecules (interacting with both the MHC and the peptide it presents), T lymphocytes are said to be **MHC restricted**.

Several types of T lymphocytes exist, with various functions. Important subpopulations of T cells include the following:

- **Helper T cells** (Th cells) are characterized by **CD4**, the coreceptor with the TCR for binding MHC class II molecules and the peptides they are presenting (Figure 14–6a). Activated by such binding, helper T cells greatly assist immune responses by producing cytokines that promote differentiation of B cells into plasma cells, activate macrophages to become phagocytic, activate cytotoxic T lymphocytes (CTLs), and induce many parts of an inflammatory reaction. Some specifically activated helper T cells persist as long-lived memory helper T cells, which allow a more rapid response if the antigen appears again later.
- **CTLs** are **CD8⁺**. Their TCRs together with CD8 coreceptors bind specific antigens on foreign cells or virus-infected cells displayed by MHC class I molecules (Figure 14–6b). In the presence of interleukin-2 (IL-2) from helper T cells, cytotoxic T cells that have recognized such antigens are activated and proliferate. Also called killer T cells, they attach to the cell sources of the antigens and remove them by releasing perforins and granzymes, which trigger apoptosis. This represents **cell-mediated immunity** and its mechanism is largely similar to that of NK cells. Activation of cytotoxic T cells also results in a population of memory cytotoxic T cells.
- **Regulatory T cells** (T$_{regs}$ or suppressor T cells) are CD4⁺CD25⁺ and serve to inhibit specific immune responses. These cells, also identified by the presence of the Foxp3 transcription factor, play crucial roles in allowing immune tolerance, maintaining unresponsiveness to self-antigens and suppressing excessive immune responses. These cells produce **peripheral tolerance**, which acts to supplement the central tolerance that develops in the thymus.
- **γδ T lymphocytes** represent a smaller subpopulation whose TCRs contain γ (gamma) and δ (delta) chains

instead of α and β chains. The γδ T cells migrate to the epidermis and mucosal epithelia, becoming largely intraepithelial, and do not recirculate to secondary lymphoid organs. They function in many ways like cells of innate immunity, in the front lines against invading microorganisms.

B Lymphocytes

In B lymphocytes the surface receptors for antigens are monomers of IgM or IgD, with each B cell covered by about 150,000 such **B-cell receptors (BCRs)** (Figure 14–5b). BCRs bind an antigen, which may be free in solution, on an exposed part of an infectious agent, or already bound to antibodies, and the surface complexes then undergo endocytosis. Degraded in endosomes, peptides from the antigens are presented on MHC class II molecules of the B cell. A helper T cell then binds this B cell and activates it further with a cytokine, inducing recombination in the immunoglobulin genes and stimulating several cycles of cell proliferation (see Figure 14–6c).

In all secondary lymphoid tissues B lymphocytes interact with scattered **follicular dendritic cells (FDCs)**, which have long filamentous processes. Unlike other dendritic cells, FDCs are mesenchymal in origin and their function does not involve MHC class II molecules. Surfaces of these cells are covered with antibody-antigen complexes bound to receptors for complement proteins and for immunoglobulin Fc regions, causing B cells to attach, become activated, and aggregate as a small **primary lymphoid nodule** (or follicle). With the help of adjacent T cells, these B cells now form a much larger and more prominent **secondary lymphoid nodule** (Figure 14–7).

Secondary nodules are characterized by a lightly stained **germinal center** filled with large lymphoblasts (or centroblasts) undergoing immunoglobulin gene recombination, rapid proliferation, and quality control. Growth of activated B cells in germinal centers is exuberate and very rapid, causing naive, nonproliferating B cells to be pushed aside and produce the more darkly stained peripheral **mantle** (Figure 14–7). After 2 to 3 weeks of proliferation, most cells of the germinal center and mantle are dispersed and the structure of the secondary lymphoid nodule is gradually lost.

Most of these new, specific B lymphocytes differentiate into **plasma cells** secreting antibodies that will bind the same epitope recognized by the activated B cell. Because the antibodies specified by B cells circulate in lymph and blood throughout the body, B cells are said to provide **humoral**

FIGURE **14–6** Activation of lymphocytes.

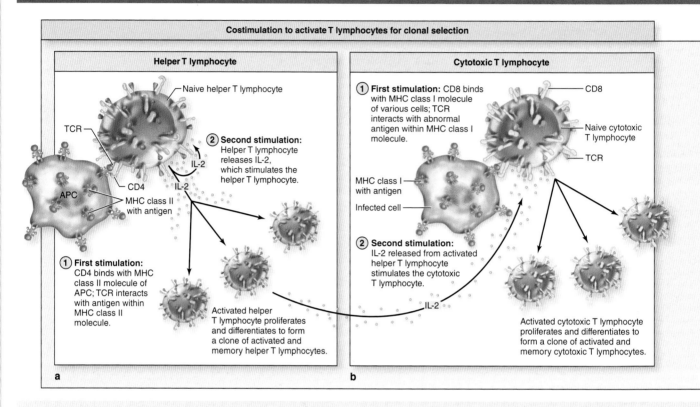

Lymphocyte activation requires costimulation of at least two receptors and causes cell proliferation that produces many effector cells and a smaller population of memory cells. **(a)** The TCR and CD4 proteins of a **helper T cell** bind antigens presented on MHC class II

molecules and with interleukin-2 (IL-2) stimulation, the lymphocyte is activated and proliferates. **(b) Cytotoxic T lymphocytes**, or CTLs, recognize and bind abnormal peptides on MHC class I molecules, and triggered by IL-2 from helper T cells the CTLs proliferate.

immunity. As with activated T cells, some of the newly formed B cells remain as long-lived **memory B cells**. Formation of long-lived memory lymphocytes is a key feature of adaptive immunity, which allows a very rapid response upon subsequent exposure to the same antigen.

❯ THYMUS

While immature B lymphocytes emerge from the bone marrow, the primary or central lymphoid organ in which T cells are produced is the **thymus**, a bilobed structure in the mediastinum (Figure 14-8). A main function of the thymus is induction of **central tolerance**, which along with regulatory T cells prevents autoimmunity. The organ originates from the embryo's third pair of pharyngeal pouches (endoderm), with precursor lymphoblasts circulating from the bone marrow to invade and proliferate in this unique thymic epithelium during its development. Fully formed and functional at birth, the thymus remains large and very active in T-cell production until puberty during which it normally undergoes **involution**, with decreasing lymphoid tissue mass and cellularity and reduced T cell output (Figure 14–8).

> **❯❯ MEDICAL APPLICATION**
>
> Failure of the third (and fourth) pharyngeal pouches to develop normally in the embryo leads to **DiGeorge syndrome**, characterized by **thymic hypoplasia** (or aplasia). Lacking many or all thymic epithelial cells, such individuals cannot produce T lymphocytes properly and have severely depressed cell-mediated immunity.

The thymus has a vascularized connective tissue capsule that extends septa into the parenchyma, dividing the organ into many incompletely separated lobules. Each lobule has an outer darkly basophilic **cortex** surrounding a more lightly stained **medulla**. The staining differences reflect the much greater density of lymphoblasts and small lymphocytes in the cortex than the medulla (Figure 14–8b).

The thymic cortex contains an extensive population of T lymphoblasts (or **thymocytes**), some newly arrived via venules, located among numerous macrophages and associated with the unique **thymic epithelial cells (TECs)** that have certain features of both epithelial and reticular cells. These

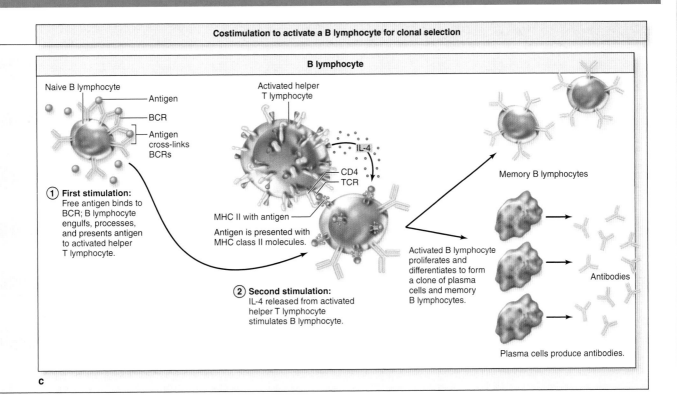

Costimulation to activate a B lymphocyte for clonal selection

B lymphocyte

Naive B lymphocyte

— Antigen

— BCR

— Antigen cross-links BCRs

Activated helper T lymphocyte

IL-4
CD4
TCR

MHC II with antigen

Antigen is presented with MHC class II molecules.

① **First stimulation:** Free antigen binds to BCR; B lymphocyte engulfs, processes, and presents antigen to activated helper T lymphocyte.

② **Second stimulation:** IL-4 released from activated helper T lymphocyte stimulates B lymphocyte.

Activated B lymphocyte proliferates and differentiates to form a clone of plasma cells and memory B lymphocytes.

Memory B lymphocytes

Antibodies

Plasma cells produce antibodies.

c

(c) Antigen bound to the immunoglobulin receptors on B cells (BCRs) is endocytosed, processed, and presented on MCH class II proteins to helper T cells. These then secrete IL-4 and other cytokines that stimulate gene recombination and clonal proliferation of these specific B cells. They differentiate as **plasma cells** producing antibodies against the antigen that was originally bound and processed.

cells usually have large euchromatic nuclei but are morphologically and functionally diverse. There are three major types of TECs in the cortex of the thymus:

- Squamous TECs form a layer, joined by desmosomes and occluding junctions, line the connective tissue of the capsule and septa and surround the microvasculature. This creates an isolated cortical compartment and, together with the vascular endothelial cells and pericytes, forms a **blood-thymus barrier** preventing unregulated exposure of thymocytes to antigens.
- Throughout this compartment another population of stellate TECs, with processes containing keratin tonofilaments joined by desmosomes, form a **cytoreticulum** to which macrophages and developing lymphocytes attach instead of to reticulin fibers (Figure 14–9). Importantly, these cells are APCs, expressing MHC class II molecules in addition to MHC class I. They also secrete numerous cytokines for T-cell development and other immune functions, justifying this organ's inclusion among endocrine glands.
- Other squamous cortical TECs also express MHC class II molecules but form a sheetlike structure contributing to

a functional **corticomedullary barrier** between these two regions of each lobule.

The more lightly stained thymic medulla contains fewer and larger, more mature lymphocytes. Three related types of medullary TECs form the following:

- A second layer of the boundary between cortex and medulla.
- A cytoreticulum that (1) supports T lymphocytes, dendritic cells, and macrophages (all less densely packed than in the cortex), and (2) expresses many specialized proteins specific to cells of other organs.
- Large aggregates of TECs, sometimes concentrically arranged, called **Hassall corpuscles** (Figure 14–10). Up to 100 μm in diameter, thymic corpuscles are unique to the medulla. Their cells secrete several cytokines that control activity of local dendritic cells, including factors that promote development of regulatory T cells for peripheral tolerance.

The microvasculature of the medulla is not surrounded by a tight layer of TECs, and mature T lymphocytes exit the thymus by passing through the walls of venules and efferent lymphatics in this region.

FIGURE **14–7** Lymphoid nodules (or follicles).

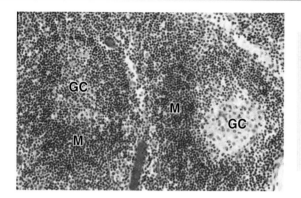

Large aggregates of B cells called lymphoid nodules or follicles tran-siently characterize all secondary lymphoid organs. Aggregates of uniform cell density and staining represent **primary nodules**, while those with larger, more euchromatic cells centrally are termed **second-ary nodules**. Here two secondary nodules can be seen, with germinal centers (**GC**) at different stages of development. Rapid proliferation of activated B lymphoblasts in the germinal center causes smaller, naive lymphocytes to be pushed aside and crowded together peripherally as the follicular mantle (**M**).

FIGURE **14–8** Thymus.

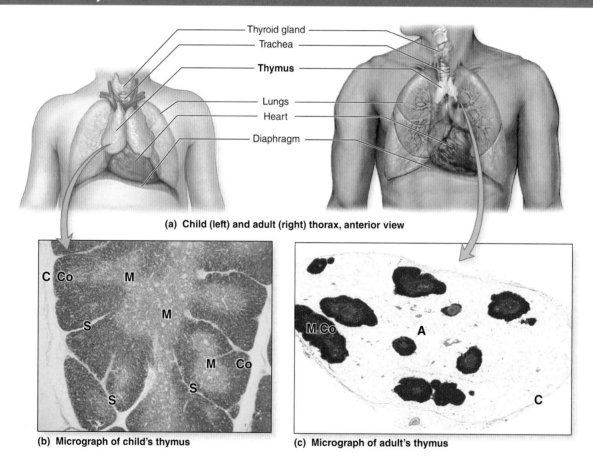

(a) Child (left) and adult (right) thorax, anterior view

(b) Micrograph of child's thymus

(c) Micrograph of adult's thymus

(a) The thymus is a bilobed organ in the mediastinum that is most active and prominent before puberty and undergoes involution with less activity in the adult.

(b) A child's thymus, showing connective tissue of the capsule (**C**) and septa (**S**) between thymic lobules, each having an outer cortex (**Co**) and incompletely separated medulla (**M**) of lymphoid tissue. (H&E; X40)

(c) After-involution the thymus shows only small regions of lym-phoid tissue, here still with cortex (**Co**) and medulla (**M**), and these are embedded in adipose tissue (**A**). Age-related thymic involution reduces production of naïve T cells and may be involved with the decline of immune function in the elderly. (H&E; X24)

FIGURE **14–9** Cortex of the thymus.

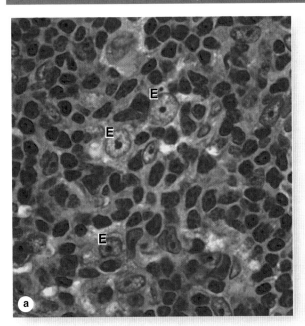

a

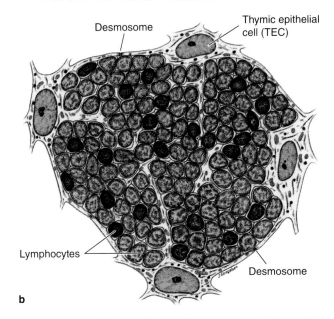

Desmosome

Thymic epithelial cell (TEC)

Lymphocytes

Desmosome

b

(a) The cortical zone of an active thymus is packed with small, highly basophilic lymphoblasts that proliferate as well as undergo positive and negative selection in that region. The lymphoblasts are supported on a meshwork (or cytoreticulum) of unusual thymic epithelial cells (**E**). (X400; PT)

(b) The epithelial reticular cells throughout the cortex are APCs and extend long processes bound together by desmosomes to make the framework, the cytoreticulum, for the lymphoblasts, having a cytoreticulum consisting of APC cellular processes. rather than the more common network of simple reticulin fibers, allows regulated specificity of lymphocyte binding via the changing antigens on MHC proteins. Some cortical epithelial cells also secrete cytokines that promote T-cell maturation.

FIGURE **14–10** Medulla of the thymus with Hassall corpuscles.

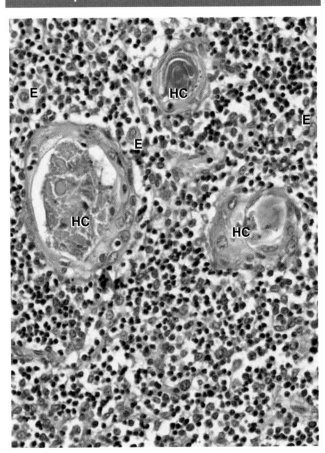

The thymic medulla contains fewer lymphocytes than the cortex, and the epithelial cells (**E**) located here have different morphology and function. The most characteristic feature of the medulla in humans is the presence of thymic (Hassall) corpuscles (**H**). These are of variable size and contain aggregates of thymic epithelial cells releasing many cytokines important within the medullary microenvironment, especially for dendritic cell activity and for the differentiation of regulatory T cells. Dendritic cells in the medulla are difficult to discern without special staining. (X200; H&E)

Role of the Thymus in T-Cell Maturation & Selection

The thymus is the site of T-lymphocyte differentiation and the selective removal of T cells reactive against self-antigens, a key part of inducing central self-tolerance.

T lymphoblasts arriving in the thymus do not yet express CD4, CD8, or a TCR. These cells populate the cortex and begin to proliferate, recombine variable regions of the TCR α and β chain genes, and then express these TCR proteins as well as both CD4 and CD8. With these key functional components in place, thymocytes begin a stringent, **two-stage selection process** of quality control, which ensures that mature T cells have TCRs that are fully functional but do not recognize and strongly bind MHC with self-antigens. The selection process

FIGURE **14–11**   **Thymic selection of functional but not self-reactive T cells.**

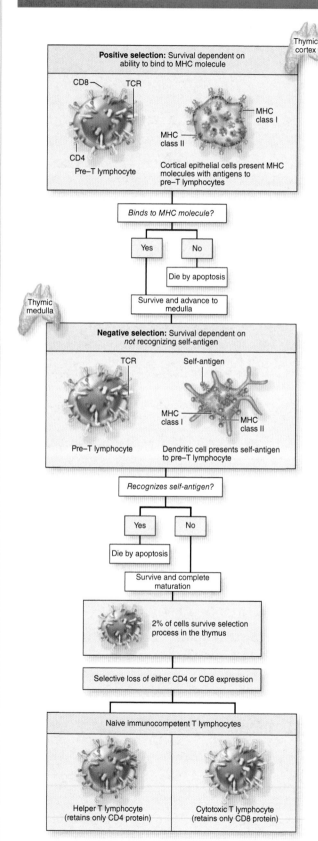

FIGURE **14–11**   **Thymic selection of functional but not self-reactive T cells.**

for each pre–T lymphocyte begins in the cortex, ends in the medulla, and lasts about 2 weeks. Key events are summarized in Figure 14–11, and a simplified account of how the process is currently understood is presented here.

TECs in the cytoreticulum of the cortex present the developing thymocytes with peptides on both MHC class I and class II proteins, which are important for development of CD8⁺ and CD4⁺ T cells, respectively. This interaction determines whether the newly made TCR proteins of these cells are functional. The cells are examined by **positive selection**, with a cell's survival depending on whether its TCRs can recognize and bind antigens on the MHC molecules properly. If the cell cannot pass this test, due, for example, to faulty gene recombination and expression of α and β chains, they are nonfunctional and completely useless. Such cells (as many as 80% of the total) undergo apoptosis and are removed by the macrophages. T cells with normal binding to antigens on either MHC class I or class II are positively selected and move to the medullary compartment (Figure 14–11).

In the medulla T cells with functional TCRs encounter antigens presented on both cytoreticular TECs and dendritic cells. Here the focus is on removing T cells whose TCRs strongly bind self-antigens, a process called **negative selection** because survival depends on a cell *not* binding to MHC molecules with such peptides (Figure 14–11). Among the peptides presented in this compartment are those from proteins specific for many tissues other than the thymus. This occurs because medullary thymic epithelial cells express high levels of the gene *Aire* (*auto*immune *re*gulator), whose product promotes expression of a multitude of such **tissue-specific antigens** in these cells. Most of these proteins are transferred to neighboring dendritic cells, which play the major role in presenting them to developing thymocytes.

T cells that strongly bind MHCs containing these self-peptides undergo apoptosis, which is important because release of any such cells from the thymus would lead to a damaging autoimmune response. Only about 2% of all developing T lymphocytes pass both the positive and negative selection tests and survive to exit the thymus as immunocompetent T cells. Depending on which class of MHC they interacted with, most of these lymphocytes will have stopped expressing *either* CD8 or CD4, and become either helper T cells or cytotoxic T cells (Figure 14–11).

Deletion of self-reactive helper and cytotoxic T lymphocytes in the thymus is the basis for the central immunotolerance produced there. Supplementing this throughout the body is the peripheral tolerance mediated by regulatory T cells, which also develop initially in the thymic medulla under the influence of cytokines from Hassall corpuscles.

Positive selection occurs in the cortex and allows survival only of T cells with functional TCRs that recognize MHC class I and class II molecules. **Negative selection** occurs in the medulla and allows survival only of T cells that do *not* tightly bind self-antigens presented on dendritic cells there.

MUCOSA-ASSOCIATED LYMPHOID TISSUE

Secondary lymphoid structures, where most lymphocytes are activated by antigen presentation, include the mucosa-associated lymphoid tissue (MALT), the lymph nodes, and the spleen.

The mucosa or inner lining of the digestive, respiratory, and genitourinary tracts is a common site of invasion by pathogens because their lumens open to the external environment. To protect against such invaders mucosal connective tissue of these tracts contains large and diffuse collections of lymphocytes, IgA-secreting plasma cells, APCs, and lymphoid nodules, all of which comprise the MALT.

Lymphocytes are also present within the epithelial lining of such mucosae. Most of the immune cells in MALT are dispersed diffusely in the connective tissue; others are found in aggregates that form large, conspicuous structures such as the **tonsils**, the **Peyer patches** in the ileum, and the **appendix**. Collectively the MALT is one of the largest lymphoid organs, containing up to 70% of all the body's immune cells. Most of the lymphocytes here are B cells; among T cells, CD4⁺ helper T cells predominate.

Tonsils are large, irregular masses of lymphoid tissue in the mucosa of the posterior oral cavity and nasopharynx where their cells encounter antigens entering the mouth and nose. Named by their location these masses are the **palatine**, **lingual**, and **pharyngeal tonsils** (Figure 14–12a). In all

FIGURE **14–12** Tonsils.

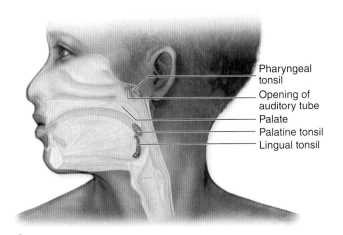

Pharyngeal tonsil
Opening of auditory tube
Palate
Palatine tonsil
Lingual tonsil

a

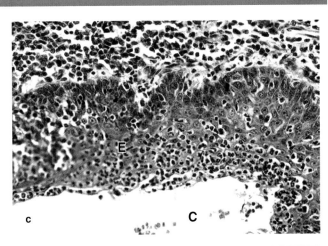

c

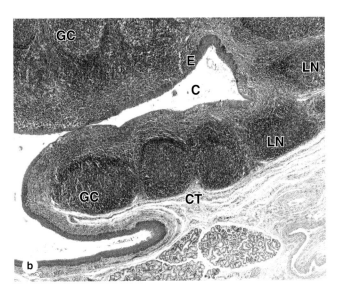

b

(a) Palatine tonsils are located in the posterior lateral walls of the oral cavity, and **lingual** tonsils are situated along the surface of the posterior third of the tongue. Both are covered with stratified squamous epithelium. The **pharyngeal tonsil** is a single medial mass situated in the posterior wall of the nasopharynx. It is usually covered by ciliated pseudostratified columnar epithelium, but areas with stratified epithelium can also be observed. Hypertrophied regions of pharyngeal tonsils resulting from chronic inflammation are called **adenoids**.

(b) A section showing several lymphoid nodules (**LN**), collectively covered by stratified squamous epithelium (**E**) on one side and a connective tissue capsule (**CT**) on the other. Some nodules show lighter staining germinal centers (**GC**). Infoldings of the mucosa in some tonsils form crypts (**C**), along which nodules are especially numerous. Lumens of crypts contain desquamated epithelial cells, live and dead lymphocytes, and bacteria. (X140; H&E)

(c) Epithelium (**E**) surrounding tonsillar crypts (**C**) often becomes infiltrated with lymphocytes and other leukocytes and can become difficult to recognize histologically. Adjacent connective tissue at the top of the photo also contains numerous lymphocytes. (X200; H&E)

(Figure 14-12b and c used with permission from Dr Paulo A. Abrahamsohn, Institute of Biomedical Sciences, University of São Paulo, Brazil.)

Masses of lymphoid nodules comprising tonsils are collected in three general locations in the wall of the pharynx.

tonsils the lymphoid tissue is closely associated with the surface epithelium. Other features include the following:

- **Palatine tonsils**, located posteriorly on the soft palate, are covered by stratified squamous epithelium. The surface area of each is enlarged with 10-20 deep invaginations or **tonsillar crypts** in which the epithelial lining is densely infiltrated with lymphocytes and other leukocytes (Figure 14–12). The lymphoid tissue is filled diffusely with lymphocytes, with many secondary lymphoid nodules around the crypts. This tissue is underlain by dense connective tissue that acts as a partial capsule.
- **Lingual tonsils** are situated along the base of the tongue, are also covered by stratified squamous epithelium with crypts, and have many of the same features as palatine tonsils but lack distinct capsules.
- The single **pharyngeal tonsil** is situated in the posterior wall of the nasopharynx, is covered by pseudostratified ciliated columnar epithelium, and has a thin underlying capsule. The mucosa with diffuse lymphoid tissue and lymphoid nodules is invaginated with shallow infoldings but lacks crypts.

> **》》MEDICAL APPLICATION**
>
> Inflammation of the tonsils, **tonsillitis**, is more common in children than adults. Chronic inflammation of the pharyngeal lymphoid tissue and tonsils of children often produces hyperplasia and enlargement of the tonsils to form "**adenoids**," which can obstruct the eustachian tube and lead to middle ear infections.

Diffuse MALT extends from the pharynx along the entire gastrointestinal tract but becomes very well-developed again in the mucosa and submucosa of the ileum. Here large aggregates of lymphoid nodules comprise the **Peyer patches**, each containing dozens of nodules with no underlying connective tissue capsule (Figure 14–13a).

The simple columnar epithelium that covers the lymphoid nodules of Peyer patches includes large epithelial **M cells** with apical microfolds rather than the brush border typical of the neighboring enterocytes. On the basal side M cells have large intracellular pockets that contain transient populations of lymphocytes and dendritic cells and open to the underlying lymphoid tissue through a unique porous or sieve-like basement membrane (Figure 14–13b). Antigens in the intestinal lumen are continuously sampled at the apical surface of these cells and transferred to the immune cells in the pockets. Lymphocytes and dendritic cells leaving the M cell pockets through the basement membrane pores interact and initiate adaptive responses to the antigens, which results in formation of the secondary lymphoid nodules. Locally produced B cells give rise to plasma cells secreting IgA, which is transported by enterocytes into the intestinal lumen to bind and neutralize potentially harmful antigens.

Another significant collection of MALT occurs in the mucosa of the **appendix**, a short, small-diameter projection from the cecum. Typically the mucosa of the appendix is almost completely filled with lymphoid tissue, effacing the glands otherwise found in the large intestine wall (Figure 14–14). The lumen contains the normal bacterial flora of the large intestine and may serve to retain some of these beneficial bacteria there during diarrheal illnesses.

》 LYMPH NODES

Lymph nodes are bean-shaped, encapsulated structures, generally only 10 mm by 2.5 cm in size, distributed throughout the body along the lymphatic vessels (see Figure 14–1). A total of 400 to 450 lymph nodes are present, in the axillae (armpits) and groin, along the major vessels of the neck, and in the thorax and abdomen, and especially in the visceral mesenteries. The nodes constitute a series of in-line filters of lymph that defend against the spread of microorganisms and tumor cells and provide enclosed environments for antigen presentation and development of plasma cells secreting non-IgA antibodies. Before merging with the bloodstream, all lymph is filtered and has antibodies added by at least one lymph node.

Embedded in loose connective tissue, a lymph node has a convex surface where **afferent lymphatics** enter and a concave depression, the **hilum**, where an **efferent lymphatic** leaves and where an artery, vein, and nerve penetrate the organ (Figure 14–15). A dense connective tissue capsule surrounds the lymph node, extending trabeculae internally through which the blood vessels branch. Valves in the lymphatics ensure unidirectional lymph flow.

The most abundant cells of lymph nodes are lymphocytes of all types, plasma cells, dendritic cells, macrophages, and other APCs. FDCs are present within lymphoid nodules. All of these cells are arranged in a stroma of reticulin fibers and reticular cells to form three major regions within each lymph node: an outer **cortex** containing the nodules; a deeper extension of cortex called the **paracortex**, which lacks nodules; and a **medulla** with prominent draining sinusoids adjacent to the hilum (Figures 14–15 and 14–16). Unlike the thymus these regions of lymph nodes are not compartmentalized by epithelium.

The **cortex** includes the following components:

- A **subcapsular sinus**, immediately inside the capsule, receives lymph from the afferent lymphatics (Figure 14–17). From this space **cortical sinuses** (or trabecular sinuses) branch internally among the lymphoid nodules along trabeculae. These sinuses are lined by a very thin, discontinuous endothelium penetrated by reticulin fibers and processes of dendritic cells. Lymph containing antigens, lymphocytes, and APCs passes through these sinuses and percolates easily into the surrounding lymphoid tissue.
- **Lymphoid nodules**, with or without germinal centers, consist largely of developing B lymphocytes and occupy

FIGURE **14–13** Peyer's patch and M cells.

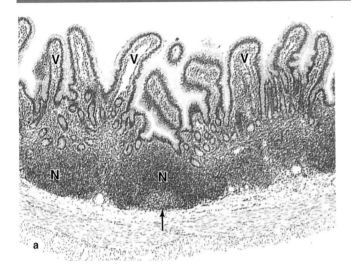

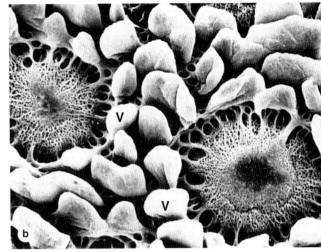

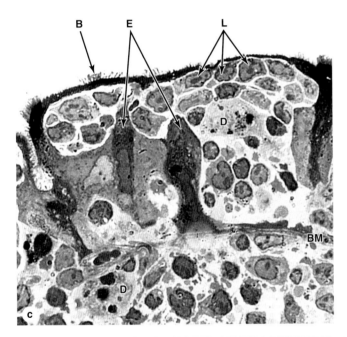

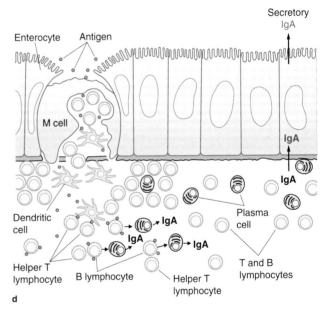

Peyer's patches are very large clusters of lymphoid follicles located in the wall of the ileum which allow close monitoring of microorganisms in the gut.

(a) A section through a Peyer patch shows a few lymphoid nodules (**N**), some with germinal centers (arrow). The mucosa of the small intestine is folded into many projecting villi (**V**). (X100; H&E)

(b) With the surface epithelial cells removed, scanning electron microscopy (SEM) shows typical basement membrane over the villi (**V**) but reveals a highly porous covering over lymphoid nodules of the Peyer patch. This sieve-like basement membrane facilitates interactions between immune cells and M cells in the epithelium over the nodules.

(Used with permission from Dr Samuel G. McClugage, Department of Cell Biology and Anatomy, Louisiana State University Health Sciences Center, New Orleans, LA.)

(c) The TEM shows that the epithelium directly over a Peyer patch lymphoid nodule has unique cells called **M (microfold) cells** with

short apical folds but no brush border. The basal surface of M cells forms a large intracellular pocket that harbors a transient population of T and B lymphocytes (**L**) and dendritic cells (**D**) which move through the openings in the basement membrane (**BM**). Darker cytoplasm of adjacent enterocytes (**E**) with brush borders (**B**) is also seen. (X1000)

(Used with permission from Dr Marian R. Neutra, Children's Hospital, Harvard Medical School, Boston, MA.)

(d) A summary diagram showing that antigens in the gut lumen are bound by **M cells** and undergo transcytosis into their intraepithelial pockets where **dendritic cells** take up the antigen, process it, and present it to **T helper cells**. B lymphocytes stimulated by the Th cells differentiate into **plasma cells** secreting IgA antibodies. The IgA is transported into the gut lumen where it binds its antigen on the surface of microorganisms, neutralizing potentially harmful invaders before they penetrate the mucosa.

FIGURE **14–14** Appendix.

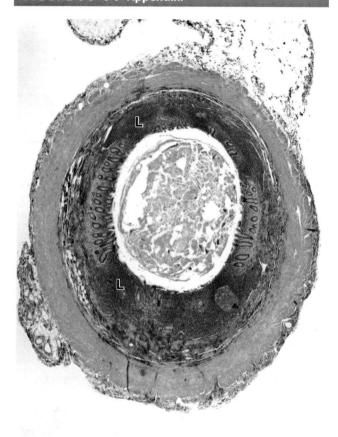

A blind evagination off the cecum, the appendix is a significant part of the MALT with its lamina propria and submucosa filled with lymphocytes and lymphoid follicles (**L**). The small lumen contains a sample of the microbial flora of the intestine, along with undigested material. (X20; H&E)

much of the cortex not filled with helper T lymphocytes (Figures 14–16 and 14–17). Each nodule is organized around the long, interdigitating processes of follicular dendritic cells (FDCs), but these are not readily seen by routine light microscopy. Numerous macrophages are also present for removal of newly formed defective B cells which undergo apoptosis.

The region between the cortex and medulla, the **paracortex** does not have precise boundaries but can be distinguished from the outer cortex by its lack of nodules (Figure 14–16). Unlike the superficial cortex, the paracortex contains lymphoid tissue rich in T cells distinguishable by immunohistochemistry (Figure 14–18).

Also located mainly in the paracortex, specialized postcapillary venules called **high endothelial venules (HEVs)** represent an important entry point for most (90%) circulating

lymphocytes into lymph nodes. Endothelial cells of these vessels become unusually enlarged or cuboidal and express specific apical surface glycoproteins that mediate the tethering and diapedesis of B and T cells from the blood into the paracortex of the lymph node (Figure 14–19). HEVs also occur in the large accumulations of MALT discussed previously, but are less well-characterized in those tissues.

The **medulla** of a lymph node has two major components (Figures 14–16 and 14–20):

- **Medullary cords** are branched cordlike masses of lymphoid tissue extending from the paracortex. They contain T and B lymphocytes and many plasma cells.
- **Medullary sinuses** are dilated spaces lined by discontinuous endothelium that separate the medullary cords. As shown in Figure 14–20, the lumens of medullary sinuses include a meshwork of processes from reticular cells, which represent a final lymph filter. These sinuses contain many macrophages and sometimes neutrophils if the lymph node is draining an infected region. They are continuous with the cortical sinuses and converge at the hilum as the efferent lymphatic vessel (Figure 14–15).

>> **MEDICAL APPLICATION**

Metastatic cancer cells detached from a primary tumor can enter lymphatics and are carried to nearby lymph nodes, especially the **sentinel lymph node** that is the first one downstream of the region with the tumor. Cells from well-established tumors are often immunosuppressive themselves and may continue growth as a **secondary tumor** within lymph nodes. During cancer surgery lymph nodes in the lymphatics draining the tumor area are examined by pathologists for the presence of cancer cells. The presence of such metastatic cells in lymph nodes is a key determinant in most **staging systems** for various types of cancer and an important prognostic indicator.

Role of Lymph Nodes in the Immune Response

The lymph arriving at a lymph node contains antigens free in solution or bound to antibodies or complement, still on microorganisms, or already internalized and transported by APCs. If draining from an infected or inflamed region, lymph may also contain microorganisms and cytokines. Antigens not yet phagocytosed will be internalized by APCs in the lymph nodes and presented on MHC class II molecules.

Circulating B and T lymphocytes traffic from node to node, entering via the lymph or HEVs, where B cells contact antigens on FDCs and T cells sample antigens presented on dendritic cells and other APCs. Lymphocytes whose receptors recognize such antigens will be activated. B cells will proliferate rapidly in germinal centers of follicles with the help

FIGURE **14–15** Lymph node.

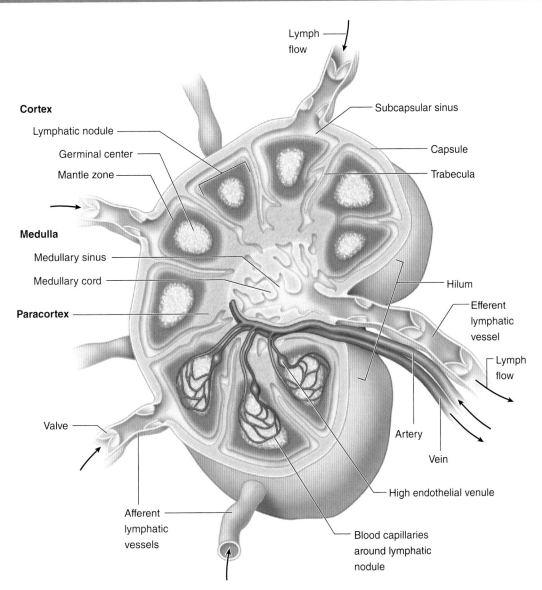

Lymph nodes are small encapsulated structures positioned along lymphatic vessels to filter lymph and facilitate antibody production. Valves in the lymphatic vessels ensure the one-way flow of lymph, indicated by arrows. The three major regions of a lymph node include the outer **cortex** receiving lymph from the afferent lymphatics, an inner **paracortex** where most lymphocytes enter via high endothelial venules (HEVs), and a central **medulla** with sinuses converging at the efferent lymphatic.

of Th cells, often enlarging the entire lymph node. Activated cytotoxic T cells in the paracortex proliferate to a much lesser extent without forming follicles.

Many newly made B cells, now activated against a specific antigen, differentiate as plasma cells and move to the medulla or to downstream sites beyond the lymph node where they produce antibodies. Specific Th cells, CTLs, and T_{regs} also recirculate in the efferent lymph and with the antibodies spread the immune defenses against those microorganisms throughout the body. Both B and T memory cells also move elsewhere in the body, providing long-lived protection and proliferating more rapidly upon subsequent exposure to their specific antigen.

❱❱ MEDICAL APPLICATION

Neoplastic proliferation of lymphocytes, producing a malignant **lymphoma**, may occur diffusely but is often located in one or more lymph nodes. Such growth can completely obliterate the normal architecture of the node and convert it to an enlarged, encapsulated structure filled with lymphocytes, a condition called **lymphadenopathy**.

FIGURE **14–16** Regions of a lymph node.

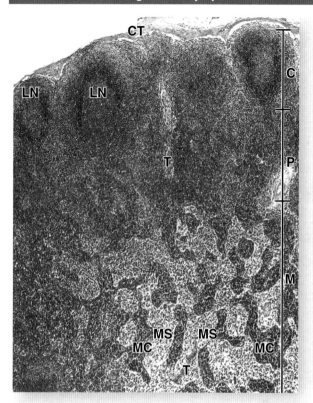

A low-magnification section of a lymph node shows the three functional regions: the cortex (**C**), the paracortex (**P**), and the medulla (**M**). Connective tissue of the capsule (**CT**) completely surrounds each lymph node and extends as several trabeculae (**T**) throughout the lymphoid tissue. Major spaces for lymph flow are present in this tissue under the capsule and along the trabeculae. A changing population of immune cells is suspended on reticular fibers throughout the cortex, paracortex, and medulla. Lymphoid nodules (**LN**) are normally restricted to the cortex, and the medulla is characterized by sinuses (**MS**) and cords (**MC**) of lymphoid tissue. (X40; H&E)

(Used with permission from Dr Paulo A. Abrahamsohn, Institute of Biomedical Sciences, University of São Paulo, Brazil.)

FIGURE **14–17** Lymph node cortex.

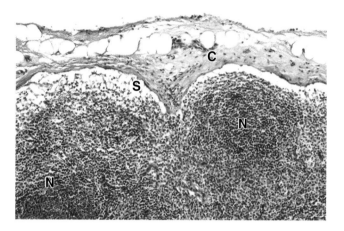

The outer regions on the convex sides of a lymph node include the capsule (**C**), subcapsular sinuses (**S**), and diffuse lymphoid tissue with lymphoid nodules (**N**). Afferent lymphatic vessels (which are only rarely shown well in sections) penetrate this capsule, dumping lymph into the sinus where its contents are processed by lymphocytes and APCs. (X140; H&E)

(Used with permission from Dr Paulo A. Abrahamsohn, Institute of Biomedical Sciences, University of São Paulo, Brazil.)

› SPLEEN

The spleen contains the largest single accumulation of lymphoid tissue in the body and is the only lymphoid organ involved in filtration of blood, making it an important organ in defense against blood-borne antigens. It is also the main site of old erythrocyte destruction. As is true of other secondary lymphoid organs, the spleen is a production site of antibodies and activated lymphocytes, which here are delivered directly into the blood.

Located high in the left upper quadrant of the abdomen and typically about 12 × 7 × 3 cm in size, the spleen's volume varies with its content of blood and tends to decrease very slowly after puberty. The organ is surrounded by a capsule of dense connective tissue from which emerge trabeculae to penetrate the parenchyma or **splenic pulp** (Figure 14–21). Large trabeculae originate at the hilum, on the medial surface of the spleen, and carry branches of the splenic artery, vein, lymphatics, and nerves into the splenic pulp.

Functions of Splenic White & Red Pulp

The spleen is filled with reticular tissue containing reticular cells and fibers, many lymphocytes and other blood cells, macrophages, and APCs. This splenic pulp has two components: the **white pulp** (20% of the spleen) and the **red pulp** (Figure 14–21). The small masses of white pulp consist of **lymphoid nodules** and the **periarteriolar lymphoid sheaths (PALS)**, while the red pulp consists of blood-filled **sinusoids** and **splenic cords**.

As expected of an organ where the blood is monitored immunologically, the splenic microvasculature contains unique regions shown schematically in Figure 14–22. Branching from the hilum, small **trabecular arteries** leave the trabecular connective tissue and enter the parenchyma as arterioles enveloped by the PALS, which consists primarily of T cells with some macrophages, DCs, and plasma cells as part of the white pulp. Surrounded by the PALS, these vessels are known as **central arterioles** (Figure 14–23). B cells located within the PALS

FIGURE **14–18** Lymph node cortex and paracortex.

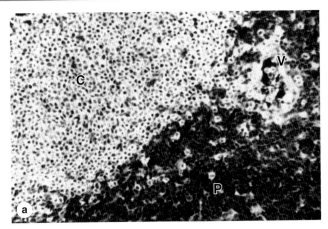
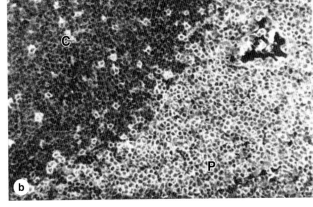

The region just inside the cortex is called the **paracortex**. Although most lymphocytes in the cortex are B cells, many located in nodules, the lymphocytes of the paracortex are largely T cells. This separation is indicated in the fluorescence micrographs here using immunohistochemistry on adjacent sections of lymph node. **(a)** Antibody against a B-cell surface marker labels nearly all the lymphocytes in the cortex (**C**), as well as many cells around an HEV (**V**)

in the paracortex, but few cells in the paracortex proper (**P**). **(b)** Stained with an antibody against a T-cell marker, the paracortex is heavily labeled, but only a few cells in the cortex (**C**) are stained, possibly T helper cells. (X200)

(*Used with permission from I. L. Weissman, Stanford University School of Medicine, Palo Alto, CA.*)

FIGURE **14–19** High endothelial venules.

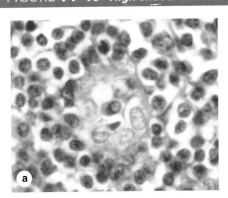
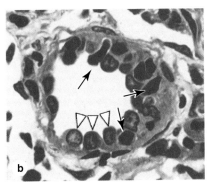
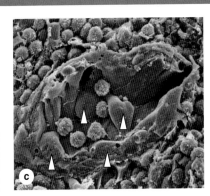

HEVs are found in the paracortex of lymph nodes, as shown, as well as in tonsils and Peyer patches. Their endothelial cells are unusually shaped but generally cuboidal and facilitate rapid translocation of lymphocytes from blood into the lymphoid tissue. L-selectin on the lymphocytes recognizes sugar-rich ligands on the surfaces of these endothelial cells and, as a consequence, the lymphocytes stop there. Integrins promote adhesion between lymphocytes and the endothelial cells, and the lymphocytes cross the vessel wall into the lymph node parenchyma.

(a) HEVs can be difficult to identify in H&E-stained paraffin sections. (X400; H&E)

(b) Plastic sections more clearly reveal the HEVs (arrowheads) and the lymphocytes passing between them (arrows). (X400; PT).

(c) SEM of a sectioned HEV shows five typical lymphocytes adhering to endothelial cells (arrowheads) before migrating between them and joining other lymphocytes in the surrounding paracortex. (X500)

(*Figure 14–19c reproduced, with permission, from Fujita T. Prog Clin Biol Res 1989;295:493.*)

FIGURE **14–20** Lymph node medulla.

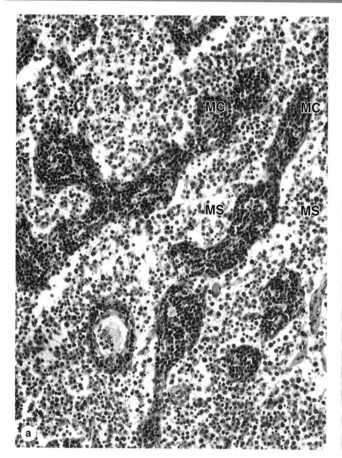

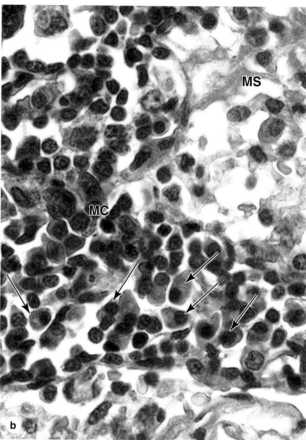

(a) The medulla of a lymph node consists mainly of the medullary sinuses (**MS**) separated by intervening medullary cords (**MC**). Lymphocytes and plasma are abundant and predominate in number over other cell types. A blood vessel within a medullary cord is also seen. (X200; H&E)

(b) Higher magnification of a medullary cord (**MC**) shows plasma cells (arrows) with spherical, eccentric nuclei and much more

cytoplasm than lymphocytes. Efferent lymph is rich in newly synthesized antibodies. A medullary sinus (**MS**) with a meshwork of eosinophilic processes from surrounding reticular cells is also seen. (X400; H&E)

(Used with permission from Dr Paulo A. Abrahamsohn, Institute of Biomedical Sciences, University of São Paulo, Brazil.)

FIGURE **14–21** Spleen.

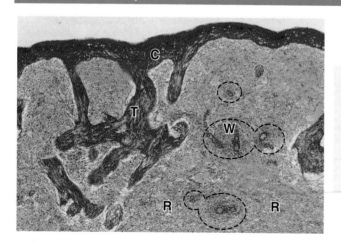

The capsule (**C**) of the spleen connects to trabeculae (**T**) extending into the pulp-like interior of the organ. The red pulp (**R**) occupies most of the parenchyma, with white pulp (**W**) restricted to smaller areas, mainly around the central arterioles. Names of these splenic areas refer to their color in the fresh state: red pulp is filled with blood cells of all types, located both in cords and sinuses; white pulp is lymphoid tissue. Large blood vessels and lymphatics enter and leave the spleen at a hilum. (X20; Picro-Sirius-hematoxylin)

FIGURE **14–22** Blood flow in the spleen.

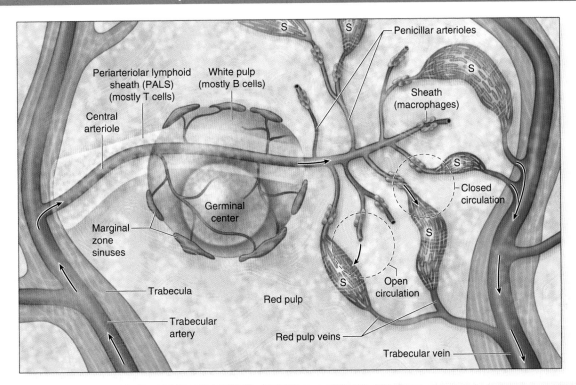

Schematic view of the blood circulation and the structure of the spleen, from the trabecular artery to the trabecular vein. Small branches of these arteries are called **central arterioles** and become enclosed within a sheath of lymphoid cells, the **periarteriolar lymphoid sheath (PALS)** in white pulp. B cells in these sheaths can form nodules as the largest masses of white pulp, and around these nodules are located the marginal zone sinuses.

Emerging from the white pulp, the central arteriole branches as the penicillar arterioles, which lead to sheathed capillaries. From these, blood flows into either a **closed circulation** passing directly into splenic sinuses (**S**) or an **open circulation**, being dumped from the vasculature into the lymphoid tissue of the red pulp's splenic cords. From there viable blood cells reenter the vasculature through the walls of the sinuses.

FIGURE **14–23** White pulp of the spleen.

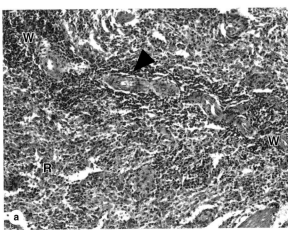

The splenic white pulp consists of lymphoid tissue surrounding the central arterioles as the PALS and the nodules of proliferating B cells in this sheath. **(a)** Longitudinal section of white pulp (**W**) in a PALS surrounding a central arteriole (arrowhead). Surrounding the PALS is much red pulp (**R**).

(b) A large nodule with a germinal center forms in the PALS and the central arteriole (arrowhead) is displaced to the nodule's periphery. Small vascular sinuses can be seen at the margin between white (**W**) and red (**R**) pulp. (Both X20; H&E)

(*Used with permission from Dr Paulo A. Abrahamsohn, Institute of Biomedical Sciences, University of São Paulo, Brazil.*)

may be activated by a trapped antigen from the blood and form a temporary lymphoid nodule like those of other secondary lymphoid organs (Figure 14–23b). In growing nodules the arteriole is pushed to an eccentric position but is still called the central arteriole. These arterioles send capillaries throughout the white pulp and to small sinuses in a peripheral marginal zone of developing B cells around each lymphoid nodule (Figure 14–22).

Each central arteriole eventually leaves the white pulp and enters the red pulp, losing its sheath of lymphocytes and branching as several short straight **penicillar arterioles** that continue as capillaries (Figure 14–22). Some of these capillaries are sheathed with APCs for additional immune surveillance of blood.

The red pulp is composed almost entirely of **splenic cords** (of Billroth) and **splenic sinusoids** and is the site where effete RBCs in blood are removed (Figure 14–24). The splenic cords contain a network of reticular cells and fibers filled with T and B lymphocytes, macrophages, other leukocytes, and red blood cells. The splenic cords are separated by the sinusoids (Figure 14–25). Unusual elongated endothelial cells called **stave cells** line these sinusoids, oriented parallel to the blood flow and sparsely wrapped in reticular fibers and highly discontinuous basal lamina (Figure 14–26).

Blood flow through the splenic red pulp can take either of two routes (Figure 14–22):

■ In the **closed circulation**, capillaries branching from the penicillar arterioles connect directly to the sinusoids and the blood is always enclosed by endothelium.

■ In the **open circulation**, capillaries from about half of the penicillar arterioles are uniquely *open-ended*, dumping blood into the stroma of the splenic cords. In this route plasma and all the formed elements of blood must reenter the vasculature by passing through narrow slits between the stave cells into the sinusoids. These small openings present no obstacle to platelets, to the motile leukocytes, or to thin flexible erythrocytes. However stiff or effete, swollen RBCs at their normal life span of 120 days are blocked from passing between the stave cells and undergo selective removal by macrophages (Figure 14–24).

Removal of defective RBCs and recycling of their iron are major functions of the red pulp. Iron released from hemoglobin during the degradation of RBCs is stored by macrophages within complexes of ferritin proteins or bound to transferrin, returned to the circulation, and reused primarily for erythropoiesis. Iron-free heme is either bound to its transport protein, hemopexin, or is metabolized to bilirubin and excreted in the bile by liver cells. After surgical removal of the spleen (splenectomy), the number of abnormal erythrocytes in the circulation increases although most such cells are then removed by macrophages in sinusoids of the bone marrow and liver.

From the sinusoids blood proceeds to small red pulp veins that converge as the **trabecular veins** (Figure 14–22), which in turn form the splenic vein. The trabecular veins lack significant smooth muscle and resemble endothelium-lined channels hollowed out in the trabecular connective tissue.

Important aspects of the major lymphoid organs (thymus, MALT, lymph nodes, and spleen) are summarized and compared in Table 14–4.

FIGURE 14–24 Erythrocyte removal by splenic macrophages.

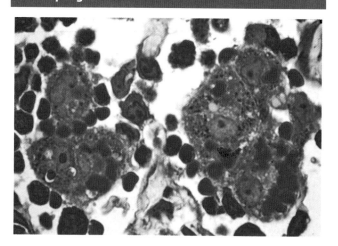

A micrograph showing five macrophages in a splenic cord engaged in phagocytosis of effete erythrocytes. (X400; PT)

>> MEDICAL APPLICATION

Enlargement of the spleen, **splenomegaly**, can occur from a variety of causes, including lymphoma or other malignant growth, infections such as mononucleosis, or sickle cell disease and other types of anemia. The splenic capsule is relatively thin, and an enlarged spleen is susceptible to traumatic rupture, a potentially life-threatening occurrence due to loss of blood into the abdominal cavity. Such rupture may require prompt surgical removal of the spleen, **splenectomy**, after which most functions of the organ are carried out by other lymphoid organs, with erythrocyte removal occurring in the liver and bone marrow.

FIGURE **14–25**   Red pulp of the spleen.

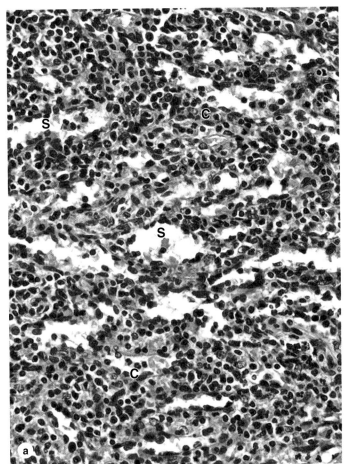

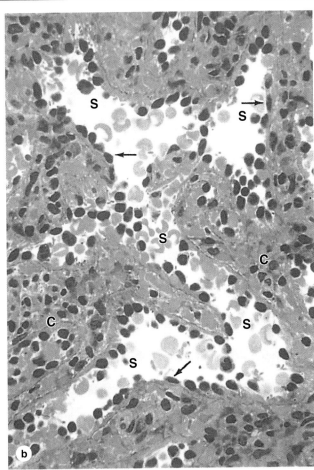

(a) The splenic red pulp is composed entirely of sinusoids (**S**) and splenic cords (**C**), both of which contain blood cells of all types. The cords, often called **cords of Billroth**, are reticular tissue rich in macrophages and lymphocytes. (X140; H&E)

(b) Higher magnification shows that the sinusoids (**S**) are lined by endothelial cells (arrows) with large nuclei bulging into the

sinusoidal lumens. The unusual endothelial cells are called **stave cells** and have special properties that allow separation of healthy from effete red blood cells in the splenic cords (**C**). (X200; H&E)

(*Figure 14-25a and b, used with permission from Dr Paulo A. Abrahamsohn, Institute of Biomedical Sciences, University of São Paulo, Brazil.*)

TABLE **14–4**   **Important histologic comparisons of the major lymphoid organs.**

	Thymus	**MALT**	**Lymph Nodes**	**Spleen**
Cortex/medulla	Distinctly present	Absent	Present	Absent
Lymphoid nodules	Absent	Present	Present (in cortex only)	Present (in white pulp only)
Lymphatic vessels	No afferents; few efferents in septa	No afferents; efferents present	Afferents at capsule, emptying into subcapsular sinus; efferent at hilum	No afferents; efferents in trabeculae
Unique features	Hassall (thymic) corpuscles in medulla; epithelial- reticular cells in cortex and medulla	Crypts lined by surface mucosa in tonsils; epithelial M cells in mucosa over Peyer patches	Thin paracortical region between cortex and medulla, with high endothelial venules (HEV); medullary cords and sinuses	Minor white pulp component, with central arterioles; major red pulp component, with many sinusoids

FIGURE **14–26** Structure and function of splenic sinusoids.

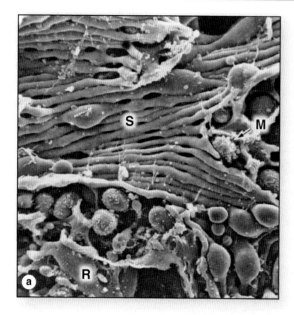

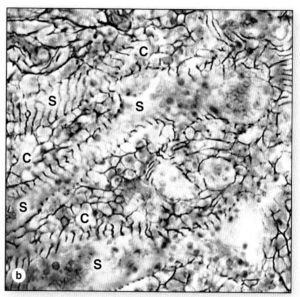

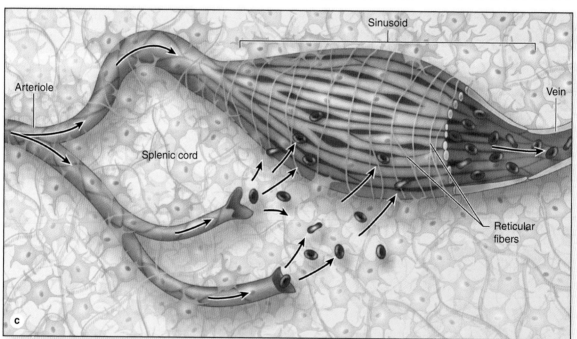

The endothelial stave cells that line the sinusoids in red pulp are long cells oriented lengthwise along the sinusoids. **(a)** SEM clearly shows the parallel alignment of the stave cells (**S**), as well as many macrophages (**M**) in the surrounding red pulp (**R**). (X500)

   (*Figure 14-26a, reproduced with permission from Fujita T. Prog Clin Biol Res 1989;295:493*)

(b) A silver-stained section of spleen shows dark reticular fibers in the splenic cords (**C**) and wrapped around the sinusoids (**S**). The sinusoidal basement membrane is highly discontinuous and open to the passage of blood cells. (X400)

(c) Diagram showing these components of splenic sinusoids, with the structures resembling a loosely organized wooden barrel. In the open circulation mode of blood flow, blood cells dumped into the splenic cords move under pressure or by their own motility through the spaces between **stave cells**, reentering the vasculature and soon leaving the spleen via the splenic vein. Cells that cannot move between the stave cells, mainly effete erythrocytes, are removed by macrophages.

The Immune System & Lymphoid Organs SUMMARY OF KEY POINTS

Basic Immunology

- **Innate immunity** is present from birth and involves leukocytes (mainly granulocytes), and proteins such as defensins, complement, lysozyme, and interferons; **adaptive immunity** develops more slowly and is based on antigen presentation to **lymphocytes**.
- Immune cells communicate with one another and regulate one another's activities via polypeptide hormones called **cytokines**.
- **Antigens** are the regions of macromolecules, usually proteins, that are recognized by lymphocytes to elicit a specific immune response against them.
- **Antibodies** are immunoglobulins produced by plasma cells after a progenitor B cell is activated by a specific antigen and rearranges its immunoglobulin genes so the antibody matches the antigen.
- Surfaces of all nucleated cells bear fragments of their constituent proteins on **major histocompatibility complex (MHC) class I** molecules.
- Only **antigen-presenting cells (APCs)**, mostly derived from monocytes, also present fragments of endocytosed foreign (usually from microorganisms) proteins on surface **MHC class II** molecules.

Lymphocyte Origins and Differentiation

- Lymphocytes originate in the **primary lymphoid organs**: bone marrow for **B lymphocytes** and the thymus for **T lymphocytes**.
- B cells produce antibodies for **humoral immunity**; T cells function in **cell-mediated immunity**.
- T cells develop receptors (**TCRs**), usually containing α and β chains, that bind antigen along with another surface protein designated by a **CD ("cluster of differentiation") numbering system**.
- Important classes of T cells include **CD4⁺ T helper cells**; **CD8⁺ cytotoxic T cells**; **CD4⁺CD25⁺ regulatory T cells**; and γδ **T cells**, which have those TCR chains and are mainly in epithelia.
- **B-cell receptors (BCRs)** are IgM or IgD antibodies on the cell surface that bind specific antigens whenever they contact them.
- B and T cells are often activated, proliferate, and begin to function in the **secondary lymphoid organs**: the **lymph nodes**, all **mucosa-associated lymphoid tissue (MALT)**, and the **spleen**.
- In these organs lymphocytes are distributed within a meshwork of reticulin produced by fibroblastic **reticular cells**, and most APCs are **dendritic cells** with many processes.
- In secondary lymphoid tissues, BCRs bind antigen not presented in MHC class II molecules of another cell, the **follicular dendritic cell (FDC)**.
- With cytokines from helper T cells, a FDC-activated B cell proliferates clonally to produce temporarily a large **lymphoid nodule** (or follicle), which develops a pale **germinal center**.
- From lymphoid nodules cells produced there disperse as **plasma cells**, various T cells, and **B and T memory cells** that respond and proliferate quickly if their specific antigen reappears.

Thymus

- T lymphoblasts, or **thymocytes**, attach in the **thymus** to a **cytoreticulum** composed of interconnected **thymic epithelial cells (TECs)**.
- The thymic epithelial cells (TECs) also secrete many **cytokines**, **compartmentalize** the thymus into a **cortex** and a **medulla**, and in the cortex surround blood vessels in the **blood-thymus barrier**.
- Developing T cells with nonfunctional TCRs are detected and removed in the thymic cortex by a process of **positive selection**; cells with functional TCRs move into the thymic medulla.

- In the thymic medulla T cells whose TCRs bind strongly to "**self-proteins**," including proteins of many nonthymus cell types made by thymic epithelial cells expressing the *Aire* gene, are induced to undergo apoptosis there in a process of **negative selection**.
- This two-stage thymic selection leads to **central immune tolerance**, producing functional T cells that do not bind to proteins of the host.
- **Peripheral immune tolerance** occurs throughout the body when specific immune reactions are suppressed by **regulatory T cells** that also originate largely in the thymic medulla.
- Regulatory T cells form in the thymus upon interacting with dendritic cells presenting self-antigens in a process promoted by cytokines from thymic epithelial cell (TEC) aggregates called **Hassall corpuscles**, found only in the thymic medulla.

Mucosa-Associated Lymphoid Tissue (MALT)

- **MALT** is found in the mucosa of most tracts but is concentrated in the **palatine**, **lingual** and **pharyngeal tonsils**, **Peyer patches**, and the **appendix**.
- Unlike MALT, **lymph nodes** are completely **encapsulated** and occur along the lymphatic vessels; each has several **afferent lymphatics** and one **efferent lymphatic**.

Lymph Nodes

- Each lymph node **filters lymph** and **provides a site for B-cell activation** and differentiation to antibody-secreting **plasma cells**.
- A lymph node has three functional but not physically separate compartments: an outer **cortex**, a underlying **paracortex**, and an inner **medulla** adjacent to the hilum and efferent lymphatic.
- Lymphatics enter at the **cortex** of a node, where **B cells** encounter antigens, proliferate in **lymphoid nodules**, and then move into the deeper regions of the lymph node.
- Most lymphocytes enter at the **paracortex** of the lymph node via **high endothelial venules (HEVs)** located there only; most lymphocytes in this region are **T helper cells**.
- The medulla has **medullary cords** containing reticular fibers with many plasma cells, macrophages, and other leukocytes; between the cords are lymph-filled **medullary sinuses** that converge at the efferent lymphatic.

Spleen

- The **spleen** is a large lymphoid organ without a cortex/medulla structure; instead, it has two intermingled but functionally different regions: **white pulp** and **red pulp**.
- **White pulp**, only 20% of the spleen, is **secondary lymphoid tissue** associated with small central arterioles that are also enclosed by **periarteriolar lymphoid sheaths (PALS)** of T cells.
- **Red pulp**, which filters blood, removes defective erythrocytes, and recycles hemoglobin iron, consists of **splenic cords** with macrophages and blood cells of all kinds and **splenic sinusoids**.
- The **splenic sinusoids** are lined by unusual endothelial cells called **stave cells** that are elongated and aligned parallel to the blood flow, with open slits between the cells.
- Blood flow in red pulp is either a **closed circulation**, moving from capillaries into the venous sinusoids, or an **open circulation**, with capillaries opening directly into the splenic cords.
- **Blood filtration** in the open circulation involves interaction with splenic cord **macrophages** that remove old, swollen RBCs unable to slip between stave cells to reenter the venous blood flow.

The Immune System & Lymphoid Organs ASSESS YOUR KNOWLEDGE

1. Which function is carried out by all lymphoid tissues and organs?
   a. Filtration of lymph
   b. Filtration of blood
   c. Extramedullary hemopoiesis
   d. Production of lymphocytes
   e. Destruction of old erythrocytes

2. Which structure is partly encapsulated and covered by nonkeratinized stratified squamous epithelium?
   a. Appendix
   b. Lymph node
   c. Palatine tonsil
   d. Peyer's patch
   e. Thymic (Hassal's) corpuscle

3. Which cell type gives rise to both memory and effector cells and is primarily associated with humoral immunity?
   a. B lymphocyte
   b. NK cell
   c. Macrophage
   d. T lymphocyte
   e. Reticular cell

4. Recycling of iron and heme, the major complex containing iron, occurs most actively in which lymphoid organ(s)/tissue(s)?
   a. Lymph nodes
   b. Peyer's patches
   c. Tonsils
   d. Spleen
   e. Lymphatic vessels

5. Which description is true of all secondary (peripheral) lymphoid organs?
   a. Capable of antigen-independent lymphopoiesis
   b. Contain crypts
   c. Contain epithelial-reticular cells
   d. Lack connective tissue capsules
   e. Contain lymphoid nodules

6. Which structure would be most heavily labeled by an immunohistochemical method targeting the CD8 surface antigen?
   a. Germinal centers
   b. Paracortex
   c. Peyer's patch
   d. Sheathed arterioles
   e. Splenic cords

7. A baby is born with a cleft palate and a condition called DiGeorge syndrome, which involves failure of third and fourth pharyngeal pouch derivatives to develop properly. The palate defect is corrected surgically, but regarding the pharyngeal pouch defect the parents are advised that the growing child may expect which of the following health problems?
   a. Insufficient B-cell production by lymph nodes in the head and neck
   b. Inability to secrete IgA
   c. Excessive numbers of circulating but defective erythrocytes
   d. Increased oral infections due lack of palatine and pharyngeal tonsils
   e. Conditions related to autoimmunity

8. Many immune-related cellular activities are often impaired in aged patients. Which lymphoid organ(s) normally develop less functionality and increasing amounts of adipose tissue with age?
   a. Axillary lymph nodes
   b. Lingual tonsils
   c. Thymus
   d. Splenic white pulp
   e. Splenic red pulp

9. A 12-year-old African-American girl presents with anemia and a large percentage of her peripheral erythrocytes appear sickle-shaped. Genetic testing reveals homozygosity for sickle cell disease. In which of the following sites will the abnormal RBCs be removed from the circulation?
   a. Thymic cortex
   b. Periarteriolar lymphoid sheaths of splenic white pulp
   c. Medullary sinuses of lymph nodes
   d. Thymic medulla
   e. Splenic cords (of Billroth)

10. A 6-year-old boy is brought to the clinic where his mother reports that was bitten by a neighbor's dog two days earlier. The child's right hand is lacerated between the thumb and index finger and this area is inflamed but healing. The doctor's examination reveals small but painless swellings beneath the skin inside the right elbow and arm pit and he explains to the mother that these are active lymph nodes enlarged in response to the infection in the hand. What has produced the swelling?
   a. Increased flow of lymph through the nodes' afferent lymphatics
   b. Formation of germinal centers for B-cell proliferation in each node's cortex
   c. Arrival of antigen-presenting cells in each node's medulla
   d. Enlargement and increased activity of the nodes' high endothelial venules
   e. Increased thickness of each node's paracortex

CHAPTER

15 Digestive Tract

The digestive system consists of the **digestive tract**—oral cavity, esophagus, stomach, small and large intestines, and anus—and its associated glands—salivary glands, liver, and pancreas (Figure 15–1). Also called the **gastrointestinal (GI) tract** or **alimentary canal**, its function is to obtain from ingested food the molecules necessary for the maintenance, growth, and energy needs of the body. During digestion proteins, complex carbohydrates, nucleic acids, and fats are broken down into their small molecule subunits that are easily absorbed through the small intestine lining. Most water and electrolytes are absorbed in the large intestine. In addition, the inner layer of the entire digestive tract forms an important protective barrier between the content of the tract's lumen and the internal milieu of the body's connective tissue and vasculature.

Structures within the digestive tract allow the following:

- **Ingestion**, or introduction of food and liquid into the oral cavity,
- **Mastication**, or chewing, which divides solid food into digestible pieces,
- **Motility**, muscular movements of materials through the tract,
- **Secretion** of lubricating and protective mucus, digestive enzymes, acidic and alkaline fluids, and bile,
- **Hormone release** for local control of motility and secretion,
- **Chemical digestion** or enzymatic degradation of large macromolecules in food to smaller molecules and their subunits,
- **Absorption** of the small molecules and water into the blood and lymph,
- **Elimination** of indigestible, unabsorbed components of food.

GENERAL STRUCTURE OF THE DIGESTIVE TRACT

All regions of the GI tract have certain structural features in common. The GI tract is a hollow tube with a lumen of variable diameter and a wall made up of four main layers: the **mucosa, submucosa, muscularis,** and **serosa**. Figure 15–2 shows a general overview of these four layers; key features of each layer are summarized here.

- The **mucosa** consists of an **epithelial lining**; an underlying **lamina propria** of loose connective tissue rich in blood vessels, lymphatics, lymphocytes, smooth muscle cells, and often containing small glands; and a thin layer of smooth muscle called the **muscularis mucosae** separating mucosa from submucosa and allowing local movements of the mucosa. The mucosa is also frequently called a **mucous membrane**.
- The **submucosa** contains denser connective tissue with larger blood and lymph vessels and the **submucosal (Meissner) plexus** of autonomic nerves. It may also contain glands and significant lymphoid tissue.
- The thick **muscularis** (or muscularis externa) is composed of smooth muscle cells organized as two or more sublayers.

FIGURE **15–1** The digestive system.

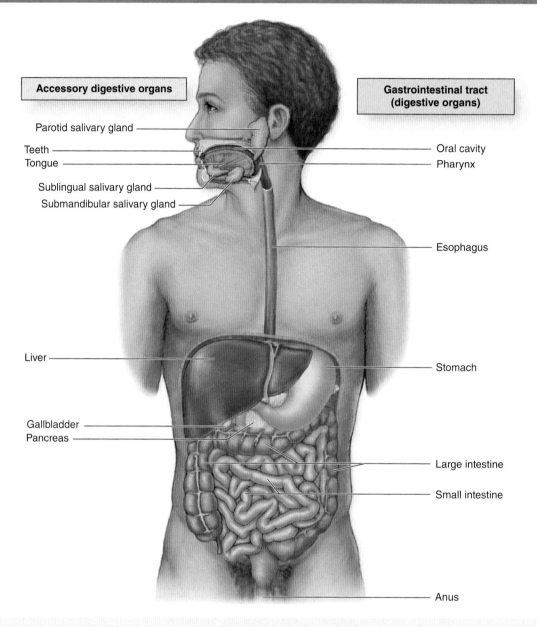

Accessory digestive organs

- Parotid salivary gland
- Teeth
- Tongue
- Sublingual salivary gland
- Submandibular salivary gland
- Liver
- Gallbladder
- Pancreas

Gastrointestinal tract (digestive organs)

- Oral cavity
- Pharynx
- Esophagus
- Stomach
- Large intestine
- Small intestine
- Anus

The digestive system consists of the tract from the mouth (oral cavity) to the anus, as well as the digestive glands emptying into this tract, primarily the salivary glands, liver, and pancreas. These accessory digestive glands are described in Chapter 16.

In the internal sublayer (closer to the lumen), the fiber orientation is generally circular; in the external sublayer it is longitudinal. The connective tissue between the muscle sublayers contains blood and lymph vessels, as well as the **myenteric (Auerbach) nerve plexus** of many autonomic neurons aggregated into small ganglia and interconnected by pre- and postganglionic nerve fibers. This and the submucosal plexus together comprise the **enteric nervous system** of the digestive tract. Contractions of the muscularis, which mix and propel the luminal contents forward, are generated and coordinated by the myenteric plexus.

- The **serosa** is a thin layer of loose connective tissue, rich in blood vessels, lymphatics, and adipose tissue, with a simple squamous covering epithelium or **mesothelium**. In the abdominal cavity, the serosa is continuous with **mesenteries**, thin membranes covered by mesothelium on both sides that support the intestines. Mesenteries are continuous with the **peritoneum**, a serous membrane that lines that cavity. In places where the digestive tract is not suspended in

FIGURE **15–2** Major layers and organization of the digestive tract.

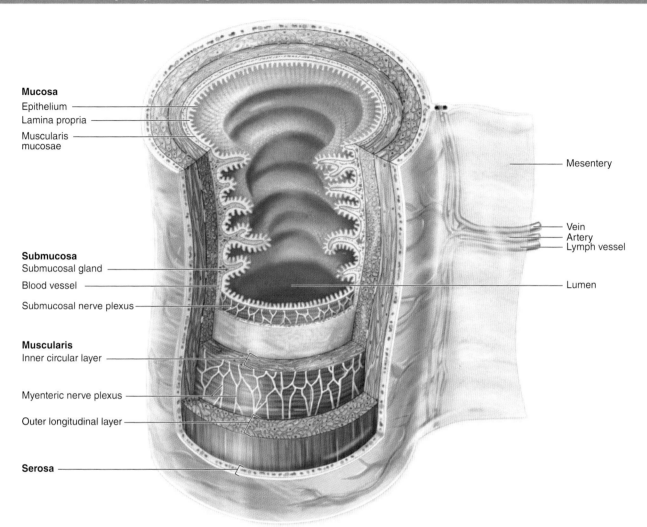

Mucosa
Epithelium
Lamina propria
Muscularis
mucosae

Submucosa
Submucosal gland
Blood vessel
Submucosal nerve plexus

Muscularis
Inner circular layer

Myenteric nerve plexus

Outer longitudinal layer

Serosa

Mesentery

Vein
Artery
Lymph vessel

Lumen

Diagram showing the structure of the small intestine portion of the digestive tract, with the four main layers and their major components listed on the left. The stomach, small intestine, and large intestine are suspended by mesenteries that are the sites of nerves, blood vessels and lymphatics from the stomach and intestines.

a cavity but bound directly to adjacent structures, such as in the esophagus (Figure 15–1), the serosa is replaced by a thick **adventitia**, a connective tissue layer that merges with the surrounding tissues and lacks mesothelium.

The numerous free immune cells and lymphoid nodules in the mucosa and submucosa constitute the MALT described in Chapter 14. The digestive tract normally contains thousands of microbial species, including both useful inhabitants of the gut as well as potential pathogens ingested with food and drink. The mucosa-associated immune defense system provides an essential backup to the thin physical barrier of the epithelial lining. Located just below the epithelium, the lamina propria is rich with macrophages and lymphocytes, many for production of IgA antibodies. Such antibodies undergo transcytosis into the intestinal lumen bound to the secretory protein produced by the epithelial cells. This IgA complex resists proteolysis by digestive enzymes and provides important protection against specific viral and bacterial pathogens.

❯❯ MEDICAL APPLICATION

In diseases such as **Hirschsprung disease** (congenital aganglionic megacolon) or **Chagas disease** (**trypanosomiasis**, infection with the protozoan *Trypanosoma cruzi*), plexuses in the digestive tract's enteric nervous system are absent or severely injured, respectively. This disturbs digestive tract motility and produces dilations in some areas. The rich autonomic innervation of the enteric nervous system also provides an anatomic explanation of the well-known actions of emotional stress on the stomach and other regions of the GI tract.

〉 ORAL CAVITY

The oral cavity (Figure 15–1) is lined with stratified squamous epithelium, which may be keratinized, partially keratinized, or nonkeratinized depending on the location. Epithelial differentiation, keratinization, and the interface between the epithelium and lamina propria are similar to those features in the epidermis and dermis and are discussed more extensively with skin (see Chapter 18). Like the keratinized surface cells of epidermis, the flattened superficial cells of the oral epithelium undergo continuous desquamation, or loss at the surface. Unlike those of the epidermis, the shed cells of the nonkeratinized or parakeratinized oral epithelium retain their nuclei.

〉〉 MEDICAL APPLICATION

Viral infections with herpes simplex 1 cause death of infected epithelial cells that can lead to vesicular or ulcerating lesions of the oral mucosa or skin near the mouth. In the oral cavity such areas are called **canker sores**, and on the skin they are usually called **cold sores** or **fever blisters**. Such lesions, often painful and clustered, occur when the immune defenses are weakened by emotional stress, fever, illness, or local skin damage, allowing the virus, present in the local nerves, to move into the epithelial cells.

The keratinized cell layers resist damage from abrasion and are best developed in the **masticatory mucosa** on the gingiva (gum) and hard palate. The lamina propria in these regions rests directly on the periosteum of underlying bone. Nonkeratinized squamous epithelium predominates in the **lining mucosa** over the soft palate, cheeks, the floor of the mouth, and the **pharynx** (or throat), the posterior region of the oral cavity leading to the esophagus. Lining mucosa overlies a thick submucosa containing many minor salivary glands, which secrete continuously to keep the mucosal surface wet, and diffuse lymphoid tissue. Throughout the oral cavity, the epithelium contains transient antigen-presenting cells and rich sensory innervation.

The well-developed core of striated muscle in the **lips**, or labia, (Figure 15–3) makes these structures highly mobile for ingestion, speech, and other forms of communication. Both lips have three differently covered surfaces:

- The internal mucous surface has lining mucosa with a thick, nonkeratinized epithelium and many minor labial salivary glands.
- The red **vermilion zone** of each lip is covered by very thin keratinized stratified squamous epithelium and is transitional between the oral mucosa and skin. This region lacks salivary or sweat glands and is kept moist with saliva from the tongue. The underlying connective tissue is very rich in both sensory innervation and capillaries, which impart the pink color to this region.

FIGURE 15–3 Lip.

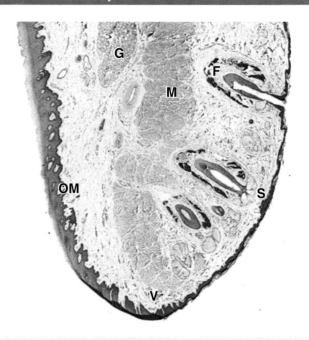

Low-magnification micrograph of a lip section showing one side covered by typical oral mucosa (**OM**), the opposite side covered by skin (**S**) containing hair follicles (**F**) and associated glands. Between the oral portion of the lips and normal skin is the vermilion zone (**V**), where epidermis is very thin, lightly keratinized, and transparent to blood in the rich microvasculature of the underlying connective tissue. Because this region lacks the glands for oil and sweat, it is prone to excessive dryness and chapping in cold, dry weather. Internally, the lips contain much striated muscle (**M**) and many minor salivary glands (**G**). (X10; H&E)

- The outer surface has thin skin, consisting of epidermal and dermal layers, sweat glands, and many hair follicles with sebaceous glands.

Tongue

The tongue is a mass of striated muscle covered by mucosa, which manipulates ingested material during mastication and swallowing. The muscle fibers are oriented in all directions, allowing a high level of mobility. Connective tissue between the small fascicles of muscle is penetrated by the lamina propria, which makes the mucous membrane strongly adherent to the muscular core. The lower surface of the tongue is smooth, with typical lining mucosa. The dorsal surface is irregular, having hundreds of small protruding **papillae** of various types on its anterior two-thirds and the massed lingual tonsils on the posterior third, or root of the tongue (Figure 15–4). The papillary and tonsillar areas of the lingual surface are separated by a V-shaped groove called the **sulcus terminalis**.

The lingual papillae are elevations of the mucous membrane that assume various forms and functions. There are four types (Figure 15–4):

FIGURE 15–4 Tongue, lingual papillae, and taste buds.

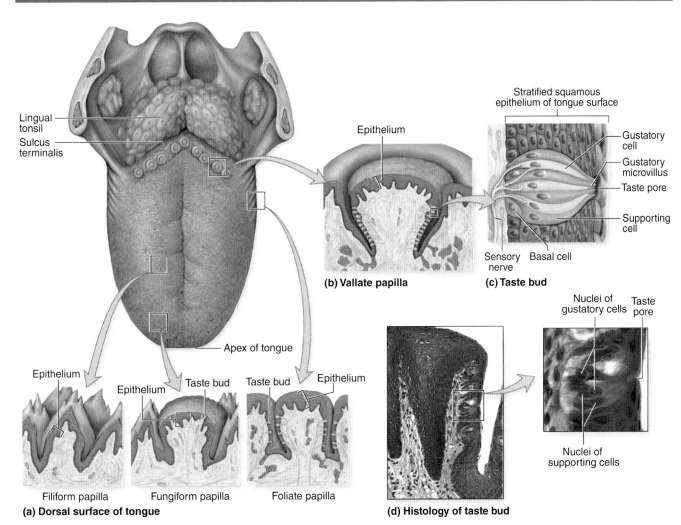

Lingual tonsil
Sulcus terminalis

Epithelium

Stratified squamous epithelium of tongue surface

Gustatory cell
Gustatory microvillus
Taste pore
Supporting cell

Sensory nerve   Basal cell

(b) Vallate papilla   **(c) Taste bud**

Apex of tongue

Epithelium
Epithelium   Taste bud   Taste bud   Epithelium

Filiform papilla   Fungiform papilla   Foliate papilla

(a) Dorsal surface of tongue

Nuclei of gustatory cells   Taste pore

Nuclei of supporting cells

(d) Histology of taste bud

On its dorsal surface **(a)**, the posterior third of the tongue has the **lingual tonsils** and the anterior portion has numerous **lingual papillae** of four types. Pointed **filiform papillae** provide friction to help move food during chewing. Ridge-like **foliate papillae** on the sides of the tongue are best developed in young children. **Fungiform papillae** are scattered across the dorsal surface, and 8-12 large **vallate papillae (b)** are present in a V-shaped line near the terminal sulcus. **Taste buds** are present on fungiform and foliate papillae but are much more abundant on vallate papillae.

(c) Diagram of a single taste bud shows the **gustatory (taste) cells**, the **supporting cells** whose function is not well understood, and the basal **stem cells**. Microvilli at the ends of the gustatory cells project through an opening in the epithelium, the **taste pore**. Afferent sensory axons enter the basal end of taste buds and synapse with the gustatory cells. In the stratified squamous epithelium of the tongue surface, taste buds form as distinct clusters of cells that are recognizable histologically even at low magnification **(d)**. At higher power the taste pore may be visible, as well as the elongated nuclei of gustatory and supporting cells. (140X and 500X; H&E)

- **Filiform papillae** (Figure 15–5) are very numerous, have an elongated conical shape, and are heavily keratinized, which gives their surface a gray or whitish appearance. They provide a rough surface that facilitates movement of food during chewing.
- **Fungiform papillae** (Figure 15–5) are much less numerous, lightly keratinized, and interspersed among the filiform papillae. They are mushroom-shaped

with well-vascularized and innervated cores of lamina propria.
- **Foliate papillae** consist of several parallel ridges on each side of the tongue, anterior to the sulcus terminalis, but are rudimentary in humans, especially older individuals.
- **Vallate** (or **circumvallate**) **papillae** (Figure 15–5) are the largest papillae, with diameters of 1-3 mm. Eight to

FIGURE **15–5** **Lingual papillae.**

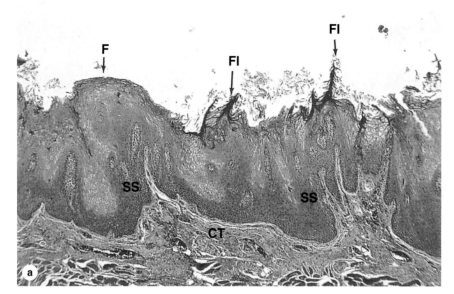
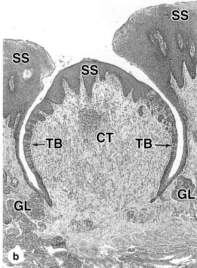

(a) Section of the dorsal surface of tongue showing both filiform (**FI**) and fungiform papillae (**F**). Both types are elevations of the connective tissue (**CT**) covered by stratified squamous epithelium (**SS**), but the filiform type is pointed and heavily keratinized while the fungiform type is mushroom-shaped, lightly keratinized, and has a few taste buds.

(b) Micrograph shows a single very large vallate papilla with two distinctive features: many taste buds (**TB**) around the sides and several small salivary glands (**GL**) emptying into the cleft or moat formed by the elevated mucosa surrounding the papilla. These glands continuously flush the cleft, renewing the fluid in contact with the taste buds. (Both X20; H&E)

twelve vallate papillae are normally aligned just in front of the terminal sulcus. Ducts of several small, serous **salivary (von Ebner) glands** empty into the deep, moatlike groove surrounding each vallate papilla. This provides a continuous flow of fluid over the taste buds that are abundant on the sides of these papillae, washing away food particles so that the taste buds can receive and process new gustatory stimuli. Secretions from these and other minor salivary glands associated with taste buds contain a lipase that prevents the formation of a hydrophobic film on these structures that would hinder gustation.

Taste buds are ovoid structures within the stratified epithelium on the tongue's surface, which sample the general chemical composition of ingested material (Figures 15–4 and 15–5). Approximately 250 taste buds are present on the lateral surface of each vallate papilla, with many others present on fungiform and foliate (but not the keratinized filiform) papillae. They are not restricted to papillae and are also widely scattered elsewhere on the dorsal and lateral surfaces of the tongue, where they are also continuously flushed by numerous minor salivary glands.

A taste bud has 50-100 cells, about half of which are elongated **gustatory (taste) cells**, which turn over with a 7- to 10-day life span. Other cells present are slender **supportive cells**, immature cells, and slowly dividing basal **stem cells** that

give rise to the other cell types. The base of each bud rests on the basal lamina and is entered by afferent sensory axons that form synapses with the gustatory cells. At the apical ends of the gustatory cells, microvilli project toward a 2-μm-wide opening in the structure called the **taste pore**. Molecules (tastants) dissolved in saliva contact the microvilli through the pore and interact with cell surface taste receptors (Figure 15–4).

Taste buds detect at least five broad categories of tastants: sodium ions (salty); hydrogen ions from acids (sour); sugars and related compounds (sweet); alkaloids and certain toxins (bitter); and amino acids such as glutamate and aspartate (umami; Jap. *umami*, savory). Salt and sour tastes are produced by ion channels and the other three taste categories are mediated by G-protein–coupled receptors. Receptor binding produces depolarization of the gustatory cells, stimulating the sensory nerve fibers that send information to the brain for processing. Conscious perception of tastes in food requires olfactory and other sensations in addition to taste bud activity.

Teeth

In the adult human there are normally 32 **permanent teeth**, arranged in two bilaterally symmetric arches in the maxillary and mandibular bones (Figure 15–6a). Each quadrant has eight teeth: two incisors, one canine, two premolars, and three permanent molars. Twenty of the permanent teeth are preceded

FIGURE **15–6** Teeth.

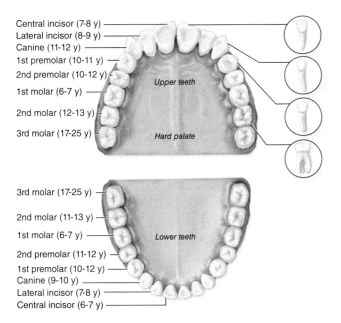

(a) Permanent teeth

Central incisor (7-8 y)
Lateral incisor (8-9 y)
Canine (11-12 y)
1st premolar (10-11 y)
2nd premolar (10-12 y)
1st molar (6-7 y)
2nd molar (12-13 y)
3rd molar (17-25 y)

Upper teeth

Hard palate

3rd molar (17-25 y)
2nd molar (11-13 y)
1st molar (6-7 y)
2nd premolar (11-12 y)
1st premolar (10-12 y)
Canine (9-10 y)
Lateral incisor (7-8 y)
Central incisor (6-7 y)

Lower teeth

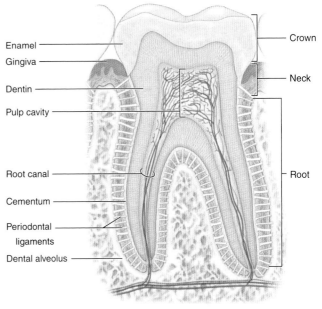

(b) Molar

Enamel
Gingiva
Dentin
Pulp cavity
Root canal
Cementum
Periodontal ligaments
Dental alveolus

Crown
Neck
Root

All teeth are similar embryologically and histologically.

(a) The dentition of the permanent teeth is shown, as well as the approximate age at eruption for each tooth.

(b) Diagram of a molar's internal structure is similar to that of all teeth, with an enamel-covered **crown**, cementum-covered **roots** anchoring the tooth to alveolar bone of the jaw, and a slightly constricted **neck** where the **enamel** and **cementum** coverings meet at the gingiva. Most of the roots and neck consists of **dentin**. A **pulp cavity** extends into the neck and is filled with well-vascularized, well-innervated mesenchymal connective tissue. Blood vessels and nerves enter the tooth through apical foramina at the root tips. **Periodontal ligaments** hold the tooth to bone of the jaw.

by **primary teeth** (deciduous or milk teeth) that are shed; the others are permanent molars with no deciduous precursors. Each tooth has a **crown** exposed above the gingiva, a constricted **neck** at the gum, and one or more **roots** that fit firmly into bony sockets in the jaws called **dental alveoli** (Figure 15–6b).

The crown is covered by very hard, acellular **enamel** and the roots by a bone-like tissue called **cementum**. These two coverings meet at the neck of the tooth. The bulk of a tooth is composed of another calcified material, **dentin**, which surrounds an internal **pulp cavity** (Figure 15–6b). **Dental pulp** is highly vascular and well-innervated and consists largely of loose, mesenchymal connective tissue with much ground substance, thin collagen fibers, fibroblasts, and mesenchymal stem cells. The pulp cavity narrows in each root as the **root canal**, which extends to an opening (**apical foramen**) at the tip of each root for the blood vessels, lymphatics, and nerves of the pulp cavity. The **periodontal ligaments** are fibrous connective tissue bundles of collagen fibers inserted into both the cementum and the alveolar bone.

Dentin

Dentin is a calcified tissue harder than bone, consisting of 70% hydroxyapatite. The organic matrix contains type I collagen and proteoglycans secreted from the apical ends of **odontoblasts**, tall polarized cells derived from the cranial neural crest that line the tooth's pulp cavity (Figure 15–7a). Mineralization of the **predentin** matrix secreted by odontoblasts involves matrix vesicles in a process similar to that occurring in osteoid during bone formation (see Chapter 8).

Long apical **odontoblast processes** extend from the odontoblasts within **dentinal tubules** (Figure 15–7b) that penetrate the full thickness of the dentin, gradually becoming longer as the dentin becomes thicker. Along their length, the processes extend fine branches into smaller lateral branches of the tubules (Figure 15–7c). The odontoblast processes are important for the maintenance of dentin matrix. Odontoblasts continue predentin production into adult life, gradually reducing the size of the pulp cavity, and are stimulated to repair dentin if the tooth is damaged.

FIGURE **15–7** Dentin and odontoblasts.

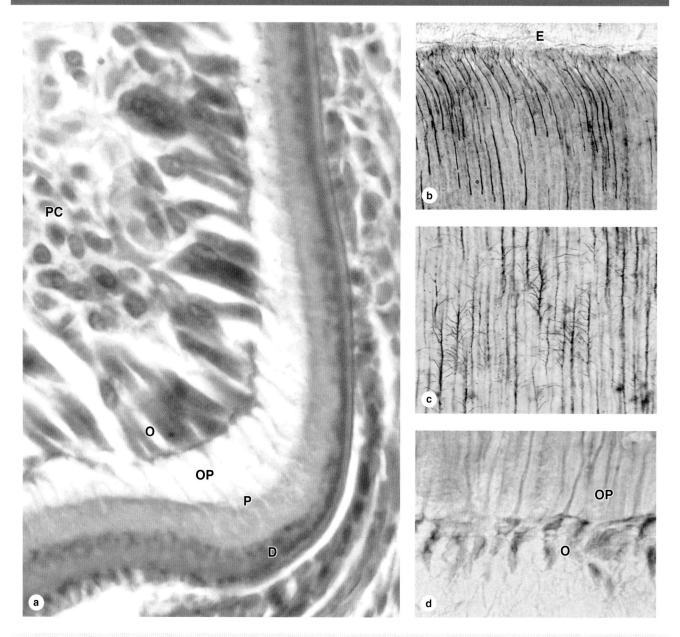

(a) Odontoblasts (**O**) are long polarized cells derived from mesenchyme of the developing pulp cavity (**PC**). Odontoblasts are specialized for collagen and GAG synthesis and are bound together by junctional complexes as a layer, with no basal lamina, so that a collagen-rich matrix called predentin (**P**) is secreted only from their apical ends at the dentinal surface. Within approximately 1 day of secretion, predentin mineralizes to become dentin (**D**) as hydroxyapatite crystals form in a process similar to that occurring in osteoid of developing bones (see Chapter 8). In this process the collagen is masked, and calcified matrix becomes much more acidophilic and stains quite differently than that of predentin. When predentin secretion begins, an apical extension from each cell, the odontoblast process (**OP**), forms and is surrounded by new matrix. As the dentin-predentin layer thickens, these processes lengthen. When tooth formation is complete, odontoblasts persist and

their processes are maintained in canals called **dentinal tubules** that run through the full thickness of the dentin. (X400; Mallory trichrome)

(b) Odontoblast processes can be silver-stained and shown to branch near the junction of dentin with enamel (**E**) and along their length closer to their source (**c**), with the lateral branches occupying smaller canaliculi within dentin. (Both X400; Silver)

(d) These odontoblast process (**OP**) connections to the odontoblasts (**O**), shown with stained nuclei here, are important for the maintenance of dentin in adult teeth. (X400; Mallory trichrome)

(*Figure 15-7b, c, and d, used with permission from M. F. Santos, Department of Histology and Embryology, Institute of Biomedical Sciences, University of São Paulo, Brazil.*)

Teeth are sensitive to stimuli such as cold, heat, and acidic pH, all of which can be perceived as pain. Pulp is highly innervated, and unmyelinated nerve fibers extend into the dental tubules along with odontoblast processes near the pulp cavity (Figure 15–8). Such stimuli can affect fluid inside the dentinal tubules, stimulating these nerve fibers and producing tooth sensitivity.

Enamel

Enamel is the hardest component of the human body, consisting of 96% calcium hydroxyapatite and only 2%-3% organic material including very few proteins and no collagen. Other ions, such as fluoride, can be incorporated or adsorbed by the hydroxyapatite crystals; enamel containing fluorapatite is more resistant to acidic dissolution caused by microorganisms, hence the addition of fluoride to toothpaste and water supplies.

Enamel consists of uniform, interlocking columns called **enamel rods** (or prisms), each about 5 µm in diameter and surrounded by a thinner layer of other enamel. Each rod extends through the entire thickness of the enamel layer, which averages 2 mm. The precise, interlocked arrangement of the enamel rods is crucial for enamel's hardness and resistance to great pressures during mastication.

In a developing tooth bud, the matrix for the enamel rods is secreted by tall, polarized cells, the **ameloblasts** (Figure 15–9a), which are part of a specialized epithelium in the tooth bud called the **enamel organ**. The apical ends of the ameloblasts face those of the odontoblasts producing predentine (Figure 15–10). An apical extension from each ameloblast, the **ameloblast** (or **Tomes**) **process**, contains numerous secretory granules with the proteins of the enamel matrix. The secreted matrix undergoes very rapid mineralization. Growth of the hydroxyapatite crystals to produce each elongating enamel rod is guided by a small (20 kDa) protein **amelogenin**, the main structural protein of developing enamel.

Ameloblasts are derived from the ectodermal lining of the embryonic oral cavity, while odontoblasts and most tissues of the pulp cavity develop from neural crest cells and mesoderm, respectively. Together, these tissues produce a series of 52 tooth buds in the developing oral cavity, 20 for the primary teeth and 32 for the secondary or permanent teeth. Primary teeth complete development and begin to erupt about 6 months after birth. Development of the secondary tooth buds arrests at the "bell stage," shown in Figure 15–10a, until about 6 years of age, when these teeth begin to erupt as the primary teeth are shed.

FIGURE **15–8** Ultrastructure of dentinal tubule.

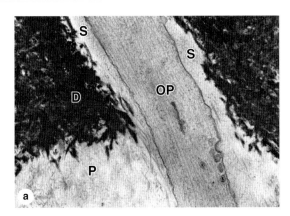
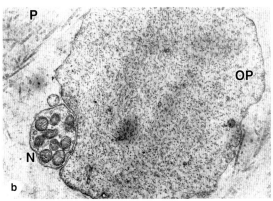

(a) TEM shows the calcification of dentin (**D**) at its border with not-yet calcified predentin (**P**). An odontoblast process (**OP**) with microtubules and a few secretory vesicles is seen in the fluid-filled space (**S**) in the dentinal tubule. A process extends from each odontoblast, and the tubules continue completely across the dentin layer. (X32,000)

(b) TEM cross section of an odontoblast process (**OP**) near predentin (**P**) shows its close association with an unmyelinated nerve fiber (**N**) extending there from fibers in the pulp cavity. These nerves respond to various stimuli, such as cold temperatures, reaching the nerve fibers through the dentinal tubules. (X61,000)

FIGURE **15–9** Ameloblasts and enamel.

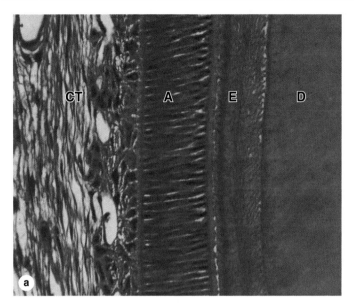

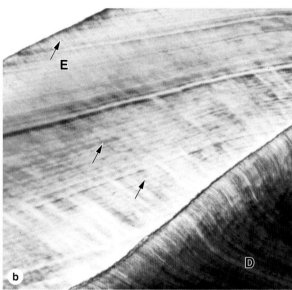

(a) In a section of tooth bud ameloblasts (**A**) are tall polarized cells whose apical ends initially contact dentin (**D**). Ameloblasts are joined to form a cell layer surrounded basally by connective tissue (**CT**). As odontoblasts secrete predentin, ameloblasts secrete a matrix lacking collagens, but rich in proteins such as amelogenin that quickly initiate calcium hydroxyapatite formation to make enamel (**E**), the hardest material in the body. Enamel forms a layer but consists of enamel rods or prisms, solidly fused together by more enamel. Each enamel rod represents the product of one ameloblast. No cellular processes occur in enamel, and the layer of ameloblasts surrounding the developing crown is completely lost during tooth eruption. Teeth that have been decalcified for histologic sectioning typically lose their enamel layer completely. (X400; H&E)

(b) Micrograph of a thin preparation of a tooth prepared by grinding. Fine, long tubules can be observed in the dentin (**D**), and rods aligned the same way can be very faintly observed (arrows) in the enamel (**E**). The more prominent lines that cross enamel diagonally represent incremental growth lines produced as the enamel matrix is secreted cyclically by the ameloblast layer. (X400; Unstained)

>> **MEDICAL APPLICATION**

Periodontal diseases include **gingivitis**, inflammation of the gums, and **periodontitis**, which involves inflammation at deeper sites, both of which are caused most commonly by bacterial infections with poor oral hygiene. Chronic periodontitis weakens the periodontal ligament and can lead to loosening of the teeth. The depth of the gingival sulcus, measured during clinical dental examinations, is an important indicator of potential periodontal disease.

Periodontium

The **periodontium** comprises the structures responsible for maintaining the teeth in the maxillary and mandibular bones, and includes the **cementum**, the **periodontal ligament**, and the **alveolar bone** with the associated **gingiva** (Figure 15–6b; Figure 15–11).

Cementum covers the dentin of the root and resembles bone, but it is avascular. It is thickest around the root tip where **cementocytes** reside in lacunae with processes in canaliculi, especially near the cementum surface. Although less labile than bone, cementocytes maintain their surrounding matrix and react to stresses by gradually remodeling.

The **periodontal ligament** is fibrous connective tissue with bundled collagen fibers (Sharpey fibers) binding the cementum and the alveolar bone (Figure 15–11). Unlike typical ligaments, it is highly cellular and has a rich supply of blood vessels and nerves, giving the periodontal ligament sensory and nutritive functions in addition to its role in supporting the tooth. It permits limited movement of the tooth within the alveolus and helps protect the alveolus from the recurrent pressure exerted during mastication. Its thickness (150-350 μm) is fairly uniform along the root but decreases with aging.

The **alveolar bone** lacks the typical lamellar pattern of adult bone but has osteoblasts and osteocytes engaging in continuous remodeling of the bony matrix. It is surrounded by the periodontal ligament, which serves as its periosteum. Collagen fiber bundles of the periodontal ligament penetrate this bone, binding it to the cementum (Figure 15–11c).

Around the peridontium the keratinized oral mucosa of the **gingiva** is firmly bound to the periosteum of the maxillary and mandibular bones (Figure 15–11). Between the enamel and the gingival epithelium is the **gingival sulcus**, a groove

FIGURE **15–10** Tooth formation.

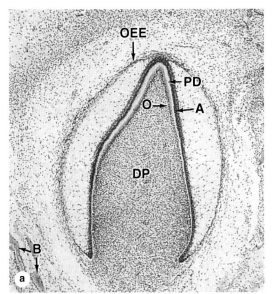

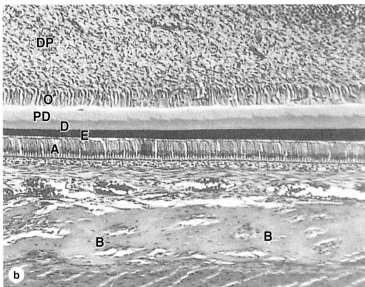

Tooth formation begins in the embryo when ectodermal epithelium lining the oral cavity grows into the underlying mesenchyme of the developing jaws. At a series of sites corresponding to each future tooth, these epithelial cells proliferate extensively and become organized as **enamel organs**, each shaped like a wine glass with its stem initially still attached to the oral lining. **Ameloblasts** form from the innermost layer of cells in the enamel organ. Mesenchymal cells inside the concave portion of the enamel organ include neural crest cells that differentiate as the layer of **odontoblasts** with their apical ends in contact with the apical ends of the ameloblasts.

(a) A section of enamel organ in which production of dentin and enamel has begun. The ameloblast layer (**A**) is separated from the outer enamel epithelium (**OEE**) by a thick intervening region rich in GAGs but with few, widely separated cells. Surrounding the enamel organ is mesenchyme, some parts of which begin to undergo intramembranous bone formation (**B**). Inside the cavity of each enamel organ, mesenchymal cells comprise the dental papilla (**DP**), in which the outermost cells are the layer of

odontoblasts (**O**) facing the ameloblasts. These two cell layers begin to move apart as the odontoblasts begin to produce the layer of predentin (**PD**). Contact with dentin induces each ameloblast to begin secretion of an enamel rod. More slowly, calcifying interprismatic enamel fuses all the enamel rods into a very strong, solid mass. (X20; H&E)

(b) Detail of an enamel organ showing the layers of predentin (**PD**) and dentin (**D**) and a layer of enamel (**E**), along with the organized cell layers that produced this material. Odontoblasts (**O**) are in contact with the very cellular mesenchyme of the dental papilla (**DP**) that will become the pulp cavity. Ameloblasts (**A**) are prominent in the now much thinner enamel organ, which is very close to developing bone (**B**). Enamel formation continues until shortly before tooth eruption; formation of dentin continues after eruption until the tooth is fully formed. Odontoblasts persist around the pulp cavity, with processes penetrating the dental layer, producing factors to help maintain dentin. Mesenchymal cells immediately around the enamel organ differentiate into the cells of cementum and other periodontal tissues. (X120; H&E)

up to 3 mm deep surrounding the neck (Figure 15–11a). A specialized part of this epithelium, the **junctional epithelium**, is bound to the tooth enamel by means of a cuticle, which resembles a thick basal lamina to which the epithelial cells are attached by numerous hemidesmosomes.

▷ ESOPHAGUS

The **esophagus** is a muscular tube, about 25-cm long in adults, which transports swallowed material from the pharynx to the stomach. The four layers of the GI tract (Figure 15–12) first become well-established and clearly seen in the esophagus. The esophageal mucosa has nonkeratinized stratified squamous epithelium, and the submucosa contains small mucus-secreting glands, the **esophageal glands**, which lubricate

and protect the mucosa (Figure 15–13a). Near the stomach the mucosa also contains groups of glands, the **esophageal cardiac glands**, which secrete additional mucus.

❯❯ MEDICAL APPLICATION

The lubricating mucus produced in the esophagus offers little protection against acid that may move there from the stomach. Such movement can produce **heartburn** or **reflux esophagitis**. An incompetent inferior esophageal sphincter may result in chronic heartburn, which can lead to erosion of the esophageal mucosa or **gastroesophageal reflux disease (GERD)**. Untreated GERD can produce metaplastic changes in the stratified squamous epithelium of the esophageal mucosa, a condition called **Barrett esophagus**.

FIGURE **15–11** Periodontium.

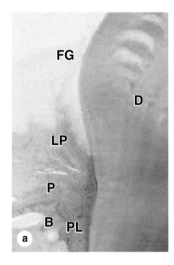
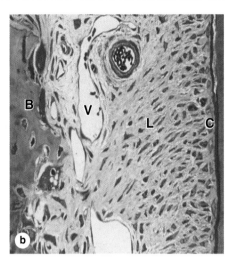
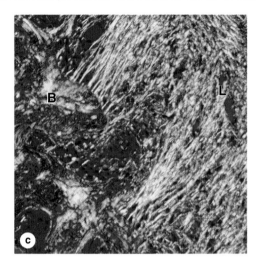

The periodontium of each tooth consists of the **cementum**, **periodontal ligament**, **alveolar bone**, and **gingiva**.

(a) Micrograph of decalcified tooth shows the gingiva. The free gingiva (**FG**) is against the dentin (**D**), with little of the gingival sulcus apparent. Gingiva stratified squamous epithelium over connective tissue of the lamina propria (**LP**). The connective tissue is continuous with that of the periosteum (**P**) covering the alveolar bone (**B**) and with the periodontal ligament (**PL**). (X10; H&E)

(b) Micrograph shows the periodontal ligament (**L**) with its many blood vessels (**V**) and insertions into the alveolar bone (**B**). This ligament serves as the periosteum of the alveolar in tooth sockets and is also continuous with developing layers of cementum (**C**) that covers the dentin. Cementum forms a thin layer of bone-like material secreted by large, elongated cells called **cementoblasts**. (X100; H&E)

(c) Polarizing light micrograph shows the continuity of collagen fibers in alveolar bone (**B**), with the bundles in the periodontal ligament (**L**). (X200; Picrosirius in polarized light)

FIGURE **15–12** Esophagus.

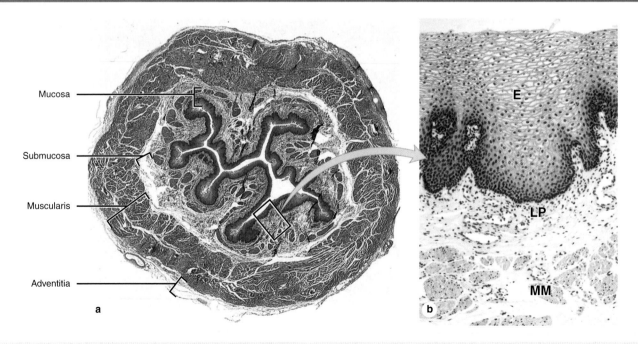

(a) In cross section the four major layers of the GI tract are clearly seen. The esophageal mucosa is folded longitudinally, with the lumen largely closed. (X10; H&E)

(b) Higher magnification of the mucosa shows the stratified squamous epithelium (**E**), the lamina propria (**LP**) with scattered lymphocytes, and strands of smooth muscle in the muscularis mucosae (**MM**). (X65; H&E)

FIGURE 15–13 Esophagus.

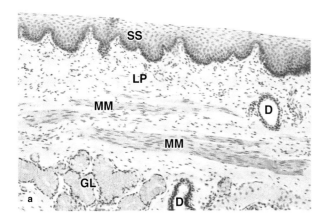
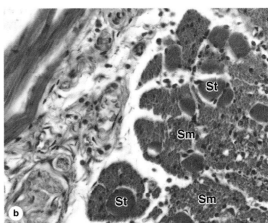

(a) Longitudinal section of esophagus shows mucosa consisting of nonkeratinized stratified squamous epithelium (**SS**), lamina propria (**LP**), and smooth muscles of the muscularis mucosae (**MM**). Beneath the mucosa is the submucosa containing esophageal mucous glands (**GL**) that empty via ducts (**D**) onto the luminal surface. (X40; H&E)

(b) Transverse section showing the muscularis halfway along the esophagus reveals a combination of large skeletal or striated muscle fibers (**St**) and smooth muscle fibers (**Sm**) in the outer layer, which is cut transversely here. This transition from muscles under voluntary control to the type controlled autonomically is important in the swallowing mechanism. (X200; H&E)

Swallowing begins with voluntary muscle action but finishes with involuntary peristalsis. In approximately the upper one-third of the esophagus, the muscularis is exclusively skeletal muscle like that of the tongue. The middle portion of the esophagus has a combination of skeletal and smooth muscle fibers (Figure 15–13b), and in the lower third the muscularis is exclusively smooth muscle. Only the distal 1-2 cm of the esophagus, in the peritoneal cavity, is covered by serosa; the rest is enclosed by the loose connective tissue of the adventitia, which blends into the surrounding tissue.

❯ STOMACH

The stomach is a greatly dilated segment of the digestive tract whose main functions are:

- To continue the digestion of carbohydrates initiated by the amylase of saliva,
- To add an acidic fluid to the ingested food and mixing its contents into a viscous mass called **chyme** by the churning activity of the muscularis,
- To begin digestion of triglycerides by a secreted lipase
- To promote the initial digestion of proteins with the enzyme **pepsin**.

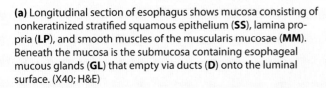

❯❯ MEDICAL APPLICATION

For various reasons, including autoimmunity, parietal cells may be damaged to the extent that insufficient quantities of intrinsic factor are secreted and **vitamin B$_{12}$** is not absorbed adequately. This vitamin is a cofactor required for DNA synthesis; low levels of vitamin B$_{12}$ can reduce proliferation of erythroblasts, producing **pernicious anemia**.

Four major regions make up the stomach: the cardia, fundus, body, and pylorus (Figure 15–14a). The **cardia** is a narrow transitional zone, 1.5-3 cm wide, between the esophagus and the stomach; the **pylorus** is the funnel-shaped region that opens into the small intestine. Both these regions are primarily involved with mucus production and are similar histologically. The much larger **fundus** and **body** regions are identical in microscopic structure and are the sites of gastric glands releasing acidic gastric juice. The mucosa and submucosa of the empty stomach have large, longitudinally directed folds called **rugae**, which flatten when the stomach fills with food. The wall in all regions of the stomach is made up of all four major layers (Figures 15–14c and 15–15).

❯❯ MEDICAL APPLICATION

Gastric and duodenal ulcers are painful erosive lesions of the mucosa that may extend to deeper layers. Such ulcers can occur anywhere between the lower esophagus and the jejunum, and their causes include bacterial infections with *Helicobacter pylori*, effects of nonsteroidal anti-inflammatory drugs, overproduction of HCl or pepsin, and lowered production or secretion of mucus or bicarbonate.

Mucosa

Changing abruptly at the esophagogastric junction (Figures 15–14b), the mucosal surface of the stomach is a simple columnar epithelium that invaginates deeply into the lamina propria. The invaginations form millions of **gastric pits**, each with an opening to the stomach lumen (see Figure 15–14; Figure 15–16). The **surface mucous cells** that line the lumen

FIGURE **15–14** Stomach.

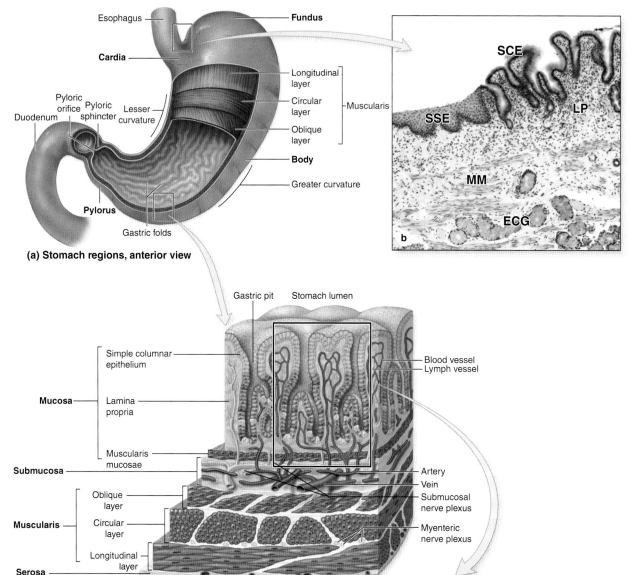

Esophagus — **Fundus**

Cardia —

Pyloric orifice
Duodenum — Pyloric sphincter — Lesser curvature
Pylorus

Gastric folds

(a) Stomach regions, anterior view

Longitudinal layer
Circular layer — Muscularis
Oblique layer
Body
Greater curvature

SCE
SSE
LP
MM
ECG
b

Gastric pit   Stomach lumen

Mucosa — Simple columnar epithelium

Lamina propria

Muscularis mucosae

Submucosa —

Muscularis — Oblique layer
Circular layer
Longitudinal layer

Serosa —

Blood vessel
Lymph vessel

Artery
Vein
Submucosal nerve plexus
Myenteric nerve plexus

(c) Stomach wall, sectional view

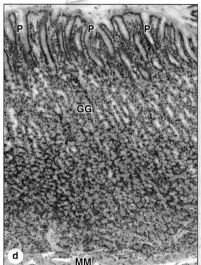

P   P   P

GG

d

MM

The stomach is a muscular dilation of the digestive tract where mechanical and chemical digestion occurs.

(a) The major stomach regions are the **cardia, fundus, body**, and **pylorus**, all with longitudinal gastric folds, or rugae. The muscularis has three layers.

(b) At the esophagogastric junction, stratified squamous epithelium (**SSE**) lining the esophagus is abruptly replaced by simple columnar epithelium (**SCE**) of the stomach. Also seen here are the mucous esophageal cardiac glands (**ECG**) beneath the lamina propria (**LP**) and muscularis mucosae (**MM**). (X60; H&E)

(c) The mucosa of the stomach wall contains invaginations called **gastric pits** that lead into **gastric glands**. These structures are lined by simple columnar epithelium containing five functional cell types.

(d) A histologic section of the gastric mucosa shows the gastric pits (**P**) and glands (**GG**) surrounded by cells of the lamina propria. The underlying muscularis mucosae (**MM**) is also seen. (X60; H&E)

FIGURE **15–15** Wall of the stomach with rugae.

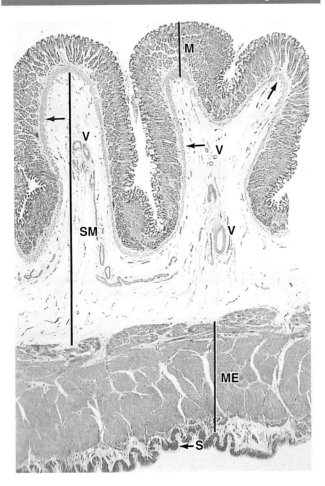

A low-magnification micrograph of the stomach wall at the
fundus shows the relative thickness of the four major layers:
the mucosa (**M**), the submucosa (**SM**), the muscularis externa
(**ME**), and the serosa (**S**). Two rugae (folds) cut transversely and
consisting of mucosa and submucosa are included. The mucosa
is packed with branched tubular glands penetrating the full
thickness of the lamina propria so that this sublayer cannot be
distinguished at this magnification. The **muscularis mucosae**
(arrows), immediately beneath the basal ends of the gastric
glands, is shown. The submucosa is largely loose connective tissue, with blood vessels (**V**) and lymphatics. (X12; H&E)

FIGURE **15–16** Gastric pits and glands.

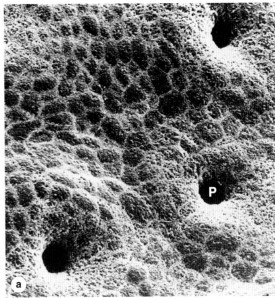

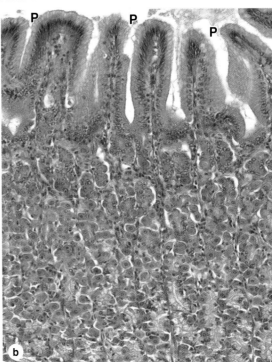

(a) SEM of the stomach lining cleared of its mucous layer
reveals closely placed gastric pits (**P**) surrounded by polygonal
apical ends of surface mucous cells. (X600)

(b) A section of the same lining shows that these surface
mucous cells are part of a simple columnar epithelium continuous with the lining of the pits (**P**). Each pit extends into the
lamina propria and then branches into several tubular glands.
These glands coil and fill most of the mucosa. Around the various cells of the closely-packed gastric glands are cells, capillaries, and small lymphatics of the connective tissue lamina
propria. (X200; H&E)

and gastric pits secrete a thick, adherent, and highly viscous
mucous layer that is rich in bicarbonate ions and protects the
mucosa from both abrasive effects of intraluminal food and
the corrosive effects of stomach acid.

The gastric pits lead to long, branched, tubular glands that
extend through the full thickness of the lamina propria. **Stem
cells** for the epithelium that lines the glands, pits, and stomach
lumen are found in a narrow segment (isthmus) between each
gastric pit and the gastric glands. The pluripotent stem cells
divide asymmetrically, producing progenitor cells for all the
other epithelial cells. Some of these move upward to replace
surface mucous cells, which have a turnover time of 4-7 days.

Other progenitor cells migrate more deeply and differentiate into the secretory cells of the glands that turn over much more slowly than the surface mucous cells.

The vascularized **lamina propria** that surrounds and supports the gastric pits and glands contains smooth muscle fibers, lymphoid cells, capillaries, and lymphatics. Separating the mucosa from the underlying submucosa is a layer of smooth muscle, the **muscularis mucosae** (Figure 15–15).

In the fundus and body the **gastric glands** themselves fill most of the mucosa, with several such glands formed by branching at the isthmus or neck of each gastric pit. Secretory epithelial cells of the gastric glands are distributed unevenly and release products that are key to the stomach's functions.

These cells are of four major types and important properties of each are as follows:

- **Mucous neck cells** are present in clusters or as single cells among the other cells in the necks of gastric glands and include many progenitor and immature surface mucous cells (Figure 15–17). Less columnar than the surface mucous cells lining the gastric pits, mucous neck cells are often distorted by neighboring cells, but they have round nuclei and apical secretory granules. Their mucus secretion is less alkaline than that of the surface epithelial mucous cells.
- **Parietal** (oxyntic) **cells** produce hydrochloric acid (HCl) and are present among the mucous neck cells

FIGURE 15–17 Gastric glands.

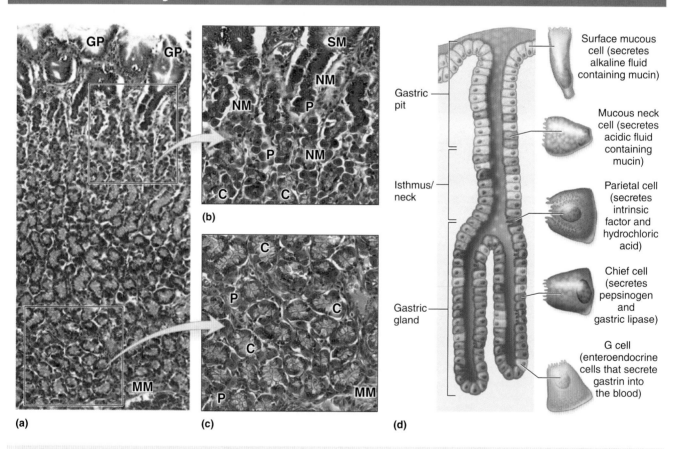

Throughout the **fundus** and **body** regions of the stomach, the gastric pits lead to gastric glands with various cell types.

(a) The long, coiled gastric glands penetrate the complete thickness of the mucosa, from the gastric pits (**GP**) to the muscularis mucosae (**MM**).

(b) In the neck of a gastric gland, below the surface mucous cells (**SM**) lining the gastric pit, are small mucous neck cells (**MN**), scattered individually or clustered among parietal cells (**P**) and stem cells that give rise to all epithelial cells of the glands. The numerous parietal cells (**P**) are large distinctive cells often bulging from the tubules, with central nuclei surrounded by intensely eosinophilic cytoplasm with unusual ultrastructure. These cells produce HCl, and the numerous mitochondria required for this process

cause the eosinophilia. Chief cells (**C**) begin to appear in the neck region. Around these tubular glands are various cells and microvasculature in connective tissue.

(c) Near the muscularis mucosae (**MM**), the bases of these glands contain fewer parietal cells (**P**) but many more zymogenic chief cells (**C**). Chief cells are found in clusters, with basal nuclei and basophilic cytoplasm. From their apical ends chief cells secrete pepsinogen, the zymogen precursor for the major protease pepsin. Zymogen granules are often removed or stain poorly in routine preparations. (Both X200; H&E)

(d) Diagram showing general morphology and functions of major gastric gland cells.

and throughout deeper parts of the gland. They are large cells, usually appearing rounded or pyramidal, each with one (sometimes two) central round nucleus. The cytoplasm is intensely eosinophilic due to the high density of mitochondria (Figure 15–17). A

striking ultrastructural feature of an active parietal cell is a deep, circular invagination of the apical plasma membrane to form an **intracellular canaliculus** with a large surface area produced by thousands of microvilli (Figure 15–18). As shown in Figure 15–19,

FIGURE **15–18** Ultrastructure of parietal cells.

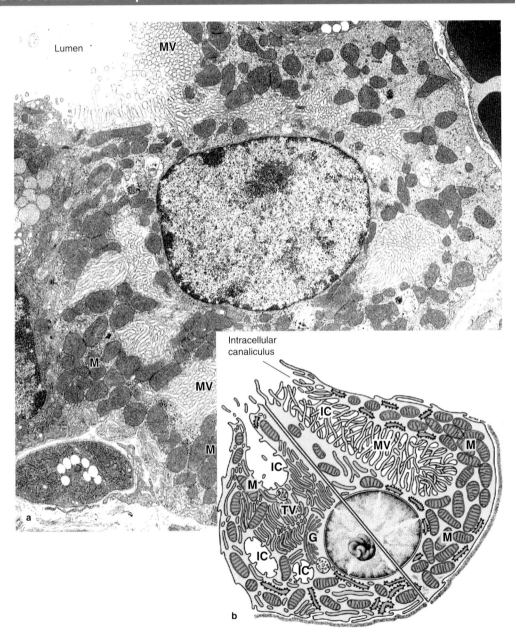

(a) A TEM of an active parietal cell shows abundant microvilli (**MV**) protruding into the intracellular canaliculi, near the lumen and deep in the cell. The cytoplasm contains numerous mitochondria (**M**). (X10,200)

 (*Figure 15–18a, used with permission from Dr Susumu Ito, Department of Cell Biology, Harvard Medical School, Boston, MA.*)

(b) Composite diagram of a parietal cell shows the ultrastructural differences between a resting cell (left) and an active cell (right). In

the resting cell a number of tubular vesicles (**TV**) are seen below the apical plasmalemma (left), but the cell has few microvilli and only short intracellular canaliculi (**IC**) among the mitochondria (**M**) and Golgi vesicles (**G**). When stimulated to produce HCl (right), the tubular vesicles fuse with the cell membrane to form large intracellular canaliculi (**IC**) and microvilli (**MV**), thus providing a generous increase in the surface of the cell membrane for diffusion and ion pumps. Prolonged activity may produce more mitochondria.

FIGURE **15–19** Synthesis of HCl by parietal cells.

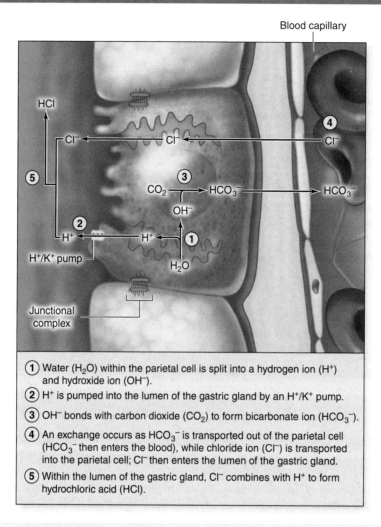

Blood capillary

(1) Water (H_2O) within the parietal cell is split into a hydrogen ion (H^+) and hydroxide ion (OH^-).

(2) H^+ is pumped into the lumen of the gastric gland by an H^+/K^+ pump.

(3) OH^- bonds with carbon dioxide (CO_2) to form bicarbonate ion (HCO_3^-).

(4) An exchange occurs as HCO_3^- is transported out of the parietal cell (HCO_3^- then enters the blood), while chloride ion (Cl^-) is transported into the parietal cell; Cl^- then enters the lumen of the gastric gland.

(5) Within the lumen of the gastric gland, Cl^- combines with H^+ to form hydrochloric acid (HCl).

The main steps in the synthesis of HCl at parietal cells are indicated here. Conversion of H_2O and CO_2 to HCO_3^- (**bicarbonate ion**) and H^+ (**a proton**) is catalyzed by the enzyme carbonic anhydrase. Active transport is used to pump H^+ into canaliculi in exchange for K^+ and to discharge HCO_3^- by an antiport at the basal cell domain in exchange for Cl^-. The Cl^- ions diffuse from the cell into the canaliculi and combine with protons in the lumen of the gastric gland to form HCl.

Basally released bicarbonate ions enter the local interstitial fluid and microvasculature, helping to maintain the neutral pH of the mucosa. Other HCO_3^- is taken up by surface mucous cells and used to raise the pH of mucus.

carbonic anhydrase catalyzes the conversion of cytoplasmic water and CO_2 into HCO_3^- and H^+. The HCO_3^- is transported from the basal side of the cell and H^+ is pumped from the cell apically, along with Cl^-. In the lumen the H^+ and Cl^- ions combine to form HCl. While the gastric secretion becomes highly acidic, the mucosa itself remains at a more neutral pH partly because of the bicarbonate released into the lamina propria. The abundant mitochondria provide energy primarily for operating the cells' ion pumps.

Parietal cells also secrete **intrinsic factor**, a glycoprotein required for uptake of vitamin B_{12} in the small intestine.

Parietal cell secretory activity is stimulated both by parasympathetic innervation and by paracrine release of histamine and the polypeptide **gastrin** from enteroendocrine cells.

- **Chief (zymogenic) cells** predominate in the lower regions of the gastric glands (Figure 15–17) and have all the characteristics of active protein-secreting cells. Ultrastructurally chief cells show abundant RER and numerous apical secretory granules (Figure 15–20). The granules contain inactive enzyme **pepsinogens**, precursors which are converted in the acid environment of the stomach into active **pepsins** (Gr. *peptein*, to digest). Pepsins are endoproteinases with broad specificity and maximal activity at a pH between 1.8 and 3.5. Pepsins initiate the hydrolysis of ingested protein in the stomach. Chief cells also produce **gastric lipase**, which digests many lipids.

FIGURE **15–20** Ultrastructure of parietal, chief, and enteroendocrine cells.

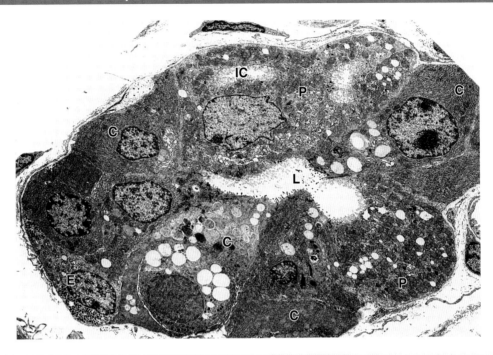

TEM of a transversely sectioned gastric gland shows the ultra-structure of three major cell types. Parietal cells (**P**) contain abundant mitochondria and intracellular canaliculi (**IC**). Also shown are chief cells (**C**), which have extensive rough ER and apical secretory granules near the lumen (**L**). An enteroendocrine cell (**E**) shows dense basal secretory granules and is a closed-type enteroendocrine cell; that is, it has no contact with the gland's lumen and secretes product in an endocrine/paracrine manner. (X1200)

■ **Enteroendocrine cells** are scattered epithelial cells in the gastric mucosa with endocrine or paracrine functions. In the fundus small enteroendocrine cells secreting **serotonin** (5-hydroxytryptamine) are found at the basal lamina of the gastric glands (Figure 15–20). In the pylorus other enteroendocrine cells are located in contact with the glandular lumens, including **G cells** producing the peptide **gastrin**.

Various enteroendocrine cells secreting different hormones, usually peptides, are also found in the intestinal mucosa and are of major importance for function of the digestive tract. Important examples are summarized in Table 15–1. Seldom seen by routine light microscopy, these cells can be visualized by TEM tissue treatment with chromium or silver salts. This provided the alternative names **enterochromaffin (EC) cells** and **argentaffin cells**, respectively. Now usually visualized immunohistochemically using antibodies against their product, they are named with the initial letter of the main hormone they produce (Table 15–1). Most of these cells process amines and are also collectively called **APUD cells** for their "amine precursor uptake and decarboxylation" activity. All such cells are more generally considered part of the **diffuse neuroendocrine system (DNES)**, which is discussed further in Chapter 20.

> **》 MEDICAL APPLICATION**
>
> Tumors called **carcinoids**, which arise from enteroendocrine EC cells, are responsible for the clinical symptoms caused by overproduction of serotonin. Serotonin increases gut motility, and chronic high levels of this hormone/neurotransmitter can produce mucosal vasoconstriction and tissue damage.

Upon stimulation, these cells release their hormone products that then exert paracrine (local) or endocrine (systemic) effects via the vasculature. Cells of the digestive tract DNES fall into two classes: a "closed" type, in which the cellular apex is covered by neighboring epithelial cells (Figure 15–20), and an "open" type, in which the constricted apical end of the cell contacts the lumen and bears chemoreceptors that sample the lumen's contents. Effects of the hormones include regulation of peristalsis and tract motility; secretion of digestive enzymes, water, and electrolytes; and the sense of being satiated after eating.

In the cardia and pylorus regions of the stomach, the mucosa also contains tubular glands, with long pits, branching into coiled secretory portions, called **cardiac glands** and **pyloric glands** (Figure 15–21). These glands lack both parietal and chief cells, primarily secreting abundant mucus.

TABLE 15–1	Principal enteroendocrine cells in the gastrointestinal tract.			
			Major Action	
Cell Type	**Major Location**	**Hormone Produced**	**Promotes**	**Inhibits**
D cells	Pylorus, duodenum, and pancreatic islets	Somatostatin		Secretion from other DNES cells nearby
EC cells	Stomach, small and large intestines	Serotonin and substance P	Increased gut motility	
G cells	Pylorus	Gastrin	Gastric acid secretion	
I cells	Small intestine	Cholecystokinin (CCK)	Pancreatic enzyme secretion, gallbladder contraction	Gastric acid secretion
K cells	Duodenum and jejunum	Gastric inhibitory polypeptide (GIP)		Gastric acid secretion
L cells	Ileum and colon	Glucagon-like peptide (GLP-1)	Insulin secretion	Gastric acid secretion Sense of hunger
L cells	Ileum and colon	Peptide YY	H_2O and electrolyte absorption in large intestine	Gastric acid secretion
Mo cells	Small intestine	Motilin	Increased gut motility	
N cells	Ileum	Neurotensin		Gastric acid secretion
S cells	Small intestine	Secretin	Pancreatic and biliary bicarbonate and water secretion	Gastric acid secretion Stomach emptying

Other Layers

The other major layers of the stomach wall are summarized in Figures 15–14 and 15–15. In all stomach regions the **submucosa** is composed of connective tissue with large blood and lymph vessels and many lymphoid cells, macrophages, and mast cells. The **muscularis** has three poorly defined layers of smooth muscle: an outer longitudinal layer, a middle circular layer, and an innermost oblique layer. Rhythmic contractions of the muscularis thoroughly mix ingested food and chyme with mucus, HCl, and digestive enzymes from the gastric mucosa. At the pylorus the middle layer is greatly thickened to form the **pyloric sphincter**. The stomach is covered by a thin **serosa**.

⟩ SMALL INTESTINE

The small intestine is the site where the digestive processes are completed and where the nutrients (products of digestion) are absorbed by cells of the epithelial lining. The small intestine is relatively long—approximately 5 m—and consists of three segments: the **duodenum**, **jejunum**, and **ileum**. These segments have most histologic features in common and are discussed together.

⟩⟩ MEDICAL APPLICATION

Leiomyomas, benign tumors of smooth muscle cells, are the most common type of tumor in the stomach and small intestine and may become large. Autopsy records suggest that the muscularis of the stomach may include leiomyomas in up to 50% of the population older than 50 years.

Mucosa

Viewed macroscopically, the lining of the small intestine shows a series of permanent circular or semilunar folds (**plicae circulares**), consisting of mucosa and submucosa (Figures 15–22a and 15–23), which are best developed in the jejunum. Densely covering the entire mucosa of the small intestine are short (0.5-1.5 mm) mucosal outgrowths called **villi** that project into the lumen (Figure 15–22). These finger- or leaflike projections are covered by a simple columnar epithelium of absorptive cells called **enterocytes**, with many interspersed **goblet cells**. Each villus has a core of loose connective tissue that extends from the lamina propria and contains fibroblasts, smooth muscle fibers, lymphocytes and plasma cells, fenestrated capillaries, and a central lymphatic called a **lacteal**.

⟩⟩ MEDICAL APPLICATION

Celiac disease (celiac sprue) is a disorder of the small intestine mucosa that causes **malabsorption** and can lead to damage or destruction of the villi. The cause of celiac disease is an immune reaction against gluten or other proteins in wheat and certain other types of grain. The resulting inflammation affects the enterocytes, leading to reduced nutrient absorption.

Between the villi are the openings of short tubular glands called **intestinal glands** or **crypts** (or **crypts of Lieberkühn**) and the epithelium of each villus is continuous with that of the intervening glands (Figure 15–22c). The epithelium of the

FIGURE **15–21** Pyloric glands.

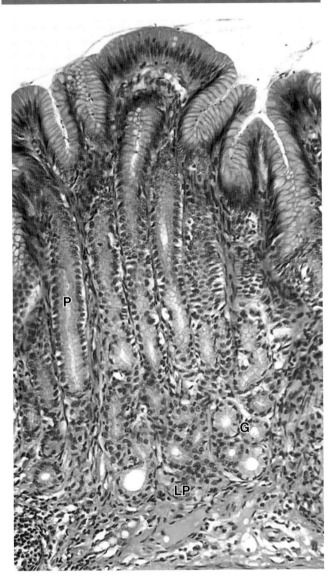

The pyloric region of the stomach has deep gastric pits (**P**) leading to short, coiled pyloric glands (**G**) in the lamina propria (**LP**). Cardial glands are rather similar histologically and functionally. Cells of these glands secrete mucus and lysozyme primarily, with a few enteroendocrine G cells also present. The glands and pits are surrounded by cells of the lamina propria connective tissue containing capillaries, lymphatics and MALT. Immediately beneath the glands is the smooth muscle layer of the muscularis mucosae. (X140; H&E)

ordered region called the **striated** (or **brush**) **border**. Ultrastructurally the striated border is seen to be a layer of densely packed **microvilli** covered by **glycocalyx** through which nutrients are taken into the cells (Figures 15–22e and 15–24c). As discussed in Chapter 4, each microvillus is a cylindrical protrusion of the apical cytoplasm approximately 1-µm tall and 0.1 µm in diameter containing actin filaments and enclosed by the cell membrane. Each enterocyte has an average of 3000 microvilli and each 1 mm^2 of mucosal surface contains about 200 million of these structures. Microvilli, villi, and the plicae circulares all greatly increase the mucosal surface area in contact with nutrients in the lumen, which is an important feature in an organ specialized for nutrient absorption. It is estimated that plicae increase the intestinal surface area 3-fold, the villi increase it 10-fold, and the microvilli increase it another 20-fold, resulting in a total absorptive area of over 200 m^2 in the small intestine!

The mechanism of nutrient absorption varies with the different molecules produced by digestion. Disaccharidases and peptidases secreted by the enterocytes and located within the glycocalyx hydrolyze disaccharides and dipeptides into monosaccharides and amino acids. These are easily absorbed by active transport and immediately released again from the enterocytes for uptake by the capillaries. Digestion of fats by gastric and pancreatic lipases produces lipid subunits, including glycerol, monoglycerides, and fatty acids, which are emulsified by bile salts into small (2 nm) micelles from which lipids enter enterocytes by passive diffusion and membrane transporters. These lipids are reesterified into triglycerides in the enterocyte smooth ER and then complexed with apoproteins in the Golgi apparatus, producing **chylomicrons** that are discharged from the cell's basolateral surface for uptake by the lacteal. Figure 15–25 summarizes basic aspects of lipid absorption.

- **Goblet cells** are interspersed among the absorptive enterocytes (Figures 15–22d and 15–24a, b). They secrete glycoprotein mucins that are then hydrated to form mucus, whose main function is to protect and lubricate the lining of the intestine.
- **Paneth cells**, located in the basal portion of the intestinal crypts below the stem cells, are exocrine cells with large, eosinophilic secretory granules in their apical cytoplasm (Figure 15–26). Paneth cell granules release lysozyme, phospholipase A_2, and hydrophobic peptides called **defensins**, all of which bind and break down membranes of microorganisms and bacterial cell walls. Paneth cells have an important role in innate immunity and in regulating the microenvironment of the intestinal crypts.
- **Enteroendocrine cells** are present in varying numbers throughout the length of the small intestine, secreting various peptide hormones (Table 15–1). Many of these are of the "open" type, in which the constricted apical

intestinal glands includes differentiating cells and pluripotent stem cells for all the cell types of the small intestine. These include the following:

- **Enterocytes**, the absorptive cells, are tall columnar cells, each with an oval nucleus located basally (Figure 15–24). The apical end of each enterocyte displays a prominent

FIGURE **15–22** Absorptive surface of the small intestine.

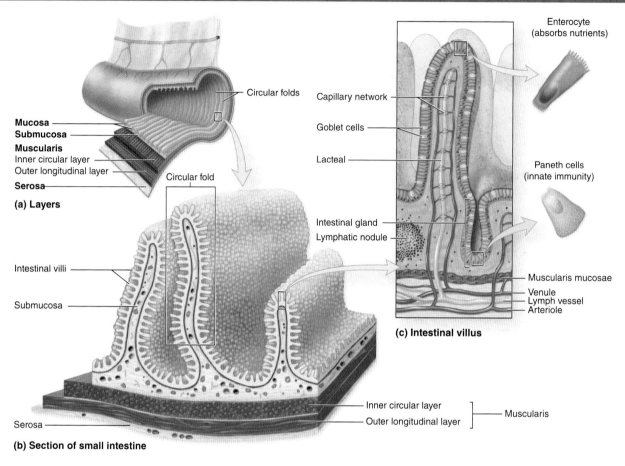

Mucosa
Submucosa
Muscularis
Inner circular layer
Outer longitudinal layer
Serosa

Circular folds

(a) Layers

Circular fold

Intestinal villi

Submucosa

Serosa

Inner circular layer
Outer longitudinal layer
Muscularis

(b) Section of small intestine

Enterocyte
(absorbs nutrients)

Capillary network

Goblet cells

Lacteal

Intestinal gland
Lymphatic nodule

Paneth cells
(innate immunity)

Muscularis mucosae
Venule
Lymph vessel
Arteriole

(c) Intestinal villus

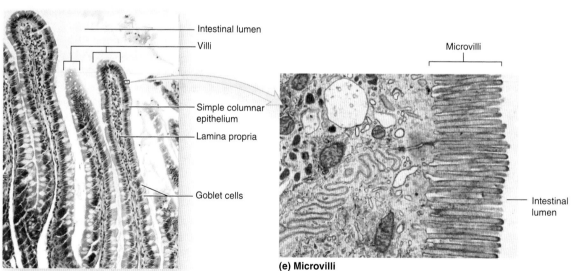

Intestinal lumen
Villi

Simple columnar
epithelium

Lamina propria

Goblet cells

Microvilli

Intestinal
lumen

(d) Intestinal villi

(e) Microvilli

(a) The **mucosa** and **submucosa** are the inner two of the gut's four concentric layers.

(b) They form circular folds or **plicae circulares**, which increase the absorptive area.

(c) They are lined by a dense covering of fingerlike projections called **villi**. Internally each villus contains lamina propria connective tissue with microvasculature and lymphatics called **lacteals**.

(d) Villi are covered with a simple columnar epithelium composed of absorptive enterocytes and goblet cells. (X70; H&E)

(e) At the apical cell membrane of each enterocyte are located dense **microvilli**, which serve to increase greatly the absorptive surface of the cell. (X18,000; TEM)

FIGURE **15–23** Circular folds (plicae circulares) of the jejunum.

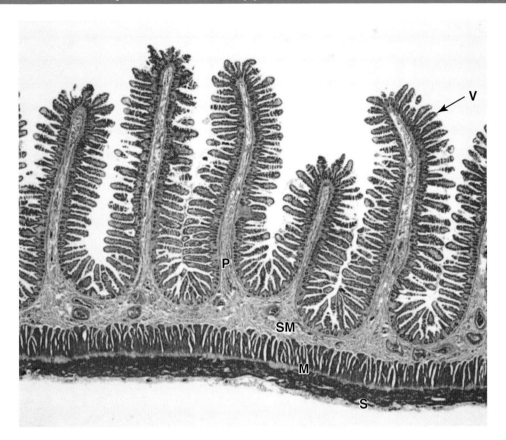

The mucosa and submucosa (**SM**) of the small intestine form distinct projecting folds called plicae (**P**), which encircle or spiral around the inner circumference and are best developed in the jejunum. On each fold the mucosa forms a dense covering of projecting structures called **villi** (**V**). In this longitudinal section the two layers of the muscularis (**M**) are clearly distinguished. The inner layer has smooth muscle encircling the submucosa; the outer layer runs lengthwise just inside the serosa (**S**), the gut's outer layer. This arrangement of smooth muscle provides for strong peristaltic movement of the gut's contents. (X12; Masson trichrome)

end of the cell contacts the intestinal lumen and has chemoreceptors similar to those of taste buds, sampling levels of certain nutrients such as sugars to regulate hormone release basally (Figure 15–27).

■ **M (microfold) cells** are specialized epithelial cells in the mucosa of the ileum overlying the lymphoid follicles of Peyer patches. As discussed in Chapter 14, these cells are characterized by the presence of basal membrane invaginations or pockets containing many intraepithelial lymphocytes and antigen-presenting cells (see Figure 14–13). M cells selectively endocytose antigens and transport them to the underlying lymphocytes and dendritic cells, which then migrate to lymph nodes for an appropriate immune response.

Other Layers

Along the entire small intestine loose connective tissue of the mucosal lamina propria contains extensive blood and lymph microvasculature, nerve fibers, smooth muscle cells, and diffuse lymphoid tissue. The lamina propria penetrates the core of each intestinal villus, bringing with it microvasculature, lymphatics, and nerves (Figures 15–22c and 15–28). Smooth muscle fibers extending from the muscularis mucosae produce rhythmic movements of the villi that increase the absorption efficiency. Fibers of the muscularis mucosae also produce local movements of plicae circulares that help propel lymph from the lacteals into submucosal and mesenteric lymphatics.

The submucosa has larger blood and lymph vessels and the diffuse, interconnected neurons of the **submucosal (Meissner) nerve plexus**. The proximal part of the duodenum has in the submucosa and mucosa large clusters of branched tubular mucous glands, the **duodenal** (or **Brunner) glands**, with small excretory ducts opening among the intestinal crypts (Figure 15–29). Mucus from these glands is distinctly alkaline (pH 8.1-9.3), which neutralizes chyme entering the duodenum from the pylorus, protecting the mucous membrane, and

FIGURE **15–24** Cells covering the villi.

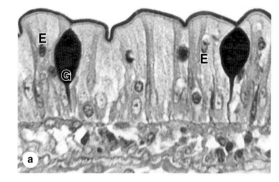

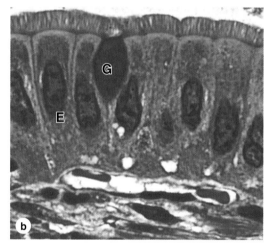

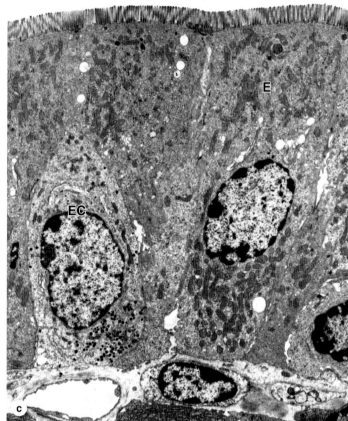

(a) The columnar epithelium that covers intestinal villi consists mainly of the tall absorptive enterocytes (**E**). The apical ends of these cells are joined and covered by a **brush border** of microvilli. Covered by a coating of glycoproteins, the brush border, along with the mucus-secreting goblet cells (**G**), stains with carbohydrate staining methods. Other cells of the epithelium are scattered enteroendocrine cells, which are difficult to identify in routine preparations, and various immune cells such as intraepithelial lymphocytes. The small spherical nuclei of lymphocytes can be seen between the enterocytes. (X250; PAS-hematoxylin)

(b) At higher magnification individual **microvilli** of enterocytes are better seen and the striated appearance of the border is apparent. (X500)

(c) TEM shows microvilli and densely packed mitochondria of enterocytes (**E**), and enteroendocrine cells (**EC**) with basal secretory granules can be distinguished along the basal lamina. (X2500)

bringing the intestinal contents to the optimum pH for pancreatic enzyme action. In the ileum both the lamina propria and submucosa contain well-developed mucosa-associated lymphoid tissue (MALT), consisting of the large lymphoid nodule aggregates known as **Peyer patches** underlying the epithelial M cells.

The muscularis is well developed in the small intestine, composed of an internal circular layer, an external longitudinal layer, and between them the neurons of the **myenteric (Auerbach) nerve plexus** which produce peristalsis (Figure 15–30). Neurons of both the submucosal and myenteric plexuses are largely autonomic and collectively make up the large enteric nervous system. The small intestine is covered by a thin serosa with mesothelium continuous with that of mesenteries (Figure 15–22a).

> **>> MEDICAL APPLICATION**
>
> Crohn disease is a chronic inflammatory bowel disease that occurs most commonly in the ileum or colon, resulting from a poorly understood combination of immune, environmental, and genetic factors. Excessive lymphocytic activity and inflammation occur in any or all layers of the tract wall, producing pain, localized bleeding, malabsorption, and diarrhea.

> LARGE INTESTINE

The large intestine or bowel, which absorbs water and electrolytes and forms indigestible material into feces, has the following regions: the short **cecum**, with the **ileocecal valve** and

FIGURE **15–25** Lipid absorption and processing by enterocytes.

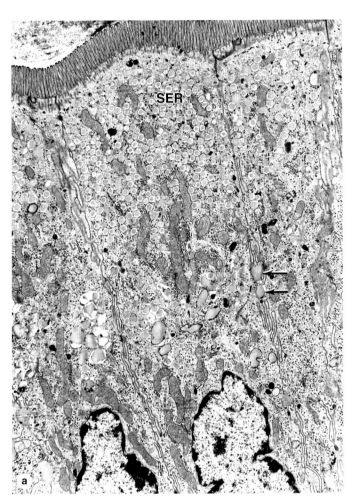

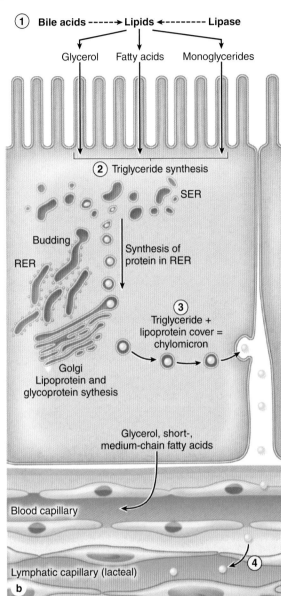

(a) TEM shows that enterocytes involved in lipid absorption accumulate many small lipid droplets in vesicles of the smooth ER (**SER**). These vesicles fuse near the nucleus, forming larger globules that are moved laterally and cross the cell membrane to the extracellular space (**arrows**) for eventual uptake by lymphatic capillaries (lacteals) in the lamina propria. (X3000)

(*Figure 15–25a, used with permission from Dr Robert R. Cardell, Jr, Department of Cancer and Cell Biology, University of Cincinnati College of Medicine, Cincinnati, OH.*)

(b) Diagram showing lipid processing by enterocytes. Ingested fats are emulsified by **bile acids** to form a suspension of lipid droplets from which lipids are digested by **lipases** to produce glycerol, fatty acids, and monoglycerides (**1**). The products of hydrolysis diffuse passively across the microvilli membranes and are collected in the cisternae of the smooth ER, where they are resynthesized as **triglycerides (2)**. Processed through the RER and Golgi, these triglycerides are surrounded by a thin layer of proteins and packaged in vesicles containing **chylomicrons** (0.2-1 μm in diameter) of lipid complexed with protein (**3**). Chylomicrons are transferred to the lateral cell membrane, secreted by exocytosis, and flow into the extracellular space in the direction of the lamina propria, where most enter the lymph in **lacteals (4)**.

FIGURE **15–26** Intestinal crypts or glands, with Paneth cells.

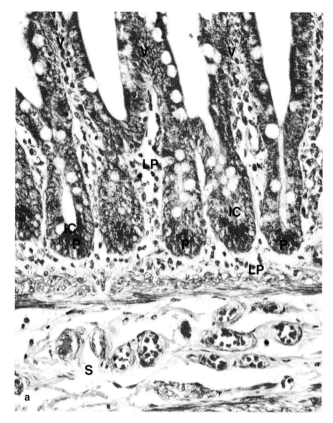

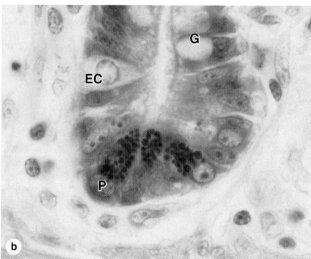

(a) Between villi (**V**) throughout the small intestine, the covering epithelium invaginates into the lamina propria (**LP**) to form short tubular glands called intestinal glands or intestinal crypts (**IC**). The lining near the openings of the crypts contains a population of stem cells for the entire epithelial lining of the small intestine. Daughter cells slowly move with the growing epithelium out of the crypts, differentiating as **goblet cells, enterocytes,** and **enteroendocrine cells.** These cells continue to move up each villus and within a week are shed at the tip, with billions shed

throughout the small intestine each day. At the base of the crypts are many Paneth cells (**P**) with an innate immune function. The submucosa (**S**) has many lymphatics draining lacteals. (X200; H&E)

(b) Higher magnification at the base of an intestinal gland shows the typical eosinophilic granules of Paneth cells (**P**), along with an open-type enteroendocrine cell (**EC**) and a differentiating goblet cell (**G**). (X400; H&E)

the **appendix**; the ascending, transverse, descending, and sigmoid **colon**; and the **rectum**, where feces is stored prior to evacuation (Figure 15–31). The mucosa lacks villi and except in the rectum has no major folds. Less than one-third as long as the small intestine, the large intestine has a greater diameter (6-7 cm). The wall of the colon is puckered into a series of large sacs called **haustra** (L. sing. *haustrum*, bucket, scoop).

The mucosa of the large bowel is penetrated throughout its length by tubular **intestinal glands**. These and the intestinal lumen are lined by goblet and absorptive cells, with a small number of enteroendocrine cells (Figures 15–32 and 15–33). The columnar absorptive cells or **colonocytes** have irregular microvilli and dilated intercellular spaces indicating active fluid absorption (Figure 15–33d). Goblet cells producing lubricating mucus become more numerous along the length of

the colon and in the rectum. Epithelial stem cells are located in the bottom third of each gland.

The lamina propria is rich in lymphoid cells and in lymphoid nodules that frequently extend into the submucosa (Figure 15–32). The richness in MALT is related to the large bacterial population of the large intestine. The appendix has little or no absorptive function but is a significant component of MALT (see Chapter 14).

The muscularis of the colon has longitudinal and circular layers but differs from that of the small intestine, with fibers of the outer layer gathered in three separate longitudinal bands called **teniae coli** (L., ribbons of the colon) (Figure 15–32a). Intraperitoneal portions of the colon are covered by serosa, which is characterized by small, pendulous protuberances of adipose tissue.

FIGURE **15–27** Enteroendocrine cell.

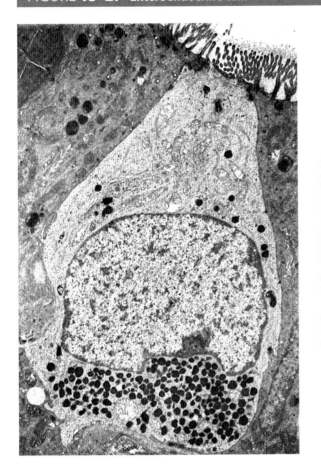

TEM of an open-type enteroendocrine cell in the epithelium of the duodenum shows microvilli at its apical end in contact with the lumen. The microvilli have components of nutrient-sensing and signal transduction systems similar in some components to those of taste bud gustatory cells. Activation of these cells by nutrients triggers the release at the basolateral membranes of peptide factors, including satiation peptides, which diffuse through extracellular fluid to enter capillaries (endocrine) or to bind receptors on nearby nerve terminals, smooth muscle fibers, or other cells (paracrine). Hormones from the various enteroendocrine cells act in a coordinated manner to control gut motility, regulate secretion of enzymes, HCl, bile and other components for digestion, and produce the sense of satiety in the brain. (X4500)

(*Used with permission from A.G.E. Pearse, Department of Histochemistry, Royal Postgraduate Medical School, London, UK.*)

FIGURE **15–28** Microvasculature, lymphatics, and muscle in villi.

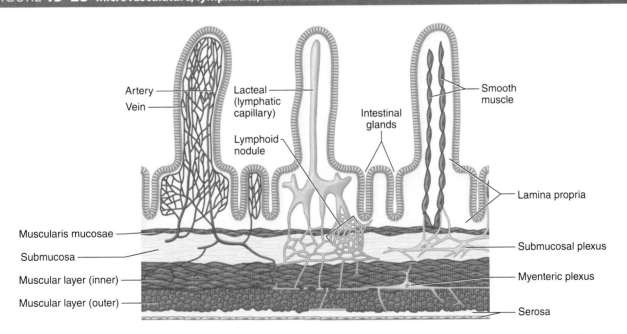

The villi of the small intestine contain blood microvasculature (left), lymphatic capillaries called lacteals (center), and both innervation and smooth muscle fibers (right).

FIGURE **15–29** Duodenal (Brunner) glands.

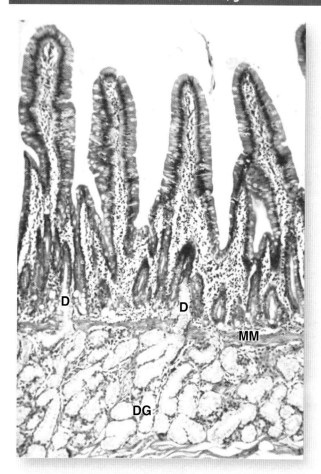

Concentrated in the upper duodenum are large masses of compound tubular **mucous glands**, the duodenal glands (**DG**), with many lobules that occupy much of the submucosa and may extend above the muscularis mucosae (**MM**) into the mucosa. Many small excretory ducts (**D**) extend from these lobules through the lamina propria and empty into the lumen among the small intestinal crypts. Alkaline mucus from duodenal glands neutralizes the pH of material entering the duodenum and supplements the mucus from goblet cells in lubricating and protecting the lining of the small intestine. (X100; H&E)

≫ MEDICAL APPLICATION

Colorectal cancer is an **adenocarcinoma** that develops initially from benign **adenomatous polyps** in the mucosal epithelium. Such polyps usually occur in epithelium of the rectum, sigmoid colon, or distal descending colon and are more common in individuals with low-fiber diets, which reduce the bulk of fecal material, and this in turn prolongs contact of the mucosa with toxins in feces. Screens for colorectal cancer include **sigmoidoscopy** or **colonoscopy** to see polyps and tests for **fecal occult blood** resulting from mucosal bleeding as an adenocarcinoma invades more deeply into the mucosa.

FIGURE **15–30** Small intestine muscularis and myenteric plexus.

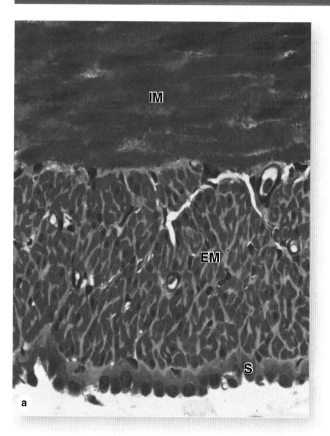

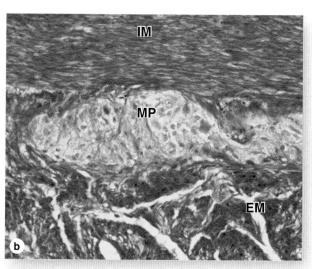

(a) Transverse sections of the small intestinal wall show the orientation of the internal (**IM**) and external (**EM**) smooth muscle layers. The inner layer is predominantly circular while the outer layer is longitudinal. The serosa (**S**) is a thin connective tissue covered here by a mesothelium of cuboidal or squamous cells. (X200; PT)

(b) Between the internal and external layers of muscularis (**IM** and **EM**) are ganglia of pale-staining neurons and other cells of the myenteric plexus (**MP**). (X100; H&E)

FIGURE **15–31** Large intestine.

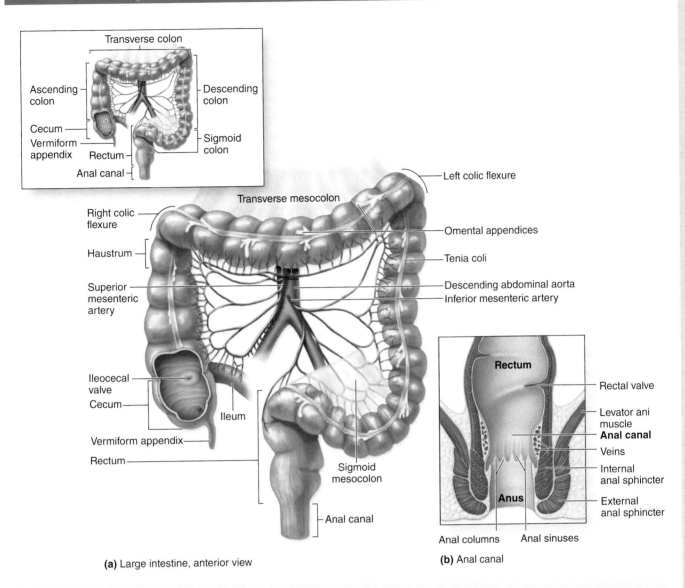

(a) Large intestine, anterior view

(b) Anal canal

As shown at the top, the large intestine consists of the cecum; the ascending, transverse, descending, and sigmoid regions of the colon; and the rectum.

(a) Anterior view of the large intestine with the proximal end exposed shows the **ileocecal valve** at its attachment to the ileum, along with the sac called the **cecum** and its extension, the **appendix**. The mucosa has only shallow plicae and no villi. The muscularis has two layers, but the outer longitudinal layer consists only of three distinct bundles of muscle fibers called **teniae coli** that produce the haustra in the colon wall. The serosa of the colon is continuous with that of the supporting mesenteries and displays a series of suspended masses of adipose tissue called **omental appendages**.

(b) At the distal end of the rectum, the anal canal, the mucosa, and submucosa are highly vascularized, with venous sinuses, and are folded as a series of longitudinal folds called **anal columns** (of Morgagni) with intervening **anal sinuses**. Fecal material accumulates in the rectum is eliminated by muscular contraction, including action of an **internal anal sphincter** continuous with the circular layer of the muscularis and an **external sphincter** of striated (voluntary) muscle.

FIGURE 15–32 Wall of the large intestine.

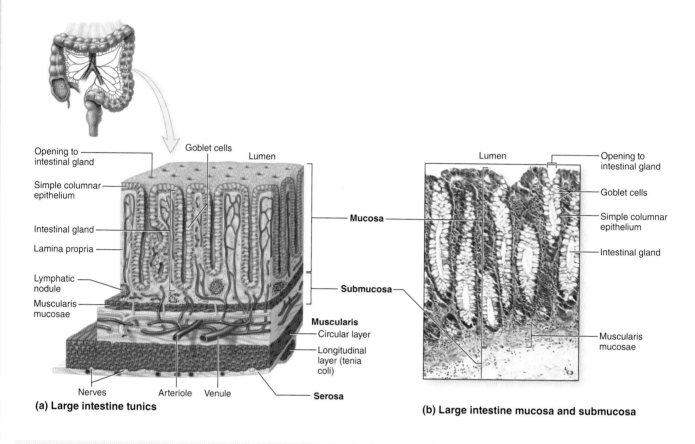

(a) Large intestine tunics

(b) Large intestine mucosa and submucosa

(a) Diagram shows the wall of the large intestine composed of the four typical layers. The **submucosa** is well vascularized. The **muscularis** has a typical inner circular layer, but the outer longitudinal muscle is only present in three equally spaced bands, the **teniae coli**.

(b) The **mucosa** is occupied mostly by tubular **intestinal glands** extending as deep as the muscularis mucosae and by lamina propria rich in MALT. (X80; H&E)

❯❯ MEDICAL APPLICATION

Swollen blood vessels in the mucosa or submucosa of the anal canal can cause a painful disorder called **hemorrhoids**. This common condition typically results from a low-fiber diet, **constipation**, prolonged sitting, or straining at defecation, conditions that produce increased pressure on these blood vessels.

The distal end of the GI tract is the **anal canal**, 3-4 cm long. At the **rectoanal junction** the simple columnar mucosal lining of the rectum is replaced by stratified squamous epithelium (Figure 15–34). The mucosa and submucosa of the anal canal form several longitudinal folds, the **anal columns** (Figure 15–31b), in which the lamina propria and submucosa include sinuses of the rectal venous plexus. Near the anus the circular layer of the rectum's muscularis forms the

internal anal sphincter. Defecation involves the action of voluntary muscle comprising the **external anal sphincter** (Figure 15–31b).

Table 15–2 summarizes the histologic features distinguishing each region of the digestive tract.

❯❯ MEDICAL APPLICATION

Herniation or outpocketing of the mucosa and submucosa of the colon can occur between the teniae coli, forming bulges (diverticula) and a condition called **diverticulosis**. This disorder can result from structural defects in the colon wall or from high intraluminal pressure or **constipation**. Fecal material can become immobilized in the diverticula and cause localized inflammation or **diverticulitis**.

FIGURE **15–33** Colon mucosa.

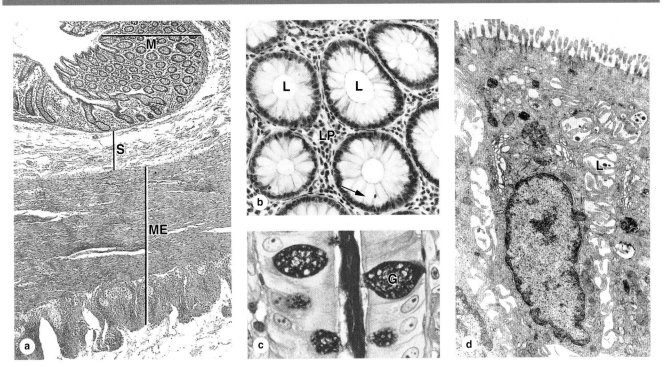

(a) Transverse section of the colon shows the muscularis externa (**ME**), including a **tenia coli** cut transversely in the lower part of the figure, the submucosa (**S**), the mucosa (**M**) filled with **tubular intestinal glands**. Some of these glands are cut longitudinally, but most seen here are cut transversely. (X14; H&E)

(b) Transversely cut glands are seen to consist of simple columnar epithelium surrounded by a tubular lumen (**L**) and embedded in lamina propria (**LP**) with many free lymphocytes. Lymphocytes can also be seen penetrating the epithelium (**arrow**). (X200; H&E)

(c) Longitudinal section of one intestinal gland stained for glyco-proteins shows **mucus** in the lumen and two major cell types in the epithelium: goblet cells (**G**) and the neighboring columnar cells specialized for water absorption. (X400; PAS)

(d) TEM of the absorptive cells, or **colonocytes**, reveals short **microvilli** at their apical ends and dilated **intercellular spaces** with interdigitating leaflets of cell membrane (**L**), a sign of active water transport. The absorption of water is passive, following the active transport of sodium from the basolateral surfaces of the epithelial cells. (X2500)

FIGURE **15–34** Mucosa of the rectoanal junction.

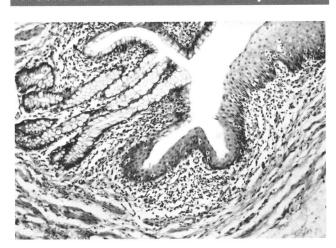

The simple columnar epithelium with tubular **intestinal glands** in the rectum (left side of photo) changes abruptly to stratified squamous epithelium in the **anal canal** (right side of photo), as seen in this longitudinal section. The connective tissue of the lamina propria is seen to contain many free lymphocytes. (X40; H&E)

TABLE 15–2	Summary of distinguishing digestive tract features, by region and layers.			
Region and Subdivisions	Mucosa (Epithelium, Lamina Propria, Muscularis Mucosae)	Submucosa (With Submucosal Plexuses)	Muscularis (Inner Circular and Outer Longitudinal Layers, With Myenteric Plexuses Between Them)	Adventitia/Serosa
Esophagus (upper, middle, lower)	Nonkeratinized **stratified squamous epithelium**; **cardiac glands** at lower end	Small **esophageal glands** (mainly mucous)	Both layers **striated muscle** in upper region; both layers **smooth muscle** in lower region; **smooth and striated muscle** fascicles mingled in middle region	Adventitia, except at lower end with serosa
Stomach (cardia, fundus, body, pylorus)	**Surface mucous cells** and **gastric pits** leading to **gastric glands** with **parietal and chief cells**, (in the fundus and body) or to mucous **cardiac glands** and **pyloric glands**	No distinguishing features	**Three indistinct layers** of smooth muscle (inner oblique, middle circular, and outer longitudinal)	Serosa
Small intestine (duodenum, jejunum, ileum)	**Plicae circulares**; **villi**, with **enterocytes** and **goblet cells**, and **crypts/glands** with **Paneth cells** and **stem cells**; **Peyer patches** in ileum	**Duodenal (Brunner) glands** (entirely mucous); possible extensions of Peyer patches in ileum	No distinguishing features	Mainly serosa
Large intestine (cecum, colon, rectum)	**Intestinal glands** with **goblet cells** and **absorptive cells**	No distinguishing features	Outer longitudinal layer separated into three bands, the **teniae coli**	Mainly serosa, with adventitia at rectum
Anal canal	**Stratified squamous epithelium**; longitudinal **anal columns**	Venous sinuses	Inner circular layer thickened as **internal sphincter**	Adventitia

Digestive Tract SUMMARY OF KEY POINTS

Oral Cavity

- The **oral cavity** is lined primarily by **mucosa** with **nonkeratinized stratified squamous epithelium**, with **keratinized stratified squamous epithelium** on the **hard palate** and **gingiva**.
- The dorsal surface mucosa of the tongue has projecting **lingual papillae** of four types: **filiform** papillae with keratinized epithelium and nonkeratinized **foliate**, **fungiform**, and large **vallate** papillae.
- All lingual papillae, except the filiform type, have epithelial **taste buds** on their sides, with chemosensory **gustatory cells** with synapses to basal sensory innervation, **support cells**, and an apical **taste pore**.
- Each tooth has **enamel** covering its **crown** and **neck** and a vascularized, innervated central **pulp cavity** within the **dentin** that makes up the **roots** and extends into the neck.
- Enamel calcifies as parallel **enamel rods** in a process guided by the protein **amelogenin** after secretion from columnar epithelial cells called **ameloblasts** in the **enamel organ** of the embryonic **tooth bud**.

- **Predentin** is secreted as elongated **dentinal tubules** from tall **odontoblasts** that line the pulp cavity and persist in the fully formed tooth, with apical **odontoblast processes** extending between the tubules.
- The **periodontium** of each tooth consists of a thin layer of bone-like **cementum** surrounding dentin of the roots and the **periodontal ligament** binding the cementum to **alveolar bone** on the jaw socket.

Layers of the Digestive Tract

- From the esophagus to the rectum, the digestive tract has **four major layers**: a lining **mucosa**, a **submucosa**, a **muscularis**, and an outermost **adventitia** or mesothelium-covered **serosa**.
- The **mucosa** varies regionally along the tract but always consists of a lining **epithelium** on a **lamina propria** of loose connective tissue and smooth muscle fibers extending from **muscularis mucosae** layer.

Esophagus

- The mucosa of the **esophagus** has **nonkeratinized stratified squamous epithelium**; its muscularis is striated at its superior end with smooth muscle at its inferior end, with mixed fiber types in the middle.
- Most of the outer layer of the esophagus is **adventitia**, merging with other tissues of the mediastinum.
- At the **esophagogastric junction**, stratified squamous epithelium changes abruptly to **simple columnar epithelium** invaginating into the lamina propria as many branched tubular glands.

Stomach

- The **stomach** has four major regions: the superior **cardia** and inferior **pylorus**, which are rather similar histologically, and the intervening **fundus** and **body**, which are also similar.
- The mucosa of the stomach fundus and body is penetrated by numerous **gastric pits**, which are lined like the stomach lumen with **surface mucous cells** and which lead into branching **gastric glands**.
- The surface mucous cells secrete a thick layer of **viscous mucus with bicarbonate ions**, which protects these cells and the underlying lamina propria.
- The **gastric glands** are lined by epithelium with four **major cell types**, as well as their pluripotent **stem cells** that are located in the narrow neck regions of these glands:
    - **Mucous neck cells** include immature precursors of the surface mucous cells but produce less alkaline mucus while migrating up into the gastric pits.
    - **Parietal cells** are large cells with many mitochondria and **large intracellular canaliculi** for production of HCl in the gastric secretion; they also secrete **intrinsic factor** for vitamin B_{12} uptake.
    - **Chief (zymogenic) cells**, clustered mainly in the lower half of the gastric glands, secrete the protein **pepsinogen** that is activated by the low pH in the lumen to form the major protease **pepsin**.
    - **Enteroendocrine cells** are scattered epithelial cells of the **diffuse neuroendocrine system**, which release **peptide hormones** to regulate activities of neighboring tissues during food digestion.
- The mucosa of the stomach cardiac and pyloric regions has branching **cardial and pyloric glands** that consist almost entirely of columnar **mucous cells**, lacking parietal and chief cells.

Small Intestine

- The **small intestine** has three regions: the **duodenum** with large mucous glands in the submucosa called **duodenal glands**; the **jejunum**; and the **ileum** with the large mucosal and submucosal **Peyer patches**.
- In all regions of small intestine the mucosa has millions of projecting **villi**, with simple columnar epithelium over cores of lamina propria, and intervening simple tubular **intestinal glands** (or crypts).
- Stem cells in these glands produce the columnar epithelial cells of villi, mainly **goblet cells** and **enterocytes** for nutrient absorption, as well as defensin-producing **Paneth cells** deep in the glands.
- **Sugars** and **amino acids** produced by final steps of digesting carbohydrates and polypeptides in the **glycocalyx** undergo transcytosis through **enterocytes** for uptake by **capillaries**.
- Products of **lipid** digestion associate with bile salts, are taken up by enterocytes, and are converted to **triglycerides** and **lipoproteins** for release as **chylomicrons** and uptake by a lymphatic called a **lacteal** in the core of each villus.
- **Smooth muscle** of the lamina propria and **muscularis mucosae**, under the control of the autonomic **submucosal (Meissner) plexus**, moves the villi and helps propel lymph through the lacteals.
- **Smooth muscle** in the **inner circular layer** and the **outer longitudinal layer** of the muscularis, under the control of the autonomic **myenteric (Auerbach) plexus**, produces strong **peristalsis**.

Large Intestine

- The **large intestine** has three major regions: the short **cecum**, with the appendix; the long **colon**, with its ascending, transverse, descending, and sigmoid portions; and the **rectum**.
- Along its entire length, the mucosa of the large intestine has millions of short simple tubular **intestinal glands**, lined by lubricant **goblet cells** and **absorptive cells** for the uptake of water and electrolytes.
- The **muscularis** of the colon has its outer longitudinal layer subdivided into three bands of smooth muscle called **teniae coli**, which act in the peristaltic movement of feces to the rectum.

Anal Canal

- At the **anal canal** the simple columnar epithelium lining the rectum shifts abruptly to **stratified squamous epithelium** of the skin at the **anus**.
- Near the anus the circular layer of the rectum's muscularis forms the **internal anal sphincter**, with further control exerted by **striated muscle** of the **external anal sphincter**.

Digestive Tract ASSESS YOUR KNOWLEDGE

1. In which of the following structures of the oral cavity would taste buds be localized in the highest concentration?
   a. Fungiform papillae
   b. Gingiva
   c. Filiform papillae
   d. Ventral surface of the tongue
   e. Vallate papillae

2. Certain antibiotic therapies slow the replacement of the cells lining the small intestine. This may cause the loss of what tissue type?
   a. Ciliated pseudostratified columnar epithelium
   b. Simple cuboidal epithelium
   c. Simple columnar epithelium
   d. Pseudostratified columnar epithelium with stereocilia
   e. Stratified squamous, nonkeratinized epithelium

3. The teniae coli of the large intestine represent an organ-specific specialization of which layer of the intestinal tract wall?
   a. Epithelium
   b. Lamina propria
   c. Muscularis mucosa
   d. Muscularis externa
   e. Serosa

4. Which of the following would most likely result from a reduction in the number of Paneth cells?
   a. Thinning of the glycocalyx
   b. Reduced breakdown of fats
   c. Elevated levels of undigested proteins
   d. Decreased mucus in the intestine
   e. Increased number of intestinal bacteria

5. A medical student on a rotation in the pathology laboratory is given an unlabeled microscope slide with tissue provided by a gastroenterologist from a cancer patient she is attending. The mucosa and submucosa are poorly preserved, with only the thick muscularis well-stained, showing striated fibers. The slide most likely shows a biopsy of which region of the GI tract?
   a. Pyloric sphincter
   b. Esophagus
   c. Colon
   d. Corpus of the stomach
   e. Ileum

6. Diarrhea may result if which of the following organs fails to carry out its role in absorbing water from the feces?
   a. Anal canal
   b. Cecum
   c. Colon
   d. Jejunum
   e. Duodenum

7. Which of the following is true of the absorptive cells of the small intestine?
   a. Also called enteroendocrine cells
   b. Have many microvilli covering their basal surfaces
   c. Absorb lipids by active transport
   d. Synthesize triglycerides from absorbed lipids
   e. Undergo mitosis at tips of villi and are sloughed off into crypts

8. A 52-year-old man is diagnosed with a carcinoid after an appendectomy. The enteroendocrine cells producing this disorder differ from goblet cells in which of the following?
   a. The direction of release of secretion
   b. The use of exocytosis for release of secretory product
   c. Their presence in the small and large intestines
   d. The origin from a crypt stem cell
   e. Their location in a simple columnar epithelium

9. A 14-month-old girl is brought to the pediatric dentistry clinic because her erupted deciduous teeth are opalescent with fractured and chipped surfaces. X-rays reveal bulb-shaped crowns, thin roots, and enlarged central cavities. Tissue immediately surrounding one tooth's central cavity is biopsied and prepared for histology, which reveals irregular, widely spaced tubules. Which of the following applies to this irregular tissue layer?
   a. It has a composition similar to that of bone and is produced by cells similar in appearance to osteocytes.
   b. It is formed on a noncollagenous matrix that is resorbed after mineralization by the same cells that secreted it.
   c. It contains abundant nerves, microvasculature, and loose connective tissue.
   d. It consists of mineralized collagen secreted by cells derived from the neural crest.
   e. It is the site of inflammation in diabetic patients and is sensitive to vitamin C deficiency.

10. A 39-year-old woman presents with dyspnea, fatigue, pallor, tachycardia, anosmia, and diarrhea. Laboratory results are: hematocrit 32% (normal 36.1%-44.3%), MCV 102 fL (normal 78-98 fL), 0.3% reticulocytes (normal 0.5%-2.0%), 95 pg/mL vitamin B_{12} (normal 200-900 pg/mL), and an abnormal stage I of the Schilling test. Autoantibodies are detected against a cell type located in one region of the GI tract. In which regions would those cells be found?
    a. Esophagus
    b. Body of the stomach
    c. Pyloric region of the stomach
    d. Cardiac region of the stomach
    e. Duodenum

16 Organs Associated with the Digestive Tract

The organs associated with the digestive tract include the major salivary glands, the pancreas, the liver, and the gallbladder. Products of these organs facilitate transport and digestion of food within the gastrointestinal tract. The main functions of the salivary glands are to moisten and lubricate ingested food and the oral mucosa, to initiate the digestion of carbohydrates and lipids with amylase and lipase, and to secrete innate immune components such as lysozyme and lactoferrin.

The pancreas secretes digestive enzymes that act in the small intestine and hormones important for the metabolism of the absorbed nutrients. Bile, whose components are necessary for digestion and absorption of fats, is made in the liver but stored and concentrated in the gallbladder. The liver also plays a major role in carbohydrate and protein metabolism, inactivates many toxic substances and drugs, and synthesizes most plasma proteins and factors necessary for blood coagulation.

› SALIVARY GLANDS

Exocrine glands in the mouth produce saliva, which has digestive, lubricating, and protective functions. With a normal pH of 6.5-6.9, saliva also has an important buffering function and in some species is also important for evaporative cooling. There are three pairs of large salivary glands: the **parotid**, **submandibular**, and **sublingual glands** (Figure 16–1), in addition to the numerous minor or intrinsic salivary glands located throughout most of the oral mucosa which secrete about 10% of the total saliva volume.

›› MEDICAL APPLICATION

Inadequate saliva production, leading to **dry mouth** or **xerostomia**, can be caused by various factors affecting the major salivary glands, such as **mumps** viral infection, radiation of the glands, or the normal side effect of drugs such as **antihistamines**.

A connective tissue capsule surrounds each major salivary gland. The parenchyma of each consists of secretory units on a branching duct system arranged in lobules, separated by septa of connective tissue. The secretion of each gland is either serous, seromucous, or mucous, depending on its content of the glycoprotein mucin. Saliva from the parotids is serous and watery. The submandibular and sublingual glands produce a seromucous secretion, while that of the minor glands is mostly mucous. Saliva is modified by the cells of the duct system draining the secretory units, with much Na^+ and Cl^- reabsorbed while certain growth factors and digestive enzymes are added.

Three epithelial cell types comprise the salivary secretory units:

- **Serous cells** are polarized protein-secreting cells, usually pyramidal in shape, with round nuclei, well-stained RER, and apical secretory granules (Figures 16–2 through 16–4). Joined apically by tight and adherent junctions, serous cells form a somewhat spherical unit called an **acinus** (L. grape), with a very small central lumen (Figure 16–2). Serous acinar cells secrete enzymes and other proteins.

- **Mucous cells** are somewhat more columnar in shape, with more compressed basal nuclei (Figures 16–2 and 16–4). Mucous cells contain apical granules with hydrophilic mucins that provide lubricating properties in saliva but cause poor cell staining in routine preparations (Figure 16–5). Mucous cells are most often organized as cylindrical **tubules** rather than acini. Mixed salivary glands have tubuloacinar secretory units with both serous and mucous secretion.

- **Myoepithelial cells**, described in Chapter 4, are found inside the basal lamina surrounding acini, tubules, and the proximal ends of the duct system (Figures 16–2 and 16–4). These small, flattened cells extend several

FIGURE **16–1** Major salivary glands.

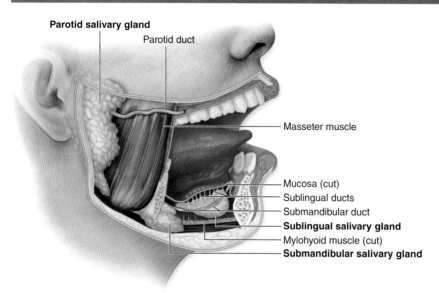

Parotid salivary gland
Parotid duct
Masseter muscle
Mucosa (cut)
Sublingual ducts
Submandibular duct
Sublingual salivary gland
Mylohyoid muscle (cut)
Submandibular salivary gland

There are three bilateral pairs of major salivary glands, the **parotid**, **submandibular**, and **sublingual glands**, which together produce about 90% of saliva. Their locations, relative sizes, and excretory ducts are shown here. These glands plus microscopic minor salivary glands located throughout the oral mucosa produce 0.75-1.50 L of saliva daily.

FIGURE **16–2** Epithelial components of a submandibular gland lobule.

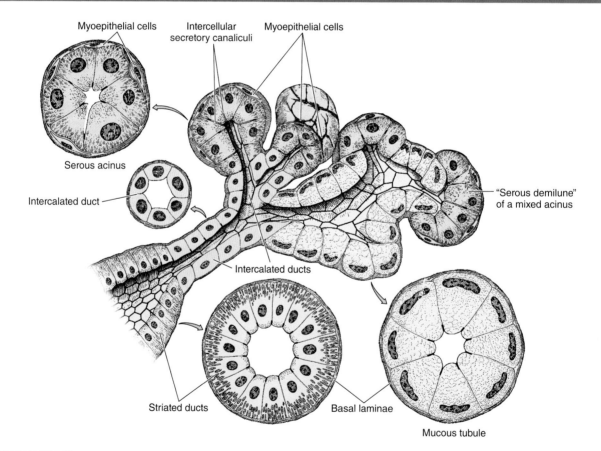

Myoepithelial cells
Intercellular secretory canaliculi
Myoepithelial cells
Serous acinus
Intercalated duct
Intercalated ducts
"Serous demilune" of a mixed acinus
Striated ducts
Basal laminae
Mucous tubule

The secretory portions are composed of pyramidal serous (violet) and mucous (tan) cells. **Serous acini** consist of typical protein-secreting cells with rounded nuclei, basal accumulation of RER, and apical ends filled with secretory granules. The cells of **mucous tubules** have flattened, basal nuclei with condensed chromatin. In the submandibular gland mixed tubuloacinar secretory units also occur, combining short mucous tubules with distal clusters of serous cells called "**serous demilune**."The short **intercalated ducts** are lined with low cuboidal epithelium. The **striated ducts** consist of columnar cells with characteristics of ion-transporting cells: basal membrane invaginations with mitochondrial accumulations. **Myoepithelial cells** are shown around the serous acini.

FIGURE **16–3** Parotid gland.

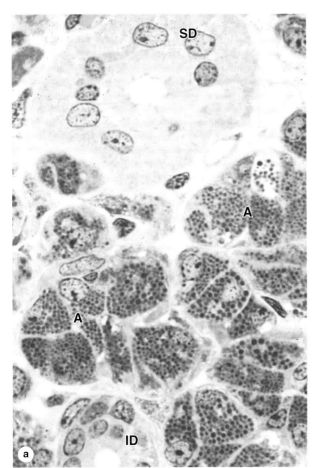

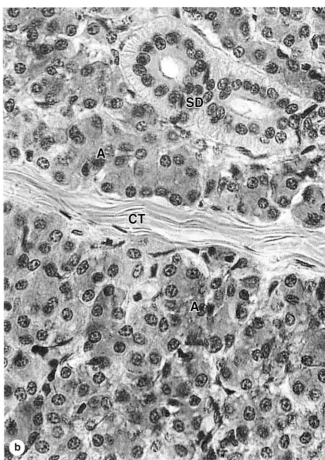

The large parotid gland consists entirely of serous acini with cells producing amylase and other proteins for storage in secretory granules.

(a) Micrograph of a parotid gland shows densely packed serous acini (**A**) with ducts. Secretory granules of serous cells are clearly

shown in this plastic section, as well as an intercalated duct (**ID**) and striated duct (**SD**), both cut transversely. (X400; PT)

(b) Striations of a duct (**SD**) are better seen here, along with a septum (**CT**) and numerous serous acini (**A**). The connective tissue often includes adipocytes. (X200; H&E)

contractile processes around the associated secretory unit or duct and their activity is important for moving secretory products into and through the ducts.

≫ MEDICAL APPLICATION

Excessive saliva production, or **sialorrhea**, is associated with the autonomic activity of **nausea**, inflammation within the oral cavity, and **rabies** viral infection.

In the **intralobular duct system**, secretory acini and tubules empty into short **intercalated ducts**, lined by cuboidal epithelial cells, and several of these ducts join to form a

striated duct (Figure 16–2). The more columnar striated duct cells have many infoldings of their basolateral membrane, all aligned with numerous mitochondria that, by light microscopy, appear as faint basal striations radiating toward the nuclei (Figure 16–6). Striated ducts reabsorb Na$^+$ ions from the initial secretion and their folded cell membranes present a large surface area with ion transporters, facilitating rapid ion transcytosis and making the secretion slightly hypotonic.

Plasma cells in the connective tissue surrounding the small intralobular ducts release **IgA**, which forms a complex with the secretory component synthesized by the epithelial cells of the serous acini and intralobular ducts. Transferred into the saliva, the IgA complex released into the saliva provides defense against specific pathogens in the oral cavity.

FIGURE **16–4** **Ultrastructure of serous and mucous cells.**

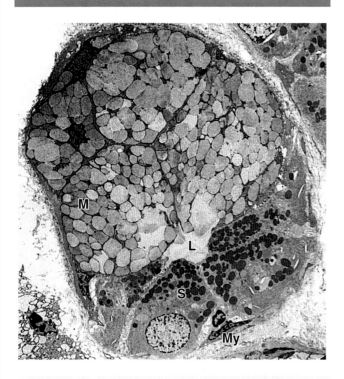

A micrograph of a mixed acinus from a submandibular gland shows both serous and mucous cells surrounding the small lumen (**L**). Mucous cells (**M**) have large, hydrophilic granules like those of goblet cells, while serous cells (**S**) have small, dense granules. Small myoepithelial cells (**My**) extend contractile processes around each acinus. (X2500)

(*Used with permission from Dr John D. Harrison, King's College London Dental Institute, London, UK.*)

Ducts from each lobule converge and drain into interlobular **excretory ducts** with increasing size and thicker connective tissue layers. The lining of these ducts is unusual, combining various epithelial types, including simple cuboidal or columnar, stratified cuboidal or columnar, and pseudostratified epithelia, distributed in no apparent pattern. These atypical epithelia may reflect their composition of cells with many diverse functions, including cells for ion reabsorption, cells for secretion of mucin and other proteins, enteroendocrine cells, and basal stem cells, all in highly branched ducts of small diameter. Before emptying into the oral cavity, the main duct of each gland is lined with nonkeratinized stratified squamous epithelium.

Vessels and nerves enter the large salivary glands at a hilum and gradually branch into the lobules. A rich vascular and nerve plexus surrounds the secretory and duct components of each lobule. The capillaries surrounding the secretory units provide fluid important for saliva production, which is stimulated by the autonomic nervous system. Parasympathetic

stimulation, usually elicited through the smell or taste of food, provokes a copious watery secretion with relatively little organic content. Sympathetic stimulation inhibits such secretion and produces the potential for dry mouth often associated with anxiety.

Features specific to each group of major salivary glands include the following:

- **Parotid glands**, located in each cheek near the ear, are branched acinar glands with exclusively serous acini (Figure 16–3). Serous cells of parotid glands secrete abundant **α-amylase** that initiates hydrolysis of carbohydrates and proline-rich proteins with antimicrobial and other protective properties.
- **Submandibular glands**, which produce two-thirds of all saliva, are branched tubuloacinar glands, having primarily serous acini, but with many mixed tubuloacinar secretory units (Figures 16–4 and 16–5a). Within the mixed units grouped serous cells occur distally on short mucous tubules and often assume a crescent-shaped arrangement called a **serous demilune** (Figure 16–5a). In addition to α-amylase and proline-rich proteins, serous cells of the submandibular gland secrete **lysozyme** for hydrolysis of bacterial walls.
- **Sublingual glands**, the smallest of the major glands, are also considered branched tubuloacinar glands, but here secretory tubules of mucous cells predominate and the main product of the gland is mucus (Figure 16–5b). The few serous cells present add amylase and lysozyme to the secretion.

As described in Chapter 15, small, nonencapsulated salivary glands are distributed throughout the oral mucosa and submucosa with short ducts to the oral cavity. These minor salivary glands are usually mucous, except for the small serous glands at the bases of circumvallate papillae. Plasma cells releasing IgA are also common within the minor salivary glands.

› PANCREAS

The **pancreas** is a mixed exocrine-endocrine gland that produces both digestive enzymes and hormones. It is an elongated retroperitoneal organ, with a large head near the duodenum and more narrow body and tail regions that extend to the left (Figure 16–7). The pancreas has a thin capsule of connective tissue, from which septa extend to cover the larger vessels and ducts and to separate the parenchyma into lobules (Figure 16–8). The secretory acini are surrounded by a basal lamina that is supported only by a delicate sheath of reticular fibers with a rich capillary network. Endocrine function of the pancreas involves primarily smaller cells similar to enteroendocrine cells located in variously sized clusters called the **pancreatic islets** (islets of Langerhans). These are described with the endocrine organs in Chapter 20.

FIGURE **16–5** Submandibular gland and sublingual gland.

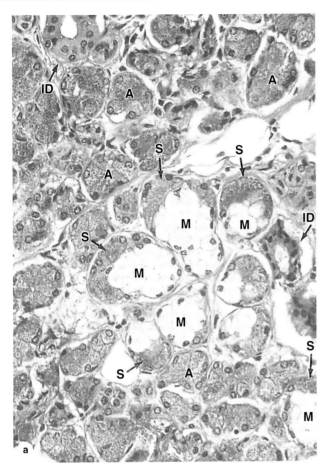

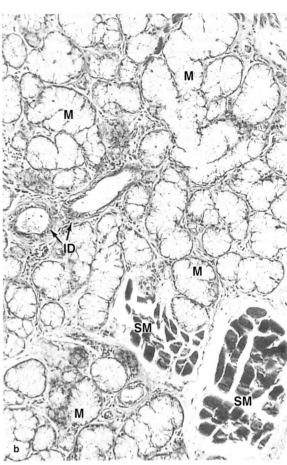

(a) The submandibular gland is a mixed serous and mucous gland (serous cells predominate), and shows well-stained serous acini (**A**) and "serous demilunes" (**S**) and pale-staining mucous cells (**M**) grouped as tubules in this tubuloacinar gland. (The crescent-shaped "serous demilunes" arise at least in part artifactually due to disproportionate swelling of the adjacent mucous cells during

slide preparation.) Small intralobular ducts (**ID**) drain each lobule. (X340; H&E)

(b) The sublingual gland is a mixed but largely mucous gland with a tubuloacinar arrangement of poorly stained mucous cells (**M**). Small intralobular ducts (**ID**) are seen in connective tissue, as well as small fascicles of lingual striated muscle (**SM**). (X140; H&E)

›› MEDICAL APPLICATION

Pancreatic cancer, which is usually a carcinoma of duct cells, can arise anywhere in the gland but occurs most often in the head of the organ near the duodenum. The tumor is usually asymptomatic until growth and metastasis are well advanced, leading to the low rate of early detection and subsequent high rate of mortality. Metastasis may be facilitated by the relatively sparse connective tissue around the ducts and vasculature of the pancreas.

The digestive enzymes are produced by cells of serous acini in the larger exocrine portion of the pancreas (Figure 16–9a). This somewhat resembles the parotid gland histologically, although the pancreas lacks striated ducts and the parotid glands lack islets of endocrine tissue. Each pancreatic acinus consists of several serous cells surrounding a very small lumen, without myoepithelial cells (Figure 16–9). The acinar cells are polarized, with round basal nuclei, and numerous zymogen granules apically, typical of protein-secreting cells (Figure 16–10).

Each acinus is drained by a short **intercalated duct** of simple squamous or low cuboidal epithelium. The initial cells of these small ducts extend into the lumen of the acinus as small pale-staining **centroacinar cells** that are unique to the pancreas. Cells of the intercalated ducts secrete a large volume of fluid, rich in HCO_3^- (bicarbonate ions), which alkalinizes and transports hydrolytic enzymes produced in the acini.

FIGURE 16–6   Striated ducts.

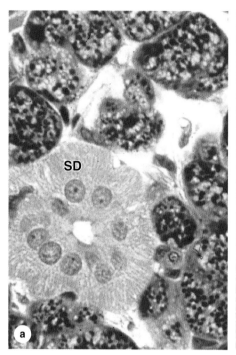

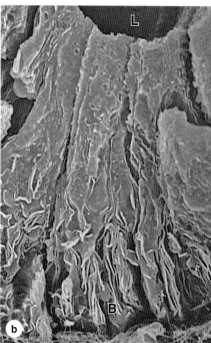

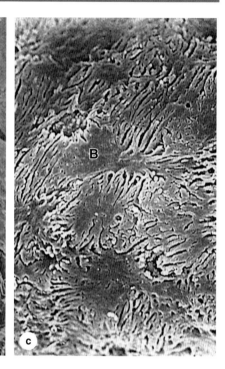

(a) A striated duct (**SD**) shows very faint striations in the basal half of the columnar cells, which represent mitochondria located in the folds of the lateral cell membrane. (X200; H&E)

(b) SEM indicates that the apical ends of the cells are joined together near the small lumen (**L**), with interdigitating folds of cell membrane best developed at the basal end (**B**). (X4000)

(c) SEM shows the bases (**B**) of several such cells with the basal lamina removed, revealing the interlocking of folded membrane between neighboring cells. Mitochondria within the folds supply energy for rapid ion uptake from saliva. (X4000)

The intercalated ducts merge with **intralobular ducts** and larger **interlobular ducts**, which have increasingly columnar epithelia before joining the main pancreatic duct that runs the length of the gland.

The exocrine pancreas secretes approximately 1.5 L of alkaline pancreatic juice per day and delivers it directly into the duodenum where the HCO_3^- ions neutralize the acidic chyme entering there from the stomach and establish the pH for optimal activity of the pancreatic enzymes. These digestive enzymes include several **proteases, α-amylase, lipases,** and **nucleases (DNAase** and **RNAase).** The proteases are secreted as inactive zymogens (**trypsinogen, chymotrypsinogen,** proelastase, kallikreinogen, and procarboxipeptidases). Trypsinogen is cleaved and activated by enteropeptidases in the duodenum, generating trypsin that activates the other proteases in a cascade. Pancreatic tissue is protected against autodigestion by the following:

- Restricting protease activation to the duodenum,
- Trypsin inhibitor, which is copackaged in the secretory granules with trypsinogen,

- The higher pH in the acini and duct system due to HCO_3^- secreted by the centroacinar and intercalated duct cells, which helps keep all the enzymes inactive.

❯❯ MEDICAL APPLICATION

In **acute pancreatitis**, the proenzymes may be activated and digest pancreatic tissues, leading to very serious complications. Possible causes include infection, gallstones, alcoholism, drugs, and trauma. **Chronic pancreatitis** can produce progressive fibrosis and loss of pancreatic function.

Exocrine secretion in the pancreas is regulated mainly through two polypeptide hormones produced by enteroendocrine cells of the small intestine:

- **Cholecystokinin (CCK)** stimulates enzyme secretion by the acinar cells.
- **Secretin** promotes water and HCO_3^- secretion by the duct cells.

FIGURE **16–7** Pancreas and duodenum.

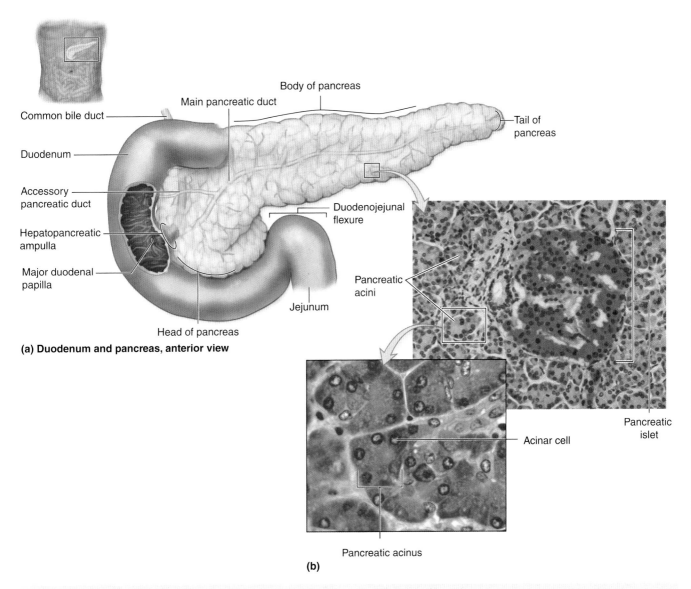

(a) Duodenum and pancreas, anterior view

(b)

(a) The main regions of the pancreas are shown in relation to the two pancreatic ducts and the duodenum.

(b) Micrographs show a pancreatic islet and several pancreatic acini. (X75 and X200; H&E)

Autonomic (parasympathetic) nerve fibers also stimulate secretion from both acinar and duct cells.

❯❯ MEDICAL APPLICATION

In the normal liver most dense connective tissue is found only in the portal areas, surrounding the blood vessels and bile ductule. In liver **cirrhosis**, which occurs late in **chronic liver disease**, fibrosis and proliferation of fibroblasts and hepatic stellate cells occur beyond the portal areas. The excessive connective tissue may disrupt the normal hepatic architecture and interfere with liver function.

❯ LIVER

The **liver** is the largest internal organ, in adults averaging about 1.5 kg or 2% of the body weight. Located in the right upper quadrant of the abdomen just below the diaphragm (see Figure 15–1), the liver has major left and right lobes with two smaller inferior lobes, most of which are covered by a thin capsule and mesothelium of the visceral peritoneum. The capsule thickens at the hilum (or porta hepatis) on the inferior side, where the **dual blood supply** from the **hepatic portal vein** and **hepatic artery** enters the organ and where the hepatic vein, lymphatics, and common hepatic (bile) duct exit.

FIGURE **16–8** Pancreas.

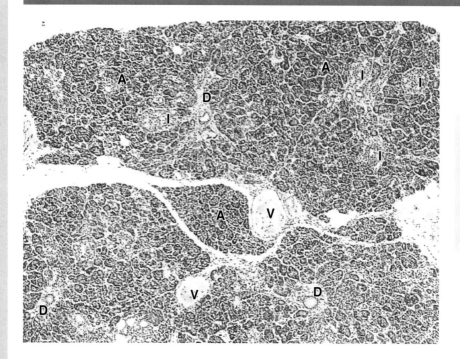

Low-power view of pancreas includes several islets (**I**) surrounded by many serous acini (**A**). The larger intralobular ducts (**D**) are lined by simple columnar epithelium. The ducts and blood vessels (**V**) are located in connective tissue, which also provides a thin capsule to the entire gland and thin septa separating the lobules of secretory acini. (X20; H&E)

FIGURE **16–9** Pancreatic acini.

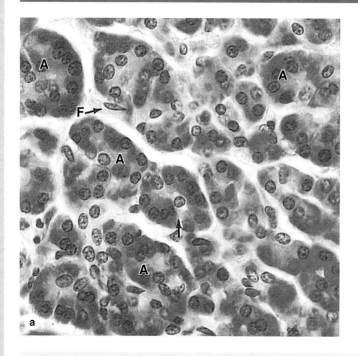

Centroacinar cells

Basal lamina

Intercalated duct

Zymogen granules

Acinar cells

b

(a) Micrograph of exocrine pancreas shows the serous, enzyme-producing cells arranged in small acini (**A**) with very small lumens. Acini are surrounded by only small amounts of connective tissue with fibroblasts (**F**). Each acinus is drained by an intercalated duct with its initial cells, the centroacinar cells (**arrow**), inserted into the acinar lumen. (X200; H&E)

(b) The diagram shows the arrangement of cells more clearly. Under the influence of secretin, the **centroacinar** and **intercalated duct cells** secrete a copious HCO_3^--rich fluid that hydrates, flushes, and alkalinizes the enzymatic secretion of the acini.

FIGURE **16–10** Pancreatic acinar cell ultrastructure.

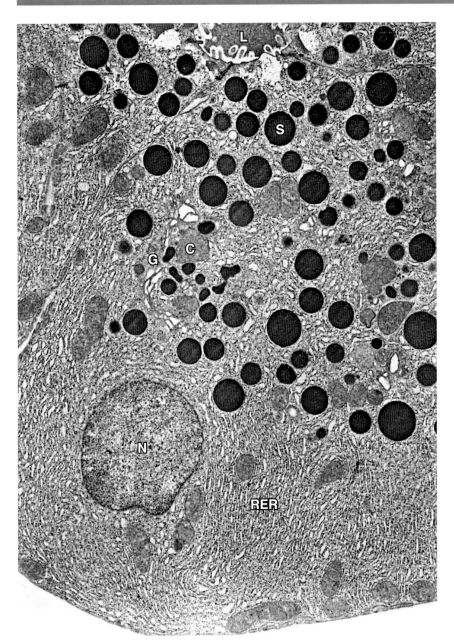

TEM of a pancreatic acinar cell shows its pyramidal shape and the round, basal nucleus (**N**) surrounded by cytoplasm packed with cisternae of rough ER (**RER**). The Golgi apparatus (**G**) is situated at the apical side of the nucleus and is associated with condensing vacuoles (**C**) and numerous secretory granules (**S**) with zymogen. The small lumen (**L**) of the acinus contains proteins recently released from the cell by exocytosis. Exocytosis of digestive enzymes from secretory granules is promoted by CCK, released by enteroendocrine cells of the duodenum when food enters that region from the stomach. (X8000)

The main digestive function of the liver is production of **bile**, a complex substance required for the emulsification, hydrolysis, and uptake of fats in the duodenum. The liver is also the major interface between the digestive system and the blood, as the organ in which nutrients absorbed in the small intestine are processed before distribution throughout the body. About 75% of the blood entering the liver is nutrient-rich (but O_2-poor) blood from the portal vein arising from the stomach, intestines, and spleen; the other 25% comes from the hepatic artery and supplies the organ's O_2.

Hepatocytes (Gr. *hepar*, liver), the key cells of this organ, are among the most functionally diverse cells of the body.

In addition to an exocrine function in the secretion of bile components, hepatocytes and other liver cells process the contents of blood, with many specific functions:

- Synthesis and endocrine secretion into the blood of the major **plasma proteins**, including albumins, fibrinogen, apolipoproteins, transferrin, and many others
- Conversion of amino acids into glucose (**gluconeogenesis**);
- Breakdown (**detoxification**) and conjugation of ingested toxins, including many drugs;
- Amino acid **deamination**, producing **urea** removed from blood in kidneys;

- **Storage of glucose** in glycogen granules and **triglycerides** in small lipid droplets;
- **Storage of vitamin A** (in **hepatic stellate cells**) and other fat-soluble vitamins;
- Removal of effete erythrocytes (by specialized macrophages, or **Kupffer cells**);
- **Storage of iron** in complexes with the protein **ferritin**.

Hepatocytes & Hepatic Lobules

The liver's unique histologic organization and microvasculature allow hepatocytes to perform their diverse metabolic,

exocrine, and endocrine functions. Hepatocytes are large cuboidal or polyhedral epithelial cells, with large, round central nuclei and eosinophilic cytoplasm rich in mitochondria. The cells are frequently binucleated and about 50% of them are polyploid, with two to eight times the normal chromosome number.

The liver parenchyma is organized as thousands of small (~0.7 × 2 mm) **hepatic lobules** in which hepatocytes form hundreds of irregular plates arranged radially around a small **central vein** (Figures 16–11 through 16–13). The hepatocyte plates are supported by a delicate stroma of reticulin fibers (Figure 16–13b). Peripherally each lobule has three to

---

FIGURE 16–11 Liver.

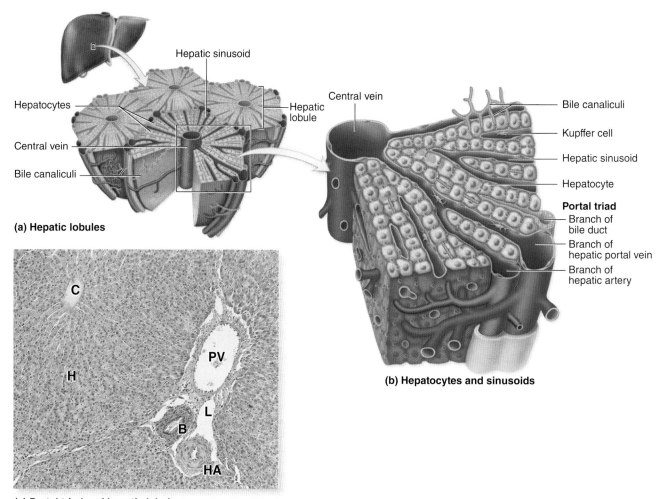

(a) Hepatic lobules

Hepatic sinusoid
Hepatocytes
Central vein
Bile canaliculi
Hepatic lobule
Central vein

(b) Hepatocytes and sinusoids

Bile canaliculi
Kupffer cell
Hepatic sinusoid
Hepatocyte
Portal triad
Branch of bile duct
Branch of hepatic portal vein
Branch of hepatic artery

(c) Portal triad and hepatic lobule

The liver, a large organ in the upper right quadrant of the abdomen, immediately below the diaphragm, is composed of thousands of polygonal structures called **hepatic lobules**, which are the basic functional units of the organ.

(a) Diagram showing a small **central vein** in the center of a hepatic lobule and several sets of blood vessels at its periphery. The peripheral vessels are grouped in connective tissue of the **portal tracts** and include a branch of the portal vein, a

branch of the hepatic artery, and a branch of the bile duct (the **portal triad**).

(b) Both blood vessels in this triad branch as **sinusoids**, which run between plates of **hepatocytes** and drain into the central vein.

(c) Micrograph of a lobule shows the central vein (**C**), plates of hepatocytes (**H**), and in an adjacent portal area a small lymphatic (**L**) and components of the portal triad: a portal venule (**PV**), hepatic arteriole (**HA**), and bile ductule (**B**). (X220; H&E)

FIGURE **16–12** Hepatic lobule.

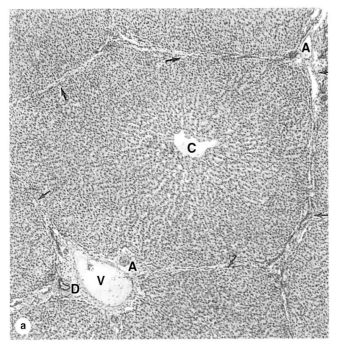

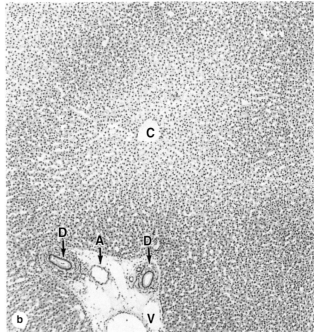

Cut transversely, hepatic lobules are polygonal units show-ing plates of epithelial cells called **hepatocytes** radiating from a central venule (**C**). **(a)** Hepatic lobules of some mam-mals, such as the pig, are delimited on all sides by connective tissue.

(b) In humans these lobules have much less connective tissue and their boundaries are more difficult to distinguish. In both cases peripheral connective tissue of portal areas contains the portal triad: small bile ductules (**D**), venule (**V**) branches of the portal vein, and arteriole (**A**) branches of the hepatic artery. (Both X150; H&E)

six portal areas with more fibrous connective tissue, each of which contains three interlobular structures that comprise the **portal triad** (Figures 16–11 and 16–13d):

- A **venule** branch of the portal vein, with blood rich in nutrients but low in O_2,
- An **arteriole** branch of the hepatic artery that supplies O_2,
- One or two small **bile ductules** of cuboidal epithelium, branches of the bile conducting system.

Most of the peripheral portal areas also contain lymphat-ics and nerve fibers and in some species (eg, pigs) extend thin sheets of fibrous connective tissue completely around the lob-ules, making individual lobules easier to distinguish than in humans (Figure 16–12b).

Between all of the anastomosing plates of hepatocytes of a hepatic lobule are important vascular **sinusoids** that emerge from the peripheral branches of the portal vein and hepatic artery and converge on the lobule's **central vein** (Figures 16–11 through 16–13c). The venous and arterial blood mixes in these irregular hepatic sinusoids. The anastomosing sinusoids have thin, discontinuous linings of fenestrated endothelial cells surrounded by sparse basal lamina and reticular fibers. The discontinuities and fenestrations allow plasma to fill a narrow

perisinusoidal space (or **space of Disse**) and directly bathe the many irregular microvilli projecting from the hepatocytes into this space (Figure 16–14). This direct contact between hepatocytes and plasma facilitates most key hepatocyte func-tions that involve uptake and release of nutrients, proteins, and potential toxins.

Two other functionally important cells are found with the sinusoids of hepatic lobules:

- Numerous specialized **stellate macrophages**, usually called **Kupffer cells**, are found within the sinusoid lin-ing (Figure 16–15). These cells recognize and phagocy-tose aged erythrocytes, freeing heme and iron for reuse or storage in ferritin complexes. Kupffer cells are also antigen-presenting cells and remove any bacteria or debris present in the portal blood.
- In the perisinusoidal space are **hepatic stellate cells** (or **Ito cells**) with small lipid droplets that store vitamin A and other fat-soluble vitamins (Figure 16–15b). These mesenchymal cells, which are difficult to see in routine preparations, also produce extracellular matrix (ECM) components (becoming myofibroblasts after liver injury) and cytokines that help regulate Kupffer cell activity.

FIGURE **16–13** Hepatic lobule microvasculature.

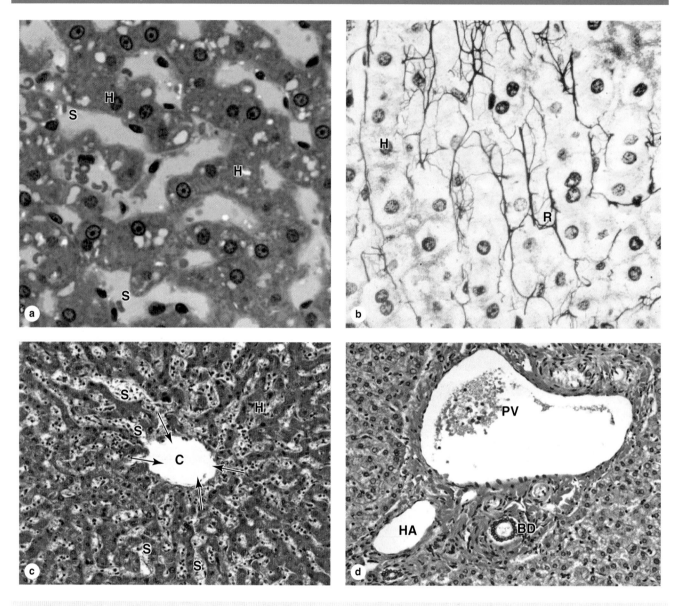

(a) Hepatocytes (**H**) are polygonal epithelial cells that form branching, irregular plates separated by venous sinusoids (**S**). (H&E X400)

(b) Reticulin (collagen type III) fibers (**R**) running along the plates of hepatocytes (**H**), supporting these and the intervening sinusoids. Most connective tissue in the liver is found in the septa and portal tracts. (X400; Silver)

(c) With plates of hepatocytes (**H**) appearing to radiate from it, the central vein (**C**) of the lobule has more collagen than the smaller

sinusoids (**S**) that drain into it from all directions (arrows). (X200; Mallory trichrome)

(d) Peripheral portal areas contain more connective tissue and are the sites of the portal triad: a portal venule (**PV**), an arteriole branching off the hepatic artery (**HA**), and one or two bile ductules (**BD**). (X200; H&E)

The endothelium of the central vein in the middle of each hepatic lobule is supported by a very thin layer of fibrous connective tissue (Figure 16–13c). Central venules from each lobule converge into larger veins, which eventually form two or more large **hepatic veins** that empty into the inferior vena cava.

Blood always flows from the periphery to the center of each hepatic lobule. Consequently, oxygen and metabolites, as well as all other toxic or nontoxic substances absorbed in the intestines, reach the lobule's peripheral cells first and then the more central cells. This direction of blood flow partly explains why the properties and function of the periportal hepatocytes

FIGURE **16–14** Ultrastructure of hepatocytes, perisinusoidal space, and bile canaliculi.

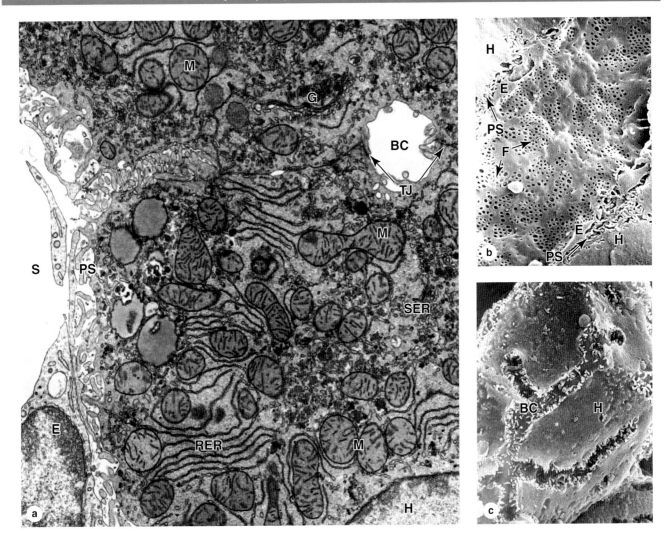

(a) TEM of hepatocytes shows small bile canaliculi (**BC**) between tight junctions (**TJ**) joining two cells. A hepatocyte nucleus (**H**) is in the lower right corner, surrounded by small tubular vesicles of smooth ER (**SER**), much rough ER (**RER**), many mitochondria (**M**), small electron-dense glycogen granules, and Golgi complexes (**G**). Between the hepatocytes and the fenestrated endothelial cell (**E**) of the sinusoid (**S**) is the very small perisinusoidal space (**PS**) almost filled with **microvilli**. (X9500)

(Figure 16–14a, used with permission from Douglas L. Schmucker, Department of Anatomy, University of California, San Francisco, CA.)

(b) SEM of the luminal surface of the endothelium lining a hepatic sinusoid shows grouped fenestrations (**F**). At the border are seen

cut edges of endothelial cells (**E**) in this discontinuous sinusoid and hepatocytes (**H**). Between these two cells is the thin perisinusoidal space (**PS**), into which project **microvilli** from the hepatocytes surface. (X6500)

(Figure 16–14b, used with permission from Eddie Wisse, Electron Microscopy Unit, Department of Pathology, University of Maastricht, Maastricht, the Netherlands.)

(c) SEM of hepatocytes (**H**) broken apart from one another reveals the length of a bile canaliculus (**BC**) along the cell's surface. Such canaliculi run between the cells of the hepatocyte plates in the hepatic lobules and carry bile toward the portal areas where the canaliculi join cuboidal bile ductules. (X8000)

differ from those of the centrolobular cells. Hepatocytes near the portal areas can rely on aerobic metabolism and are often more active in protein synthesis, while the more central cells are exposed to lower concentrations of nutrients and oxygen and are more involved with detoxification and glycogen metabolism.

While the sinusoidal (basolateral) domains of hepatocytes process nutrients and other blood components and

secrete the plasma proteins, the smaller apical surfaces of the hepatocytes form **bile canaliculi** and are involved in exocrine secretion of bile (Figures 16–14 and 16–16). Within the hepatic plates hepatocytes adhere firmly with desmosomes and junctional complexes. The apical surfaces of two adherent hepatocytes are grooved and juxtaposed to form the canaliculus, sealed by tight junctions, into which bile components are secreted (Figure 16–14). These canaliculi are elongated spaces

FIGURE 16–15   **Hepatic sinusoids.**

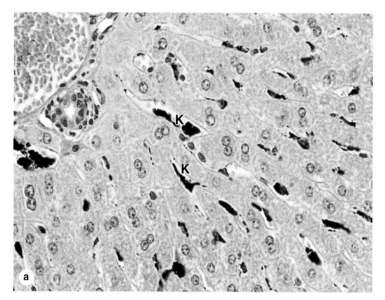

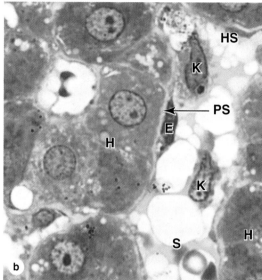

In the endothelial lining of the hepatic sinusoids are numerous specialized stellate macrophages or **Kupffer cells** that detect and phagocytose effete erythrocytes.

(a) Kupffer cells (**K**) are seen as black cells in a liver lobule from a rat injected with particulate India ink. (X200; H&E)

(b) In a plastic section, Kupffer cells (**K**) are seen in the sinusoid (**S**) between two groups of hepatocytes (**H**). They are larger than the

flattened endothelial cells (**E**). Between the endothelium and the hepatocytes is a very thin space called the perisinusoidal space (**PS**) of Disse, in which are located small hepatic stellate cells (**HS**), or Ito cells, that maintain the very sparse ECM of this compartment and also store vitamin A in small lipid droplets. These cells are numerous but are difficult to demonstrate in routine histologic preparations. (X750; PT)

(total length > 1 km) with lumens only 0.5-1µm in diameter with large surface areas due to the many short microvilli from the constituent hepatocytes (Figures 16–14 and 16–16).

The bile canaliculi form a complex anastomosing network of channels through the hepatocyte plates that end near the portal tracts (Figures 16–11b and 16–17). The bile flow therefore progresses in a direction opposite to that of the blood, that is, from the center of the lobule to its periphery. Bile canaliculi are the smallest branches of the biliary tree or bile conducting system. They empty into bile **canals of Hering** (Figure 16–17) composed of cuboidal epithelial cells called **cholangiocytes**. The short bile canals quickly merge in the portal areas with the **bile ductules** lined by cuboidal or columnar cholangiocytes and with a distinct connective tissue sheath. Bile ductules gradually merge, enlarge, and form right and left **hepatic ducts** leaving the liver.

Into the canaliculi hepatocytes continuously secrete bile, a mixture of **bile acids** (organic acids such as cholic acid), **bile salts** (the deprotonated forms of bile acids), electrolytes, fatty acids, phospholipids, cholesterol, and **bilirubin**. Some bile components are synthesized in hepatocyte SER, but most are taken up from the perisinusoidal space; all are quickly secreted

into the bile canaliculi (Figure 16–16). Bile acids/salts have an important function in emulsifying the lipids in the duodenum, promoting their digestion and absorption.

Bilirubin is a pigmented breakdown product of heme that is released from splenic macrophages primarily, but also from Kupffer cells, and carried to hepatocytes bound to albumen. Released into the duodenum with bile, bilirubin is converted by intestinal bacteria into other pigmented products, some of which are absorbed in the intestinal mucosa to be processed and excreted again in the liver or excreted into urine by the kidneys. These bilirubin-related compounds give feces and urine their characteristic colors.

>> **MEDICAL APPLICATION**

The **fibrosis** characteristic of **cirrhosis** produces connective tissue that can fill the perisinusoidal space and interfere with metabolic exchange between the hepatocytes and the sinusoids. Blockage of hepatocyte secretion into the blood can result in **clotting disorders**, **hypoalbuminemia**, and other medical problems.

FIGURE **16–16** Hepatocyte ultrastructure and major functions.

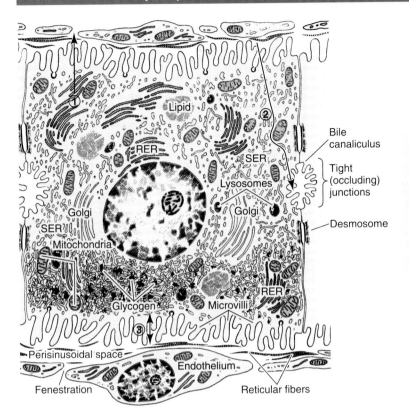

A diagram of hepatocyte cytoplasmic organization, with major functions localized. (**1**) RER is primarily engaged in synthesis of **plasma proteins** for release into the perisinusoidal space. (**2**) Potentially toxic compounds, bilirubin (bound to albumin) and bile acids are taken up from the perisinusoidal space, processed by enzymes in the tubulovesicular system of the SER, and secreted into the **bile canaliculi**. (**3**) Glucose is taken up from the perisinusoidal space and stored in **glycogen granules**, with the process reversed when glucose is needed.

FIGURE **16–17** Bile ductules.

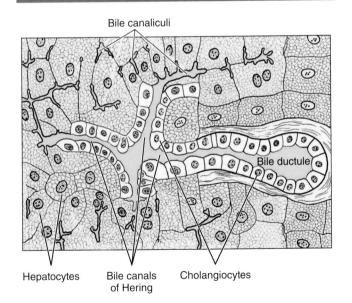

Near the periphery of each hepatic lobule, many bile canaliculi join with the much larger bile canals of Hering, which are lined by cuboidal epithelial cells called **cholangiocytes**. These canals soon join the bile ductules in the portal areas and drain into the biliary tree.

Structure & Function in the Liver

As mentioned previously, hepatocytes are highly versatile cells with diverse functions that are reflected in their structure (Figure 16–16). Abundant **rough ER** is focused on synthesis of plasma proteins and causes cytoplasmic basophilia, which is often more pronounced in hepatocytes near the portal areas (Figure 16–12). Abundant **smooth ER**, distributed more evenly throughout the cytoplasm, contains the enzyme systems for the biotransformation or detoxification of substances in blood, which are then usually excreted with bile. These include enzymes responsible for oxidation, methylation, and conjugation of steroids, barbiturates, antihistamines, anticonvulsants, and other drugs. Under some conditions prolonged presence of drugs can lead to increased amounts of SER in hepatocytes, thus improving the liver's detoxification capacity. Other SER enzymes (glucuronosyl transferases) conjugate bilirubin to glucuronate, rendering it more soluble and facilitating its excretion in bile.

Glycogen granules and small **lipid droplets** in hepatocytes, and very small electron-dense **ferritin complexes** (hemosiderin) primarily in the Kupffer cells, respectively mediate temporary storage of glucose, triglycerides, and iron.

Hepatocyte **peroxisomes** are also abundant and important for oxidation of excess fatty acids, catalase-mediated breakdown of the hydrogen peroxide generated by fatty acid oxidation (by means of catalase activity), and conversion of excess purines to uric acid. Many Golgi complexes are also present, involved in synthesis of both plasma proteins and bile components. The numerous mitochondria provide energy for all these activities (Figure 16–16).

The different categories of hepatocyte functions—including secretion of proteins into blood, the exocrine secretion of bile, and the removal of diverse small compounds from blood—have led to three ways of considering liver lobule structure, which are summarized in Figure 16–18.

■ The classic **hepatic lobule** (Figure 16–18a), with blood flowing past hepatocytes from the portal areas to a central venule, emphasizes the endocrine function of the structure producing factors for uptake by plasma.

■ The concept of **portal lobules** of hepatocytes is more useful when considering the exocrine function of these cells, that is, bile secretion. The portal area has the bile ductule at the center, and bile, moving in the opposite direction as the blood, flows toward it from all the surrounding hepatocytes. The tissue draining bile into

FIGURE 16–18 Concepts of structure-function relationships in liver.

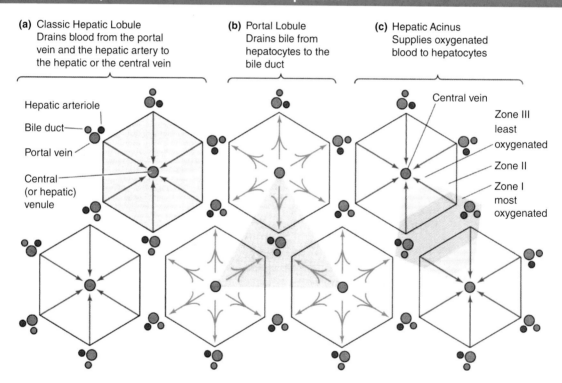

(a) Classic Hepatic Lobule
Drains blood from the portal vein and the hepatic artery to the hepatic or the central vein

(b) Portal Lobule
Drains bile from hepatocytes to the bile duct

(c) Hepatic Acinus
Supplies oxygenated blood to hepatocytes

Hepatic arteriole
Bile duct
Portal vein
Central (or hepatic) venule

Central vein
Zone III least oxygenated
Zone II
Zone I most oxygenated

Studies of liver microanatomy, physiology, and pathology have given rise to three related ways to view the liver's organization, which emphasize different aspects of hepatocyte activity.

(a) The **classic lobule** concept offers a basic understanding of the structure-function relationship in liver organization and emphasizes the endocrine function of hepatocytes as blood flows past them toward the central vein.

(b) The **portal lobule** emphasizes the hepatocytes' exocrine function and the flow of bile from regions of three classic lobules toward the bile duct in the portal triad at the center here. The area drained by each bile duct is roughly triangular.

(c) The **hepatic acinus** concept emphasizes the different oxygen and nutrient contents of blood at different distances along the

sinusoids, with blood from each portal area supplying cells in two or more classic lobules. Major activity of each hepatocyte is determined by its location along the oxygen/nutrient gradient: periportal cells of zone I get the most oxygen and nutrients and show metabolic activity generally different from the pericentral hepatocytes of zone III, exposed to the lowest oxygen and nutrient concentrations. Many pathologic changes in the liver are best understood from the point of view of liver acini.

(Used with permission from Boron WF, Boulpaep EL. Medical Physiology: A Cellular and Molecular Approach. Philadelphia, PA: Saunders Elsevier, 2005.)

each portal area duct is roughly triangular in shape, with the central veins of three classic lobules at its angles (Figure 16–18b).

- The **hepatic acinus**, a third way of viewing liver cells, emphasizes the nature of the blood supply to the hepatocytes and the oxygen gradient from the hepatic artery branch to the central vein. In a liver acinus hepatocytes make up an irregular oval or diamond-shaped area extending from two portal triads to the two closest central veins (Figure 16–18c). Periportal hepatocytes nearest the hepatic arteriole, comprising zone I in the acinus, get the most oxygen and nutrients and can most readily carry out functions requiring oxidative metabolism such as protein synthesis. Hepatocytes in zone III, near the central vein, get the least oxygen and nutrients. They are the preferential sites of glycolysis, lipid formation, and drug biotransformations and are the first hepatocytes to undergo fatty accumulation and ischemic necrosis. In the intervening zone II, hepatocytes have an intermediate range of metabolic functions between those in zones I and III. The major activities in any given hepatocyte result from the cell adapting to the microenvironment produced by the contents of the blood to which it is exposed.

❯❯ MEDICAL APPLICATION

An important function of hepatocyte SER is the conjugation of hydrophobic (water-insoluble), yellow bilirubin by glucuronosyl transferases to form water-soluble, nontoxic bilirubin glucuronide, which is excreted into the bile canaliculi. When bilirubin glucuronide is not formed or excreted properly, various diseases characterized by jaundice can result.

A frequent cause of jaundice in newborns is an underdeveloped state of the hepatocyte SER (neonatal hyperbilirubinemia). A treatment in these cases is exposure to blue light from ordinary fluorescent tubes, which transforms unconjugated bilirubin into a water-soluble photoisomer that can be excreted by the kidneys.

Unlike the salivary glands and pancreas, the liver has a strong capacity for **regeneration** despite its normal slow rate of cell renewal. Hepatocyte loss from the action of toxic substances triggers mitosis in the remaining healthy hepatocytes in a process of **compensatory hyperplasia** that maintains the original tissue mass. Surgical removal of a liver portion produces a similar response in the hepatocytes of the remaining lobe(s). The regenerated liver tissue is usually well organized, with the typical lobular arrangement, and replaces the functions of the destroyed tissue. This regenerative capacity is important clinically because one major liver lobe can sometimes be donated by a living relative for surgical transplant and full liver function restored in both donor and recipient.

Besides proliferation of existing hepatocytes, a role for **liver stem cells** in regeneration has been shown in some experimental models. Such cells, often called **oval cells**, are present among cholangiocytes of the bile canals near portal areas and produce progenitor cells for both hepatocytes and cholangiocytes.

❯❯ MEDICAL APPLICATION

Most malignant tumors of the liver derive from hepatocytes or cholangiocytes of the hepatic ducts. The pathogenesis of **liver carcinoma** is associated with a variety of acquired disorders, such as **chronic viral hepatitis** (B or C) and cirrhosis.

❯ BILIARY TRACT & GALLBLADDER

The bile produced by the hepatocytes flows through the **bile canaliculi, bile ductules**, and **bile ducts**. These structures gradually merge, forming a converging network that ultimately forms the **common hepatic duct** that joins the **cystic duct** from the gallbladder and continues to the duodenum as the **common bile duct** (Figure 16–19).

The hepatic, cystic, and common bile ducts are lined with a mucous membrane having a simple columnar epithelium of cholangiocytes. The lamina propria and submucosa are relatively thin, with mucous glands in some areas of the cystic duct, and surrounded by a thin muscularis. This muscle layer becomes thicker near the duodenum and finally, in the duodenal papilla, forms a sphincter that regulates bile flow into the small bowel.

The gallbladder is a hollow, pear-shaped organ (Figure 16–19) attached to the lower surface of the liver, capable of storing 30-50 mL of bile that is concentrated during storage. The wall of the gallbladder consists of a mucosa composed of simple columnar epithelium and lamina propria, a thin muscularis with bundles of muscle fibers oriented in several directions, and an external adventitia or serosa (Figure 16–20a). The mucosa has numerous folds that are particularly evident when the gallbladder is empty.

The lining epithelial cells of the gallbladder have prominent mitochondria, microvilli, and large intercellular spaces, all indicative of cells actively transporting water, in this case for concentrating bile (Figure 16–20b). The mechanism for this includes activity of Na^+ pumps in the basolateral membranes, followed by passive movement of water from the bile. To move stored bile into the duodenum, contraction of the gallbladder muscularis is induced by cholecystokinin (CCK) released from enteroendocrine cells of the small intestine. Release of CCK is, in turn, stimulated by the presence of ingested fats in the small intestine. Gallbladder removal due to obstruction or chronic inflammation leads to the direct flow of bile from liver to gut, with few major consequences on digestion.

FIGURE 16–19 Biliary tract and gallbladder.

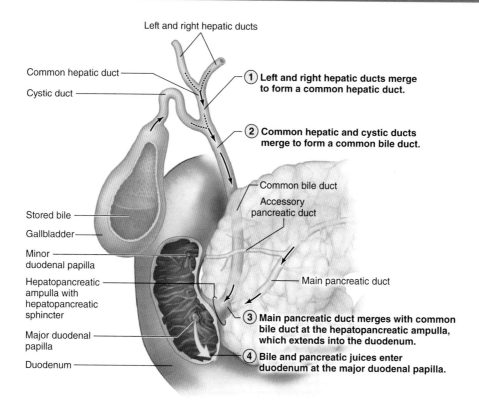

Left and right hepatic ducts

Common hepatic duct

Cystic duct

① Left and right hepatic ducts merge to form a common hepatic duct.

② Common hepatic and cystic ducts merge to form a common bile duct.

Common bile duct

Accessory pancreatic duct

Stored bile

Gallbladder

Minor duodenal papilla

Hepatopancreatic ampulla with hepatopancreatic sphincter

Main pancreatic duct

③ Main pancreatic duct merges with common bile duct at the hepatopancreatic ampulla, which extends into the duodenum.

Major duodenal papilla

④ Bile and pancreatic juices enter duodenum at the major duodenal papilla.

Duodenum

Bile leaves the liver in the left and right hepatic ducts, which merge to form the common hepatic duct, which connects to the cystic duct serving the gallbladder. The latter two ducts merge to form a common bile duct. All these ducts carrying bile are lined by cuboidal or low columnar cells called **cholangiocytes**, similar to those of the small bile ductules in the liver.

The main pancreatic duct merges with the common bile duct at the hepatopancreatic ampulla, which enters the wall of the duodenum at a major papilla (of Vater); the accessory pancreatic duct enters the duodenum at a minor papilla. Bile and pancreatic juices are mixed before release into the duodenal lumen.

❯❯ MEDICAL APPLICATION

Reabsorption of water from bile in the gallbladder is involved in the formation of **gallstones** in the lumen of the gallbladder or biliary ducts, a condition called **cholelithiasis**.

This disorder usually originates with bile that already contains excessive amounts of normal bile components. Supersaturation of cholesterol in bile can lead to the formation of **cholesterol stones**, the most common form. Brown or black **pigment stones** can form when bile contains excessive amounts of unconjugated bilirubin, which can result from chronic hemolysis associated with disorders such as sickle cell anemia. Gallstones can lead to biliary obstruction or more commonly to inflammation in **acute or chronic cholecystitis**.

Organs Associated with the Digestive Tract SUMMARY OF KEY POINTS

Salivary Glands

- **Salivary glands** have secretory units of either **protein-secreting serous cells**, usually organized in round or oval **acini**, or of **mucin-secreting mucous cells** in elongated **tubules**.
- **Parotid glands** have only serous acini; **sublingual glands** are mixed but have primarily mucous tubules, some with **serous demilunes**; **submandibular glands** are also mixed but have mainly serous acini.
- Salivary secretory units are drained by simple cuboidal **intercalated ducts** that merge as simple columnar **striated ducts**, which merge as larger interlobular or **excretory ducts**.

- Cells of **striated ducts** have mitochondria-lined, basolateral membrane folds specialized for electrolyte reabsorption from the secretion; **excretory ducts** are unusual in having **stratified cuboidal or columnar** cells.

Pancreas

- **Pancreatic islets** of endocrine cells are embedded in exocrine **serous acinar tissue**, which comprises most of the pancreas and in which the cells secrete hydrolytic **digestive enzymes** for delivery to the duodenum.

FIGURE **16–20** Gallbladder.

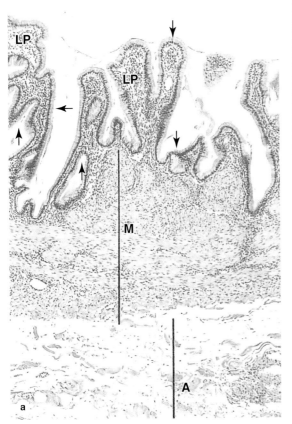
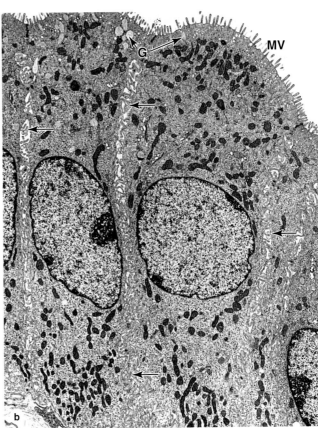

The gallbladder is a saclike structure that stores and concentrates bile, and releases it into the duodenum after a meal.
(a) Its wall consists largely of a highly folded mucosa, with a simple columnar epithelium (arrows) overlying a typical lamina propria (**LP**); a muscularis (**M**) with bundles of muscle fibers oriented in all directions to facilitate emptying of the organ; and an external adventitia (**A**) where it is against the liver and a serosa where it is exposed. (X60; H&E)

(b) TEM of the epithelium shows cells specialized for water uptake across apical microvilli (**MV**) and release into the intercellular spaces (**arrows**) along the folded basolateral cell membranes. From these spaces water is quickly removed by capillaries in the lamina propria. Abundant mitochondria provide the energy for this pumping process. Scattered apical secretory granules (**G**) contain mucus. (X5600)

■ Each pancreatic **acinar cell** is pyramidal, with **secretory (zymogen) granules** in the narrow apical end and Golgi complexes, much rough ER, and a large nucleus at the basal end.

■ **Intercalated ducts** draining pancreatic acini, including their initial **centroacinar cells** that insert into the acinar lumen, **secrete bicarbonate** ions (HCO_3^-) to neutralize chyme entering the duodenum from the stomach.

Liver

■ Liver **hepatocytes** are large epithelial cells with large central nuclei (polyploid and often binucleated), much smooth and rough ER, and many small Golgi complexes.

■ **Hepatocytes** have many functions, including endocrine (**plasma protein** secretion), exocrine (**bile** secretion), glucose storage (**glycogen granules**), and **detoxification** (using SER and peroxisomes).

■ In the liver, **hepatocytes** are organized into irregular plates to form polygonal **hepatic lobules** in which the hepatocyte plates radiate toward a small **central vein**.

■ Each hepatic lobule is surrounded by sparse connective tissue that is more abundant in the **portal areas** at the corners.

■ **Portal areas** or tracts contain a small lymphatic and the **portal triad**: a **portal venule** branch from the portal vein, a **hepatic arteriole** branch of the hepatic artery, and a **bile ductule** branch of the biliary tree.

■ In the lobules the portal venule and hepatic arteriole both branch into irregular **sinusoids** between the hepatic plates where the nutrient-rich and O_2-rich blood mixes, flows past hepatocytes, and drains to the central vein.

■ The endothelium of the hepatic **sinusoids** is **discontinuous and fenestrated**; between it and the hepatocytes is the **perisinusoidal space (of Disse)** where exchange occurs between the hepatocytes and blood plasma.

■ The sinusoidal endothelium includes many specialized **stellate macrophages** or **Kupffer cells**, which recognize and remove effete erythrocytes, **releasing iron and bilirubin** for uptake by hepatocytes.

■ Also present in the perisinusoidal spaces are **hepatic stellate cells (or Ito cells)** containing many small lipid droplets for **storage of vitamin A** and other fat-soluble vitamins.

■ Between adherent hepatocytes in the hepatic plates are grooves called **bile canaliculi**, sealed by tight junctions, into which hepatocytes secrete water and **bile components**, including **bilirubin** and **bile acids**.

■ In each hepatic lobule, all bile canaliculi converge on the **bile canals (of Hering)**, which join the **bile ductules** in the portal areas and eventually all merge to form the left and right **hepatic ducts.**

Biliary Tract and Gallbladder

■ All bile-conducting ducts after the bile canaliculi are lined by simple cuboidal or columnar cells called **cholangiocytes.**

■ The common hepatic duct leads to the cystic duct that carries bile to the **gallbladder** for temporary **bile storage** and concentration.

■ The mucosa of the **gallbladder** has many **folds** with a large surface area, a well-vascularized lamina propria, and a lining of **columnar cholangiocytes** specialized for water uptake from bile.

■ Contraction of the gallbladder **muscularis** sends bile to the duodenum via the **common bile duct** and is induced by **cholecystokinin (CCK)** from enteroendocrine cells in the duodenum when food is present.

Organs Associated with the Digestive Tract ASSESS YOUR KNOWLEDGE

1. In a liver biopsy from a long-time drug user which of the following hepatocyte organelles would be expected to be more extensive than normal?
   a. Rough endoplasmic reticulum
   b. Golgi apparatus
   c. Lysosomes
   d. Peroxisomes
   e. Smooth endoplasmic reticulum

2. Which description is true of pancreatic zymogens?
   a. Are packaged for secretion in the SER
   b. Are synthesized on free ribosomes
   c. Are inactive until they reach the duodenal lumen
   d. Are stored in the basal cytoplasm of acinar cells
   e. Are produced by cuboidal cells lining the pancreatic duct

3. Which process increases in response to parasympathetic stimulation of the salivary glands?
   a. Volume of secretion
   b. Cell division in secretory acini
   c. Mucus content of saliva
   d. Inorganic salts content of saliva
   e. Cell division in interlobular ducts

4. Which feature is unique to the exocrine pancreas?
   a. Insulin-secreting β cells
   b. Centroacinar cells
   c. Predominately serous secretory cells
   d. Striated interlobular ducts
   e. Striated intralobular ducts

5. Which description is true of the bile canaliculi?
   a. Are bordered directly by endothelial cells
   b. Are part of the portal triad
   c. Are surrounded by the hepatic sinusoids
   d. Lumens are entirely sealed by junctional complexes
   e. Normally contain some blood plasma

6. Which description is true of the gallbladder?
   a. Dilutes bile
   b. Absorbs bile
   c. Secretes mucus
   d. Has a thick submucosa
   e. Is covered entirely by serosa

7. Which description is true for the hepatic space of Disse?
   a. Is surrounded by the hepatic sinusoid
   b. Contents flow toward the central vein
   c. Is directly contacted by hepatocytes
   d. Lumen sealed by junctional complexes
   e. Contents empty into canals of Hering lined by cholangiocytes

8. A 50-year-old woman presents to the family medicine clinic. She admits to drinking a six-pack of beer each day with a little more intake on weekends. Laboratory tests show elevated alanine aminotransferase/serum glutamic oxaloacetic transaminase (AST/SGOT). Her sclerae appear jaundiced and her serum bilirubin is 2.5 mg/dL (normal 0.3-1.9 mg/dL). A biopsy shows hepatic fibrosis with significant loss of normal lobular structure. Jaundice is most likely to result when the proper location or orientation of what hepatic structures is disrupted?
   a. Central veins
   b. Spaces of Disse
   c. Kupffer cells
   d. Hepatocytes
   e. Merging sinusoids

9. A 48-year-old woman is referred to an allergy and rheumatology specialist with itching eyes, dryness of the mouth, difficulty swallowing, loss of the sense of taste, hoarseness, fatigue, and swollen parotid glands. She reports increasing joint pain over the past 2 years. She complains of frequent mouth sores. Laboratory tests show a positive antinuclear antibody (ANA) and rheumatoid factor (RF) levels of 70 U/mL (normal < 60 U/mL) by the nephelometric method. A parotid gland biopsy shows inflammatory infiltrates in the interlobular connective tissue with damage to the acinar cells and striated ducts. In this case, resorption of which of the following will be most altered by destruction of those ducts?
   a. Na^+
   b. H_2O
   c. HCO_3^-
   d. Cl^-
   e. Ca^{2+}

10. A young child presents with hepatomegaly and renomegaly, failure to thrive, stunted growth, and hypoglycemia. A deficiency in glucose-6-phosphatase is identified and the diagnosis of von Gierke disease is made. What cellular structures would be expected to accumulate in hepatocytes during progression of this disorder?
    a. Chylomicrons
    b. Glycogen granules
    c. Mitochondria
    d. Zymogen granules
    e. Ribosomes

Answers: 1e, 2c, 3a, 4b, 5d, 6c, 7c, 8d, 9a, 10b

The respiratory system provides for exchange of O_2 and CO_2 to and from the blood. Respiratory organs include the **lungs** and a branching system of bronchial tubes that link the sites of gas exchange with the external environment. Air is moved through the lungs by a ventilating mechanism, consisting of the thoracic cage, intercostal muscles, diaphragm, and elastic components of the lung tissue. The system can be divided anatomically into the upper and lower respiratory tracts (Figure 17–1). Functionally the system has two components:

- The **conducting portion**, which consists of the nasal cavities, pharynx, larynx, trachea, bronchi (Gr. *bronchos*, windpipe), bronchioles, and terminal bronchioles
- The **respiratory portion**, where the system's main function of gas exchange occurs, consisting of respiratory bronchioles, alveolar ducts, and alveoli.

Alveoli, the cellular sites of the exchange of O_2 and CO_2 between inspired air and blood, are small, air-filled, saclike structures that make up most of the lung structure.

The conducting portion cleans and humidifies inspired air and provides conduits for air movement to and from alveoli. To ensure an uninterrupted supply of air, a combination of cartilage, collagen and elastic fibers, and smooth muscle provides the conducting portion with rigid structural support and the necessary flexibility and extensibility.

› NASAL CAVITIES

The left and right nasal cavities each have two components: the external, dilated **vestibule** and the internal **nasal cavity**. Skin of the nose enters the **nares** (nostrils) partway into the vestibule and includes sweat glands, sebaceous glands, and coarse, moist **vibrissae** (hairs) that filter out particulate material from inspired air. Within the vestibule, the epithelium loses its keratinized nature and undergoes a transition to typical pseudostratified columnar epithelium which also lines the nasal cavities.

The nasal cavities lie within the skull as two cavernous chambers separated by the osseous **nasal septum**. Extending from each lateral wall are three bony shelflike projections (Figure 17–1) called **conchae**, or turbinate bones. The mucosa covering these and other parts of the nasal cavity walls has a lamina propria with important roles in conditioning inhaled air. A complex vasculature with loops of capillaries near the epithelial surface carries blood in a general direction counter to the flow of inspired air and releases heat to warm that air while it is humidified by water secreted from small seromucous glands. The thin mucus layer produced by these glands and goblet cells serves to trap particulate and gaseous air impurities that are then removed. The secretions also contain immunoglobulin A (IgA) from plasma cells in the lamina propria.

Epithelium on the middle and inferior conchae is respiratory epithelium; the roof of the nasal cavities and the superior conchae are covered with specialized olfactory epithelium.

Respiratory Epithelium

Most of the nasal cavities and conducting portion of the system is lined with mucosa having ciliated pseudostratified columnar epithelium, described with epithelia in Chapter 4 and

FIGURE **17-1** Anatomy of the respiratory system.

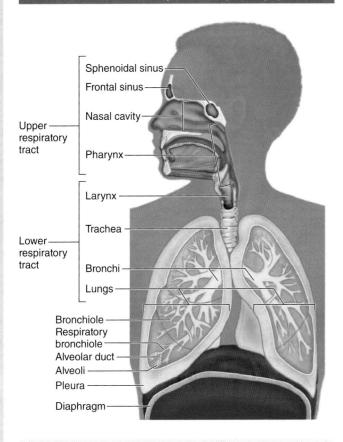

Sphenoidal sinus
Frontal sinus
Nasal cavity
Upper respiratory tract
Pharynx
Larynx
Trachea
Lower respiratory tract
Bronchi
Lungs
Bronchiole
Respiratory bronchiole
Alveolar duct
Alveoli
Pleura
Diaphragm

Anatomically, the respiratory tract has upper and lower parts. Histologically and functionally, the respiratory system has a **conducting portion**, which consists of all the components that condition air and bring it into the lungs, and a **respiratory portion**, where gas exchange actually occurs, consisting of respiratory bronchioles, alveolar ducts, and alveoli in the lungs. Portions of two sets of paranasal sinuses are also shown here.

commonly known as **respiratory epithelium** (Figure 17–2). This epithelium has five major cell types, all of which contact an unusually thick basement membrane:

- **Ciliated columnar cells** are the most abundant, each with 250-300 cilia on its apical surface (Figure 17–2).
- **Goblet cells** are also numerous and predominate in some areas (Figure 17–2), with basal nuclei and apical domains filled with granules of mucin glycoproteins.
- **Brush cells** are a much less numerous, columnar cell type, in which a small apical surface bears sparse, blunt microvilli (Figure 17–2c). Brush cells are **chemosensory receptors** resembling gustatory cells, with similar signal transduction components and synaptic contact with afferent nerve endings on their basal surfaces.
- **Small granule cells** (or Kulchitsky cells) are difficult to distinguish in routine preparations, but possess numerous dense core granules 100-300 nm in diameter. Like enteroendocrine cells of the gut, they are part of the

diffuse neuroendocrine system (DNES; see Chapter 20). Like brush cells, they represent only about 3% of the cells in respiratory epithelium.
- **Basal cells** are mitotically active stem and progenitor cells that give rise to the other epithelial cell types.

❯❯ MEDICAL APPLICATION

The chronic presence or accumulation of toxins that occur with heavy cigarette smoking or industrial air pollution affects the respiratory epithelium beginning in the nasal cavities. Immobilization of the cilia causes failure to clear mucus containing filtered material and exacerbates the problem, leading eventually to the likelihood of **squamous metaplasia** of the epithelium. A change from pseudostratified ciliated columnar to stratified squamous epithelium can occur, particularly in the mucosa of bronchi. This can produce precancerous **cell dysplasia** in this tissue.

Olfactory Epithelium

The olfactory chemoreceptors for the sense of smell are located in the **olfactory epithelium**, a specialized region of the mucous membrane covering the superior conchae at the roof of the nasal cavity. In adult humans, it is about 10 cm² in area and up to 100 μm in thickness. This thick, pseudostratified columnar epithelium has three major cell types (Figure 17–3):

- **Olfactory neurons** are bipolar neurons present throughout this epithelium. Their nuclei form an irregular row near the middle of this thick epithelium. The apical (luminal) pole of each olfactory cell is its dendrite end and has a knoblike swelling with about a dozen basal bodies, from which long cilia project into the overlying aqueous layer. These cilia have nonmotile axonemes and collectively provide a large surface for transmembrane chemoreceptors. The receptors respond to odoriferous substances by generating an action potential along the axons extending from the basal ends of these neurons. The axons leave the epithelium and unite in the lamina propria as very small nerves that then pass to the brain through foramina in the **cribriform plate** of the ethmoid bone (Figure 17–3). There they form the olfactory nerve (cranial nerve I) and eventually synapse with neurons in the olfactory bulb of the brain.
- **Supporting cells** are columnar, with narrow bases and broad, cylindrical apexes containing the nuclei and extending microvilli into the fluid layer. Well-developed junctional complexes bind the supporting cells to the olfactory cells. The supportive role of these cells is not well understood, but they express abundant ion channels that help maintain a microenvironment conducive to olfactory function and survival.
- **Basal cells** are small, spherical or cone-shaped cells near the basal lamina. These are the stem cells for the other two types, replacing the olfactory neurons every 2-3 months and support cells less frequently.

FIGURE **17–2** Respiratory epithelium.

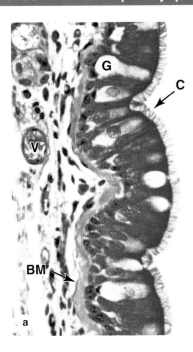
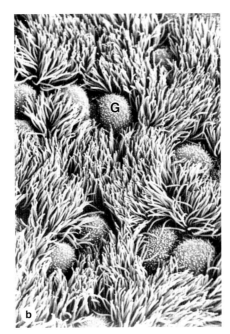
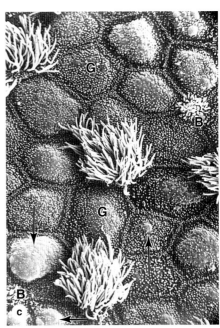

Respiratory epithelium is the classic example of **pseudostratified ciliated columnar epithelium**.

(a) Details of its structure vary in different regions of the respiratory tract, but it usually rests on a very thick basement membrane (**BM**) and has several cell types, some columnar, some basal, and all contacting the basement membrane. Ciliated columnar cells are most abundant, with hundreds of long robust cilia (**C**) on each of their bulging apical ends that provide a lush cover of cilia on the luminal surface. Most of the small rounded cells at the basement membrane are stem cells and their differentiating progeny, which together make up about 30% of the epithelium. Mucus-secreting goblet cells (**G**) and intraepithelial lymphocytes and dendritic cells are also present in respiratory epithelium. The lamina propria is well-vascularized (**V**). (X400; Mallory trichrome)

(b) Scanning electron microscopy (SEM) shows the luminal surface of goblet cells (**G**) among the numerous ciliated cells.

(c) SEM of another region shows that goblet cells (**G**) predominate in some areas, with subsurface accumulations of mucus evident in some (**arrows**). The film of mucus traps most airborne dust particles and microorganisms, and the ciliary movements continuously propel the sheet of mucus toward the pharynx for elimination. Other columnar cells, representing only about 3% of the cells in respiratory epithelium, are brush cells (**B**) with small apical surfaces bearing a tuft of short, blunt microvilli. (Both X3000)

(*Figure 17–2b, c, used with permission from Andrews, P. M. (1974), A scanning electron microscopic study of the extrapulmonary respiratory tract. Am. J. Anat., 139: 399–423. doi: 10.1002/aja.1001390308.*)

The lamina propria of the olfactory epithelium possesses large serous glands, the **olfactory glands** (of Bowman), which produce a constant flow of fluid surrounding the olfactory cilia and facilitating the access of new odoriferous substances.

❯❯ MEDICAL APPLICATION

The loss or reduction of the ability to smell, **anosmia** or **hyposmia**, respectively, can be caused by traumatic damage to the ethmoid bone that severs olfactory nerve axons or by damage to the olfactory epithelium caused by intranasal drug use.

The olfactory neurons are the best-known neurons to be replaced regularly because of **regenerative activity** of the epithelial stem cells from which they arise. For this reason, loss of the sense of smell due to toxic fumes or physical injury to the olfactory mucosa itself is usually temporary.

Paranasal Sinuses

The **paranasal sinuses** are bilateral cavities in the frontal, maxillary, ethmoid, and sphenoid bones of the skull (Figure 17–1). They are lined with a thinner respiratory epithelium having fewer goblet cells. The lamina propria contains only a few small glands and is continuous with the underlying periosteum. The paranasal sinuses communicate with the nasal cavities through small openings; mucus produced there is moved into the nasal passages by the activity of the ciliated epithelial cells.

❯❯ MEDICAL APPLICATION

Sinusitis is an inflammatory process of the sinuses that may persist for long periods of time, mainly because of obstruction of drainage orifices. Chronic sinusitis and **bronchitis** are components of **primary ciliary dyskinesia**, or Kartagener syndrome, an inherited genetic disorder characterized by defective ciliary action.

FIGURE 17–3 Olfactory mucosa.

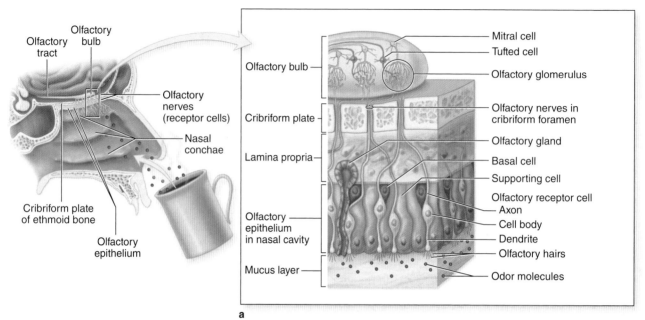

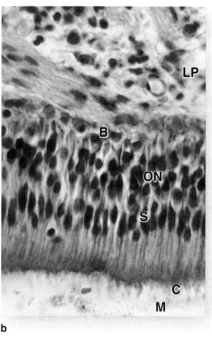

(a) The olfactory mucosa covers the superior conchae bilaterally and sends axons from throughout its entire 10 cm² area to the brain via small openings in the cribriform plate of the ethmoid bone. It is a pseudostratified epithelium, containing basal stem cells and columnar support cells in addition to the bipolar olfactory neurons. The dendrites of these neurons are at the luminal ends and have cilia specialized with many membrane receptors for odor molecules. Binding such ligands causes depolarization which passes along basal axons to the olfactory bulb of the brain.

(b) Only a thin basement membrane separates the olfactory basal cells (**B**) from the underlying lamina propria (**LP**). Nuclei of the bipolar olfactory neurons (**ON**) lie in the middle of the pseudostratified olfactory epithelium, with a zone of supporting cell (**S**) nuclei above it. At the apical end of the cells are the nonmotile cilia (**C**), or olfactory hairs, and a layer of mucus (**M**). (X200; H&E)

› PHARYNX

The nasal cavities open posteriorly into the nasopharynx, the first part of the **pharynx**. The nasopharynx is continuous caudally with the oropharynx (throat), the posterior part of the oral cavity leading to the larynx and esophagus (Figure 17–1). Unlike the stratified squamous epithelium of the oropharynx, the nasopharynx lining is respiratory epithelium, and its mucosa contains the medial pharyngeal tonsil and the openings of the two auditory tubes which connect to each middle ear cavity.

› LARYNX

The **larynx** is a short (4 cm × 4 cm) passage for air between the pharynx and the trachea (Figure 17–1). Its rigid wall is reinforced by hyaline cartilage (in the thyroid, cricoid, and the inferior arytenoid cartilages) and smaller elastic cartilages (in the epiglottis, cuneiform, corniculate, and the superior arytenoid cartilages), all of which are connected by ligaments. In addition to maintaining an open airway, movements of these cartilages by skeletal muscles participate in sound production during phonation.

The **epiglottis**, a flattened structure projecting from the upper rim of the larynx, serves to prevent swallowed food or fluid from entering that passage. Its upper, or lingual, surface has stratified squamous epithelium; at variable points on its laryngeal surface this epithelium undergoes a transition to ciliated pseudostratified columnar (respiratory) epithelium. Mixed mucous and serous glands are found in the lamina propria beneath the epithelium.

Below the epiglottis and vestibule of the larynx, the mucosa projects bilaterally into the lumen with two pairs of folds separated by a narrow space or ventricle (Figure 17–4). The upper pair, the immovable **vestibular folds**, is partly covered with typical respiratory epithelium overlying numerous seromucous glands and occasional lymphoid nodules. The lower pair of folds, the **vocal folds** (or cords), have features important for phonation or sound production:

- Each is covered with nonkeratinized stratified squamous epithelium that protects the mucosa from abrasion and desiccation from rapid air movement.

FIGURE **17–4** Larynx.

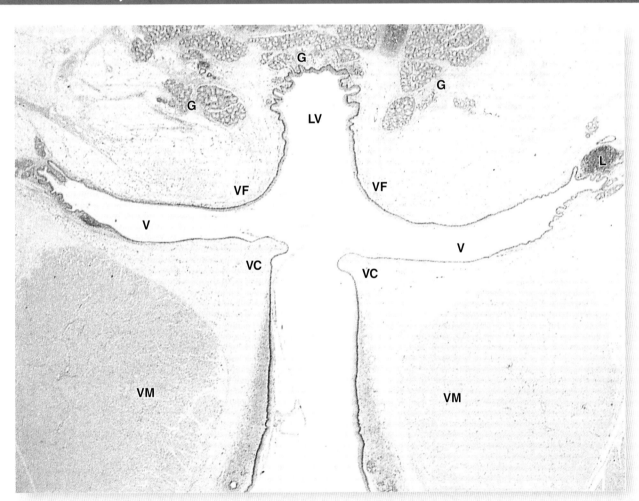

The larynx is a short air passage between the pharynx and trachea. Its wall contains skeletal muscles and pieces of cartilage, all of which make the larynx specialized for sound production, or phonation. This low-power micrograph shows the laryngeal vestibule (**LV**), which is surrounded by seromucous glands (**G**). The lateral walls of this region bulge as a pair of vestibular folds (**VF**). These also contain seromucous glands and areolar tissue with MALT, often with lymphoid nodules (**L**) and are largely covered by respiratory epithelium, with regions near the epiglottis having stratified squamous epithelium.

Below each large vestibular fold is a narrow space or ventricle (**V**), below which is another pair of lateral folds, the vocal folds or cords (**VC**). These are covered by stratified squamous epithelium and project more sharply into the lumen. Each contains a large striated vocalis muscle (**VM**) and nearer the surface a small ligament, which is cut transversely and therefore difficult to see here. Variable tension of these ligaments caused by the muscles produces different sounds as air is expelled across the vocal cords. All the structures and spaces above these folds add resonance to the sounds, assisting phonation. (X15; H&E)

- A dense regular bundle of elastic connective tissue, the vocal ligament, supports the free edge of each vocal fold.
- Deep to the mucosa are large bundles of striated fibers comprising the **vocalis muscle** which allow each vocal fold to be moved.

During phonation muscles of the larynx draw the paired vocal folds together (adduction), which narrows the opening between them, the **rima glottidis**, and air expelled from the lungs causes the adducted vocal folds (cords) to vibrate and produce sound.

The pitch and other qualities of the sound are altered by changing the tension on the vocal folds, the width of the rima glottidis, the volume of air expelled, etc. The vestibular folds and ventricles, along with other structures and spaces higher in the respiratory tract, contribute to the resonance of sound produced in the larynx. Speech is produced when sounds made in the larynx are modified by movements of the pharynx, tongue, and lips. The larynx is larger and the vocal folds longer in males than in females after puberty, causing men's voices typically to have a deeper range than women's voices.

> **》》 MEDICAL APPLICATION**
>
> Inflammation of the larynx, or **laryngitis**, is typically due to viral infection and is usually accompanied by **edema** or swelling of the organ's lamina propria. This changes the shape of the vocal folds or other parts of the larynx, producing **hoarseness** or complete **loss of voice**. Croup is a similar syndrome in young children in which edema of the laryngeal mucosa is accompanied by both hoarseness and **coughs** that typically are loud and harsh. Benign reactive polyps, called **singer's nodules**, are frequent in the stratified squamous epithelium of the true vocal cords, affecting the voice.

》 TRACHEA

The **trachea**, 10-12 cm long in adults, is lined with typical respiratory mucosa in which the lamina propria contains numerous seromucous glands producing watery mucus (Figure 17–5). A series with about a dozen C-shaped rings of hyaline cartilage between the submucosa and adventitia reinforces the wall and keeps the tracheal lumen open (Figure 17–6). The open ends of the cartilage rings are on the posterior surface, against the esophagus, and are bridged by a bundle of smooth muscle called the **trachealis muscle** and a sheet of fibroelastic tissue attached to the perichondrium.

The trachealis muscle relaxes during swallowing to facilitate the passage of food by allowing the esophagus to bulge into the lumen of the trachea, with the elastic layer preventing excessive distention of the lumen. The muscle strongly contracts in the cough reflex to narrow the tracheal lumen and provide for increased velocity of the expelled air and better loosening of material in the air passage.

FIGURE **17–5** Tracheal wall.

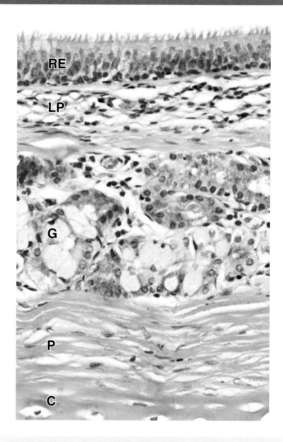

The trachea is lined by typical respiratory epithelium (**RE**) underlain by connective tissue of the lamina propria (**LP**) and seromucous glands (**G**) in the lamina propria and submucosa. Adjacent to the submucosa are the C-shaped rings of hyaline cartilage (**C**) covered by perichondrium (**P**). (X50; H&E)

Major features of all upper respiratory tract structures are summarized in Table 17–1.

> **》》 MEDICAL APPLICATION**
>
> **Coughing** is a reflex action produced most often by viral infection or other irritation of the trachea or other region of the respiratory tract. A persistent **dry cough**, in which no mucus (phlegm) is produced, can be treated by **cough suppressants** that act on the brain stem and vagus nerve, while **productive coughs** are often treated with **expectorants** that help loosen mucus covering the respiratory mucosa.

》 BRONCHIAL TREE & LUNG

The trachea divides into two **primary bronchi** that enter each lung at the hilum, along with arteries, veins, and lymphatic vessels. After entering the lungs, the primary bronchi course

FIGURE **17–6** Bronchial tree.

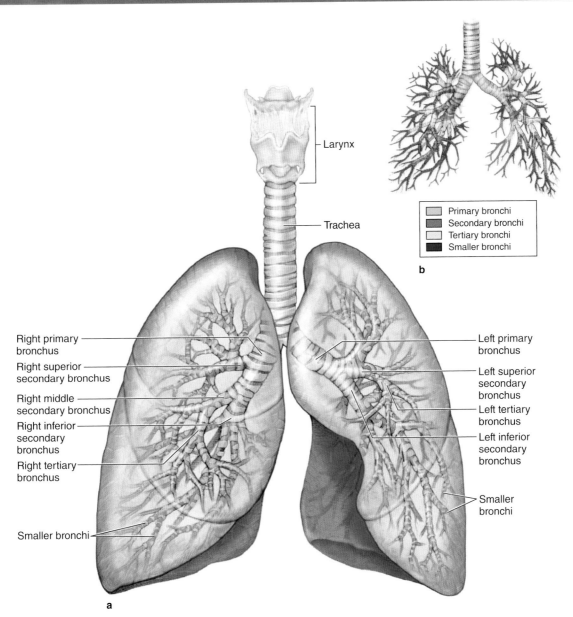

Larynx

Trachea

Primary bronchi
Secondary bronchi
Tertiary bronchi
Smaller bronchi

b

Right primary
bronchus

Right superior
secondary bronchus

Right middle
secondary bronchus

Right inferior
secondary
bronchus

Right tertiary
bronchus

Smaller bronchi

Left primary
bronchus

Left superior
secondary
bronchus

Left tertiary
bronchus

Left inferior
secondary
bronchus

Smaller
bronchi

a

The **trachea** bifurcates as right and left **primary bronchi** that enter the hilum on the posterior side of each **lung** along with the pulmonary vessels, lymphatics, and nerves.

(a) Within each lung, bronchi subdivide further to form the **bronchial tree**, the last component of the air conducting system.

(b) The small diagram shows the color-coded major branches of the bronchial tree.

downward and outward, giving rise to three **secondary (lobar) bronchi** in the right lung and two in the left lung (Figure 17–6), each of which supplies a pulmonary lobe. These lobar bronchi again divide, forming **tertiary (segmental) bronchi**. Each of the tertiary bronchi, together with the smaller branches it supplies, constitutes a **bronchopulmonary segment**—approximately 10%-12% of each lung with its own connective tissue capsule and blood supply. The existence of such lung segments facilitates the specific surgical resection of diseased lung tissue without affecting nearby healthy tissue.

The tertiary bronchi give rise to smaller and smaller bronchi, whose terminal branches are called **bronchioles**. Each bronchiole enters a pulmonary lobule, where it branches to form five to seven **terminal bronchioles**. The pulmonary lobules are each pyramid-shaped, with the apex aimed at the pulmonary hilum, and each is delineated by a thin layer of connective tissue, which in adults is frequently incomplete.

TABLE 17–1	Histologic features of the upper respiratory tract, larynx, and trachea.			
Region	Epithelium	Glands	Musculoskeletal Support	Other Features and Major Functions
Vestibules of nasal cavities	Stratified squamous, keratinized to nonkeratinized	Sebaceous and sweat glands	Hyaline cartilage	Vibrissae (stiff hairs) and moisture both filter and humidify air
Most areas of nasal cavities	Respiratory	Seromucous glands	Bone and hyaline cartilage	Rich vasculature and glands warm, humidify, and clean air
Superior areas of nasal cavities	Olfactory, with bipolar neurons	Serous (Bowman) glands	Bone (ethmoid)	Solubilize and detect odorant molecules in air
Nasopharynx and posterior oropharynx	Respiratory and stratified squamous	Seromucous glands	Bone and skeletal muscle	Conduct air to larynx; pharyngeal and palatine tonsils
Larynx	Respiratory and stratified squamous	Mucous glands, smaller seromucous glands	Elastic and hyaline cartilage, ligaments, skeletal muscle	Site for phonation; epiglottis closes while swallowing
Trachea	Respiratory	Mainly mucous glands, some serous or mixed glands	C-shaped rings of hyaline cartilage, with smooth (trachealis) muscle in posterior opening of each	Conduct air to primary bronchi entering lungs; some MALT

Moving through the smaller bronchi and bronchioles toward the respiratory portion, the histological organization of both the epithelium and the underlying lamina propria gradually becomes more simplified (Table 17–2).

Bronchi

Each primary bronchus branches repeatedly, with each branch becoming progressively smaller until it reaches a diameter of 1-2 mm. The mucosa of the larger bronchi is structurally similar to the tracheal mucosa except for the organization of cartilage and smooth muscle (Figure 17–7). In the primary bronchi most cartilage rings completely encircle the lumen, but as the bronchial diameter decreases, cartilage rings are gradually replaced with smaller isolated plates of hyaline cartilage. Small mucous and serous glands are abundant, with ducts opening into the bronchial lumen. The lamina propria also contains crisscrossing bundles of spirally arranged smooth muscle and elastic fibers (Figures 17–7 and 17–8), which become more prominent in the smaller bronchial branches. Contraction of this muscle layer is responsible for the folded appearance of the bronchial mucosa observed histologically in cross sections.

Numerous lymphocytes are found both within the lamina propria and among the epithelial cells. Lymphatic nodules are

TABLE 17–2	Features of airways within the lungs.		
Region of Airway	Epithelium	Muscle and Skeletal Support	Other Features and Major Functions
Bronchi	Respiratory	Prominent spiral bands of smooth muscle; irregular hyaline cartilage plates	Repeated branching; conduct air deeper into lungs
Bronchioles	Simple ciliated cuboidal to columnar, with exocrine club cells	Prominent circular layer of smooth muscle; no cartilage	Conduct air; important in bronchoconstriction and bronchodilation
Terminal bronchioles	Simple cuboidal, ciliated cells and club cells	Thin, incomplete circular layer of smooth muscle; no cartilage	Conduct air to respiratory portions of lungs; exocrine club cells with several protective and surfactant functions
Respiratory bronchioles	Simple cuboidal, ciliated cells and club cells, with scattered alveoli	Fewer smooth muscle fibers, mostly around alveolar openings	Conduct air deeper, with some gas exchange, and protective and surfactant functions of club cells
Alveolar ducts and sacs	Simple cuboidal between many alveoli	Bands of smooth muscle around alveolar openings	Conduct air, with much gas exchange
Alveoli	Types I and II alveolar cells (pneumocytes)	None (but with network of elastic and reticular fibers)	Sites of all gas exchange; surfactant from type II pneumocytes; dust cells

FIGURE **17–7** Tertiary (segmental) bronchus.

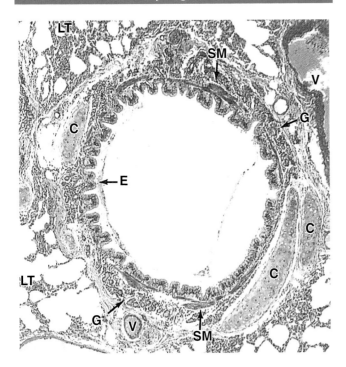

In a cross section of a large bronchus, the lining of respiratory epithelium (**E**) and the mucosa are folded due to contraction of its smooth muscle (**SM**). At this stage in the bronchial tree, the wall is also surrounded by many pieces of hyaline cartilage (**C**) and contains many seromucous glands (**G**) in the submucosa which drain into the lumen. In the connective tissue surrounding the bronchi can be seen arteries and veins (**V**), which are also branching as smaller and smaller vessels in the approach to the respiratory bronchioles. All bronchi are surrounded by distinctive lung tissue (**LT**) showing the many empty spaces of pulmonary alveoli. (X56; H&E)

present, especially at the branching points of the bronchial tree. Like the smooth muscle and elastic fibers, mucosa-associated lymphoid tissue (MALT) also becomes relatively more abundant as bronchi become smaller and the cartilage and other connective tissue are reduced.

Bronchioles

Bronchioles are the intralobular airways with diameters of 1 mm or less, formed after about the tenth generation of branching; they lack both mucosal glands and cartilage, although dense connective tissue is associated with the smooth muscle (Figure 17–9). In the larger bronchioles, the epithelium is still ciliated pseudostratified columnar, but this decreases in height and complexity to become ciliated simple columnar or simple cuboidal epithelium in the smallest **terminal bronchioles**, which are the last parts of the air conducting system. The ciliated epithelial lining of bronchioles begins the **mucociliary**

apparatus or escalator, important in clearing debris and mucus by moving it upward along the bronchial tree and trachea.

The cuboidal epithelium of terminal bronchioles consists largely of **club cells** or bronchiolar exocrine cells (previously called Clara cells), with nonciliated, dome-shaped apical ends containing secretory granules (Figure 17–10). These exocrine cells have various functions, including the following:

- **Secretion of surfactant** lipoproteins and mucins in the fluid layer on the epithelial surface,
- **Detoxification** of inhaled xenobiotic compounds by enzymes of the SER,
- Secretion of **antimicrobial peptides** and cytokines for local immune defense.

Also present in the terminal bronchiole epithelium are chemosensory **brush cells** and **DNES small granule cells** like those of the epithelium higher in the respiratory system. A small population of stem cells provides for replacement of the other bronchiolar cell types.

The bronchiolar lamina propria still contains elastic fibers and smooth muscle, producing folds in the mucosa. Muscular contraction in both the bronchi and the bronchioles is controlled primarily by nerves of the autonomic nervous system.

FIGURE **17–8** Bronchial wall.

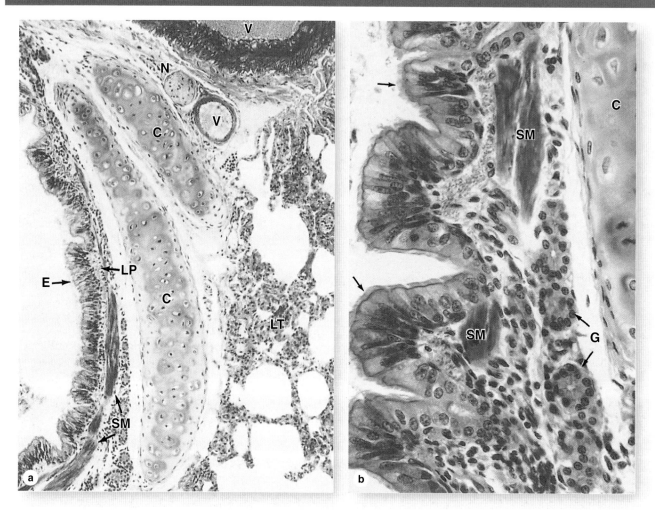

(a) The epithelial lining (**E**) of bronchi is mainly pseudostratified ciliated columnar cells with a few goblet cells. The lamina propria (**LP**) contains the distinct layer of smooth muscle (**SM**) surrounding the entire bronchus. The submucosa is the site of the supporting cartilage (**C**) and the adventitia includes blood vessels (**V**) and nerves (**N**). Lung tissue (**LT**) directly surrounds the adventitia of bronchi. (X140; H&E)

(b) In the smaller bronchi the epithelium is primarily of columnar cells with cilia (arrows), with fewer goblet cells. The lamina propria has both smooth muscle (**SM**) and small serous glands (**G**) near cartilage (**C**). (X400; H&E)

Respiratory Bronchioles

Each terminal bronchiole subdivides into two or more **respiratory bronchioles** that include saclike **alveoli** and represent, therefore, the first-part of this system's respiratory region (Figure 17–11). The respiratory bronchiolar mucosa resembles that of the terminal bronchioles, except for having a few openings to alveoli where gas exchange occurs. Smooth muscle and elastic connective tissue comprise the lamina propria and the epithelium consists of club cells, with simple squamous cells at the alveolar openings and extending into the alveolus. Proceeding distally along the respiratory bronchioles, alveoli become more numerous and closer together.

>> **MEDICAL APPLICATION**

Obstruction of the air supply in bronchi due to excess mucus or to aspirated material can lead to collapse of pulmonary lobules as circulating blood absorbs gases from the affected alveoli. This condition, called **atelectasis**, is normally reversible when the blockage is relieved but, if persistent, can cause fibrosis and loss of respiratory function.

Alveolar Ducts

Distal ends of respiratory bronchioles branch into tubes called **alveolar ducts** that are completely lined by the openings of alveoli (Figures 17–11 and 17–12). Both the alveolar ducts

FIGURE **17–9** Bronchioles.

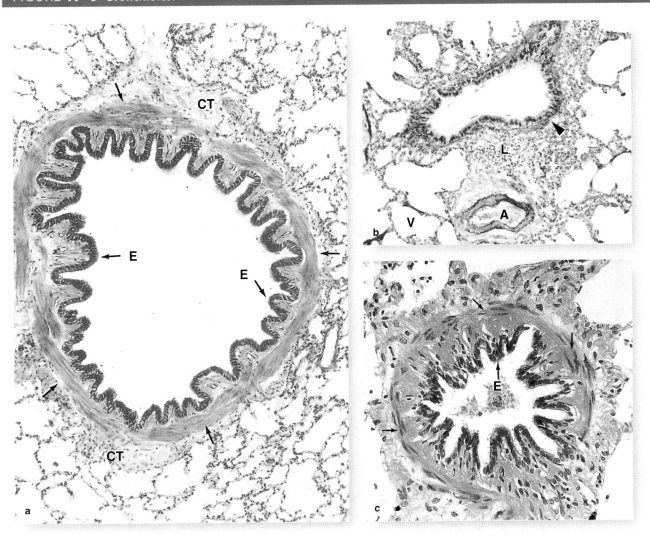

The smallest branches of the bronchial tree are the bronchioles, which lack supporting cartilage and glands.

(a) A large bronchiole has the characteristically folded respiratory epithelium (**E**) and prominent smooth muscle (arrows), but it is supported only by fibrous connective tissue (**CT**). (X140; H&E)

(b) Staining for elastic fibers reveals the high elastic content of the smooth muscle (**arrowhead**) associated with the muscle of a smaller bronchiole in which the epithelium is simple columnar but still ciliated. Darkly stained elastic fibers are also present in the tunica media of a large arteriole (**A**) nearby and to a lesser extent in the accompanying venule (**V**). The connective tissue includes many lymphocytes (**L**) of diffuse MALT and lymphoid nodules. (X180; Aldehyde fuchsin)

(c) In very small bronchioles the epithelium (**E**) is reduced to simple cuboidal cells with cilia. Several layers of smooth muscle cells (**arrows**) comprise a high proportion of the wall. (X300; H&E)

and the alveoli themselves are lined with extremely attenuated squamous cells. In the thin lamina propria, a strand of smooth muscle cells surrounds each alveolar opening and a network of elastic and collagen fibers supports both the duct and its alveoli.

Larger clusters of alveoli called **alveolar sacs** form the ends of alveolar ducts distally and occur occasionally along their length (Figures 17–11 and 17–12). The lamina propria is now extremely thin, consisting essentially of a web of **elastic** and **reticular fibers** that encircles the alveolar openings and closely surrounds each alveolus. Prominent in this sparse connective tissue, a network of **capillaries** also surrounds each alveolus.

FIGURE **17–10**   Terminal bronchiole and exocrine bronchiolar cells.

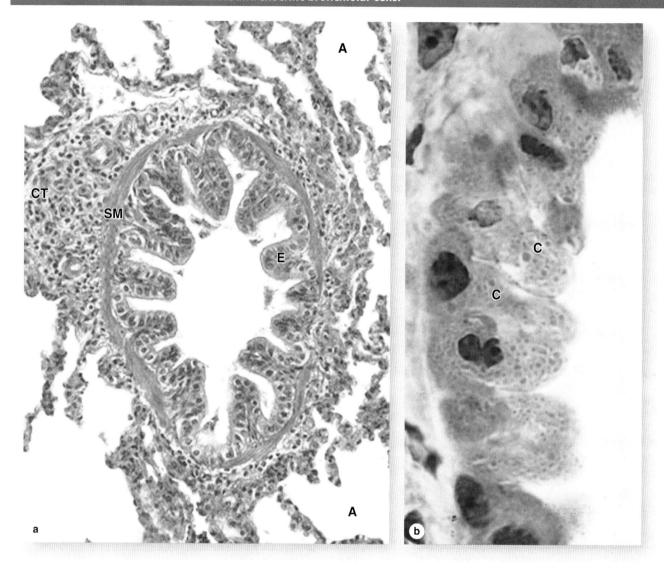

The last parts of the air conducting system before the sites of gas exchange appear are called the **terminal bronchioles**.

(a) A terminal bronchiole has a mucosa with nonciliated cuboidal or low columnar epithelium (**E**), surrounded by only one or two layers of smooth muscle (**SM**) embedded in connective tissue (**CT**). Alveoli (**A**) are seen in the surrounding lung tissue. (X300; PT)

(b) Most of the epithelium consists of exocrine **club cells** (**C**) with bulging domes of apical cytoplasm contain granules, as shown here in a plastic section. These cells have several important

functions. They secrete components of surfactant which reduces surface tension and helps prevent collapse of the bronchioles. The P450 enzyme system of their smooth ER detoxifies potentially harmful compounds in air. In other defensive functions the cells also produce the secretory component for the transfer of IgA into the bronchiolar lumen; lysozyme and other enzymes active against bacteria and viruses; and several cytokines that regulate local inflammatory responses. Also included among the cuboidal cells are stem cells that give rise to all of the cells within the bronchiolar epithelium. (X500; PT)

❯❯ MEDICAL APPLICATION

Diffuse alveolar damage or **adult respiratory distress syndrome** can be produced by various types of injuries to the alveolar epithelial and the capillary endothelial cells. Common causes of such injuries include viral and bacterial

respiratory tract infections; inhalation of **toxic gases**, chemicals, or air with **excessive oxygen**; and **fat embolism syndrome**, in which adipocytes enter the blood during surgery, circulate, and later block the capillary beds. With removal of the initiating factors, normal alveolar wall components can often be restored and at least partial function restored.

FIGURE 17–11 Pulmonary circulation, terminal and respiratory bronchioles, and alveoli.

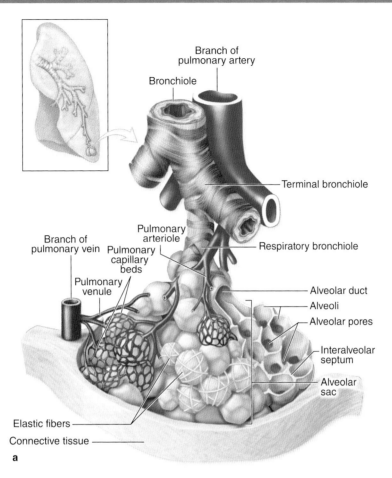

Branch of pulmonary artery

Bronchiole

Terminal bronchiole

Pulmonary arteriole

Respiratory bronchiole

Branch of pulmonary vein

Pulmonary capillary beds

Pulmonary venule

Alveolar duct

Alveoli

Alveolar pores

Interalveolar septum

Alveolar sac

Elastic fibers

Connective tissue

a

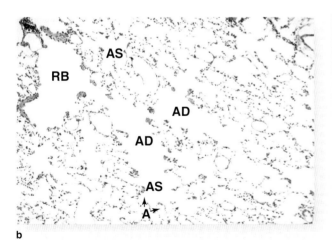

b

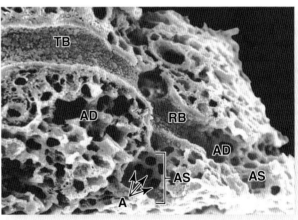

c

Terminal bronchioles branch into **respiratory bronchioles**, which then branch further into **alveolar ducts** and individual **alveoli**. Respiratory bronchioles are similar in most respects to terminal bronchioles except for the presence of scattered alveoli along their length.

(a) The diagram shows this branching relationship, as well as the pulmonary blood vessels that travel with the bronchioles and the dense layer of branching capillaries and elastic fibers that surround each alveolus.

(b) The micrograph shows the branching nature of the air passages in two dimensions: respiratory bronchiole (**RB**), alveolar ducts (**AD**), alveolar sacs (**AS**), and individual alveoli (**A**). (X60; H&E)

(c) SEM shows in three dimensions the relationship of a terminal bronchiole (**TB**), respiratory bronchiole (**RB**), alveolar duct (**AD**), alveolar sacs (**AS**), and individual alveoli (**A**). (X180)

FIGURE 17–12 Respiratory bronchioles, alveolar ducts, and alveoli.

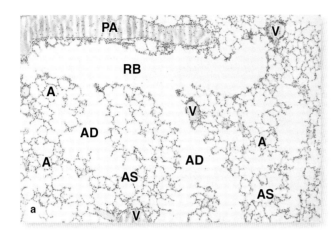
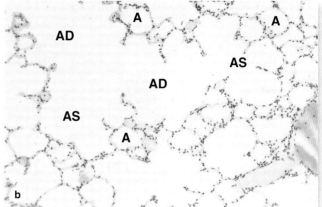

Lung tissue has a spongy structure because of the abundant air passages and pockets called alveoli.

(a) Typical section of lung tissue includes many bronchioles, some of which are respiratory bronchioles (**RB**) cut lengthwise, and shows the branching continuity with alveolar ducts (**AD**) and sacs (**AS**). Respiratory bronchioles still have a layer of smooth muscle and some regions of cuboidal epithelium, but alveolar ducts consist of a linear series of alveoli, each with smooth muscle fibers around the opening. Individual alveoli (**A**) all open to the sacs or ducts. The respiratory bronchiole runs along a

thin-walled branch of the pulmonary artery (**PA**), while branches of the pulmonary vein (**V**) course elsewhere in the parenchyma. (X14; H&E)

(b) Higher magnification shows the relationship of the many rounded, thin-walled alveoli (**A**) to alveolar ducts (**AD**). Alveolar ducts end in two or more clusters of alveoli called alveolar sacs (**AS**). Smooth muscle around the alveolar openings appears as eosinophilic "knobs" between adjacent alveoli. The alveoli here that do not show openings to the ducts or the sacs have their connections in planes of adjacent sections. (X140; H&E)

Alveoli

Alveoli are saclike evaginations, each about 200 µm in diameter, from the respiratory bronchioles, alveolar ducts, and alveolar sacs. Along with the airways, alveoli are responsible for the spongy structure of the lungs (Figures 17–11 and 17–12). Each adult lung has approximately 200 million alveoli with a total internal surface area of 75 m². Each alveolus resembles a small rounded pouch open on one side to an alveolar duct or alveolar sac. Air in these structures exchanges O_2 and CO_2 with the blood in surrounding capillaries, through thin specialized alveolar walls that enhance diffusion between the external and internal environments.

Between neighboring alveoli lie thin **interalveolar septa** consisting of scattered fibroblasts and sparse extracellular matrix (ECM), notably elastic and reticular fibers, of connective tissue. The arrangement of elastic fibers enables alveoli to expand with inspiration and contract passively with expiration; reticular fibers prevent both collapse and excessive distention of alveoli. The interalveolar septa are vascularized with the richest capillary networks in the body (Figure 17–11).

The densely anastomosing pulmonary capillaries within the interalveolar septa are supported by the meshwork of reticular and elastic fibers in the alveolar walls. Air in the alveoli is separated from capillary blood by three components referred

to collectively as the respiratory membrane or **blood-air barrier** (Figures 17–13 through 17–15):

- Two or three highly attenuated, thin cells lining the alveolus,
- The fused basal laminae of these cells and the endothelial cells of capillaries, and
- The thin capillary endothelial cells.

The total thickness of these layers varies from 0.1 to 1.5 µm. Macrophages and other leukocytes can also be found within the septa (Figures 17–13 and 17–14). **Alveolar pores** (of Kohn), ranging 10-15 µm in diameter, penetrate the interalveolar septa (Figure 17–13) and connect neighboring alveoli that open to different bronchioles. Such pores equalize air pressure in these alveoli and permit collateral circulation of air if a bronchiole becomes obstructed.

O_2 from the alveolar air diffuses through the blood-air barrier into the capillary blood and binds **hemoglobin** in erythrocytes; CO_2 diffuses into the alveolar air from the pulmonary blood. Most CO_2 arrives in the lungs as part of H_2CO_3 inside erythrocytes and is liberated through the action of carbonic anhydrase.

Capillary endothelial cells of alveolar walls are very thin, but continuous (not fenestrated) (Figure 17–15).

FIGURE **17–13** Alveoli and the blood-air barrier.

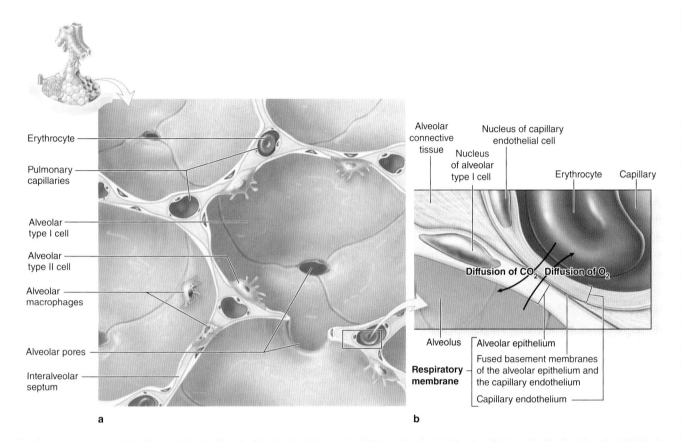

Gas exchange between air and blood occurs at a membranous barrier between each alveolus and the capillaries surrounding it.

(a) This diagram shows the relationship between capillaries in the interalveolar septa and the saclike alveoli. The alveolar pores (of Kohn) allow air pressure to equilibrate and air to circulate between alveoli on different ducts or sacs if the local airway becomes blocked.

(b) The air-blood barrier consists of an alveolar type I cell, a capillary endothelial cell, and their fused basement membranes. Oxygen diffuses from alveolar air into capillary blood and carbon dioxide moves in the opposite direction. The inner lining of alveoli is covered by a layer of surfactant (not shown) which lowers fluid surface tension and helps prevent collapse of alveoli.

Perinuclear clustering of most organelles allows the rest of the cell to become very thin and demonstrate highly efficient gas exchange. Ultrastructurally the most prominent features in the flattened portions of the cell are numerous pinocytotic vesicles.

Type I alveolar cells (or **type I pneumocytes**) are also extremely attenuated cells that line the alveolar surfaces. Type I cells constitute the alveolar side of the blood-air barrier and make up about 95% of the alveolar lining; type II alveolar cells (described below) make up the remainder. Type I cells are so thin that the TEM was needed to demonstrate that all alveoli have such linings (Figure 17–15). Organelles here are also grouped around the nucleus, reducing the thickness of the remaining cytoplasm at the blood-air barrier to as little as 25 nm. Pinocytotic vesicles in the attenuated cytoplasm may play a role in the turnover of surfactant and the removal

of small particulate contaminants from the outer surface. In addition to desmosomes, all type I epithelial cells have tight junctions that prevent the leakage of tissue fluid into the alveolar air space (Figure 17–16).

Type II alveolar cells (type II pneumocytes or septal cells) are cuboidal cells that bulge into the air space, interspersed among the type I alveolar cells and bound to them with tight junctions and desmosomes (Figure 17–16). Type II cells often occur in groups of two or three at points where two or more alveolar walls unite. These epithelial cells rest on the same basal lamina and have the same origin as the type I cells that line most of the alveolus. Type II cell nuclei are rounded and may have nucleoli, and their cytoplasm is typically lightly stained with many vesicles.

Many vesicles of type II alveolar cells are structures termed **lamellar bodies**, which TEM reveals to be membrane-bound

FIGURE **17–14** Alveolar walls.

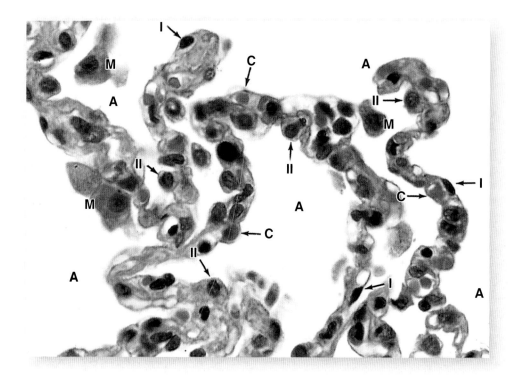

The septa between alveoli (**A**) contain several cell types. As seen here, the capillaries (**C**) include erythrocytes and leukocytes. The alveoli are lined mainly by squamous type I alveolar cells (**I**), which line almost the entire alveolus surface and across which gas exchange occurs. Type II alveolar cells line a bit of each alveolus and are large rounded cells, often bulging into the alveolus (**II**). These type II cells have many functions of club cells, including production of surfactant. Also present are alveolar macrophages (**M**), sometimes called **dust cells**, which may be in the alveoli or in the interalveolar septa. (X450; H&E)

granules 100-400 nm in diameter that contain closely stacked parallel membrane lamellae (Figures 17–16 and 17–17). Lamellar bodies contain a variety of lipids, phospholipids, and proteins that are continuously synthesized and released at the apical cell surface. The secreted material acts as **pulmonary surfactant** by spreading over the entire inner alveolar surface as a complex film of phospholipids and lipoproteins over a thin aqueous phase at the cell membranes. The surfactant film lowers surface tension at the air-epithelium interface, which helps prevent alveolar collapse at exhalation and allows alveoli to be inflated with less inspiratory force, easing the work of breathing.

Critical components of the surfactant layer produced by type II alveolar cells include the phospholipid **dipalmitoylphosphatidylcholine (DPPC)**, cholesterol, and four surfactant proteins (Figure 17–17). Surfactant protein A (SP-A), an abundant hydrophilic glycoprotein, and SP-D are important for innate immune protection within alveoli. SP-B and SP-C are hydrophobic membrane proteins required for the proper orientation of DPPC in the surfactant film lining the alveolus.

The surfactant layer turns over constantly, being gradually removed by pinocytosis in both types of alveolar cells and by macrophages. In fetal development, surfactant appears in the last weeks of gestation as type II cells differentiate and form lamellar bodies. Lack of adequate surfactant is a major cause of respiratory distress in premature neonates.

>> **MEDICAL APPLICATION**

Infant respiratory distress syndrome, the leading cause of death in premature babies, is due to incomplete differentiation of type II alveolar cells and a resulting deficit of surfactant and difficulty in expanding the alveoli in breathing. Treatment involves insertion of an endotracheal tube to provide both **continuous positive airway pressure (CPAP)** and **exogenous surfactant**, either synthesized chemically or purified from lungs of cattle.

Alveolar macrophages, also called **dust cells**, are found in alveoli and in the interalveolar septum (Figures 17–13 and 17–14). Tens of millions of monocytes migrate daily from the microvasculature into the lung tissue, where they phagocytose erythrocytes lost from damaged capillaries and airborne particulate matter that has penetrated as far as alveoli.

FIGURE **17–15** Ultrastructure of the blood-air barrier.

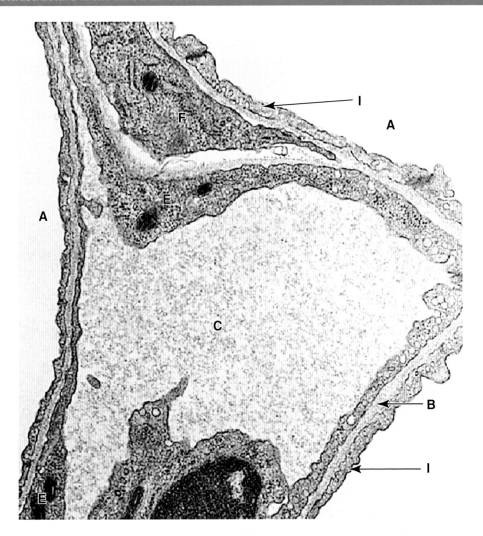

TEM of a capillary (**C**) in an interalveolar septum shows areas for gas exchange between blood and air in three alveoli (**A**). The endothelium is continuous but extremely thin, and its basal lamina (**B**) fuses with that of the type I alveolar cells (**I**) and type II cells. A fibroblast (**F**) can be seen in the septum, and the thickened nuclear regions of two endothelial cells (**E**) are also included. The nucleus at the bottom belongs to an endothelial cell or a circulating leukocyte. (X17,000)

Active alveolar macrophages can often be distinguished from type II pneumocytes because they are slightly darker due to their content of dust and carbon from air and complexed iron (hemosiderin) from erythrocytes (Figure 17–14). Filled macrophages have various fates: most migrate into bronchioles where they move up the mucociliary apparatus for removal into the esophagus; others exit the lungs in the lymphatic drainage; and some remain in the interalveolar septa connective tissue for years.

Alveolar lining fluids are also removed via the conducting passages as a result of the ciliary activity. As the secretions pass up through the airways, they combine with bronchial mucus to form **bronchoalveolar fluid**, which helps remove particulate components from inspired air. The bronchoalveolar fluid is bacteriostatic, containing lysozyme and other protective agents produced by club cells, type II alveolar cells, and alveolar macrophages.

Important histologic features and major functions at each level of airways in lungs are summarized in Table 17–2.

>> **MEDICAL APPLICATION**

In congestive heart failure, the lungs become congested with blood, and erythrocytes pass into the alveoli, where they are phagocytized by alveolar macrophages. In such cases, these macrophages are called **heart failure cells** when present in the lung and sputum; they are identified by a positive histochemical reaction for iron pigment (hemosiderin).

FIGURE **17–16** Ultrastructure of type II alveolar cells.

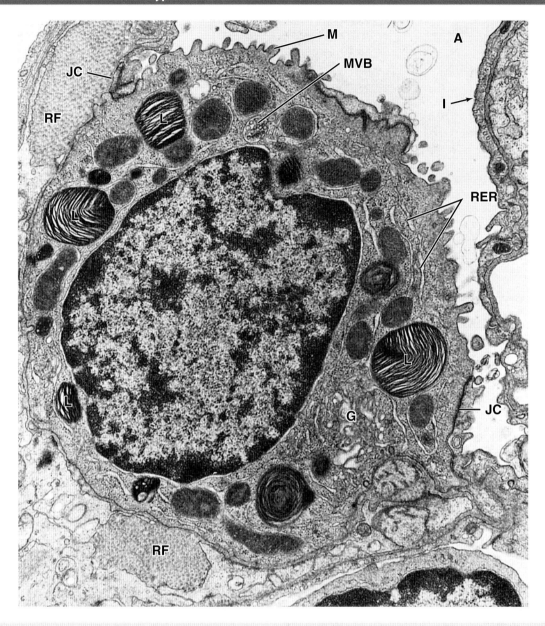

TEM of a type II alveolar cell protruding into the alveolar lumen shows unusual lamellar bodies (**L**) that contain newly synthesized pulmonary surfactant after processing of its components in rough ER (**RER**) and the Golgi apparatus (**G**). Smaller multivesicular bodies (**MVB**) with intralumenal vesicles are also present. Short microvilli (**M**) cover the apical cell surface at the alveolus (**A**) lumen.

The type II cell is attached via junctional complexes (**JC**) to the neighboring type I cell (**I**). Reticular fibers (**RF**) are prominent in the ECM. (X30,000)

(*Used with permission from Dr Mary C. Williams, Pulmonary Center Boston University School of Medicine, MA.*)

Regeneration in the Alveolar Lining

Inhalation of toxic gases or similar materials can kill type I and II cells lining pulmonary alveoli. Death of alveolar cells stimulates mitotic activity in the remaining type II cells, the progeny of which become progenitors for both cell types. The normal turnover rate of type II cells is estimated to be 1% per day and results in a continuous renewal of both alveolar cells.

With increased toxic stress, stem cells are also stimulated to divide and give rise to new alveolar cell progenitors.

❯ LUNG VASCULATURE & NERVES

Blood circulation in the lungs includes both the **pulmonary circulation**, carrying O_2-depleted blood for gas exchange,

FIGURE **17–17** Type II alveolar cell function.

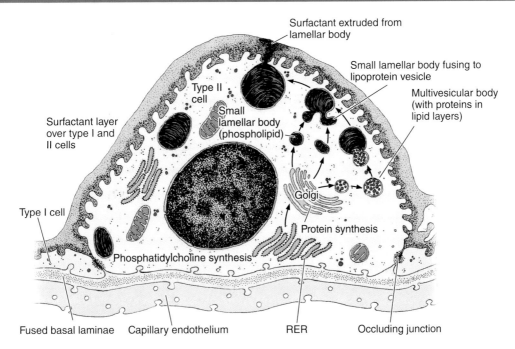

The diagram illustrates **surfactant production** by a type II cell. Surfactant contains protein-lipid complexes synthesized initially in the ER and Golgi apparatus, with further processing and storage in large organelles called **lamellar bodies**. Also present are smaller multivesicular bodies that form when membrane components of an early endosome are sorted, invaginate, and pinch off into smaller vesicles inside the endosome's lumen. In surfactant-producing cells the vesicles in multivesicular bodies are added to the lamellar bodies. Surfactant is secreted continuously by exocytosis and forms an oily film containing phospholipids and surfactant proteins.

and **bronchial circulation**, carrying O₂-rich blood. The pulmonary arteries and veins are both relatively thin-walled as a result of the low pressures (25 mm Hg systolic, 5 mm Hg diastolic) within the pulmonary circuit. Entering the lung at the hilum, the pulmonary artery branches and accompanies the bronchial tree (Figures 17–11 and 17–12), with its branches sharing the connective tissue of the bronchi and bronchioles. Beginning at the level of the respiratory bronchioles, branches of this artery give rise to dense capillary networks that surround each alveolus in the interalveolar septa.

❯❯ MEDICAL APPLICATION

Emphysema, a chronic lung disease most commonly caused by cigarette smoking, involves dilation and permanent enlargement of the bronchioles leading to pulmonary acini. Emphysema is accompanied by loss of cells in the alveoli and other parts of the airway walls, leading to an irreversible loss of respiratory function. Any type of infection in the respiratory regions of the lung produces the local inflammatory condition called **pneumonia**.

Venules arising from the capillary networks are found singly in the lung parenchyma, somewhat removed from the airways (Figures 17–11 and 17–12), supported by a thin covering of connective tissue. After small pulmonary veins leave a lobule, they follow the bronchial tree toward the hilum.

Bronchial arteries also enter the lung hilum and branch along with the bronchial tree, distributing blood to most of the lung down to the level of the respiratory bronchioles, at which point they anastomose with small branches of the pulmonary artery and mix their blood with that entering the capillary networks drained by pulmonary venules.

The lymphatic vessels originate in the connective tissue of bronchioles. They follow the bronchioles, bronchi, and pulmonary vessels and all drain into lymph nodes in the region of the hilum. This deep network of lymphatics parallels a superficial network draining areas near the lung surfaces located in the visceral pleura. Lymphatic vessels are not found in the terminal portions of the bronchial tree beyond the alveolar ducts.

Both parasympathetic and sympathetic autonomic fibers innervate the lungs and control reflexes regulating smooth muscle contractions which determine the diameters of the airways. General visceral afferent fibers, carrying poorly localized pain sensations, are also present. The nerves are found primarily in the connective tissue surrounding the larger elements of the bronchial tree and exit the lung at the hilum.

› PLEURAL MEMBRANES

The lung's outer surface and the internal wall of the thoracic cavity are covered by a serous membrane called the **pleura** (Figure 17–18). The membrane attached to lung tissue is called the **visceral pleura** and the membrane lining the thoracic walls is the **parietal pleura**. The two layers are continuous at the hilum and are both composed of simple squamous mesothelial cells on a thin connective tissue layer containing collagen and elastic fibers. The elastic fibers of the visceral pleura are continuous with those of the pulmonary parenchyma.

The narrow **pleural cavity** (Figure 17–18) between the parietal and visceral layers is entirely lined with mesothelial cells that normally produce a thin film of serous fluid that acts as a lubricant, facilitating the smooth sliding of one surface over the other during respiratory movements.

In certain pathologic states, the pleural cavity may contain liquid or air. Like the walls of the peritoneal and pericardial cavities, the serosa of the pleural cavity is water-permeable and fluid exuded from blood plasma commonly accumulates (as a pleural effusion) in this cavity during inflammation and other abnormal conditions.

›› MEDICAL APPLICATION

The condition **pneumothorax** is a partially or completely **collapsed lung** caused by air trapped in the pleural cavity, typically resulting from blunt or penetrating trauma to the chest and producing shortness of breath and hypoxia. Inflammation of the pleura, a condition called either **pleuritis** or **pleurisy**, is most commonly caused by an acute viral infection or pneumonia. **Pleural effusion** or fluid buildup in the pleural cavity produces shortness of breath and can be one result of inflamed pleura.

› RESPIRATORY MOVEMENTS

During inhalation, contraction of the external intercostal muscles elevates the ribs and contraction of the diaphragm lowers the bottom of the thoracic cavity, increasing its diameter and resulting in pulmonary expansion. The bronchi and bronchioles increase in diameter and length during inhalation. The respiratory portion also enlarges, mainly as a result of expansion of the alveolar ducts. Individual alveoli enlarge only slightly. The elastic fibers of the pulmonary parenchyma are stretched by this expansion. During exhalation, the lungs retract passively because of muscle relaxation and the elastic fibers' return to the unstretched condition.

›› MEDICAL APPLICATION

Lung cancer is one of the most common forms of this disease. **Squamous cell carcinoma**, which is closely correlated with a history of smoking, arises most often from epithelial cells of segmental bronchi. **Adenocarcinoma**, the most

FIGURE **17–18** Pleura.

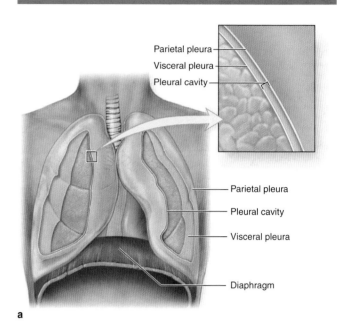

a

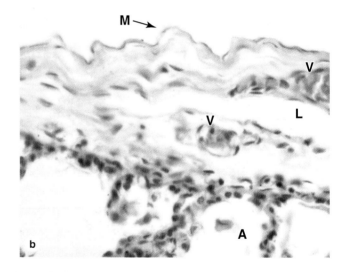

b

The pleura are serous membranes (serosa) associated with each lung and thoracic cavity. **(a)** The diagram shows the **parietal pleura** lining the inner surface of the thoracic cavity and the **visceral pleura** covering the outer surface of the lung. Between these layers is the narrow space of the pleural cavity. **(b)** Both layers are similar histically and consist of a simple squamous mesothelium (**M**) on a thin layer of connective tissue, as shown here for visceral pleura covering alveoli (**A**). The connective tissue is rich in both collagen and elastic fibers and contains both blood vessels (**V**) and lymphatics (**L**). (X140; H&E)

common lung cancer in nonsmokers, usually arises from epithelial cells more peripherally, in bronchioles and alveoli. **Small cell carcinoma**, a less common but highly malignant form of lung cancer, develops after neoplastic transformation of small granule Kulchitsky cells in bronchial respiratory epithelium.

The Respiratory System SUMMARY OF KEY POINTS

- The function of the respiratory system is to provide **oxygen to the blood**, with a secondary function of sound production in the larynx.
- The respiratory system consists of an air **conducting region** (the upper respiratory tract in the head, as well as the larynx, trachea, bronchi, and most bronchioles) and a **respiratory region** with **alveoli**.

Nasal Cavities

- The left and right **nasal cavities** of the upper respiratory tract have **vestibules** where air enters and three projections called **conchae** from their medial walls, which create turbulence in inspired air.
- Moist **vibrissae** in the vestibular openings, the **nares** or nostrils, filter some material from inspired air.
- Deeper areas of the vestibules and the floor, lateral walls, and most of the conchae of the nasal cavities themselves are lined by **respiratory epithelium: pseudostratified ciliated columnar epithelium**.
- **Respiratory epithelium** includes **goblet cells** secreting mucus, **ciliated columnar cells** sweeping the mucus along the surface, chemosensory **brush cells**, scattered **endocrine cells**, and basal **stem cells**.
- The roof and part of the superior concha in each nasal cavity are covered by **olfactory epithelium**, which is pseudostratified epithelium containing bipolar **olfactory neurons**, **support cells**, and **stem cells**.
- The mucosa of the **nasal cavities** and **nasopharynx** also contains a rich **vasculature** and many **seromucous glands**, which help warm, humidify, and clean inspired air.

Larynx, Trachea, and Bronchial Tree

- Within the lumen of the **larynx**, bilateral projecting **vocal folds** (or cords) can be placed under variable tension by the underlying **vocalis muscles** and caused to vibrate by expelled air, producing sounds.
- The **trachea** is completely lined by **respiratory epithelium** and is supported by C-shaped rings of **hyaline cartilage**, with smooth **trachealis muscles** in the posterior opening of the rings.
- Left and right **primary bronchi** enter the two lungs and bifurcate repeatedly as **secondary**, **tertiary**, and **smaller segmental bronchi** with the lung tissue as the **bronchial tree**.

- **Bronchi** and their branches are lined by **respiratory mucosa**, with prominent spiraling bands of **smooth muscle** and increasingly smaller pieces of **hyaline cartilage**.
- Branches of the bronchial tree with diameters of 1 mm or less are generally called **bronchioles**, which are lined by **simple columnar** or **cuboidal ciliated cells**, with circular smooth muscle but **no cartilage**.
- **Terminal bronchioles** are the last branches to lack alveoli and are lined by **simple cuboidal epithelium** consisting mainly of **club cells**, which have **innate immune** and **surfactant secretory functions**.

Respiratory Region

- Terminal bronchioles subdivide into two or three **respiratory bronchioles**, lined by simple cuboidal epithelium and interrupted by scattered squamous evaginations called **alveoli**, the sites of gas exchange.
- A respiratory bronchiole leads to an **alveolar duct**, which is lined by a continuous series of alveoli and which ends in a cluster of alveoli called the **alveolar sac**.
- All alveoli are surrounded by sparse connective tissue in **interalveolar septa** that consist primarily of **elastic and reticular fibers** and a dense **capillary network**.
- The wall of each alveolus consists of **alveolar cells**, or **pneumocytes**, of two types: extremely **thin type I alveolar cells** and **cuboidal type II alveolar cells** with **surfactant** secreting and **innate immune** properties.
- **Type II alveolar cells** are characterized ultrastructurally by unique cytoplasmic **lamellar bodies**, large granules with closely stacked layers of membrane involved in **surfactant synthesis**.
- The **blood-air barrier** allowing gas exchange at each alveolus consists of the **thin type I alveolar cell**, the **thin capillary endothelial cells**, and **the fused basal laminae** of these two cells.
- The **surfactant material** secreted by exocrine club cells and type II alveolar cells is an oily mixture of cholesterol, **phospholipids** and **surfactant proteins**, which forms a film and **lowers surface tension** in alveoli.
- Each lung is covered by **visceral pleura**, a layer of thin connective tissue and mesothelium, and is continuous with **parietal pleura**, a similar tissue layer that lines the pleural cavity.

The Respiratory System ASSESS YOUR KNOWLEDGE

1. Which of the following components increase(s) as a proportion of the respiratory tract wall from trachea to alveoli?
   a. Cilia
   b. Elastic fibers
   c. Smooth muscle
   d. Cartilage
   e. Goblet cells

2. Air moving rapidly across the vocal cords and causing them to vibrate and produce sound is contacting what type of epithelium?
   a. Pseudostratified ciliated
   b. Stratified squamous keratinized
   c. Stratified squamous nonkeratinized
   d. Simple squamous
   e. Simple cuboidal

3. Which structural feature distinguishes between terminal and respiratory bronchioles?
   a. Alveoli
   b. Cilia
   c. Exocrine bronchiolar cells
   d. Mucous glands in lamina propria
   e. Smooth muscle

4. Which of the following features distinguishes a bronchus within a lung from the primary bronchi?
   a. Glands in the submucosa
   b. Pseudostratified ciliated columnar epithelium
   c. Smooth muscle in the wall
   d. Irregular plates of cartilage
   e. Goblet cells in the epithelium

5. Which feature involved in protection of the respiratory tract is absent from the digestive tract?
   a. Goblet cells
   b. Cilia
   c. Lymphoid nodules
   d. Secretory IgA
   e. Tight junctions

6. Which of the following is true of pulmonary surfactant?
   a. Secreted by type I pneumocytes
   b. Forms layer rich in phospholipid overlying a thin aqueous phase
   c. Prevents alveolar collapse by increasing surface tension
   d. Does not affect bacterial survival
   e. Is secreted by goblet cells

7. The pulmonary (functional) and bronchial (nutritive) arterial systems enter the lungs separately at the hilus but anastomose into a single system at which level?
   a. Bronchi
   b. Larynx
   c. Terminal bronchioles
   d. Segmental bronchioles
   e. Respiratory bronchioles

8. After 35 weeks of gestation, a 5 lb 5 oz girl is born to a 30-year-old gravid 2, para 2 (G2P2) woman. The infant has rapid and labored breathing that is viewed as transient tachypnea of the newborn. The infant's 1- and 5-min APGAR scores are 8 and 9, respectively. She has respiratory distress, with a normal pulse and no heart murmurs. She is transported to the neonatal intensive care unit with worsening tachypnea. In this infant which of the following is likely to be involved?
   a. Failure of the type I pneumocytes to form complete blood-air barriers
   b. Absence of elastic fibers from the bronchiolar walls and interalveolar septa
   c. Failure of type II pneumocytes to complete differentiation and become fully functional
   d. Failure of type II pneumocyte progenitors to proliferate adequately during gestation
   e. Inadequate development of the parietal and visceral pleura

9. A teenage girl presents at the ER with paroxysms of dyspnea, cough, and wheezing. Her parents indicate that she had these "attacks" during the past winter and that they have worsened and become more frequent during the spring. Which of the following cell types and their location is correctly matched to a function it may perform in this patient's disorder?
   a. Cilia in alveolar ducts, enhanced mucociliary transport
   b. Plasma cells in bronchus-associated lymphoid tissue (BALT), bronchoconstriction
   c. Eosinophils in BALT, bronchodilation
   d. Goblet cells in bronchioles, hyposecretion
   e. Mast cells in BALT, bronchoconstriction and edema

10. A 28-year-old man is diagnosed with a testicular germ cell tumor. The tumor is surgically removed and he begins chemotherapy with cisplatin, etoposide, and bleomycin. Bleomycin chemotherapy is known to affect the lung blood-air barrier. Which of the following best describes the structural site of those effects?
   a. Fused basal laminae of epithelial and endothelial cells
   b. Alveolar pores of Kohn
   c. Alveolar macrophages in interalveolar septa
   d. Type II pneumocytes linked by junctional complexes
   e. Smooth muscle cells of the pulmonary and bronchial arteries

Answers: 1b, 2c, 3a, 4d, 5b, 6b, 7e, 8c, 9e, 10a

18 Skin

The skin is the largest single organ of the body, typically accounting for 15%-20% of total body weight and, in adults, presenting 1.5-2 m² of surface to the external environment. Also known as the **integument** (L. *integumentum*, covering) or **cutaneous layer**, the skin is composed of the **epidermis**, an epithelial layer of ectodermal origin, and the **dermis**, a layer of mesodermal connective tissue (Figure 18–1). At the irregular junction between the dermis and epidermis, projections called **dermal papillae** interdigitate with invaginating **epidermal ridges** to strengthen adhesion of the two layers. Epidermal derivatives include hairs, nails, and sebaceous and sweat glands. Beneath the dermis lies the **subcutaneous tissue** or **hypodermis** (Gr. *hypo*, under + *derma*, skin), a loose connective tissue layer usually containing pads of adipocytes. The subcutaneous tissue binds the skin loosely to the underlying tissues and corresponds to the superficial fascia of gross anatomy.

The specific functions of the skin fall into several broad categories.

- **Protective:** It provides a physical barrier against thermal and mechanical insults such as friction and against most potential pathogens and other material. Microorganisms that do penetrate skin alert resident lymphocytes and antigen-presenting cells (APCs) in skin and an immune response is mounted. The dark pigment melanin in the epidermis protects cell nuclei from ultraviolet (UV) radiation. Skin is also a permeability barrier against excessive loss or uptake of water, which has allowed for terrestrial life. Skin's selective permeability allows some lipophilic drugs such as certain steroid hormones and medications to be administered via skin patches.
- **Sensory:** Many types of sensory receptors allow skin to constantly monitor the environment, and various skin

mechanoreceptors help regulate the body's interactions with physical objects.
- **Thermoregulatory:** A constant body temperature is normally easily maintained thanks to the skin's insulating components (eg, the fatty layer and hair on the head) and its mechanisms for accelerating heat loss (sweat production and a dense superficial microvasculature).
- **Metabolic:** Cells of skin synthesize vitamin D_3, needed in calcium metabolism and proper bone formation, through the local action of UV light on the vitamin's precursor. Excess electrolytes can be removed in sweat, and the subcutaneous layer stores a significant amount of energy in the form of fat.
- **Sexual signaling:** Many features of skin, such as pigmentation and hair, are visual indicators of health involved in attraction between the sexes in all vertebrate species, including humans. The effects of sex pheromones produced by the apocrine sweat glands and other skin glands are also important for this attraction.

The dermal-epidermal interdigitations are of the peg-and-socket variety in most skin (Figure 18–1), but they occur as well-formed ridges and grooves in the thick skin of the palms and soles, which is more subject to friction. These ridges and the intervening sulci form distinctive patterns unique for each individual, appearing as combinations of loops, arches, and whorls, called dermatoglyphs, also known as fingerprints and footprints. Skin is elastic and can expand rapidly to cover swollen areas and, like the gut lining, is self-renewing throughout life. In healthy individuals injured skin is repaired rapidly. The molecular basis of skin healing is increasingly well understood and provides a basis for better understanding of repair and regeneration in other organs.

FIGURE **18–1**   **Layers and appendages of skin.**

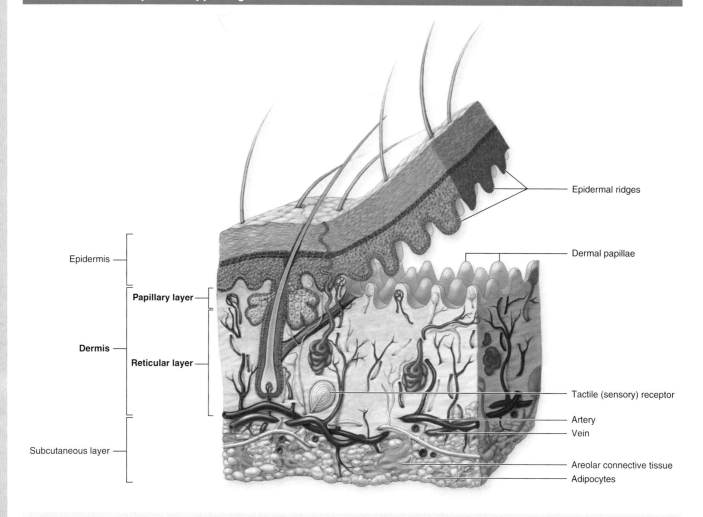

Epidermis

Papillary layer

Dermis

Reticular layer

Subcutaneous layer

Epidermal ridges

Dermal papillae

Tactile (sensory) receptor

Artery

Vein

Areolar connective tissue

Adipocytes

Diagrammatic overview of skin, showing the major layers and epidermal appendages (hair follicles, sweat, and sebaceous glands), the vasculature, and the major sensory receptors.

› EPIDERMIS

The epidermis consists mainly of a stratified squamous keratinized epithelium composed of cells called **keratinocytes**. There are also three much less abundant epidermal cell types: pigment-producing **melanocytes**, antigen-presenting **Langerhans cells**, and tactile epithelial cells called **Merkel cells** (Figure 18–2).

The epidermis forms the major distinction between **thick skin** (Figure 18–2a), found on the palms and soles, and **thin skin** (Figure 18–3) found elsewhere on the body. The designations "thick" and "thin" refer to the thickness of the epidermal layer, which alone varies from 75 to 150 μm for thin skin and from 400 to 1400 μm (1.4 mm) for thick skin. Total skin thickness (epidermis plus dermis) also varies according to the site. For example, full skin on the back is about 4-mm thick, whereas that of the scalp is about 1.5-mm thick. Like all

epithelia, the stratified squamous epidermis lacks microvasculature, its cells receiving nutrients and O_2 by diffusion from the dermis.

From the dermis, the epidermis consists of four layers of keratinocytes (or five layers in thick skin, Figure 18–2):

- The **basal layer (stratum basale)** is a single layer of basophilic cuboidal or columnar cells on the basement membrane at the dermal-epidermal junction (Figures 18–2 and 18–3). Hemidesmosomes in the basal cell membranes join these cells to the basal lamina, and desmosomes bind the cells of this layer together in their lateral and upper surfaces. The stratum basale is characterized by intense mitotic activity and contains, along with the deepest part of the next layer, progenitor cells for all the epidermal layers. In addition to the basal stem cells for keratinocytes found here, a niche for such cells also occurs in the hair

FIGURE **18–2** Layers (strata) of epidermis in thick skin.

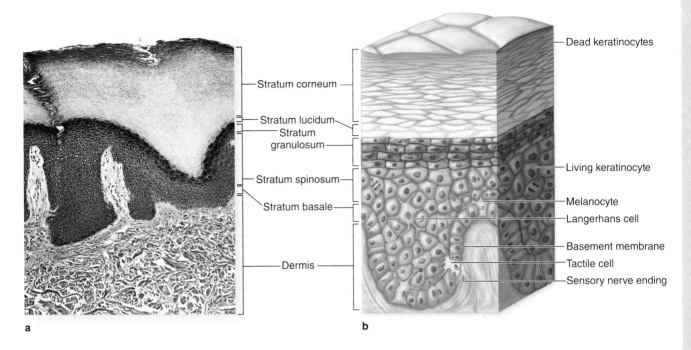

a

b

(a) Micrograph shows the sequence of the epidermal layers in thick skin and the approximate sizes and shape of keratinocytes in these layers. Also shown are the coarse bundles of collagen in the dermis and on the far left, the duct from a sweat gland entering the epidermis from a dermal papilla and coiling to a surface pore through all the strata. (X100; H&E)

(b) Diagram illustrating the sequence of the epidermal layers also indicates the normal locations of three important nonkeratinocyte cells in the epidermis: melanocytes, a Langerhans cell, and a tactile Merkel cell.

follicle sheaths that are continuous with the epidermis. The human epidermis is renewed about every 15-30 days, depending on age, the region of the body, and other factors. An important feature of all keratinocytes in the stratum basale is the cytoskeletal **keratins**, intermediate filaments about 10 nm in diameter. During differentiation, the cells move upward and the amount and types of keratin filaments increase until they represent half the total protein in the superficial keratinocytes.

❯❯ MEDICAL APPLICATION

Friction blisters are lymph-filled spaces created between the epidermis and dermis of thick skin by excessive rubbing, as with ill-fitting shoes or hard use of the hands. If continued, such activity produces protective thickening and hardening of the outer cornified epidermal layers, seen as **corns** and **calluses.**

▪ The **spinous layer (stratum spinosum)** is normally the thickest layer, especially in the epidermal ridges (Figures 18–2 and 18–3), and consists of generally polyhedral cells having central nuclei with nucleoli and cytoplasm actively synthesizing keratins. Just above the basal layer, some cells may still divide and this combined

zone is sometimes called the **stratum germinativum**. The keratin filaments assemble here into microscopically visible bundles called **tonofibrils** that converge and terminate at the numerous desmosomes holding the cell layers together. The cells extend slightly around the tonofibrils on both sides of each desmosome (and the extensions elongate if the cells shrink slightly during histologic processing), leading to the appearance of many short "spines" or prickles at the cell surfaces (Figure 18–4). The epidermis of thick skin subject to continuous friction and pressure (such as the foot soles) has a thicker stratum spinosum with more abundant tonofibrils and desmosomes.

❯❯ MEDICAL APPLICATION

In adults, one-third of all cancers originate in the skin. Most of these derive from cells of the basal or spinous layers, producing, respectively, basal cell carcinomas and squamous cell carcinomas. Fortunately, both types of tumors can be diagnosed and excised early and consequently are rarely lethal. Skin cancer shows an increased incidence in fair-skinned individuals residing in regions with high amounts of solar radiation.

FIGURE **18–3** Layers of epidermis in thin skin.

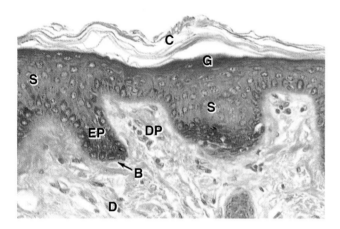

The interface between dermis and epidermis in thin skin is held together firmly by interlocking epidermal ridges or pegs (**EP**) and dermal papillae (**DP**). The dermis (**D**) of thin skin is more cellular and well vascularized than that of thick skin, with elastin and less coarse bundles of collagen. The epidermis usually shows only four layers in thin skin: the one-cell thick stratum basale (**B**) containing most mitotic cells; the stratum spinosum (**S**) where synthesis of much keratin and other proteins takes place; the stratum granulosum (**G**); and the stratum corneum (**C**), consisting of dead squames composed mostly of keratin. (X240; H&E)

- The **granular layer (stratum granulosum)** consists of three to five layers of flattened cells, now undergoing the terminal differentiation process of **keratinization**. Their cytoplasm is filled with intensely basophilic masses (Figures 18–2 and 18–3; Figure 18–5) called **keratohyaline granules**. These are dense, non–membrane-bound masses of **filaggrin** and other proteins associated with the keratins of tonofibrils, linking them further into large cytoplasmic structures.

  Characteristic features in cells of the granular layer also include Golgi-derived **lamellar granules**, small ovoid (100 × 300 nm) structures with many lamellae containing various lipids and glycolipids. Among the last activities of the keratinocytes, the lamellar granules undergo exocytosis, producing a lipid-rich, impermeable layer around the cells. This material forms a major part of the skin's barrier against water loss. Formation of this barrier, which appeared first in ancestral reptiles, was a key evolutionary process that permitted animals to develop on land. Together, keratinization and production of the lipid-rich layer also have a crucial sealing effect in skin, forming the barrier to penetration by most foreign materials.

- The **stratum lucidum**, found only in thick skin, consists of a thin, translucent layer of flattened eosinophilic keratinocytes held together by desmosomes (Figures 18–2 and 18–5). Nuclei and organelles have been lost, and the cytoplasm consists almost exclusively of packed keratin filaments embedded in an electron-dense matrix.

- The **stratum corneum** (Figures 18–2 and 18–3) consists of 15-20 layers of squamous, keratinized cells filled with birefringent filamentous keratins. Keratin filaments contain at least six different polypeptides with molecular

FIGURE **18–4** Keratinocytes of the stratum spinosum.

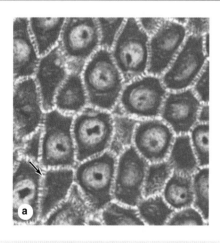

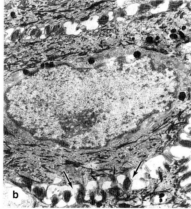

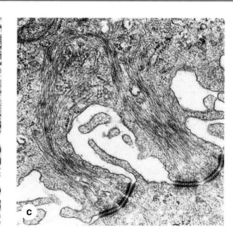

(a) A section of epidermal stratum spinosum of thick skin, showing cells with numerous short cytoplasmic projections (**arrow**). (X400; PT)

(b) TEM of a single spinous keratinocyte with surface projections (**arrows**). (8400)

(c) Detail of the desmosomes joining two cells showing intermediate filaments associated with desmosomes. (X40,000)

FIGURE **18–5** **Stratum granulosum and stratum lucidum: thick skin.**

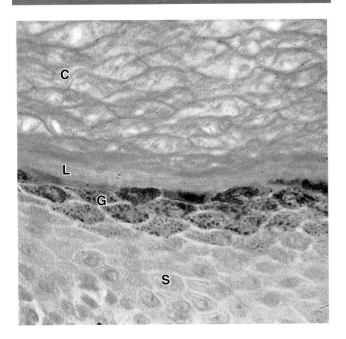

In keratinocytes moving upward from the stratum spinosum (**S**), differentiation proceeds with the cells becoming filled with numerous large, amorphous masses of protein called **keratohyaline granules**.

Cells with these basophilic granules make up the stratum granulosum (**G**), where keratin filaments are cross-linked with filaggrin and other proteins from these granules to produce tight bundles filling the cytoplasm and flattening the cells. Smaller organelles called **lamellar granules** undergo exocytosis in this layer, secreting a lipid-rich layer around the cells which makes the epidermis impermeable to water. Together, the lipid envelope and the keratin-filled cells determine most of the physical properties of the epidermis.

The cells leaving the stratum granulosum, still bound together by desmosomes, undergo terminal differentiation and in thick skin appear as a dense, thin layer called the **stratum lucidum** (**L**). Here proteins are dispersed through the tonofibril bundles, giving this layer a regular, "clear" appearance. In the most superficial stratum corneum (**C**), the cells have lost nuclei and cytoplasm, consisting only of flattened, keratinized structures called **squames** bound by hydrophobic, lipid-rich intercellular cement. At the surface they are worn away (thick skin) or flake off (thin skin). (X560; H&E)

masses ranging from 40 to 70 kDa, synthesized during cell differentiation in the immature layers. As they form, keratin tonofibrils become heavily massed with filaggrin and other proteins in keratohyaline granules. By the end of keratinization, the cells contain only amorphous, fibrillar proteins with plasma membranes surrounded by the lipid-rich layer. These fully keratinized or cornified cells called **squames** are continuously shed at the epidermal surface as the desmosomes and lipid-rich cell envelopes break down.

Important features of the epidermal strata are summarized in Table 18–1.

Melanocytes

The color of the skin is the result of several factors, the most important of which are the keratinocytes' content of **melanin** and **carotene** and the number of blood vessels in the dermis.

Eumelanins are brown or black pigments produced by the **melanocyte** (Figures 18–6 and 18–7), a specialized cell of the epidermis found among the cells of the basal layer and in hair follicles. The similar pigment found in red hair is called **pheomelanin** (Gr. *phaios*, dusky + *melas*, black). Melanocytes are neural crest derivatives that migrate into the embryonic epidermis' stratum basale, where eventually one melanocyte accumulates for every five or six basal keratinocytes (600-1200/mm^2 of skin). They have pale-staining, rounded cell bodies attached by hemidesmosomes to the basal lamina, but lacking attachments to the neighboring keratinocytes. Several long irregular cytoplasmic extensions from each melanocyte cell body penetrate the epidermis, running between the cells of the basal and spinous layers and terminating in invaginations of 5-10 keratinocytes. Ultrastructurally a melanocyte has numerous small mitochondria, short cisternae of RER, and a well-developed Golgi apparatus (Figure 18–6).

The first step in melanin synthesis is catalyzed by **tyrosinase**, a transmembrane enzyme in Golgi-derived vesicles (Figure 18–7). Tyrosinase activity converts tyrosine into **3,4-dihydroxyphenylalanine (DOPA)**, which is then further transformed and polymerized into the different forms of melanin. Melanin pigment is linked to a matrix of structural proteins and accumulates in the vesicles until they form mature elliptical granules about 1-μm long called **melanosomes** (Figure 18–7).

Melanosomes are then transported via kinesin to the tips of the cytoplasmic extensions. The neighboring keratinocytes phagocytose the tips of these dendrites, take in the melanosomes, and transport them by dynein toward their nuclei. The melanosomes accumulate within keratinocytes as a supranuclear cap that prior to keratinization absorbs and scatters sunlight, protecting DNA of the living cells from the ionizing, mutagenic effects of UV radiation.

Although melanocytes produce melanosomes, the keratinocytes are the melanin depot and contain more of this

TABLE 18–1	Summary of skin and subcutaneous layers.	

Layer	Specific Layer	Description
Epidermis 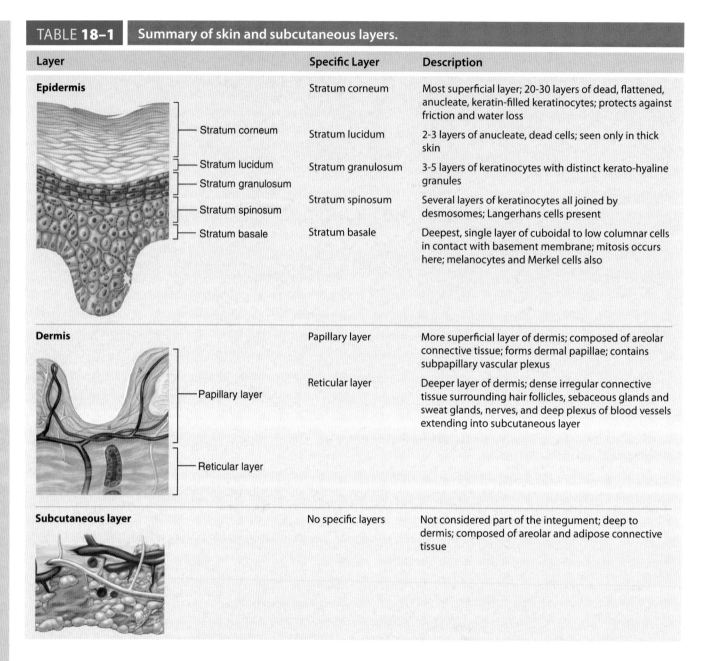 Stratum corneum Stratum lucidum Stratum granulosum Stratum spinosum Stratum basale	Stratum corneum	Most superficial layer; 20-30 layers of dead, flattened, anucleate, keratin-filled keratinocytes; protects against friction and water loss
	Stratum lucidum	2-3 layers of anucleate, dead cells; seen only in thick skin
	Stratum granulosum	3-5 layers of keratinocytes with distinct kerato-hyaline granules
	Stratum spinosum	Several layers of keratinocytes all joined by desmosomes; Langerhans cells present
	Stratum basale	Deepest, single layer of cuboidal to low columnar cells in contact with basement membrane; mitosis occurs here; melanocytes and Merkel cells also
Dermis Papillary layer Reticular layer	Papillary layer	More superficial layer of dermis; composed of areolar connective tissue; forms dermal papillae; contains subpapillary vascular plexus
	Reticular layer	Deeper layer of dermis; dense irregular connective tissue surrounding hair follicles, sebaceous glands and sweat glands, nerves, and deep plexus of blood vessels extending into subcutaneous layer
Subcutaneous layer	No specific layers	Not considered part of the integument; deep to dermis; composed of areolar and adipose connective tissue

pigment than the cells that make it. One melanocyte plus the keratinocytes into which it transfers melanosomes make up an **epidermal-melanin unit**. The density of such units in skin is similar in all individuals. Melanocytes of people with ancestral origins near the equator, where the need for protection against the sun is greatest, produce melanin granules more rapidly and accumulate them more abundantly in keratinocytes.

❯❯ MEDICAL APPLICATION

Albinism is a congenital disorder producing skin **hypopigmentation** due to a defect in tyrosinase or some other component of the melanin-producing pathway. An acquired condition called **vitiligo** involves skin **depigmentation**, often only in affected patches, due to the loss or decreased activity of melanocytes. The causes of melanocyte loss are not clear, but they may include environmental, genetic, or autoimmune conditions.

In regions with much less sunlight such as northern Europe, the small amount of UV radiation penetrating dark skin barely sustains adequate vitamin D_3 synthesis. Individuals with ancestry there have one or more genetic polymorphisms that affect steps in melanin formation and cause more lightly pigmented keratinocytes that have increased UV penetration and vitamin D_3 synthesis.

Darkening of the skin, or tanning, after exposure to solar radiation at wavelengths of 290-320 nm is a two-step process. A physicochemical reaction darkens preexisting melanin. At the same time, paracrine factors secreted by keratinocytes experiencing increased UV radiation accelerate melanin synthesis and its accumulation in the epidermis.

FIGURE **18–6** Melanocytes.

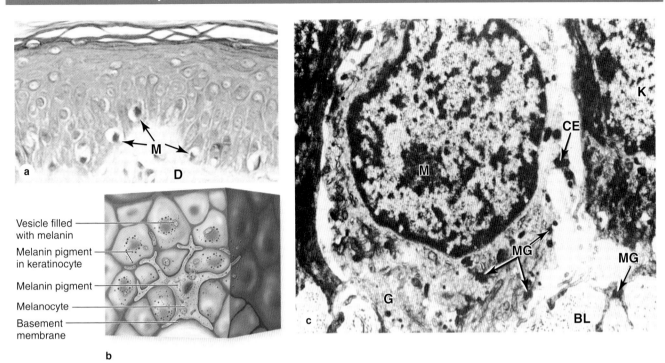

Melanocytes are located in the epidermal basal layer and synthesize **melanin granules** and transfer them into neighboring **keratinocytes**. Transfer occurs through many long, branching melanocyte processes that extend into the spinous layer and are not seen in routine microscopy.

(a) In light microscopy melanocytes (**M**) typically appear as rounded, pale-staining or clear cells just above the dermis (**D**). Melanocytes are difficult to distinguish from Merkel cells by routine microscopy. Langerhans cells are also rounded, poorly stained cells but are typically located more superficially than melanocytes, in the stratum spinosum. (X400; H&E)

(b) Diagram of a melanocyte shows the irregular cytoplasmic processes between neighboring keratinocytes for transfer of melanin to those cells.

(c) Ultrastructurally, a melanocyte is located on the basal lamina (**BL**) and has well-developed Golgi complexes (**G**) producing the vesicles in which melanin is synthesized. As they fill, these vesicles become melanin granules (**MG**), which accumulate at the tips of the dendritic cytoplasmic extensions (**CE**) before transfer to keratinocytes (**K**). (X14,000)

>> MEDICAL APPLICATION

Melanocytes can normally proliferate in skin to produce **moles**, or **benign melanocytic nevi** of various types. Changes in the size or appearance of moles are sometimes indicative of dysplasia that can progress further to **malignant melanoma**. Dividing rapidly, malignantly transformed melanocytes often penetrate the basal lamina, enter the dermis, and metastasize by invading blood and lymphatic vessels.

Langerhans Cells

Antigen-presenting cells (APCs) called **Langerhans cells**, derived from monocytes, represent 2%-8% of the cells in epidermis and are usually most clearly seen in the spinous layer. Cytoplasmic processes extend from these dendritic cells between keratinocytes of all the layers, forming a fairly dense network in the epidermis (Figure 18–8). Langerhans cells bind, process, and present antigens to T lymphocytes in the same manner as immune dendritic cells in other organs (see Chapter 14).

Microorganisms cannot penetrate the epidermis without alerting these dendritic cells and triggering an immune response. Langerhans cells, along with more scattered epidermal lymphocytes and other APCs in the dermis, comprise a major component of the skin's adaptive immunity.

Because of its location, the skin is continuously in close contact with many antigenic molecules. Various epidermal features participate in both innate and adaptive immunity (see Chapter 14), providing an important immunologic component to the skin's overall protective function.

Merkel Cells

Merkel cells, or **epithelial tactile cells**, are sensitive mechanoreceptors essential for light touch sensation. Joined by desmosomes to keratinocytes of the basal epidermal layer, Merkel cells resemble the surrounding cells but with few, if any, melanosomes. They are abundant in highly sensitive skin like that of fingertips and at the bases of some hair follicles. Merkel cells originate from the same stem cells as keratinocytes and are characterized

FIGURE **18–7** Melanosome formation.

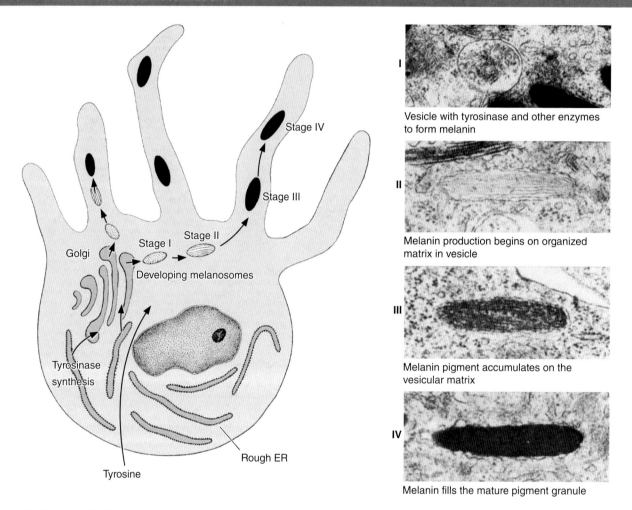

I

Vesicle with tyrosinase and other enzymes to form melanin

II

Melanin production begins on organized matrix in vesicle

III

Melanin pigment accumulates on the vesicular matrix

IV

Melanin fills the mature pigment granule

The diagram of a melanocyte shows the main features of melanin formation. The granules containing melanin mature through four stages that are characterized ultrastructurally, as shown on the right. **Tyrosinase** is synthesized in the rough ER, processed through the Golgi apparatus, and accumulates in vesicles that also have a fine granular matrix of other proteins (**stage I** melanosomes). Melanin synthesis begins in the ovoid **stage II** melanosomes, in which the matrix has been organized into parallel filaments on which polymerized melanin is deposited and accumulates in **stage III**. A mature melanin granule (**stage IV**) has lost tyrosinase and other activities and has the internal matrix completely filled with melanin. The mature granules are ellipsoid, approximately 0.5 by 1 μm in size, and visible by light microscopy.

Melanin granules are transported to the tips of the processes of melanocyte and are then transferred to the neighboring keratinocytes of the basal and spinous layers. In keratinocytes the melanin granules are transported to a region near the nucleus, where they accumulate as a supranuclear cap shading the DNA against the harmful effects of UV radiation.

by small, Golgi-derived dense-core neurosecretory granules containing peptides (Figure 18–9). The basolateral surfaces of the cells contact expanded terminal discs of unmyelinated sensory fibers penetrating the basal lamina (Figure 18–9).

❯❯ MEDICAL APPLICATION

Merkel cells are of clinical importance because Merkel cell carcinomas, though uncommon, are very aggressive and difficult to treat. Merkel cell carcinoma is 40 times less common than malignant melanoma but has twice the mortality of that disease.

❯ DERMIS

The **dermis** is the layer of connective tissue (Figures 18–1 and 18–2) that supports the epidermis and binds it to the subcutaneous tissue (hypodermis). The thickness of the dermis varies with the region of the body and reaches its maximum of 4 mm on the back. The surface of the dermis is very irregular and has many projections (dermal papillae) that interdigitate with projections (epidermal pegs or ridges) of the epidermis (Figure 18–1), especially in skin subject to frequent pressure, where they reinforce the dermal-epidermal junction.

FIGURE **18–8** Langerhans cells.

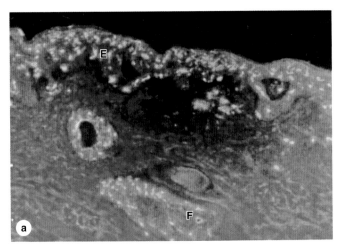

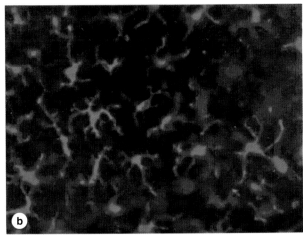

Langerhans cells are dendritic APCs of the epidermis where they comprise an important defense against pathogens and environmental insults. Like other APCs, they develop in the bone marrow, move into the blood circulation, and finally migrate into stratified squamous epithelia where they are difficult to identify in routinely stained sections.

(a) Section of immunostained skin shows Langerhans cells (yellow) abundant in hair follicles (**F**), where many microorganisms live, and throughout the epidermis (**E**). Keratin of the epidermis and follicles is stained green. (X40) Antibodies against langerin/CD207 and keratin.

(b) Face-on view of an epidermal sheet stained using the same antibody showing the network of Langerhans cells among the other epidermal cells, which detects invading microorganisms. After sampling the invaders' antigens, Langerhans cells leave the epidermis and travel to the nearest lymph node to elicit lymphocytes that can mount a collective immune response. (X200; Anti-langerin/CD207)

(Reproduced, with permission, from Romani N, et al. Acta Path Micro Immunol Scandinavica. 2003;111:725.)

FIGURE **18–9** Merkel or tactile cell.

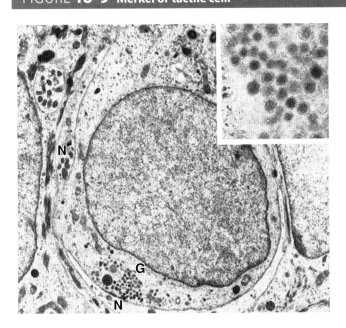

Merkel cells in the basal epidermal layer have high tactile sensitivity and function as mechanoreceptors. This TEM of a Merkel cell shows a mass of dense-core cytoplasmic granules (**G**) near the basolateral cell membrane, which is in direct contact with the expanded, disc-like ending of a sensory nerve (**N**). (X14,000; Inset: Granules are similar in morphology and content to the granules of many neuroendocrine cells. X61,500)

A **basement membrane** always occurs between the stratum basale and the dermis, and follows the contour of the interdigitations between these layers. Nutrients for keratinocytes diffuse into the avascular epidermis from the dermal vasculature through the basement membrane.

The dermis contains two sublayers with indistinct boundaries (Figure 18–1; Table 18–1):

- The thin **papillary layer**, which includes the dermal papillae, consists of loose connective tissue, with types I and III collagen fibers, fibroblasts and scattered mast cells, dendritic cells, and leukocytes. From this layer, anchoring fibrils of type VII collagen insert into the basal lamina, helping to bind the dermis to the epidermis.
- The underlying **reticular layer** is much thicker, consists of dense irregular connective tissue (mainly bundles of type I collagen), with more fibers and fewer cells than the papillary layer. A network of elastic fibers is also present (Figure 18–10), providing elasticity to the skin. Between the collagen and elastic fibers are abundant proteoglycans rich in dermatan sulfate.

Both dermal regions contain a rich network of blood and lymphatic vessels. Nutritive vessels form two major plexuses (Table 18–1):

- Between the papillary and reticular dermal layers lies the microvascular **subpapillary plexus**, from which capillary branches extend into the dermal papillae and form a rich, nutritive capillary network just below the epidermis.

FIGURE 18–10 **Elastic fibers of dermis.**

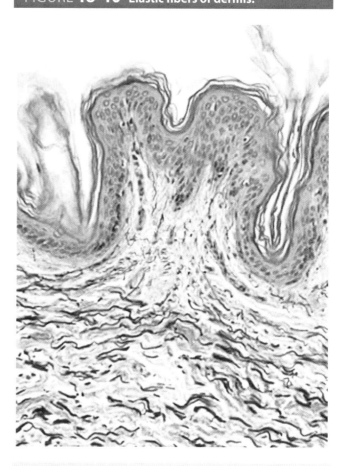

A section of thin skin stained for elastic fibers shows the extensive distribution of these darkly stained fibers among the eosinophilic collagen bundles. In the dermal papillary layer, the diameter of fibers decreases as they approach the epidermis and insert into the basement membrane. (X100; Weigert elastic stain)

- A deep plexus with larger blood and lymphatic vessels lies near the interface of the dermis and the subcutaneous layer.

In addition to the nutritive function, dermal vasculature has a thermoregulatory function, which involves numerous **arteriovenous anastomoses** or shunts (see Chapter 11) located between the two major plexuses. The shunts decrease blood flow in the papillary layer to minimize heat loss in cold conditions and increase this flow to facilitate heat loss when it is hot, thus helping maintain a constant body temperature. Lymphatic vessels begin in the dermal papillae and converge to form two plexuses located with the blood vessels.

The dermis is also richly innervated. Sensory afferent nerve fibers form a network in the papillary dermis and around hair follicles, ending at epithelial and dermal receptors shown in Figure 18–11. Autonomic effector nerves to dermal sweat glands and smooth muscle fibers in the skin of some

FIGURE **18–11** Tactile receptors.

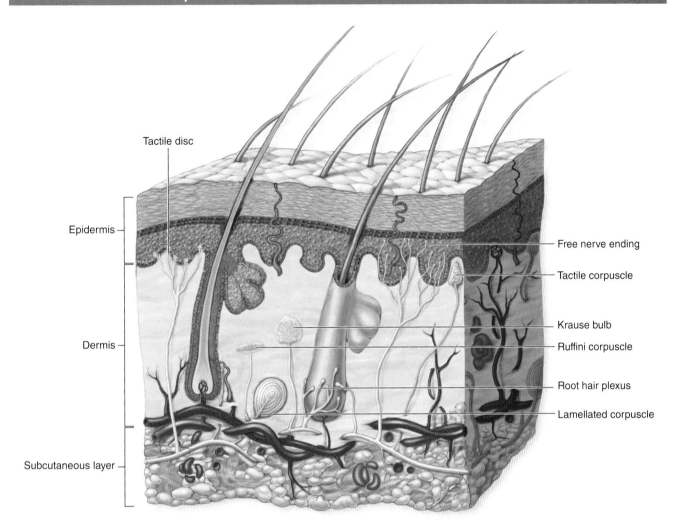

Tactile disc

Epidermis

Dermis

Subcutaneous layer

Free nerve ending

Tactile corpuscle

Krause bulb

Ruffini corpuscle

Root hair plexus

Lamellated corpuscle

Skin contains several types of **sensory receptors**, with or without capsules of collagen and modified Schwann cells. Most are difficult to see in routine preparations. In the epidermis are **free nerve endings** and tactile discs of nerve fibers associated with **Merkel cells** in the basal layer. Both have unencapsulated nerve fibers, as does the **root hair plexus** around the bases of hair follicles in the dermis. They detect light touch or movement of hair, although epidermal free nerve endings also detect pain and temperature extremes.

More complex, encapsulated tactile receptors are located in the dermis and hypodermis, and include **Meissner corpuscles** for light touch, **lamellated (pacinian) corpuscles** detecting pressure and high-frequency vibration, **Krause end bulbs** for low-frequency vibrations/movements, and **Ruffini corpuscles** detecting tissue distortion. The latter two receptors are less widely distributed in skin and less commonly seen.

areas are postganglionic fibers of sympathetic ganglia; no parasympathetic innervation is present.

❱ SUBCUTANEOUS TISSUE

The **subcutaneous layer** (see Figure 18–1) consists of loose connective tissue that binds the skin loosely to the subjacent organs, making it possible for the skin to slide over them. This layer, also called the **hypodermis** or **superficial fascia**, contains adipocytes that vary in number in different body regions

and vary in size according to nutritional state. The extensive vascular supply at the subcutaneous layer promotes rapid uptake of insulin or drugs injected into this tissue.

❱ SENSORY RECEPTORS

With its large surface and external location, the skin functions as an extensive receiver for various stimuli from the environment. Diverse sensory receptors are present in skin, including both simple nerve endings with no Schwann cell or collagenous coverings and more complex structures with sensory fibers enclosed by glia and delicate connective tissue capsules (Figure 18–11). The unencapsulated receptors include the following:

- The **Merkel cells**, each associated with expanded nerve endings (Figure 18–9), which function as tonic receptors for sustained light touch and for sensing an object's texture.
- **Free nerve endings** in the papillary dermis and extending into lower epidermal layers, which respond primarily to high and low temperatures, pain, and itching, but also function as tactile receptors.

- **Root hair plexuses**, a web of sensory fibers surrounding the bases of hair follicles in the reticular dermis that detects movements of the hairs.

The encapsulated receptors are all phasic mechanoreceptors, responding rapidly to stimuli on the skin. Four are recognized in human skin, although only the first two are seen in routine preparations:

- **Meissner corpuscles** are elliptical structures, 30-75 μm by 50-150 μm, consisting of sensory axons winding among flattened Schwann cells arranged perpendicular to the epidermis in the dermal papillae (Figure 18–12a).

FIGURE **18–12**   **Meissner and lamellated (pacinian) corpuscles.**

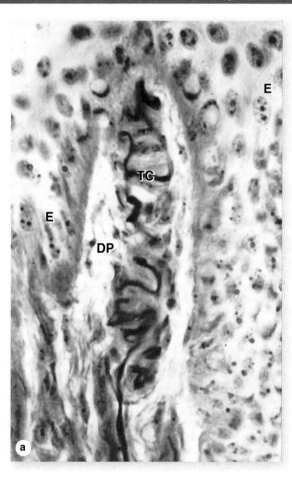
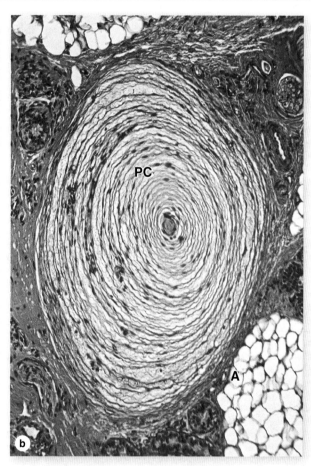

Micrographs show the two most commonly seen sensory receptors of skin.

(a) Meissner tactile corpuscles (**TC**) are specialized to detect light touch and are frequently located in dermal papillae (**DP**), partially surrounded by epidermis (**E**). They are elliptical, approximately 150-μm long, with an outer capsule (from the perineurium) and thin, stacked inner layers of modified Schwann cells, around which course nerve fibers. (X400; H&E)

(b) Lamellated (pacinian) corpuscles (**PC**) detect coarse touch or pressure and are large oval structures, frequently 1 mm in length,

found among adipose tissue (**A**) deep in the reticular dermis or in the subcutaneous tissue. Here the outer connective tissue capsule surrounds 15-50 thin, concentric layers of modified Schwann cells, each separated by slightly viscous interstitial fluid. Several axons enter one end of the corpuscle and lie in the cylindrical, inner core of the structure. Movement or pressure of this corpuscle from any direction displaces the inner core, leading to a nerve impulse. (X40; H&E)

They initiate impulses when light-touch or low-frequency stimuli against skin temporarily deform their shape. They are numerous in the fingertips, palms, and soles but decline slowly in number during aging after puberty.

- **Lamellated (pacinian) corpuscles** are large oval structures, approximately 0.5 mm by 1 mm, found deep in the reticular dermis and hypodermis, with an outer capsule and 15-50 thin, concentric lamellae of flattened Schwann cells and collagen surrounding a highly branched, unmyelinated axon (Figure 18–12b). Lamellated corpuscles are specialized for sensing coarse touch, pressure (sustained touch), and vibrations, with distortion of the capsule amplifying a mechanical stimulus to the axonal core where an impulse is initiated. Pacinian corpuscles are also found in the connective tissue of organs located deep in the body, including the wall of the rectum and urinary bladder, where they also produce the sensation of pressure when the surrounding tissue is distorted.
- **Krause end bulbs** are simpler encapsulated, ovoid structures, with extremely thin, collagenous capsules penetrated by a sensory fiber. They are found primarily in the skin of the penis and clitoris where they sense low-frequency vibrations.
- **Ruffini corpuscles** have collagenous, fusiform capsules anchored firmly to the surrounding connective tissue, with sensory axons stimulated by stretch (tension) or twisting (torque) in the skin.

❯ HAIR

Hairs are elongated keratinized structures that form within epidermal invaginations, the **hair follicles** (Figure 18–13). The color, size, shape, and texture of hairs vary according to age, genetic background, and region of the body. All skin has at least minimal hair except the glabrous skin of the palms, soles, lips, glans penis, clitoris, and labia minora. The face has about 600 hairs/cm² and the remainder of the body has about 60/cm². Hairs grow discontinuously, with periods of growth followed by periods of rest, and this growth does not occur synchronously in all regions of the body or even in the same area.

The growing hair follicle has a terminal dilation called a **hair bulb** (Figure 18–13a). A **dermal papilla** inserts into the base of the hair bulb and contains a capillary network required to sustain the hair follicle. Keratinocytes continuous with those of the basal epidermis cover the dermal papilla. These cells form the matrix of the elongating **hair root**; the part of a hair extending beyond the skin surface is the **hair shaft**.

The keratinocytes of the hair bulb are generally similar to those in the basal and spinous layers of epidermis. They divide rapidly in the region immediately around the dermal papilla and then undergo keratinization, melanin accumulation, and terminal differentiation. Melanocytes in the hair bulb matrix transfer melanosomes into the epithelial cells that will later differentiate to form the hair. Unlike the epidermis in which all keratinocytes give rise to the stratum corneum, cells in the hair root matrix differentiate with variable amounts and types of keratin. The keratin of hair is harder and more compact than that of the stratum corneum, maintaining its structure as the hair shaft much longer.

In most thick hairs large, vacuolated, and moderately keratinized cells form the central **medulla** of the hair root (Figures 18–13b and 18–14). Heavily keratinized, densely packed cells make up the **cortex** around the medulla. The most peripheral cells of the hair root comprise the **cuticle**, a thin layer of heavily keratinized, squamous cells covering the cortex (Figures 18–13c and 18–14).

The outermost cells of the hair bulb are continuous with the epithelial root sheath, in which two layers can be recognized. The **internal root sheath** completely surrounds the initial part of the hair root but degenerates above the level of the attached sebaceous glands. The **external root sheath** covers the internal sheath and extends all the way to the epidermis, where it is continuous with the basal and spinous layers. Separating the hair follicle from the dermis is an acellular hyaline layer, the thickened basement membrane called the **glassy membrane** (Figure 18–14b). The surrounding dermis forms a connective tissue sheath.

The **arrector pili muscle**, a small bundle of smooth muscle cells, extends from the midpoint of the fibrous sheath to the dermal papillary layer (Figure 18–13a). Contraction of these muscles pulls the hair shafts to a more erect position, usually when it is cold in an effort to trap a layer of warm air near the skin. In regions where hair is fine, contraction of arrector pili muscles is seen to produce tiny bumps on the skin surface ("goose bumps") where each contracting muscle distorts the attached dermis.

As mentioned earlier hairs grow asynchronously, cyclically, and at different rates in different regions of the body. The hair growth cycle has three major phases:

- A generally long period of mitotic activity and growth **(anagen)**
- A brief period of arrested growth and regression of the hair bulb **(catagen)**
- A final long period of inactivity **(telogen)** during which the hair may be shed

FIGURE 18–13 Hair.

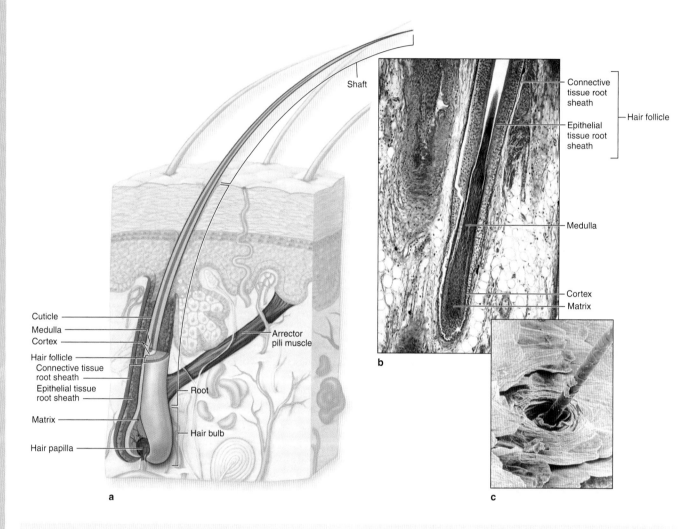

a

b

c

All types of body hair have a similar composition and form in hair follicles derived from the epidermis but extending deep into the dermis.

(a) The diagram shows major parts of a hair and its follicle, including vascularized, nutritive hair dermal **papilla** and the **arrector pili muscle** that pulls the hair erect.

(b) A longitudinal section of a hair root and bulb shows the **matrix**, **medulla**, and **cortex** in the root and the surrounding

epithelial and connective tissue sheaths. Cells of the hair bulb matrix proliferate, take up melanin granules, and undergo keratinization to differentiate as the three concentric layers of the hair. (X70; H&E)

(c) The outermost layer of the hair is the thin **cuticle**, composed of shingle-like cells, shown in this SEM of a hair shaft emerging at the stratum corneum. (X260)

At the beginning of the next anagen phase, epidermal stem cells located in small bulge of the external root sheath near the arrector pili muscle produce progenitor cells for the matrix of a new hair bulb. Hair growth on the face and pubis is strongly influenced by sex hormones, especially androgens, and begins at puberty.

is not well understood. Arresting mitotic activity in the hair matrix during cancer **chemotherapy** disrupts both the function and the structural integrity of hair follicles and usually leads to rapid, reversible alopecia.

❯❯ MEDICAL APPLICATION

Loss of hair to produce **baldness** or **alopecia** results from a complex combination of genetic and hormonal factors that

❯ NAILS

A similar process of keratinization also produces the **nails**, which are hard plates of keratin on the dorsal surface of each distal phalanx (Figure 18–15). The proximal part of the nail is the

FIGURE **18–14** Layers of a hair and its follicle.

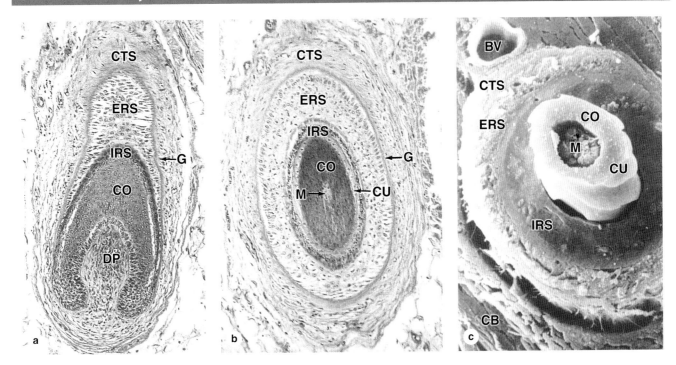

(a) The base of a hair follicle sectioned obliquely shows the vascularized dermal papilla (**DP**) continuous with the surrounding connective tissue sheath (**CTS**). The papilla is enclosed by the deepest part of the epithelial sheath, which is continuous with both the internal root sheath (**IRS**) and external root sheath (**ERS**). Both of these layers are in turn continuous with the stratified epidermis. Just outside the ERS is the glassy membrane (**G**) that is continuous with the basement membrane of the epidermis. The epithelial cells (keratinocytes) around the papilla proliferate and differentiate as the root of the hair itself. Above the papilla, only the cortex (**CO**) of the hair is clearly seen in this section. (X140; H&E)

(b) A hair root sectioned more transversely shows the same layers of the follicular sheath, but the layers of the hair root are now seen

to include the medulla (**M**), cortex (**CO**), and cuticle (**CU**). Other labels are like those of part **(a)**. (X140; H&E)

(c) SEM of a similar specimen gives a different perspective on these layers, including the shingle-like nature of the thin cuticle surface (**CU**), and the small blood vessel (**BV**) and collagen bundles (**CB**) near the surrounding connective tissue sheath (**CTS**). Other labels are like those of **(a)**. (X2600)

(Figure 18–14c, used with permission from Kessel RG, Kardon RH. Tissues and Organs: A Text-Atlas of Scanning Electron Microscopy. San Francisco, CA: W.H. Freeman & Co.; 1979.)

nail root and is covered by a fold of skin, from which the epidermal stratum corneum extends as the **cuticle**, or **eponychium**. The **nail plate** is bound to a bed of epidermis, the **nail bed**, which contains only the basal and spinous epidermal layers.

The nail root forms from the **nail matrix** in which cells divide, move distally, and become keratinized in a process somewhat similar to hair formation but without keratohyaline granules. The nail root matures and hardens as the nail plate (Figure 18–15). Continuous growth in the matrix pushes the nail plate forward over the nail bed (which makes no contribution to the plate) at a rate of about 3 mm/mo for fingernails and 1 mm/mo for toenails. The distal end of the plate becomes free of the nail bed at the epidermal fold called the **hyponychium**. The nearly transparent nail plate and the thin epithelium of the nail bed provide a useful window on the amount of oxygen in the blood by showing the color of blood in the dermal vessels.

❯ SKIN GLANDS

Sebaceous Glands

Sebaceous glands are embedded in the dermis over most of the body, except in the thick, glabrous skin of the palms and soles. There is an average of about 100 such glands per square centimeter of skin, but the frequency increases to 400-900/cm^2 in the face and scalp. Sebaceous glands are branched acinar glands with several acini converging at a short duct that usually empties into the upper portion of a hair follicle (Figure 18–16). A hair follicle and its associated sebaceous glands make up a **pilosebaceous unit**. The stem cell niche of the follicle's bulge region also forms the progenitor cells of the associated sebaceous glands. In certain hairless regions, such as the penis, clitoris, eyelids, and nipples, sebaceous ducts open directly onto the epidermal surface.

FIGURE **18–15** Nails.

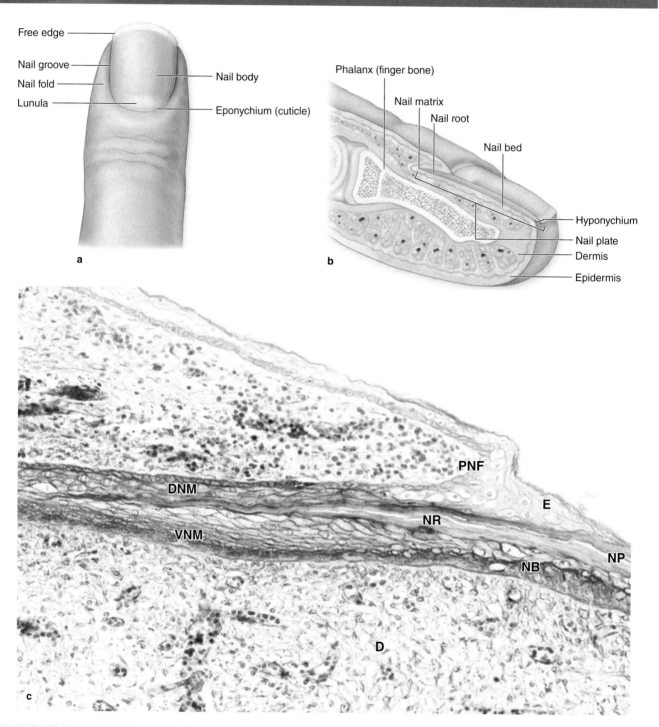

a

Free edge
Nail groove
Nail fold
Lunula
Nail body
Eponychium (cuticle)

b

Phalanx (finger bone)
Nail matrix
Nail root
Nail bed
Hyponychium
Nail plate
Dermis
Epidermis

c

PNF
DNM
E
NR
VNM
NB
NP
D

Nails are hard, keratinized derivatives formed in a process similar to that of the stratum corneum and hair.

(a) Surface view of a finger shows the nail's major parts, including the crescent-shaped white area called the **lunula**, which derives its color from the opaque nail matrix and immature nail plate below it.

(b) A diagrammatic sagittal section includes major internal details of the growing nail and the hyponychium where the free end of the nail plate is bound to epidermis.

(c) A sagittal section from a finger shows the proximal nail fold (**PNF**) and its epidermal extension, the eponychium (**E**) or cuticle.

The nail root (**NR**), the most proximal region of the nail plate (**NP**), is formed like the hair root by a matrix of proliferating, differentiating keratinocytes. These cells make up the dorsal nail matrix (**DNM**) and ventral nail matrix (**VNM**), which contribute keratinized cells to the nail root. The mature nail plate remains attached to the nail bed (**NB**), which consists of basal and spinous epidermal layers over dermis (**D**), but is pushed forward on this bed by continuous growth in the nail matrix. (X100; Mallory trichrome)

FIGURE **18–16** Glands of skin.

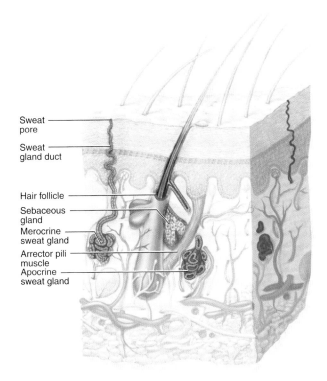

Sweat pore

Sweat gland duct

Hair follicle

Sebaceous gland

Merocrine sweat gland

Arrector pili muscle

Apocrine sweat gland

Skin includes three major types of exocrine glands. **Sebaceous glands** are usually part of a pilosebaceous unit with a hair follicle and secrete oily sebum into the space around the hair root. Thermoregulatory **eccrine sweat glands** empty their secretion onto the skin surface via sweat pores. **Apocrine sweat glands** secrete a more protein-rich sweat into the follicles of hair in skin of the axillae and perineum.

The acini of sebaceous glands are the classic example of holocrine secretion. They have a basal layer of flattened epithelial cells on the basal lamina, which proliferate and are displaced centrally, undergoing terminal differentiation as large, lipid-producing **sebocytes** filled with small fat droplets (Figure 18–17). Their nuclei shrink and undergo autophagy along with other organelles, and near the duct the cells disintegrate, releasing the lipids as the main secretory product. This product, called **sebum**, gradually covers the surfaces of both the epidermis and hair shafts.

Sebum is a complex mixture of lipids that includes wax esters, squalene, cholesterol, and triglycerides that are hydrolyzed by bacterial enzymes after secretion. Secretion from sebaceous glands increases greatly at puberty, stimulated primarily by testosterone in men and by ovarian and adrenal androgens in women. Sebum helps maintain the stratum corneum and hair shafts and exerts weak antibacterial and antifungal properties.

Sweat Glands

Sweat glands develop as long epidermal invaginations embedded in the dermis (Figure 18–1). There are two types of sweat glands, eccrine and apocrine, with distinct functions, distributions, and structural details.

Eccrine sweat glands (Figures 18–16 and 18–18) are widely distributed in the skin and are most numerous on the foot soles (620/cm^2). Collectively the 3 million eccrine sweat glands of the average person approximately equal the mass of a kidney and produce as much as 10 L/d, a secretory rate far exceeding that of other exocrine glands. Sweating is a physiologic response to increased body temperature during physical exercise or thermal stress and is the most effective means of temperature regulation of humans.

Both the secretory components and ducts of eccrine sweat glands are coiled and have small lumens. The **secretory part** is generally more pale-staining than the ducts and consists of an unusual stratified cuboidal epithelium with three cell types (Figure 18–18b):

- Pale-staining **clear cells** located on the basal lamina produce the sweat, having abundant mitochondria and microvilli to provide large surface areas. Interstitial fluid from the capillary-rich dermis around the gland is transported through the clear cells, either directly into the gland's lumen or into intercellular canaliculi that open to the lumen.
- **Dark cells** filled with strongly eosinophilic granules line most of the lumen and do not contact the basal lamina (Figure 18–18). The granules undergo merocrine secretion to release a poorly understood mixture of glycoproteins with bactericidal activity.
- **Myoepithelial cells** on the basal lamina (Figure 18–18b) contract to move the watery secretion into the duct.

The **ducts** of eccrine sweat glands have two layers of more acidophilic cells filled with mitochondria and having cell membranes rich in Na$^+$, K$^+$-ATPase. These duct cells absorb Na$^+$ ions from the secreted water to prevent excessive loss of this electrolyte. At the epidermis each duct merges with the stratum basale and sweat flow continues in a spiraling channel through the five epidermal strata to an excretory **sweat pore** in the skin surface (Figures 18–2a and 18–16). Sweat quickly evaporates upon release, cooling the skin and the blood

FIGURE **18–17** Sebaceous glands.

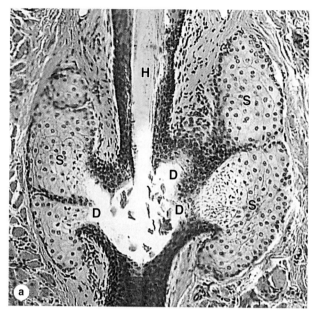

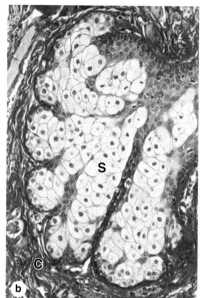

Sebaceous glands secrete a complex, oily mixture of lipids called **sebum** into short ducts that in most areas open into hair follicles. Sebum production is the classic example of holocrine secretion, in which the entire cell dies and contributes to the secretory product.

(a) A section of a pilosebaceous unit shows acini composed of large sebocytes (**S**), which undergo terminal differentiation by

filling with small lipid droplets and then disintegrating near the ducts (**D**) opening at the hair (**H**). shaft (X122; H&E)

(b) A micrograph shows the gland's capsule (**C**) and differentiates sebocytes (**S**) at higher magnification. Proliferation of the small progenitor cells just inside the capsule continuously forces sebum into the ducts; myoepithelial cells are not present. (X400; H&E)

present there. Eccrine sweat glands also function as auxiliary excretory organs, eliminating small amounts of nitrogenous waste and excess salts.

Apocrine sweat glands are largely confined to skin of the axillary and perineal regions. Their development depends on sex hormones and is not complete and functional until after puberty. The secretory components of apocrine glands have much larger lumens than those of the eccrine glands (Figure 18–19) and consist of simple cuboidal, eosinophilic cells with numerous secretory granules that also undergo exocytosis. Thus the glands are misnamed: their cells show merocrine, not apocrine, secretion.

The ducts of apocrine glands are similar to those of the eccrine glands, but they usually open into hair follicles at the epidermis (Figure 18–16) and may contain the protein-rich product. The slightly viscous secretion is initially odorless but may acquire a distinctive odor as a result of bacterial activity. The production of pheromones by apocrine glands is well established in many mammals and is likely in humans, although in a reduced or vestigial capacity. Apocrine sweat glands are innervated by adrenergic nerve endings, whereas eccrine sweat glands receive cholinergic fibers.

❯❯ MEDICAL APPLICATION

The sweat of infants with cystic fibrosis (CF) is often salty and is commonly taken as indicative of this genetic disease. CF patients have defects in a transmembrane conductance regulator (CFTR) of epithelial cells that lead to disruptive accumulations of thick mucus in the respiratory and digestive tracts. Failure to remove salt from sweat is related to the same genetic defect.

❯ SKIN REPAIR

Skin has a good capacity for repair, which is important in this exposed and easily damaged organ. The process of **cutaneous wound healing**, whether initiated surgically or accidentally, involves several overlapping stages that vary in duration with the size of the wound (Figure 18–20). In the first phase blood from cut vessels coagulates in the wound, releasing polypeptide growth factors and chemokines from the disintegrating platelets. Neutrophils and macrophages undergo diapedesis locally and remove bacteria and debris from the wound.

FIGURE **18–18** Eccrine sweat glands.

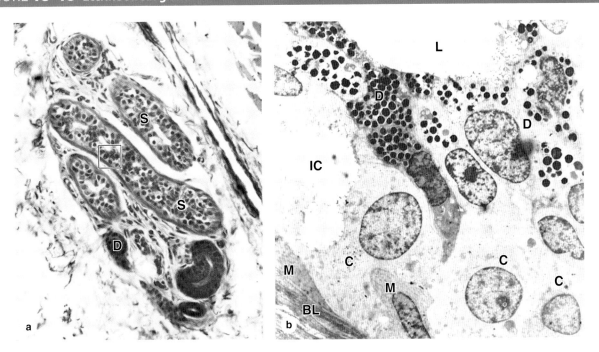

(a) Histologically eccrine glands have small lumens in the secretory components (**S**) and ducts (**D**), both of which have an irregular stratified cuboidal appearance. Both clear and acidophilic cells are seen in the stratified cuboidal epithelium of the secretory units. The box indicates an area with such cells like that shown ultrastructurally in part (b). (X200; Mallory trichrome)

(b) TEM of these important thermoregulatory structures reveals three cell types in their secretory portions. Myoepithelial cells (**M**) are present at the basal lamina (**BL**) to propel sweat into the duct. Irregular pyramidal cells called **dark cells** (**D**) border the lumen (**L**) and are filled with the electron-dense, eosinophilic secretory granules that release bactericidal peptides and other components of innate immunity. Columnar or cuboidal clear cells (**C**) on the basal lamina rapidly transport water from interstitial fluid in the capillary-rich dermis directly into the lumen or into intercellular canaliculi (**IC**) continuous with the lumen. Na⁺ ions are recovered from this fluid through the action of cells in the ducts. (X6500)

FIGURE **18–19** Apocrine sweat glands.

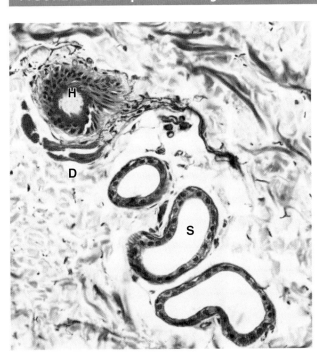

The secretory portions (**S**) of apocrine sweat glands have lumens that are much larger than those of eccrine sweat glands. The ducts (**D**) of apocrine glands also differ from those of eccrine glands in opening into hair follicles (**H**) rather than to the epidermal surface. (X200; Mallory trichrome)

FIGURE **18–20** Major stages of cutaneous wound healing.

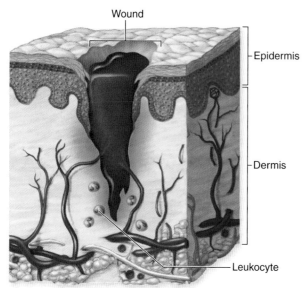

Wound
— Epidermis
— Dermis
— Leukocyte

① Cut blood vessels bleed into the wound.

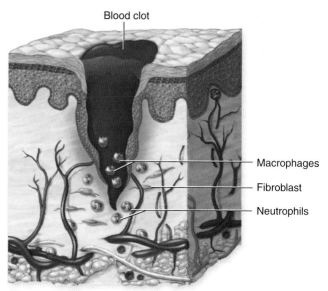

Blood clot
— Macrophages
— Fibroblast
— Neutrophils

② Blood clot forms, and leukocytes clean wound.

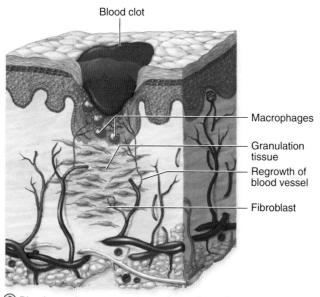

Blood clot
— Macrophages
— Granulation tissue
— Regrowth of blood vessel
— Fibroblast

③ Blood vessels regrow, and granulation tissue forms.

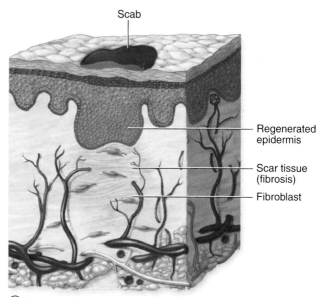

Scab
— Regenerated epidermis
— Scar tissue (fibrosis)
— Fibroblast

④ Epithelium regenerates, and connective tissue fibrosis occurs.

Skin repair occurs in overlapping stages shown here schematically. The process begins with blood quickly clotting at the wound site, releasing platelet-derived growth factors and other substances (**1**). Macrophages and neutrophils enter the wound as inflammation begins, and epithelial cells from the cut edges of the stratum basale begin to migrate beneath and through the blood clot (**2**).

Under the influence of growth factors and hydrolytic enzymes released in part from macrophages, fibroblasts proliferate and produce much new collagen to form "granulation tissue" containing many new, growing capillaries (**3**). The epidermis gradually reestablishes continuity over the wound site, but excessive collagen usually remains in the dermis as scar tissue (**4**).

These are major events of **inflammation** that typically lasts 2-3 days.

Before this phase is completed, **epithelialization** begins as cells of the epidermal basal layer remove their desmosomes and hemidesmosomes and migrate laterally beneath the blood clot that becomes an increasingly desiccated eschar, or scab. If much epidermis has been lost, new cells may migrate from the bulge region of surviving hair follicles. Growth of epidermal cells and fibroblasts is stimulated by several different growth factors released from macrophages and other cells and from their binding sites in ECM proteoglycans. Matrix metalloproteinases and other proteases from the migrating cells and macrophages facilitate cell migration.

Proliferating fibroblasts and newly sprouted capillaries produce new collagen-rich, well-vascularized tissue in the dermis called **granulation tissue**, which gradually replaces the blood clot (Figure 18–20). In the final stage the epidermis reestablishes continuity but has lost the ability to form new hair or glands. The granulation tissue undergoes **remodeling** and a more normal vasculature is reestablished. Collagen bundles and fibroblasts in this new connective tissue are at least initially much more abundant and disorganized than in uninjured skin, producing **scar tissue** at the wound site.

Skin SUMMARY OF KEY POINTS

- Skin consists mainly of a superficial stratified squamous epithelium, the **epidermis**, and a thicker layer of connective tissue, the **dermis**, which overlies a **subcutaneous hypodermis**.

Epidermis

- The epidermis consists of **keratinocytes** that undergo a terminal differentiation process called **keratinization** in a series of steps that form distinct epidermal strata or layers.
- The **stratum basale** is one layer of mitotically active cuboidal cells attached by **hemidesmosomes** and integrins to the basement membrane and to each other by **desmosomes**.
- The **stratum spinosum** has several layers of polyhedral cells attached to each other by desmosomes at the tips of short projections containing **bundled keratin, or tonofibrils**.
- The **stratum granulosum** is a thinner layer of keratinocytes, now flattened and filled densely with **keratohyaline granules** containing **filaggrin** and other proteins binding the tonofibrils.
- The superficial **stratum corneum** protects against water loss, friction, and microbial invasion, and consists of flattened, terminally differentiated cells, or **squames**, which are slowly lost.
- The epidermis-dermis interface is enlarged and strengthened by interdigitating **epidermal ridges** or **pegs** and **dermal papillae** in which microvasculature also supplies nutrients and O_2 for the epidermis.
- **Melanocytes** in the basal epidermis synthesize dark melanin pigment in **melanosomes** and transport these to adjacent keratinocytes, which accumulate them to protect nuclear DNA from UV damage.
- Antigen-presenting cells (APCs) called **Langerhans cells** form a network through the epidermis, intercepting and sampling microbial invaders before moving to lymph nodes in an adaptive immune response.

Dermis

- The **dermis** has two major layers: a superficial **papillary layer** or loose connective tissue with a microvascular plexus, and a thicker dense irregular **reticular layer** containing larger blood vessels.

Cutaneous Sensory Receptors

- Sensory receptors in the epidermis include **free nerve endings**, which detect pain and temperature extremes, and basal **Merkel cells**, light-touch (tactile) receptors associated with sensory fibers.
- Other cutaneous sensory structures include **Meissner corpuscles**, encapsulated elliptical mechanoreceptors that surround sensory axons and also detect light touch.
- Deeper in the dermis and subcutaneous layer are **lamellated or pacinian corpuscles**, which are ovoid and much larger than Meissner corpuscles, for detection of pressure or firm touch.

Epidermal Appendages

- **Hairs** form in **hair follicles**, in which keratinocytes comprising the matrix of the deep **hair bulb** proliferate rapidly and undergo keratinization to form the medulla, cortex, and cuticle of a **hair root**.
- A large dermal **hair papilla** penetrates the base of the hair bulb, and its vasculature supplies nutrients and O_2 for proliferating and differentiating cells.
- The growing **hair root** is surrounded by **internal and external root sheaths** continuous with the epidermis, a **glassy membrane** formed in part by the basal lamina, and a **connective tissue sheath**.
- **Nails** are formed in a manner similar to hairs: keratinocytes proliferate in the matrix of the **nail root** and differentiate with the formation of **hard keratin** as a growing **nail plate** with edges covered by skin folds.
- **Sebaceous glands** produce **sebum** by terminal differentiation of **sebocytes**, the classic example of **holocrine secretion**, secreting this oily substance onto hair in the follicles or **pilosebaceous units**.
- **Eccrine sweat glands** in the dermis produce sweat that is mostly water onto the skin surface, where its evaporation provides an important mechanism for cooling the body.
- **Apocrine sweat glands** are restricted to skin of the axillae and perineum, have much wider lumens than eccrine glands, develop after puberty, and secrete protein-rich sweat onto the hair of hair follicles.

Skin ASSESS YOUR KNOWLEDGE

1. Which of the following components of the epidermis provides sealant between adjacent cells?
   a. Keratohyaline granules
   b. Glycolipids and lipids
   c. Keratin
   d. Desmosomes
   e. Adherent junctions

2. Which cells derive from precursors originating in the bone marrow and function as antigen-presenting cells in skin?
   a. Keratinocytes
   b. Langerhans cells
   c. Melanocytes
   d. Merkel cell
   e. Arrector pili

3. Cells responsible for producing the pigment for dark hair are located in which of following?
   a. The cortex of the hair shaft
   b. Throughout the hair shaft
   c. The internal root sheath of the hair
   d. The dermal papilla of the hair bulb
   e. The hair matrix (zone of dividing and differentiating cells) of the hair bulb

4. Which of the following separates the hair follicle from the connective tissue of the dermis?
   a. External root sheath
   b. Internal root sheath
   c. Glassy membrane
   d. Hair cuticle
   e. Medulla

5. Which structure typifies reticular dermis but not papillary dermis?
   a. Capillaries
   b. Dense irregular connective tissue
   c. Meissner tactile corpuscles
   d. Sweat gland ducts
   e. Type I collagen fibers

6. Which of the following best characterizes sebaceous glands?
   a. Its duct drains onto the skin surface.
   b. It releases its contents via holocrine secretion.
   c. It primarily secretes water and salts.
   d. Its secretory units are supplied by adrenergic stimulation.
   e. It is located typically in the reticular dermis.

7. Myoepithelial cells aid in the secretory process of which of the following?
   a. Melanocytes
   b. Sebaceous glands
   c. Keratinocytes of the granular layer
   d. Eccrine sweat glands
   e. Apocrine sweat glands

8. A 52-year-old woman presents with severe blistering over her buttocks. Analysis of her serum demonstrates the presence of antibodies which by imunohistochemical techniques stain material located at the basement membrane of the epidermis in a biopsy of her skin. The underlying biological mechanism of her skin disorder involves an abnormality in which of the following structures?
   a. Macula adherens
   b. Gap junctions
   c. Hemidesmosomes
   d. Zonula occludens (tight junctions)
   e. Zonula adherens

9. A 64-year-old woman, who has always been proud of her suntanned, healthy look, is referred to a dermatologist with a blue-violet, painless, 1.5-cm lump in the skin of her left shoulder. The lump is firm and cannot be moved, and has grown very rapidly over the past few weeks. The mass is removed surgically and the pathologist diagnoses it as a Merkel cell carcinoma. If the UV radiation to which her skin was exposed affected the Merkel cells, what other cell type sharing the same specific epidermal layer might also be affected?
   a. Fibroblasts of the papillary layer
   b. Keratinocytes of the stratum granulosum
   c. Cells of tactile (Meissner) corpuscles
   d. Keratinized epithelial cells
   e. Basal stem cells for keratinocytes

10. A 37-year-old woman presents with a suspected Schwannoma. The radiology report indicates "a soft tissue mass to the right of L1 at the level of the L1 to L2 neural foramen." The neurologist presses the base of a vibrating 128 cps tuning fork to the skin of the patient's right and left thighs and asks her to describe the sensation. She asks the patient to close her eyes and then to tell her whether the tuning fork is vibrating or not. With that instrument the doctor is primarily testing the function of which of the following sensory receptors?
    a. Lamellated (Pacinian) corpuscles
    b. Kraus end bulbs
    c. Meissner corpuscles
    d. Merkel cells
    e. Free nerve endings

Answers: 1b, 2b, 3e, 4c, 5b, 6b, 7d, 8c, 9e, 10a

19 The Urinary System

The urinary system consists of the paired kidneys and ureters, the bladder, and the urethra. This system's primary role is to ensure optimal properties of the blood, which the kidneys continuously monitor. This general role of the kidneys involves a complex combination of renal functions:

- Regulation of the balance between water and electrolytes (inorganic ions) and the acid-base balance;
- Excretion of metabolic wastes along with excess water and electrolytes in urine, the kidneys' excretory product which passes through the ureters for temporary storage in the bladder before its release to the exterior by the urethra;
- Excretion of many bioactive substances, including many drugs;
- Secretion of **renin**, a protease important for regulation of blood pressure by cleaving circulating angiotensinogen to angiotensin I;
- Secretion of **erythropoietin**, a glycoprotein growth factor that stimulates erythrocyte production in red marrow when the blood O_2 level is low;
- Conversion of the steroid prohormone vitamin D, initially produced in the skin, to the active form (**1,25-dihydroxyvitamin D$_3$** or calcitriol); and
- Gluconeogenesis during starvation or periods of prolonged fasting, making glucose from amino acids to supplement this process in the liver.

› KIDNEYS

Approximately 12-cm long, 6-cm wide, and 2.5-cm thick in adults, each kidney has a concave medial border, the **hilum**—where nerves enter, the ureter exits, and blood and lymph vessels enter and exit—and a convex lateral surface, both covered by a thin fibrous capsule (Figure 19–1). Within the hilum the upper end of the ureter expands as the **renal pelvis** and divides into two or three **major calyces**. Smaller branches,

the minor calyces, arise from each major calyx. The area surrounding the renal pelvis and calyces contains adipose tissue.

The parenchyma of each kidney has an outer **renal cortex**, a darker stained region with many round corpuscles and tubule cross sections, and an inner **renal medulla** consisting mostly of aligned linear tubules and ducts (Figure 19–1). The renal medulla in humans consists of 8-15 conical structures called **renal pyramids**, all with their bases meeting the cortex (at the corticomedullary junction) and separated from each other by extensions of the cortex called renal columns. Each pyramid plus the cortical tissue at its base and extending along its sides constitutes a **renal lobe**. Parallel ducts and tubules extending from the medulla into the cortex comprise the medullary rays; these plus their associated cortical tissue are considered renal lobules. The tip of each pyramid, called the **renal papilla**, projects into a minor calyx that collects urine formed by tubules in one renal lobe (Figure 19–1).

Kidneys each contain 1-4 million functional units called **nephrons** (Figure 19–2), each consisting of a corpuscle and a long, simple epithelial renal tubule with three main parts along its length. The major divisions of each nephron are:

- **Renal corpuscle**, an initial dilated part enclosing a tuft of capillary loops and the site of blood filtration, always located in the cortex;
- **Proximal tubule**, a long convoluted part, located entirely in the cortex, with a shorter straight part that enters the medulla;
- **Loop of Henle** (or nephron loop), in the medulla, with a **thin descending** and a **thin ascending** limb;
- **Distal tubule**, consisting of a thick straight part ascending from the loop of Henle back into the cortex and a convoluted part completely in the cortex; and
- **Connecting tubule**, a short minor part linking the nephron to collecting ducts.

FIGURE **19–1** Kidney.

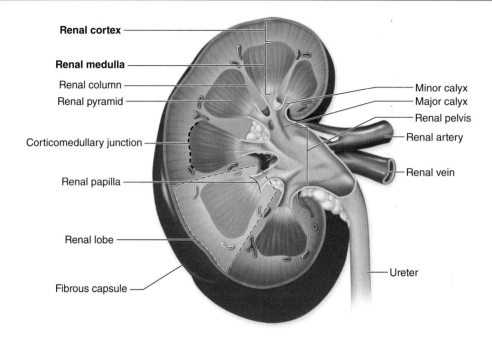

Each kidney is bean-shaped, with a concave **hilum** where the ureter and the renal artery and vein enter. The ureter divides and subdivides into several **major and minor calyces**, around which is located the renal sinus containing adipose tissue.

Attached to each minor calyx is a **renal pyramid**, a conical region of medulla delimited by extensions of **cortex**. The cortex and hilum are covered with a fibrous capsule.

Connecting tubules from several nephrons merge to form **collecting tubules** that then merge as larger **collecting ducts**. These converge in the renal papilla, where they deliver urine to a minor calyx. **Cortical nephrons** are located almost completely in the cortex while **juxtamedullary nephrons** (about one-seventh of the total) lie close to the medulla and have long loops of Henle.

❯ BLOOD CIRCULATION

As expected for an organ specialized to process the blood, the kidney vasculature is large, well-organized, and closely associated with all components of the nephron. Blood vessels of the kidneys are named according to their locations or shapes (Figure 19–3).

Each kidney's **renal artery** divides into two or more segmental arteries at the hilum. Around the renal pelvis,

these arteries branch further as the **interlobar arteries**, which extend between the renal pyramids toward the corticomedullary junction (Figure 19–3). Here the interlobar arteries divide again to form the **arcuate arteries** that run in an arc along this junction at the base of each renal pyramid. Smaller **interlobular arteries** (or cortical radial arteries) radiate from the arcuate arteries, extending deeply into the cortex.

From the interlobular arteries arise the microvascular **afferent arterioles**, which divide to form a plexus of capillary loops called the **glomerulus**, each of which is located within a renal corpuscle where the blood is filtered (Figures 19–3 and 19–4). Blood leaves the glomerular capillaries, not via venules, but via **efferent arterioles**, which at once branch again to form another capillary network, usually the **peritubular capillaries** profusely distributed throughout the cortex. From the juxtaglomerular corpuscles near the medulla, efferent arterioles do not form peritubular capillaries, but instead branch repeatedly to form parallel tassel-like bundles of capillary loops called the **vasa recta** (L. *recta*, straight) that penetrate deep into the medulla in association with the loops of Henle and collecting ducts. Collectively, the cortex receives over 10 times more blood than the medulla.

Blood leaves the kidney in veins that follow the same courses as arteries and have the same names (Figure 19–3). The outermost peritubular capillaries and capillaries in the

FIGURE 19–2 A nephron and its parts.

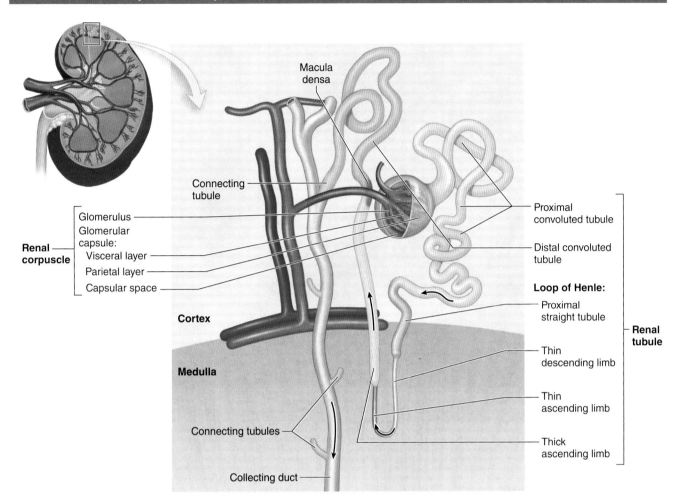

Each kidney contains 1-4 million functional units called **nephrons**. Each nephron originates in the cortex, at the **renal corpuscle** surrounding a small tuft of glomerular capillaries. Extending from the corpuscle is the long **proximal convoluted tubule** which leads to the short **proximal straight tubule** that enters the outer medulla. This tubule continues as the **thin descending limb** and the **thin ascending limb** of the nephron's loop of Henle in the medulla. The loop of Henle ends with a **thick ascending limb**, a straight tubule that reenters the cortex and ends at its thickened **macula densa** area where it contacts the arterioles entering the glomerulus. Beyond the macula densa this tubule is the **distal convoluted tubule**, the end of which is the short **connecting tubule**. Connecting tubules from many nephrons merge into cortical collecting tubules and a **collecting duct** that transports urine to the calyx.

kidney capsule converge into small stellate veins that empty into the interlobular veins.

There are many different **glomerular diseases** involving the renal corpuscles, with different causes calling for different treatments. Accurate diagnoses of such disorders by pathologists require sampling of the cortex and may involve examination of the renal corpuscles by immunofluorescence light microscopy or even by TEM.

> RENAL FUNCTION: FILTRATION, SECRETION, & REABSORPTION

All the major functions of the kidneys—the removal of metabolic wastes and excess water and electrolytes from blood—are performed by various specialized epithelial cells of the nephrons and collecting systems. Renal function involves specific activities:

■ **Filtration**, by which water and solutes in the blood leave the vascular space and enter the lumen of the nephron

FIGURE **19–3** Blood supply to the kidneys.

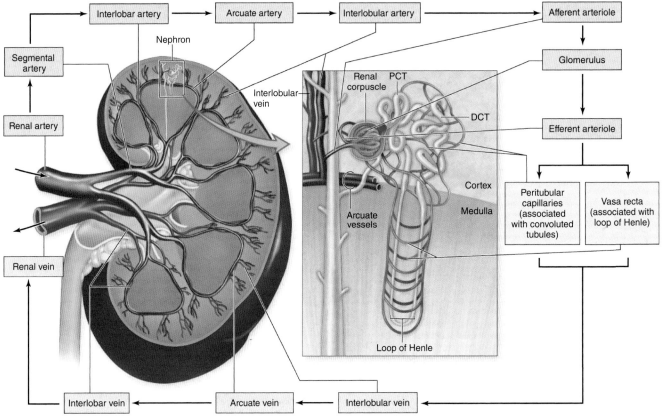

A coronal view of a kidney (left) shows the major blood vessels, with their names. An expanded diagram (right) includes the microvascular components extending into the cortex and medulla from the interlobular vessels. Pink boxes indicate vessels with arterial blood and light blue indicates the venous return. The intervening lavender boxes and vessels indicate capillaries where specific reabsorbed substances reenter the blood.

■ Tubular **secretion**, by which substances move from epithelial cells of the tubules into the lumens, usually after uptake from the surrounding interstitium and capillaries

■ Tubular **reabsorption**, by which substances move from the tubular lumen across the epithelium into the interstitium and surrounding capillaries.

Along the length of the nephron tubule and collecting system, the filtrate receives various secreted molecules while others are reabsorbed and then enters the minor calyces as **urine** and undergoes **excretion**.

The number of nephrons decreases slightly in older adults, a process accelerated by high blood pressure. If a kidney is donated for transplant (unilateral nephrectomy), the remaining kidney undergoes compensatory growth with cellular hypertrophy in the proximal parts of the nephron tubules and an increase in the rate of filtration, which allow normal renal function to continue.

〉〉 MEDICAL APPLICATION

Inflammation within the glomeruli, or **glomerulonephritis**, which can range from acute or chronic, usually stems from humoral immune reactions. Varieties of this condition involve the deposition of circulating antibody-antigen complexes within glomeruli or circulating antibodies binding to either glomerular antigens or extraneous antigens deposited in the glomeruli. Regardless of the source the accumulating immune complexes can then elicit a local inflammatory response.

Renal Corpuscles & Blood Filtration

At the beginning of each nephron is a renal corpuscle, about 200 μm in diameter and containing a tuft of glomerular capillaries, surrounded by a double-walled epithelial capsule called the **glomerular (Bowman) capsule** (Figures 19–2 and 19–5).

FIGURE **19–4** Microvasculature of the renal cortex.

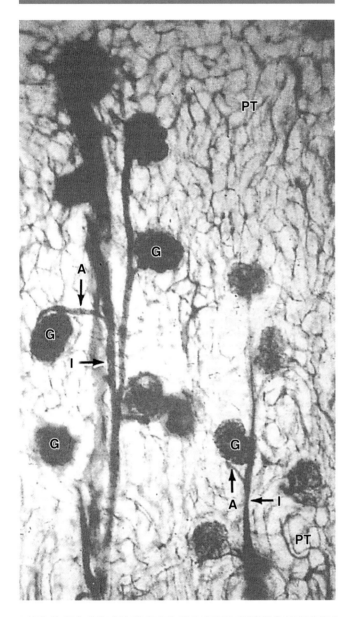

Cortical vasculature is revealed in a section of the kidney with the renal artery injected with carmine dye before fixation. Small interlobular arteries (**I**) branch from the arcuate arteries and radiate out through the cortex giving off the afferent arterioles (**A**) that bring blood to the glomerular capillaries. Each glomerulus (**G**) contains a mass of capillary loops that drain into an efferent arteriole. These then branches as a large, diffuse network of peritubular capillaries (**PT**) throughout the cortex. (X125)

The internal or **visceral layer** of this capsule closely envelops the glomerular capillaries, which are finely fenestrated. The outer **parietal layer** forms the surface of the capsule. Between the two capsular layers is the **capsular** (or **urinary) space**, which receives the fluid filtered through the capillary wall and visceral layer. Each renal corpuscle has a **vascular pole**,

where the afferent arteriole enters and the efferent arteriole leaves, and a **tubular pole**, where the proximal convoluted tubule (PCT) begins (Figure 19–5).

The outer parietal layer of a glomerular capsule consists of a simple squamous epithelium supported externally by a basal lamina. At the tubular pole, this epithelium changes to the simple cuboidal epithelium that continues and forms the proximal tubule (Figure 19–5).

The visceral layer of a renal corpuscle consists of unusual stellate epithelial cells called **podocytes** (Figures 19–5c, d), which together with the capillary endothelial cells compose the apparatus for renal filtration. From the cell body of each podocyte several **primary processes** extend and curve around a length of glomerular capillary. Each primary process gives rise to many parallel, interdigitating secondary processes or **pedicels** (L. *pedicellus*, little foot; Figures 19–5c, d). The pedicels cover much of the capillary surface, in direct contact with the basal lamina (Figures 19–5c and 19–6).

Between the interdigitating pedicels are elongated spaces, or filtration slit pores, 25- to 30-nm wide (Figures 19–5c and 19–6). Spanning adjacent pedicels and bridging the slit pores are zipper-like **slit diaphragms** (Figure 19–6). Slit diaphragms are modified and specialized occluding or tight junctions composed of **nephrins**, other proteins, glycoproteins, and proteoglycans important for renal function. Projecting from the cell membrane on each side of the filtration slit, these polyanionic glycoproteins and proteoglycans interact to form a series of openings within the slit diaphragm, with a surface that is negatively charged.

Between the highly fenestrated endothelial cells of the capillaries and the covering podocytes is the thick (300-360 nm) glomerular basement membrane (GBM) (Figure 19–6). This membrane is the most substantial part of the filtration barrier that separates the blood from the capsular space and forms by fusion of the capillary- and podocyte-produced basal laminae. Laminin and fibronectin in this fused basement membrane bind integrins of both the podocyte and endothelial cell membranes, and the meshwork of cross-linked type IV collagen and large proteoglycans restricts passage of proteins larger than about 70 kDa. Smaller proteins that are filtered from plasma are degraded, and the amino acids reabsorbed in the proximal tubule. Polyanionic GAGs in the glomerular membrane are abundant and their negative charges, like those of the slit diaphragms, tend to restrict filtration of organic anions.

Filtration, therefore, occurs through a structure with three parts:

- The fenestrations of the capillary endothelium, which blocks blood cells and platelets
- The thick, combined basal laminae, or GBM, which restricts large proteins and some organic anions
- The filtration slit diaphragms between pedicels, which restrict some small proteins and organic anions

Normally about 20% of the blood plasma entering a glomerulus is filtered into the capsular space. The initial

FIGURE **19–5** Renal corpuscles.

← Flow of blood	
← Flow of filtrate	

Parietal layer of glomerular capsule

Capsular space

Tubular pole

Vascular pole

Afferent arteriole

Proximal convoluted tubule

Juxtaglomerular apparatus:
Juxtaglomerular cell
Macula densa

Glomerulus

Podocyte of visceral layer of glomerular capsule

Distal tubule

Efferent arteriole

Pedicel

Endothelium of glomerulus

(a) Renal corpuscle

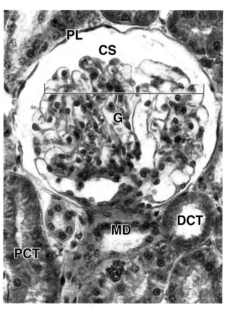

(b) Histology of renal corpuscle

Visceral layer of glomerular capsule

Pedicels   Filtration slits   Podocyte cell body

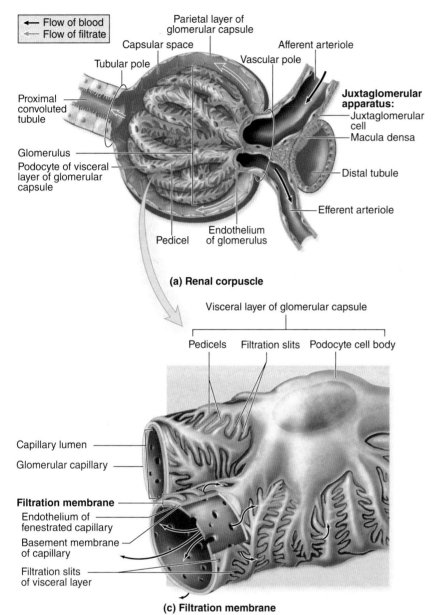

Capillary lumen

Glomerular capillary

Filtration membrane
Endothelium of fenestrated capillary
Basement membrane of capillary
Filtration slits of visceral layer

(c) Filtration membrane

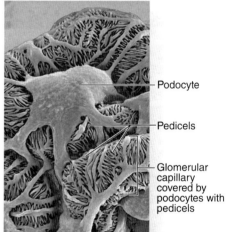

Podocyte

Pedicels

Glomerular capillary covered by podocytes with pedicels

(d) Podocytes

(a) The renal corpuscle is a small mass of capillaries called the **glomerulus** housed within a bulbous glomerular capsule. The internal lining of the capsule is composed of complex epithelial cells called **podocytes**, which cover each capillary, forming slit-like spaces between interdigitating processes called **pedicels**. Blood enters and leaves the glomerulus through the afferent and efferent arterioles, respectively.

(b) The micrograph shows the major histologic features of a renal corpuscle. The glomerulus (**G**) of capillaries is surrounded by the capsular space (**CS**) covered by the simple squamous parietal layer (**PL**) of Bowman capsule. Near the corpuscle is that nephron's

macula densa (**MD**) and sections of proximal convoluted tubules (**PCT**) and distal convoluted tubules (**DCT**). (H&E; X300)

(c) Filtrate is produced in the corpuscle when blood plasma is forced under pressure through the **capillary fenestrations**, across the filtration **membrane** or GBM surrounding the capillary, and through the **filtration slit** diaphragms located between the podocyte pedicels.

(d) The scanning electron microscopy (SEM) shows the distinctive appearance of podocytes and their pedicel processes that cover glomerular capillaries. (X800)

FIGURE **19–6** Glomerular filtration barrier.

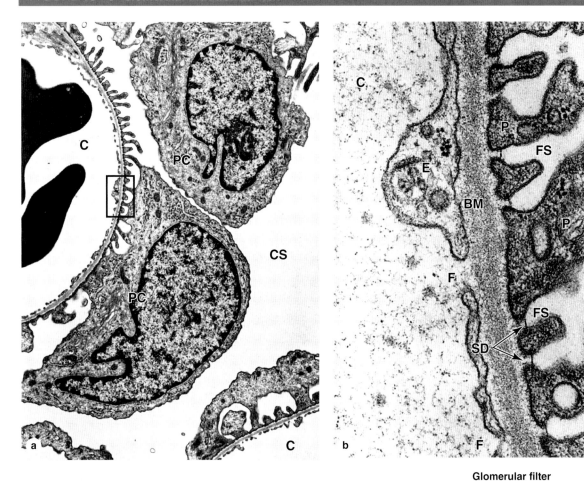

The glomerular filtration barrier consists of three layered components: the fenestrated **capillary endothelium**, the **glomerular basement membrane (GBM)**, and **filtration slit** diaphragms between pedicels. The major component of the filter is formed by fusion of the basal laminae of a podocyte and a capillary endothelial cell.

(a) TEM shows cell bodies of two podocytes (**PC**) and the series of pedicels on the capillary (**C**) basement membrane separated by the filtration slit diaphragms. Around the capillaries and podocytes is the capsular space (**CS**) into which the filtrate enters. The enclosed area is shown in part **(b)**. (X10,000)

(b) At higher magnification, both the fenestrations (**F**) in the endothelium (**E**) of the capillary (**C**) and the filtration slits (**FS**) separating the pedicels (**P**) are clearly seen on the two sides of the thick, fused basement membrane (**BM**). Thin slit diaphragms (**SD**) bridge the slits between pedicels. (X45,750)

(c) Diagram shows the three parts of the glomerular filter and their major functions.

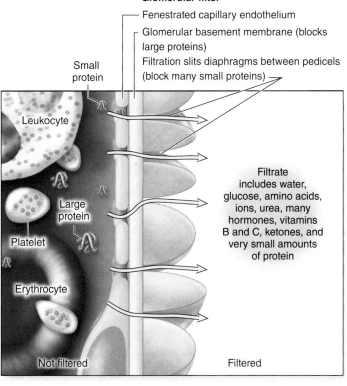

Glomerular filter

— Fenestrated capillary endothelium

— Glomerular basement membrane (blocks large proteins)

— Filtration slits diaphragms between pedicels (block many small proteins)

Small protein

Leukocyte

Large protein

Platelet

Erythrocyte

Not filtered

Filtrate includes water, glucose, amino acids, ions, urea, many hormones, vitamins B and C, ketones, and very small amounts of protein

Filtered

(c) Substances filtered by filtration membrane

glomerular filtrate has a chemical composition similar to that of plasma except that it contains very little protein. The glomerular filter blocks filtration of most plasma proteins, but smaller proteins, including most polypeptide hormones, are removed into the filtrate.

❯❯ MEDICAL APPLICATION

In diseases such as diabetes mellitus and glomerulonephritis, the glomerular filter is altered and becomes much more permeable to proteins, with the subsequent release of protein into the urine (**proteinuria**). Proteinuria is an indicator of many potential kidney disorders.

Capillaries of each glomerulus have a total length of approximately 1 cm and are uniquely situated between two arterioles—afferent and efferent—the muscle of which allows increased hydrostatic pressure in these vessels, favoring movement of plasma across the glomerular filter. The glomerular filtration rate (GFR) is constantly regulated by neural and hormonal inputs affecting the degree of constriction in each of these arterioles. The total glomerular filtration area of an adult has been estimated at 500 cm² and the average GFR at 125 mL/min or 180 L/d. Because the total amount of circulating plasma averages 3 L, it follows that the kidneys typically filter the entire blood volume 60 times every day.

In addition to capillary endothelial cells and podocytes, renal corpuscles also contain **mesangial cells** (Gr. *mesos*, in the midst + *angion*, vessel), most of which resemble vascular pericytes in having contractile properties and producing components of an external lamina. Mesangial cells are difficult to distinguish in routine sections from podocytes, but often stain more darkly. They and their surrounding matrix comprise the mesangium (Figure 19–7), which fills interstices between capillaries that lack podocytes. Functions of the mesangium include the following:

- Physical support of capillaries within the glomerulus
- Adjusted contractions in response to blood pressure changes, which help maintain an optimal filtration rate
- Phagocytosis of protein aggregates adhering to the glomerular filter, including antibody-antigen complexes abundant in many pathological conditions
- Secretion of several cytokines, prostaglandins, and other factors important for immune defense and repair in the glomerulus

Proximal Convoluted Tubule

Cells in many parts of the nephron tubule and collecting system reabsorb water and electrolytes, but other activities are restricted mainly to specific tubular regions. Table 19–1 summarizes major functions of parts within nephrons and collecting ducts, along with the histologic features involved in these activities.

FIGURE **19–7** Mesangium.

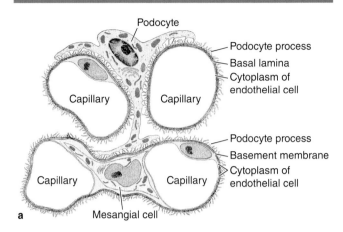

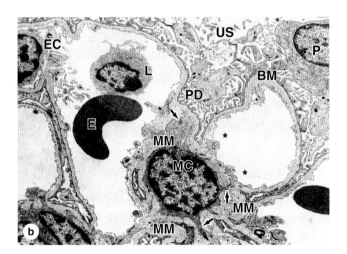

(a) Diagram shows that mesangial cells in renal corpuscles are located between capillaries and cover those capillary surface not covered by podocyte processes.

(b) The TEM shows one mesangial cell (**MC**) and the surrounding mesangial matrix (**MM**). This matrix appears similar to and in many places continuous with basement membrane (**BM**) and supports capillaries where podocytes are lacking. Mesangial cells extend contractile processes (arrows) along capillaries that help regulate blood flow in the glomerulus. Some mesangial processes appear to pass between endothelial cells (**EC**) into the capillary lumen (asterisks) where they may help remove or endocytose adherent protein aggregates. The capillary at the left contains an erythrocyte (**E**) and a lymphocyte (**L**). Podocytes (**P**) and their pedicels (**PD**) open to the urinary space (**US**) and associate with the capillary surfaces not covered by mesangial cells. (X3500)

At the tubular pole of the renal corpuscle, the simple squamous epithelium of the capsule's parietal layer is continuous with the simple cuboidal epithelium of the **proximal convoluted tubule (PCT)** (Figures 19–8 and 19–9). These long, tortuous tubules fill most of the cortex. PCT cells are specialized for both reabsorption and secretion. Over half of the water and electrolytes, and all of the organic nutrients

TABLE **19–1**	Histologic features and major functions of regions within renal tubules.		
Region of Tubule	**Histological Features**	**Locations**	**Major Functions**
PCT	Simple cuboidal epithelium; cells well-stained, with numerous mitochondria, prominent basal folds and lateral interdigitations; long microvilli, lumens often occluded	Cortex	Reabsorption of all organic nutrients, all proteins, most water and electrolytes; secretion of organic anions and cations, H^+, and NH_4^+
Loop of Henle			
Thin limbs	Simple squamous epithelium; few mitochondria	Medulla	Passive reabsorption of Na^+ and Cl^-
TAL	Simple cuboidal epithelium; no microvilli, but many mitochondria	Medulla and medullary rays	Active reabsorption of various electrolytes
DCT	Simple cuboidal epithelium; cells smaller than in PCT, short microvilli and basolateral folds, more empty lumens	Cortex	Reabsorption of electrolytes
Collecting system			
Principal cells	Most abundant, cuboidal to columnar; pale-staining, distinct cell membranes	Medullary rays and medulla	Regulated reabsorption of water & electrolytes; regulated secretion of K^+
Intercalated cells	Few and scattered; slightly darker staining	Medullary rays	Reabsorption of K^+ (low-K^+ diet); help maintain acid-base balance

DCT, distal convoluted tubule; PCT, proximal convoluted tubule; TAL, thick ascending limb.

(glucose, amino acids, vitamins, etc), filtered from plasma in the renal corpuscle are normally reabsorbed in the PCT. These molecules are transferred directly across the tubular wall for immediate uptake again into the plasma of the peritubular capillaries.

Transcellular **reabsorption** involves both active and passive mechanisms, with the cells having a large variety of transmembrane ion pumps, ion channels, transporters, enzymes, and carrier proteins. Water and certain solutes can also move passively between the cells (paracellular transport) along osmotic gradients through leaky apical tight junctions.

Small proteins in the filtrate are either reabsorbed by receptor-mediated endocytosis and degraded in the cuboidal cells, or degraded by peptidases on the luminal surface. In both cases the amino acids are released at the basolateral cell surfaces for uptake by capillaries.

Conversely, organic anions and cations not filtered in the renal corpuscle (because of the polyanions in the filter or binding to plasma proteins) may be released in the peritubular capillaries, taken up by the cells of the proximal tubules and undergo **secretion** into the filtrate (Table 19–1). Organic anion and cation transporters allow the kidneys to dispose of such substances at a higher rate than by glomerular filtration alone. Because these molecules include important substances (such as bile salts, creatinine, etc) and many antibiotics and other drugs, this process of tubular secretion is of great pharmacologic importance as a key mechanism of drug clearance.

The cells of the proximal tubules have central nuclei and very acidophilic cytoplasm (Figures 19–8 and 19–9) because of the abundant mitochondria. The cell apex has very many long microvilli that form a prominent **brush border** in the lumen that facilitates reabsorption (Figures 19–8 through 19–10). Because the cells are large, each transverse section of a PCT typically contains only three to five nuclei. In routine histologic preparations, the long brush border may be disorganized and give the lumens a fuzz-filled appearance. Peritubular capillaries are abundant in the sparse surrounding connective tissue interstitium, which fills only about 10% of the cortex (Figure 19–8).

Ultrastructurally the apical cytoplasm of these cells has numerous pits and **vesicles** near the bases of the microvilli, indicating active endocytosis and pinocytosis (Figure 19–10). These vesicles contain the small, reabsorbed proteins that will be degraded in lysosomes, with the amino acids released to the circulation. Proximal tubular cells also have many long basal **membrane invaginations** and lateral interdigitations with neighboring cells (Figure 19–10). Both the brush border and the basolateral folds contain the many types of transmembrane proteins that mediate tubular reabsorption and secretion. Long mitochondria concentrated along the basal invaginations (Figure 19–10) supply ATP locally for the membrane proteins involved in active transport. Because of the extensive interdigitations of the lateral membranes, discrete limits between cells of the proximal tubule are difficult to see with the light microscope.

Besides their major roles in reabsorption and secretion, cells of the proximal tubule also perform hydroxylation of vitamin D and release to the capillaries. Moreover, fibroblastic

FIGURE **19–8** Renal cortex: proximal and distal convoluted tubules.

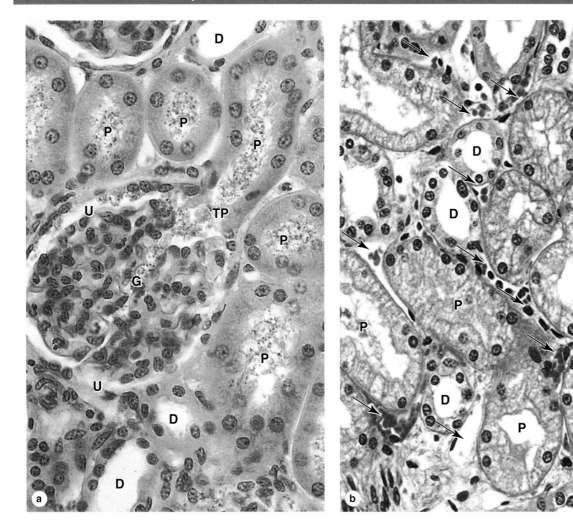

(a) The micrograph shows the continuity at a renal corpuscle's tubular pole (**TP**) between the simple cuboidal epithelium of a proximal convoluted tubule (**P**) and the simple squamous epithelium of the capsule's parietal layer. The urinary space (**U**) between the parietal layer and the glomerulus (**G**) drains into the lumen of the proximal tubule. The lumens of the proximal tubules appear filled, because of the long microvilli of the brush border

and aggregates of small plasma proteins bound to this structure. By contrast, the lumens of distal convoluted tubules (**D**) appear empty, lacking a brush border and protein.

(b) Here the abundant peritubular capillaries and draining venules (arrows) that surround the proximal (**P**) and distal (**D**) convoluted tubules are clearly seen. (Both X400; H&E)

interstitial cells in cortical areas near the proximal tubules produce erythropoietin, the growth factor secreted in response to a prolonged decrease in local oxygen concentration.

›› MEDICAL APPLICATION

Diabetic glomerulosclerosis, the thickening and loss of function in the GBM produced as part of the systemic microvascular sclerosis in diabetes mellitus, is the leading cause of (irreversible) **end-stage kidney disease** in the United States. Treatment requires either a **kidney transplant** or regular **artificial hemodialysis**.

Loop of Henle

The PCT continues with the much shorter proximal straight tubule that enters the medulla and continues as the nephron's **loop of Henle** (Figure 19–2). This is a U-shaped structure with a **thin descending limb** and a **thin ascending limb**, both composed of simple squamous epithelia. The straight part of the proximal tubule has an outer diameter of about 60 μm, but it narrows abruptly to about 30 μm in the thin limbs of the loop. The wall of the thin segments consists only of squamous cells with few organelles (indicating a primarily passive role in transport) and the lumen is prominent (Figures 19–9 and 19–11). The thin ascending limb of the loop becomes the

FIGURE **19–9** Convoluted tubules, nephron loops, and collecting ducts.

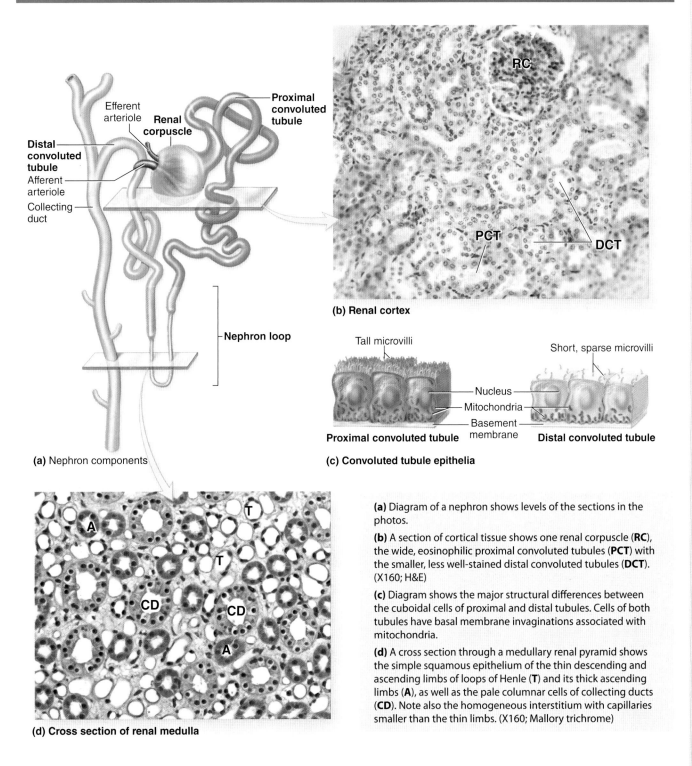

(a) Nephron components

(b) Renal cortex

(c) Convoluted tubule epithelia

(d) Cross section of renal medulla

(a) Diagram of a nephron shows levels of the sections in the photos.

(b) A section of cortical tissue shows one renal corpuscle (**RC**), the wide, eosinophilic proximal convoluted tubules (**PCT**) with the smaller, less well-stained distal convoluted tubules (**DCT**). (X160; H&E)

(c) Diagram shows the major structural differences between the cuboidal cells of proximal and distal tubules. Cells of both tubules have basal membrane invaginations associated with mitochondria.

(d) A cross section through a medullary renal pyramid shows the simple squamous epithelium of the thin descending and ascending limbs of loops of Henle (**T**) and its thick ascending limbs (**A**), as well as the pale columnar cells of collecting ducts (**CD**). Note also the homogeneous interstitium with capillaries smaller than the thin limbs. (X160; Mallory trichrome)

thick ascending limb (TAL), with simple cuboidal epithelium and many mitochondria again, in the outer medulla and extends as far as the macula densa near the nephron's glomerulus.

The loops of Henle and surrounding interstitial connective tissue are involved in further adjusting the salt content of the filtrate. Cuboidal cells of the loops' TALs actively transport sodium and chloride ions out of the tubule against a concentration gradient into the hyaluronate-rich interstitium, making that compartment hyperosmotic. This causes water to be withdrawn passively from the thin descending part of

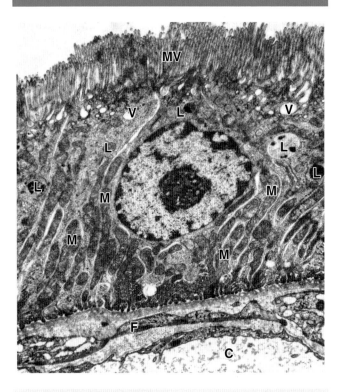

FIGURE **19–10** **Ultrastructure of proximal convoluted tubule (PCT) cells.**

TEM reveals important features of the cuboidal cells of the proximal convoluted epithelium: the long, dense apical microvilli (**MV**), the abundant endocytotic pits and vesicles (**V**) in the apical regions near lysosomes (**L**). Small proteins brought into the cells nonspecifically by pinocytosis are degraded in lysosomes and the amino acids released basally. Apical ends of adjacent cells are sealed with zonula occludens, but the basolateral sides are characterized by long invaginating folds of membrane along which many long mitochondria (**M**) are situated. Water and the small molecules released from the PCTs are taken up immediately by the adjacent peritubular capillaries (**C**). Between the basement membranes of the tubule and the capillary shown here is an extension of a fibroblast (**F**). (X10,500)

the loop, thus concentrating the filtrate. The thin ascending limbs reabsorb sodium chloride (NaCl) but are impermeable to water. The countercurrent flow of the filtrate (descending, then immediately ascending) in the two parallel thin limbs establishes a gradient of osmolarity in the interstitium of the renal pyramids, an effect that is "multiplied" at deeper levels in the medulla. Countercurrent blood flow in the descending and ascending loops of the vasa recta helps maintain the hyperosmotic interstitium. The interstitial osmolarity at the pyramid tips is about four times that of the blood. The **countercurrent multiplier system** established by the nephron loop and vasa recta is an important aspect of renal physiology in humans

Sickle cell nephropathy, one of the most common problems caused by sickle cell disease, occurs when the affected erythrocytes sickle in the vasa recta, because of the low oxygen tension there. The nephropathy results from **renal infarcts**, usually within the renal papillae or pyramids.

Distal Convoluted Tubule & Juxtaglomerular Apparatus

The ascending limb of the nephron is straight as it enters the cortex and forms the macula densa, and then becomes tortuous as the **distal convoluted tubule (DCT)** (Figure 19–2). Much less tubular reabsorption occurs here than in the proximal tubule. The simple cuboidal cells of the distal tubules differ from those of the proximal tubules in being smaller and having no brush border and more empty lumens (Figure 19–9). Because distal tubule cells are flatter and smaller than those of the proximal tubule, more nuclei are typically seen in sections of distal tubules than in those of proximal tubules (Figure 19–8). Cells of the DCT also have fewer mitochondria than cells of proximal tubules, making them less acidophilic (Figure 19–9). The rate of Na^+ absorption here is regulated by **aldosterone** from the adrenal glands.

Where the initial, straight part of the distal tubule contacts the arterioles at the vascular pole of the renal corpuscle of its parent nephron, its cells become more columnar and closely packed, forming the **macula densa** (L. thicker spot). This is part of a specialized sensory structure, the **juxtaglomerular apparatus (JGA)** that utilizes feedback mechanisms to regulate glomerular blood flow and keep the rate of glomerular filtration relatively constant. The JGA is shown in Figures 19–5 and 19–12. Cells of the macula densa typically have apical nuclei, basal Golgi complexes, and a more elaborate and varied system of ion channels and transporters. Adjacent to the macula densa, the tunica media of the afferent arteriole is also modified. The smooth muscle cells are modified as **juxtaglomerular granular (JG) cells**, with a secretory phenotype including more rounded nuclei, rough ER, Golgi complexes, and granules with the protease **renin** (Figures 19–5 and 19–12). Also at the vascular pole are **lacis cells** (Fr. *lacis*, lacework), which are extraglomerular mesangial cells that have many of the same supportive, contractile, and defensive functions as these cells inside the glomerulus.

Basic functions of the JGA in the autoregulation of the GFR and in controlling blood pressure include the following activities. Elevated arterial pressure increases glomerular capillary blood pressure, which increases the GFR. Higher GFR leads to higher luminal concentrations of Na^+ and Cl^- in the TAL of the nephron, which are monitored by cells of the macula densa. Increased ion levels in the lumen cause these cells to release ATP, adenosine, and other vasoactive compounds

FIGURE **19–11**   Renal medulla: nephron loops and collecting ducts.

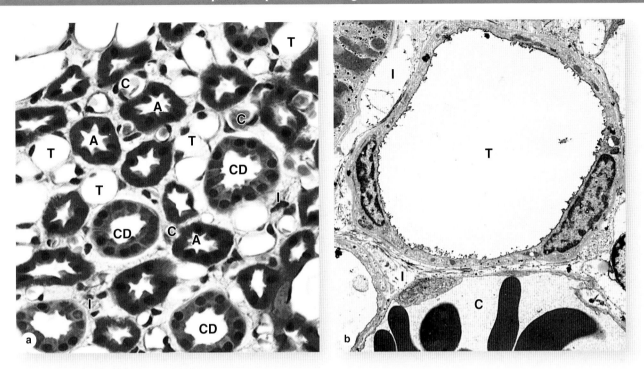

(a) A micrograph of a medullary renal pyramid cut transversely shows closely packed cross sections of the many nephron loops' thin descending and ascending limbs (**T**) and thick ascending limbs (**A**), intermingled with parallel vasa recta capillaries containing blood (**C**) and collecting ducts (**CD**). All these structures are embedded in the interstitium (**I**) that contains sparse myofibroblast-like cells in a matrix very rich in hydrophilic hyaluronate. The specialized nature of the interstitial tissue helps maintain the osmolarity gradient established by differential salt and water transport across the wall of the nephron loop which is required

to concentrate urine and conserve body water. (X400; Mallory trichrome)

(b) The TEM reveals the slightly fibrous nature of the interstitium (**I**) and shows that the simple squamous epithelium of the thin limbs (**T**) is slightly thicker than that of the nearby vasa recta capillaries (**C**). (X3300)

(Figure 19–11b, used with permission from Dr Johannes Rhodin, Department of Cell Biology and Anatomy, University of South Florida College of Medicine, Tampa.)

that trigger contraction of the afferent arteriole, which lowers glomerular pressure and decreases the GFR. This lowers tubular ion concentrations, which turns off the release of vasoconstrictors from the macula densa.

Decreased arterial pressure leads to increased autonomic stimulation to the JGA as a result of baroreceptor function, including local baroreceptors in the afferent arteriole, possibly the JG cells themselves. This causes the JG cells to release renin, an aspartyl protease, into the blood. There renin cleaves the plasma protein **angiotensinogen** into the inactive decapeptide **angiotensin I**. Angiotensin-converting enzyme (ACE) on lung capillaries clips this further to **angiotensin II**, a potent vasoconstrictor that directly raises systemic blood pressure and stimulates the adrenals to secrete **aldosterone**. Aldosterone promotes Na⁺ and water reabsorption in the distal convoluted and connecting tubules, which raises blood

volume to help increase blood pressure. The return of normal blood pressure turns off secretion of renin by JG cells.

Collecting Ducts

The last part of each nephron, the **connecting tubule**, carries the filtrate into a collecting system that transports it to a minor calyx and in which more water is reabsorbed if needed by the body. As shown in Figures 19–13, a connecting tubule extends from each nephron and several join together in the cortical medullary rays to form **collecting ducts** of simple cuboidal epithelium and an average diameter of 40 μm. In the medulla these merge further, forming larger and straighter collecting ducts with increasingly columnar cells and overall diameters reaching 200 μm (Figures 19–11 and 19–14). Approaching the apex of each renal pyramid, several medullary collecting ducts

FIGURE **19–12** Juxtaglomerular apparatus (JGA).

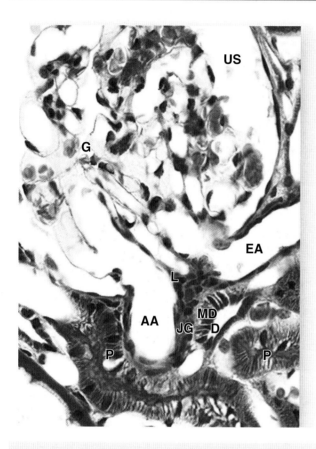

The JGA forms at the point of contact between a nephron's distal tubule (**D**) and the vascular pole of its glomerulus (**G**). At that point cells of the distal tubule become columnar as a thickened region called the macula densa (**MD**). Smooth muscle cells of the afferent arteriole's (**AA**) tunica media are converted from a contractile to a secretory morphology as juxtaglomerular granule cells (**JG**). Also present are lacis cells (**L**), which are extraglomerular mesangial cells adjacent to the macula densa, the afferent arteriole, and the efferent arteriole (**EA**). In this specimen the lumens of proximal tubules (**P**) appear filled and the urinary space (**US**) is somewhat swollen. (X400; Mallory trichrome)

merge again to form each **papillary duct** (or duct of Bellini), which deliver urine directly into the minor calyx (Figure 19–13). Running parallel with the descending and ascending limbs of the loops of Henle and vasa recta, medullary collecting ducts lie in the area with very high interstitial osmolarity (Figures 19–2 and 19–11).

Collecting tubules and ducts are composed mainly of pale-staining **principal cells** with few organelles, sparse microvilli, and unusually distinct cell boundaries (Figure 19–14). Ultrastructurally the principal cells can be seen to have basal membrane infoldings, consistent with their role in ion transport, and a primary cilium among the microvilli. The medullary collecting ducts are the final site of water reabsorption from the filtrate. Principal cells are particularly rich in **aquaporins**, the integral membrane pore proteins that function as specific channels for water molecules, but here most aquaporins are sequestered in membranous cytoplasmic vesicles.

Antidiuretic hormone (ADH), released from the posterior pituitary gland as the body becomes dehydrated, makes collecting ducts more permeable to water and increases the rate at which water molecules are pulled osmotically from the filtrate. Upon binding, ADH receptors on the basolateral cell surface stimulate the movement and insertion of vesicles with aquaporins into the apical (luminal) membranes, increasing the number of membrane channels and water movement through the cells. The high osmolarity of the interstitium draws water passively from the collecting ducts, concentrating the filtrate. The water thus saved immediately enters the blood in the vasa recta.

Scattered among the principal cells are variably darker **intercalated cells** with more abundant mitochondria and projecting apical folds. Intercalated cells, some of which also occur in the DCTs, help maintain acid-base balance by secreting either H^+ (from type A or α intercalated cells) or HCO_3^- (from type B or β intercalated cells).

Histologic features and major functions of the nephron's parts and collecting ducts are summarized in Table 19–1.

>> **MEDICAL APPLICATION**

A common problem involving the ureters is their obstruction by **renal calculi** (**kidney stones**) formed in the renal pelvis or calyces, usually from calcium salts (oxalate or phosphate) or uric acid. While urate stones are usually smooth and small, calcium stones can become large and irritate the mucosa. Most kidney stones are asymptomatic, but besides causing an obstruction that can lead to renal problems, movement of stones from the renal pelvis into the ureter can cause extreme pain on the affected side of the body. Problems caused by such stones can be corrected by either surgical removal of the stone or its disintegration using focused ultrasonic shock waves in a procedure called **lithotripsy**, although this treatment can cause significant renal damage.

› URETERS, BLADDER, & URETHRA

Urine is transported by the **ureters** from the renal pelvis to the **urinary bladder** where it is stored until emptying by micturition via the **urethra**. The walls of the ureters are similar to that of the calyces and renal pelvis, with mucosal, muscular, and adventitial layers and becoming gradually thicker closer to the bladder. The mucosa of these organs is lined by the uniquely stratified **urothelium** or transitional epithelium introduced in **Chapter 4** (Figures 19–15 and 19–16). Cells of this epithelium are organized as three layers:

- A single layer of small basal cells resting on a very thin basement membrane;
- An intermediate region containing from one to several layers of cuboidal or low columnar cells; and

FIGURE **19–13** Fluid transport in the urinary system.

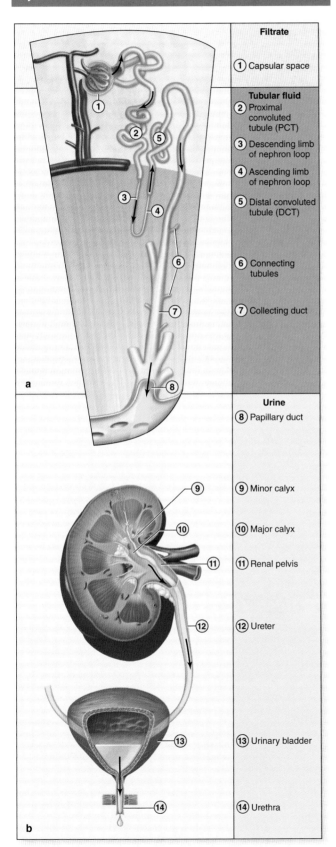

Filtrate

① Capsular space

Tubular fluid

② Proximal convoluted tubule (PCT)

③ Descending limb of nephron loop

④ Ascending limb of nephron loop

⑤ Distal convoluted tubule (DCT)

⑥ Connecting tubules

⑦ Collecting duct

Urine

⑧ Papillary duct

⑨ Minor calyx

⑩ Major calyx

⑪ Renal pelvis

⑫ Ureter

⑬ Urinary bladder

⑭ Urethra

FIGURE **19–14** Collecting ducts.

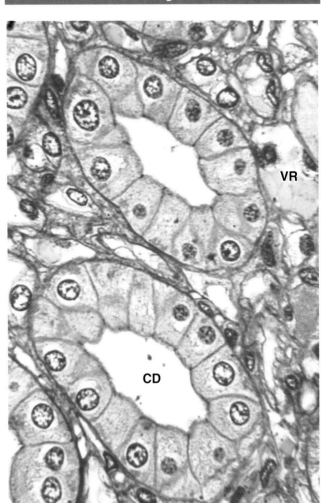

Pale-staining columnar **principal cells**, in which ADH-regulated aquaporins of the cell membrane allow more water reabsorption, are clearly seen in these transversely sectioned collecting ducts (**CD**), surrounded by interstitium with vasa recta (**VR**). (X600; PT)

▪ A superficial layer of large bulbous or elliptical **umbrella cells**, sometimes binucleated, which are highly differentiated to protect the underlying cells against the potentially cytotoxic effects of hypertonic urine.

The thick muscularis of the ureters moves urine toward the bladder by peristaltic contractions and produces prominent mucosal folds when the lumen is empty (Figure 19–16).

(Left) (a) Diagram of a nephron and collecting system shows the flow of filtrate. **(b)** Upon delivery at a minor calyx, filtrate is no longer modified by reabsorption or secretion and is called **urine**. It flows passively into the renal pelvis but moves by peristalsis along the ureters for temporary storage in the urinary bladder, which is emptied through the urethra.

FIGURE **19–15** Renal papilla, collecting ducts, and minor calyx.

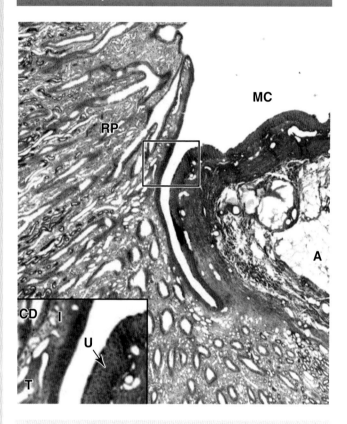

FIGURE **19–15** Renal papilla, collecting ducts, and minor calyx.

A sagittal section of a renal papilla shows numerous collecting ducts (also called the **ducts of Bellini** at this level) converging at the end of the renal papilla (**RP**) where they empty into the minor calyx (**MC**). The mucosa of the calyx contains dense connective tissue stained blue here and adipose tissue (**A**). The ducts are embedded in interstitial tissue that also contains thin limbs of the nephron loops (X50; Mallory trichrome). Inset: An enlarged area shows the columnar epithelium of the collecting ducts (**CD**), the interstitium (**I**) and thin limbs (**T**), and the protective urothelium (**U**) that lines the minor calyx. (X200; Mallory trichrome)

>> **MEDICAL APPLICATION**

Bacterial infections of the urinary tract can lead to inflammation of the renal pelvis and calyces, or **pyelonephritis**. In acute pyelonephritis bacteria often move from one or more minor calyx into the associated renal papilla, causing accumulation of neutrophils in the collecting ducts.

Umbrella cells are especially well developed in the bladder (Figure 19–17) where contact with urine is greatest. These cells, up to 100 μm in diameter, have extensive intercellular junctional complexes surrounding unique apical membranes. Most of the apical surface consists of **asymmetric unit membranes** in which regions of the outer lipid layer appear ultrastructurally to be twice as thick as the inner leaflet. These regions are composed of lipid rafts containing mostly integral membrane proteins called **uroplakins** that assemble into paracrystalline arrays of stiffened plaques 16 nm in diameter. The abundant membranous plaques, together with the tight junctions, allow this epithelium to serve as an osmotic barrier protecting its cells and the cells of surrounding tissues from hypertonic urine and preventing dilution of the stored urine.

Plaques are hinged together by more narrow regions of typical membrane. When the bladder is emptied, not only does the mucosa fold extensively, but individual umbrella cells decrease their apical surface area by folding the membrane at the hinge regions and internalizing the folded plaques in discoidal vesicles. As the bladder fills again these vesicles rejoin the apical membrane, increasing its surface area as the tight junctions are reorganized and the cells become less bulbous. The thickness of the full bladder's urothelium is half that of the empty bladder (5-7 cell layers vs. 2-3 layers), apparently the result of the intermediate cells being pushed and pulled laterally to accommodate the increased volume of urine.

Urothelium is surrounded by a folded lamina propria and submucosa, followed by a dense sheath of interwoven smooth muscle layers and adventitia (Figures 19–16 and 19–17). Urine is moved from the renal pelvises to the bladder by peristaltic contractions of the ureters.

The bladder's lamina propria and dense irregular connective tissue of the submucosa are highly vascularized. The bladder in an average adult can hold 400-600 mL of urine, with the urge to empty appearing at about 150-200 mL. The muscularis consists of three poorly delineated layers, collectively called the **detrusor muscle**, which contract to empty the bladder (Figure 19–17). Three muscular layers are seen most distinctly at the neck of the bladder near the urethra (Figure 19–17). The ureters pass through the wall of the bladder obliquely, forming a valve that prevents the backflow of urine into the ureters as the bladder fills. All the urinary passages are covered externally by an adventitial layer, except for the upper part of the bladder that is covered by serous peritoneum.

>> **MEDICAL APPLICATION**

Cystitis, or inflammation of the bladder mucosa, is the most frequent problem involving this organ. Such inflammation is common during urinary tract infections, but it can also be caused by immunodeficiency, urinary catheterization, radiation, or chemotherapy. Chronic cystitis can cause an **unstable urothelium**, with **benign urothelial changes** involving hyperplasia or metaplasia. Bladder cancer is usually some form of **transitional cell carcinoma** arising from unstable urothelium.

The **urethra** is a tube that carries the urine from the bladder to the exterior (Figure 19–18). The urethral mucosa has prominent longitudinal folds, giving it a distinctive appearance in cross section. In men, the two ducts for sperm transport during

FIGURE **19–16**   Ureters.

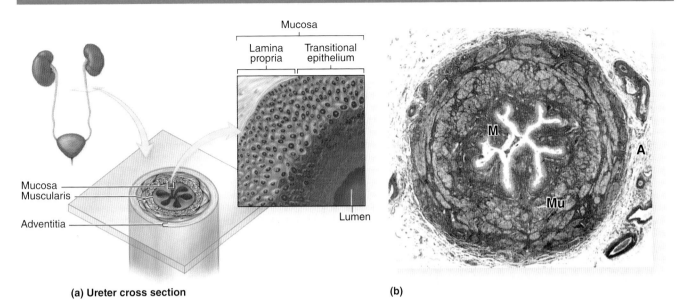

(a) Ureter cross section

(b)

(a) Diagram of a ureter in cross section shows a characteristic pattern of longitudinally folded mucosa, surrounded by a thick muscularis that moves urine by regular waves of peristalsis. The lamina propria is lined by a unique stratified epithelium called **transitional epithelium** or **urothelium** that is resistant

to the potentially deleterious effects of contact with hypertonic urine.

(b) Histologically the muscularis (**Mu**) is much thicker than the mucosa (**M**) and adventitia (**A**). (X18; H&E)

FIGURE **19–17**   Bladder wall and urothelium.

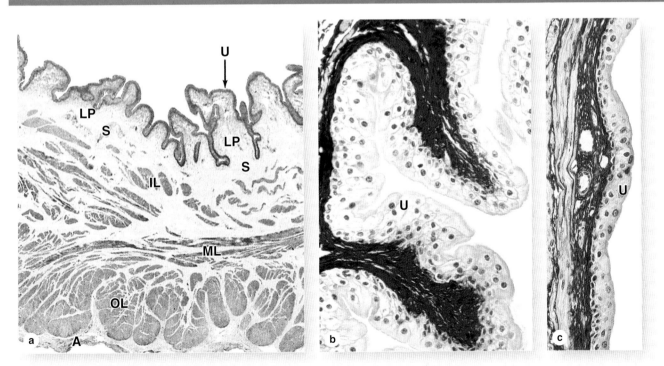

(a) In the neck of the bladder, near the urethra, the wall shows four layers: the mucosa with urothelium (**U**) and lamina propria (**LP**); the thin submucosa (**S**); inner, middle, and outer layers of smooth muscle (**IL**, **ML**, and **OL**); and the adventitia (**A**). (X15; H&E)

(b) When the bladder is empty, the mucosa is highly folded and the urothelium (**U**) has bulbous umbrella cells. (X250; PSH)

(c) When the bladder is full, the mucosa is pulled smooth, the urothelium (**U**) is thinner, and the umbrella cells are flatter. (X250; H&E)

FIGURE **19–18** Urethra.

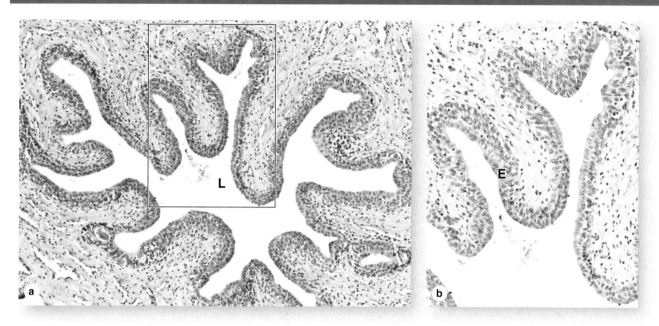

The urethra is a fibromuscular tube that carries urine from the bladder to the exterior of the body.

(a) A transverse section shows that the mucosa has large longitudinal folds around the lumen (**L**). (X50; H&E)

(b) A higher magnification of the enclosed area shows the unusual stratified columnar nature of the urethral epithelium

(**E**). This thick epithelial lining varies between stratified columnar in some areas and pseudostratified columnar elsewhere, but it becomes stratified squamous at the distal end of the urethra. (X250; H&E)

ejaculation join the urethra at the prostate gland (see Chapter 21). The male urethra is longer and consists of three segments:

- The **prostatic urethra**, 3-4 cm long, extends through the prostate gland and is lined by urothelium.
- The **membranous urethra**, a short segment, passes through an **external sphincter** of striated muscle and is lined by stratified columnar and pseudostratified columnar epithelium.
- The **spongy urethra**, about 15 cm in length, is enclosed within erectile tissue of the penis (see Chapter 21) and is lined by stratified columnar and pseudostratified columnar epithelium (Figure 19–18), with stratified squamous epithelium distally.

In women, the urethra is exclusively a urinary organ. The female urethra is a 3- to 5-cm-long tube, lined initially with transitional epithelium which then transitions to nonkeratinized stratified squamous epithelium continuous with that of the skin at the labia minora. The middle part of the urethra in both sexes is surrounded by the external striated muscle sphincter.

›› MEDICAL APPLICATION

Urinary tract infections, usually involving coliform bacteria or *Chlamydia*, often produce **urethritis** and in women often lead to cystitis because of the short urethra. Such infections are usually accompanied by a persistent or more frequent urge to urinate, and urethritis may produce pain or difficulty during urination (**dysuria**).

The Urinary System   SUMMARY OF KEY POINTS

Kidney

- Each kidney has a thick outer **cortex**, surrounding a **medulla** that is divided into 8-12 renal pyramids; each pyramid and its associated cortical tissue comprises a **renal lobe**.
- The apical papilla of each renal pyramid inserts into a **minor calyx**, a subdivision of two or three **major calyces** extending from the **renal pelvis**.
- The **ureter** carries urine from the renal pelvis and exits the **renal hilum**, where the **renal artery and vein** are also located.

Renal Vasculature

- Renal arteries branch to form smaller arteries between the renal lobes, with **interlobular arteries** entering the cortex to form the microvasculature; venous branches parallel the arterial supply.
- In the cortex **afferent arterioles** enter capillary clusters called **glomeruli**, which are drained by **efferent arterioles**, instead of venules, an arrangement that allows higher hydrostatic pressure in the capillaries.
- The **efferent arterioles** from cortical glomeruli branch diffusely as **peritubular capillaries**, while those from juxtamedullary glomeruli branch as long microvascular loops called **vasa recta** in the medulla.

Nephrons

- Functional units of the kidney are the **nephrons**, numbering about 1 million, each with a **renal corpuscle** and a long **renal tubule**, and a system of **collecting ducts**.
- The renal corpuscle has a simple squamous parietal layer of the **glomerular (Bowman) capsule**, continuous with the proximal tubule, and a specialized visceral layer of **podocytes** surrounding the glomerular capillaries.
- **Podocytes** extend large primary processes that curve around a capillary and extend short, interdigitating secondary processes or **pedicels**, between which are narrow spaces called **slit pores**.
- The elevated pressure in the capillaries forces water and small solutes of blood plasma through the **glomerular filter** into the **capsular** (or **urinary**) **space** inside the glomerular capsule.
- In each glomerulus the filter has three parts: the finely **fenestrated capillary endothelium**; the thick (330 nm) **fused basal laminae** of type IV collagen and other proteins produced by the endothelial cells and podocytes; and the **slit pores** between the pedicels, covered by thin filtration **slit diaphragms**.
- From the renal corpuscle, filtrate enters the long **nephron tubule** that extends through both the cortex and medulla, with epithelial cells for both **reabsorption** and **secretion** of substances into the filtrate.
- The first tubular part, the **proximal convoluted tubule (PCT)**, is mainly cortical, has **simple cuboidal** cells with **long microvilli**

in the lumen, **abundant mitochondria**, and large, interdigitating **basolateral folds**.

- In the **PCT**, all glucose and other organic nutrients, all small proteins and peptides (which are degraded to amino acids), and much water and electrolytes are reabsorbed from the filtrate and transferred to the peritubular capillaries.
- From the PCT filtrate flows into the **loop of Henle**, located in the medulla, which has squamous **thin descending** and **ascending** limbs; the latter extends as a **thick ascending limb (TAL)** back into the cortex.
- In the cortex the **TAL** (also known as the **distal straight tubule**) contacts the arterioles at the vascular pole of its parent renal corpuscle and there thickens focally as the **macula densa**.
- Tall epithelial cells of the **macula densa** and specialized smooth muscle cells in the adjacent afferent arteriole called **juxtaglomerular cells**, which secrete **renin**, comprise a juxtaglomerular apparatus (JGA) that is an important regulator of blood pressure.
- Beyond the macula densa, the tubule continues as the **distal convoluted tubule (DCT)**, where electrolyte levels of the filtrate are adjusted further and which lead to short **connecting tubules**.
- **Connecting tubules** from several nephrons join to form the cortical **collecting ducts**, of simple cuboidal epithelium, which enter the medulla in parallel with the loops of Henle and vasa recta and become larger with more columnar cells.

Urinary Tract

- **Principal cells** of the **collecting ducts** are pale-staining, with relatively few mitochondria and distinct cell membranes that are rich in **aquaporins** (water channels) for passive water reabsorption.
- The largest collecting ducts deliver filtrate into the **minor calyces**, where it undergoes no further modification and is called **urine**.
- The calyces, renal pelvis, ureters, and urinary bladder are lined by **urothelium**, or transitional epithelium, which protects underlying cells from hypertonic, potentially toxic effects of urine.
- Large, bulbous superficial cells of the urothelium, called **umbrella cells**, have apical membranes consisting of hinged regions with dense plaques of **uroplakin** proteins that protect the cytoplasm.
- As the **urinary bladder** fills its highly folded mucosa unfolds, the urothelium gets somewhat thinner by cell movements, and the hinged membrane plaques of umbrella cells partially unfold.
- The **urethra** drains the bladder and in both genders is lined initially by urothelium, followed (in males) by alternating **stratified columnar** and **pseudostratified columnar epithelium** and distally by stratified squamous epithelium.
- In males the urethra has three regions: the **prostatic urethra** in the prostate gland; the short **membranous urethra** passing through the urogenital diaphragm, and the long **penile urethra**.

The Urinary System ASSESS YOUR KNOWLEDGE

1. Blood in the renal arcuate arteries flows next into which vessels?
   a. Afferent arterioles
   b. Efferent arterioles
   c. Glomerular capillaries
   d. Interlobar arteries
   e. Interlobular arteries

2. Which cell type comprises the visceral layer of Bowman capsule?
   a. Endothelial cells
   b. Juxtaglomerular cells
   c. Mesangial cells
   d. Podocytes
   e. Extraglomerular mesangial (or Lacis) cells

3. Which type of epithelium lines the thick ascending limb of the loop of Henle?
   a. Pseudostratified columnar
   b. Simple columnar
   c. Simple cuboidal
   d. Simple squamous
   e. Transitional (urothelium)

4. Which cell is a modified smooth muscle cell that secretes renin?
   a. Macula densa cells
   b. Mesangial cells
   c. Podocytes
   d. Juxtaglomerular cells
   e. Endothelial cells

5. Epithelial cell membrane domains containing many stiffened plaques of protein are an important feature in which part of the urinary system?
   a. Juxtaglomerular apparatus
   b. Bladder mucosa
   c. Collecting ducts
   d. Renal pyramids
   e. Membranous urethra

6. An immunohistochemical technique using antibodies against aquaporins to stain a section of kidney would be expected to stain cells in which structures most intensely?
   a. Collecting ducts
   b. Lining of the major and minor calyces
   c. Proximal convoluted tubules
   d. Distal convoluted tubules
   e. Glomeruli

7. What type of epithelium lines the prostatic urethra?
   a. Simple columnar
   b. Pseudostratified columnar
   c. Stratified squamous
   d. Simple squamous
   e. Transitional (urothelium)

8. A 14-year-old patient presents in the nephrology clinic with fatigue, malaise, anorexia, abdominal pain, and fever. She reports a loss of 6 lb in the past 2 months. Serum gamma globulin as well as the immunoglobulins IgG, IgA, and IgM are all elevated. Her serum creatine is 1.4 mg/dL (normal 0.6-1.2 mg/dL) and urinalysis of glucose and protein are 2+ on a dipstick test, confirmed by laboratory at 8.0 g/dL and 0.95 g/dL, respectively. A renal biopsy is prepared for light microscopy and an infiltrate containing lymphocytes, plasma cells, and eosinophils is found among tubules having cells with prominent brush borders. Which one of the following statements correctly pertains to these epithelial cells?
   a. Impermeable to water despite presence of ADH
   b. The primary site for the reduction of the tubular fluid volume
   c. The site of the countercurrent multiplier
   d. The site of action of aldosterone
   e. Indirectly involved in the release of renin

9. A 45-year-old man presents with nephrolithiasis or kidney stones. The process of calcium oxalate stone formation as seen in this patient begins with Randall plaques found in the basement membranes of which one of the following structures found only in the renal medulla?
   a. Proximal convoluted tubules
   b. Distal convoluted tubules
   c. Thin loops of Henle
   d. Afferent arterioles
   e. Collecting ducts

10. A 15-year-old male presents with hematuria, hearing loss, lens dislocation, and the onset of cataracts. Genetic analysis reveals a mutation in the COL4A5 gene. Transmission EM examination of a renal biopsy confirms that the disorder has affected a component of the renal corpuscles in which damage disrupts normal glomerular filtration. Which one of the following structures would most likely be abnormal in the TEM of this patient's biopsy?
    a. Pedicels
    b. Filtration slits
    c. Slit diaphragms
    d. Glomerular basement membranes
    e. Fenestrated endothelium of glomerular capillaries

20 Endocrine Glands

Secretory cells of endocrine glands release their products, signaling molecules called **hormones**, into the neighboring vascularized compartment for uptake by capillaries and distribution throughout the body. There is no secretory duct as in exocrine glands. Endocrine cells are typically epithelial, at least in origin, and aggregated as cords or clusters. Besides the specialized endocrine glands discussed in this chapter, many other organs specialized for other functions, such as the heart, thymus, gut, kidneys, testis, and ovaries, contain various endocrine cells (Figure 20–1).

Distribution by the circulation allows hormones to act on target cells with receptors for those hormones at a distance from the site of their secretion. As discussed briefly in Chapter 2, other endocrine cells produce hormones that act on target cells only a short distance away. This may involve **paracrine** secretion, with localized dispersal in interstitial fluid or through short loops of blood vessels, as when gastrin made by pyloric G cells reaches target cells in the fundic glands, or **juxtacrine** secretion, in which a signaling molecule remains on the secreting cell's surface or adjacent extracellular matrix and affects target cells when the cells make contact. Juxtacrine signaling is particularly important in embryonic and regenerative tissue interactions. In **autocrine** secretion, cells may produce molecules that act on themselves or on cells of the same type. For example, insulin-like growth factor (IGF) produced by several cell types may act on the same cells that produced it. Endocrine glands are often also target organs for other hormones that can establish a feedback mechanism to control hormone secretion and keep blood hormonal levels within strict limits.

Hormones, like neurotransmitters, are frequently hydrophilic molecules such as proteins, glycoproteins, peptides, or modified amino acids with receptors on the surface of target cells. Alternatively, hydrophobic steroid and thyroid hormones must circulate on transport proteins but can diffuse through the cell membranes and activate cytoplasmic receptors in target cells (see Chapter 2).

▶ PITUITARY GLAND (HYPOPHYSIS)

The **pituitary gland**, or **hypophysis** (Gr. *hypo*, under + *physis*, growth), weighs about 0.5 g in adults and has dimensions of about $10 \times 13 \times 6$ mm. It lies below the brain in a small cavity on the sphenoid bone, the sella turcica (Figure 20–2). The pituitary is formed in the embryo partly from the developing brain and partly from the developing oral cavity (Figure 20–3). The neural component is the neurohypophyseal bud growing down from the floor of the future diencephalon as a stalk (or infundibulum) that remains attached to the brain. The oral component arises as an outpocketing of ectoderm from the roof of the primitive mouth and grows cranially, forming a structure called the **hypophyseal (Rathke) pouch**. The base of this pouch eventually constricts and separates from the pharynx. Its anterior wall then thickens greatly, reducing the pouch's lumen to a small fissure (Figure 20–3).

Because of its dual origin, the pituitary actually consists of two glands—the posterior **neurohypophysis** and the anterior **adenohypophysis**—united anatomically but with different functions. The neurohypophysis retains many histologic features of brain tissue and consists of a large part, the **pars nervosa**, and the smaller **infundibulum** stalk attached to the hypothalamus at the median eminence (Figures 20–2 and 20–4). The **adenohypophysis**, derived from the oral ectoderm, has

FIGURE **20–1** **Locations of the major endocrine glands.**

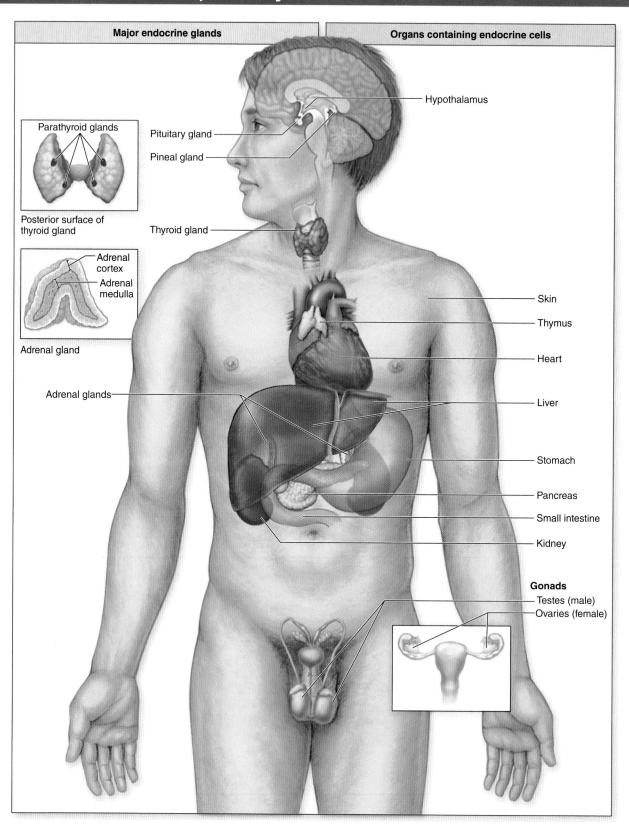

Major endocrine glands

Organs containing endocrine cells

Parathyroid glands

Posterior surface of thyroid gland

Adrenal cortex
Adrenal medulla

Adrenal gland

Adrenal glands

Hypothalamus

Pituitary gland
Pineal gland

Thyroid gland

Skin
Thymus
Heart
Liver

Stomach
Pancreas
Small intestine
Kidney

Gonads
Testes (male)
Ovaries (female)

In addition to the major endocrine glands shown at the left here, there are widely distributed endocrine cells as well as various other tissues in organs (right) throughout the body with endocrine functions. Not shown are adipocytes, which exert important endocrine functions, and the many tissues in which paracrine signalling is important.

FIGURE **20–2** Pituitary gland.

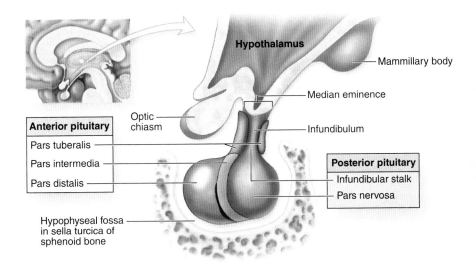

The pituitary gland is composed of an anterior part and a posterior part that is directly attached to the hypothalamus region of the brain by an infundibular stalk. The gland occupies a fossa of the sphenoid bone called the **sella turcica** (L. Turkish saddle).

FIGURE **20–3** Formation of the pituitary gland.

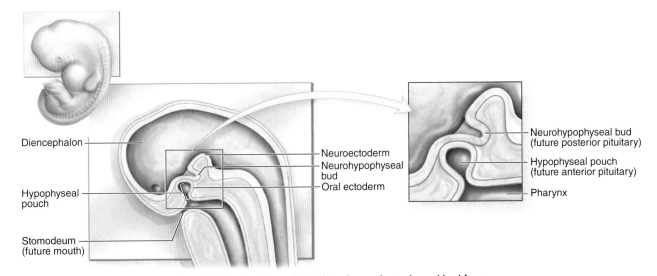

(a) Week 3: Hypophyseal pouch and neurohypophyseal bud form

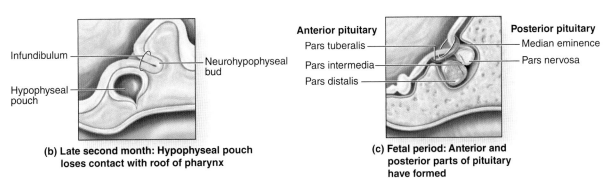

(b) Late second month: Hypophyseal pouch loses contact with roof of pharynx

(c) Fetal period: Anterior and posterior parts of pituitary have formed

The pituitary gland forms from two separate embryonic structures.

(a) During the third week of development, a hypophyseal pouch (or Rathke pouch, the future anterior pituitary) grows from the roof of the pharynx, while a neurohypophyseal bud (future posterior pituitary) forms from the diencephalon.

(b) By late in the second month, the hypophyseal pouch detaches from the roof of the pharynx and merges with the neurohypophyseal bud. (c) During the fetal period, the anterior and posterior parts of the pituitary complete development.

FIGURE **20–4** **Pituitary gland.**

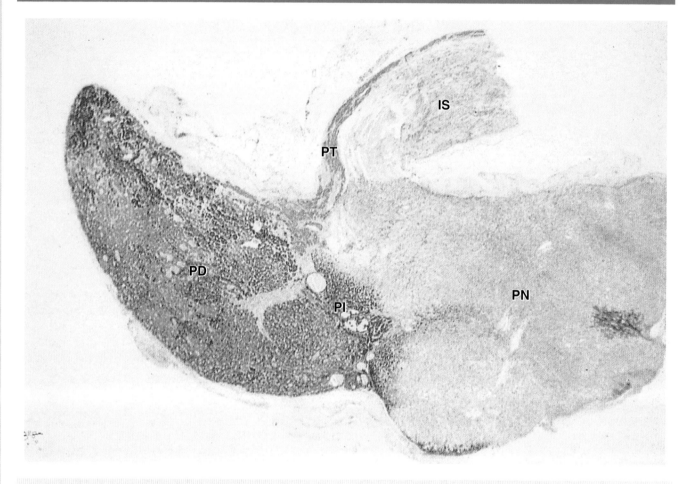

Histologically the two parts of the pituitary gland reflect their origins, as seen in this low-magnification section of an entire gland. The infundibular stalk (**IS**) and pars nervosa (**PN**) of the neurohypophysis resemble CNS tissue, while the adenohypophysis' pars distalis (**PD**), pars intermediate (**PI**), and pars tuberalis (**PT**) are typically glandular in their level of staining. (X30; H&E)

three parts: a large **pars distalis** or **anterior lobe**; the **pars tuberalis**, which wraps around the infundibulum; and the thin **pars intermedia** adjacent to the posterior pars nervosa (Figures 20–2 and 20–4).

The Hypothalamic-Hypophyseal Tract & Blood Supply

The pituitary gland's neural connection to the brain and its blood supply are both of key importance for its function (Figures 20–4 and 20–5). Embryologically, anatomically, and functionally, the pituitary gland is connected to the hypothalamus at the base of the brain. In addition to the vascular portal system carrying small regulatory peptides from the hypothalamus to the adenohypophysis, a bundle of axons called the **hypothalamic-hypophyseal tract** courses into the neurohypophysis from two important hypothalamic nuclei.

The peptide hormones ADH (antidiuretic hormone) and oxytocin are synthesized by large neurons in the **supraoptic** and the **paraventricular nuclei**, respectively. Both hormones undergo axonal transport and accumulate temporarily in the axons of the hypothalamic-hypophyseal tract before their release and uptake by capillaries branching from the inferior arteries.

The blood supply derives from two groups of vessels coming off the internal carotid artery and drained by the hypophyseal vein. The **superior hypophyseal arteries** supply the median eminence and the infundibular stalk; the **inferior hypophyseal arteries** provide blood mainly for the neurohypophysis. The superior arteries divide into a **primary plexus** of fenestrated capillaries that irrigate the stalk and median eminence. These capillaries then rejoin to form venules that branch again as a larger secondary capillary plexus in the adenohypophysis (Figure 20–5). These vessels make up the

FIGURE **20–5** The hypothalamic-hypophyseal tract and portal system.

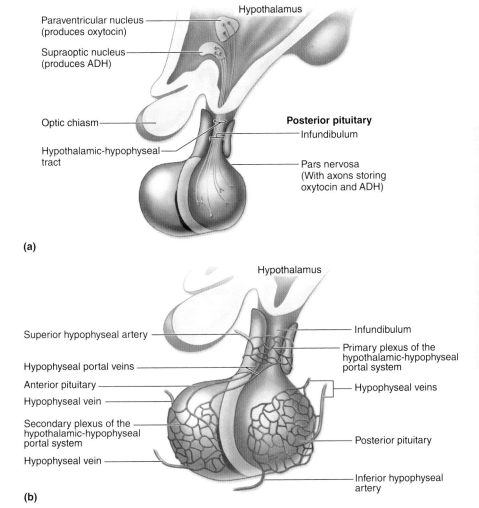

(a)

(b)

(a) The **hypothalamic-hypophyseal tract** consists of axons extending from the hypothalamic supraoptic and paraventricular nuclei, through the infundibulum and into the pars nervosa of the posterior pituitary, where peptide hormones they carry are released for capillary uptake.
(b) The **hypothalamic-hypophyseal portal system**, with blood from the superior hypophyseal artery, consists of two capillary networks connected by the hypophyseal portal vein. The primary plexus surrounds the infundibulum and median eminence, and the second is throughout the pars distalis and drains into the hypophyseal veins.

hypothalamic-hypophyseal portal system that has great importance because it carries neuropeptides from the median eminence the short distance to the adenohypophysis where they either stimulate or inhibit hormone release by the endocrine cells there.

Adenohypophysis (Anterior Pituitary)

The three parts of the adenohypophysis are derived embryonically from the hypophyseal pouch.

Pars Distalis

The **pars distalis** accounts for 75% of the adenohypophysis and has a thin fibrous capsule. The main components are cords of well-stained endocrine cells interspersed with fenestrated capillaries and supporting reticular connective tissue (Figures 20–4 and 20–6). Common stains suggest two broad groups of cells in the pars distalis with different staining affinities: **chromophils**

and **chromophobes**. Chromophils are secretory cells in which hormone is stored in cytoplasmic granules. They are also called **basophils** and **acidophils**, based on their affinities for basic and acidic dyes, respectively (Figure 20–6).

Subtypes of basophilic and acidophilic cells are identified by their granular morphology in the TEM or more easily by immunohistochemistry (Figure 20–7). Specific cells are usually named according to their hormone's target cells (Table 20–1). Acidophils secrete either growth hormone (somatotropin) or prolactin and are called **somatotrophs** and **lactotrophs** (or somatotropic cells and lactotropic cells), respectively. The basophilic cells are the **corticotrophs, gonadotrophs,** and **thyrotrophs**, with target cells in the adrenal cortex, gonads, and thyroid gland, respectively. Somatotrophs typically constitute about half the cells of the pars distalis in humans, with thyrotrophs the least abundant.

With two exceptions, each type of anterior pituitary cell makes one kind of hormone (Table 20–1). Gonadotrophs

FIGURE 20–6 Pars distalis: Acidophils, basophils, and chromophobes.

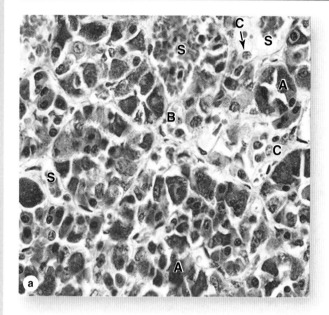

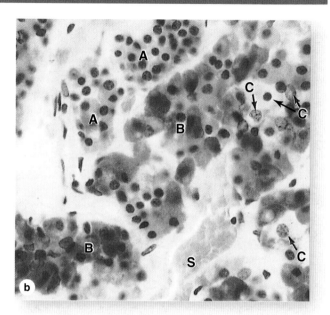

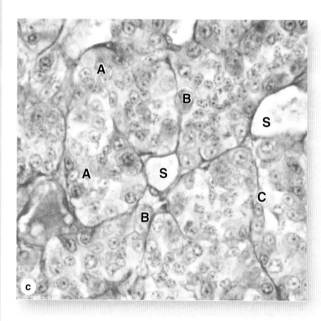

(a, b) Most general staining methods simply allow the parenchymal cells of the pars distalis to be subdivided into acidophil cells (**A**), basophils (**B**), and chromophobes (**C**) in which the cytoplasm is poorly stained. Also shown are capillaries and sinusoids (**S**) in the second capillary plexus of the portal system. Cords of acidophils and basophils vary in distribution and number in different regions of the pars distalis, but are always closely associated with microvasculature that carries off secreted hormones into the general circulation. (X400; H&E)

(c) The same area is seen after staining with Gomori trichrome. (X400)

secrete two different glycoproteins: **follicle-stimulating hormone (FSH)** and **luteinizing hormone (LH**; called interstitial cell-stimulating hormone [ICSH] in men). The main protein synthesized in corticotrophs is pro-opiomelanocortin (POMC), which is cleaved posttranslationally into the polypeptide hormones **adrenocortical trophic hormone (ACTH)** and **β-lipotropin (β-LPH)**. Hormones produced by the pars distalis have widespread functional activities. They regulate almost all other endocrine glands, ovarian function and sperm production, milk production, and the metabolism of muscle, bone, and adipose tissue (Table 20–1; Figure 20–8).

Chromophobes stain weakly, with few or no secretory granules, and also represent a heterogeneous group,

including stem and undifferentiated progenitor cells as well as any degranulated cells present.

Pars Tuberalis

The **pars tuberalis** is a smaller funnel-shaped region surrounding the infundibulum of the neurohypophysis (Figures 20–2 and 20–4). Most of the cells of the pars tuberalis are gonadotrophs.

Pars Intermedia

A narrow zone lying between the pars distalis and the pars nervosa, the **pars intermedia** contains basophils

FIGURE 20–7 Ultrastructure and immunohistochemistry of somatotropic cells.

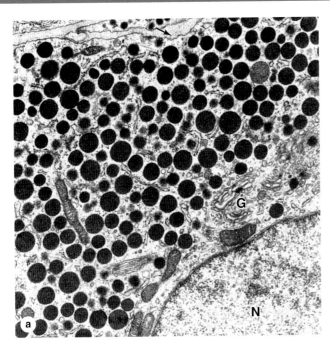
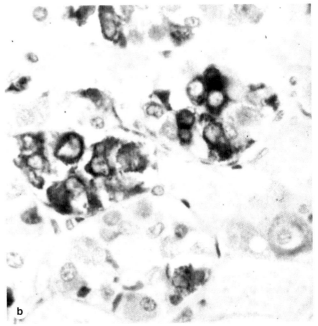

(a) Ultrastructurally, cytoplasm of all chromophil cells is shown to have well-developed Golgi complexes (**G**), euchromatic nuclei (**N**), and cytoplasm filled with secretory granules, as seen here in a somatotroph, the most common acidophil. The **arrow** indicates the cell membrane. Specific chromophils are more easily identified using immunohistochemistry and antibodies against the hormone products. (X10,000)

(b) The micrograph shows somatotrophs stained using an antibody against somatotropin. (X400; Hematoxylin counterstain)

TABLE 20–1 Major cell types of the anterior pituitary and their major functions.

Cell Type	% of Total Cells	Hormone Produced	Major Function
Somatotrophs	50	Somatotropin (growth hormone, GH), a 22-kDa protein	Stimulates growth in epiphyseal plates of long bones via insulin-like growth factors (IGFs) produced in liver
Lactotrophs (or mammotrophs)	15-20	Prolactin (PRL), a 22.5-kDa protein	Promotes milk secretion
Gonadotrophs	10	Follicle-stimulating hormone (FSH) and luteinizing hormone (LH; interstitial cell-stimulating hormone [ICSH] in men), both 28-kDa glycoprotein dimers, secreted from the same cell type	FSH promotes ovarian follicle development and estrogen secretion in women and spermatogenesis in men; LH promotes ovarian follicle maturation and progesterone secretion in women and interstitial cell androgen secretion in men
Thyrotrophs	5	Thyrotropin (TSH), a 28-kDa glycoprotein dimer	Stimulates thyroid hormone synthesis, storage, and liberation
Corticotrophs	15-20	Adrenal corticotropin (ACTH), a 4-kDa polypeptide	Stimulates secretion of adrenal cortex hormones
		Lipotropin (LPH)	Helps regulate lipid metabolism

FIGURE **20–8** Hormones of the pars distalis and their targets.

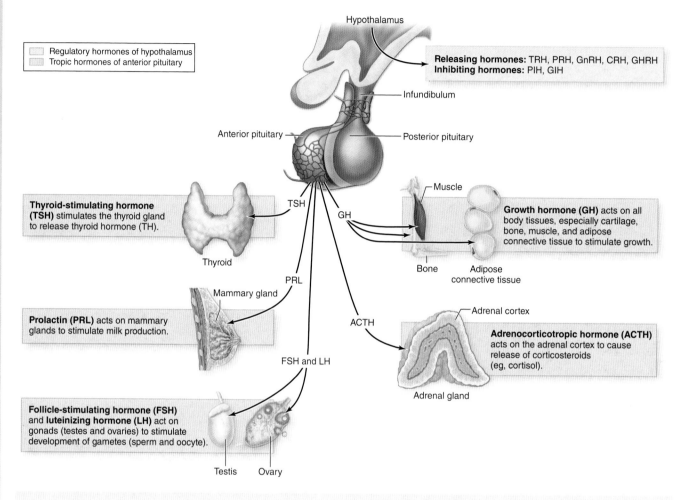

The anterior pituitary secretes six major tropic hormones controlling the activities of their target organs. Release of these hormones is regulated primarily by hypothalamic factors carried by the hypothalamic-hypophyseal blood supply.

(corticotrophs), chromophobes, and small, colloid-filled cysts derived from the lumen of the embryonic hypophyseal pouch (Figure 20–9). Best-developed and active during fetal life, corticotrophs of the pars intermedia express POMC but cleave it differently from cells in the pars distalis, producing mainly smaller peptide hormones, including two forms of melanocyte-stimulating hormone (MSH), γ-LPH, and β-endorphin. MSH increases melanocyte activity, but the overall functional significance of the pars intermedia remains uncertain.

Control of Hormone Secretion in the Anterior Pituitary

The activities of the cells of the anterior pituitary are controlled primarily by peptide-related **hypothalamic hormones** produced by small neurons near the third ventricle, discharged from axons in the median eminence, and transported by capillaries of the portal system throughout the anterior pituitary. As shown in Table 20–2, most of these hormones are **releasing hormones** that stimulate secretion by specific anterior pituitary cells. Two of the hypothalamic factors, however, are **inhibiting hormones** that block hormone secretion in specific cells of the adenohypophysis (Table 20–2). Because of the strategic position of the hypothalamic neurons and the control they exert on the adenohypophysis and therefore on many bodily functions, many sensory stimuli coming to the brain or arising within the CNS can affect pituitary function and then also quickly affect activities of many other organs and tissues.

Another mechanism controlling activity of anterior pituitary cells is **negative feedback** by hormones from the target organs on secretion of the relevant hypothalamic factors

All these mechanisms allow the fine tuning of hormone secretion by cells of the anterior pituitary.

Neurohypophysis (Posterior Pituitary)

The neurohypophysis consists of the pars nervosa and the infundibular stalk (Figures 20–2 and 20–4) and, unlike the adenohypophysis, does not contain the cells that synthesize its two hormones. It is composed of neural tissue, containing some 100,000 unmyelinated axons of large secretory neurons with cell bodies in the supraoptic and paraventricular nuclei of the hypothalamus (Figure 20–5). Also present are highly branched glial cells called **pituicytes** that resemble astrocytes and are the most abundant cell type in the posterior pituitary (Figure 20–11).

The secretory neurons have all the characteristics of typical neurons, including the ability to conduct an action potential, but have larger-diameter axons and well-developed synthetic components related to the production of the 9-amino acid peptide hormones **antidiuretic hormone (ADH)**—also called arginine vasopressin—and **oxytocin**. Transported axonally into the pars nervosa, these hormones accumulate in axonal dilations called **neurosecretory bodies** or **Herring bodies**, visible in the light microscope as faintly eosinophilic structures (Figure 20–11). The neurosecretory bodies contain membrane-enclosed granules with either oxytocin or ADH bound to 10-kDa carrier proteins called **neurophysin I and II**, respectively. The hormone-neurophysin complex is synthesized as a single protein and then cleaved to produce the peptide hormone and its binding protein. Nerve impulses along the axons trigger the release of the peptides from the neurosecretory bodies for uptake by the fenestrated capillaries of the pars nervosa, and the hormones are then distributed to the general circulation. Axons from the supraoptic and paraventricular nuclei mingle in the neurohypophysis but are mainly concerned with ADH and oxytocin secretion, respectively.

ADH is released in response to increased blood tonicity, sensed by osmoreceptor cells in the hypothalamus, which then stimulate ADH synthesis in supraoptic neurons. ADH increases the permeability of the renal collecting ducts to water (see Chapter 19) so that more water is reabsorbed from the filtrate in these tubules and osmotic balance of body fluids is restored (Table 20–3).

FIGURE **20–9** Pars intermedia.

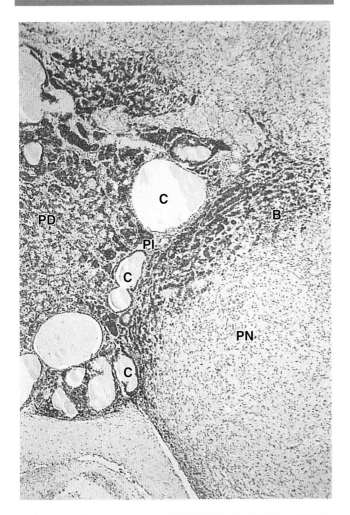

The pars intermedia (**PI**) is a narrow region lying between the pars distalis (**PD**) and the pars nervosa (**PN**), with many of its basophils (**B**) often invading the latter. Remnants of the embryonic hypophyseal pouch's lumen are usually present in this region as colloid-filled cysts (**C**) of various sizes. Function of this region in humans is not clear. (X56; H&E)

and on hormone secretion by the relevant pituitary cells. Figure 20–10 illustrates this mechanism, using the thyroid as an example, and shows the complex chain of events that begins with the action of neural stimuli in the hypothalamus and ends with the effects of hormones from the pituitary's target organs.

Finally, hormone secretion in the anterior pituitary is affected by other hormones from outside the feedback loop or even outside the major target tissues. Examples include the polypeptide ghrelin produced mainly in the stomach mucosa, which also acts as a releasing hormone for somatotropin secretion, and oxytocin, released in the posterior pituitary during breast-feeding, which increases secretion of prolactin.

TABLE **20–2**	Hypothalamic hormones regulating cells of the anterior pituitary.	
Hormone	**Chemical Form**	**Functions**
Thyrotropin-releasing hormone (TRH)	3-amino acid peptide	Stimulates release of thyrotropin (TSH)
Gonadotropin-releasing hormone (GnRH)	10-amino acid peptide	Stimulates the release of both follicle-stimulating hormone (FSH) and luteinizing hormone (LH)
Somatostatin	14-amino acid peptide	Inhibits release of both somatotropin (GH) and TSH
Growth hormone–releasing hormone (GHRH)	40- or 44-amino acid polypeptides (2 forms)	Stimulates release of GH
Dopamine	Modified amino acid	Inhibits release of prolactin (PRL)
Corticotropin-releasing hormone (CRH)	41-amino acid polypeptide	Stimulates synthesis of pro-opiomelanocortin (POMC) and release of both β-lipotropin (β-LPH) and corticotropin (ACTH)

FIGURE **20–10** Negative feedback loops affecting anterior pituitary secretion.

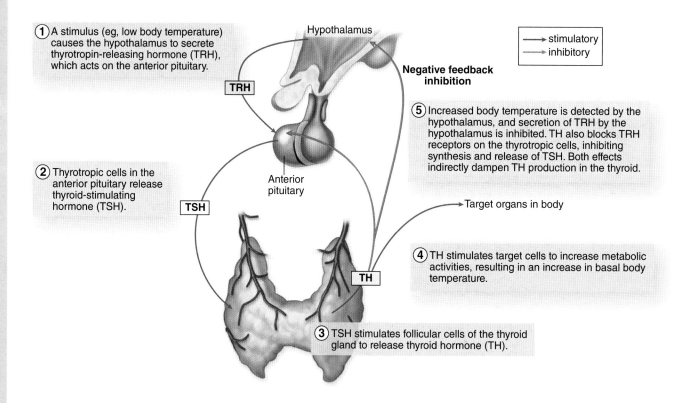

① A stimulus (eg, low body temperature) causes the hypothalamus to secrete thyrotropin-releasing hormone (TRH), which acts on the anterior pituitary.

② Thyrotropic cells in the anterior pituitary release thyroid-stimulating hormone (TSH).

Hypothalamus

Negative feedback inhibition

→ stimulatory
→ inhibitory

⑤ Increased body temperature is detected by the hypothalamus, and secretion of TRH by the hypothalamus is inhibited. TH also blocks TRH receptors on the thyrotropic cells, inhibiting synthesis and release of TSH. Both effects indirectly dampen TH production in the thyroid.

Anterior pituitary

Target organs in body

④ TH stimulates target cells to increase metabolic activities, resulting in an increase in basal body temperature.

③ TSH stimulates follicular cells of the thyroid gland to release thyroid hormone (TH).

Relationship between the hypothalamus, the anterior pituitary, and its target organs is shown, using the thyroid as an example. Hypothalamic thyrotropin-releasing hormone (**TRH**) stimulates secretion of thyroid-stimulating hormone or thyrotropin (**TSH**), which stimulates synthesis and secretion of thyroid hormone (**TH**). In addition to their effects on target organs, TH inhibits TSH secretion from the pars distalis and TRH secretion from the hypothalamus by negative feedback.

FIGURE **20–11** Pars nervosa: neurosecretory bodies and pituicytes.

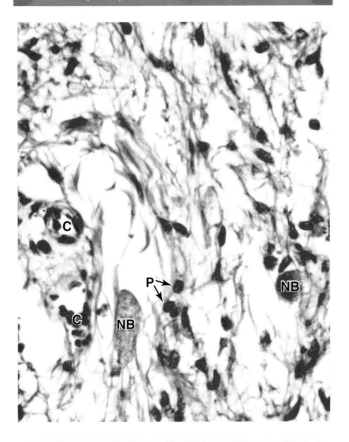

The pars nervosa of the posterior pituitary consists of modified neural tissues containing unmyelinated axons supported and ensheathed by glia cells called **pituicytes** (**P**), the most numerous cell present. The axons run from the supraoptic and paraventricular hypothalamic nuclei, and have swellings called **neurosecretory** (**Herring**) **bodies** (**NB**) from which either oxytocin or vasopressin is released upon neural stimulation. The released hormones are picked up by capillaries (**C**) for distribution. (X400; H&E)

TABLE **20–3**	Hormones of the posterior pituitary.
Hormone	**Function**
Vasopressin/antidiuretic hormone (ADH)	Increases water permeability of renal collecting ducts
Oxytocin	Stimulates contraction of mammary gland myoepithelial cells and uterine smooth muscle

❯❯ MEDICAL APPLICATION

Posterior pituitary function can be adversely affected by heritable mutations in the gene for vasopressin (ADH)-neurophysin, by compression from a tumor in adjacent tissues, and

by **head trauma**. By lowering levels of vasopressin, such conditions can produce diabetes insipidus, a disorder characterized by inability to concentrate urine, which leads to frequent urination (**polyuria**) and increased thirst (**polydipsia**).

Oxytocin stimulates contraction of uterine smooth muscle during childbirth and the myoepithelial cells in mammary gland alveoli (Table 20–3). A nursing infant induces oxytocin secretion by stimulating sensory tracts that act on the hypothalamus in a neurohormonal reflex causing rapid ejection of milk. Oxytocin also produces psychological effects, such as promotion of pair bonding behavior.

❯ ADRENAL GLANDS

The **adrenal** (or suprarenal) **glands** are paired organs lying near the superior poles of the kidneys, embedded in the pararenal adipose tissue and fascia (Figures 20–1 and 20–12). They are flattened structures with a half-moon shape, about 4-6-cm long, 1-2-cm wide, and 4-6-mm thick in adults. Together, they

FIGURE **20–12** Location and blood supply of the adrenal glands.

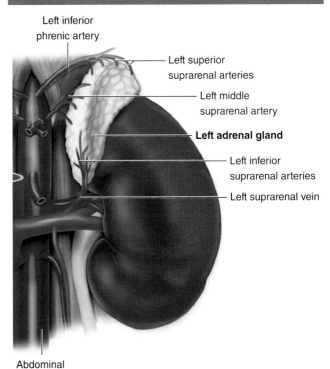

The paired adrenal glands are located at the superior pole of each kidney and each consists of an outer cortex that produces a variety of steroid hormones and an inner medulla that produces epinephrine and norepinephrine. This anterior view of the left adrenal gland and kidney shows the blood vessels supplying these glands.

weigh about 8 g, but their weight and size vary with the age and physiologic condition of the individual. Adrenal glands are each covered by a dense connective tissue capsule that sends thin trabeculae into the gland's parenchyma. The stroma consists mainly of reticular fibers supporting the secretory cells and microvasculature. Each gland has two concentric regions: a yellowish **adrenal cortex** and a reddish-brown central **adrenal medulla**.

The adrenal cortex and medulla can be considered two different organs with distinct embryonic origins, functions, and morphologic characteristics that become united during embryonic development. The cortex arises from mesoderm and the medulla from the neural crest. The general histologic appearance of the adrenal gland is typical of an endocrine gland in which cells of both cortex and medulla are grouped in cords along wide capillaries.

The adrenal gland lacks a hilum; superior, middle, and inferior suprarenal arteries arising from larger abdominal arteries penetrate the capsule independently (Figure 20–12) and branch immediately to form a subcapsular arterial plexus. From this plexus arterioles for the adrenal cortex and medulla emerge separately to form rich networks of fenestrated capillaries and sinusoids. Cortical capillaries irrigate endocrine cells of the cortex and then drain into the microvasculature of the medulla. The adrenal medulla thus has a dual blood supply: arterial blood from the medullary arterioles and venous blood from capillaries of the cortex. Venous drainage from the glands occurs via the suprarenal veins (Figure 20–12).

Adrenal Cortex

Cells of the adrenal cortex have characteristic features of steroid-secreting cells: acidophilic cytoplasm rich in lipid droplets, with central nuclei. Ultrastructurally their cytoplasm shows an exceptionally profuse smooth ER (SER) of interconnected tubules, which contain the enzymes for cholesterol synthesis and conversion of the steroid prohormone pregnenolone into specific active steroid hormones. The mitochondria are often spherical, with tubular rather than shelflike cristae (Figure 20–13). These mitochondria not only synthesize ATP but also contain the enzymes for converting cholesterol to pregnenolone and for some steps in steroid synthesis. The function of steroid-producing cells involves close collaboration between SER and mitochondria.

Steroid hormones are not stored in granules like proteins nor undergo exocytosis. As small lipid-soluble molecules, steroids diffuse freely from cells through the plasma membrane.

The adrenal cortex has three concentric zones in which the cords of epithelial steroid-producing cells are arranged somewhat differently and which synthesize different classes of steroid hormones (Figure 20–14):

- The **zona glomerulosa**, immediately inside the capsule and comprising about 15% of the cortex, consists of closely packed, rounded or arched cords of columnar or pyramidal cells with many capillaries (Figure 20–15). The steroids made by these cells

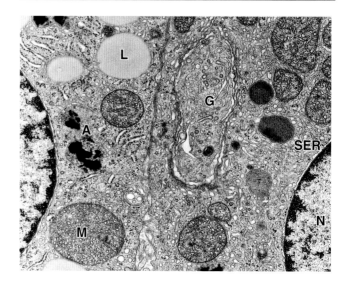

FIGURE 20–13 Ultrastructure of cortical adrenalocytes.

TEM of two adjacent steroid-secreting cells from the zona fasciculate shows features typical of steroid-producing cells: lipid droplets (**L**) containing cholesterol esters, mitochondria (**M**) with tubular and vesicular cristae, abundant **SER**, and autophagosomes (**A**), which remove mitochondria and SER between periods of active steroid synthesis. Also seen are the euchromatic nuclei (**N**), a Golgi apparatus (**G**), RER, and lysosomes. (X25,700)

are called **mineralocorticoids** because they affect uptake of Na⁺, K⁺, and water by cells of renal tubules. The principal product is **aldosterone**, the major regulator of salt balance, which acts to stimulate Na⁺ reabsorption in the distal convoluted tubules (see Chapter 19). Aldosterone secretion is stimulated primarily by angiotensin II and also by an increase in plasma K⁺ concentration, but only weakly by ACTH.

- The middle **zona fasciculata**, occupies 65%-80% of the cortex and consists of long cords of large polyhedral cells, one or two cells thick, separated by fenestrated sinusoidal capillaries (Figure 20–15). The cells are filled with lipid droplets and appear vacuolated in routine histologic preparations. These cells secrete **glucocorticoids**, especially **cortisol**, which affect carbohydrate metabolism by stimulating gluconeogenesis in many cells and glycogen synthesis in the liver. Cortisol also suppresses many immune functions and can induce fat mobilization and muscle proteolysis. Secretion is controlled by ACTH with negative feedback proportional to the concentration of circulating glucocorticoids (Figure 20–10). Small amounts of weak androgens are also produced here.

- The innermost **zona reticularis** comprises about 10% of the cortex and consists of smaller cells in a network of irregular cords interspersed with wide capillaries (Figure 20–15). The cells are usually more heavily

FIGURE **20–14** **Adrenal gland.**

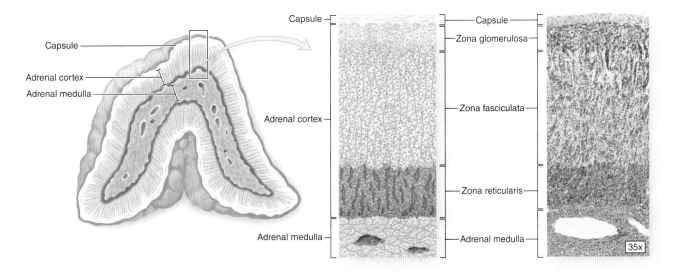

Inside the capsule of each adrenal gland is an adrenal cortex, formed from embryonic mesodermal cells, which completely surrounds an innermost adrenal medulla derived embryologically from neural crest cells. Both regions are very well vascularized with fenestrated sinusoidal capillaries. Cortical cells are arranged as three layers: the zona glomerulosa near the capsule, the zona fasciculata (the thickest layer), and the zona reticularis.

stained than those of the other zones because they contain fewer lipid droplets and more lipofuscin pigment. Cells of the zona reticularis also produce cortisol but primarily secrete the **weak androgens**, including **dehydroepiandrosterone (DHEA)** that is converted to testosterone in both men and women. Secretion by these cells is also stimulated by ACTH with regulatory feedback.

❯❯ MEDICAL APPLICATION

Addison disease or **adrenal cortical insufficiency** is a disorder, usually autoimmune in origin, which causes degeneration in any layer of adrenal cortex, with concomitant loss of glucocorticoids, mineralocorticoids, or androgen production.

Adrenal Medulla

The adrenal medulla is composed of large, pale-staining polyhedral cells arranged in cords or clumps and supported by a reticular fiber network (Figure 20–16). A profuse supply of sinusoidal capillaries intervenes between adjacent cords and a few parasympathetic ganglion cells are present. Medullary parenchymal cells, known as **chromaffin cells**, arise from neural crest cells, as do the postganglionic neurons of sympathetic and parasympathetic ganglia. Chromaffin cells can be considered modified sympathetic postganglionic neurons, lacking axons and dendrites and specialized as secretory cells.

❯❯ MEDICAL APPLICATION

In the adrenal medulla, benign pheochromocytomas periodically secrete high levels of catecholamines that cause swings in blood pressure between hypertension and hypotension.

Unlike cells of the adrenal cortex, chromaffin cells contain many electron-dense granules, 150-350 nm in diameter, for storage and secretion of catecholamines, either **epinephrine** or **norepinephrine**. The granules of epinephrine-secreting cells are less electron-dense and generally smaller than those of norepinephrine-secreting cells (Figure 20–16). Both catecholamines, together with Ca^{2+} and ATP, are bound in granular storage complexes with 49-kDa proteins called **chromogranins**.

Norepinephrine-secreting cells are also found in paraganglia (collections of catecholamine-secreting cells adjacent to the autonomic ganglia) and in various viscera. The conversion of norepinephrine to epinephrine (adrenalin) occurs only in chromaffin cells of the adrenal medulla. About 80% of the catecholamine secreted from the adrenal is epinephrine.

Medullary chromaffin cells are innervated by preganglionic sympathetic neurons, which trigger epinephrine and norepinephrine release during stress and intense emotional reactions. Epinephrine increases heart rate, dilates bronchioles, and dilates arteries of cardiac and skeletal muscle. Norepinephrine constricts vessels of the digestive system and skin, increasing blood flow to the heart, muscles, and brain. Both

FIGURE **20–15** Adrenal cortex.

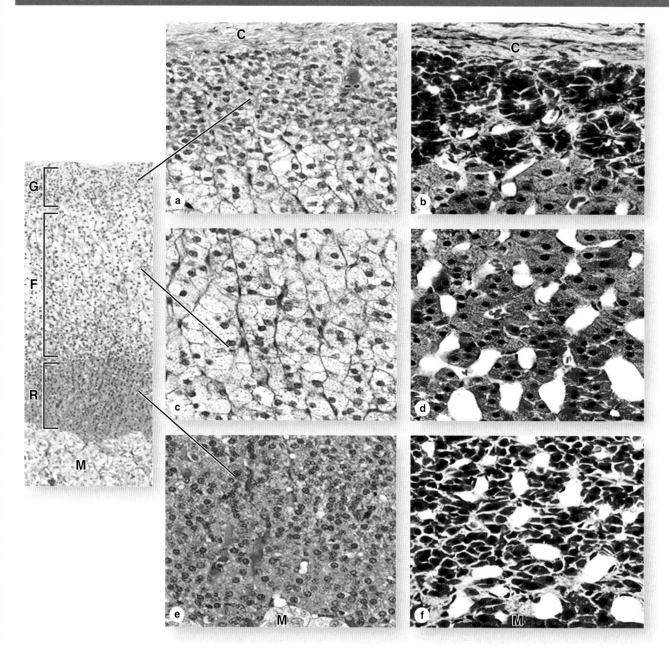

The steroid-secreting cells of the adrenal cortex are arranged differently to form three fairly distinct concentric layers, the zonae glomerulosa (**G**), fasciculata (**F**), and reticularis (**R**), surrounding the medulla (**M**). As with all endocrine glands, the layers of the adrenal cortex all contain a rich microvasculature. Shown here are sections from two adrenal glands, stained with H&E (left) and Mallory trichrome, in which the sparse collagen appears blue (right).

(a, b) Immediately beneath the capsule (**C**), the **zona glomerulosa** consists of rounded clusters of columnar or pyramidal cells principally secreting the mineral corticoid aldosterone. Blood-filled regions are parts of the subcapsular arterial plexus.

(c, d) The thick middle layer, the **zona fasciculata**, consists of long cords of large, spongy-looking cells mainly secreting glucocorticoids such as cortisol.

(e, f) Cells of the innermost **zona reticularis**, next to the medulla (**M**), are small, have fewer lipid droplets and are therefore better stained, arranged in a close network and secrete mainly sex steroids, including the androgen precursor DHEA. Cells of all the layers are closely associated with capillaries and sinusoids.

Left: (X20); a–f: (X200)

| FIGURE **20–16** Adrenal medulla. |

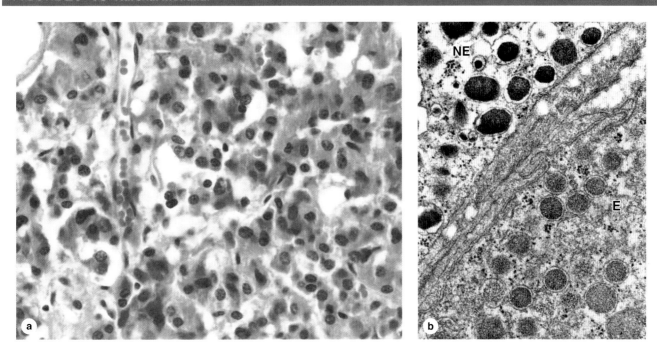

The hormone-secreting cells of the adrenal medulla are chromaffin cells, which resemble sympathetic neurons.

(a) The micrograph shows that they are large pale-staining cells, arranged in cords interspersed with wide capillaries. Faintly stained cytoplasmic granules can be seen in most chromaffin cells. (X200; H&E)

(b) TEM reveals that the granules of norepinephrine-secreting cells (**NE**) are more electron-dense than those of cells secreting epinephrine (**E**), which is due to the chromogranins binding the catecholamines. Most of the hormone produced is epinephrine, which is only made in the adrenal medulla. (X33,000)

hormones stimulate glycogen breakdown, elevating blood glucose levels. Together these effects augment the capability for defensive reactions or escape of stressors, the fight-or-flight response. During normal activity the adrenal medulla continuously secretes small quantities of these hormones.

❯ PANCREATIC ISLETS

The **pancreatic islets** (islets of Langerhans) are compact spherical or ovoid masses of endocrine cells embedded within the acinar exocrine tissue of the pancreas (Figure 20–17). Most islets are 100-200 μm in diameter and contain several hundred cells, but some have only a few cells. The pancreas has more than 1 million islets, mostly in the gland's narrow tail region, but they only constitute 1%-2% of the organ's total volume. A very thin reticular capsule surrounds each islet, separating it from the adjacent acinar tissue. Pancreatic islets have the same embryonic origin as the pancreatic acinar tissue: in epithelial outgrowths from endoderm of the developing gut.

The cells of islets are polygonal or rounded, smaller, and more lightly stained than the surrounding acinar cells, arranged in cords separated by fenestrated capillaries (Figure 20–17).

Routine stains or trichrome stains show that most islet cells are acidophilic or basophilic with fine cytoplasmic granules (Figure 20–17). Ultrastructural features are those of active polypeptide-secreting cells, with secretory granules that vary

in size, morphology, and electron density from cell to cell. The major islet cells are most easily identified and studied by immunohistochemistry:

- α or A cells secrete primarily **glucagon** and are usually located peripherally.
- β or B cells produce **insulin** (L. *insula*, island), are the most numerous, and are located centrally.
- δ or D cells, secreting **somatostatin**, are scattered and much less abundant.

FIGURE **20–17** Pancreatic islets.

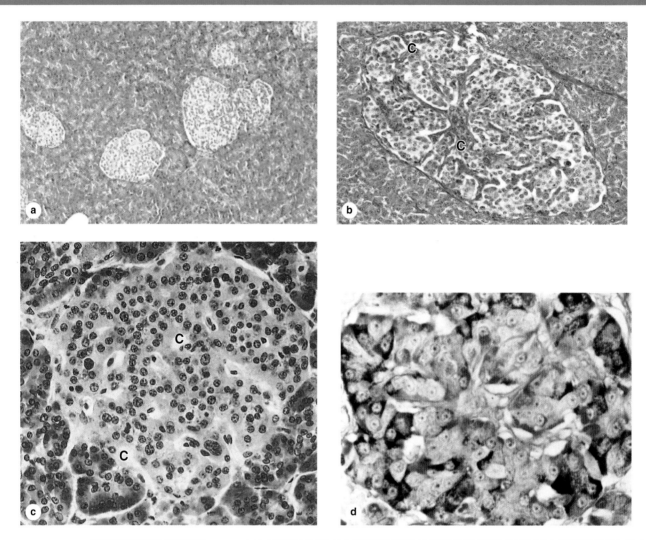

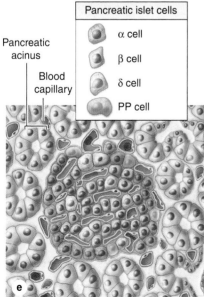

Pancreatic islets are clumped masses of pale-staining endocrine cells embedded in the exocrine acinar tissue of the pancreas.

(a) The islets are clusters of cells smaller and lighter staining than cells of the surrounding tissue. (X12.5; H&E)

(b) At higher magnification an islet's capillary system can be seen. Several arterioles enter each islet, branch into fenestrated capillaries (**C**) among the peripheral islet cells, then converge centrally before leaving the islet as efferent capillaries carrying blood to the acini surrounding the islet. This local vascular system allows specific islet hormones to help control secretion of other islet cells and the neighboring acini. (X40; H&E)

(c) With H&E staining all cells of an islet appear similar, although differences in cell size and basophilia may be apparent. Capillaries (**C**) are also apparent. (X55; H&E)

(d) An islet prepared with a modified aldehyde fuchsin stain shows that granules in the peripheral α cells are a deep brownish purple and the central β cells granules are brownish orange. Reticulin connective tissue of the islet capsule and along the capillaries stains green in this preparation. Immunohistochemistry with antibodies against the various islet polypeptide hormones allows definitive identification of each islet cell type. (X300; Modified aldehyde fuchsin and light green)

(Figure 20-17d, used with permission from Dr Arthur A. Like, Department of Pathology, University of Massachusetts Medical School, Worcester.)

(e) The diagram shows the four major islet hormones and the cells secreting them: α cells making glucagon, β cells making insulin, δ cells making somatostatin, and PP cells making pancreatic polypeptide.

TABLE **20–4**	Major cell types and hormones of pancreatic islets.			
Cell Type	**Quantity (%)**	**Hormone Produced**	**Hormone Structure and Size**	**Hormone Function**
α	~20	Glucagon	Polypeptide; 3500 Da	Acts on several tissues to make energy stored in glycogen and fat available through glycogenolysis and lipolysis; increases blood glucose content
β	~70	Insulin	Dimer of α and β chains with S-S bridges; 5700-6000 Da	Acts on several tissues to cause entry of glucose into cells and promotes decrease of blood glucose content
δ or D	5-10	Somatostatin	Polypeptide; 1650 Da	Inhibits release of other islet cell hormones through local paracrine action; inhibits release of GH and TSH in anterior pituitary and HCl secretion by gastric parietal cells
PP	Rare	Pancreatic polypeptide	Polypeptide; 4200 Da	Stimulates activity of gastric chief cells; inhibits bile secretion, pancreatic enzyme and bicarbonate secretion, and intestinal motility

A minor fourth cell type, more common in islets located within the head of the pancreas, are **PP cells**, which secrete **pancreatic polypeptide**. Table 20–4 summarizes the types, quantities, and main functions of the major pancreatic hormones. Pancreatic islets also normally contain a few enterochromaffin cells, like those of the digestive tract, which are also scattered in the pancreatic acini and ducts and which secrete other hormones affecting the digestive system.

Activity of α and β cells is regulated largely by blood glucose levels above or below the normal level of 70 mg/dL. Increased glucose levels stimulate β cells to release insulin and inhibit α cells from releasing glucagon; decreased glucose levels stimulate α cells to release glucagon. Opposing actions of these hormones help to precisely control blood glucose concentration, an important factor in homeostasis (Table 20–4).

These hormones and somatostatin from the δ cells also act in a paracrine manner to affect hormone release within an islet as well as activity of the neighboring acinar cells. Sympathetic and parasympathetic nerve endings are closely associated with about 10% of α, β, and δ cells and can also function as part of the control system for insulin and glucagon secretion. Gap junctions transfer the autonomic neural stimulus to the other cells. Sympathetic fibers increase glucagon release and inhibit insulin release; parasympathetic fibers increase secretion of both glucagon and insulin.

▶ DIFFUSE NEUROENDOCRINE SYSTEM

The enterochromaffin cells scattered in both the islets and small ducts of the pancreas are similar to those of the digestive tract (see Chapter 15). Collectively these dispersed cells,

as well as similar cells in the respiratory mucosa, make up the **diffuse neuroendocrine system (DNES)**. Like the pancreatic islets, most of these cells are derived from endodermal cells of the embryonic gut or bronchial buds. These secretory cells are considered neuroendocrine because they produce many of the same polypeptides and neurotransmitter-like molecules, such as serotonin (5-hydroxytryptamine), also released by neurosecretory cells in the CNS. Several such cells, along with their hormones and major functions, are summarized in Table 15–1 with the digestive system. Most of these hormones are polypeptides and act in a paracrine manner, affecting primarily the activities of neighboring contractile cells and secretory cells (both exocrine and endocrine). Enteroendocrine cells of the stomach and small bowel are shown ultrastructurally in Figures 15–20, 15–24c, and 15–27.

Many cells of the DNES are stained by solutions of chromium salts and have therefore been called **enterochromaffin cells**, while those staining with silver nitrate are sometimes called **argentaffin cells**. DNES cells secreting serotonin or certain other amine derivatives demonstrate amine precursor uptake and decarboxylation and are often referred to acronymically as **APUD cells**. Such names are still widely used, but, as indicated in Table 15–1, they have been largely replaced by letter designations like those used for pancreatic islet cells. Whatever name is used, cells of the DNES are highly important due to their role in regulating motility and secretions of all types within the digestive system.

▶ THYROID GLAND

The thyroid gland, located anterior and inferior to the larynx, consists of two lobes united by an isthmus (Figure 20–18). It originates in early embryonic life from the foregut endoderm

FIGURE 20–18 Thyroid gland.

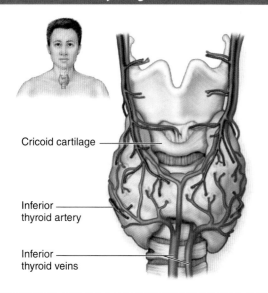

Cricoid cartilage

Inferior thyroid artery

Inferior thyroid veins

The thyroid is a highly vascular, butterfly-shaped gland surrounding the anterior surface of the trachea just below the larynx.

near the base of the developing tongue. It synthesizes the thyroid hormones **thyroxine** (tetra-iodothyronine or **T4**) and **tri-iodothyronine** (**T3**), which help control the basal metabolic rate in cells throughout the body, as well as the polypeptide hormone **calcitonin**.

The parenchyma of the thyroid is composed of millions of rounded epithelial **thyroid follicles** of variable diameter, each with simple epithelium and a central lumen densely filled with gelatinous acidophilic **colloid** (Figure 20–19). The thyroid is the only endocrine gland in which a large quantity of secretory product is stored. Moreover, storage is outside the cells, in the colloid of the follicle lumen, which is also unusual. There is sufficient hormone in follicles to supply the body for up to 3 months with no additional synthesis. Thyroid colloid contains the large glycoprotein **thyroglobulin** (660 kDa), the precursor for the active thyroid hormones.

The thyroid gland is covered by a fibrous capsule from which septa extend into the parenchyma, dividing it into lobules and carrying blood vessels, nerves, and lymphatics. Follicles are densely packed together, separated from one another only by sparse reticular connective tissue (Figure 20–19), although this stroma is very well vascularized with fenestrated capillaries for transfer of released hormone to the blood.

The follicular cells, or **thyrocytes**, range in shape from squamous to low columnar (Figure 20–19), their size and other features varying with their activity that is controlled by thyroid-stimulating hormone (TSH) from the anterior pituitary. Active glands have more follicles of low columnar epithelium; glands with mostly squamous follicular cells are hypoactive.

Thyrocytes have apical junctional complexes and rest on a basal lamina (Figure 20–20). The cells exhibit organelles indicating active protein synthesis and secretion, as well as phagocytosis and digestion. The nucleus is generally round and central. Basally the cells are rich in rough ER and apically, facing the follicular lumen, are Golgi complexes, secretory granules, numerous phagosomes and lysosomes, and microvilli.

Another endocrine cell type, the **parafollicular cell**, or **C cell**, is also found inside the basal lamina of the follicular epithelium or as isolated clusters between follicles (Figure 20–20). Derived from the neural crest, parafollicular cells are usually somewhat larger than follicular cells and stain less intensely. They have a smaller amount of rough ER, large Golgi complexes, and numerous small (100-180 nm in diameter) granules containing calcitonin (Figure 20–20). Secretion of calcitonin is triggered by elevated blood Ca^{2+} levels and it inhibits osteoclast activity, but this function in humans is less important than the roles of parathyroid hormone and vitamin D in the regulation of normal calcium homeostasis.

❯❯ MEDICAL APPLICATION

Chronic dietary iodine deficiencies inhibit thyroid hormone production, causing thyrotropic cells of the anterior pituitary gland to produce excess TSH. This leads to excessive growth of thyroid follicles and enlargement of the thyroid gland, a condition known as **goiter**.

Production of Thyroid Hormone & Its Control

Production, storage, and release of thyroid hormones involve an unusual, multistage process in the thyrocytes, with both an exocrine phase and an endocrine phase. Both phases are promoted by TSH and occur in the same cell, as summarized in Figure 20–21. The major activities of this process include the following:

1. The **production of thyroglobulin**, which is similar to that in other glycoprotein-exporting cells, with synthesis in the rough ER and glycosylation in the Golgi apparatus. Thyroglobulin has no hormonal activity itself but contains 140 tyrosyl residues critical for thyroid hormone synthesis. The glycoprotein is released as an exocrine product from apical vesicles of thyrocytes into the follicular lumen.

2. The **uptake of iodide** from blood by Na/I symporters (NIS) in the thyrocytes' basolateral cell membranes, which allows for 30-fold concentration of dietary iodide in thyroid tissue relative to plasma. Decreased levels of circulating iodide trigger synthesis of NIS, increasing iodide uptake and compensating for the lower plasma concentration. An apical iodide/chloride transporter (also called **pendrin**) pumps I⁻ from thyrocytes into the colloid.

FIGURE **20–19** Thyroid follicular cells and parafollicular cells.

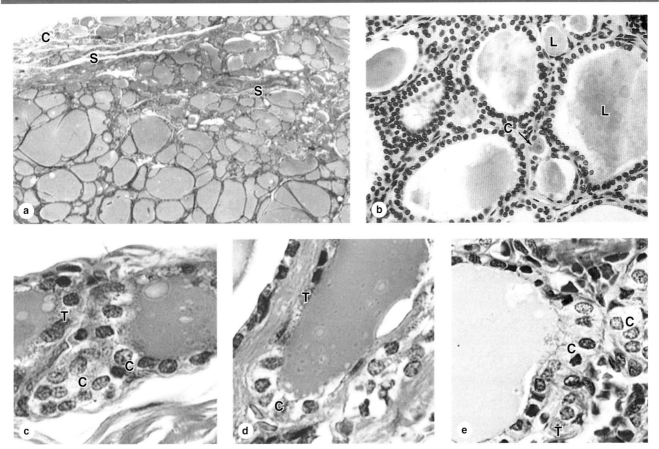

(a) A low-power micrograph of thyroid gland shows the thin capsule (**C**), from which septa (**S**) with the larger blood vessels, lymphatics, and nerves enter the gland. The parenchyma of the organ is distinctive, consisting of colloid-filled epithelial follicles of many sizes. The lumen of each follicle is filled with a lightly staining colloid of a large gelatinous protein called **thyroglobulin**. (X12; H&E)

(b) The lumen (**L**) of each follicle is surrounded by a simple epithelium of thyrocytes in which the cell height ranges from squamous to low columnar. Also present are large pale-staining parafollicular or C cells (**C**) that secrete calcitonin, a polypeptide involved with calcium metabolism. (X200; H&E)

(c-e) C cells may be part of the follicular epithelium or present singly or in groups outside of follicles. Thyrocytes (**T**) can usually be distinguished from parafollicular C cells (**C**) by their smaller size and darker staining properties. Unlike thyrocytes, C cells seldom vary in their size or pale staining characteristics. C cells are somewhat easier to locate in or between small follicles. c and d: (X400;H&E); e: (X400; Mallory trichrome)

3. **Iodination of tyrosyl residues** in thyroglobulin with either one or two atoms occurs in the colloid after oxidation of iodide to iodine by membrane-bound thyroid peroxidase on the microvilli surfaces of thyrocytes.

4. **Formation of T_3 and T_4** (also called thyroxine) occurs as two iodinated tyrosines, still part of colloidal thyroglobulin, which are covalently conjugated in coupling reactions.

5. **Endocytosis of iodinated thyroglobulin** by the thyrocytes involves both fluid-phase pinocytosis and receptor-mediated endocytosis. The endocytic vesicles fuse with lysosomes, and the thyroglobulin is thoroughly degraded by lysosomal proteases, freeing active thyroid hormone as both T_3 and T_4.

6. **Secretion of T_4 and T_3** at the basolateral domains of thyrocytes occurs in an endocrine manner: both molecules are immediately taken up by capillaries.

Nearly all of both thyroid hormones are carried in blood tightly bound to thyroxine-binding globulin or albumen. T_4 is the more abundant compound, constituting 90% of the circulating thyroid hormone. Both molecules bind the same intracellular receptors of target cells, but T_3 is 2- to 10-fold more active than T_4. The half-life of T_3 is 1.5 days in comparison with a week for T_4. Both thyroid hormones increase the number and size of mitochondria and stimulate mitochondrial protein synthesis, helping to enhance metabolic activity.

FIGURE **20–20** Ultrastructure of thyroid follicular and parafollicular cells.

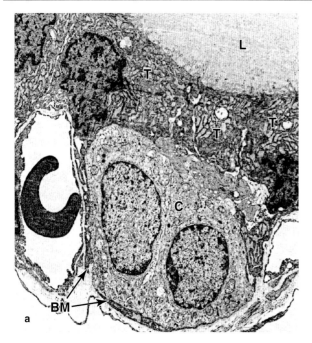

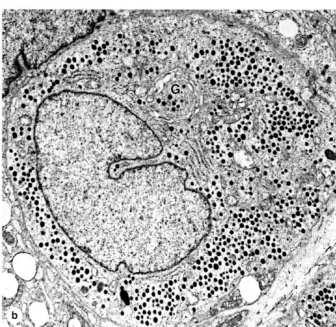

(a) TEM of the follicular epithelium shows pseudopodia and micro-villi extending from the follicular thyrocytes (**T**) into the colloid of the lumen (**L**). The cells have apical junctional complexes, much RER, well-developed Golgi complexes, and many lysosomes. Inside the basement membrane (**BM**) of the follicle, but often not contacting the colloid in the lumen, are occasional C cells (**C**). To the left and right of the two C cells seen here are capillaries intimately associated with the follicular cells, but outside the basement membrane. (X2000)

(b) A TEM of a C cell, with its large Golgi apparatus (**G**), extensive RER, and cytoplasm filled with small secretory granules containing calcitonin. (X5000)

The major regulator of the anatomic and functional state of thyroid follicles is TSH (thyrotropin) from the anterior pituitary (Figure 20–8). With TSH receptors abundant on the basal cell membrane of thyrocytes, this tropic hormone increases cell height in the follicular epithelium and stimulates all stages of thyroid hormone production and release. Thyroid hormones inhibit the release of TSH, maintaining levels of circulating T_4 and T_3 within the normal range (Figure 20–10). Secretion of TSH in the pituitary is also increased by exposure to cold and decreased by heat and stressful stimuli.

manifested by tiredness, weight gain, intolerance of cold, and decreased ability to concentrate.

❯ PARATHYROID GLANDS

The **parathyroid glands** are four small ovoid masses—each 3×6 mm—with a total weight of about 0.4 g. They are located on the back of the thyroid gland, usually embedded in the larger gland's capsule (Figure 20–22). The microvasculature of each arises from the inferior thyroid arteries. Each parathyroid gland is contained within a thin capsule from which septa extend into the gland. A sparse reticular stroma supports dense elongated clusters of secretory cells.

The parathyroid glands are derived from the embryonic pharyngeal pouches—the superior glands from the fourth pouch and the inferior glands from the third pouch. Their migration to the developing thyroid gland is sometimes misdirected so that the number and locations of the glands are somewhat variable. Up to 10% of individuals may have parathyroid tissue attached to the thymus, which originates from the same pharyngeal pouches.

❯❯ MEDICAL APPLICATION

Graves disease is an autoimmune disorder in which antibodies produce chronic stimulation of the follicular cells and release of thyroid hormones (**hyperthyroidism**), which causes a hypermetabolic state marked by weight loss, nervousness, sweating, heat intolerance, and other features. **Hypothyroidism**, with reduced thyroid hormone levels, can be caused by local inflammation (**thyroiditis**) or inadequate secretion of TSH by the anterior pituitary gland and is often

FIGURE 20–21 Thyrocyte activities in thyroid hormone synthesis.

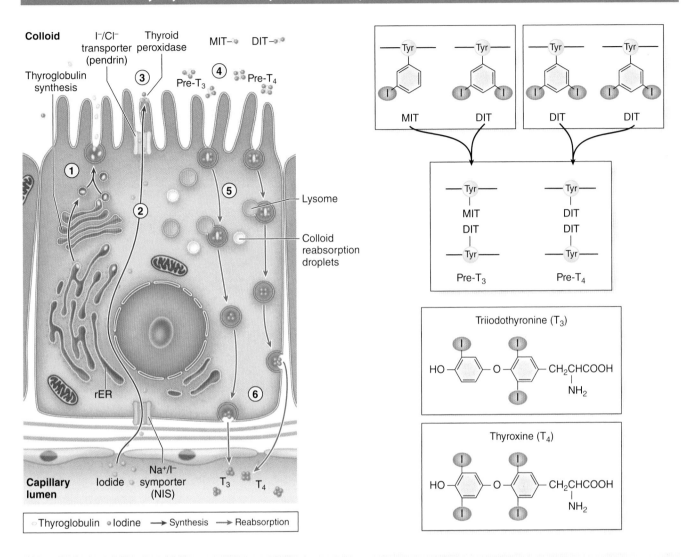

The diagram shows the multistep process by which thyroid hormones are produced via the stored thyroglobulin intermediate. In an exocrine phase of the process, the glycoprotein **thyroglobulin** is made and secreted into the follicular lumen and iodide is pumped across the cells into the lumen. In the lumen **tyrosine residues** of thyroglobulin are iodinated, to form monoiodotyrosine (MIT) or diiodotyrosine (DIT), and then covalently coupled to form

T_3 and T_4 still within the glycoprotein. The iodinated thyroglobulin is then endocytosed by thyrocytes and degraded by lysosomes, releasing **free active T_3 and T_4** to the adjacent capillaries in an endocrine manner. Detailed steps are given in the text. Both phases are promoted by TSH and may occur simultaneously in the same cell.

The endocrine cells of the parathyroid glands, called **principal (chief) cells**, are small polygonal cells with round nuclei and pale-staining, slightly acidophilic cytoplasm (Figure 20–23). Irregularly shaped cytoplasmic granules contain the polypeptide **parathyroid hormone (PTH)**, an important regulator of blood calcium levels. PTH has three major targets:

■ Osteoblasts respond to PTH by producing an osteoclast-stimulating factor that increases the number and activity of osteoclasts. The resulting resorption of the calcified

bone matrix and release of Ca^{2+} increase the concentration of circulating Ca^{2+}, which suppresses PTH production. The effect of PTH on blood levels of Ca^{2+} is thus opposite to that of calcitonin.

■ In the distal convoluted tubules of the renal cortex, PTH stimulates Ca^{2+} reabsorption (and inhibits phosphate reabsorption in the proximal tubules).

■ PTH also indirectly increases the Ca^{2+} absorption in the small intestine by stimulating vitamin D activation.

FIGURE **20–22** Parathyroid glands.

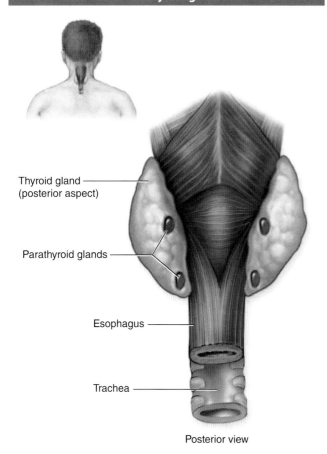

Thyroid gland
(posterior aspect)

Parathyroid glands

Esophagus

Trachea

Posterior view

The parathyroid glands are four small nodules normally embedded in the capsule on the posterior surface of the thyroid gland.

FIGURE **20–23** Parathyroid principal cells.

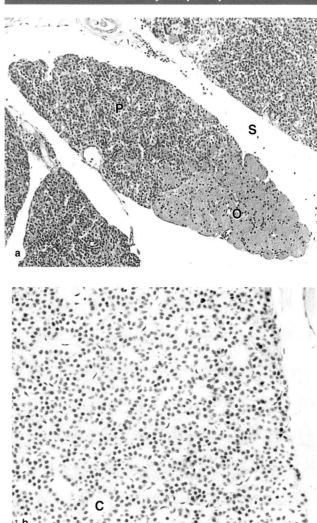

(a) A small lobe of parathyroid gland, surrounded by connective tissue septa (**S**), shows mainly densely packed cords of small principal cells (**P**). Older parathyroid glands show increasing numbers of much larger and acidophilic nonfunctional oxyphil cells (**O**) that may occur singly or in clumps of varying sizes. (X60; H&E)

(b) Higher magnification shows that principal cells have round central nuclei and pale-staining cytoplasm. Cords of principal cells secreting PTH surround capillaries (**C**). (X200; H&E)

With increasing age, many secretory cells are replaced with adipocytes, which may constitute more than 50% of the gland in older people.

Much smaller populations of **oxyphil cells**, often clustered, are sometimes also present in parathyroid glands, more commonly in older individuals. These are much larger than the principal cells and are characterized by very acidophilic cytoplasm filled with abnormally shaped mitochondria. Some oxyphil cells show low levels of PTH synthesis, suggesting that these cells are transitional derivatives of principal cells.

〉〉 MEDICAL APPLICATION

In **hypoparathyroidism**, diminished secretion of PTH can cause bones to become more mineralized and denser and striated muscle to exhibit abnormal contractions due to inadequate calcium ion concentrations. Excessive **PTH** produced in **hyperparathyroidism** stimulates osteoclast number and activity, leading to increased levels of blood calcium that can be deposited pathologically in cartilage, arteries, or the kidneys.

〉 PINEAL GLAND

The **pineal gland**, also known as the **epiphysis cerebri**, regulates the daily rhythms of bodily activities. A small, pine cone-shaped organ, approximately 5-8 mm by 3-5 mm, the pineal gland develops from neuroectoderm in the posterior wall of the third ventricle and remains attached to the brain by a short stalk. The pineal gland is covered by connective tissue of the pia mater, from which septa containing small blood vessels emerge and subdivide variously sized lobules.

Prominent and abundant secretory cells called **pinealocytes** have slightly basophilic cytoplasm and irregular euchromatic nuclei (Figure 20–24). Ultrastructurally pinealocytes are seen to have secretory vesicles, many mitochondria, and long cytoplasmic processes extending to the vascularized septa, where they end in dilatations near capillaries, indicating an endocrine function. These cells produce **melatonin**, a low-molecular-weight tryptophan derivative. Unmyelinated sympathetic nerve fibers enter the pineal gland and end among pinealocytes, with some forming synapses.

Melatonin release from pinealocytes is promoted by darkness and inhibited by daylight. The resulting diurnal fluctuation in blood melatonin levels induces rhythmic changes in the activity of the hypothalamus, pituitary gland, and other endocrine tissues that characterize the circadian (24 hours, day/night) rhythm of physiological functions and behaviors. In humans and other mammals, the cycle of light and darkness is detected within the retinas and transmitted to the pineal via the retinohypothalamic tract, the suprachiasmatic nucleus, and the tracts of sympathetic fibers entering the pineal. The pineal gland acts, therefore, as a neuroendocrine transducer, converting sensory input regarding light and darkness into variations in many hormonal functions.

The pineal gland also has interstitial glial cells that are modified **astrocytes**, staining positively for glial fibrillary acidic protein, which represent about 5% of the cells. These

FIGURE **20–24** Pineal gland.

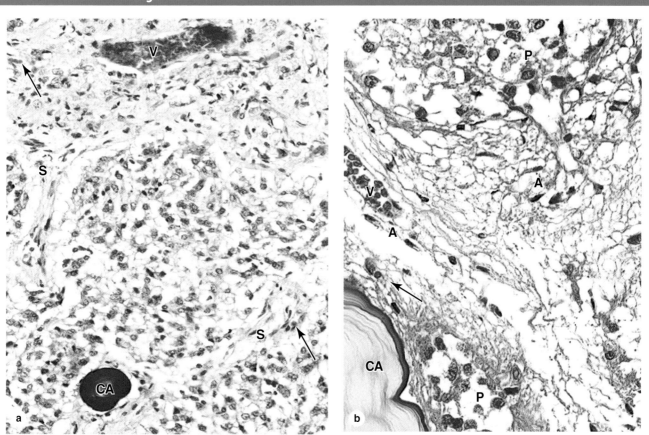

(a) The micrograph shows a group of pinealocytes surrounded by septa (**S**) containing venules (**V**) and capillaries (arrows). Also seen is an extracellular mineral deposit called a **corpus arenaceum** (**CA**) of unknown physiologic significance but an excellent marker for the pineal. (X200; H&E)

(b) At higher magnification the numerous large pinealocytes (**P**) with euchromatic nuclei can be compared to much fewer astrocytes (**A**) that have darker, more elongated nuclei and are located mainly within septa and near small blood vessels (**V**). Capillaries

(arrow) are not nearly as numerous as in other endocrine glands. At the lower left is a part of a very large corpus arenaceum (**CA**), the calcified structures also known as **brain sand**. Along the septa run unmyelinated tracts of sympathetic fibers, associated indirectly with photoreceptive neurons in the retinas and running to the pinealocytes to stimulate melatonin release in periods of darkness. Levels of circulating melatonin are one factor determining the diurnal rhythms of hormone release and physiologic activities throughout the body. (X400; H&E)

have elongated nuclei more heavily stained than those of pinealocytes and are usually found in perivascular areas and between the groups of pinealocytes.

A characteristic feature of the pineal gland is the presence of variously sized concretions of calcium and magnesium salts called **corpora arenacea**, or brain sand, formed by mineralization of extracellular protein deposits. Such concretions appear during childhood and gradually increase in number and size with age, with no apparent effect on the gland's function.

>> **MEDICAL APPLICATION**

Densely calcified corpora arenacea can be used as landmarks for the midline location of the pineal gland in various **radiological examinations** of the brain. **Tumors** originating from pinealocytes are very rare, but they can be either benign or highly malignant.

Table 20–5 summarizes the major endocrine cells, hormones, and functions of the adrenal gland, pancreatic islets, thyroid, parathyroid, and pineal glands.

TABLE 20–5	Cells, important hormones, and functions of other major endocrine organs.		
Gland	**Endocrine Cells**	**Major Hormones**	**Major Functions**
Adrenal glands: Cortex	Cells of zona glomerulosa	Mineralocorticoids	Stimulate renal reabsorption of water and Na^+ and secretion of K^+ to maintain salt and water balance
	Cells of zona fasciculata	Glucocorticoids	Influence carbohydrate metabolism; suppress immune cell activities
	Cells of zona reticularis	Weak androgens	Precursors for testosterone or estrogen
Adrenal glands: Medulla	Chromaffin cells	Epinephrine	Increases heart rate and blood pressure
		Norepinephrine	Constricts vessels; increases heart rate and blood pressure
Pancreatic islets	α Cells	Glucagon	Raises blood glucose levels
	β Cells	Insulin	Lowers blood glucose levels
	δ Cells	Somatostatin	Inhibits secretion of insulin, glucagon, and somatotropin
	PP cells	Pancreatic polypeptide	Inhibits secretion of pancreatic enzymes and HCO_3^-
Thyroid glands	Follicular cells	Thyroid hormones (T_3 and T_4)	Increases metabolic rate
	Parafollicular or C cells	Calcitonin	Lowers blood Ca^{2+} levels by inhibiting osteoclast activity
Parathyroid glands	Chief cells	Parathyroid hormone (PTH)	Raises blood Ca^{2+} levels by stimulating osteoclast activity
Pineal gland	Pinealocytes	Melatonin	Regulates circadian rhythms

Endocrine Glands SUMMARY OF KEY POINTS

Pituitary Gland

- The **pituitary gland** has two major parts: the **posterior** part called the **pars nervosa** develops as a downgrowth of the developing brain and is attached in the **hypothalamus** by the **infundibulum**.
- The anterior pituitary includes the large **pars distalis**, the **pars tuberalis** that surrounds the infundibulum, and the thin **pars intermedia** adjacent to the pars nervosa.
- Blood vessels of the **hypothalamic hypophyseal portal system** are important in carrying peptide factors from hypothalamic neurons to cells of the anterior pituitary where they control cell secretion.
- This portal system includes a **primary capillary plexus** in the infundibulum and lower hypothalamus and a **secondary plexus** in the pars distalis, connected by **portal veins** and draining to the **hypophyseal vein**.
- Endocrine cells of the anterior pituitary can be called **acidophils**, **basophils**, or **chromophobes** based on their general staining properties; the latter lack secretory granules and stain poorly.
- Acidophils and basophils can be identified as to which pituitary hormone they produce using **immunohistochemistry** and antibodies against specific hormones.
  - Acidophils are primarily **somatotrophs** producing **somatotropin (growth hormone)**, or **lactotrophs** (or mammotrophs) producing **prolactin (PRL)**.
  - Basophils include **gonadotrophs** producing **follicle-stimulating hormone (FSH)** and **luteinizing hormone (LH)** and **thyrotrophs** making **thyroid-stimulating hormone (TSH)**.
  - A third type of basophil is the corticotroph, synthesizing **proopiomelanocortin (POMC)** that is cleaved by proteases to make **adrenocorticotropic hormone (ACTH)** and **β-lipotropic hormone (LPH)**.
- The **hypothalamic hypophyseal tract** includes bundles of nerve axons that run from **supraoptic nucleus (SON)** and **paraventricular nucleus (PVN)** in the hypothalamus to the posterior pituitary.
- Hypothalamic neurons in the SON synthesize **vasopressin/antidiuretic hormone (ADH)** and those of the PVN synthesize oxytocin, both of which are stored in axonal dilations called **Herring bodies** before release to capillaries in the posterior pituitary.

Adrenal Glands

- The **adrenal cortex** of each adrenal gland consists of three concentric zones, all histologically distinct but with cells producing steroid hormones and all drained by the same system of capillaries.

- The most superficial **zona glomerulosa** has round clusters of cells producing **mineralocorticoids** such as **aldosterone** regulating electrolyte levels.
- The wider, middle **zona fasciculata** has elongated strands of cells producing **glucocorticoids** such as **cortisol**, which regulates several aspects of carbohydrate metabolism.
- The innermost zona reticularis has a network of cells making the **weak androgen dehydroepiandrosterone (DHEA)** that is converted to testosterone in men and women.
- The **adrenal medulla** contains neural crest-derived **chromaffin cells** synthesizing either **epinephrine** or **norepinephrine** that regulate the stress response.

Pancreatic Islets

- **Pancreatic islets of Langerhans** are small clusters of pale cells embedded within the exocrine tissue; they have primarily **β cells making insulin, α cell (glucagon), and δ cells (somatostatin)**.

Thyroid Gland

- The **thyroid gland** consists mainly of spherical **follicles** composed of simple epithelium of **thyrocytes** surrounding a lumen with **thyroglobulin**.
- **Thyroglobulin** is a large glycoprotein in which tyrosine residues are **iodinated** as precursors of the thyroid hormones **thyroxine** and **triiodothyronine**, which are released when the precursor is degraded.
- Thyroid follicles also contain a few smaller and paler cells called **parafollicular cells**, or **C cells**, which produce the polypeptide hormone **calcitonin**.

Parathyroid Glands

- On the posterior side of the thyroid glands are four small **parathyroid glands**, each consisting primarily of **principal cells** that secrete **parathyroid hormone (PTH)** that raises blood Ca^{2+} levels.

Pineal Gland

- The **pineal gland** develops from embryonic neuroectoderm, remains attached to the brain, and contains modified neurons called **pinealocytes** that secrete the amine **melatonin**.
- Pineal landmarks are the concretions called **corpora arenacea** (brain sand); neural connections from the retina to **pinealocytes** allow diurnal secretion of melatonin and rhythms in physiological activities.

Endocrine Glands ASSESS YOUR KNOWLEDGE

1. Which of the following accurately describes glucocorticoids?
   a. Include the steroid hormone aldosterone
   b. Are produced in response to stimulation by ACTH
   c. Are produced primarily by the zone glomerulosa
   d. Typically enhance the immune response
   e. Include the steroid hormone dehydroepiandrosterone (DHEA)

2. Pregnant women who have begun labor but in whom this process is no longer progressing are often given an IV injection of Pitocin to stimulate uterine contractions and facilitate parturition. Pitocin is a trade name of a hormone produced in what endocrine tissue?
   a. Pars distalis
   b. Ovarian follicles
   c. Pars nervosa
   d. Placenta
   e. Pars tuberalis

3. What hormone is produced in response to decreased blood calcium levels?
   a. Pancreatic polypeptide
   b. β-endorphin
   c. Somatostatin
   d. Calcitonin
   e. Parathyroid hormone

4. Addison disease (or adrenal cortex insufficiency) is a disorder, usually autoimmune in origin, which can cause degeneration and cell loss in the adrenal glands. Fludrocortisone is a mineralocorticoid used to treat Addison patients. Response to this drug indicates that which region of the adrenal glands was involved in the disease?
   a. Medulla
   b. Zona glomerulosa
   c. Zona reticularis
   d. Macula densa
   e. Zona fasciculate

5. A glucagonoma is a malignant tumor consisting of what cells?
   a. A or α cells
   b. B or β cells
   c. Chromophils
   d. D or δ cells
   e. Mucous cells

6. Secretion, chemical modification and storage, reuptake, and digestion of a protein occur in epithelial cells of what endocrine tissue?
   a. Neurohypophysis
   b. Adrenal medulla
   c. Adenohypophysis
   d. Thyroid gland
   e. Neuroendocrine cells in the duodenum

7. Secretion in what neuroendocrine cell is controlled directly by neural activity and involves a hormone that generally slows metabolic activity at night?
   a. Pituicyte
   b. Melanocyte
   c. Herring body of the neurohypophysis
   d. Chromaffin cell
   e. Pinealocyte

8. Some mammalian endocrine tissues or cells can be experimentally transplanted to other well-vascularized sites (such as the oral mucosa) in genetically similar hosts and the tissue's function continues normally and with proper regulation. The pars distalis is not a good candidate for such transplantation studies for which one of the following reasons?
   a. More severe rejection of neurally-related tissue occurs compared with other endocrine organs.
   b. Its hormonal source is unavailable after its axonal connections to the hypothalamus are disrupted.
   c. Its cells stop functioning when separated from the hypothalamo-hypophyseal portal system.
   d. Neogenesis of blood vessels into this tissue will not occur at the transplant site.
   e. The vascular wall of the superior hypophyseal arteries is unique.

9. A 45-year-old corporate executive presents with the primary complaint of "always being tired." She comments that she has been tired for 4 months even though she is sleeping more. She complains of being unable to finish chores at home and "dragging at work." She indicates that she is often constipated and is intolerant of cold. She is continuously turning the thermostats in the house and office to higher temperatures, to the dismay of family and coworkers. She also complains that her skin is very dry; use of lotions and creams have not helped the dryness. A biopsy of her thyroid gland shows dense lymphocytic infiltration with germinal centers throughout the parenchyma and a battery of further tests is carried out, which also suggest thyroiditis. Which of the following results of blood tests would be most likely in this patient?
   a. Elevated TSH levels
   b. Elevated T_3 and T_4 levels
   c. Autoantibodies to the thyroid hormone receptor
   d. Elevated calcitonin levels
   e. Elevated glucocorticoid levels

10. A 9-year-old girl, the youngest of four daughters, is taken to the pediatrician by her mother, who indicates that for at least 4 months the child has seemed "hyperactive," unable to sleep soundly because "she says her room is too hot," and no longer able to concentrate in school. Upon questioning, the mother also remembers that her daughter's periods also began within the past few months. Blood tests indicate high levels of estrogen-related hormones and cortisol. Which of the following tentative diagnoses is consistent with all of these symptoms?
   a. Graves disease, caused by antibodies stimulating the TSH receptor
   b. A defect in the paraventricular nucleus
   c. Excessively active cells of the adrenal cortex zona glomerulosa
   d. A benign tumor involving cells in the adenohypophysis
   e. A disabling mutation in the gene for thyroblobulin

CHAPTER 21

The Male Reproductive System

The male reproductive system consists of the testes, genital ducts, accessory glands, and penis (Figure 21–1). Testes produce sperm but also contain endocrine cells secreting hormones such as testosterone, which drives male reproductive physiology. Testosterone is important for spermatogenesis, sexual differentiation during embryonic and fetal development, and control of gonadotropin secretion in the pituitary. A metabolite of testosterone, dihydrotestosterone, also begins to act on many tissues during puberty (eg, male accessory glands and hair follicles).

The genital ducts and accessory glands produce secretions required for sperm activity and contract to propel spermatozoa and the secretions from the penile urethra. These secretions provide nutrients for spermatozoa while they are confined to the male reproductive tract. Spermatozoa and the secretions of the accessory glands make up the **semen** (L. seed), which is introduced into the female reproductive tract by the penis.

❯ TESTES

Each **testis** (or testicle) is surrounded by a dense connective tissue capsule, the **tunica albuginea**, which thickens on the posterior side to form the **mediastinum testis**. From this fibrous region, septa penetrate the organ and divide it into about 250 pyramidal compartments or **testicular lobules** (Figures 21–2 and 21–3). Each lobule contains sparse connective tissue with endocrine **interstitial cells** (or **Leydig cells**) secreting testosterone, and one to four highly convoluted **seminiferous tubules** in which sperm production occurs.

The testes develop retroperitoneally in the dorsal wall of the embryonic abdominal cavity and are moved during fetal development to become suspended in the two halves of the scrotal sac, or scrotum, at the ends of the spermatic cords (Figure 21–2). During migration from the abdominal cavity, each testis carries with it a serous sac, the **tunica vaginalis**, derived from the peritoneum. This tunic consists of an outer parietal layer lining the scrotum and an inner visceral layer, covering the tunica albuginea on the anterior and lateral sides of the testis (Figure 21–2).

Having evolved in cold-blooded animals, certain molecular events in the process of sperm formation cannot occur at the core body temperature of 37°C. A permissive temperature of about 34°C is maintained in the scrotal sac by various mechanisms. Each testicular artery is surrounded by a rich **pampiniform venous plexus** containing cooler blood from the testis, which draws heat from the arterial blood by a countercurrent heat-exchange system. Evaporation of sweat from the scrotum also contributes to heat loss. Relaxation or contraction of the thin dartos muscle of the scrotum and the cremaster muscles of the spermatic cords move the testes away from or closer to the body, respectively, allowing further control on testicular temperature.

FIGURE **21-1** The male reproductive system.

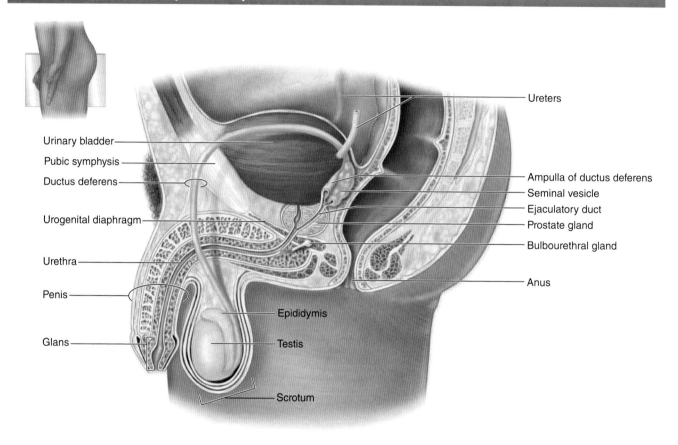

The diagram shows the locations and relationships of the testes, epididymis, glands, and the ductus deferens running from the scrotum to the urethra. The ductus deferens is located along the anterior and superior sides of the bladder as a result of the testes descending into the scrotum from the abdominal cavity during fetal development.

Interstitial Tissue

The interstitial tissue of the testis between the seminiferous tubules consists of sparse connective tissue containing fibroblasts, lymphatics, and blood vessels including fenestrated capillaries. During puberty **interstitial cells**, or **Leydig cells**, develop as large round or polygonal cells with central nuclei and eosinophilic cytoplasm rich in small lipid droplets (Figures 21–2b and 21–4). These cells produce the steroid hormone **testosterone**, which promotes development of the secondary male sex characteristics. Testosterone is synthesized by enzymes present in the smooth ER and mitochondria similar to the system in adrenal cortical cells.

Testosterone secretion by interstitial cells is triggered by the pituitary gonadotropin, **luteinizing hormone (LH)**, which is also called **interstitial cell stimulating hormone (ICSH)**. Testosterone synthesis thus begins at puberty, when the hypothalamus begins producing gonadotropin-releasing hormone. In the late embryonic testes gonadotropin from the placenta stimulates interstitial cells to synthesize the testosterone needed for development of the ducts and glands of the male reproductive system. These fetal interstitial cells are very active during the third and fourth months of pregnancy, then regress and become quiescent cells resembling fibroblasts until

FIGURE **21–2** Testes and seminiferous tubules.

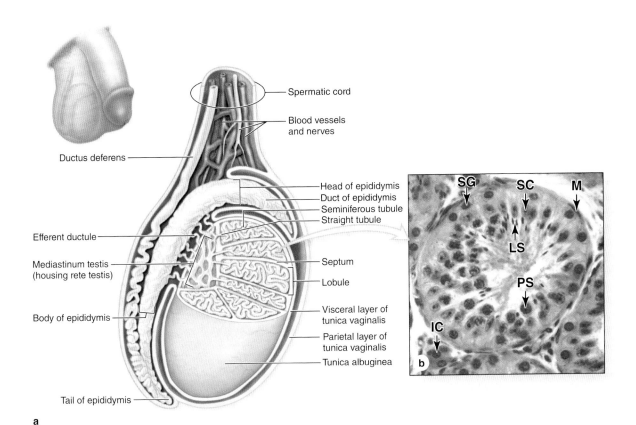

The anatomy of a testis is shown. **(a)** The diagram shows a partially cutaway sagittal section of the testis. **(b)** A seminiferous tubule cross section shows spermatogonia (**SG**) near the periphery, near nuclei of Sertoli cells (**SC**), primary spermatocytes (**PS**), and late spermatids (**LS**) near the lumen, with interstitial cells (**IC**) in the surrounding connective tissue. (X400; H&E)

puberty when they resume testosterone synthesis in response to the pituitary gonadotropin.

❯❯ MEDICAL APPLICATION

Both **interstitial cell tumors** and **Sertoli cell tumors** are rare. Most (95%) **testicular cancer** involves **germ cell tumors**, which only appear after puberty and are much more likely to develop in men with untreated cryptorchidism.

Seminiferous Tubules

Sperm are produced in the seminiferous tubules at a rate of about 2×10^8 per day in the young adult. Each testis has from 250 to 1000 such tubules in its lobules, and each tubule measures 150-250 μm in diameter and 30-70 cm in length. The combined length of the tubules of one testis totals about 250 m.

Each tubule is actually a loop linked by a very short, narrower segment, the **straight tubule**, to the **rete testis**, a labyrinth of epithelium-lined channels embedded in the mediastinum testis (Figures 21–2a and 21–3). About 10-20 **efferent ductules** connect the rete testis to the head of the **epididymis** (Figure 21–2a).

Each seminiferous tubule is lined with a complex, specialized stratified epithelium called **germinal** or **spermatogenic epithelium** (Figure 21–2b). The basement membrane of this epithelium is covered by fibrous connective tissue, with an innermost layer containing flattened, smooth muscle-like **myoid cells** (Figure 21–2b), which allow weak contractions of the tubule. The germinal epithelium consists of two types of cells:

- Large nondividing **Sertoli cells** (Figure 21–4), which physically and metabolically support developing sperm cell precursors
- Dividing cells of the **spermatogenic lineage** (Figure 21–5a)

FIGURE **21–3** Lobules converging at rete testis.

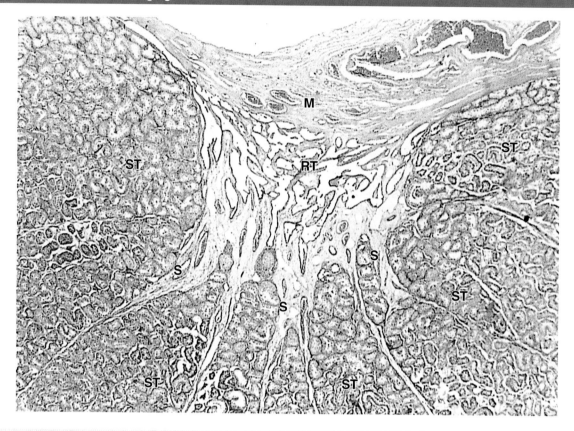

The dense capsule of the testis, the tunica albuginea, thickens on the posterior side as the mediastinum (**M**) testis, from which many thin septa (**S**) subdivide the organ into about 250 lobules. Each lobule contains one to four convoluted seminiferous tubules (**ST**)

in a sparse connective tissue interstitium. Each tubule is a loop attached by means of a short straight tubule to the rete testis (**RT**), a maze of channels embedded in the mediastinum testis. From the rete testis the sperm move into the epididymis. (X60; H&E)

The cells of the spermatogenic lineage, comprising four or more concentric layers of cells in the germinal epithelium, develop from progenitor cells to fully formed sperm cells over a period of approximately 10 weeks. As shown in Figure 21–5 **spermatogenesis**, the first part of sperm production, involves mainly mitosis and meiosis and is followed by **spermiogenesis**, the final differentiation process occurring in the haploid male germ cells.

Spermatogenesis

Spermatogenesis begins at puberty with proliferation of stem and progenitor cells called **spermatogonia** (Gr. *sperma* + *gone*, generation), small round cells about 12 μm in diameter. These cells occupy a basal niche in the epithelial wall of the tubules, next to the basement membrane and closely associated with Sertoli cell surfaces (Figures 21–5, 21–6 and 21–7).

Different stages of spermatogonia development can be recognized by subtle changes in shape and staining properties of their nuclei. Spermatogonia with dark, ovoid nuclei act as stem cells, dividing infrequently and giving rise both

to new stem cells and to cells with more pale-staining, ovoid nuclei that divide more rapidly as transit amplifying (progenitor) cells (Figure 21–7). These **type A spermatogonia** each undergo several unique clonal divisions that leave most of the cells interconnected as a syncytium. These become **type B spermatogonia**, which have more spherical and pale nuclei.

Each type B spermatogonium then undergoes a final mitotic division to produce two cells that grow in size and become **primary spermatocytes**, which are spherical cells with euchromatic nuclei (Figures 21–6 and 21–7). Primary spermatocytes replicate their DNA, so each chromosome consists of duplicate chromatids, and enter meiosis, during which homologous chromosomes come together in synapsis, DNA recombination occurs, and two rapid cell divisions produce haploid cells (see Chapter 3).

The primary spermatocyte has 46 (44 + XY) chromosomes, the diploid number, and a DNA content of 4N. (The letter N denotes either the haploid number of chromosomes, 23 in humans, or the amount of DNA in this set.) Soon after their formation, these cells enter the first meiotic prophase

FIGURE **21–4** Interstitial cells and Sertoli cells.

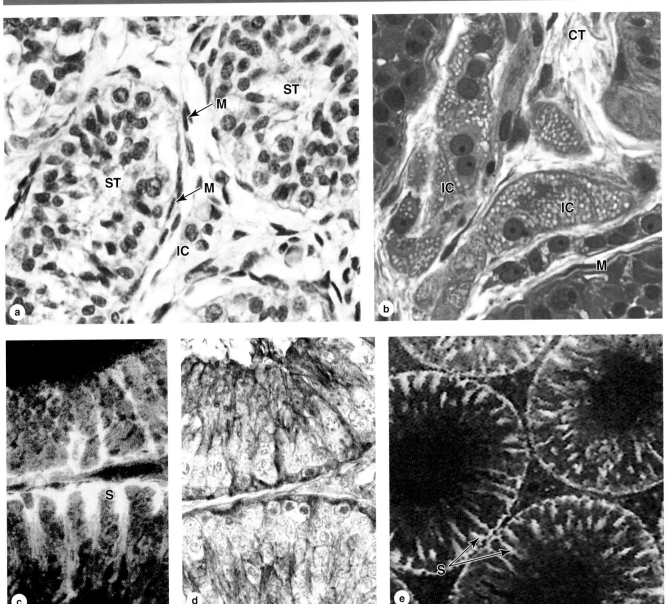

(a) Seminiferous tubules (**ST**) are surrounded by stroma containing many interstitial cells (**IC**), typically located near capillaries, which secrete androgens. The seminiferous tubule wall consists of a unique germinal epithelium composed of columnar Sertoli cells and dividing spermatogenic stem cells. Seen around the seminiferous tubules are myoid cells (**M**) with elongated nuclei, the contractions of which help move fluid and mature sperm in the tubules. (X400; H&E)

(b) A plastic section shows lipid droplets filling the cytoplasm of the clumped interstitial cells (**IC**), or Leydig cells, in the connective tissue (**CT**) between tubules. Such cytoplasm is typical of steroid-secreting endocrine cells and here indicates cells actively secreting testosterone. The epithelium of a nearby seminiferous tubule is immediately surrounded by myoid cells (**M**). (X400; PT)

(c) Immunohistochemistry of a seminiferous tubule wall with antibodies against prosaposin, a glycoprotein abundant in Sertoli cells. The yellow fluorescent stain indicates the tall columnar shape of Sertoli cells (**S**) and the dendritic nature of their apical ends. Sertoli cells support all spermatogenic cells physically and metabolically, phagocytize debris, and have endocrine roles affecting spermatogenesis and fetal development of the male reproductive tract. (400X; Immunofluorescence)

(d) Immunohistochemistry of the seminiferous tubule with the same antibodies labeled with peroxidase shows the tall Sertoli cells in brown. This bright-field preparation also shows the close association of the Sertoli cells with the numerous spermatogenic cells, whose nuclei are stained with hematoxylin. (400X; Immunoperoxidase & hematoxylin)

(e) Lower magnification of the same preparation as part c shows the distribution and density of Sertoli cells (**S**) in the seminiferous tubules. (200X; Immunofluorescence)

(Figure 21–4c-e, used with permission from Dr. Richard Sharpe and Chris McKinnell, University of Edinburgh, MRC Centre for Reproductive Health, UK.)

FIGURE **21–5** Spermatogenesis and spermiogenesis.

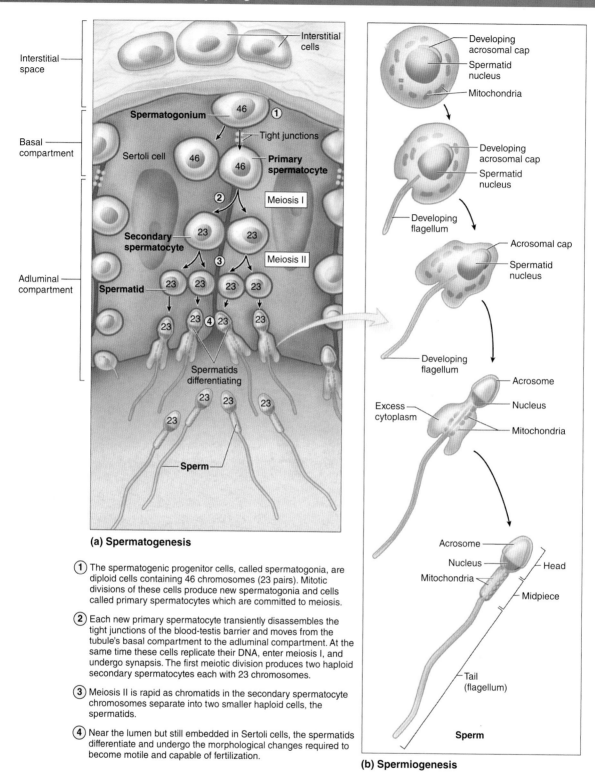

(a) Spermatogenesis

① The spermatogenic progenitor cells, called spermatogonia, are diploid cells containing 46 chromosomes (23 pairs). Mitotic divisions of these cells produce new spermatogonia and cells called primary spermatocytes which are committed to meiosis.

② Each new primary spermatocyte transiently disassembles the tight junctions of the blood-testis barrier and moves from the tubule's basal compartment to the adluminal compartment. At the same time these cells replicate their DNA, enter meiosis I, and undergo synapsis. The first meiotic division produces two haploid secondary spermatocytes each with 23 chromosomes.

③ Meiosis II is rapid as chromatids in the secondary spermatocyte chromosomes separate into two smaller haploid cells, the spermatids.

④ Near the lumen but still embedded in Sertoli cells, the spermatids differentiate and undergo the morphological changes required to become motile and capable of fertilization.

(b) Spermiogenesis

(a) The diagram shows two large, columnar Sertoli cells with their surfaces binding many germ cells in various stages of spermatogenesis. Near the basement membrane are **spermatogonia,** which divide by mitosis to produce both more spermatogonia and also **primary spermatocytes** that undergo meiosis to produce secondary spermatocytes and then haploid spermatids that differentiate as sperm. Newly formed spermatocytes temporarily disassemble the tight junctions between Sertoli cells that act as the "blood-testis barrier" in order to move from the **basal compartment** to the **adluminal compartment** of the tubule.

(b) Spermiogenesis is the process of cell differentiation by which spermatids become sperm. The major changes that occur during spermiogenesis are shown here. These involve flattening of the nucleus, formation of an **acrosome** that resembles a large lysosome, growth of a **flagellum** (tail) from the basal body, reorganization of the mitochondria in the **midpiece** region, and shedding of unneeded cytoplasm.

FIGURE **21–6** Seminiferous tubules: Sertoli cells and spermatogenesis.

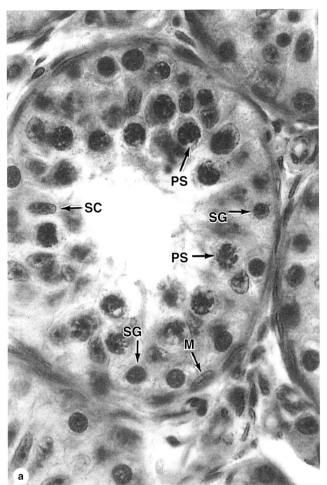

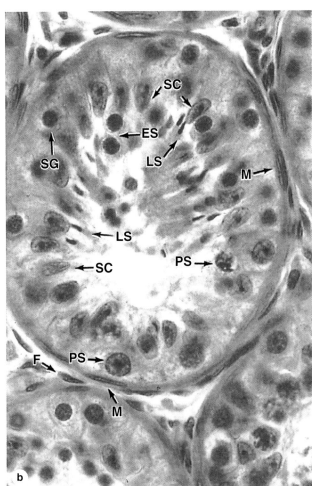

(a, b) In these cross sections of seminiferous tubules, most of their cell types can be seen. Outside the tubules are myoid cells (**M**) and fibroblasts (**F**). Inside near the basement membrane are many prominent spermatogonia (**SG**), small cells that divide mitotically but give rise to a population that enters meiosis. The primary spermatocytes (**PS**) remain for 3 weeks in prophase of the first meiotic division during which recombination occurs. Primary spermatocytes are the largest spermatogenic cells and are usually abundant at all levels between the basement membrane and the lumen.

Each divides to form two secondary spermatocytes, which are rarely seen because they undergo the second meiotic division almost immediately to form two haploid spermatids. Newly formed round, early spermatids (**ES**) differentiate and lose volume in becoming late spermatids (**LS**) and finally highly specialized sperm cells. All stages of spermatogenesis and spermiogenesis occur with the cells intimately associated with the surfaces of adjacent Sertoli cells (**SC**) that perform several supportive functions. (Both X750; H&E)

that lasts about 3 weeks. Most spermatocytes seen in sections of testis are in this phase of meiosis. The primary spermatocytes are the largest cells of the spermatogenic lineage and are characterized by the presence of partially condensed chromosomes in various stages of synapsis and recombination (Figure 21–6).

Homologous chromosomes separate in the first meiotic division, which produces smaller cells called **secondary spermatocytes** (Figures 21–5a and 21–7) with only 23 chromosomes (22 + X or 22 + Y), but each still consists of two chromatids so the amount of DNA is 2N (see Chapter 3). Secondary spermatocytes are rare in testis sections because

they are very short-lived cells that remain in interphase only briefly and quickly undergo the second meiotic division.

Division of each secondary spermatocyte separates the chromatids of each chromosome and produces two haploid cells called **spermatids** each of which contains 23 chromosomes (Figures 21–5a, 21–6, and 21–7). Because no S phase (DNA replication) occurs between the first and second meiotic divisions, the amount of DNA per cell is reduced by half when the chromatids separate and the cells formed are haploid (1N). With fertilization, a haploid ovum and sperm produced by meiosis unite and the normal diploid chromosome number is restored.

FIGURE **21–7** Clonal nature of spermatogenesis.

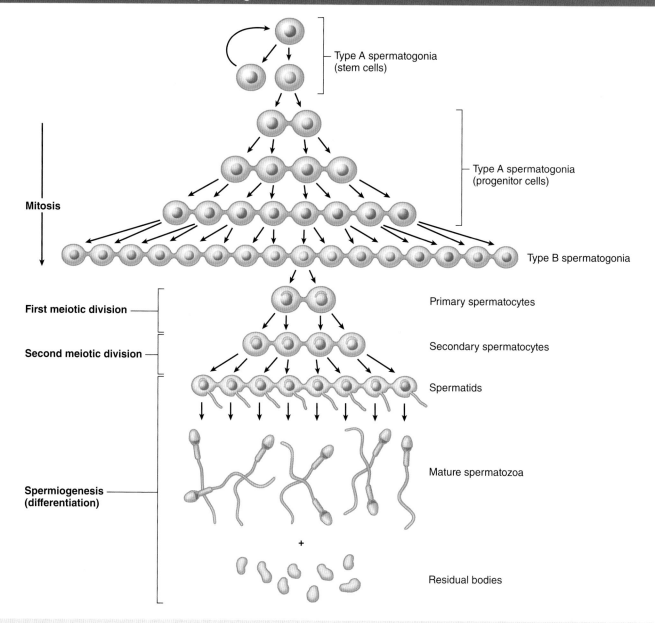

A subpopulation called **type A spermatogonia** act as stem cells, dividing to produce new stem cells and other type A spermatogonia that undergo transit amplification as progenitor cells. Mitosis in these cells occurs with incomplete cytokinesis, leaving the cytoplasm of most or all of these cells connected by cytoplasmic bridges. Type A spermatogonia divide mitotically two or three more times, then differentiate as **type B spermatogonia** that undergo a final round of mitosis to form the cells that then enter meiosis and become **primary spermatocytes** (only two are shown), still interconnected. The intercellular bridges persist during the **first and second meiotic divisions**, but they are lost as the haploid **spermatids** differentiate into **sperm** (spermiogenesis) and shed excess cytoplasm as **residual bodies**. The interconnections of the cells allow free cytoplasmic communication among the cells and help coordinate their progress through meiosis and spermiogenesis.

The Clonal Nature of Male Germ Cells

The stem cells produced by mitotic divisions of spermatogonia remain as separate cells. However, all subsequent divisions of the daughter cells that become transit amplifying progenitor cells have incomplete cytokinesis after telophase and the cells remain attached to one another by **intercellular bridges** of cytoplasm (Figure 21–7). These allow free cytoplasmic communication among the cells during their remaining mitotic and meiotic divisions. Although some cells degenerate without completing spermatogenesis and some cells may separate, clones of approximately a hundred cells may remain linked through meiosis. The complete significance of this **spermatogenic syncytium** is not clear, but the cytoplasmic bridges allow the haploid cells to be supplied with products of the complete diploid genome, including proteins and RNA encoded by genes on the X or Y chromosome missing in their haploid nuclei. The germ cells finally become separated from one another during differentiation (Figure 21–7).

The cellular events and changes between the final mitoses of spermatogonia and the formation of spermatids take about 2 months. The spermatogenic cells are not randomly distributed in the spermatogenic epithelium. Cells at different stages of development are typically grouped together along the tubule, with the intercellular bridges helping to coordinate their divisions and differentiation.

Spermiogenesis

Spermiogenesis, the final phase of sperm production, is the temperature-sensitive process by which spermatids differentiate into **spermatozoa**, which are highly specialized to deliver male DNA to the ovum. No cell division occurs during this process, and as with spermatogenesis the cells involved remain associated with Sertoli cells.

The haploid spermatids are small (7-8 μm in diameter) cells near the lumen of the seminiferous tubules (Figures 21–5a and 21–6b). Spermiogenesis includes formation of the **acrosome** (Gr. *akron*, extremity + *soma*, body), condensation and elongation of the nucleus, development of the **flagellum** (L, whip), and the loss of much of the cytoplasm. The end result is the mature spermatozoon, which is released from the Sertoli cell surface into the tubule's lumen. Spermiogenesis is commonly divided into four phases:

■ In the **Golgi phase** the cytoplasm contains a prominent Golgi apparatus near the nucleus, mitochondria, paired centrioles, and free ribosomes. Small **proacrosomal vesicles** from the Golgi apparatus coalesce as a single membrane-limited **acrosomal cap** close to one end of the nucleus (Figures 21–5b and 21–8). The centrioles migrate to a position farthest from the acrosomal cap and one acts as a basal body, organizing the axoneme of the flagellum which is structurally and functionally similar to that of a cilium (see Chapter 2).

FIGURE **21–8** **Differentiating spermatid.**

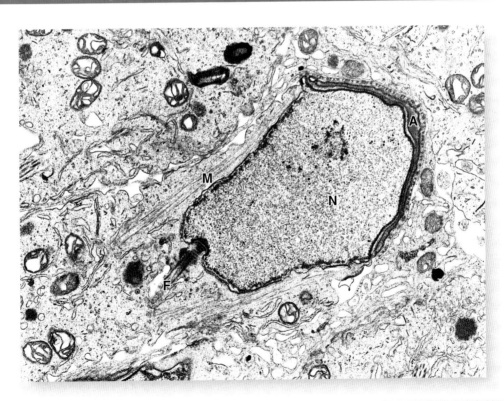

The head of a late spermatid during spermiogenesis is completely enclosed by Sertoli cell cytoplasm. As shown by TEM, the sperm nucleus (**N**) is half covered by the thin Golgi-derived acrosomal cap (**A**). The flagellum (**F**) can be seen emerging from a basal body near the nucleus on the side opposite the acrosome. A perinuclear bundle of microtubules and actin filaments called the **manchette** (**M**) transports vesicles, mitochondria, and keratins into position as the spermatid elongates. (X7500)

- In the **cap phase** the acrosomal cap spreads over about half of the condensing nucleus (Figures 21–5b and 21–8). The acrosome is a specialized type of lysosome containing hydrolytic enzymes, mainly hyaluronidase and a trypsin-like protease called **acrosin**. These enzymes are released when a spermatozoon encounters an oocyte and the acrosomal membrane fuses with the sperm's plasma membrane. They dissociate cells of the corona radiata and digest the zona pellucida, both structures that surround the egg (see Chapter 22). This process, the **acrosomal reaction**, is one of the first steps in fertilization.

- In the **acrosome phase** the **head** of the developing sperm, containing the acrosome and the condensing nucleus, remains embedded in the Sertoli cell while the growing axoneme extends into the lumen of the tubule (Figure 21–6b). Nuclei become more elongated and very highly condensed, with the histones of nucleosomes replaced by small basic peptides called **protamines**. Flagellum growth continues as the tail and mitochondria aggregate around its proximal region to form a thickened **middle piece** where the ATP for flagellar movements is generated (Figure 21–5).

- In the **maturation phase** of spermiogenesis, unneeded cytoplasm is shed as a **residual body** from each spermatozoon and remaining intercellular bridges are lost. Mature but not yet functional or mobile sperm (Figure 21–5) are released into the lumen of the tubule.

❯❯ MEDICAL APPLICATION

Decreased semen quality, which is frequently idiopathic (arising from unknown causes), is a major cause of **male infertility**. Common features of poor semen quality include **oligospermia** (ejaculate volume > 2 mL), sperm cell density less than 10-20 million/mL, abnormal sperm morphology, and flagellar defects that impair sperm motility.

Sertoli Cells

The **Sertoli cells**, named after Enrico Sertoli (1842-1910) who first demonstrated their physiologic significance, are tall "columnar" epithelial cells that nourish the spermatogenic cells and divide the seminiferous tubules into two (basal and adluminal) compartments (Figure 21–4c-e). All cells of the spermatogenic lineage are closely associated with the extended surfaces of Sertoli cells and depend on them for metabolic and physical support.

Sertoli cells adhere to the basal lamina and their apical ends extend to the lumen, as shown immunohistochemically in Figure 21–4c-e. In routine preparations the outlines of Sertoli cells surrounding the spermatogenic cells are very poorly defined (Figures 21–6 and 21–8). Each Sertoli cell supports 30-50 developing germ cells. Ultrastructurally Sertoli cells are seen to contain abundant SER, some rough ER, well-developed

Golgi complexes, numerous mitochondria, and lysosomes (Figure 21–8). Their nuclei are typically ovoid or triangular, euchromatic, and have a prominent nucleolus, features that allow Sertoli cells to be distinguished from the neighboring germ cells (Figure 21–6).

Important in Sertoli cell function are elaborate tight occluding junctions between their basolateral membranes that form a **blood-testis barrier** within the seminiferous epithelium (Figure 21–5a). The tightest blood-tissue barrier in mammals, this physical barrier is one part of a system that prevents autoimmune attacks against the unique spermatogenic cells, which first appear after the immune system is mature and central self-tolerance is well established.

Spermatogonia lie in a **basal compartment** of the tubule, below the tight junctions and not sealed off from the vascularized interstitial tissue containing lymphocytes and other immune cells. Newly formed primary spermatocytes temporarily disassemble the adhesion molecules of the local occluding junctions and move into the tubule's **adluminal compartment** while still adhering to Sertoli cells (Figure 21–5a). Like the spermatogonia, all spermatocytes and spermatids lie within invaginations of the Sertoli cells surfaces. Adluminal migration occurs without compromising the blood-testis barrier, which is all the more impressive when one remembers that the germ cells remain linked by intercellular bridges. Sertoli cells are also connected and coupled ionically by gap junctions, which may help regulate the transient changes in the occluding junctions and synchronize activities in the spermatogenic cells. As the flagellar tails of the spermatids develop, they appear as tufts extending from the apical ends of the Sertoli cells.

Related to their role in establishing the blood-testis barrier, Sertoli cells have three general functions:

- **Support, protection, and nutrition of the developing spermatogenic cells:** Because spermatocytes, spermatids, and developing sperm are isolated from plasma proteins and nutrients by the blood-testis barrier, they depend on Sertoli cells for production or transport into the lumen of metabolites and nutritive factors such as the iron-transport protein transferrin. Thus, while protecting spermatogenic cells from circulating immune components, Sertoli cells supply many plasma factors needed for cell growth and differentiation.

- **Exocrine and endocrine secretion:** Sertoli cells continuously release into the seminiferous tubules water that carries new sperm out of the testis. Production of nutrients and **androgen-binding protein (ABP)**, which concentrates testosterone to a level required for spermiogenesis, is promoted by follicle-stimulating hormone (FSH). As endocrine cells, they secrete the 39-kDa glycoprotein **inhibin**, which feeds back on the anterior pituitary gland to suppress FSH synthesis and release. In the fetus Sertoli cells also secrete a 140-kDa glycoprotein called **müllerian-inhibiting substance (MIS)** that causes regression of the embryonic müllerian (paramesonephric) ducts; in the absence of MIS these ducts persist and become parts of the female reproductive tract.

- **Phagocytosis**: During spermiogenesis, excess cytoplasm shed as residual bodies is phagocytosed and digested by Sertoli cell lysosomes. No proteins from sperm normally pass back across the blood-testis barrier.

›› MEDICAL APPLICATION

Acute or chronic inflammation of the testis, **orchitis**, frequently involves the ducts connecting this organ to the epididymis. Common forms of orchitis are produced by infective agents and occur secondarily to a **urinary tract infection** or a **sexually transmitted pathogen** such as *Chlamydia* or *Neisseria gonorrhoeae* entering the testis from the epididymis or via the lymphatics.

Acute **epididymitis** is a result of **sexually transmitted infections** such as gonorrhea or *Chlamydia* infection and causes intrascrotal pain and tenderness. Persistent inflammation of the epididymis, such as that associated with **gonorrhea** infections, includes massive invasion by leukocytes into the infected duct, stimulating fibrosis that obstructs the epididymis and is a common cause of **male infertility**.

›› INTRATESTICULAR DUCTS

The intratesticular ducts are the **straight tubules** (or tubuli recti), the **rete testis**, and the **efferent ductules** (Figure 21–2), all of which carry spermatozoa and liquid from the seminiferous tubules to the duct of the epididymis (Table 21–1).

The loops of seminiferous tubules join the rete testis by the short straight tubules, which are lined initially only by Sertoli cells (Figure 21–9). These empty into the rete testis, an interconnected network of channels lined with cuboidal epithelium and supported by connective tissue of the mediastinum (Figure 21–9).

The rete testis drains into about 20 efferent ductules lined by an unusual epithelium in which groups of nonciliated cuboidal cells alternate with groups of taller ciliated cells and give the tissue a characteristic scalloped appearance (Figure 21–10). The nonciliated cells absorb some of the fluid secreted by the Sertoli cells of seminiferous tubules. This absorption and the ciliary activity create a fluid flow that carries sperm passively out of the testis toward the epididymis. A thin layer of circularly oriented smooth muscle cells in the walls of efferent ductules aids the movement of sperm into the duct of the epididymis.

TABLE 21–1	Summary of histology and functions of male genital ducts.			
	Location	Epithelium	Support Tissues	Function(s)
Seminiferous tubules	Testicular lobules	Spermatogenic, with Sertoli cells and germ cells	Myoid cells and loose connective tissue	Produce sperm
Straight tubules (tubuli recti)	Periphery of the mediastinum testis	Sertoli cells in proximal portions, simple cuboidal in distal portions	Connective tissue	Convey sperm into the rete testis
Rete testis	In mediastinum testis	Simple cuboidal cells with microvilli and single cilia	Dense irregular connective tissue	Channels with sperm from all seminiferous tubules
Efferent ductules	From rete testis to head of epididymis	Alternating patches of simple cuboidal nonciliated and simple columnar ciliated	Thin circular layer of smooth muscle and vascular loose connective tissue	Absorb most fluid from seminiferous tubules; convey sperm into the epididymis
Epididymal duct	Head, body, and tail of the epididymis	Pseudostratified columnar, with small basal cells and tall principal cells bearing long stereocilia	Circular smooth muscle initially, with inner and outer longitudinal layers in the tail	Site for sperm maturation and short-term storage; expels sperm at ejaculation
Ductus (vas) deferens	Extends from epididymis to ejaculatory ducts in prostate gland	Pseudostratified columnar, with fewer stereocilia	Fibroelastic lamina propria and three very thick layers of smooth muscle	Carries sperm by rapid peristalsis from the epididymis to the ejaculatory ducts
Ejaculatory ducts	In prostate, formed by union of ductus deferens and ducts of the seminal vesicles	Pseudostratified and simple columnar	Fibroelastic tissue and smooth muscle of the prostate stroma	Mix sperm and seminal fluid; deliver semen to urethra, where prostatic secretion is added

FIGURE **21–9** Seminiferous tubules, straight tubules and rete testis.

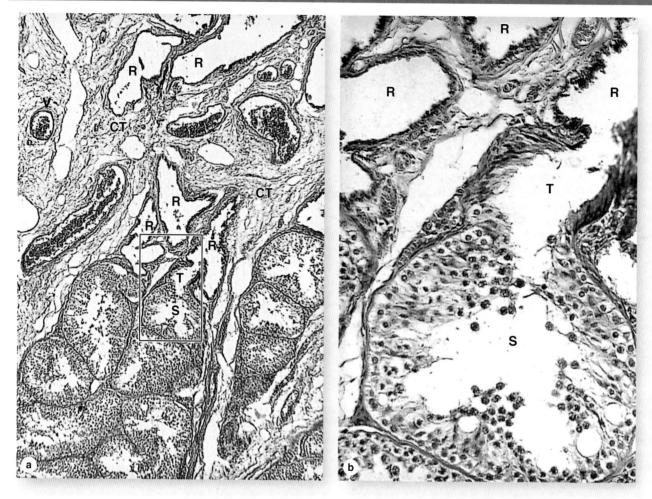

(a) The seminiferous tubules (**S**) drain into short, much narrower straight tubules (**T**), which connect to the rete testis (**R**), a network of channels embedded along with blood vessels (**V**) in the connective tissue (**CT**) of the mediastinum testis. (X120; H&E)

(b) At higher magnification the enclosed portion of part **a** shows the transition from wide seminiferous tubule (**S**) to the straight

tubule (**T**). Initially the straight tubule wall has only tall Sertoli cells devoid of germ cells. The wall becomes a simple cuboidal epithelium near its connection to the rete testis (**R**), which is also lined with simple cuboidal epithelium. (X300; H&E)

❯ EXCRETORY GENITAL DUCTS

The excretory genital ducts are those of the **epididymis**, the **ductus** (or **vas**) **deferens**, and the **urethra**. They transport sperm from the scrotum to the penis during ejaculation.

Epididymis

The long, coiled duct of the epididymis, surrounded by connective tissue, lies in the scrotum along the superior and posterior sides of each testis (Figure 21–2). About 4-5 m in length, it includes a **head** region where the efferent ductules enter, a **body**, and a **tail** opening into the ductus deferens. Passage of

sperm through the duct of the epididymis normally takes 2-4 weeks, during which spermatozoa undergo maturation and acquire the ability to fertilize. Important changes within sperm while passing through the epididymis include:

- Development of the competence for independent forward motility,
- Maturation of the acrosome, and
- Biochemical and organizational changes within the cell membrane.

Fluid within the epididymis contains glycolipid "decapacitation factors" that bind the plasma membranes of sperm and block acrosomal reactions and fertilizing ability until these

FIGURE **21–10**   Rete testis and efferent ductules.

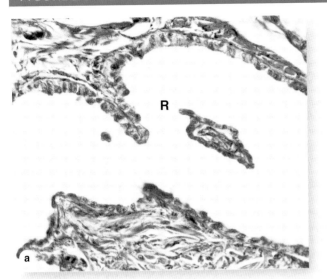

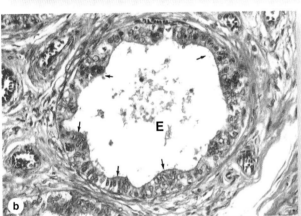

(a) Rete testis (**R**) channels have simple cuboidal epithelium and are usually embedded in dense connective tissue. (X350; Mallory trichrome)

(b) The efferent ductules (**E**) that drain the rete testis have a lining with a characteristic scalloped appearance in section, consisting of patches of simple cuboidal cells with water-absorbing microvilli alternating with patches of taller cells with cilia (**arrows**). (X350; H&E)

factors are removed as part of the **capacitation** process in the female reproductive tract.

The epididymal duct is lined with pseudostratified columnar epithelium consisting of columnar **principal cells**, with characteristic long **stereocilia**, and small round stem cells (Figure 21–11). The principal cells secrete glycolipids and glycoproteins, but also absorb most of the remaining water and remove residual bodies or other debris not removed earlier by Sertoli cells. The duct epithelium is surrounded by a few layers of smooth muscle cells, arranged as inner and outer longitudinal layers as well as a circular layer in the tail of the epididymis.

At ejaculation peristaltic contractions of this muscle move the sperm rapidly along the duct and empty the epididymal tail and distal body regions.

Ductus or Vas Deferens

From the epididymis the **ductus** (or **vas**) **deferens**, a long straight tube with a thick, muscular wall and a relatively small lumen, continues toward the prostatic urethra where it empties (Figure 21–1). As shown in Figure 21–12, its mucosa is slightly folded longitudinally, the lamina propria contains many elastic fibers, and the epithelial lining is pseudostratified with some cells having sparse stereocilia. The very thick muscularis consists of longitudinal inner and outer layers and a middle circular layer. The muscles produce strong peristaltic contractions during ejaculation, which rapidly move sperm along this duct from the epididymis.

The ductus (vas) deferens forms part of the spermatic cord, which also includes the testicular artery, the pampiniform plexus, and nerves (Figure 21–2). Following the general path along which the embryonic testes descend, each ductus passes over the urinary bladder where it enlarges as an **ampulla** (L. a small bottle) where the epithelium is thicker and more extensively folded (Figure 21–13). Within the prostate gland, the ends of the two ampullae merge with the ducts of the two seminal vesicles, joining these ducts to form the **ejaculatory ducts** which open into the **prostatic urethra**.

The histology of the intratesticular and excretory ducts is summarized in Table 21–1.

>> **MEDICAL APPLICATION**

The accessibility of the ductus (vas) deferens in the spermatic cords allows for the most common surgical method of male contraception: **vasectomy**. In this procedure a very small incision is made through the scrotal skin near the two ducts and each vas is exposed, cut, and the two ends (or only the end leading to the abdomen) are cauterized and tied.

After vasectomy sperm are still produced, but they degenerate and are removed by macrophages in the epididymis (and in the scrotal sac if the short portion of the vas is left open-ended.) Inflammatory and other changes occur in the mucosa of the epididymis, but serious adverse effects of vasectomy are usually minimal. A vasectomy may be reversed by surgically reconnecting the two ends of each ductus deferens. However, even successful surgery very often fails to restore fertility, due to incomplete sperm maturation in the epididymis changed by postvasectomy inflammation.

▸ ACCESSORY GLANDS

The accessory glands of the male reproductive tract produce secretions that are mixed with sperm during ejaculation to produce semen and that are essential for reproduction.

FIGURE **21–11** Epididymis.

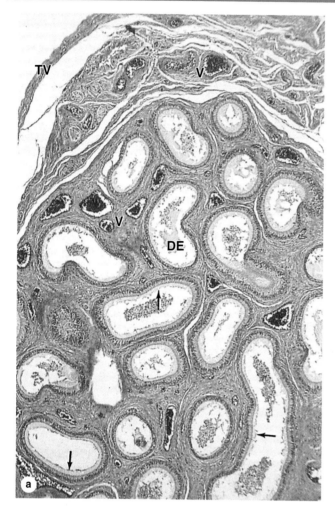

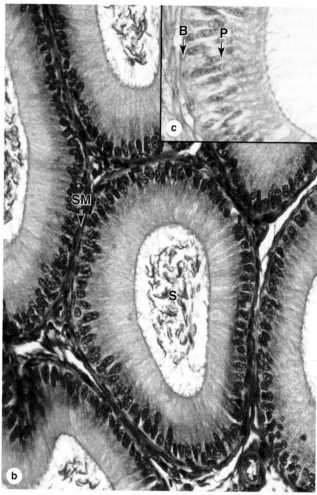

(a) The long, coiled duct of the epididymis (**DE**), where sperm undergo maturation and short-term storage, is enclosed by connective tissue with many blood vessels (**V**) and covered by a capsule and the tunica vaginalis (**TV**). The duct is lined by a pseudostratified columnar epithelium with long stereocilia (**arrows**). (X140; H&E)

(b) The columnar epithelium of the epididymal duct is surrounded by a thin circular layer of smooth muscle (**SM**) cells and its lumen

contains sperm (**S**). The smooth muscle becomes thicker and a longitudinal layer develops in the body and tail of the epididymis. (X400; H&E)

(c) The inset photo shows the epithelium with tall principal cells (**P**) with stereocilia and fewer small basal stem cells (**B**). Intraepithelial lymphocytes are also commonly seen in the epididymal duct. (X500; H&E)

The accessory genital glands are the **seminal vesicles** (or glands), the **prostate gland**, and the **bulbourethral glands** (Figure 21–13).

Seminal Vesicles

The two **seminal vesicles** consist of highly tortuous tubes, each about 15-cm long, enclosed by a connective tissue capsule. The unusual mucosa of the tube displays a great number of thin, complex folds that fill much of the lumen (Figure 21–14). The folds are lined with simple or pseudostratified columnar epithelial cells rich in secretory granules. The lamina propria contains elastic fibers and is surrounded by smooth muscle with inner

circular and outer longitudinal layers that empty the gland during ejaculation. The seminal vesicles are exocrine glands in which production of their viscid, yellowish secretion depends on testosterone. Fluid from seminal vesicles typically makes up about 70% of the ejaculate and its components include the following:

- **Fructose**, a major energy source for sperm, as well as inositol, citrate, and other metabolites;
- **Prostaglandins**, which stimulate activity in the female reproductive tract; and
- **Fibrinogen**, which allows semen to coagulate after ejaculation.

FIGURE **21–12** Ductus deferens.

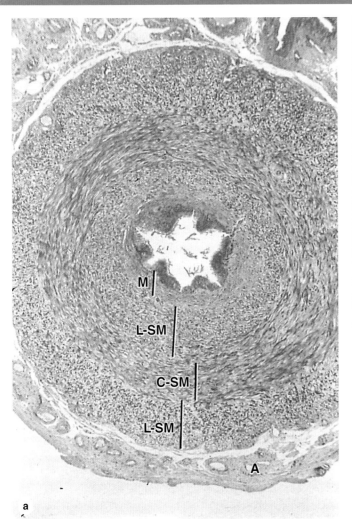

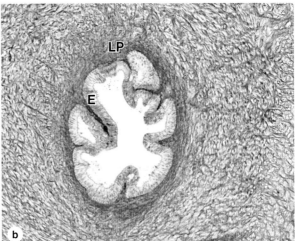

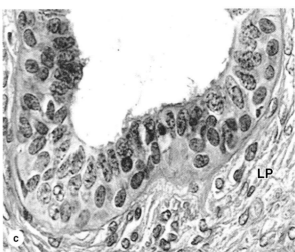

(a) A cross section of the vas deferens shows the mucosa (**M**), a thick muscularis with inner and outer layers of longitudinal smooth muscle (**L-SM**) and an intervening layer of circular smooth muscle (**C-SM**), and an external adventitia (**A**). The muscularis is specialized for powerful peristaltic movement of sperm at ejaculation. (X60; H&E)

(b) The lamina propria (**LP**) is rich in elastic fibers and the thick epithelial lining (**E**) shows longitudinal folds (X150; Mallory trichrome).

(c) Higher magnification of the mucosa shows that the epithelium is pseudostratified with basal cells and many columnar cells, some with stereocilia. (X400; H&E)

Prostate Gland

The **prostate gland** is a dense organ that surrounds the urethra below the bladder. It is approximately 2 cm × 3 cm × 4 cm in size and weighs about 20 g. The prostate is a collection of 30-50 tubuloacinar glands embedded in a dense fibromuscular stroma in which smooth muscle contracts at ejaculation (Figure 21–13b). Ducts from individual glands may converge but all empty directly into the prostatic urethra, which runs through the center of the prostate. As shown in Figure 21–15, the glands are arranged in three major zones around the urethra:

- The **transition zone** occupies only about 5% of the prostate volume, surrounds the superior portion of

the urethra, and contains the periurethral **mucosal glands**.
- The **central zone** comprises 25% of the gland's tissue and contains the periurethral **submucosal glands** with longer ducts.
- The **peripheral zone**, with about 70% of the organ's tissue, contains the prostate's **main glands** with still longer ducts (Figure 21–16).

The tubuloacinar glands of the prostate are all lined by a simple or pseudostratified columnar epithelium and produce fluid that contains various glycoproteins, enzymes, and small molecules such as prostaglandins and is stored until ejaculation. A clinically important product of the prostate is

FIGURE **21–13** Accessory glands of the male reproductive tract.

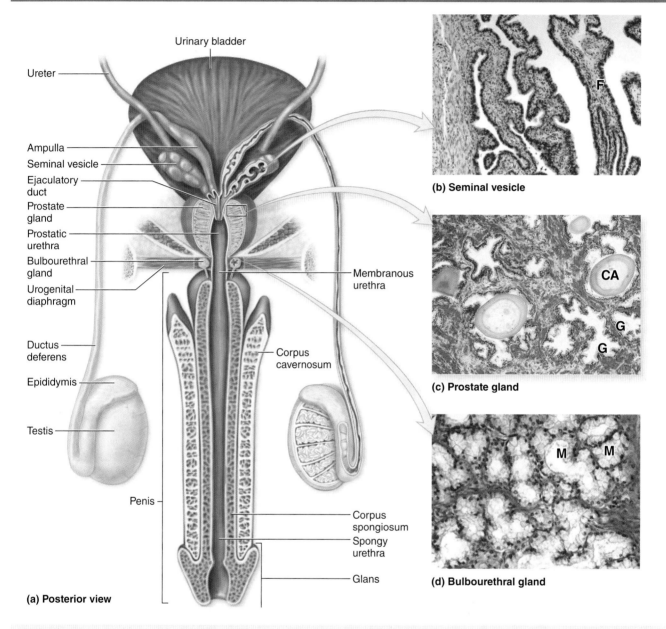

(a) Posterior view

Labels in diagram:
Urinary bladder
Ureter
Ampulla
Seminal vesicle
Ejaculatory duct
Prostate gland
Prostatic urethra
Bulbourethral gland
Urogenital diaphragm
Ductus deferens
Epididymis
Testis
Penis
Membranous urethra
Corpus cavernosum
Corpus spongiosum
Spongy urethra
Glans

(b) Seminal vesicle
(c) Prostate gland
(d) Bulbourethral gland

(a) Three sets of glands connect to the ductus deferens or urethra: the paired **seminal vesicles**, the **prostate**, and the paired **bulbourethral glands**. The first two types of glands contribute the major volume to semen and the latter produces a secretion that lubricates the urethra before ejaculation.

(b) Seminal vesicles have a characteristic thin mucosal folds (**F**) with a large surface area.

(c) The prostate gland has many individual tubuloacinar glands (**G**), some containing concretions called corpora amylacea (**CA**).

(d) The small, paired bulbourethral glands consist mainly of mucous acini (**M**). (b,c,d: X80; H&E)

prostate-specific antigen (PSA), a 34-kDa serine protease that helps liquefy coagulated semen for the slow release of sperm after ejaculation. Small amounts of PSA also leak normally into the prostatic vasculature; elevated levels of circulating PSA indicate abnormal glandular mucosa typically due to prostatic carcinoma or inflammation.

Small spherical concretions, 0.2-2 mm in diameter and often partially calcified, are normally present in the lumens of many prostatic tubuloacinar glands (Figure 21–16). These concretions, called **corpora amylacea**, containing primarily deposited glycoproteins and keratan sulfate, may become more numerous with age but seem to have no physiologic or

FIGURE 21–14 Seminal vesicles.

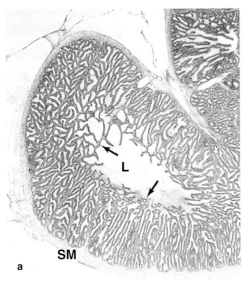
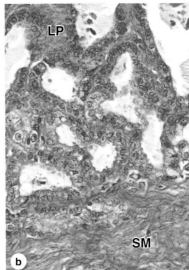
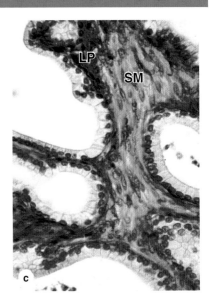

The seminal vesicles are paired exocrine glands that secrete most seminal fluid, including sperm nutrients. **(a)** A low-power micrograph shows that each consists of a coiled duct with a mucosa organized with many thin folds (**arrows**) in the lumen (**L**) and surrounded by two layers of smooth muscle (**SM**). (X20; H&E)

(b, c) The mucosal folds include smooth muscle (**SM**), a thin lamina propria (**LP**), and a simple columnar epithelium of principal secretory cells. (Both: X300. b: H&E; c: PSH)

FIGURE 21–15 Organization of the prostate gland.

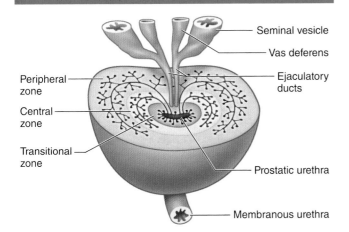

The prostate consists of 30-50 branched tubuloacinar glands organized into three layers, shown here schematically. Around the prostatic urethra is the **transition zone** containing mucosal glands. Surrounding most of that zone is the intermediate **central zone**, which contains the submucosal glands. The outermost and largest layer is the **peripheral zone**, which contains the most numerous main glands.

clinical significance. The prostate is surrounded by a fibroelastic capsule, from which septa extend and divide the gland into indistinct lobes. Like the seminal vesicles, the prostate's structure and function depend on the level of testosterone.

❯❯ MEDICAL APPLICATION

The prostate gland is prone to three common problems: (1) **chronic prostatitis**, usually involving bacteria or other infectious agents; (2) **nodular hyperplasia** or **benign prostatic hypertrophy**, occurring mainly in the periurethral mucosal glands where it often leads to compression of the urethra and problems with urination; and (3) **prostate cancer** (adenocarcinoma), the most common cancer in nonsmoking men, occurring mainly in glands of the peripheral zone.

Bulbourethral Glands

The paired round **bulbourethral glands** (or Cowper glands), 3-5 mm in diameter, are located in the urogenital diaphragm (Figure 21–13) and empty into the proximal part of the penile urethra. Each gland has several lobules with tubuloacinar secretory units surrounded by smooth muscle cells and lined by a mucus-secreting simple columnar epithelium that is also

FIGURE **21-16** Prostate gland.

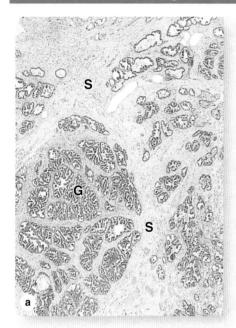
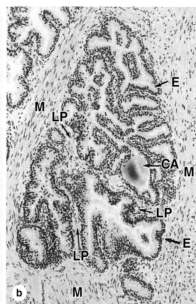
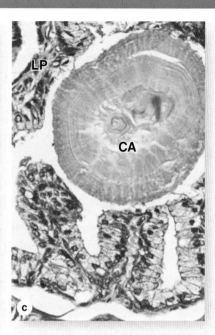

(a) The prostate has a dense fibromuscular stroma (**S**) in which are embedded a large number of small tubuloalveolar glands (**G**). (X20; H&E)

(b) A micrograph of one prostatic gland, showing a corpus amylaceum (**CA**) concretion and the secretory epithelium (**E**)

surrounded by a thin lamina propria (**LP**) and thick smooth muscle (**M**). (X122; H&E)

(c) Higher magnification shows the lamellar nature of a corpus amylaceum (**CA**) and the pseudostratified columnar epithelium underlain by lamina propria (**LP**). (X300; Mallory trichrome)

testosterone-dependent (Figure 21–13d). During erection the bulbourethral glands, as well as numerous, very small, and histologically similar urethral glands along the penile urethra, release a clear mucus-like secretion that coats and lubricates the urethra in preparation for the imminent passage of sperm.

❯ PENIS

The penis consists of three cylindrical masses of erectile tissue, plus the penile urethra, surrounded by skin (see Figure 21–1). Two of the erectile masses—the **corpora cavernosa**—are dorsal; the ventral **corpus spongiosum** surrounds the urethra (Figure 21–17). At its end the corpus spongiosum expands, forming the **glans** (Figure 21–13a). Most of the penile urethra is lined with pseudostratified columnar epithelium. In the glans, it becomes stratified squamous epithelium continuous with that of the thin epidermis covering the glans surface. Small mucus-secreting **urethral glands** are found along the length of the penile urethra. In uncircumcised men the glans is covered by the

FIGURE **21-17** Structure of the penis.

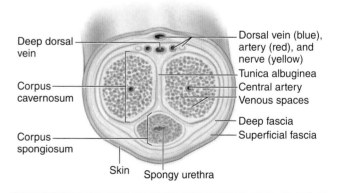

A diagram of the penis in transverse section shows the relationships of the three **erectile bodies,** the **tunica albuginea,** and **major blood vessels**. Compare this section to the longitudinal diagram in Figure 21–13.

FIGURE **21–18** Penis.

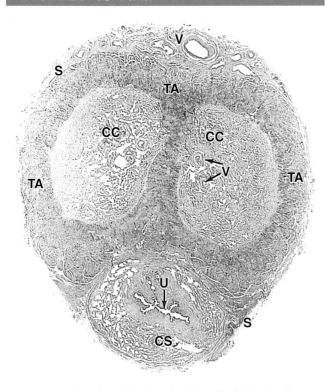

The corpus spongiosum (**CS**) is on the ventral side of the penis and surrounds the urethra (**U**). Two corpora cavernosa (**CC**) occupy most of the dorsal side and are surrounded by dense, fibrous tunica albuginea (**TA**). Along the dorsal side run the major blood vessels (**V**) and deep in each corporal mass of erectile tissue are smaller blood vessels (**V**), including the central arteries. The penis is covered loosely by skin (**S**), which distally forms the large foreskin fold and becomes thin over the glans. (X15; H&E)

prepuce or foreskin, a retractable fold of thin skin with sebaceous glands on the internal surface.

The corpora cavernosa are each surrounded by a dense fibroelastic layer, the **tunica albuginea** (Figures 21–17 and 21–18). All three erectile tissues consist of many venous **cavernous spaces** lined with endothelium and separated by trabeculae with smooth muscle and connective tissue continuous with the surrounding tunic (Figure 21–19).

Central arteries in the corpora cavernosa branch to form nutritive arterioles and small coiling **helicine arteries**, which lead to the cavernous vascular spaces of erectile tissue. Arteriovenous shunts are present between the central arteries and the dorsal veins.

Penile **erection** involves blood filling the cavernous spaces in the three masses of erectile tissue. Triggered by external stimuli to the CNS, erection is controlled by autonomic nerves in these vascular walls. Parasympathetic stimulation relaxes the trabecular smooth muscle and dilates the helicine arteries, allowing increased blood flow and filling the cavernous spaces. This enlarges the corpora cavernosa and causes them to compress the dorsal veins against the dense tunica albuginea, which blocks the venous outflow and produces tumescence and rigidity in the erectile tissue. Beginning at ejaculation, sympathetic stimulation constricts the helicine arteries and trabecular muscle, decreasing blood flow into the spaces, lowering the pressure there, and allowing the veins to drain most blood from the erectile tissue.

〉〉 MEDICAL APPLICATION

At the beginning of an erection acetylcholine from parasympathetic nerves causes the vascular endothelial cells of the helicine arteries and cavernous tissue to release nitric oxide (NO). NO diffuses into the surrounding smooth muscle cells and activates guanylate cyclase to produce cyclic GMP, which causes these cells to relax and promotes blood flow for the erection.

Erectile dysfunction, or **impotence**, can result from diabetes, anxiety, vascular disease, or nerve damage during prostatectomy. The drug **sildenafil** may alleviate the problem by inhibiting the phosphodiesterase that degrades cyclic GMP in the smooth muscle cells of helicine arteries and erectile tissue. The subsequent higher level of cGMP promotes relaxation of these cells and enhances the neural effect to produce or maintain an erection.

The Male Reproductive System SUMMARY OF KEY POINTS

Testes

- In each **testis** approximately each of 250 **lobules** contains one or more very long, convoluted **seminiferous tubules** in a sparse, vascular stroma containing testosterone-producing **interstitial cells (of Leydig)**.
- **Seminiferous tubules** consist of spermatogenic epithelium containing columnar **Sertoli cells**, each of which supports and nourishes many germ cells embedded at its surface.
- **Tight junctions** between Sertoli cells establish two compartments within seminiferous tubules: a **basal compartment** with spermatogonia and an **adluminal compartment** with spermatocytes and spermatids.

- **Sertoli cells** also produce **androgen-binding protein**, which concentrates testosterone, **phagocytose** shed debris from differentiating spermatids, and **secrete fluid** that carries sperm along the tubules.
- Stem cells called **spermatogonia** undergo mitosis and give rise to **primary spermatocytes**, which undergo a first meiotic division to form haploid **secondary spermatocytes**.
- After a very short interval, **secondary spermatocytes** undergo the second meiotic division to produce small, round **spermatids**, which differentiate while still associated with Sertoli cells.
- A spermatid undergoes **spermiogenesis** by greatly condensing its **nucleus**, forming a long **flagellum** with a surrounding mitochondrial **middle piece**, and forming a perinuclear **acrosomal cap**.

FIGURE 21–19   Penile urethra and erectile tissue.

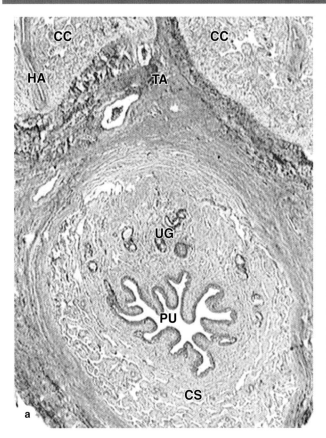

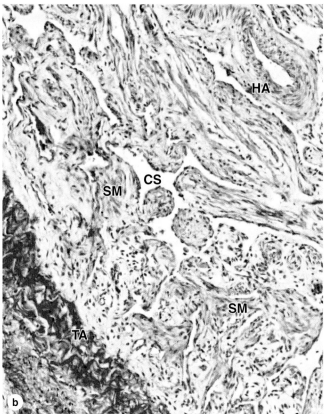

(a) The corpus spongiosum (**CS**) surrounds the longitudinally folded wall of the penile urethra (**PU**). Small urethral glands (**UG**) with short ducts to the urethra release mucus during erection, supplementing the similar secretion from the bulbourethral glands. The two dorsal corpora cavernosa (**CC**) are ensheathed by dense, fibrous tunica albuginea (**TA**) and in one here a small helicine artery (**HA**) is shown. (X100; H&E)

(b) A higher magnification of erectile tissue is shown with a small portion of tunica albuginea (**TA**) and fibrous, elastic connective tissue containing smooth muscle (**SM**) and many small, cavernous spaces (**CS**) lined by vascular endothelium. Very little blood normally passes through this vasculature due to constriction of the helicine arteries (**HA**) serving them. During erection the vascular smooth muscle relaxes, allowing rapid blood flow into the cavernous spaces, filling them and causing compression of their venous drainage, which makes the masses of erectile tissue much larger and turgid. (X200; H&E)

- From the seminiferous tubules, sperm enter the short **straight tubules** that lead to channels of the **rete testis** in the mediastinum testis, then move via 15 or 20 **efferent ductules** where fluid is absorbed.

Excretory Genital Ducts

- From efferent ductules sperm move into and through **epididymal ducts** in the head, body, and tail of **epididymis**, also located in the scrotum, for their final maturation and short storage prior to ejaculation.
- The **epididymis** is lined by **pseudostratified columnar epithelium** containing **principal cells** that have long **stereocilia**; along its length the number of surrounding **smooth muscle** layers increases.
- At ejaculation the body and tail of the **epididymis** are emptied into the **ductus (vas) deferens**, also lined by **pseudostratified columnar epithelium** with three thick layers of **smooth muscle** for peristalsis.

Accessory Glands

- Most semen is produced by the paired **seminal vesicles** (glands), characterized by numerous thin folds of secretory mucosa; a duct from each seminal vesicle joins a ductus deferens as an **ejaculatory duct**.

- Inside the medial **prostate gland** the two **ejaculatory ducts** and dozens of ducts from **tubuloacinar prostatic glands** merge with the **urethra** that transports semen through the penis.
- **Semen** components from the seminal vesicles nourish sperm (eg, **fructose**), activate ciliary and muscle activity in the female reproductive tract (**prostaglandins**), and control the product's viscosity for sperm release.

Penis

- The **penis** contains two dorsal **corpora cavernosa** and a periurethral **corpus spongiosum**, all composed of vascular **cavernous tissue** and small amounts of surrounding **smooth muscle** and **helicine arteries**.
- For erection **parasympathetic** stimulation relaxes muscle of the small **helicine arteries** and adjacent tissues, allowing vessels of the **cavernous tissue** to fill with blood; the enlarging corpora compress the venous drainage, producing further enlargement and turgidity in the three corpora masses.
- The **sympathetic** stimulation at ejaculation constricts blood flow through the **helicine** arteries, allowing blood to empty from the cavernous tissues.

The Male Reproductive System ASSESS YOUR KNOWLEDGE

1. Which of the following accurately describes spermiogenesis?
   a. Occurs before puberty
   b. Involves stem cells, meiosis, and spermatogenesis
   c. Involves cytodifferentiation of early spermatids
   d. Occurs in diploid cells
   e. Results in the formation of primary spermatocytes

2. A man with a pituitary gonadotrophic tumor causing hyposecretion of follicle-stimulating hormone (FSH) is most likely to exhibit which condition?
   a. No symptoms, since he has no ovarian follicles
   b. Loss of libido (sex drive)
   c. Low serum testosterone levels
   d. Low sperm count
   e. Prostatic hypertrophy

3. Interstitial cells of Leydig have an important function in male gamete production. Because of this function, which of the following organelles is abundant within these cells?
   a. Lysosomes
   b. Smooth endoplasmic reticulum
   c. Peroxisomes
   d. Polyribosomes
   e. Golgi complexes

4. While studying a germ cell line developed from a patient's testicular biopsy, the researcher notes that colchicine-treated cells blocked in metaphase have 46 chromosomes. From which of the following regions of the male genital tract would you expect these cells to have originated?
   a. Within the rete testis
   b. At the basal lamina of the seminiferous tubule
   c. In the middle region of the germinal epithelium
   d. Within the adluminal compartment of the seminiferous tubule
   e. Within the straight tubules

5. Which of the following organs is normally characterized by the accumulation of corpora amylacea with increasing age?
   a. Prostate
   b. Seminal vesicles
   c. Bulbourethral (Cowper) glands
   d. Epididymis
   e. Ductus (vas) deferens

6. Within the male reproductive tract, stereocilia project from cells lining which of the following regions?
   a. Rete testis
   b. Seminiferous tubules
   c. Ampulla of the ductus deferens
   d. Epididymis
   e. Penile urethra

7. As sperm pass through the male genital ducts, proteins and low molecular weight products are added from several sources producing semen. Which of the following provides a nutritive, fructose-rich secretion?
   a. Interstitial cells of Leydig
   b. Bulbourethral (Cowper) glands
   c. Prostate gland
   d. Epididymis
   e. Seminal vesicles

8. A 20-year-old man contracts cholera during a long tour of military duty in a remote, completely undeveloped region. After a 5-day period of severe diarrhea and treatment, he gradually recovers and slowly returns to work. He is married 3 years later but after a few years of trying to conceive a child, semen analysis reveals that his sperm are few in number and malformed, and blood tests show a high titer of antibodies against sperm antigens. The causative agent of cholera, *Vibrio cholera*, secretes a toxin that interferes with tight (occluding) junctions. What cells in the gonad are the likely target of this toxin in the male reproductive system?
   a. Spermatogonia
   b. Sertoli cells
   c. Myoid cells
   d. Interstitial cells of Leydig
   e. Differentiating spermatozoa

9. A 29-year-old man presents with testicular pain and a burning sensation during urination. Tests reveal the presence of *Neisseria gonorrhea* and penicillin is prescribed. Gonorrhea often produces acute or chronic inflammation of the testes and frequently involves the channels that connect the testis to the epididymis. What is the name of these channels?
   a. The mediastinum testis
   b. The rete testis
   c. Efferent ductules
   d. The straight tubules (tubuli recti)
   e. The seminiferous tubules

10. A 39-year-old man undergoing an extensive series of tests for infertility is found to have a genetic mutation that prevents formation of a functional synaptonemal complex during meiosis, causing almost complete failure of sperm formation. Which cells would be directly affected by this mutation?
    a. Primary spermatocytes
    b. Spermatogonia
    c. Secondary spermatocytes
    d. Spermatids
    e. Cells undergoing spermiogenesis

Answers: 1c, 2d, 3b, 4b, 5a, 6d, 7e, 8b, 9c, 10a

22 The Female Reproductive System

The female reproductive system consists of the paired ovaries and oviducts (or uterine tubes), the uterus, the vagina, and the external genitalia (Figure 22–1). This system produces the female gametes (**oocytes**), provides the environment for fertilization, and holds the embryo during its complete development through the fetal stage until birth. As with male gonads, the ovaries produce steroidal sex hormones that control organs of the reproductive system and influence other organs. Beginning at **menarche**, when the first menses occurs, the reproductive system undergoes monthly changes in structure and function that are controlled by neurohormonal mechanisms. **Menopause** is a variably timed period during which the cyclic changes become irregular and eventually disappear. In the postmenopausal period the reproductive organs slowly involute. Although the mammary glands do not belong to the genital system, they are included here because they undergo changes directly connected to the functional state of the reproductive organs.

❯ OVARIES

Ovaries are almond-shaped bodies approximately 3-cm long, 1.5-cm wide, and 1-cm thick. Each ovary is covered by a simple cuboidal epithelium, the **surface** (or germinal) **epithelium**, continuous with the mesothelium and overlying a layer of dense connective tissue capsule, the **tunica albuginea**,

like that of the testis. Most of the ovary consists of the **cortex**, a region with a stroma of highly cellular connective tissue and many **ovarian follicles** varying greatly in size after menarche (Figure 22–1). The most internal part of the ovary, the **medulla**, contains loose connective tissue and blood vessels entering the organ through the hilum from mesenteries suspending the ovary (Figures 22–1 and 22–2). There is no distinct border between the ovarian cortex and medulla.

Early Development of the Ovary

In the first month of embryonic life, a small population of **primordial germ cells** migrates from the yolk sac to the gonadal primordia. There the cells divide and differentiate as **oogonia**. In developing ovaries of a 2-month embryo, there are about 600,000 oogonia that produce more than 7 million by the fifth month. Beginning in the third month, oogonia begin to enter the prophase of the first meiotic division but arrest after completing synapsis and recombination, without progressing to later stages of meiosis (see Chapter 3). These cells arrested in meiosis are called **primary oocytes** (Gr. *oon*, egg + *kytos*, cell). Each primary oocyte becomes surrounded by flattened support cells called **follicular cells** to form an ovarian follicle. By the seventh month of development, most oogonia have transformed into primary oocytes within follicles. Many primary oocytes, however, are lost through a slow, continuous degenerative process called **atresia**, which

FIGURE **22–1** The female reproductive system and overview of ovary.

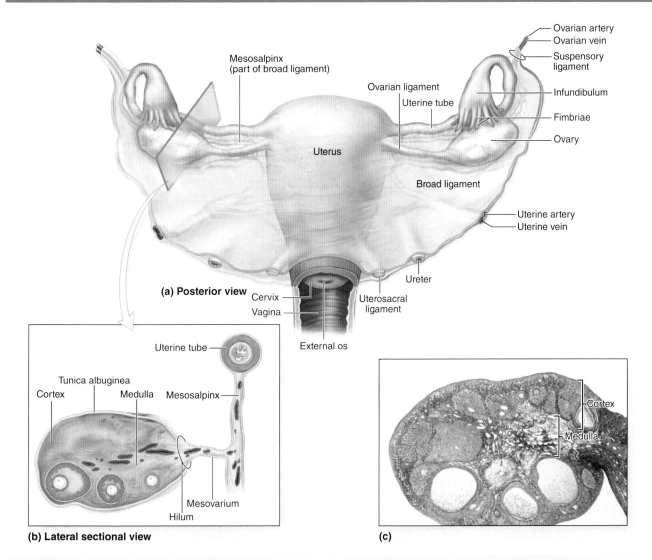

(a) Posterior view

(b) Lateral sectional view

(c)

(a) The diagram shows the internal organs of the female reproductive system, which includes as the principal organs the **ovaries**, **uterine tubes**, **uterus**, and **vagina**. **(b)** A lateral sectional view of an ovary shows the ovary and the relationship of its main supporting mesenteries, the mesovarium and the mesosalpinx of the broad ligament. **(c)** A sectioned ovary, indicating the medulla and cortex, with follicles of several different sizes in the cortex. (X15; H&E)

continues through a woman's reproductive life. At puberty the ovaries contain about 300,000 oocytes. Because generally only one oocyte resumes meiosis with ovulation during each menstrual cycle (average duration, 28 days) and the reproductive life of a woman lasts about 30 to 40 years, only about 450 oocytes are liberated from ovaries by ovulation. All others degenerate through atresia.

Ovarian Follicles

An ovarian follicle consists of an oocyte surrounded by one or more layers of epithelial cells within a basal lamina.

The follicles that are formed during fetal life—**primordial follicles**—consist of a primary oocyte enveloped by a single layer of the flattened follicular cells (Figure 22–2b; Figures 22–3 and 22–4). These follicles occur in the superficial ovarian cortex. The oocyte in the primordial follicle is spherical and about 25 μm in diameter, with a large nucleus containing chromosomes in the first meiotic prophase. The organelles tend to be concentrated near the nucleus and include numerous mitochondria, several Golgi complexes, and extensive RER. The basal lamina surrounds the follicular cells, marking a clear boundary between the follicle and the vascularized stroma.

FIGURE 22–2 Follicle development and changes within the ovary.

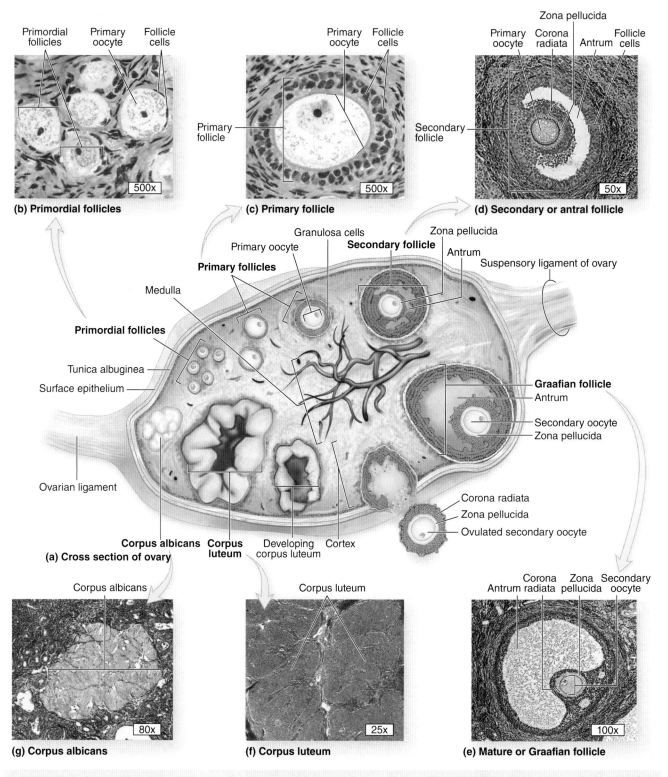

Primordial follicles · Primary oocyte · Follicle cells

(b) Primordial follicles
500x

Primary oocyte · Follicle cells
Primary follicle

(c) Primary follicle
500x

Zona pellucida
Primary oocyte · Corona radiata · Antrum · Follicle cells
Secondary follicle

(d) Secondary or antral follicle
50x

Granulosa cells · Primary oocyte · **Secondary follicle** · Zona pellucida · Antrum
Primary follicles
Medulla
Suspensory ligament of ovary
Primordial follicles
Tunica albuginea
Surface epithelium
Graafian follicle
Antrum
Secondary oocyte
Zona pellucida
Ovarian ligament
Corona radiata
Zona pellucida
Ovulated secondary oocyte
Corpus albicans **Corpus luteum** · Developing corpus luteum · Cortex

(a) Cross section of ovary

Corpus albicans

(g) Corpus albicans
80x

Corpus luteum

(f) Corpus luteum
25x

Antrum · Corona radiata · Zona pellucida · Secondary oocyte

(e) Mature or Graafian follicle
100x

The ovary produces both oocytes and sex hormones. A diagram of a sectioned ovary **(a)** shows the **different stages of follicle maturation**, **ovulation**, and **corpus luteum formation and degeneration**. All of the stages and structures shown in this diagram actually would appear at different times during the ovarian cycle and do not occur simultaneously. Follicles are arranged here for easy comparisons. The **primordial follicles** shown are greatly enlarged. The histologic sections identify primordial follicles **(b)**, a primary follicle **(c)**, a secondary follicle **(d)**, and a large vesicular follicle **(e)**. After ovulation, the portion of the follicle left behind forms the corpus luteum **(f)**, which then degenerates into the corpus albicans **(g)**. (All H&E)

FIGURE 22–3 Stages of ovarian follicles, from primordial to mature.

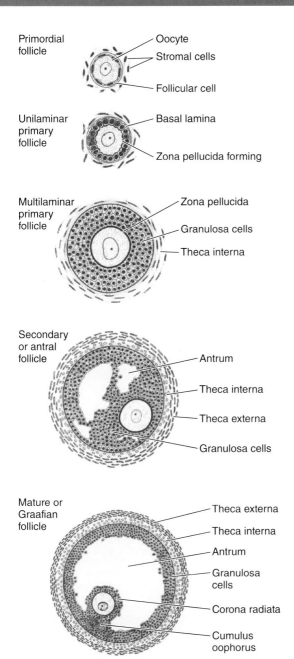

Diagrams of sectioned ovarian follicles show the changing size and morphology of follicular/granulosa cells at each stage and the disposition of the surrounding thecal cells. However, the relative proportions of the follicles are not maintained in the series of drawings: mature follicles are much larger relative to the early follicles. Deep within each follicle is a single large, growing oocyte with a large nucleus and prominent nucleolus. Follicular or granulosa cells around the oocyte support that cell's rapid growth.

FIGURE 22–4 Primordial follicles.

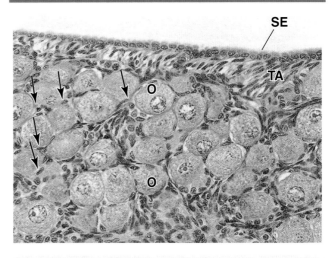

The cortical region of an ovary is surrounded by the surface epithelium (**SE**), a mesothelium with usually cuboidal cells. This layer is sometimes called the germinal epithelium because of an early erroneous view that it was the source of oogonia precursor cells. Underlying the epithelium is a connective tissue layer, the tunica albuginea (**TA**). Groups of primordial follicles, each formed by an oocyte (**O**) surrounded by a layer of flat epithelial follicular cells (**arrows**), are present in the ovarian connective tissue (stroma). (X200; H&E)

Follicular Growth & Development

Beginning in puberty with the release of follicle-stimulating hormone (FSH) from the pituitary, a small group of primordial follicles each month begins a process of follicular growth. This involves growth of the oocyte, proliferation and changes in the follicular cells, as well as proliferation and differentiation of the stromal fibroblasts around each follicle. Selection of the primordial follicles that undergo growth and recruitment early in each cycle and of the dominant follicle destined to ovulate that month both involve complex hormonal balances and subtle differences among follicles in FSH receptor numbers, aromatase activity, estrogen synthesis, and other variables.

Prompted by FSH, an oocyte grows most rapidly during the first part of follicular development, reaching a diameter of about 120 μm. Oocyte differentiation includes the following:

■ Growth of the cell and nuclear enlargement.
■ Mitochondria becoming more numerous and uniformly distributed.
■ RER becoming much more extensive and Golgi complexes enlarging and moving peripherally.
■ Formation of specialized secretory granules called **cortical granules** containing various proteases. These lie just inside the oocyte's plasma membrane and undergo exocytosis early in fertilization.

Follicular cells undergo mitosis and form a simple cuboidal epithelium around the growing oocyte. The follicle is now

FIGURE 22–5 Primary follicles: unilaminar and multilaminar.

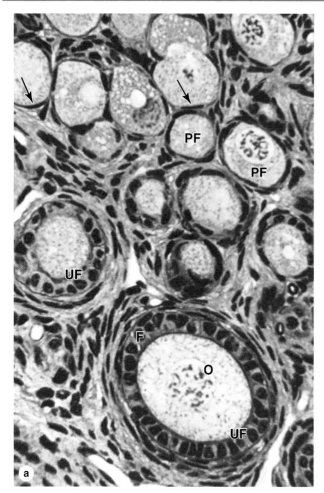

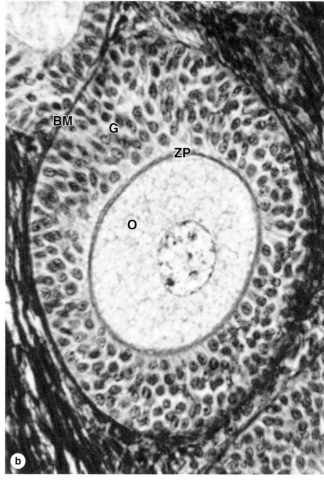

(a) A micrograph of ovarian cortex shows several primordial follicles (**PF**) and their flattened follicle cells (**arrows**), and two unilaminar primary follicles (**UF**) in which the follicle cells (**F**) form a single cuboidal layer around the large primary oocyte (**O**). (X200; PT)

(b) This micrograph taken at the same magnification shows a larger multilayered primary follicle. Follicle cells are now active granulosa cells (**G**) and have proliferated to form several layers. Between them and the oocyte (**O**) is the 5- to 10-μm-thick zona pellucida (**ZP**), a glycoprotein layer produced by the oocyte that is required for sperm binding and fertilization. The primary oocyte is now a very large cell. With this stain, the basement membrane (**BM**) that separates the follicle from the surrounding stroma can also be seen. (X200; PSH)

called a **unilaminar primary follicle** (Figures 22–3 and 22–5a). The follicular cells continue to proliferate, forming a stratified follicular epithelium, the **granulosa**, in which the cells communicate through gap junctions. Follicular cells are now termed **granulosa cells** and the follicle is a **multilaminar primary follicle** (Figures 22–3 and 22–5b) still surrounded by a basement membrane.

Between the oocyte and the first layer of granulosa cells of the growing primary follicle, extracellular material accumulates as the **zona pellucida**, 5-10-μm thick and containing four glycoproteins secreted by the oocyte (Figures 22–5b and 22–6). The zona pellucida components ZP3 and ZP4 are important sperm receptors, binding specific proteins on the sperm surface and inducing acrosomal activation. Filopodia of granulosa cells and microvilli of the oocyte penetrate the

zona pellucida, allowing communication between these cells via gap junctions.

›› MEDICAL APPLICATION

Growing primary follicles can become involved in **polycystic ovary syndrome** (**PCOS**) that is characterized by enlarged ovaries with numerous cysts and an anovulatory state (with no follicles completing maturation successfully.) The clinical presentation of this disorder is variable and the etiology is unclear, although increased androgen production by the ovaries or adrenals is likely involved. PCOS is a common cause of **infertility** in women.

FIGURE 22–6 Ultrastructure of primary follicle and zona pellucida.

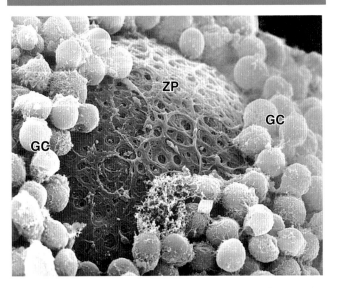

An SEM of a fractured primary follicle shows the oocyte surrounded by granulosa cells (**GC**) of the corona radiata. Between the very large oocyte surface and the granulosa cells is a layer of extracellular material, the zona pellucida (**ZP**), which contains four related glycoproteins that bind sperm and form an irregular meshwork. (X3000)

Stromal cells immediately outside each growing primary follicle differentiate to form the **follicular theca** (Gr. *theca*, outer covering). This subsequently differentiates further as two distinct tissues around the follicle (see Figure 22–3; Figures 22–7 and 22–8):

■ A well-vascularized endocrine tissue, the **theca interna**, with typical steroid-producing cells secreting androstenedione. This precursor molecule diffuses into the follicle through the basement membrane, and in the granulosa cells the enzyme aromatase converts it to estradiol, an FSH-dependent function. This estrogen returns to the thecae and stroma around the follicle, enters capillaries, and is distributed throughout the body.

■ A more fibrous **theca externa** with fibroblasts and smooth muscle merges gradually with the surrounding stroma.

As the primary follicles grow, they move deeper in the ovarian cortex. Within such follicles small spaces appear between the granulosa layers as the cells secrete **follicular fluid** (or **liquor folliculi**). This fluid accumulates, the spaces enlarge and gradually coalesce, and the granulosa cells reorganize themselves around a larger cavity called the **antrum** (Figures 22–3 and 22–7a), producing follicles now called **secondary** or **antral follicles**. Follicular fluid contains the large GAG hyaluronic acid, growth factors, plasminogen,

fibrinogen, the anticoagulant heparan sulfate proteoglycan, and high concentrations of steroids (progesterone, androstenedione, and estrogens) with binding proteins.

As the antrum develops, the granulosa cells around the oocyte form a small hillock, the **cumulus oophorus**, which protrudes into the antrum (Figures 22–3 and 22–7b). Those granulosa cells that immediately surround the zona pellucida make up the **corona radiata** and accompany the oocyte when it leaves the ovary at ovulation.

The single large antrum of a **mature** or **Graafian follicle** (named after the 17th-century reproductive biologist Regnier De Graaf) rapidly accumulates more follicular fluid and expands to a diameter of 2 cm or more. A mature follicle forms a bulge at the ovary surface that is visible with ultrasound imaging. The granulosa layer becomes thinner at this stage because its cells do not multiply in proportion to the growth of the antrum. A mature follicle has thick thecal layers and normally develops from a primordial follicle over a period of about 90 days.

Follicular Atresia

Most ovarian follicles undergo the degenerative process called **atresia**, in which follicular cells and oocytes die and are disposed of by phagocytic cells. Follicles at any stage of development, including nearly mature follicles, may become atretic (Figure 22–9). Atresia involves apoptosis and detachment of the granulosa cells, autolysis of the oocyte, and collapse of the zona pellucida. Macrophages invade the degenerating follicle and phagocytose the apoptotic material and other debris. Although follicular atresia takes place from before birth until a few years after menopause, it is most prominent just after birth, when levels of maternal hormones decline rapidly, and during both puberty and pregnancy, when qualitative and quantitative hormonal changes occur again.

During a typical menstrual cycle, one follicle becomes dominant and develops farther than the others. The **dominant follicle** usually reaches the most developed stage of follicular growth and undergoes ovulation, while the other primary and antral follicles undergo atresia. Although their oocytes are never directly used, the large growing follicles each month produce much estrogen before becoming atretic. As described later, this estrogen stimulates preparation of the reproductive tract to transport and sustain the embryo if the oocyte from the dominant follicle is fertilized.

❯❯ MEDICAL APPLICATION

Late primary or antral follicles can produce **follicular cysts**, which are thin-walled, fluid-filled structures with both granulosa and thecal endocrine cells. Follicular cysts are common and usually benign, but can produce high estrogen levels and lead to menstrual irregularities. If cyst formation disrupts blood vessels blood enters the fluid, often rapidly, and produces a **hemorrhagic cyst**.

FIGURE 22–7   Antral follicle and mature follicle.

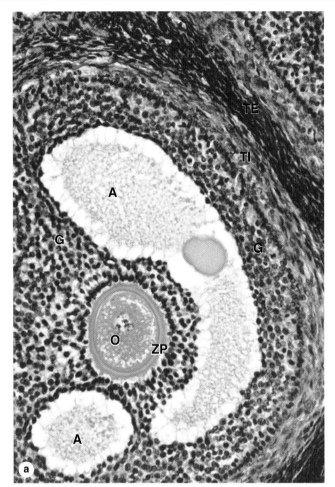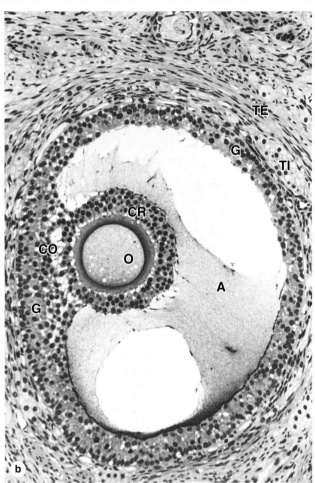

(a) An antral follicle shows the large, fluid-filled antral cavities or vesicles (**A**) that form within the granulosa layer as the cells produce follicular fluid. The oocyte (**O**) is surrounded by the zona pellucida (**ZP**) and granulosa cells (**G**), which also line the wall of the follicle. Fibroblastic cells immediately outside the growing follicles have developed as a steroid-secreting theca interna (**TI**) and a covering theca externa (**TE**). (X100; H&E)

(b) A slightly more developed preovulatory follicle shows a very large single antrum (**A**) filled with follicular fluid in which the proteins formed a thin film during fixation. The oocyte (**O**) now projects into this fluid-filled cavity, still surrounded by granulosa cells that now make up the corona radiata (**CR**). The corona radiata and oocyte are attached to the side of the follicle within a larger mass of granulosa cells called the **cumulus oophorus** (**CO**) which is continuous with the cells of the granulosa layer (**G**). Thecae interna (**TI**) and externa (**TE**) surround the whole follicle. (X100; PT)

Ovulation & Its Hormonal Regulation

Ovulation is the hormone-stimulated process by which the oocyte is released from the ovary. Ovulation normally occurs midway through the menstrual cycle, that is, around the 14th day of a typical 28-day cycle. In the hours before ovulation, the mature dominant follicle bulging against the tunica albuginea develops a whitish or translucent ischemic area, the **stigma**, in which tissue compaction has blocked blood flow. In humans usually only one oocyte is liberated during each cycle, but sometimes either no oocyte or two or more simultaneous oocytes may be expelled.

Just before ovulation the oocyte completes the first meiotic division, which it began and arrested in prophase during fetal life (Figure 22–10). The chromosomes are equally divided between the two daughter cells, but one of these retains almost all of the cytoplasm. That cell is now the **secondary oocyte** and the other becomes the **first polar body**, a very small nonviable cell containing a nucleus and a minimal amount of cytoplasm. Immediately after expulsion of the first polar body, the nucleus of the oocyte begins the second meiotic division but arrests at metaphase and never completes meiosis unless fertilization occurs (Figure 22–10).

FIGURE **22–8** Wall of antral follicle.

FIGURE **22–8** Wall of antral follicle.

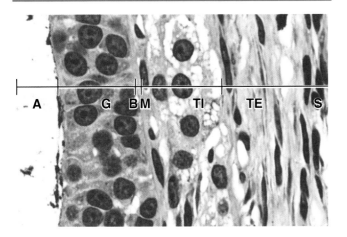

At higher magnification, a small part of the wall of an antral follicle shows the cell layers of the granulosa (**G**) next to the antrum (**A**), in which proteins have aggregated on cells in contact with the follicular fluid. The theca interna (**TI**) surrounds the follicle, its cells appearing vacuolated and lightly stained because of their cytoplasmic lipid droplets, a characteristic of steroid-producing cells. The overlying theca externa (**TE**) contains fibroblasts and smooth muscle cells and merges with the stroma (**S**). A basement membrane (**BM**) separates the theca interna from the granulosa, blocking vascularization of the latter. (X400; PT)

FIGURE **22–9** Atresia.

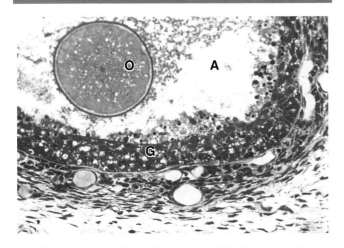

Atresia or degeneration of a follicle can begin at any stage of follicular development and is shown here in a follicle that had already developed a large antrum. Atresia is characterized by apoptosis of follicle or granulosa cells (**G**) and autolysis of the oocyte, with macrophages entering the degenerating structure to clean up debris. Many apoptotic bodies are seen loose in the antrum (**A**) here and the cells of the corona radiata have already disappeared, leaving the degenerative oocyte (**O**) free within the antrum. (X200; PT)

As mentioned before, follicular development depends on FSH from pituitary gonadotrophs, whose secretion is stimulated by gonadotropin-releasing hormone (GnRH) from the hypothalamus. Figure 22–11 summarizes the main hormonal interactions that regulate follicular growth as well as the ovulation and formation of the corpus luteum. Note that negative feedback of estrogen and progesterone on the hypothalamus and anterior pituitary is reinforced by a polypeptide hormone, **inhibin**, also produced by granulosa and luteal cells. In the days preceding ovulation, the dominant follicle secretes higher levels of estrogen which stimulate more rapid pulsatile release of GnRH from the hypothalamus.

The increased level of GnRH causes a surge of LH release from the pituitary gland that rapidly triggers a sequence of major events in and around the dominant follicle:

- **Meiosis I is completed** by the primary oocyte, yielding a secondary oocyte and the first polar body which degenerates (Figure 22–10).
- Granulosa cells are stimulated to produce much greater amounts of both **prostaglandin** and extracellular **hyaluronan**. This hydrophilic GAG loosens these cells and rapidly increases the volume, pressure, and viscosity of the follicular fluid.
- Ballooning at the stigma, the **ovarian wall weakens** as activated plasminogen (plasmin) from broken capillaries degrades collagen in the tunica albuginea and surface epithelium.
- **Smooth muscle contractions** begin in the theca externa, triggered by prostaglandins diffusing from follicular fluid.

The increasing pressure with the follicle and weakening of the wall lead to rupture of the ovarian surface at the stigma. The oocyte and corona radiata, along with follicular fluid, are expelled by the local smooth muscle contractions. The ovulated secondary oocyte adheres loosely to the ovary surface in the viscous follicular fluid and, as described later, is drawn into the opening of the uterine tube where fertilization may occur. If not fertilized within about 24 hours, the secondary oocyte begins to degenerate. Cells of the ovulated follicle that remain in the ovary redifferentiate under the influence of LH and give rise to the corpus luteum (Figure 22–11).

Corpus Luteum

After ovulation, the granulosa cells and theca interna of the ovulated follicle reorganize to form a larger temporary endocrine gland, the **corpus luteum** (L., yellowish body), in the ovarian cortex. Ovulation is followed immediately by the collapse and folding of the granulosa and thecal layers of the follicle's wall, and blood from disrupted capillaries typically accumulates as a clot in the former antrum (Figure 22–12). The granulosa is now invaded by capillaries. Cells of both the granulosa and theca interna change histologically and functionally under the influence of LH, becoming specialized for

FIGURE **22–10** Oogenesis.

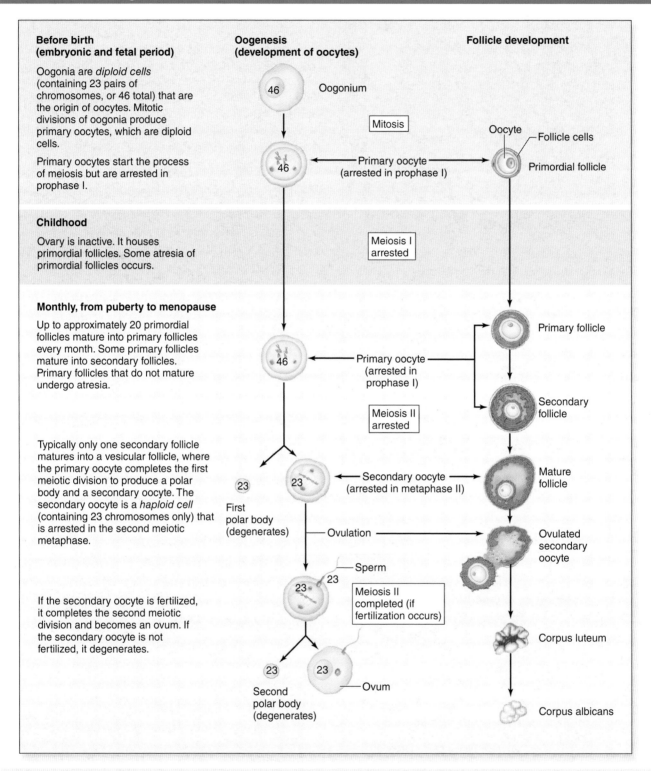

Before birth (embryonic and fetal period)

Oogonia are *diploid cells* (containing 23 pairs of chromosomes, or 46 total) that are the origin of oocytes. Mitotic divisions of oogonia produce primary oocytes, which are diploid cells.

Primary oocytes start the process of meiosis but are arrested in prophase I.

Childhood

Ovary is inactive. It houses primordial follicles. Some atresia of primordial follicles occurs.

Monthly, from puberty to menopause

Up to approximately 20 primordial follicles mature into primary follicles every month. Some primary follicles mature into secondary follicles. Primary follicles that do not mature undergo atresia.

Typically only one secondary follicle matures into a vesicular follicle, where the primary oocyte completes the first meiotic division to produce a polar body and a secondary oocyte. The secondary oocyte is a *haploid cell* (containing 23 chromosomes only) that is arrested in the second meiotic metaphase.

If the secondary oocyte is fertilized, it completes the second meiotic division and becomes an ovum. If the secondary oocyte is not fertilized, it degenerates.

Oogenesis (development of oocytes)

46 Oogonium

Mitosis

46 Primary oocyte (arrested in prophase I)

Meiosis I arrested

46 Primary oocyte (arrested in prophase I)

Meiosis II arrested

23 First polar body (degenerates)

23 Secondary oocyte (arrested in metaphase II)

Ovulation

Sperm

23 23

Meiosis II completed (if fertilization occurs)

23 Second polar body (degenerates)

23 Ovum

Follicle development

Oocyte — Follicle cells

Primordial follicle

Primary follicle

Secondary follicle

Mature follicle

Ovulated secondary oocyte

Corpus luteum

Corpus albicans

Oogenesis begins in the female fetus, with primary oocytes arresting at prophase I in primordial follicles, which remain inactive during childhood. At puberty, a population of primordial follicles begins to develop each month. Typically one per month produces a female gamete (a secondary oocyte).

FIGURE **22–11** Hormonal regulation of ovarian function.

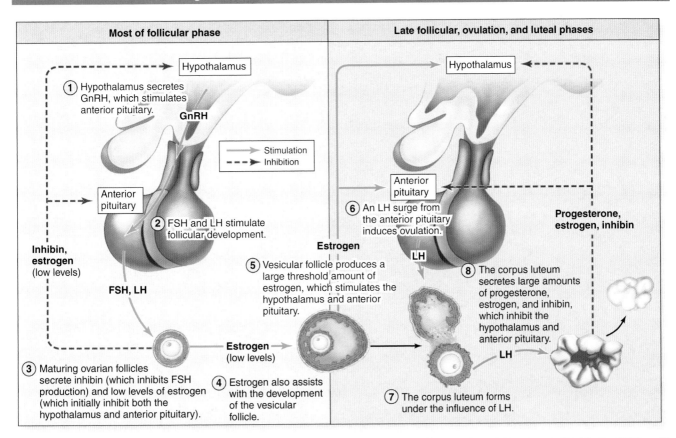

Most of follicular phase

Hypothalamus

① Hypothalamus secretes GnRH, which stimulates anterior pituitary.

GnRH

→ Stimulation
- - -→ Inhibition

Anterior pituitary

② FSH and LH stimulate follicular development.

Inhibin, estrogen (low levels)

FSH, LH

③ Maturing ovarian follicles secrete inhibin (which inhibits FSH production) and low levels of estrogen (which initially inhibit both the hypothalamus and anterior pituitary).

Estrogen (low levels)

④ Estrogen also assists with the development of the vesicular follicle.

⑤ Vesicular follicle produces a large threshold amount of estrogen, which stimulates the hypothalamus and anterior pituitary.

Late follicular, ovulation, and luteal phases

Hypothalamus

Anterior pituitary

⑥ An LH surge from the anterior pituitary induces ovulation.

Estrogen

LH

Progesterone, estrogen, inhibin

⑧ The corpus luteum secretes large amounts of progesterone, estrogen, and inhibin, which inhibit the hypothalamus and anterior pituitary.

LH

⑦ The corpus luteum forms under the influence of LH.

The **ovarian cycle** of follicular growth, followed by ovulation and luteal formation, is initiated when the hypothalamus secretes

GnRH. This stimulates the anterior pituitary to secrete FSH and LH, in the cycle depicted here.

---

more extensive production of progesterone in addition to estrogens.

Granulosa cells increase greatly in size (20-35 μm in diameter), without dividing, and eventually comprise about 80% of the corpus luteum. They are now called **granulosa lutein cells** (Figure 22–12) and have lost many features of protein-secreting cells to expand their role in aromatase conversion of androstenedione into estradiol. The former theca interna forms the rest of the corpus luteum, as **theca lutein cells** (Figure 22–12). These cells are half the size of the granulosa lutein cells and are typically aggregated in the folds of the wall of the corpus luteum, which, like all endocrine glands, becomes well vascularized. LH causes these cells to produce large amounts of progesterone as well as androstenedione.

The short-term fate of the corpus luteum depends on whether a pregnancy occurs. The ovulatory LH surge causes the corpus luteum to secrete progesterone for 10-12 days. Without further LH stimulation and in the absence of pregnancy, both major cell types of the corpus luteum cease steroid production and undergo apoptosis, with regression of the tissue. A consequence of the decreased secretion of progesterone

is menstruation, the shedding of part of the uterine mucosa. Estrogen produced by the active corpus luteum inhibits FSH release from the pituitary. However, after the corpus luteum degenerates, the blood steroid concentration decreases and FSH secretion increases again, stimulating the growth of another group of follicles and beginning the next menstrual cycle. The corpus luteum that persists for part of only one menstrual cycle is called a **corpus luteum of menstruation**. Remnants from its regression are phagocytosed by macrophages, after which fibroblasts invade the area and produce a scar of dense connective tissue called a **corpus albicans** (L., white body) (Figure 22–13).

If pregnancy occurs, the uterine mucosa must not be allowed to undergo menstruation because the embryo would be lost. To prevent the drop in circulating progesterone, trophoblast cells of the implanted embryo produce a glycoprotein hormone called **human chorionic gonadotropin (HCG)** with targets and activity similar to that of LH. HCG maintains and promotes further growth of the corpus luteum, stimulating secretion of progesterone to maintain the uterine mucosa. This **corpus luteum of pregnancy** becomes very large and

FIGURE 22–12 Corpus luteum.

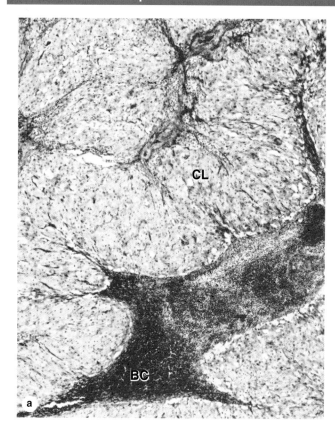

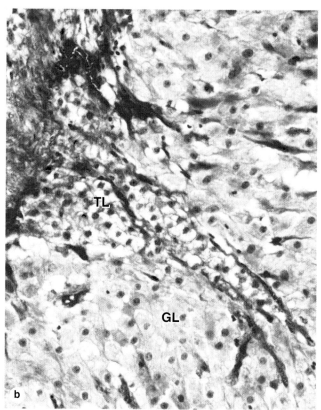

The corpus luteum is a large endocrine structure formed from the remains of the large dominant follicle after it undergoes ovulation. **(a)** A low-power micrograph shows the corpus luteum (**CL**), characterized by folds of the former granulosa that collapses as the theca externa contracts at ovulation. The former antrum often contains a blood clot (**BC**) from vessels in the thecal layers disrupted during ovulation. Cells of the granulosa and theca interna become reorganized under the influence of pituitary LH and their names are changed. (X15; H&E)

(b) Granulosa lutein cells (**GL**), seen at higher magnification here, undergo significant hypertrophy, producing most of the corpus luteum's increased size, and begin producing progesterone. The theca lutein cells (**TL**) increase only slightly in size, are somewhat darker-staining than the granulosa lutein cells, and continue to produce estrogens. Theca lutein cells, derived from the theca interna, are typically located within the folds that comprise the bulk of this tissue. (X100; H&E)

is maintained by HCG for 4-5 months, by which time the placenta itself produces progesterone (and estrogens) at levels adequate to maintain the uterine mucosa. It then degenerates and is replaced by a large corpus albicans.

〉 UTERINE TUBES

The paired **uterine tubes**, or **oviducts**, supported by ligaments and mesenteries that allow considerable mobility, each measure about 10-12 cm in length (Figure 22–14). Each opens into the peritoneal cavity near the ovary, with regions in the following sequence:

- The **infundibulum**, a funnel-shaped opening fringed with fingerlike extensions called **fimbriae** (L., fringes) next to the ovary;

- The **ampulla**, the longest and expanded region where fertilization normally occurs;
- The **isthmus**, a more narrow portion nearer the uterus; and
- The **uterine** or **intramural part**, which passes through the wall of the uterus and opens into the interior of this organ.

The wall of the oviduct consists of a folded mucosa, a thick, well-defined muscularis with interwoven circular (or spiral) and longitudinal layers of smooth muscle (Figure 22–15a), and a thin serosa covered by visceral peritoneum with mesothelium.

The numerous branching, longitudinal folds of the mucosa are most prominent in the ampulla, which in cross section resembles a labyrinth (Figure 22–14b). These mucosal folds become smaller in the regions closer to the uterus and are absent in the intramural portion of the tube.

FIGURE 22–13 Corpus albicans.

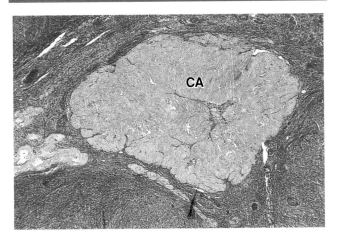

A corpus albicans (**CA**) is the scar of connective tissue that forms at the site of a corpus luteum after its involution. It contains mostly collagen, with few fibroblasts or other cells, and gradually becomes very small and lost in the ovarian stroma. Involution of the corpus luteum does not involve atresia. (X60; H&E)

Along its entire length, the mucosa is lined by simple columnar epithelium on a lamina propria of loose connective tissue (Figure 22–15b). The epithelium contains two interspersed, functionally important cell types:

- **Ciliated cells** in which ciliary movements sweep fluid toward the uterus,
- **Secretory peg cells**, nonciliated and often darker staining, often with an apical bulge into the lumen, which secrete glycoproteins of a nutritive mucus film that covers the epithelium.

Triggered primarily by estrogens, the cilia elongate and both cell types undergo hypertrophy during the follicular growth phase of the ovarian cycle and undergo atrophy with loss of cilia during the late luteal phase.

At the time of ovulation, mucosal hypertrophy and increased local blood flow have enlarged and moved the uterine tubes. The fringed infundibulum lies very close to the ovary and the fimbriae partially surround that organ. This favors the transport of the ovulated secondary oocyte into the tube. Promoted by sweeping muscular contractions of the fimbriae and ciliary activity, the oocyte enters the infundibulum and moves to the ampulla. The secretion covering the mucosa has nutritive and protective functions for both the oocyte and the sperm, including **capacitation factors** that activate sperm and make those cells able to fertilize an oocyte.

❯ MAJOR EVENTS OF FERTILIZATION

Fertilization, the union of the female and male gametes, normally occurs in the ampulla of a uterine tube, a site usually reached by only a few hundred of the millions of ejaculated sperm. Only sperm that have undergone capacitation in the female reproductive tract are capable of fertilization, a process with the following major steps:

- Upon contact with cells of the corona radiata, sperm undergo the **acrosomal reaction** in which **hyaluronidase** is released by exocytosis at multiple locations around the sperm head. This allows sperm to move more easily to the zona pellucida.
- Specific proteins on the sperm surface bind the receptors ZP3 and ZP4, activating the protease **acrosin** on the acrosomal membrane to degrade the zona pellucida locally.
- The first sperm penetrating the zona pellucida fuses with the oocyte plasmalemma and triggers Ca^{2+} release from vesicles, which induces exocytosis of proteases from the cortical granules. This **cortical reaction** quickly spreads like a wave across the entire surface of the oocyte, with the proteases converting the zona pellucida to the impenetrable **perivitelline barrier** that constitutes a permanent block to polyspermy.
- The nucleus of the secondary oocyte immediately completes meiosis II, producing a second polar body and the **female pronucleus** of the haploid **ovum**. The haploid nucleus of the single penetrating sperm head undergoes decondensation, becoming the **male pronucleus**. Fusion of the two pronuclei yields the new diploid cell, the **zygote** (Gr. *zygotos*, yoked together).

Cell division occurs while the embryo is transported by contractions of the oviduct muscularis and ciliary movements to the uterus, which takes about 5 days. This transport occurs normally in women with immotile ciliary syndrome, indicating a more important role for muscle contractions in moving the embryo.

❯❯ MEDICAL APPLICATION

Tubal ligation is a common surgical type of contraception. The uterine tube mucosa can become inflamed if infectious agents ascend from the lower genital tract, producing a condition named **salpingitis** after another name for these tubes, the **salpinges**. Mucosal damage or adhesions caused by chronic salpingitis can lead to **infertility** or an **ectopic (tubal) pregnancy** if there is blockage of oocyte or embryo transport to the uterus.

In tubal pregnancies, the lamina propria may react like the uterine endometrium and form decidual cells. But because of its small diameter and inability to expand, the tube cannot contain the growing embryo and will rupture, causing potentially fatal hemorrhage.

❯ UTERUS

As shown in Figure 22–14, the uterus is a pear-shaped organ with thick, muscular walls. Its largest part, the **body**, is entered by the left and right uterine tubes and the curved, superior

FIGURE 22–14 Uterine tubes and uterus.

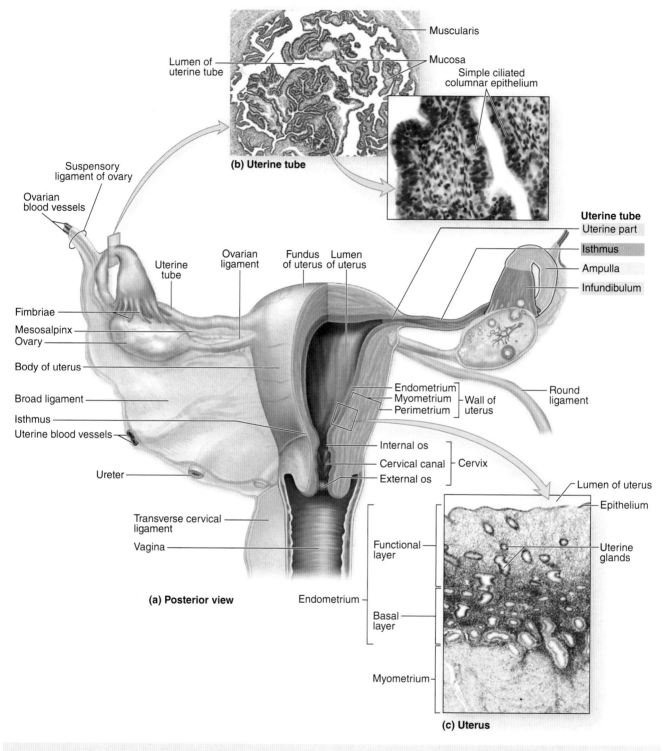

Muscularis

Lumen of uterine tube

Mucosa

Simple ciliated columnar epithelium

(b) Uterine tube

Suspensory ligament of ovary

Ovarian blood vessels

Uterine tube

Ovarian ligament

Fundus of uterus

Lumen of uterus

Uterine tube
- Uterine part
- Isthmus
- Ampulla
- Infundibulum

Fimbriae

Mesosalpinx

Ovary

Body of uterus

Broad ligament

Isthmus

Uterine blood vessels

Ureter

Endometrium
Myometrium — Wall of uterus
Perimetrium

Round ligament

Internal os

Cervical canal — Cervix

External os

Transverse cervical ligament

Vagina

(a) Posterior view

Endometrium

Lumen of uterus

Epithelium

Functional layer

Uterine glands

Basal layer

Myometrium

(c) Uterus

The uterine tubes or oviducts are paired ducts that catch the ovulated secondary oocyte, nourish both the oocyte and sperm, provide the microenvironment for fertilization, and transport the embryo undergoing cleavage to the uterus. **(a)** The diagram shows the relationship between the uterine tubes and the uterus in an intact posterior view (left) and in a cutaway view (right).

(b) Shown here is a cross section of uterine tube with a high magnification of the mucosa. (X35 and 400; H&E)

(c) Shown here is the uterine wall with the myometrium and the two layers of the endometrium. (X45; H&E)

FIGURE **22–15** Mucosa of the uterine tube wall.

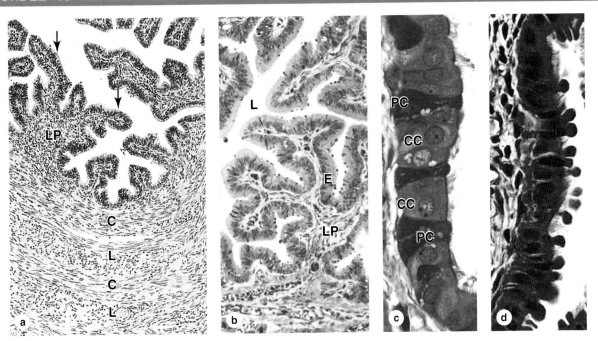

(a) A cross section of the uterine tube at the ampulla shows the interwoven circular (**C**) and longitudinal (**L**) layers of smooth muscle in the muscularis and in the complex of folded mucosa, the lamina propria (**LP**) underlying a simple columnar epithelium (arrows). (X140; H&E)

(b) The oviduct mucosa, with folds projecting into the lumen (**L**), has simple columnar epithelium (**E**) on the lamina propria (**LP**). (X200; PT)

(c, d) Higher magnification of the epithelium shows two cell types: ciliated cells (**CC**) interspersed with the secretory peg cells (**PC**), which produce the nutritive fluid covering the epithelium. These cells' histologic and functional features vary during the ovarian cycle due to hormonal fluctuations. In **(d)** the peg cells shown are at their most developed and most active state in the period shortly after ovulation when an embryo might be present. (c: X400, PT; d: X400, Mallory trichrome)

area between the tubes is called the **fundus**. The uterus narrows in the **isthmus** and ends in a lower cylindrical structure, the **cervix**. The lumen of the cervix, the **cervical canal**, has constricted openings at each end: the **internal os** (L. *os*, mouth) opens to the main uterine lumen and the **external os** to the vagina (Figure 22–14).

Supported by the set of ligaments and mesenteries also associated with the ovaries and uterine tubes (Figure 22–1), the uterine wall has three major layers (Figure 22–14):

■ An outer connective tissue layer, the **perimetrium**, continuous with the ligaments, which is adventitial in some areas, but largely a serosa covered by mesothelium;
■ A thick tunic of highly vascularized smooth muscle, the **myometrium** (Figure 22–16); and
■ A mucosa, the **endometrium**, lined by simple columnar epithelium.

These three layers are continuous with their counterparts in the uterine tubes. The thickness and structure of the endometrium is influenced cyclically by the shifting levels of ovarian hormones even more than the mucosa of the uterine tubes (Figure 22–17).

Myometrium

The **myometrium** (Gr. *myo*, muscle + *metra*, uterus), the thickest tunic of the uterus, shows bundles of smooth muscle fibers separated by connective tissue containing venous plexuses and lymphatics (Figure 22–16). The smooth muscle forms interwoven layers, with fibers of the inner and outer layers disposed generally parallel to the long axis of the organ.

During pregnancy, the myometrium goes through a period of extensive growth involving both **hyperplasia** (increasing the number of smooth muscle cells), cell hypertrophy, and increased collagen production by the muscle cells, which strengthens the uterine wall. This well-developed uterine myometrium contracts very forcefully during **parturition** to expel the infant from the uterus. After pregnancy, uterine smooth muscle cells shrink and many undergo apoptosis, with removal of unneeded collagen, and the uterus returns almost to its prepregnancy size.

Endometrium

The lamina propria or stroma of the **endometrium** contains primarily nonbundled type III collagen fibers with abundant

FIGURE **22–16** Uterus.

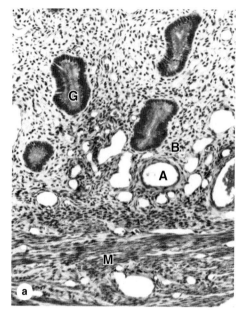
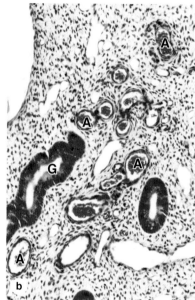
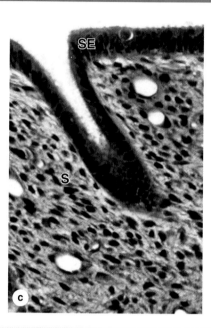

(a) The basal layer (**B**) of the endometrium, bordering the myometrium (**M**), contains the basal ends of the uterine glands (**G**) and many small arteries (**A**) embedded in a distinctive connective tissue stroma with many fibroblasts, ground substance and primarily fine type III collagen, but no adipocytes. (X100; Mallory trichrome)

(b) Superficial to the basal layer of the endometrium is its functional layer, the part that changes histologically and functionally

depending on estrogen levels. This micrograph shows only the functional layer and includes parts of the long uterine glands (**G**) as well as one spiral artery (**A**). (X100; Mallory trichrome)

(c) The surface epithelium (**SE**) lining the endometrium is simple columnar, with many cells having cilia. The underlying stroma (**S**) has an extensive microvasculature, much ground substance, and fibroblastic cells with large, active nuclei. (X400; Mallory trichrome)

fibroblasts and ground substance. Its simple columnar epithelial lining has both ciliated and secretory cells, and the latter line the numerous tubular **uterine glands** that penetrate the full thickness of the endometrium (Figures 22–16 and 22–18).

The endometrium has two concentric zones:

- The **basal layer** adjacent to the myometrium has a more highly cellular lamina propria and contains the deep basal ends of the uterine glands (Figure 22–16a).
- The superficial **functional layer** has a spongier lamina propria, richer in ground substance, and includes most of the length of the glands, as well as the surface epithelium (Figure 22–16b,c).

The functional layer undergoes profound changes during the menstrual cycles, but the basal layer remains relatively unchanged (Figure 22–17).

The blood vessels supplying the endometrium have special significance in the periodic sloughing of the functional

layer during menses. Arcuate arteries in the middle layers of the myometrium send two sets of smaller arteries into the endometrium (Figure 22–18): **straight arteries**, which supply only the basal layer, and long, progesterone-sensitive **spiral arteries**, which extend farther and bring blood throughout the functional layer. Spiral arteries branch with numerous arterioles supplying a rich, superficial capillary bed that includes many dilated, thin-walled **vascular lacunae** drained by venules.

Menstrual Cycle

Throughout the female reproductive system, estrogens and progesterone control growth and differentiation of epithelial cells and associated connective tissue. Even before birth, these cells are influenced by circulating maternal estrogen and progesterone that reach the fetus through the placenta. After menopause, diminished synthesis of these hormones results in a general involution of tissues in the reproductive tract.

FIGURE **22–17** Correlation of ovarian and menstrual cycles with levels of their controlling hormones.

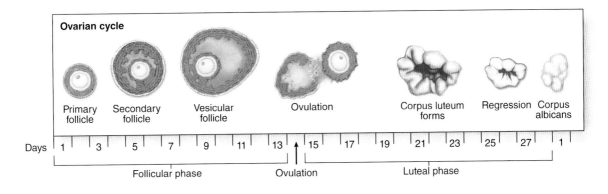

Ovarian cycle

Primary follicle | Secondary follicle | Vesicular follicle | Ovulation | Corpus luteum forms | Regression | Corpus albicans

Days 1 3 5 7 9 11 13 15 17 19 21 23 25 27 1

Follicular phase | Ovulation | Luteal phase

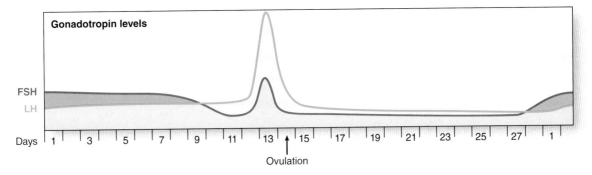

Gonadotropin levels

FSH
LH

Days 1 3 5 7 9 11 13 15 17 19 21 23 25 27 1

Ovulation

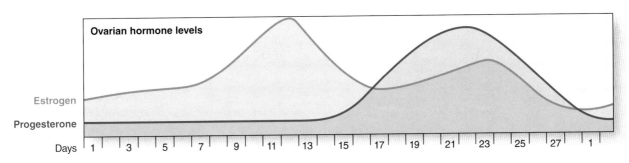

Ovarian hormone levels

Estrogen
Progesterone

Days 1 3 5 7 9 11 13 15 17 19 21 23 25 27 1

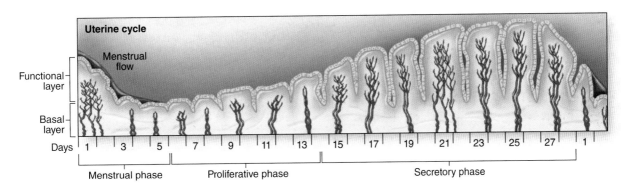

Uterine cycle

Menstrual flow

Functional layer

Basal layer

Days 1 3 5 7 9 11 13 15 17 19 21 23 25 27 1

Menstrual phase | Proliferative phase | Secretory phase

The cyclic development of **ovarian follicles** and the **corpus luteum**, controlled by the pituitary **gonadotropins** FSH and LH, lead to cyclic shifts in the levels of the major ovarian hormones: steroidal **estrogens and progesterone**. Estrogen stimulates the proliferative phase of the uterine cycle and its level peaks near the day of ovulation, which marks the midpoint of the ovarian cycle. After ovulation the corpus luteum forms and produces both progesterone and estrogens, which together promote growth and development of the endometrial **functional layer.** Unless fertilization and implantation of an embryo occur, regression of the corpus luteum leads to declining levels of the steroid hormones and failure of the new endometrial tissue to be maintained. This tissue sloughs off as the menstrual flow, the first day of which is taken to mark day 1 of both the ovarian cycle and the uterine cycle. The basal layer of endometrium is not sensitive to the loss of progesterone and is retained during menstruation, serving to regenerate the functional layer during the ensuing proliferative phase.

FIGURE 22–18 Arterial supply to the endometrium.

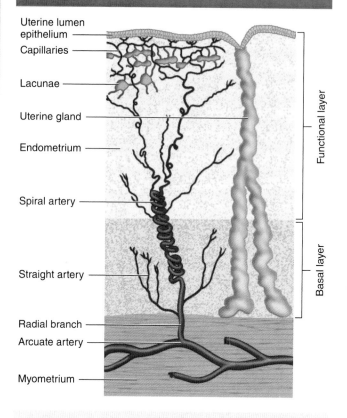

Uterine lumen epithelium
Capillaries
Lacunae
Uterine gland
Endometrium
Spiral artery
Straight artery
Radial branch
Arcuate artery
Myometrium

Functional layer
Basal layer

The **basal and functional layers** of the endometrium are supplied by different sets of small arteries emerging from the uterine arcuate arteries in the myometrium: the **straight arteries** and **spiral arteries**, respectively. The spiral arteries are uniquely sensitive to progesterone, growing rapidly in a spiral fashion as the functional layer thickens under the influence of that luteal steroid and providing blood to a microvasculature that includes many lacunae lined by thin endothelium. This blood supply brings oxygen and nutrients to cells of the functionalis and to an embryo implanting itself into that tissue. If no embryo is present to produce the gonadotropin replacing LH, the corpus luteum undergoes regression 8-10 days after ovulation. The rapid decline in the level of progesterone causes constriction of the spiral arteries and other changes that quickly lead to local ischemia in the functional layer and its separation from the basal layer during menstruation.

›› MEDICAL APPLICATION

Viable endometrial cells frequently undergo menstrual reflux into or through the uterine tubes. In some women this can lead to **endometriosis**, a disorder with pelvic pain due to endometrial tissue growing on the ovaries, oviducts, or elsewhere. Under the influence of estrogen and progesterone, the ectopic tissue grows and degenerates monthly but cannot be removed effectively from the body. In addition to pain endometriosis can produce inflammation, ovarian cysts, adhesions, and scar tissue that can cause infertility.

From puberty until menopause at about age 45-50, pituitary gonadotropins produce cyclic changes in ovarian hormone levels, which cause the endometrium to undergo cyclic modifications during the menstrual cycle (Figures 22–17 and 22–19). The duration of the menstrual cycle may be variable but averages 28 days. Because menstrual cycles are a consequence of ovarian follicle changes related to oocyte production, a woman is fertile only during the years when she is having menstrual cycles.

Day 1 of the menstrual cycle is usually taken as the day when menstrual bleeding appears. The menstrual discharge consists of degenerating endometrium mixed with blood from its ruptured microvasculature. The **menstrual period** lasts 3-4 days on average. The next phase of the cycle, the **proliferative phase**, is of variable length, 8-10 days on average, and the **secretory phase** begins at ovulation and lasts about 14 days (Figure 22–17). The cyclic structural changes occur gradually and the activities characterizing these phases overlap to some extent.

Proliferative Phase

After the menstrual phase, the uterine mucosa is relatively thin (~0.5 mm). The beginning of the **proliferative phase**, also called the **follicular or estrogenic phase**, coincides with the rapid growth of a small group of ovarian follicles growing as vesicular follicles. With development of their thecae interna, these follicles actively secrete estrogen and increase its plasma concentrations (Figure 22–17).

Estrogens act on the endometrium, inducing regeneration of the functional layer lost during menstruation. Cells in the basal ends of glands proliferate, migrate, and form the new epithelial covering over the surface exposed during menstruation. During the proliferative phase, the endometrial lining is a simple columnar surface epithelium and the uterine glands are relatively straight tubules with narrow, nearly empty lumens (Figure 22–19a,d). Mitotic figures can be found among both the epithelial cells and fibroblasts. Spiral arteries lengthen as the functional layer is reestablished and grows (Figure 22–16) and extensive microvasculature forms near the surface of the functional layer. At the end of the proliferative phase, the endometrium is 2-3 mm thick.

Secretory Phase

After ovulation, the **secretory** or **luteal phase** starts as a result of the progesterone secreted by the corpus luteum. Progesterone stimulates epithelial cells of the uterine glands that formed during the proliferative phase and these cells begin to secrete and accumulate glycogen, dilating the glandular lumens and causing the glands to become coiled (Figure 22–19b,e). The superficial microvasculature now includes thin-walled, blood-filled lacunae (Figures 22–18 and 22–19e). The endometrium reaches its maximum thickness (5 mm) during the secretory phase as a result of the accumulation of secretions and edema in the stroma.

If fertilization occurred during the day after ovulation, the embryo has been transported to the uterus by about

FIGURE **22–19** Proliferative, secretory, and premenstrual phases in the uterus.

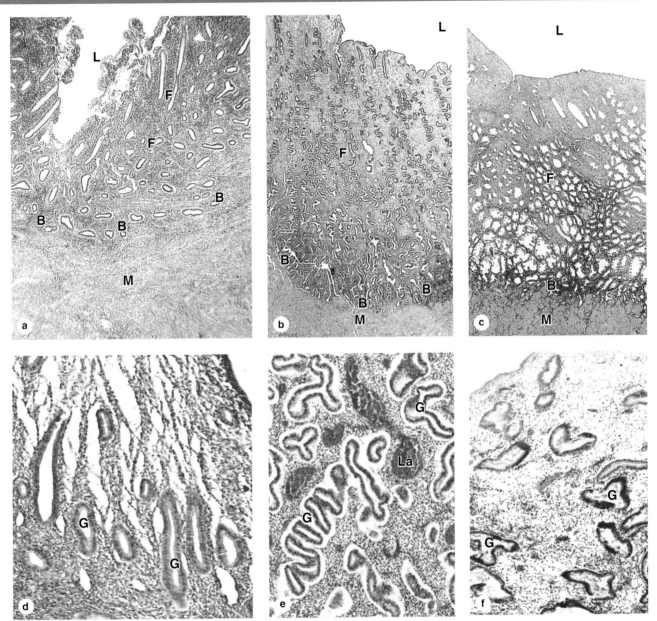

The major phases of the uterine cycle overlap but produce distinctly different and characteristic changes in the functional layer (**F**) closest to the lumen (**L**) with little effect on the basal layer (**B**) and myometrium (**M**). Characteristic features of each phase include the following. During most of the proliferative phase (**a, d**), the functional layer is still relatively thin, the stroma is more cellular, and the glands (**G**) are relatively straight, narrow, and empty.

In the secretory phase (**b, e**) the functional layer is less heavily cellular and perhaps four times thicker than the basal layer. The tubular glands have wider lumens containing secretory product

and coil tightly up through the stroma, giving a zigzag or folded appearance histologically. Superficially in the functional layer, lacunae (**La**) are widespread and filled with blood.

The short premenstrual phase (**c, f**) begins with constriction of the spiral arteries, which produces hypoxia that causes swelling and dissolution of the glands (**G**). The stroma of the peripheral functionalis is more compact and that near the basal layer typically appears more sponge-like during this time of blood stasis, apoptosis, and breakdown of the stromal matrix. (a: X20; b and c: X12; d, e, and f: X50. All H&E)

TABLE 22–1	Summary of events of the menstrual cycle.			
	Stage of Cycle			
	Proliferative	**Secretory or Luteal**		**Menstrual**
Main actions of pituitary hormones	Follicle-stimulating hormone stimulates rapid growth of ovarian follicles	Peak of luteinizing hormone at the beginning of secretory stage, secreted after estrogen stimulation, induces ovulation and development of the corpus luteum		
Main events in the ovary	Growth of ovarian follicles; dominant follicle reaches preovulatory stage	Ovulation	Development of the corpus luteum	Degeneration of the corpus luteum
Dominant ovarian hormone	Estrogens, produced by the growing follicles, act on vagina, tubes, and uterus	Progesterone, produced by the corpus luteum, acts mainly on the uterus	Progesterone production ceases	
Main events in the endometrium	Growth of the mucosa after menstruation	Further growth of the mucosa, coiling of glands, secretion		Shedding of part of the mucosa about 14 days after ovulation

5 days later and now attaches to the uterine epithelium when the endometrial thickness and secretory activity are optimal for embryonic implantation and nutrition. The major nutrient source for the embryo before and during implantation is the uterine secretion. In addition to promoting this secretion, progesterone inhibits strong contractions of the myometrium that might interfere with embryo implantation.

Menstrual Phase

When fertilization of the oocyte and embryonic implantation do not occur, the corpus luteum regresses and circulating levels of progesterone and estrogens begin to decrease 8-10 days after ovulation, causing the onset of menstruation (Figure 22–17). The drop-off in progesterone produces (1) spasms of muscle contraction in the small spiral arteries of the functional layer, interrupting normal blood flow, and (2) increased synthesis by arterial cells of prostaglandins, which produce strong vasoconstriction and local hypoxia. Cells undergoing hypoxic injury release cytokines that increase vascular permeability and immigration of leukocytes. The leukocytes release collagenase and several other matrix metalloproteinases (MMPs) that degrade basement membranes and other ECM components (Figure 22–19c,f).

The basal layer of the endometrium, not dependent on the progesterone-sensitive spiral arteries, is relatively unaffected by these activities. However, major portions of the functional layer, including the surface epithelium, most of each gland, the stroma and blood-filled lacunae, detach from the endometrium and slough away as the menstrual flow or **menses**. Arterial constriction normally limits blood loss during menstruation, but some blood does emerge from the open ends of venules. The amount of endometrium and blood lost in menstruation varies among women and in the same woman at different times.

At the end of the menstrual phase, the endometrium is usually reduced to a thin layer and is ready to begin a new cycle as its cells begin dividing to reconstitute the mucosa. Table 22–1 summarizes the main events of the menstrual cycle.

❯ EMBRYONIC IMPLANTATION, DECIDUA, & THE PLACENTA

The zygote produced by fertilization undergoes mitotic cleavages as it is moved toward the uterus, with its cells called **blastomeres** (Gr. *blastos*, germ + *meros*, part) in a compact aggregate called the **morula** (L. *morum*, mulberry). No growth occurs during the period of cell cleavage, with blastomeres becoming smaller at each division, and the morula is about the same size as the oocyte at fertilization.

About 5 days after fertilization the embryo reaches the uterine cavity, by which time blastomeres have moved to form a central cavity in the morula and the embryo enters the **blastocyst** stage of development. The blastomeres then arrange themselves as a peripheral layer called the **trophoblast** around the cavity, while a few cells just inside this layer make up the **embryoblast** or **inner cell mass** (Figure 22–20). The blastocyst remains in the lumen of the uterus for about 2 days, immersed in the endometrial glands' secretion on the mucosa.

Implantation, or nidation, involves attachment of the blastocyst to the surface epithelial cells of the endometrium and its proteolytic penetration through this epithelium into the underlying stroma (Figure 22–20), a process that lasts about 3 days. Cells of the trophoblast drive the events of implantation, during which time cells of the embryoblast rearrange around two new cavities, the **amnion** and the **yolk sac**. Where the cells lining these cavities make contact, the **bilaminar embryonic disc** develops with its **epiblast** layer

FIGURE **22–20** Embryo implantation.

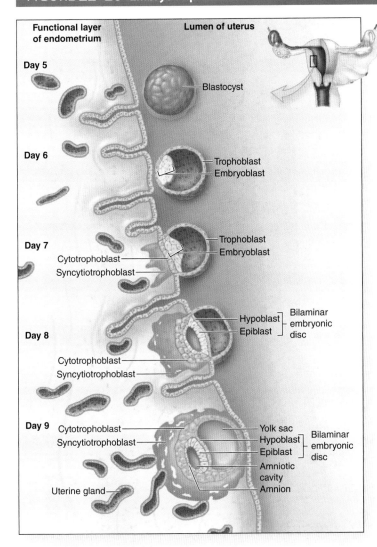

The embryo enters the uterus as a blastocyst about 5 days after ovulation or fertilization, when the uterus is in the secretory phase and best prepared for implantation. To begin implantation, receptors on cells of the outer embryonic trophoblast bind glycoprotein ligands on the endometrial epithelium. The trophoblast forms an invasive, outer syncytial layer called the **syncytiotrophoblast**. Proteases are activated and/or released locally to digest stroma components, which allows the developing embryo to embed itself within the stroma. The newly implanted embryo absorbs nutrients and oxygen from the endometrial tissue and blood in the lacunae.

continuous with the amnion and its **hypoblast** layer continuous with the yolk sac (Figure 22–20).

All parts of the embryo develop from this early embryonic disc. The yolk sac and amnion form extraembryonic structures, but only the latter persists throughout pregnancy. As shown in Figure 22–20, the trophoblast differentiates during implantation into the following:

- The **cytotrophoblast**, a layer of mitotically active cells immediately around the amnion and yolk sac, and
- The **syncytiotrophoblast**, a more superficial, nonmitotic mass of multinucleated cytoplasm which invades the surrounding stroma.

By about the ninth day after ovulation, the embryo is totally implanted in the endometrium and derives nutrients primarily from blood there. Cytotrophoblast cells synthesize anti-inflammatory cytokines to prevent an adverse uterine reaction to the implanted embryo and these are supplemented later by various embryonic factors that produce local immune tolerance for the embryo throughout the pregnancy.

The endometrial stroma undergoes histologic changes in the period following implantation. Fibroblasts become enlarged, polygonal, more active in protein synthesis, and are now called **decidual cells**. The whole endometrium is now called the **decidua** (L. *deciduus*, falling off, shedding) and includes three areas (Figure 22–21):

- The **decidua basalis** between the implanted embryo and the myometrium;
- The **decidua capsularis**, the region between the embryo and the uterine lumen which thins as the embryo gets larger; and
- The **decidua parietalis**, on the side of the uterus away from the embryo.

FIGURE 22–21 Extraembryonic membranes, decidua, and placenta.

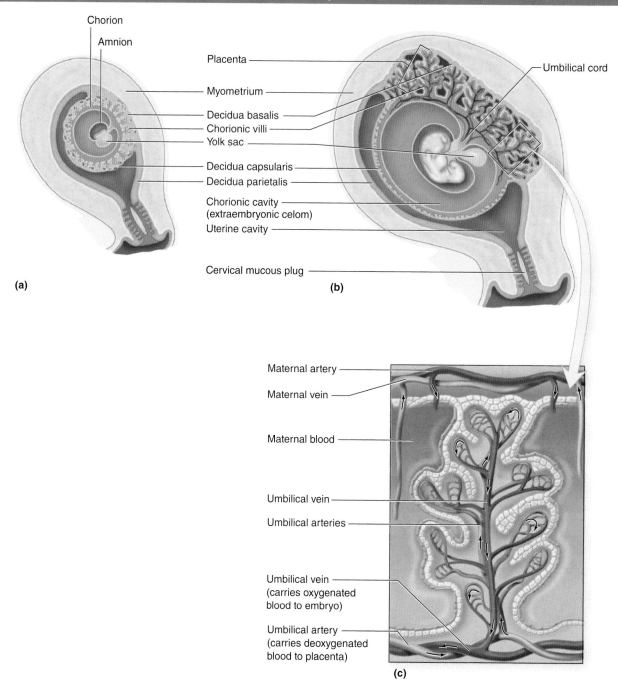

(a)

(b)

(c)

(a) The membranous extraembryonic **amnion**, **chorion**, and **yolk sac** appear during the second week of development, with the embryonic disc between the amnion and yolk sac.

(b) After implantation the endometrium changes histologically and is called the **decidua**. It develops three different regions: **decidua basalis, capsularis,** and **parietalis**. **Chorionic villi** develop most profusely in the decidua basalis, which becomes the major portion of the **placenta**.

(c) Each of the many **chorionic stem villi** in the placenta contains a branch of the umbilical artery and vein, which form loops of microvasculature into smaller villus branches. The entire stem villus is bathed in maternal blood circulated by endometrial arteries and veins.

❯❯ MEDICAL APPLICATION

The initial attachment of the embryo usually occurs on the ventral or dorsal walls of the body of the uterus. Sometimes the embryo attaches close to the internal os. In this case the placenta will be interposed between the fetus and the vagina, obstructing the passage of the fetus at parturition. This situation, called **placenta previa,** must be recognized by the physician, and the fetus must be delivered by cesarean section. Otherwise, obstructed parturition can lead to death of the fetus.

The **placenta** is the site of exchange for nutrients, wastes, O_2, and CO_2 between the mother and the fetus and contains tissues from both individuals. The embryonic part is the **chorion**, derived from the trophoblast and the maternal part is from the decidua basalis. Exchange occurs between embryonic blood in chorionic villi outside the embryo and maternal blood in lacunae of the decidua basalis. Chorionic villi of the developing placenta go through three stages:

- **Primary villi** appear 2 days after implantation as simple cords of proliferating cytotrophoblast cells covered by syncytiotrophoblast extend into lacunae containing maternal blood.
- **Secondary villi** begin to form on about the 15th day of embryonic development as the primary villi are invaded by extraembryonic mesenchyme.
- **Tertiary villi** develop within a few more days as mesenchyme in the secondary villi differentiates to form capillary loops continuous with the embryonic circulatory system.

By the end of the first month of the pregnancy, the placenta contains thousands of tertiary chorionic villi, each branching many times and each branch having one or more capillary loops (Figure 22–21c). Suspended in pools of maternal blood in the decidua, the chorionic villi provide an enormous surface area for metabolite exchange (Figure 22–22). Exchange of gases, nutrients, and wastes occurs between fetal blood in the capillaries and maternal blood bathing the villi, with diffusion occurring across the trophoblast layer and the capillary endothelium.

The placenta is also an endocrine organ, producing HCG, a lactogen, relaxin, and various growth factors, in addition to estrogen and progesterone. More detailed information on the developing embryo and placenta should be sought in embryology textbooks.

FIGURE 22–22 Term placenta.

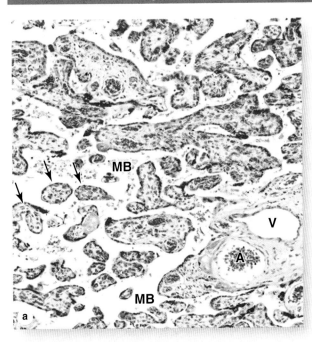
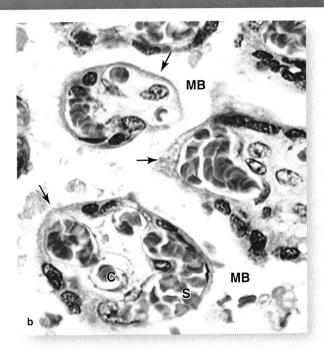

The placenta contains chorionic villi of the fetus and maternal blood pooled in the decidua. **(a)** A full-term placenta has many villus stems, containing arteries (**A**) and (**V**) of the extraembryonic vasculature, and hundreds of smaller villus branches (**arrows**) that contain connective tissue and microvasculature. Maternal blood (**MB**) fills the space around the villi. (X50; H&E)

(b) Higher magnification of villus branches surrounded by maternal blood (**MB**) each containing capillaries (**C**) or sinusoids (**S**) with fetal blood. By the end of pregnancy cytotrophoblast cells have greatly decreased in number in many areas, leaving only a thin syncytiotrophoblast and basement membrane covering the villus in these regions (**arrows**). The extraembryonic blood vessels become closely associated with these areas of thin trophoblast for maximal diffusion of material between the two pools of blood. (X400; H&E)

› CERVIX

As noted earlier the **cervix** is the lower, cylindrical part of the uterus (see Figure 22–1). The cervix differs histologically from the rest of the uterus. The **endocervical mucosa** is a simple columnar epithelium on a thick lamina propria, with many large, branched, mucus-secreting **cervical glands**. It lacks spiral arteries, does not change its 2-3 mm thickness during the ovarian cycle, and is not shed during menstruation.

The cervical region around the external os projects slightly into the upper vagina and is covered by the **exocervical mucosa** with nonkeratinized stratified squamous epithelium continuous with that of the vagina. The junction between this squamous epithelium and the mucus-secreting columnar epithelium of the endocervix occurs in the **transformation zone**, an area just outside the external os that shifts slightly with the cyclical changes in uterine size (Figure 22–23). Periodic exposure of the squamous-columnar junction to the

---

FIGURE 22–23   Cervix.

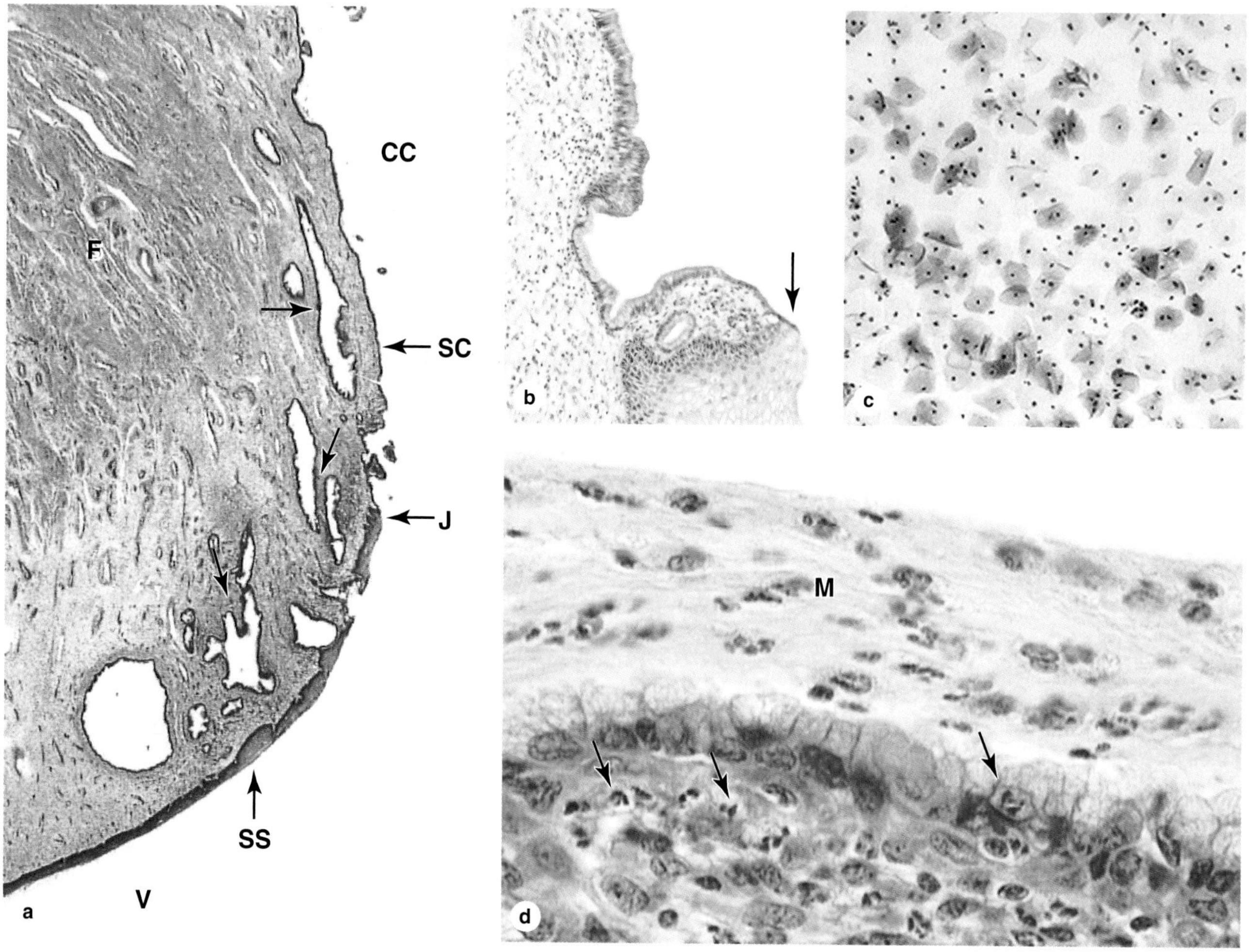

(a) The mucosa of the cervical canal (**CC**) is continuous with the endometrium and like that tissue is lined by simple columnar epithelium (**SC**). This endocervical mucosa includes many large branched cervical mucous glands (**arrows**). At the external os, the point at which the cervical canal opens into the vagina (**V**), there is an abrupt junction (**J**) between the columnar epithelium and the stratified squamous epithelium (**SS**) covering the exocervix and vagina. Deeper, the cervical wall is primarily fibromuscular tissue (**F**). (X15; H&E)

(b) The epithelial junction (**arrow**) is seen more clearly. (X50; H&E)

(c) Exfoliative cytology of epithelial cells from the exocervical mucosa in a routine cervical smear. The squamous cells, stained on

a slide by the Papanicolaou procedure using hematoxylin, orange G, and eosin, stain differently according to their content of keratins. Cells with atypical nuclei or other abnormalities can be detected by this method that is used routinely to check for cervical carcinoma. (X200; Papanicolaou stain)

(d) The endocervical mucosa is exposed to a relatively high population of microorganisms and normally has a large number of neutrophils and other leukocytes. Such cells occur in the lamina propria and epithelium (**arrows**), but they are also numerous and readily apparent in the layer of mucus (**M**) that was fixed in place here. (X400; H&E)

vaginal environment can stimulate reprogramming of epithelial stem cells, which occasionally leads to intraepithelial neoplasia at that site.

Under the influence of progesterone, the consistency of **cervical mucus** changes cyclically and plays a significant role in fertilization and early pregnancy. At ovulation, mucous secretion is abundant and watery, facilitating sperm movements into the uterus. In the luteal phase mucus is more viscous and hinders the passage of sperm. During pregnancy, the cervical glands proliferate and secrete highly viscous mucus that forms a plug in the cervical canal (Figure 22–21b).

The deeper wall of the cervix consists mainly of dense connective tissue, with much less smooth muscle than the rest of the uterus (Figure 22–23). The cervix becomes relatively rigid during pregnancy and helps retain the fetus in the uterus. Before parturition a process of **cervical effacement** occurs in which its connective tissue undergoes extensive remodeling and significant collagen removal, mediated in part by macrophages. As a result the cervix softens, the cervical canal dilates, and birth occurs more easily.

> ❯❯ **MEDICAL APPLICATION**
>
> The incidence of **cervical cancer** worldwide has been greatly reduced by widespread, routine screening by **exfoliative cytology** to examine for dysplasia of the cervical epithelium (Figure 22-23c). The test called the **Pap smear** after its developer George Papanicolaou, who introduced this diagnostic technique in the 1920s, uses cells that have been lightly scraped from cervix. Abnormal cells suggestive of precancerous changes in the epithelium are then detected microscopically. The epithelial dysplasia that precedes **squamous cell neoplasia**, the most common type of cervical cancer, occurs in metaplastic cells of the transformation zone at a mean age of 54 years. The **human papillomas virus** (HPV) is strongly implicated in the pathogenesis of this cancer.

❯ VAGINA

The wall of the **vagina** (L., *vagina*, sheath, scabbard) lacks glands and consists of a **mucosa**, a **muscular layer**, and an **adventitia**.

The epithelium of the vaginal mucosa is stratified squamous, with a thickness of 150-200 μm in adults (Figure 22–24). Stimulated by estrogens, the epithelial cells synthesize and accumulate glycogen. When the cells desquamate, bacteria metabolize glycogen to lactic acid, causing a relatively low pH within the vagina, which helps provide protection against pathogenic microorganisms. The lamina propria of the mucosa is rich in elastic fibers, with numerous narrow papillae projecting into the overlying epithelium (Figure 22–24). The mucosa normally contains lymphocytes and neutrophils in relatively large quantities.

Mucus in the vagina is produced by the cervical glands. During sexual arousal lubricating mucus is also provided by glands at the vaginal vestibule, including the paired **greater vestibular glands** (of Bartholin), which are homologous to the male bulbourethral glands.

The muscular layer of the vagina is composed mainly of two indistinct layers of smooth muscle, disposed as circular bundles next to the mucosa and as thicker longitudinal bundles near the adventitial layer (Figure 22–24). The dense connective tissue of the adventitia is rich in elastic fibers, making the vaginal wall strong and elastic while binding it the surrounding tissues. This outer layer also contains an extensive venous plexus, lymphatics, and nerves.

> ❯❯ **MEDICAL APPLICATION**
>
> **Atrophic vaginitis** involves thinning or atrophy of the vaginal epithelium caused by diminished estrogen levels and occurs most often in postmenopausal woman. This change allows the more frequent inflammation and infections characteristic of this condition. Primary **squamous cell carcinoma** of the vagina occurs rarely, with most vaginal malignancies having spread secondarily from the cervix or vulva.

❯ EXTERNAL GENITALIA

The female **external genitalia**, or **vulva**, include several structures, all covered by stratified squamous epithelium:

- The **vestibule**, a space whose wall includes the tubuloacinar vestibular glands;
- The paired **labia minora**, folds of skin lacking hair follicles but with numerous sebaceous glands;
- The paired **labia majora**, homologous and histologically similar to the skin of the scrotum; and
- The **clitoris**, an erectile structure homologous to the penis with paired corpora cavernosa.

The mucosa of these structures, abundantly supplied with sensory nerves and tactile receptors also found in skin (see Chapter 18), is important in the physiology of sexual arousal.

❯ MAMMARY GLANDS

The **mammary glands** of the breasts develop embryologically as invaginations of surface ectoderm along two ventral lines, the milk lines, from the axillae to the groin. In humans one set of glands resembling highly modified apocrine sweat glands persists on each side of the chest. Each mammary gland consists of 15-25 **lobes** of the compound tubuloalveolar type whose function is to secrete nutritive milk for newborns. Each lobe, separated from the others by dense connective tissue with much adipose tissue, is a separate gland with its own excretory

FIGURE **22–24** Vagina.

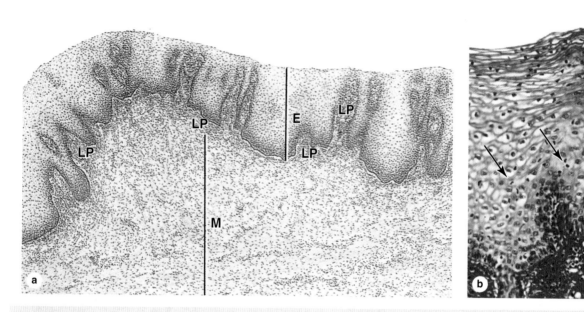

The vagina has mucosal, muscular, and adventitial layers. **(a)** The lamina propria (**L**) is highly cellular and extends narrow papillae into the thick, nonkeratinized stratified squamous epithelium (**E**). The muscular layer (**M**) has bundles of smooth muscle arranged in a circular manner near the mucosa and longitudinally near the adventitia. (X60; H&E)

(b) Higher magnification of the epithelium and lamina propria (**LP**) shows invasion of leukocytes (**arrows**) between epithelial cells from the connective tissue. (X200; PSH)

lactiferous duct (Figure 22–25). These ducts, each 2-4.5- cm long, emerge independently in the **nipple**, which has 15-25 pore-like openings, each about 0.5 mm in diameter. The histologic structure of the mammary glands varies according to sex, age, and physiologic status.

Breast Development During Puberty

Before puberty, the mammary glands in both sexes are composed only of **lactiferous sinuses** near the nipple, with very small, branching ducts emerging from these sinuses. In girls undergoing puberty, higher levels of circulating estrogens cause the breasts to grow as a result of adipocyte accumulation and elongation of the duct system.

In nonpregnant adult women each mammary gland lobe consists of many **lobules**, sometimes called **terminal duct lobular units (TDLU)**. Each lobule has several small, branching ducts, but the attached secretory units are small and rudimentary (Figure 22–25). Lactiferous sinuses are lined with stratified cuboidal epithelium, and the lining of the lactiferous ducts and terminal ducts is simple cuboidal epithelium with many myoepithelial cells. Sparse fibers of smooth muscle also encircle the larger ducts. The duct system is embedded in loose, vascular connective tissue, and a denser, less cellular connective tissue separates the lobes. In the premenstrual phase of the reproductive cycle connective tissue of the breast becomes somewhat edematous, making the breasts slightly larger.

The **areola**, or skin surrounding and covering the nipple, contains sebaceous glands and abundant sensory nerves and is continuous with the mucosa of the lactiferous sinuses. The areola contains more melanin than skin elsewhere on the breast and darkens further during pregnancy. Connective tissue of the nipple is rich in smooth muscle fibers that run parallel to the lactiferous sinuses and produce nipple erection when they contract.

Breasts During Pregnancy & Lactation

The mammary glands undergo growth during pregnancy as a result of the synergistic action of several hormones, mainly estrogen, progesterone, prolactin, and the placental lactogen. These cause cell proliferation in secretory **alveoli** at the ends of the intralobular ducts (Figures 22–25 and 22–26). The spherical alveoli are composed of cuboidal epithelium, with stellate myoepithelial cells between the secretory cells and the basal lamina. The degree of glandular development varies among lobules and even within a single lobule.

While the alveoli and duct system grow and develop during pregnancy in preparation for lactation, the stroma

FIGURE 22–25 Mammary gland.

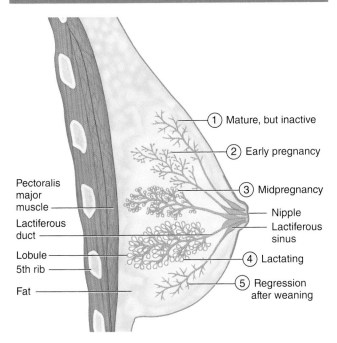

- ① Mature, but inactive
- ② Early pregnancy
- ③ Midpregnancy
- Nipple
- Lactiferous sinus
- ④ Lactating
- ⑤ Regression after weaning

Pectoralis major muscle
Lactiferous duct
Lobule
5th rib
Fat

Shown here is the sequence of changes that occur in the alveolar secretory units and duct system of mammary glands before, during, and after pregnancy and lactation. (**1**) Before pregnancy, the gland is inactive, with small ducts and only a few small secretory alveoli. (**2**) Alveoli develop and begin to grow early in a pregnancy. (**3**) By midpregnancy, the alveoli and ducts have become large and have dilated lumens. (**4**) At parturition and during the time of lactation, the alveoli are greatly dilated and maximally active in production of milk components. (**5**) After weaning, the alveoli and ducts regress with apoptotic cell death.

becomes less prominent (Figures 22–26 and 22–27). The loose connective tissue within lobules is infiltrated by lymphocytes and plasma cells, the latter becoming more numerous late in pregnancy.

Late in pregnancy the glandular alveoli and ducts are dilated by an accumulation of **colostrum**, a fluid rich in proteins and containing leukocytes, that is produced under the influence of prolactin. Immunoglobulin A (IgA) antibodies are synthesized abundantly by plasma cells and transferred into colostrum, from which passive acquired immunity is conferred on the breast-fed newborn.

Following parturition, the alveoli of mammary glands start active milk production, or **lactation**, stimulated primarily by prolactin from the anterior pituitary (see Chapter 20). Epithelial cells of the alveoli enlarge and activate various processes involved in lactation:

- Large amounts of **protein** are synthesized, packaged into secretory vesicles, and undergo **merocrine secretion**

into the lumen (Figure 22–28). Human milk contains about 1 g of protein per deciliter, including aggregated caseins (44% of the total protein), as well as soluble β-lactoglobulin and α-lactalbumin, all of which are a source of amino acids by the infant. Less abundant proteins in milk include many factors that assist digestion, several such as lactoferrin with antimicrobial activity, and various mitogenic growth factors important for gut development in the newborn.

- **Lipid droplets** form initially from short-chain fatty acids synthesized in the epithelial cells and grow by accretion of longer fatty acids and cholesterol originating from the diet or fat stores. They eventually undergo **apocrine secretion**, during which the droplets become enveloped with a portion of the apical cell membrane (see Figure 22–28). Milk contains 4 or 5 g of total fat per deciliter.
- **Lactose**, the major carbohydrate and energy source in milk, is synthesized in the Golgi apparatus and secreted with lactalbumin. Human milk contains over 7 g of lactose per deciliter, more than the combined total of proteins and lipids. Lactose is also responsible for generating the osmotic gradient that draws water and Ca^{2+} into the alveolar lumen.

Throughout lactation, secretion of proteins, membrane-bound lipid droplets, lactose, iron, and calcium is ongoing, with the products accumulating as milk in the lumens of the duct system (Figure 22–27).

❯❯ MEDICAL APPLICATION

When a woman is breast-feeding, the nursing action of the child stimulates tactile receptors in the nipple, resulting in liberation of the posterior pituitary hormone **oxytocin**. This hormone causes contraction of the smooth muscle of the lactiferous sinuses and ducts, as well as the myoepithelial cells of alveoli, resulting in the **milk**-ejection reflex. Negative emotional stimuli, such as frustration, anxiety, or anger, can inhibit the liberation of oxytocin and thus prevent the reflex.

Postlactational Regression in the Mammary Glands

When breast-feeding is stopped (weaning), most alveoli that developed during pregnancy and lactation degenerate. Epithelial cells undergo apoptosis, autophagy, or sloughing (Figure 22–29), with dead cells and debris removed by macrophages. The duct system of the gland returns to its general appearance in the inactive state (Figure 22–25). After menopause, alveoli and ducts of the mammary glands are reduced further in size and there is loss of fibroblasts, collagen, and elastic fibers in the stroma.

FIGURE 22–26 Alveolar development in the breast during pregnancy.

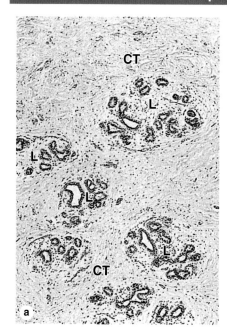

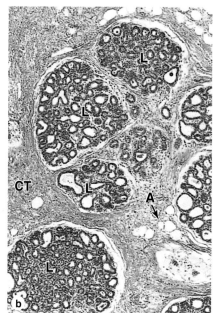

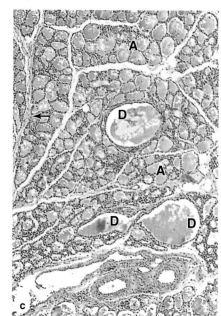

(a) The mammary glands of adult, nonpregnant women are inactive, with small ducts and few lobules (**L**) having secretory alveoli which are not well-developed. The structure with the large lumen in each lobule is part of the duct; the smaller structures are the small, undeveloped alveoli. The breasts are composed largely of connective tissue (**CT**), having considerable fat.

(b) The glands become active during pregnancy, with the duct system growing rapidly and the secretory units of each lobule becoming much larger and more extensively branched. In this micrograph adipocytes (**A**) are included, but these are only a small fraction of those present.

(c) During lactation, the lobules are greatly enlarged and the lumens of both the numerous glandular alveoli (**A**) and the excretory ducts (**D**) are filled with milk. The intralobular connective tissue is more sparse and difficult to see, except for small septa (arrows). (All X60, H&E)

❯❯ MEDICAL APPLICATION

Breast cancer is almost always derived from epithelial cells in the terminal lobules of the glands. The most common form is invasive **ductal carcinoma** in which neoplastic cells of intralobular ducts or small branches of lactiferous ducts invade the surrounding stroma, forming a fixed, palpable mass. Cell spreading (or metastasizing) from the carcinoma via the circulatory or lymphatic vessels to critical organs such as the lungs or brain is responsible for the mortality associated with breast cancer. If the treatment is **mastectomy**, axillary lymph nodes are usually also removed surgically and examined histologically for the presence of metastatic mammary carcinoma cells. Early detection (eg, through self-examination, mammography, ultrasound, and other techniques) and consequent early treatment have significantly reduced the mortality rate.

Bacterial infection of a mammary gland, or acute mastitis, may occur in the lactating or involuting breast, usually after obstruction by milk left within small components of the duct system.

FIGURE **22–27** Actively developing and lactating alveoli.

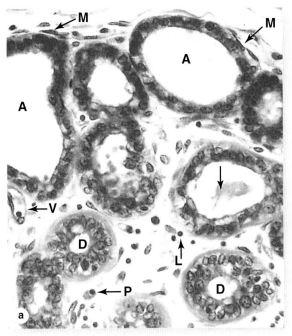

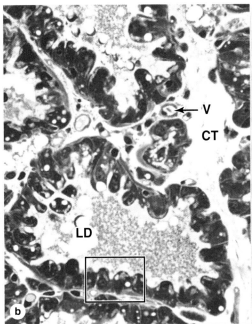

Glandular alveoli develop completely only during pregnancy and begin milk production near the end of pregnancy. **(a)** Alveoli **(A)** develop as spherical structures composed of cuboidal epithelial cells surrounded by the contractile processes of myoepithelial cells **(M)**. Development occurs at different rates throughout the breast. Late in pregnancy lymphocytes **(L)** leave venules **(V)**, accumulate in the intralobular connective tissue, and differentiate as plasma cells **(P)** secreting IgA. Intralobular ducts **(D)** are lined by epithelium

containing secretory cells, nonsecretory cells, and plasma cells; larger lumens may show milk (**arrow**). (X400; H&E)

(b) Secretory cells of the lactating gland are more columnar and contain variously sized lipid droplets, which are also visible in the milk **(LD)**. Connective tissue **(CT)** contains small blood vessels **(V)**. Secretory cells in the enclosed area are shown diagrammatically in Figure 22–28. (X400; PT)

FIGURE **22–28** Secretion in the mammary gland.

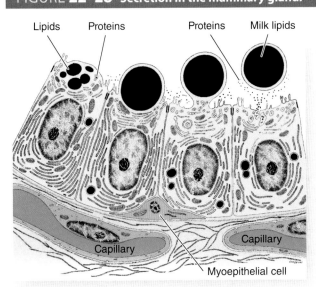

Alveolar cells of the lactating mammary gland are highly active in protein synthesis and lipid synthesis. Most proteins are packaged into secretory vesicles in the Golgi apparatus and secreted at the apical end of the cells by typical exocytosis or merocrine secretion. Lipids coalesce as free cytoplasmic droplets and eventually undergo apocrine secretion, in which they are extruded from the cell along with a portion of the apical cell membrane (and often a small amount of attached cytoplasm.) Both types of secretion are shown here in a sequence moving from left to right.

Ovaries, Follicles and Oocytes

- The female gonads, the paired **ovaries**, each have an outer **cortex** containing many hundreds of **ovarian follicles** and an inner **medulla** of dense connective tissue and large blood vessels.
- The ovary's cortex is covered by a cuboidal mesothelium, the **surface epithelium** (or germinal epithelium) that overlies a layer of connective tissue, the **tunica albuginea**.
- Before puberty all follicles are **primordial follicles**, formed in the developing fetal gonad, with each having one **primary oocyte** arrested in **meiotic prophase I** and a surrounding layer of squamous **follicular epithelial cells**.
- After puberty some primordial follicles develop each month as growing **primary follicles**, with an enlarging primary oocyte surrounded by larger epithelial cells now called **granulosa cells**.
- During follicular growth the **granulosa cells**, surrounded by a basement membrane, become stratified and actively engage in fluid secretion and steroid hormone metabolism.
- Between the oocyte and the granulosa cells a thin layer forms called the **zona pellucida**, which contains glycoproteins (**ZO proteins**) to which the sperm surface must bind to reach the oocyte at fertilization.
- **Antral** or **vesicular follicles** are larger and have developed fluid-filled spaces among their granulosa cells, but the growing oocyte is still in prophase I.
- While the primary follicle grows, mesenchymal cells immediately around it form the highly vascular layer, the **theca interna**, and a more fibrous **theca externa**, with smooth muscle cells.
- Endocrine cells of the **theca interna** secrete both **progesterone** and estrogen precursors, which are converted by **granulosa cells** into **estrogen**.
- Antral follicles continue developing as **mature, graafian follicles**, which have a large antrum filled with fluid, with the large primary oocyte enclosed by granulosa cells of the **cumulus oophorus**.
- Each month only one graafian follicle becomes a **dominant follicle** and undergoes **ovulation**; most other developing follicles arrest and degenerate with apoptosis in a process called **atresia**.

Ovulation and the Corpus Luteum

- **Ovulation** involves movement of a very large, dominant graafian follicle to the ovary surface to form a bulge, completion of meiosis I, and release of a **polar body** from the oocyte.
- Rupture of the follicle and ovarian coverings releases the **secondary oocyte**, arrested now in **metaphase II**, and a layer of attached granulosa cells that make up the **corona radiata**.
- Cells of the granulosa and thecal layers left in the ovary after ovulation are reorganized under the influence of **luteinizing hormone** (**LH**) to form the endocrine gland called the **corpus luteum**.
- The cells of the **corpus luteum** are **granulosa lutein cells**, producing estrogen and comprising 80% of the gland, and **theca lutein cells** producing progesterone.
- LH levels drop about 2 weeks after ovulation, causing the **corpus luteum** to lose activity, degenerate, and be removed by macrophages, leaving a temporary collagen-filled region called a **corpus albicans**.

Uterine Tubes or Oviducts

- The **ovulated secondary oocyte** is swept into the large open **infundibulum** of the **uterine tube**, or **oviduct**, and enters that organ's **ampulla** region where it degenerates if it is not fertilized.

- **Fertilization** involves **sperm capacitation** in the oviduct, **acrosomal activation** and sperm penetration of the corona radiata, and the following events at the oocyte surface:
  - Surface ligands on one sperm first attach to ZO proteins around the oocyte, allowing penetration.
  - **Cortical granules** of the oocyte undergo exocytosis, releasing proteases that convert the zona pellucida to the **vitelline barrier** which prevents **polyspermic fertilization**.
  - The oocyte completes meiosis II, producing the second **polar body** and the female pronucleus of the haploid **ovum**.
  - The **female pronucleus** and the **male pronucleus** from the sperm fuse, yielding a diploid cell, the **zygote**.
- The **oviduct mucosa** is **highly folded** in the ampulla region and lined by a **simple columnar epithelium** of ciliated cells and **secretory cells** producing nutritive mucus that bathes the sperm (and zygote).
- The thick oviduct **muscularis**, organized for peristaltic contractions of the tube, is mainly responsible for moving the developing embryo to the **uterus**.

Uterus

- The **uterine mucosa** or **endometrium** is lined by **simple columnar epithelium**, from which numerous large **uterine glands** extend into underlying connective tissue.
- This connective tissue is vascular, lacks fat, and consists of a highly cellular **basal layer** next to the uterine muscularis or **myometrium**, and a more superficial **functional layer**.
- Changing levels of estrogen and progesterone from the **ovarian follicles** and **corpus luteum** produce cyclic changes in the thickness, glandular activity, and vascular status of the endometrial **functional layer**.
- A **blastocyst** stage embryo arriving in the uterus attaches to the endometrial surface and implants itself into the **functional layer** when that layer's secretory activity and vascular supply are maximal.
- If no embryo implants, degeneration of the **corpus luteum** causes collapse of the progesterone-dependent **spiral arteries** that bring blood to the **functional layer**.
- Spiral artery degeneration produces ischemia in the **functional layer**, causing this layer to be sloughed off during **menses**, after which it regenerates from the **basal layer** under the influence of estrogen.
- The **placenta** consists mainly of **chorionic villi**, which form as highly vascularized projections of the extraembryonic **trophoblast** into the **vascular lacunae** of the endometrium.
- **Placental exchange** of nutrients, wastes, O_2, and CO_2 occurs between **fetal blood** in the **chorionic villi** and **maternal blood** bathing the villi in the **lacunae**.

Cervix, Vagina, and Mammary Glands

- The mucosa of the narrow, inferior end of the uterus, the **cervix**, does not change cyclically under hormone influence; at its **external os** columnar epithelium changes abruptly to stratified squamous.
- The mucosa of the **vagina** is lined by **stratified squamous epithelium**, surrounded by a muscularis.
- In the **mammary glands**, **alveolar secretory units** develop after puberty on a branching duct system with **lactiferous sinuses** converging at the **nipple**.
- **Milk secretion** (lactation), which begins in late pregnancy and continues until weaning, involves both protein exocytosis and apocrine secretion of lipid droplets.

The Female Reproductive System ASSESS YOUR KNOWLEDGE

1. Which stage of ovarian follicle development is characterized by an initial period of follicular fluid accumulation?
   a. Graafian follicle
   b. Mature follicle
   c. Primordial follicle
   d. Oocyte
   e. Secondary follicle

2. Which of the following is characteristic of granulosa lutein cells?
   a. Are a minor cell type in the corpus luteum
   b. Derive from the theca interna
   c. Contain abundant rough endoplasmic reticulum
   d. Are small and dark-staining
   e. Secrete progesterone

3. Which of the following hormones is primarily responsible for inducing ovulation?
   a. Relaxin
   b. Luteinizing hormone
   c. Progesterone
   d. Follicle-stimulating hormone
   e. Estrogen

4. Which feature is characteristic of the corpora albicans but not of atretic follicles?
   a. May contain degenerating granulosa cells floating in remnants of follicular fluid
   b. Resemble large collagenous scars
   c. Eventually removed by macrophages and replaced by stroma
   d. Are remnants of follicles that degenerate before maturation
   e. May contain degenerating oocytes

5. Endometrial glands are typically most fully developed and filled with product during which day(s) or phase of a woman's menstrual cycle?
   a. Menstrual phase
   b. Days 1-4
   c. The day ovulation occurs
   d. Proliferative phase
   e. Days 15-28

6. Which feature is characteristic of the endometrium's basal layer but not of its functional layer?
   a. Includes the uterine surface epithelium
   b. Includes connective tissue
   c. Contains cells that replace the surface epithelium after menstruation
   d. Relies solely on spiral arteries for its blood supply
   e. Undergoes cyclic thickening and shedding

7. Most lipid in milk is released from cells by which mechanism?
   a. Apocrine secretion
   b. Paracrine secretion
   c. Holocrine secretion
   d. Merocrine secretion
   e. Autocrine secretion

8. A 33-year-old woman with an average menstrual cycle of 28 days comes in for a routine Pap smear. It has been 35 days since the start of her last menstrual period, and a vaginal smear reveals clumps of basophilic cells. As her physician you suspect which of the following?
   a. She will begin menstruating in a few days.
   b. She will ovulate within a few days.
   c. Her serum progesterone levels will be found to be very low.
   d. There will be detectable levels of human choriogonadotrophin (hCG) in her serum and urine.
   e. She is undergoing early menopause.

9. A 17-year-old girl with a history of pelvic inflammatory disease presents at the emergency department with severe pain in her lower right side that came on fairly quickly. Upon questioning she replies that her last menstrual period was 6 weeks ago and that she has never missed a period before. The doctor suspects she has an ectopic pregnancy and this is quickly confirmed by ultrasound testing. The surgeon removes her right uterine tube which is inflamed, scarified, and contains the implanted embryonic tissue in the region where fertilization normally occurs. Where is this?
   a. The uterine part of the oviduct
   b. The ampulla region with highly folded mucosa
   c. The only oviduct region attached to the mesosalpinx
   d. The infundibulum region with fimbriae
   e. The isthmus region

10. A 42-year-old woman visits her physician complaining of recurrent vaginal yeast infections. The doctor explains the likelihood that the woman's vaginal lining is temporarily out of proper acid-base balance, leading to the increased susceptibility to yeast infections. The normally low pH in the vagina is maintained by which of the following?
   a. A proton pump in the epithelial cells similar to that in osteoclasts and parietal cells
   b. Secretions derived from intracellular carbonic acid
   c. Secretion of lactic acid by cells of the stratified squamous epithelium
   d. Bacterial metabolism of glycogen to produce an organic acid
   e. Synthesis and accumulation of acid hydrolases in the epithelium

23 The Eye & Ear: Special Sense Organs

Information about the external world is conveyed to the central nervous system (CNS) from **sensory receptors**. Chemoreceptive cells for the senses of taste and smell were discussed with the digestive and respiratory system (see Chapters 15 and 17, respectively) and the various mechanoreceptors that mediate the sense of touch were presented with the skin (see Chapter 18). This chapter describes the eye, both its photoreceptors and auxiliary structures, and the ear which mediates the senses of equilibrium and hearing via mechanoreceptors in the vestibulocochlear apparatus.

❯ EYES: THE PHOTORECEPTOR SYSTEM

Eyes (Figure 23–1) are highly developed photosensitive organs for analyzing the form, intensity, and color of light reflected from objects and providing the sense of sight. Protected within the orbits of the skull which also contain adipose cushions, each eyeball consists externally of a tough, fibrous globe that maintains its overall shape. Internally the eye contains transparent tissues that refract light to focus the image, a layer of photosensitive cells, and a system of neurons that collect, process, and transmit visual information to the brain.

Each eye is composed of three concentric tunics or layers (Table 23–1):

- A tough external **fibrous layer** consisting of the **sclera** and the transparent **cornea**;
- A middle **vascular layer** that includes the **choroid, ciliary body**, and **iris**; and
- An inner sensory layer, the **retina**, which communicates with the cerebrum through the posterior **optic nerve** (Figure 23–1).

Not part of these layers, the **lens** is a perfectly transparent biconvex structure held in place by a circular system of **zonular fibers** that attach it to the **ciliary body** and by close apposition to the posterior vitreous body (Figure 23–1). Partly covering the anterior surface of the lens is an opaque pigmented extension of the middle layer called the **iris**, which surrounds a central opening, the **pupil** (Figure 23–1).

Located in the anterior portion of the eye, the iris and lens are bathed in clear **aqueous humor** that fills both the **anterior chamber** between the cornea and iris and the **posterior chamber** between the iris and lens (Figure 23–1). Aqueous humor flows through the pupil that connects these two chambers.

The posterior **vitreous chamber**, surrounded by the retina, lies behind the lens and its zonular fibers and contains a large gelatinous mass of transparent connective tissue called the **vitreous body**.

Important aspects of embryonic eye formation are shown in Figure 23–2 and include the following:

- In the 4-week embryo epithelial **optic vesicles** bulge bilaterally from the forebrain, then elongate as the optic stalks bearing **optic cups** (Figure 23–2a).
- Inductive interactions between the optic cups and the overlying surface ectoderm cause the latter to invaginate and eventually detach as the initially hollow **lens vesicles** (Figure 23–2b).
- The optic stalk develops as the **optic nerve** and in an inferior groove called the **choroid fissure** encloses the **hyaloid vessels** that supply blood for the developing lens and optic cup (Figure 23–2c).
- In the ensuing weeks, head mesenchyme differentiates to form most of the tissue in the eye's two outer layers

FIGURE **23–1** Internal anatomy of the eye.

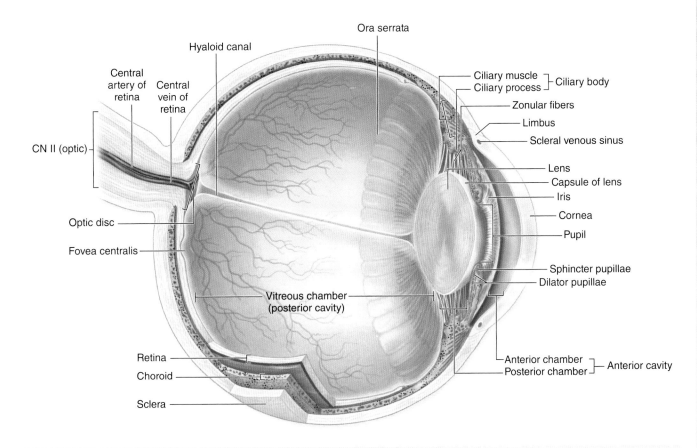

The sagittal section of an eye shows the interrelationships among the major ocular structures, the three major layers or tunics of the wall, important regions within those layers, and the refractive elements (cornea, lens, and vitreous).

and the vitreous. Ectoderm of the optic cup differentiates as the retina and surface ectoderm forms the corneal epithelium (Figure 23–2d). When the lens is fully formed, the distal hyaloid artery and vein disappear, leaving only the blood supply to the retina.

Fibrous Layer

This layer includes two major regions, the posterior sclera and anterior cornea, joined at the limbus.

Sclera

The fibrous, external layer of the eyeball protects the more delicate internal structures and provides sites for muscle insertion (Table 23–1). The white posterior five-sixths of this layer is the **sclera** (Figure 23–1), which encloses a portion of the eyeball about 22 mm in diameter in adults. The sclera averages 0.5 mm in thickness and consists mainly of dense connective tissue, with flat bundles of type I collagen parallel to the organ surface but intersecting in various directions; microvasculature is present near the outer surface.

Tendons of the **extraocular muscles** which move the eyes insert into the anterior region of the sclera. Posteriorly the sclera thickens to approximately 1 mm and joins with the epineurium covering the optic nerve. Where it surrounds the choroid, the sclera includes an inner **suprachoroid lamina**, with less collagen, more fibroblasts, elastic fibers, and melanocytes.

Cornea

In contrast to the sclera, the anterior one-sixth of the eye—the **cornea**—is transparent and completely avascular (Figure 23–1). A section of the cornea shows five distinct layers:

- An external **stratified squamous epithelium**;
- An **anterior limiting membrane (Bowman membrane)**, which is the basement membrane of the external stratified epithelium;
- The thick **stroma**;
- A **posterior limiting membrane (Descemet's membrane)**, which is the basement membrane of the endothelium; and
- An inner simple squamous **endothelium**.

TABLE **23–1**	Tunics of the eye.

	Structures	Components	Function
Fibrous Tunic (External Layer)			
	Sclera	Dense irregular connective tissue	Supports eye shape
			Protects delicate internal structures
			Extrinsic eye muscle attachment site
	Cornea	Two layers of epithelium with organized connective tissue in between	Protects anterior surface of the eye
			Refracts (bends) incoming light
Vascular Tunic or Uvea (Middle Layer)			
	Choroid	Areolar connective tissue; highly vascularized	Supplies nourishment to retina
			Pigment absorbs extraneous light
	Ciliary body	Ciliary smooth muscle and ciliary processes; covered with a secretory epithelium	Holds suspensory ligaments that attach to the lens and change lens shape for far and near vision
			Epithelium secretes aqueous humor
	Iris	Two layers of smooth muscle (sphincter pupillae and dilator pupillae) and connective tissue, with a central pupil	Controls pupil diameter and thus the amount of light entering the eye
Retina (Internal Layer)			
	Pigmented layer	Pigmented epithelial cells	Absorbs extraneous light
			Provides vitamin A for photoreceptor cells
	Neural layer	Photoreceptors, bipolar neurons, ganglion cells, and supporting Müller cells	Detects incoming light rays; light rays are converted to nerve signals and transmitted to the brain

The stratified surface epithelium is nonkeratinized, five or six cell layers thick, and comprises about 10% of the corneal thickness (Figure 23–3). The basal cells have a high proliferative capacity important for renewal and repair of the corneal surface and emerge from stem cells in the **corneoscleral limbus** that encircles the cornea. The flattened surface cells have microvilli protruding into a protective tear film of lipid, glycoprotein, and water. As another protective adaptation, the corneal epithelium also has one of the richest sensory nerve supplies of any tissue.

The basement membrane of this epithelium, often called **Bowman membrane**, is very thick (8-10 μm) and contributes to the stability and strength of the cornea, helping to protect against infection of the underlying stroma.

The **stroma**, or substantia propria, makes up 90% of the cornea's thickness and consists of approximately 60 layers of parallel collagen bundles aligned at approximately right angles to each other and extending almost the full diameter of the cornea. The uniform orthogonal array of collagen fibrils contributes to the transparency of this avascular tissue. Between the collagen lamellae are cytoplasmic extensions of flattened fibroblast-like cells called **keratocytes** (Figure 23–3). The ground substance around these cells contains proteoglycans such as lumican, with keratan sulfate and chondroitin sulfate, which help maintain the precise organization and spacing of the collagen fibrils.

>> **MEDICAL APPLICATION**

The shape or curvature of the cornea can be changed surgically to improve certain visual abnormalities involving the ability to focus. In the common ophthalmologic procedure, laser-assisted in situ keratomileusis (LASIK) surgery, the corneal epithelium is displaced as a flap and the stroma reshaped by an excimer laser which vaporizes collagen and keratocytes in a highly controlled manner with no damage to adjacent cells or ECM. After reshaping the stroma, the epithelial flap is repositioned and a relatively rapid regenerative response reestablishes normal corneal physiology. LASIK surgery is used

FIGURE **23–2** **Development of eye.**

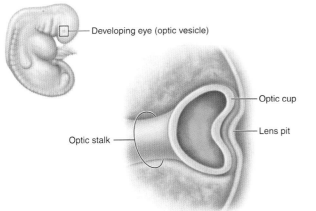

(a) **Early week 4:** Optic vesicle forms a two-layered optic cup; overlying ectoderm forms a lens pit.

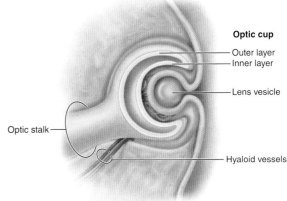

(b) **Late week 4:** Optic cup deepens and forms inner and outer layers; lens pit forms lens vesicle.

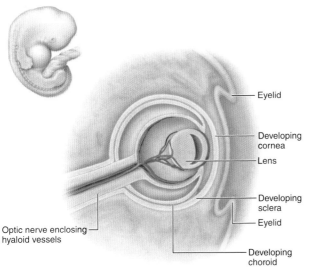

(c) **Week 6:** Lens becomes an internal structure; corneas, sclera, and choroid start to form.

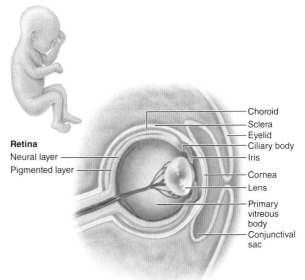

(d) **Week 20:** Three tunics of the eye have formed.

Eyes begin to form early in development as the two **optic vesicles** bulge bilaterally from the forebrain (prosencephalon). These grow, remaining connected to the developing brain by the **optic stalks**, and approach the surface ectoderm. At this point each vesicle folds in on itself to form the inner and outer layers of the **optic cup** and inducing surface ectoderm to invaginate into the cup as the **lens vesicle**, which soon detaches from the surface and lies in the opening of the optic cup.

Blood vessels, called the **hyaloid vessels**, grow along the optic stalk, enter the optic cup, and grow toward the developing lens.

Head mesenchyme associates with the developing optic cup as it forms the two major layers of the retina. The mesenchymal cells differentiate around the pigmented layer of the developing retina as the iris, ciliary body and choroid of the vascular layer, and as the more external fibrous layer. The hyaloid vessels regress, leaving a space called the **hyaloid canal**, in the vitreous body. Folds of skin develop features of the eyelids and conjunctiva, the latter developing in continuity with the surface epithelium of the cornea and sclera.

to correct myopia (near-sightedness), hyperopia (far-sightedness), or astigmatism (irregular curvature of the cornea). Corneal grafts (transplants) between unrelated individuals can usually be accomplished successfully without immune rejection due in part to this tissue's lack of both a vascular supply and lymphatic drainage and to local immune tolerance produced by ocular antigen-presenting cells and immunomodulatory factors in aqueous humor.

FIGURE 23–3 Cornea.

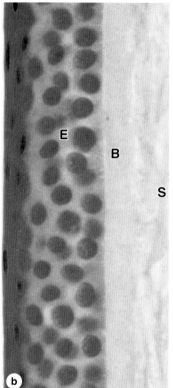

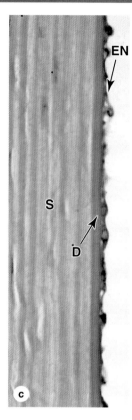

The anterior structure of the eye, the cornea has five layers.
(a) The external stratified squamous epithelium (**E**) is nonkeratinized, five or six cells thick, and densely supplied with sensory-free nerve endings that trigger the blinking reflex. The stroma (**S**) comprises approximately 90% of the cornea's thickness, consisting of some 60 layers of long type I collagen fibers arranged in a precise orthogonal array and alternating with flattened cells called **keratocytes**. The stroma is lined internally by endothelium (**EN**). (X100; H&E)

(b) The corneal epithelium (**E**) rests firmly on the thick homogeneous Bowman's membrane (**B**). The stroma (**S**) is completely avascular, and nutrients reach the keratocytes and epithelial cells by diffusion from the surrounding limbus and aqueous humor behind the cornea. (X400; H&E)

(c) The posterior surface of the cornea is covered by simple squamous epithelium (**EN**) that rests on another thick, strong layer called Descemet's membrane (**D**) adjacent to the stroma (**S**). (X400; H&E)

The posterior surface of the stroma is bounded by another thick basement membrane, called **Descemet's membrane**, which supports the internal simple squamous corneal endothelium (Figure 23–3).

This endothelium maintains Descemet's membrane and includes the most metabolically active cells of the cornea. Na⁺/K⁺ ATPase pumps in the basolateral membranes of these cells are largely responsible for regulating the proper hydration state of the corneal stroma to provide maximal transparency and optimal light refraction.

Limbus

Encircling the cornea is the **limbus**, a transitional area where the transparent cornea merges with the opaque sclera (Figures 23–1 and 23–4). Here Bowman's membrane ends and the surface epithelium becomes more stratified as the **conjunctiva** that covers the anterior part of the sclera (and lines the eyelids). As mentioned previously, epithelial stem cells located at the limbus surface give rise to rapidly dividing progenitor cells that move centripetally into the corneal epithelium. The stroma becomes vascular and less well-organized at the limbus, as the collagen bundles merge with those of the sclera.

Also at the limbus Descemet's membrane and its simple endothelium are replaced with a system of irregular endothelium-lined channels called the **trabecular meshwork** (Figure 23–5). These penetrate the stroma at the corneoscleral junction and allow slow, continuous drainage of aqueous humor from the anterior chamber. This fluid moves from these channels into the adjacent larger space of the **scleral venous sinus**, or **canal of Schlemm** (Figures 23–1, 23–4, and 23–5), which encircles the eye. From this sinus aqueous humor drains into small blood vessels (veins) of the sclera.

FIGURE 23–4   Corneoscleral junction (limbus) and ciliary body.

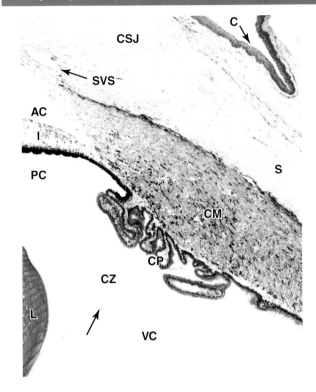

At the circumference of the cornea is the limbus or corneo-scleral junction (**CSJ**), where the transparent corneal stroma merges with the opaque, vascular sclera (**S**). The epithelium of the limbus is slightly thicker than the corneal epithelium, containing stem cells for the latter, and is continuous with the conjunctive (**C**) covering the anterior sclera and lining the eyelids. The stroma of the limbus contains the scleral venous sinus (**SVS**), or canal of Schlemm, which receives aqueous humor from an adjacent trabecular meshwork at the surface of the anterior chamber (**AC**).

Internal to the limbus, the middle layer of the eye consists of the ciliary body and its anterior extension, the iris (**I**). The thick ring of the ciliary body includes loose connective tissue containing melanocytes, smooth ciliary muscle (**CM**), numerous extensions covered by epithelium called the ciliary processes (**CP**), and the ciliary zonule (**CZ**), a system of fibrillin-rich fibers that attach to the capsule of the lens (**L**) in the center of the ciliary body. Pieces of one zonular fiber can be seen (**arrow**). Projecting into the posterior chamber (**PC**), the ciliary processes produce aqueous humor that then flows into the anterior chamber through the pupil. Behind the ciliary zonule and lens, a thin, transparent membrane (not shown) surrounds the vitreous body and separates the posterior chamber from the vitreous chamber (**VC**). (X12.5; H&E)

Vascular Layer

The eye's more vascular middle layer, known as the **uvea**, consists of three parts, from posterior to anterior: the **choroid**, the **ciliary body**, and the **iris** (Table 23–1).

Choroid

Located in the posterior two-thirds of the eye, the **choroid** consists of loose, well-vascularized connective tissue and contains numerous melanocytes (Figure 23–6). These form a characteristic black layer in the choroid and prevent light from entering the eye except through the pupil. Two layers make up the choroid (Figure 23–6):

- The inner **choroidocapillary lamina** has a rich microvasculature important for nutrition of the outer retinal layers.
- **Bruch membrane**, a thin extracellular sheet, is composed of collagen and elastic fibers surrounding the adjacent microvasculature and basal lamina of the retina's pigmented layer.

Ciliary Body

The **ciliary body**, the anterior expansion of the uvea that encircles the lens, lies posterior to the limbus (Figures 23–1 and 23–4). Like the choroid, most of the ciliary body rests on the sclera. Important structures associated with the ciliary body include the following:

- **Ciliary muscle** makes up most of the ciliary body's stroma and consists of three groups of smooth muscle fibers. Contraction of these muscles affects the shape of the lens and is important in visual accommodation (see Lens).
- **Ciliary processes** are a radially arranged series of about 75 ridges extending from the inner highly vascular region of the ciliary body. These provide a large surface area covered by a double layer of low columnar epithelial cells, the ciliary epithelium (Figure 23–7). The epithelial cells directly covering the stroma contain much melanin and correspond to the anterior projection of the pigmented retina epithelium. The surface layer of cells lacks melanin and is contiguous with the sensory layer of the retina.
- Cells of this dual epithelium have extensive basolateral folds with Na^+/K^+-ATPase activity and are specialized for secretion of **aqueous humor**. Fluid from the stromal microvasculature moves across this epithelium as aqueous humor, with an inorganic ion composition similar to that of plasma but almost no protein. As shown in Figure 23–8, aqueous humor is secreted by ciliary processes into the posterior chamber, flows through the pupil into the anterior chamber, and drains at the angle formed by the cornea and the iris into the channels of the trabecular meshwork and the scleral venous sinus, from which it enters venules of the sclera.
- The **ciliary zonule** is a system of many radially oriented fibers composed largely of fibrillin-1 and 2 produced by the nonpigmented epithelial cells on the ciliary processes. The fibers extend from grooves between the ciliary processes and attach to the surface of the lens (Figure 23–9), holding that structure in place.

FIGURE **23–5** Trabecular meshwork and scleral venous sinus.

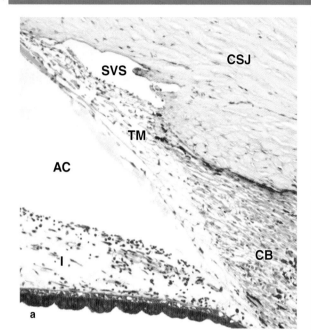

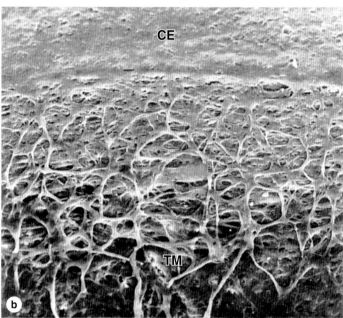

(a) At the corneoscleral junction (**CSJ**), or limbus, encircling the cornea, the posterior endothelium and its underlying Descemet's membrane are replaced by a meshwork of irregular channels lined by endothelium and supported by trabeculae of connective tissue. At the iridocorneal angle between limbus and iris (**I**), aqueous humor moves from the anterior chamber (**AC**) into channels of this

trabecular meshwork (**TM**) and is pumped by endothelial cells into the adjacent scleral venous sinus (**SVS**). (X50; H&E)

(b) Scanning electron microscopy (SEM) shows the transition from corneal endothelium (**CE**) surface to the channels of the trabecular meshwork (**TM**). (X300)

>> MEDICAL APPLICATION

Aqueous humor is produced continuously. If its drainage from the anterior chamber is impeded, typically by obstruction of the trabecular meshwork or scleral venous sinus, intraocular pressure can increase, causing the condition called **glaucoma**. Untreated glaucoma can cause pressing of the vitreous body against the retina, affecting visual function and possibly leading to neuropathy in that tissue.

When the iridocorneal angle is more narrow than usual, the thickening of the peripheral iris that occurs with dilation of the pupil can occlude the angle and obstruct drainage of aqueous humor at the trabecular meshwork. This can result in the rapid development of intraocular hypertension known as **angle closure glaucoma**, **acute glaucoma**, or **closed (narrow) angle glaucoma**. This condition usually affects both eyes and causes blurred vision, eye pain, and headache. Treatment of this type of glaucoma usually includes some form of surgical intervention.

Iris

The **iris** is the most anterior extension of the middle uveal layer which covers part of the lens, leaving a round central **pupil** (Figure 23–1). The anterior surface of the iris, exposed to aqueous humor in the anterior chamber, consists of a dense layer of fibroblasts and melanocytes with interdigitating processes and is unusual for its lack of an epithelial covering (Figure 23–10a,b). Deeper in the iris, the stroma consists of loose connective tissue with melanocytes and sparse microvasculature.

The posterior surface of the iris has a two-layered epithelium continuous with that covering the ciliary processes, but very heavily filled with melanin. The highly pigmented posterior epithelium of the iris blocks all light from entering the eye except that passing through the pupil. **Myoepithelial cells** form a partially pigmented epithelial layer and extend contractile processes radially as the very thin **dilator pupillae muscle** (Figure 23–10). Smooth muscle fibers form a circular bundle near the pupil as the **sphincter pupillae muscle**. The dilator and sphincter muscles of the iris have sympathetic and parasympathetic innervation, respectively, for enlarging and constricting the pupil.

FIGURE **23–6** Sclera, choroid, and retina.

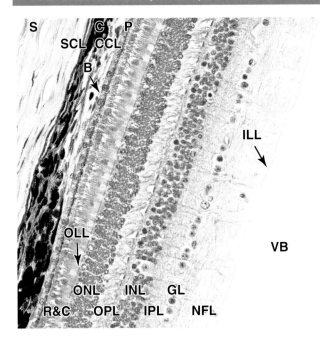

The lateral wall of an eye includes dense connective tissue of the sclera (**S**) and the loose, vascular connective tissue of the choroid (**C**). Melanocytes are prominent in the choroid, especially in its outer region, the suprachoroidal lamina (**SCL**). The choroid's inner region, the choroidocapillary lamina (**CCL**), has a rich microvasculature that helps provide O_2 and nutrients to the adjacent retina. Between the choroid and the retina is a thin layer of extracellular material known as **Bruch layer** (**B**).

The outer layer of the retina is the pigmented layer (**P**) of cuboidal epithelium containing melanin. Adjacent to this are the packed photoreceptor components of the rods and cones (**R&C**), whose cell bodies make up the outer nuclear layer (**ONL**). Junctional complexes between these cells and glia are aligned and can be seen as a thin line called the **outer limiting layer** (**OLL**). Axons of the rods and cones extend into the outer plexiform layer (**OPL**) forming synapses there with dendrites of the neurons in the inner nuclear layer (**INL**). These neurons send axons into the inner plexiform layer (**IPL**), where they synapse with dendrites of cells in the ganglionic layer (**GL**). Axons from these cells fill most of the nerve fiber layer (**NFL**) which is separated by the inner limiting layer (**ILL**) from the gelatin-like connective tissue of the vitreous body (**VB**). (X200; H&E)

Melanocytes of the iris stroma provide the color of one's eyes. In individuals with very few lightly pigmented cells in the stroma, light with a blue color is reflected back from the black pigmented epithelium on the posterior iris surface. As the number of melanocytes and density of melanin increase in the stroma, the iris color changes through various shades of green, gray, and brown. Individuals with albinism have almost no pigment and the pink color of their irises is due to the reflection of incident light from the blood vessels of the stroma.

FIGURE **23–7** Epithelium of ciliary processes.

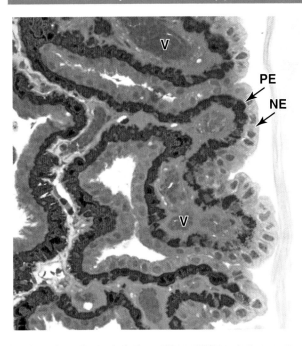

The surface epithelium of ciliary processes is a double layer of pigmented (**PE**) and nonpigmented epithelial (**NE**) low columnar or cuboidal cells. The two layers are derived developmentally from the folded rim of the embryonic optic cup, so that the exposed surface of the nonpigmented layer is actually the basal surface of the cells. No true basal lamina is present, but instead these cells produce the components that give rise to the fibers of the ciliary zonule in the embryo. Beneath the double epithelium is a core of connective tissue with many small blood vessels (**V**). Fluid from these vessels is pumped by the epithelial cells out of the ciliary processes as aqueous humor. (X200; PT)

Lens

The **lens** is a transparent biconvex structure suspended immediately behind the iris, which focuses light on the retina (Figure 23–1). Derived from an invagination of the embryonic surface ectoderm (see Figure 23–2), the lens is a unique avascular tissue and is highly elastic, a property that normally decreases with age. The lens has three principal components:

- A thick (10-20 µm), homogeneous **lens capsule** composed of proteoglycans and type IV collagen surrounds the lens (Figure 23–11) and provides the place of attachment for the fibers of the ciliary zonule (Figure 23–10). This layer originates as the basement membrane of the embryonic lens vesicle.
- A subcapsular **lens epithelium** consists of a single layer of cuboidal cells present only on the anterior surface of the lens (Figure 23–11). The epithelial cells attach basally to the surrounding lens capsule and their apical surfaces bind to the internal lens fibers. At the posterior edge of

FIGURE **23–8** **Production and removal of aqueous humor.**

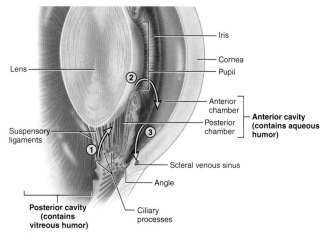

1. Aqueous humor is secreted by the ciliary processes into the posterior chamber.
2. Aqueous humor moves from the posterior chamber, through the pupil, to the anterior chamber.
3. Excess aqueous humor is resorbed via the scleral venous sinus.

Aqueous humor is a continuously flowing liquid that carries metabolites to and from cells and helps maintain an optimal microenvironment within the **anterior cavity** of the eye. As shown here, it is secreted from **ciliary processes** into the posterior chamber of the anterior cavity, flows into the anterior chamber through the **pupil**, and drains into the **scleral venous sinus (canal of Schlemm)**.

FIGURE **23–9** **Ciliary zonule fibers.**

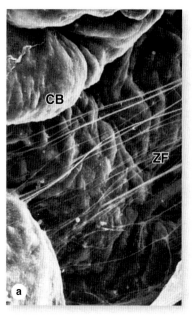
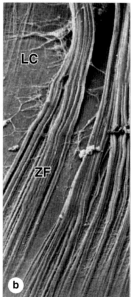

The structure of the ciliary zonule is best studied by SEM. **(a)** The surface of the ciliary body (**CB**) has projecting ciliary processes, between which emerge delicate zonular fibers (**ZF**). An array of these fibers constitutes the zonule that anchors the lens in the center of the ciliary body. (X400)

(b) Zonular fibers (**ZF**) attach tangentially to the fibrous ECM of the lens capsule (**LC**). (X500)

this epithelium, near the equator of the lens, the epithelial cells divide to provide new cells that differentiate as lens fibers. This process allows for growth of the lens and continues at a slow, decreasing rate near the equator of the lens throughout adult life.

■ **Lens fibers** are highly elongated, terminally differentiated cells that appear as thin, flattened structures (Figure 23–11). Developing from cells in the lens epithelium, lens fibers typically become 7-10-mm long, with cross-section dimensions of only 2 by 8 μm. The cytoplasm becomes filled with a group of proteins called **crystallins**, and the organelles and nuclei undergo autophagy. Lens fibers are packed tightly together and form a perfectly transparent tissue highly specialized for light refraction.

The lens is held in place by fibers of the **ciliary zonule**, which extend from the lens capsule to the ciliary body (Figures 23–1 and 23–9). Together with the ciliary muscles, this structure allows the process of visual **accommodation**, which permits focusing on near and far objects by changing the curvature of the lens (Figure 23–12). When the eye is at rest or gazing at distant objects, ciliary muscles relax and the

resulting shape of the ciliary body puts tension on the zonule fibers, which pulls the lens into a flatter shape. To focus on a close object the ciliary muscles contract, causing forward displacement of the ciliary body, which relieves some of the tension on the zonule and allows the lens to return to a more rounded shape and keep the object in focus. In the fourth decade of life **presbyopia** (Gr. *presbyter*, elder + L. *opticus*, relating to eyes) normally causes the lenses to lose elasticity and their ability to undergo accommodation.

>> **MEDICAL APPLICATION**

Presbyopia is corrected by wearing glasses with convex lenses (reading glasses). In older individuals, denaturation of crystallins commonly begins to occur in lens fibers, making them less transparent. When areas of the lens become opaque or cloudy and vision is impaired, the condition is termed a **cataract**. Causes of cataract include excessive exposure to ultraviolet light or other radiation, trauma, and as secondary effects in diseases such as diabetes mellitus and hypertension.

FIGURE **23–10** Iris.

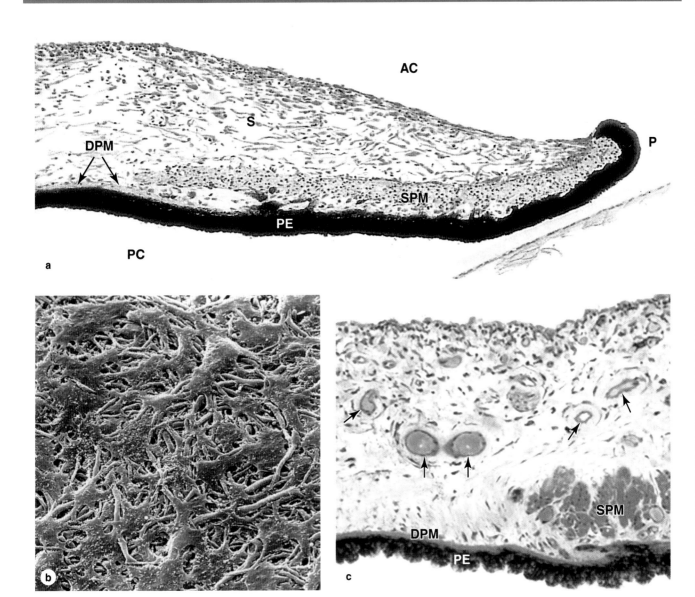

The iris regulates the amount of light to which the retina is exposed. **(a)** The low-power micrograph shows a section of the central iris, near the pupil (**P**). The anterior surface, exposed to aqueous humor in the anterior chamber (**AC**), has no epithelium and consists only of a matted layer of interdigitating fibroblasts and melanocytes. Cells of the external pigmented epithelium (**PE**) are very rich in melanin granules to protect the eye's interior from an excess of light. Cells of the other layer are myoepithelial, less heavily pigmented, and comprise the dilator pupillae muscle

(**DPM**) that extends along most of the iris. Near the pupil, fascicles of smooth muscle make up the sphincter pupillae muscle (**SPM**). The underlying stroma (**S**) contains many melanocytes with varying amounts of melanin. (X140; H&E)

(b) SEM reveals the nonepithelial anterior surface of the iris. X900. **(c)** The deep stroma also is richly vascularized (**arrows**). The myoepithelial dilator pupillae muscle (**DPM**) is more easily seen here, in relation to the sphincter pupillae muscle (**SPM**) and posterior pigmented epithelium (**PE**). (X100; PT)

In modern cataract surgery the lens is removed by aspiration of the lens substance while it is emulsified by a vibrating probe. The posterior side of the lens capsule and its inserted zonular fibers are left in place at the posterior

chamber. The concave posterior capsule, or capsular bag, is then used as the site for implantation of an acrylic intraocular lens (IOL) prosthesis. Research is under way to develop IOLs capable of natural accommodation.

FIGURE **23–11** Lens.

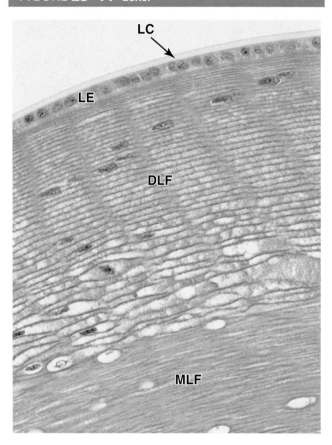

The lens is a transparent, elastic tissue that focuses light on the retina. Surrounding the entire lens is a thick, homogenous external lamina called the **lens capsule** (**LC**). The anterior surface of the lens, beneath the capsule, is covered by a simple columnar lens epithelium (**LE**). Because of its origin as an embryonic vesicle pinching off of surface ectoderm, the basal ends of the lens epithelial cells rest on the capsule and the apical regions are directed into the lens interior.

At the equator of the lens, near the ciliary zonule, the epithelial cells proliferate and give rise to cells that align parallel to the epithelium and become the lens fibers. Differentiating lens fibers (**DLF**) still have their nuclei but are greatly elongating and filling their cytoplasm with proteins called **crystallins**. The mature lens fibers (**MLF**) have lost their nuclei and become densely packed to produce a unique transparent structure. The lens is difficult to process histologically and sections usually have cracks or blebs among the lens fibers. (X200; H&E)

Vitreous Body

The **vitreous body** occupies the large vitreous chamber behind the lens (Figure 23–1). It consists of transparent, gel-like connective tissue that is 99% water (vitreous humor), with collagen fibrils and hyaluronate, contained within an external lamina called the vitreous membrane. The only cells in the vitreous body are a small mesenchymal population near the membrane called **hyalocytes**, which synthesize the hyaluronate and collagen, and a few macrophages.

FIGURE **23–12** Accommodation of the lens.

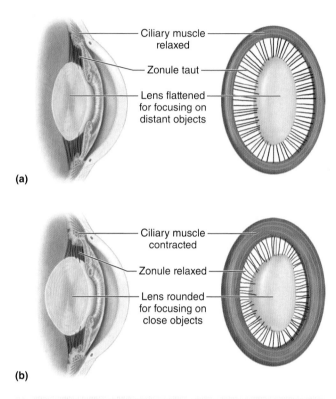

Continuous changes in the shape of the lens keep images focused on the retina. **(a)** The lens flattens for distant vision when the ciliary muscles are relaxed and the shape of the ciliary body holds the ciliary zonule taut. **(b)** To see closer objects, the ciliary muscle fibers contract, changing the shape of the ciliary body, relaxing tension on the ciliary zonule, and allowing the lens to assume the more rounded shape.

Retina

The **retina**, the innermost tunic of the eye, develops with two fundamental sublayers from the inner and outer layers of embryonic optic cup (Figure 23–2 and Table 23–1):

■ The outer **pigmented layer** is a simple cuboidal epithelium attached to Bruch's membrane and the choroido-capillary lamina of the choroid (Figure 23–6). This heavily pigmented layer forms the other part of the dual epithelium covering the ciliary body and posterior iris.

■ The inner retinal region, the **neural layer**, is thick and stratified with various neurons and photoreceptors. Although its neural structure and visual function extend anterior only as far as the **ora serrata** (Figure 23–1), this layer continues as part of the dual cuboidal epithelium that covers the surface of the ciliary body and posterior iris.

›› MEDICAL APPLICATION

The pigmented epithelium and the photoreceptor layer of the retina, derived from the two layers of the optic cup, are not firmly joined to each other. Head trauma or other conditions can cause the two layers to separate with an intervening space. In such regions of **detached retina**, the photoreceptor cells no longer have access to metabolic support from the pigmented layer and choroid and will eventually die. Prompt repositioning of the retina and reattaching it with laser surgery is an effective treatment.

Retina Pigmented Epithelium

The pigmented epithelial layer consists of cuboidal or low columnar cells with basal nuclei and surrounds the neural layer of the retina. The cells have well-developed junctional complexes, gap junctions, and numerous invaginations of the basal membranes associated with mitochondria. The apical ends of the cells extend processes and sheath-like projections that surround the tips of the photoreceptors. Melanin granules are numerous in these extensions and in the apical cytoplasm (Figure 23–13). This cellular region also contains numerous phagocytic vacuoles and secondary lysosomes, peroxisomes, and abundant smooth ER (SER) specialized for **retinal** (vitamin A) isomerization. The diverse functions of the retinal pigmented epithelium include the following:

■ The pigmented layer **absorbs scattered light** that passes through the neural layer, supplementing the choroid in this regard.

■ With many tight junctions, cells of the pigmented epithelium form an important part of the protective **blood-retina barrier** isolating retina photoreceptors from the highly vascular choroid and regulating **ion transport** between these compartments.

■ The cells play key roles in the visual cycle of **retinal regeneration**, having enzyme systems that isomerize all-*trans*-retinal released from photoreceptors and produce 11-*cis*-retinal that is then transferred back to the photoreceptors.

■ **Phagocytosis** of shed components from the adjacent photoreceptors and degradation of this material occurs in these epithelial cells.

■ Cells of pigmented epithelium **remove free radicals** by various protective antioxidant activities and **support the neural retina** by secretion of ATP, various polypeptide growth factors, and immunomodulatory factors.

Neural Retina

True to its embryonic origin, the neural retina functions as an outpost of the CNS with glia and several interconnected neuronal subtypes in well-organized strata. Nine distinct layers comprise the neural retina, described here with their functional significance.

FIGURE **23–13** Pigmented epithelium of retina.

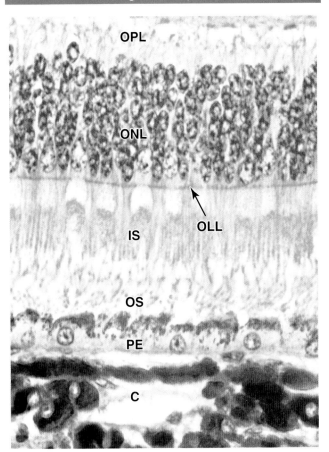

The two distinct layers of the retina are the pigmented epithelium and the photosensitive neural layer, which are derived from the outer and inner layers of the optic cup, respectively. Shown here is the interface between the two layers. The pigmented epithelium (**PE**) is of simple cuboidal cells resting on Bruch's membrane inside the choroid (**C**). Rod cells and cone cells are neurons with their nuclei collected in the outer nuclear layer (**ONL**) and with axons of one end forming synapses in an area called the outer plexiform layer (**OPL**) and modified dendrites at the other end serving as photosensitive structures. These structures have mitochondria-rich inner segments (**IS**) and photosensitive outer segments (**OS**) with stacks of folded membranes where the visual pigments are located.

The inner segments of the rod and cone cells are attached to elongated glial cells called **Müller cells**, which are modified astrocytes of the retina. The junctional complexes of these attachments can be seen in light micrographs as the outer limiting layer (**OLL**). (X500; H&E)

Three major layers contain the nuclei of the interconnected neurons (Figures 23–14b and 23–15):

■ Near the pigmented epithelium, the **outer nuclear layer (ONL)** contains cell bodies of photoreceptors (the rod and cone cells). These cells, like the pigmented epithelial cells, receive O$_2$ and nutrients by diffusion from the choroidocapillary lamina of the choroid.

FIGURE **23–14** **General organization and specialized areas of the retina.**

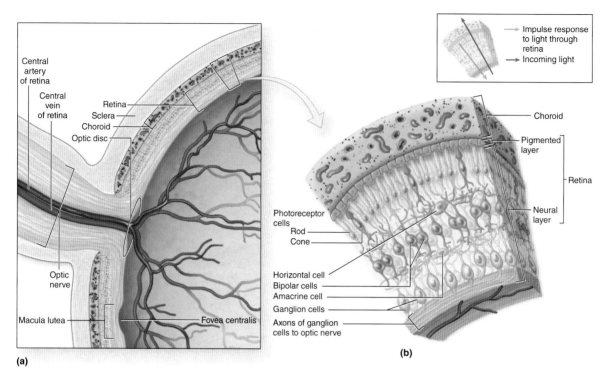

(a)

(b)

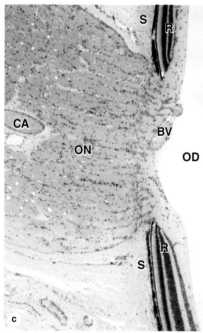

c

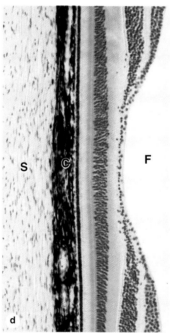

d

The retina is the thick layer of the eye immediately inside the choroid. **(a)** The central retinal artery and vein pass through the optic nerve and enter the eye at the optic disc where they divide to form smaller lateral branches in the retina's nerve fiber layer. Capillaries extend as deep as the inner nuclear layer. (Nutrients and O_2 for the outer retinal layers diffuse from capillaries in the choroid.)

(b) The layers and major neurons of the retina and their general organization are shown schematically here; the supporting Müller cells which penetrate all the neural layers are not shown.

(c) The optic disc (**OD**) at the head of the optic nerve (**ON**) is the point at which ganglionic layer axons from all regions of the retina (**R**) converge, penetrate the choroid and sclera (**S**), and leave the eye as the optic nerve to enter the brain. Blood vessels (**BV**) in the

ganglionic and nerve fiber layers converge at the optic disc to form the central artery (**CA**) and vein of the retina within the optic nerve. (X40; H&E)

(d) The fovea (F) is a small specialized area of the retina where cell bodies and axons of the ganglionic and inner layer are largely dispersed peripherally, thinning this retinal area and allowing light to hit the cones with very little light scattering. The fovea contains no rods and a greatly increased density of cones, causing the layer with their cell bodies to be slightly thicker here than elsewhere. These and other structural modifications at the fovea provide this area of the retina with the greatest visual acuity or sharpness of visual detail, but only when adequate light is present. The surrounding choroid (**C**) and sclera (**S**) are also shown. (X100; H&E)

FIGURE 23–15 Layers of the retina.

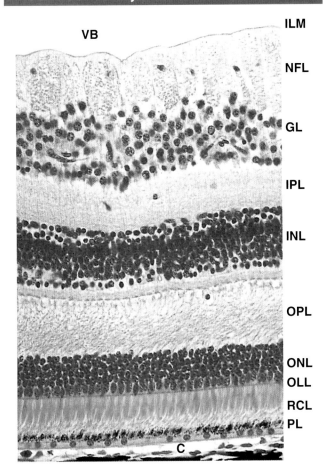

Between the vitreous body (**VB**) and the choroid (**C**), the retina can usually be seen to consist of nine neural layers and a pigmented layer. Following the path of the light, these are:

- The inner limiting membrane (**ILM**), a basement membrane covered by expanded processes of Müller cells, which are not distinguishable in routine preparations.
- The nerve fiber layer (**NFL**), containing the ganglionic cell axons that converge at the optic disc and form the optic nerve.
- The ganglionic layer (**GL**), containing cell bodies of the ganglion cells and thicker near the retina's center than its periphery.
- The inner plexiform layer (**IPL**), containing fibers and synapses of the ganglion cells and the bipolar neurons of the next layer.
- The inner nuclear layer (**INL**), with the cell bodies of several types of bipolar neurons which begin to integrate signals from the rod and cone cells.
- The outer plexiform layer (**OPL**), containing fibers and synapses of the bipolar neurons and rod and cone cells.
- The outer nuclear layer (**ONL**), with the cell bodies and nuclei of the photosensitive rod and cone cells.
- The outer limiting layer (**OLL**), a line formed by junctional complexes holding the rod and cone cells to the intervening Müller cells.
- The rod and cone layer (**RCL**), which contains the outer segments of these cells where the photoreceptors are located.
- The non-neural pigmented layer (**PL**), which has several supportive functions important for the function and maintenance of the neural retina. (X150; H&E)

- The **inner nuclear layer (INL)** contains the nuclei of various neurons, notably the bipolar cells, amacrine cells, and horizontal cells, all of which make specific connections with other neurons and integrate signals from rods and cones over a wide area of the retina.
- Near the vitreous, the **ganglionic layer (GL)** has neurons (ganglion cells) with much longer axons. These axons make up the **nerve fiber layer (NFL)** and converge to form the **optic nerve** which leaves the eye and passes to the brain. The GL is thickest near the central, macular region of the retina (Figure 23–15), but it thins peripherally to only one layer of cells (Figure 23–6).

Between the three layers with cell nuclei are two fibrous or "plexiform" regions containing only axons and dendrites connected by synapses (Figure 23–15):

- The **outer plexiform layer (OPL)** includes axons of the photoreceptors and dendrites of association neurons in the INL.
- The **inner plexiform layer (IPL)** consists of axons and dendrites connecting neurons of the INL with the ganglion cells

The rod and cone cells, named for the shape of their outer segments, are polarized neurons with their photosensitive portions aligned in the retina's **rod and cone layer (RCL)** and their axons in the OPL (Figure 23–15).

All neurons of the retina are supported metabolically by elongated, regularly arranged glial cells called **Müller cells**. With their perikarya in the INL, Müller cells extend processes that span the entire thickness of the neural retina (Figure 23–16). From these major Müller cell processes smaller lateral extensions ramify in each layer and ensheath virtually all the neuronal processes, cell bodies, and blood vessels. Müller cells are critical for retinal function, providing neurotrophic substances, removing waste products, regulating ion and water homeostasis, regulating blood flow, and maintaining a blood-inner retina barrier. Müller cells also organize two boundaries that appear as very thin retinal "layers":

- The **outer limiting layer (OLL)** is a poorly stained but well-defined series of adherent junctions (zonula adherentes) between the photoreceptors and Müller cells (Figures 23–15, 23–16, and 23–17).
- The **inner limiting membrane (ILM)** consists of terminal expansions of Müller cell processes that cover the collagenous membrane of the vitreous body and form the inner surface of the retina.

All these layers of the retina can be seen by routine light microscopy, as shown in Figures 23–6 and 23–15. It is important to note that light must pass through all the layers of the neural retina before reaching the layer of rods and cones. Branches of the central retinal artery and vein (Figure 23–1) run mainly within the nerve fiber and GLs, surrounded by perivascular feet of Müller cells and astrocytes which are located there. In some retina regions capillaries extend as deeply as

FIGURE **23–16** Müller cells.

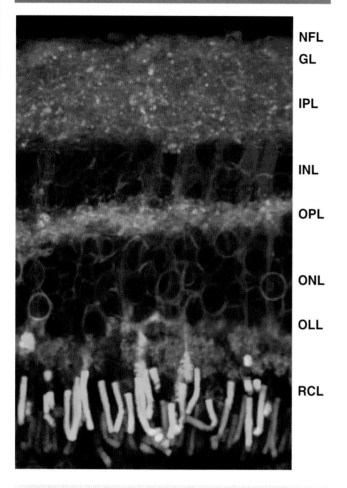

NFL
GL

IPL

INL

OPL

ONL

OLL

RCL

Müller cells are large glial cells unique to the retina, which are critical for retinal function by providing trophic and metabolic support for adjacent neurons, regulating homeostasis and synaptic activity, and helping to organize retinal components structurally. The attenuated and diffuse nature of Müller cells makes these cells difficult to study microscopically in most routine preparations of retina. This confocal light microscope image of a section of living guinea pig retina, with its layers named as in the previous figure, shows the large Müller cells in red. Axons and dendrites, abundant in the plexiform layers, and the outer segments of the photoreceptor cells appear green.

Müller cells extend through the full thickness of the retina, with fine lateral processes contacting all adjacent neuronal cell bodies, synapses, and blood vessels. At a level just beyond the layer of photoreceptor cell nuclei, Müller cells are sealed to the photoreceptor cells by zonula adherens junctions, forming the outer limiting layer (OLL). A few long Müller cell processes extend beyond this layer, into the layer of rods and cones (RCL). The nerve fiber layer (NFL) is very thin in the area shown here and the pigmented epithelium is not stained. (X400; Vital dyes Mitotracker Orange [red] & FM-43 for synaptic membranes [green])

(Used with permission from Dr. Andreas Reichenbach, Paul Flechsig Institute of Brain Research, University of Leipzig, Germany.)

the INL. A few scattered microglial cells occur throughout the neural retina.

Rod Cells

The human retina contains on average 92 million **rod cells**. They are extremely sensitive to light, responding to a single photon, and allow some vision even with light low levels, such as at dusk or nighttime. Rod cells are thin, elongated cells (50 μm × 3 μm), composed of two functionally distinct segments (Figure 23–17a). The **outer segment** is a modified primary cilium, photosensitive and shaped like a short rod; the inner segment contains glycogen, mitochondria, and polyribosomes for the cell's biosynthetic activity.

The rod-shaped segment consists mainly of 600 to 1000 flattened **membranous discs** stacked like coins and surrounded by the plasma membrane (Figure 23–17). Proteins on the cytoplasmic surface of each disc include abundant **rhodopsin** (or **visual purple**) which is bleached by light and initiates the visual stimulus. Between this outer segment and the cell's inner segment is a constriction, the **connecting stalk**, which is part of the modified primary cilium arising from a basal body (Figure 23–17).

The membranous discs form by repetitive in-folding of the plasma membrane near the connecting stalk and insertion of rhodopsin and other proteins transported there from the inner segment. In rod cells the newly assembled discs detach from the plasma membrane and are displaced distally as new discs form. Eventually the discs arrive at the end of the rod, where they are shed, phagocytosed, and digested by the cells of the pigmented epithelium (Figure 23–13). Each day approximately 90 membranous discs are lost and replaced in each rod, with the process of assembly, distal movement, and apical shedding taking about 10 days.

Cone Cells

Less numerous and less light-sensitive than rods, the 4.6 million **cone cells** in the typical human retina produce color vision in adequately bright light. There are three morphologically similar classes of cones, each containing one type of the visual pigment **iodopsin** (or photopsins). Each of the three iodopsins has maximal sensitivity to light of a different wavelength, in the red, blue, or green regions of the visible spectrum, respectively. By mixing neural input produced by these visual pigments, cones produce a color image.

Like rods cone cells (Figure 23–17a) are elongated, with outer and inner segments, a modified cilium connecting stalk, and an accumulation of mitochondria and polyribosomes. The outer segments of cones differ from those of rods in their shorter, more conical form and in the structure of their stacked membranous discs, which in cones remain as continuous invaginations of the plasma membrane along one side (Figure 23–17a). Also, newly synthesized iodopsins and other membrane proteins are distributed uniformly throughout the cone outer segment and, although iodopsin turns over, discs in cones are shed much less frequently than in rods.

FIGURE **23–17**   Rod and cone photoreceptor cells.

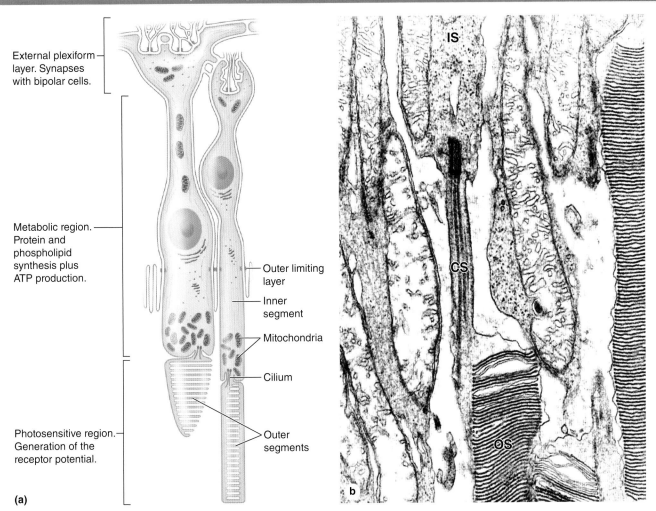

External plexiform layer. Synapses with bipolar cells.

Metabolic region. Protein and phospholipid synthesis plus ATP production.

Outer limiting layer

Inner segment

Mitochondria

Cilium

Outer segments

Photosensitive region. Generation of the receptor potential.

(a)

IS

CS

OS

b

Rod and cone cells are the two types of photoreceptors of the outer retina, sensitive to low levels of light and higher levels of light with a broader spectrum of wavelengths respectively. Both are elongated, parallel structures consisting of inner and outer segments, with the outer segments consisting largely of stacked membranous discs.

(a) Rod cells (right) and cone cells (left) all have the general shapes and important cytoplasmic features shown diagrammatically here. The membranous discs of cone outer segments are continuous with the cell membrane; those of rods are not. The line of adherent junctions attaching the rod and cone inner segments to the distal ends of the Müller glial cells (not shown) comprises the outer

limiting layer which helps anchor the photoreceptors and separates the matrix surrounding the rod and cone cell bodies from that around their metabolically unique outer segments.

(b) TEM of a sectioned retina shows two rod cell inner segments (**IS**) with mitochondria and the outer photosensitive segment (**OS**) consisting of flat, parallel membranous discs. The cell in the middle of the figure shows the connecting stalk (**CS**) between the two segments. This connection contains a basal body with a primary cilium that expands distally as the outer segment. The connecting stalks of cone cells are structurally similar. The stacked membranes of the discs in the outer segments are very distinct and electron-dense due to the high density of proteins they contain. (X24,000)

❯❯ **MEDICAL APPLICATION**

Partial **color blindness** is normally an inherited disorder due to recessive mutations in genes for one or more iodopsins or other genes required for cone function. The most common form, **red-green** color blindness, affects the cones responsible for detecting light at these two wavelengths and occurs much more frequently in men than women because many

key genes for the color sensitivity of cones are on the X chromosome. With two X chromosomes, women do not show the disability but can be carriers of the mutation.

Specialized Areas of the Retina

The **optic disc**, or blind spot, occurs at the posterior of the retina where the axons in the retina's nerve fiber layer (NFL) converge at the optic nerve head and leave the eye as the optic nerve. The optic disk is approximately 1.7 mm in diameter and lacks photoreceptors and other retinal layers except the NFL from ganglion cells. A paired central artery and vein in the optic nerve (Figure 23–14a and c) branches at the optic disc to produce microvasculature that runs through the NFL to supply the neurons and glial cells of the inner layers in all quadrants of the retina.

Near the optic disc and directly opposite the pupil, lies a specialized retinal area about 1.5 mm in diameter called the **fovea centralis** (Figure 23–14a and d) where visual acuity or sharpness is maximal. The fovea (L. *fovea*, a small pit) is a shallow depression within the retina where cells bodies of the ganglionic and inner nuclear layers are dispersed peripherally, leaving primarily cone cells (Figure 23–14d). Cone cells in the fovea are long, narrow, and closely packed. Devoid of most conducting neurons as well as capillaries, the fovea allows light to fall directly on its cones with very little light scatter. The locations and structural adaptations of the fovea centralis together account for the extremely precise visual acuity of this region.

Surrounding the fovea centralis is the macula lutea (L. *macula*, spot; *lutea*, yellow), or simply macula, 2 mm in diameter, where axons of the cone cells contain various carotenoids, giving this area its yellowish color. The carotenoids have antioxidant properties and filter potentially damaging short-wavelength light, thus helping to protect the cone cells of the fovea.

Within the GL of the entire retina a subset of ganglion cells serve as **nonvisual photoreceptors**. These neurons contain 11-*cis*-retinal bound to the protein melanopsin and serve to detect changes in light quantity and quality during each 24-hour dawn/dusk cycle. Signals from these cells pass via axons of the retinohypothalamic tract to the suprachiasmatic nuclei and the pineal gland, where they help establish the body's physiologic circadian rhythms (see Chapter 20).

Phototransduction

The stacked membranous discs of rod and cone outer segments are parallel with the retinal surface, which maximizes their exposure to light. The membranes are very densely packed with rhodopsin or one of the iodopsin proteins; one rod contains about a billion rhodopsin molecules. Each of these visual pigments contains a transmembrane protein, the **opsin**, with a small, light-sensitive **chromophore** molecule bound to it. The vitamin A derivative called **retinal** acts as the chromophore of rhodopsin in rods.

Phototransduction involves a cascade of changes in the cells triggered when light hits and activates the chromophore, a basically similar process in both rods and cones. As diagrammed for a rod in Figure 23–18, in darkness rhodopsin is not active and cation channels in the cell membrane are open. The cell is depolarized and continuously releases neurotransmitter at the synapse with the bipolar neurons. When retinal on rhodopsin absorbs a photon of light, it isomerizes within one picosecond from 11-*cis*-retinal to all-*trans*-retinal. This causes a configuration change in the opsin, which in turn activates the adjacent membrane-associated protein **transducin**,

a heterotrimeric G protein to which opsin is coupled. Transducin activity then indirectly closes cGMP-gated Na^+ channels, causing hyperpolarization which reduces the synaptic release of neurotransmitter. This change in turn depolarizes sets of bipolar neurons, which send action potentials to the ganglion cells of the optic nerve.

The conformation change induced by light in retinal which initiates this cascade of events also causes the chromophore to dissociate from the opsin, a process called **bleaching** (Figure 23–18). The free all-*trans*-retinal is transported from the rod into the adjacent pigmented epithelial cell where it is converted back to 11-*cis*-retinal, then transported back into a photoreceptor for reuse. This cycle of retinal regeneration and rhodopsin recovery from bleaching may take a minute or more and is part of the slow adaptation of the eyes that occurs when moving from bright to dim light.

›› MEDICAL APPLICATION

A leading cause of blindness in elderly individuals of developed countries is **age-related macular degeneration**, which causes blindness in the center of the visual field. Degenerative changes in the retina around the macula include depigmentation of the posterior epithelium, focal thickening of the adjacent Bruch's membrane, major changes and blood loss in the capillaries in the choroid and retina, and eventual loss of the photoreceptor cells producing blind spots. There appears to be a genetic predisposition to the disorder, along with environmental triggers such as excessive exposure to ultraviolet radiation. Progression of the disease can be slowed by laser surgery to destroy the abnormal and excessive retinal capillaries.

Accessory Structures of the Eye

Conjunctiva

The **conjunctiva** is a thin, transparent mucosa that covers the exposed, anterior portion of the sclera and continues as the lining on the inner surface of the eyelids. It consists of a stratified columnar epithelium, with numerous small goblet cells, supported by a thin lamina propria of loose vascular connective tissue (Figure 23–19). Mucous secretions from conjunctiva cells are added to the tear film that coats this epithelium and the cornea.

›› MEDICAL APPLICATION

Conjunctivitis, or **pink eye**, is a condition in which the conjunctiva is inflamed usually due to bacterial or viral infection or to allergies. The inflammation increases the discharge of mucus and enlarges the microvasculature of the sclera, causing the white sclera to have a reddish appearance. Bacterial and viral conjunctivitis are contagious but have little effect on vision.

FIGURE **23–18**   Rod cell phototransduction.

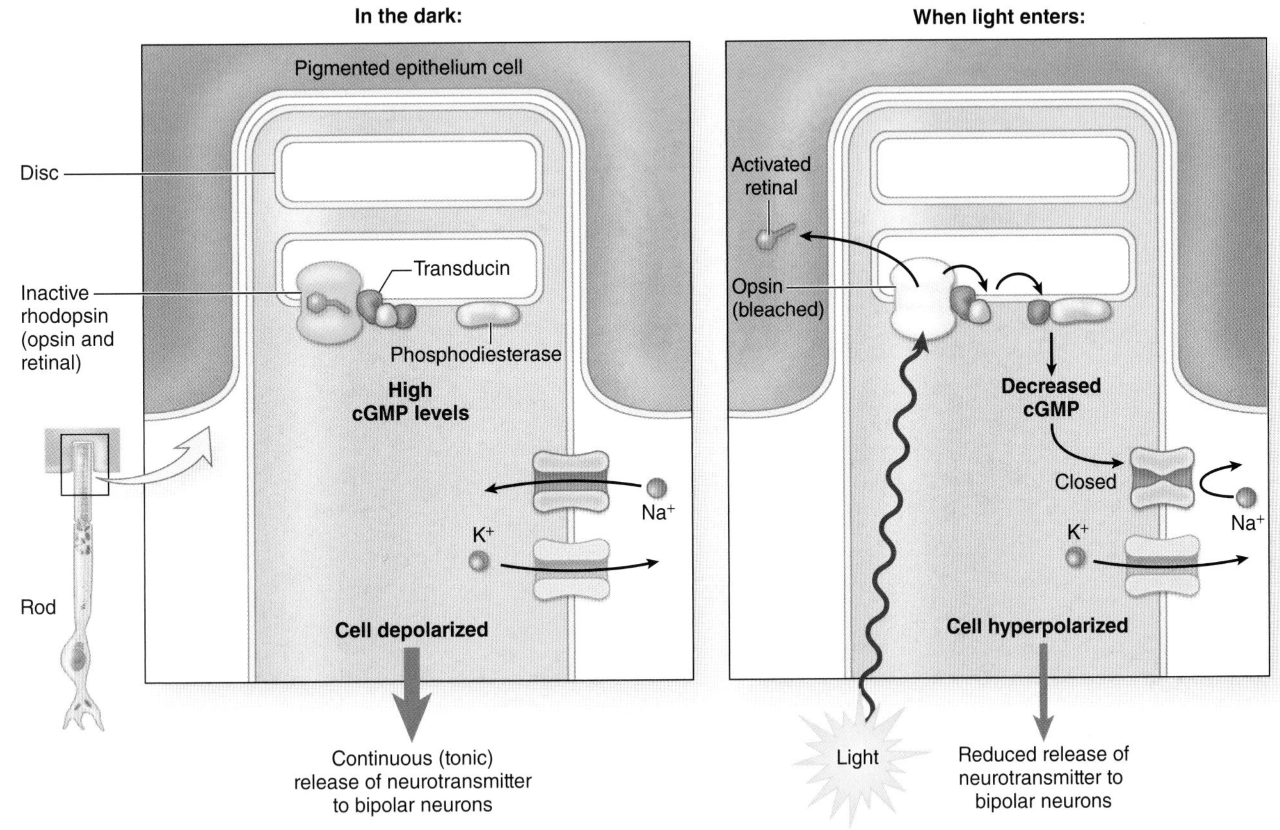

In the dark:

Pigmented epithelium cell

Disc

Inactive rhodopsin (opsin and retinal)

Transducin

Phosphodiesterase

High cGMP levels

Rod

Na⁺

K⁺

Cell depolarized

Continuous (tonic) release of neurotransmitter to bipolar neurons

When light enters:

Activated retinal

Opsin (bleached)

Decreased cGMP

Closed

K⁺

Na⁺

Cell hyperpolarized

Light

Reduced release of neurotransmitter to bipolar neurons

Phototransduction involves a series of changes in **rod and cone cells** that begins when light hits the stacked **membranous discs**. The main parts of the process are similar in both rods and cones but have been better studied in the more abundant rod cells, as shown here. Membranes of the discs are densely packed with proteins, although only one of each major type is shown here. In the dark, **rhodopsin** and its **11-*cis*-retinal** are inactive and the intracellular concentration of the second messenger cyclic GMP (**cGMP**) is high. One effect of cGMP is to keep open the abundant cGMP-gated sodium channels in the cell membrane and therefore the cell is depolarized, continuously releasing its **neurotransmitter** (glutamate) at the synapse with the bipolar neurons.

When photons of light are absorbed by the retinal of rhodopsin, the molecule isomerizes from 11-*cis*-retinal to **all-*trans*-retinal** and this change activates the opsin. This in turn activates the adjacent peripheral membrane protein **transducin**, a heterotrimeric **G protein**, allowing it to release its α subunit, which moves laterally and stimulates another membrane protein, **phosphodiesterase**, to hydrolyze cGMP. With less cGMP, many of the sodium channels now close, producing hyperpolarization of the cell which decreases the release of neurotransmitter at the synapses. This change at the synapse depolarizes sets of bipolar neurons, which then send action potentials to the various ganglion cells of the optic nerve that will allow the brain to produce an image. When retinal is activated by light, it also dissociates from rhodopsin, leaving a more pale-colored (bleached) opsin. The free retinal moves into the surrounding **pigmented epithelial cells**, where the all-*trans*-isomer is regenerated. It is then transported back into a rod or cone cell to again bind opsin and be used in another round of phototransduction.

Eyelids

Eyelids (Figure 23–19) are pliable structures containing skin, muscle, and conjunctiva that protect the eyes. The skin is loose and elastic, lacks fat, and has only very small hair follicles and fine hair, except at the distal edge, where large follicles with eyelashes are present. Associated with the follicles of eyelashes are sebaceous glands and modified apocrine sweat glands.

Beneath the skin are striated fascicles of the orbicularis oculi and levator palpebrae muscles that fold the eyelids. Adjacent to the conjunctiva is a dense fibroelastic plate called the **tarsus** that supports the other tissues. The tarsus surrounds a series of 20 to 25 large sebaceous glands, each with many acini secreting into a long central duct that opens among the eyelashes (Figure 23–19). Oils in the sebum produced by these **tarsal glands**, also called **Meibomian glands**, form a surface layer on the tear film, reducing its rate of evaporation, and help lubricate the ocular surface.

FIGURE **23–19** Eyelid.

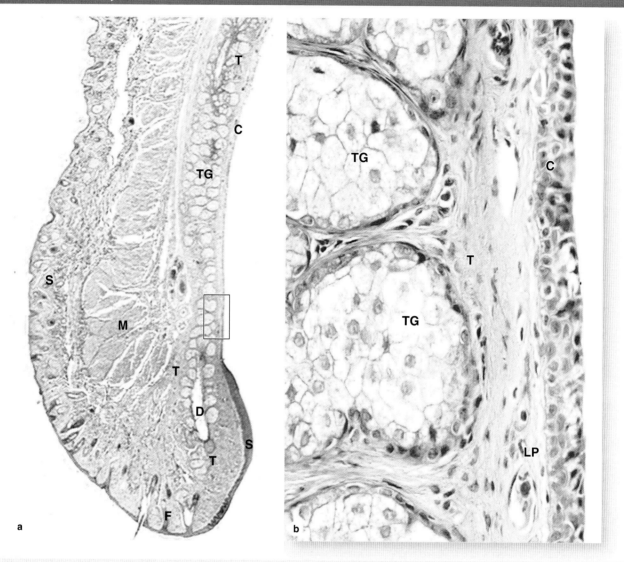

a

b

(a) The eyelid is a pliable tissue with skin (S) covering its external surface and smooth conjunctiva (C) lining its inner surface. At the outer rim of the eyelid are a series of large hair follicles (F) for the eyelashes. Associated with these hair follicles are small sebaceous glands and modified apocrine sweat glands. Internally eyelids contain fascicles of striated muscle (M) comprising the orbicularis oculi muscle and closer to the conjunctiva a thick plate of fibroelastic connective tissue called the tarsus (T). This tarsal plate provides structural support for the eyelid and surrounds a series of large sebaceous glands, the tarsal glands (TG) (aka Meibomian glands),

with acini secreting into long central ducts (D) that empty at the free edge of the eyelids. (X12.5; H&E)

(b) At higher magnification, only the inner aspect of the eyelid is seen, and it shows that the conjunctiva (C) is a mucous membrane consisting of a stratified columnar epithelium with small cells resembling goblet cells and resting on a thin lamina propria (LP). Large cells undergoing typical holocrine secretion are shown in the tarsal gland acini (TG), and the fibrous connective tissue in the tarsus (T) surrounding the acini. Sebum from these glands is added to the tear film and helps lubricate the ocular surface. (X200; H&E)

❯❯ MEDICAL APPLICATION

Infections near an opening of the tarsal gland ducts, generally caused by *Staphylococcus aureus,* are called **styes**. They are most common in infants but can occur at any age and can be quite painful. Like certain other infections, styes can occur in periods of immunosuppression caused by poor nutrition or stress.

Lacrimal Glands

The **lacrimal glands** produce fluid continuously for the tear film that moistens and lubricates the cornea and conjunctiva and supplies O_2 to the corneal epithelial cells. Tear fluid also contains various metabolites, electrolytes, and proteins of innate immunity such as lysozyme. The main lacrimal glands are located in the upper temporal portion of the orbit and have several lobes that drain through individual excretory ducts into the superior fornix, the conjunctiva-lined recess between the eyelids and the eye.

The lacrimal glands have acini composed of large serous cells filled with lightly stained secretory granules and surrounded by well-developed myoepithelial cells and a sparse, vascular stroma (Figure 23–20).

Tear film moves across the ocular surface and collects in other parts of the bilateral **lacrimal apparatus**: flowing through two small round openings (0.5 mm in diameter) to canaliculi at the medial margins of the upper and lower eyelids, then passing into the lacrimal sac, and finally draining into the nasal cavity via the nasolacrimal duct. The canaliculi are lined by stratified squamous epithelium, but the more distal sac and duct are lined by pseudostratified ciliated epithelium like that of the nasal cavity.

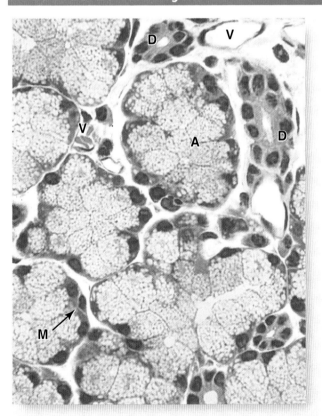

FIGURE **23–20** Lacrimal gland.

Lacrimal glands secrete most components of the tear film that moisturizes, lubricates, and helps protect the eyes. The glands have acini (**A**) composed of secretory cells filled with small, light-staining granules and myoepithelial cells (**M**). Connective tissue surrounding the acini contains blood vessels (**V**) of the microvasculature and intra- and interlobular ducts (**D**) converging as excretory ducts that empty into the superior conjunctival fornix between the upper eyelid and the eye. (X400; H&E)

❱ EARS: THE VESTIBULOAUDITORY SYSTEM

Tissues of the ear mediate the senses of equilibrium and hearing. Each ear consists of three major parts (Figure 23–21):

- The **external ear**, which receives sound waves;
- The **middle ear**, in which sound waves are transmitted from air to fluids of the internal ear via a set of small bones;
- The **internal ear**, in which these fluid movements are transduced to nerve impulses that pass via the acoustic nerve to the CNS. In addition to the auditory organ, or cochlea, the internal ear also contains the vestibular organ that allows the body to maintain equilibrium.

Most structures of the middle and internal ear develop in the embryo and are enclosed within the temporal bone as it forms from head mesenchyme.

External Ear

The **auricle**, or **pinna** (L. *pinna*, wing) is an irregular, funnel-shaped plate of elastic cartilage, covered by tightly adherent skin, which directs sound waves into the ear.

Sound waves enter the **external acoustic meatus** (L. passage), a canal lined with stratified squamous epithelium that extends from the auricle to the middle ear. Near its opening hair follicles, sebaceous glands, and modified apocrine sweat glands called **ceruminous glands** are found in the submucosa (Figure 23–22). **Cerumen**, the waxy material formed from secretions of the sebaceous and ceruminous glands, contains various proteins, saturated fatty acids, and sloughed keratinocytes and has protective, antimicrobial properties. The wall of the external auditory meatus is supported by elastic cartilage in its outer third, while the temporal bone encloses the inner part (Figure 23–21).

Across the deep end of the external acoustic meatus lies a thin, somewhat transparent sheet called the **tympanic membrane** or **eardrum**. This membrane consists of fibroelastic connective tissue covered externally with epidermis and internally by the simple cuboidal epithelium of the mucosa that lines the middle ear cavity. Sound waves cause vibrations of the tympanic membrane, which transmit energy to the middle ear (Figure 23–21).

Middle Ear

The **middle ear** contains the air-filled **tympanic cavity**, an irregular space that lies within the temporal bone between the tympanic membrane and the bony surface of the internal ear (Figure 23–21). Anteriorly, this cavity communicates with the pharynx via the **auditory tube** (also called the **eustachian** or **pharyngotympanic tube**) and posteriorly with the smaller, air-filled mastoid cavities of the temporal bone. The simple cuboidal epithelium lining the cavity rests on a thin lamina propria continuous with periosteum. Entering the auditory tube, this simple epithelium is gradually

FIGURE **23–21** Major divisions of the ear.

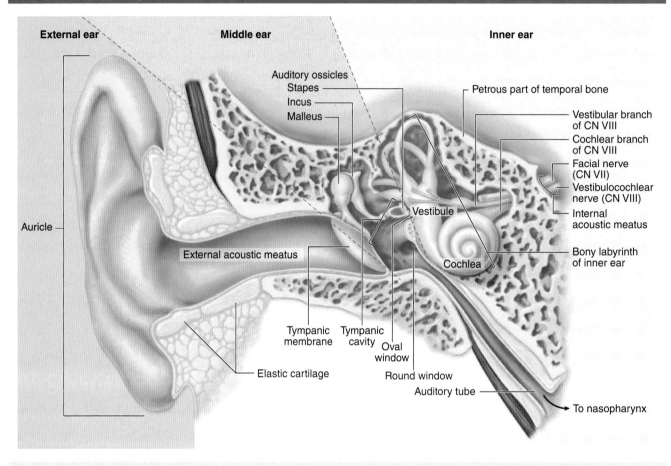

The external, middle, and internal regions of the right ear are shown here, with the major structures of each region.

replaced by the ciliated pseudostratified columnar epithelium that lines the tube. Below the temporal bone this tube is usually collapsed; swallowing opens it briefly, which serves to balance the air pressure in the middle ear with atmospheric pressure. In the medial bony wall of the middle ear are two small, membrane-covered regions devoid of bone: the **oval** and **round windows** with the internal ear behind them (Figure 23–21).

The tympanic membrane is connected to the oval window by a series of three small bones, the **auditory ossicles**, which transmit the mechanical vibrations of the tympanic membrane to the internal ear (Figure 23–23). The three ossicles are named for their shapes the **malleus, incus,** and **stapes**, from the Latin words for "hammer," "anvil," and "stirrup," respectively. The malleus is attached to the tympanic membrane and the stapes to the membrane across the oval window. The ossicles articulate at synovial joints, which along with periosteum are completely covered with simple squamous epithelium. Two small skeletal muscles, the tensor tympani and stapedius, insert into the malleus and stapes, respectively, restricting ossicle movements and protecting the

oval window and inner ear from potential damage caused by extremely loud sound.

> **›› MEDICAL APPLICATION**
>
> The middle ear tympanic cavity may show inflammation **(otitis media)** when viral or bacterial infections extend there from the upper respiratory tract via the auditory tubes. Otitis media is most common in young children, where the short auditory tubes facilitate infection of the tympanic cavity.

Internal Ear

The **internal ear** is located completely within the temporal bone, where an intricate set of interconnected spaces, the **bony labyrinth**, houses the smaller **membranous labyrinth**, a set of continuous fluid-filled, epithelium-lined tubes and chambers (Figure 23–21). The membranous labyrinth is derived from an ectodermal vesicle, the otic vesicle, which invaginates into subjacent mesenchyme during the fourth week of embryonic development, loses contact with the surface ectoderm,

FIGURE **23–22** External acoustic meatus.

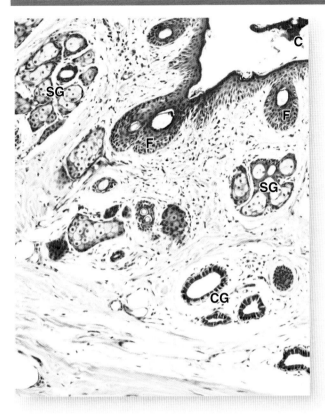

The external acoustic meatus leads from the opening in the auricle to the tympanic membrane (or eardrum). This section of the wall in the outer third of the acoustic meatus shows the lining of skin containing small hair follicles (**F**), sebaceous glands (**SG**), and modified apocrine sweat glands called **ceruminous glands** (**CG**). Secretions from these two glands form a yellowish, waxy product called **cerumen** (**C**). (X50; H&E)

and becomes embedded in rudiments of the developing temporal bone. Components of the bony and membranous labyrinths and their functions are summarized in Table 23–2.

The embryonic otic vesicle, or otocyst, forms the membranous labyrinth with its major divisions:

- Two connected sacs called the **utricle** and the **saccule**,
- Three **semicircular ducts** continuous with the utricle, and
- The **cochlear duct**, which provides for hearing and is continuous with the saccule.

Mediating the functions of the inner ear, each of these structures contains in its epithelial lining large areas with columnar mechanoreceptor cells, called **hair cells**, in specialized sensory regions:

- Two **maculae** of the utricle and saccule,
- Three **cristae ampullares** in the enlarged ampullary regions of each semicircular duct, and
- The long **spiral organ of Corti** in the cochlear duct.

FIGURE **23–23** Middle ear.

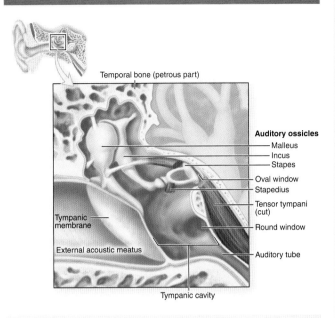

Three **auditory ossicles**, with joints and striated muscles, span the tympanic cavity of the middle ear, which is enclosed by the **temporal bone** and the **tympanic membrane**.

The entire membranous labyrinth is within the bony labyrinth, which includes the following regions (Figure 23–24):

- An irregular central cavity, the **vestibule** (L. *vestibulum*, area for entering) houses the saccule and the utricle.
- Behind this, three osseous **semicircular canals** enclose the semicircular ducts.
- On the other side of the vestibule, the **cochlea** (L. snail, screw) contains the cochlear duct (Figure 23–24). The cochlea is about 35 mm long and makes 2¾ turns around a bony core called the **modiolus** (L. hub of wheel). The modiolus contains blood vessels and surrounds the cell bodies and processes of the acoustic branch of the eighth cranial nerve in the large **spiral** or **cochlear ganglion**.

The bony and membranous labyrinths contain two different fluids (Figure 23–24). The separation and ionic differences between these fluids are important for inner ear function.

- **Perilymph** fills all regions of the bony labyrinth and has an ionic composition similar to that of cerebrospinal fluid and the extracellular fluid of other tissues, but it contains little protein. Perilymph emerges from the microvasculature of the periosteum and drains via a perilymphatic duct into the adjoining subarachnoid space. Perilymph suspends and supports the closed membranous labyrinth, protecting it from the hard wall of the bony labyrinth.
- **Endolymph** fills the membranous labyrinth and is characterized by a high-K^+ (150 mM) and low-Na^+ (16 mM)

TABLE 23–2	Structure and function of internal ear components.		
Bony Labyrinth Component (Containing Perilymph and the Membranous Labyrinth)	**Membranous Labyrinth Component (Within Bony Labyrinth and Containing Endolymph)**	**Structures With Sensory Receptors**	**Major Function**
Vestibule	Utricle, saccule	Maculae	Detect linear movements and static position of the head
Semicircular canals	Semicircular ducts	Cristae ampullares	Detect rotational movements of the head
Cochlea	Cochlear duct	Spiral organ	Detect sounds

content, similar to that of intracellular fluid. Endolymph is produced in a specialized area in the wall of the cochlear duct (described below) and drains via a small endolymphatic duct into venous sinuses of the dura mater.

Utricle and Saccule

The interconnected, membranous **utricle** and the **saccule** are composed of a very thin connective tissue sheath lined with simple squamous epithelium and are bound to the periosteum of the bony labyrinth by strands of connective tissue

FIGURE 23–24 Internal ear.

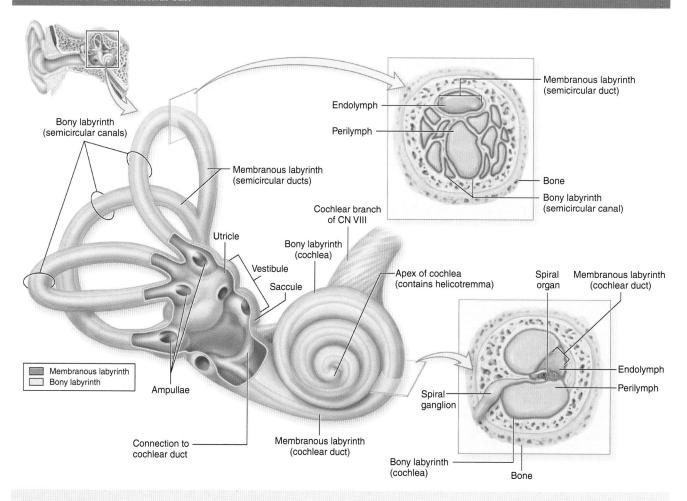

The internal ear consists of a cavity in the temporal bone, the **bony labyrinth**, which houses a fluid-filled **membranous labyrinth**. The membranous labyrinth includes the vestibular organs (the **saccule, utricle**, and **semicircular ducts**) which contribute to the sense of equilibrium and balance, and the **cochlea** for the sense of hearing.

containing microvasculature. The maculae in the walls of the utricle and saccule are small areas of columnar neuroepithelial cells innervated by branches of the vestibular nerve (Figure 23–25). The macula of the saccule lies in a plane perpendicular to that of the utricle, but both are similar histologically. Each consists of a thickening of the wall containing several thousand columnar hair cells, each with surrounding supporting cells and synaptic connections to nerve endings.

Hair cells act as mechanoelectrical transducers, converting mechanical energy into the electrical energy of nerve action potentials. Each has an apical hair bundle consisting of one rigid cilium, the **kinocilium**, up to 40-μm long, and a bundle of 30-50 rigid, unbranched **stereocilia**. The base of each stereocilia is tapered and connected to an actin-rich region of apical cytoplasm, the cuticular plate, which returns these rigid projecting structures to a normal upright position after bending. They are arranged in rows of decreasing length, with the longest adjacent to the kinocilium (Figure 23–25). The tips of the stereocilia and kinocilium are embedded in a thick, gelatinous layer of proteoglycans called the **otolithic membrane**. The outer region of this layer contains barrel-shaped crystals of $CaCO_3$ and protein called **otoliths** (or otoconia) typically 5-10 μm in diameter (Figure 23–25b).

FIGURE **23–25** Vestibular maculae.

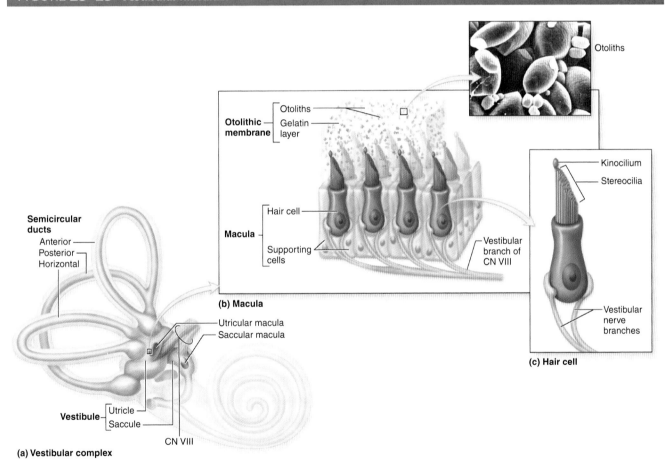

(b) Macula

(c) Hair cell

(a) Vestibular complex

(a) Two sensory areas called **maculae** occur in the **membranous labyrinth** of the vestibular **utricle** and **saccule**, both specialized for detecting gravity and endolymph movements.

(b) A more detailed diagram of a macular wall shows that it is composed of **hair cells, supporting cells**, and endings of the **vestibular branch** of the eighth cranial nerve. The apical surface of the hair cells is covered by a gelatinous otolithic layer or membrane and the basal ends of the cells have synaptic connections with the nerve fibers. The SEM shows **otoliths** embedded in this membrane. These mineralized structures make the otolithic membrane heavier

than endolymph alone, which facilitates bending of the kinocilia and stereocilia by gravity or movement of the head.

(SEM, used with permission from David J. Lim, House Ear Institute and Department of Cell & Neurobiology, University of Southern California, Los Angeles, CA.)

(c) A diagram of a single generalized hair cell shows the numerous straight **stereocilia**, which contain bundled actin, and a longer single **kinocilium**, a modified cilium whose tip may be slightly enlarged.

FIGURE **23–26** Hair cells and hair bundles.

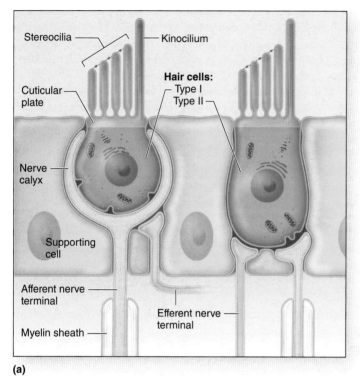

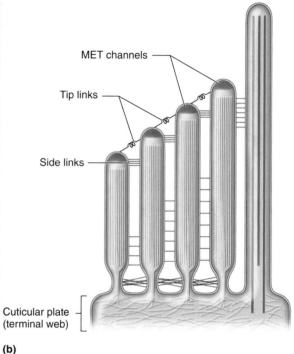

(a) This diagram shows the two types of hair cells in the maculae and cristae ampullares. Basal ends of **type I hair** cells are rounded and enclosed within a nerve calyx on the afferent fiber. **Type II hair cells** are columnar and associated with typical bouton synaptic connections to their afferents. Both types are also associated with efferent fibers.

(b) A more detailed diagram shows that **stereocilia** occur in rows of increasing height, with the tallest next to the single **kinocilium** on one side of the cell. By TEM the end of each stereocilium shows

an electron-dense region containing cation channels and proteins involved in **mechanoelectrical transduction (MET)** that converts mechanical activity of the stereocilia to electrical activity. Neighboring stereocilia are connected by proteins of various side links; the best understood of these are the **tip links** that connect the tips of stereocilia and contain very long cadherin proteins. Changes in the tension of the tip links caused by bending of the hair bundle open or close the adjacent mechanically gated K^+ channels and change the afferent synaptic activity of the hair cells.

All hair cells have basal synapses with afferent (to the brain) nerve endings but are of two types (Figure 23–26a):

- Type I hair cells have rounded basal ends completely surrounded by an afferent terminal calyx (L, cup).
- The more numerous type II hair cells are cylindrical, with bouton endings from afferent nerves.

Synaptic connections with *efferent* (from the brain) fibers are also present on hair cells of both types, or on their afferents, to modulate the sensitivity of these mechanoreceptors (Figure 23–26a). The supporting cells provide metabolic and physical support for the mechanoreceptors.

Sensory information from the utricle and saccule allows the brain to monitor the **static position** and **linear acceleration** of the head. This information, along with that provided visually and by musculoskeletal proprioceptors, is important for maintaining equilibrium and allowing the eyes to remain

fixed on the same point while moving the head. The head's position determines the position of the otolithic membrane in contact with hair cells of the two maculae. Because the otoliths are heavier than endolymph, the hair bundles are deflected by gravity when the head is not moving, when the head is tilted, and when the individual is moving in a straight line and inertia causes drag on the otolithic membrane.

Deflection or bending of the stereocilia changes the hair cells' resting potential and their rate of neurotransmitter release to the afferent nerves, which is the basis for mechanoelectrical transduction (Figure 23–26b). When the hair bundle is deflected *toward* the kinocilium, protein fibrils called **tip links** connecting the stereocilia are pulled and mechanically gated channels open to allow an influx of K^+ ions (the major cation in endolymph). The resulting depolarization of the hair cell opens voltage-gated Ca^{2+} channels in the basolateral membrane, and Ca^{2+} entry stimulates release of neurotransmitter and generates an impulse in the afferent nerve (Figure 23–27a).

When the head stops moving, the stereocilia straighten and hair cells quickly repolarize and reestablish the resting potential. Head movements that bend the stereocilia *away*

from the kinocilium cause the tip links to be slack, allowing closure of the apical cation channels and hyperpolarization of the cell. This in turn closes Ca^{2+} channels and reduces neurotransmitter release (Figure 23–27b).

FIGURE **23–27** Mechanotransduction in hair cells.

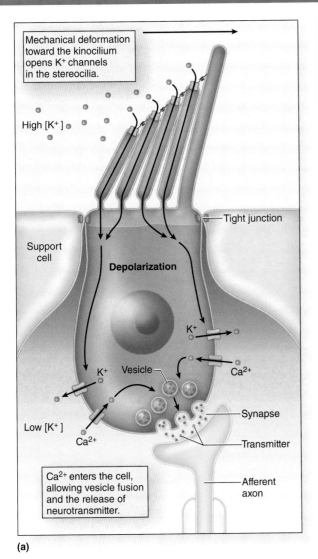

(a)

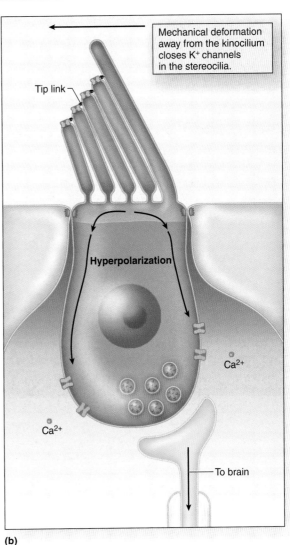

(b)

Hair cells and **supporting cells** are part of an epithelium with tight junctions. The *apical* ends of the cells are exposed to endolymph with a high concentration of K^+, and perilymph with a much lower K^+ concentration bathes their *basolateral* surface. At rest, hair cells are polarized with a small amount of K^+ entry and a low level of neurotransmitter release to afferent nerve fibers at the basal ends of the cells.

(a) As shown here, head movements that cause the stereocilia bundle to be deflected *toward* the kinocilium produce tension in the tip links, which is transduced to electrical activity by opening of adjacent cation channels. Entry of K^+ *depolarizes* the cell, opening voltage-gated basolateral Ca^{2+} channels, which stimulates release

of neurotransmitter. When this movement stops, the cells quickly repolarize.

(b) Movements in the opposite direction, *away from* the kinocilium, produce slackness on the tip links, allowing the mechanically gated apical K^+ channels to close completely, producing *hyperpolarization*, and inhibiting transmitter release. With different numbers of afferent and efferent fibers on the hair cells and with various hair cells responding differently to endolymph movements due to their positions within the maculae and cristae ampullares, the sensory information produced collectively by these cells can be processed by the vestibular regions of the brain and used to help maintain equilibrium.

Problems of the vestibular system can result in vertigo, or dizziness, a sense of bodily rotation and lack of equilibrium. Spinning the body produces **vertigo** due to overstimulation of the cristae ampullares of the semicircular ducts. Overstimulation of the maculae of the utricle caused by repetitive changes in linear acceleration and directional changes can normally lead to **motion sickness** (seasickness).

Sensory impulses from the maculae of the vestibular complex move toward the CNS via branches of the eighth cranial nerve and are interpreted along with input from the semicircular ducts and other sources to help maintain equilibrium.

The sensation of **vertigo** associated with rapid head movements can also be produced by internal ear inflammation (**vestibular neuritis**) or neurologic conditions that cause dysfunctional activity of the vestibular system. **Ménière disease** involves episodes of vertigo accompanied by hearing loss and ringing in the ears (**tinnitus**) and is caused when increased pressure within the membranous labyrinth (**endolymphatic hydrops**) leads to rupture and leakage of endolymph into the perilymph.

Semicircular Ducts

The three **semicircular ducts** extend from and return to the wall of the utricle. They lie in three different spatial planes, at approximately right angles to one another (Figure 23–24).

Each semicircular duct has one enlarged ampulla end containing hair cells and supporting cells on a crest of the wall called the **crista ampullaris** (Figure 23–28). Each crista ampullaris is perpendicular to the long axis of the duct. Cristae are histologically similar to maculae, but the proteoglycan layer called the **cupola** attached to the hair cells apically lacks otoliths and is thicker. The cupula extends completely across the ampulla, contacting the opposite nonsensory wall (-Figure 23–28).

The hair cells of the cristae ampullares act as mechanoelectrical transducers like those of the maculae in the utricle and saccule, signaling afferent axons by pulsed transmitter release determined by depolarization and hyperpolarization states (Figure 23–27). Here the mechanoreceptors detect **rotational movements** of the head as they are deflected by endolymph movement in the semicircular ducts. The cells are oriented with opposite polarity on each side of the side, so that turning the head causes hair cell depolarization on one side and hyperpolarization on the other. Neurons of the **vestibular nuclei** in the CNS receive input from the sets of semicircular ducts on each side simultaneously and interpret head rotation on the basis of the relative transmitter discharge rates of the two sides.

FIGURE 23–28 Ampullae and cristae of the semicircular ducts.

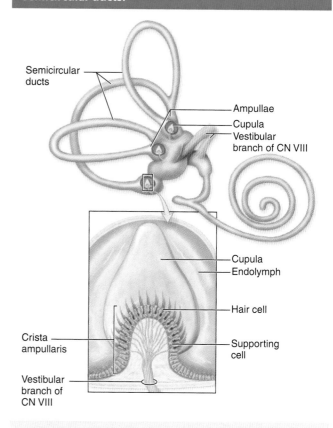

Semicircular ducts

Ampullae
Cupula
Vestibular branch of CN VIII

Cupula
Endolymph

Hair cell

Crista ampullaris

Supporting cell

Vestibular branch of CN VIII

Each of the semicircular ducts has an expanded end called the **ampulla**. The wall of each ampulla is raised as a ridge called the **crista ampullaris**, a section of which is shown here diagrammatically. Hair cells of the crista epithelium resemble the two types found in the maculae, with hair bundles projecting into a dome-shape overlying layer of proteoglycan called the **cupula**. The cupula is attached to the wall opposite the crista and is moved by endolymph movement within the semicircular duct.

Inputs from the semicircular ducts travel together with those from the utricle and saccule along the eighth cranial nerve to vestibular nuclei in the CNS. There they are interpreted together with inputs from mechanoreceptors of the musculoskeletal system to provide the basis for perceiving movement and orientation in space and for maintaining equilibrium or balance.

Brief periods of vertigo produced by sudden changes in position of head, such as standing up quickly or sitting up after lying in bed, may be examples of **benign paroxysmal positional vertigo** (**BPPV**). BPPV also results when one or more of the dense otoliths (or otoconia) detach from the otolithic membrane and move into an ampulla of the posterior semicircular duct as **canaliths** that may also adhere to the cupula. Being heavy, canaliths make that crista ampullaris gravity-sensitive. When that semicircular canal is aligned with gravity during head movements, sensory impulses are produced which the brain interprets incorrectly as continuous head rotations. A specific series of slow head movements that make up the **canalith repositioning procedure**, or **Epley maneuver**, can be used to return the otoliths to the utricle and relieve the sensation of dizziness and unsteadiness.

Cochlear Duct

The **cochlear duct**, a part of the membranous labyrinth shaped as a spiral tube, contains the hair cells and other structures that allow auditory function. Held in place within the bony cochlea, this duct is one of three parallel compartments, or scalae (L., ramps or ladders) which coil 2¾ turns within the cochlea (Figure 23–29):

- The cochlear duct itself forms the middle compartment, or **scala media**, filled with endolymph. It is continuous with the saccule and ends at the apex of the cochlea.
- The larger **scala vestibuli** contains perilymph and is separated from the scala media by the very thin **vestibular membrane (Reissner membrane)** lined on each side by simple squamous epithelium (Figure 23–30). Extensive tight junctions between cells of this membrane block ion diffusion between perilymph and endolymph.
- The **scala tympani** also contains perilymph and is separated from the scala media by the fibroelastic **basilar membrane** (Figure 23–29b).

The scalae tympani and vestibuli communicate with each other at the apex of the cochlea via a small opening called the **helicotrema**. Thus these two spaces with perilymph are actually one long tube; the scala vestibuli begins near the vestibular oval window and the scala tympani ends at the round window (Figure 23–24).

The **stria vascularis**, located in the lateral wall of the cochlear duct (scala media) (Figures 23–29b and 23–30), produces the endolymph with high levels of K⁺ that fills the entire membranous labyrinth. Stratified epithelial cells of the stria vascularis extend cytoplasmic processes and folds around the capillaries of an unusual intraepithelial plexus. K⁺ released from the capillaries is transported across tightly joined cells at the strial surface into the endolymph, which bathes the stereocilia of hair cells and produces conditions optimal for these cells' depolarization.

The **organ of Corti**, or **spiral organ**, where sound vibrations of different frequencies are detected, consists of hair cells and other epithelial structures supported by the basilar membrane (Figure 23–29). Here the sensory hair cells have precisely arranged V-shaped bundles of rigid stereocilia (Figure 23–31); each loses its single larger kinocilium during

development. Two major types of hair cells are present (Figure 23–29c, d):

- **Outer hair cells**, about 12,000 in total, occur in three rows near the saccule, increasing to five rows near the apex of the cochlea. Each columnar outer hair cell bears a V-shaped bundle of stereocilia (Figure 23–31).
- **Inner hair cells** are shorter and form a single row of about 3500 cells, each with a single more linear array of shorter stereocilia (Figure 23–31).

❯❯ MEDICAL APPLICATION

Deafness can result from many factors, which usually fall into two categories: (1) **Conductive hearing loss** involves various problems in the middle ear which can reduce conduction of vibrations by the chain of ossicles from the tympanic membrane to the oval window. A common example is otosclerosis, in which scar-like lesions develop on the bony labyrinth near the stapes which inhibit its movement of the oval window. Infection of the middle ear (otitis media) is common in young children and can reduce sound conduction due to fluid accumulation in that cavity. (2) **Sensorineural deafness** can be congenital or acquired and due to defects in any structure or cell from the cochlea to the auditory centers of the brain, but it commonly involves loss of hair cells or nerve degeneration.

Both outer and inner hair cells have synaptic connections with afferent and efferent nerve endings, with the inner row of cells more heavily innervated. The cell bodies of the afferent bipolar neurons constitute the **spiral ganglion** located in the bony core of the modiolus and (Figures 23–29 and 23–30).

Two major types of columnar supporting cells are attached to the basilar membrane in the organ of Corti (Figure 23–29c):

- Inner and outer **phalangeal cells** extend apical processes that intimately surround and support the basolateral parts of both inner and outer hair cells and the synaptic nerve endings. The apical ends of phalangeal cells are joined to those of the hair cells by tight zonulae occludens, forming an apical plate across the spiral organ through which the stereocilia bundles project into endolymph (Figure 23–31).
- **Pillar cells** are stiffened by heavy bundles of keratin and outline a triangular space, the inner tunnel, between the outer and inner complexes of hair cells and phalangeal cells. The stiff inner tunnel also plays a role in sound transmission.

On the outer hair cells the tips of the tallest stereocilia are embedded in the gel-like **tectorial membrane**, an acellular layer that extends over the organ of Corti from the connective tissue around the modiolus (Figure 23–29). The tectorial membrane consists of fine bundles of collagen (types II, V, IX, and XI) associated with proteoglycans and forms during the embryonic period from secretions of cells lining this region.

FIGURE 23–29 Cochlea and spiral organ (of Corti).

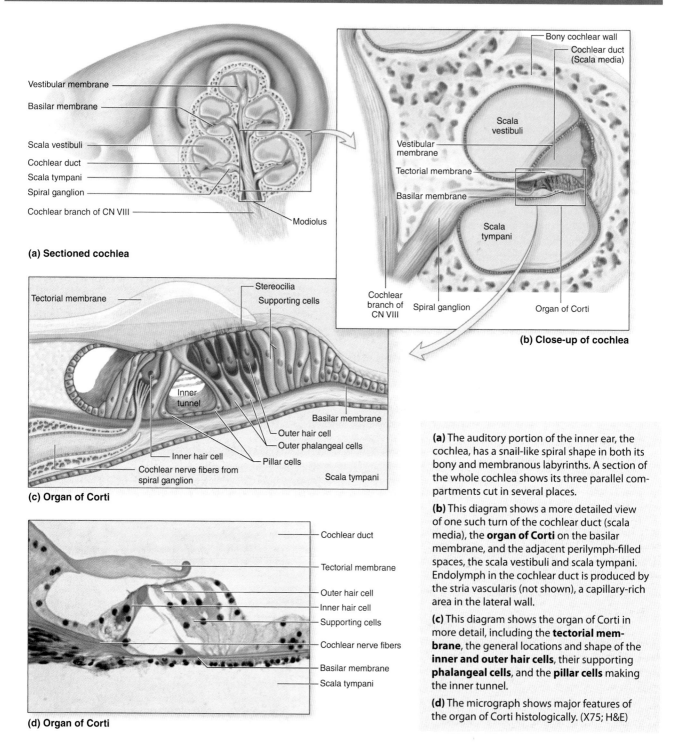

(a) Sectioned cochlea

(b) Close-up of cochlea

(c) Organ of Corti

(d) Organ of Corti

(a) The auditory portion of the inner ear, the cochlea, has a snail-like spiral shape in both its bony and membranous labyrinths. A section of the whole cochlea shows its three parallel compartments cut in several places.

(b) This diagram shows a more detailed view of one such turn of the cochlear duct (scala media), the **organ of Corti** on the basilar membrane, and the adjacent perilymph-filled spaces, the scala vestibuli and scala tympani. Endolymph in the cochlear duct is produced by the stria vascularis (not shown), a capillary-rich area in the lateral wall.

(c) This diagram shows the organ of Corti in more detail, including the **tectorial membrane**, the general locations and shape of the **inner and outer hair cells**, their supporting **phalangeal cells**, and the **pillar cells** making the inner tunnel.

(d) The micrograph shows major features of the organ of Corti histologically. (X75; H&E)

By detecting minute movements of the stereocilia, hair cells in the spiral organ of Corti act as mechanoelectrical transducers very much like those of the vestibular maculae described previously and mediate the sense of hearing. As shown in Figure 23–32, sound waves collected by the external ear cause the tympanic membrane to vibrate, which moves the chain of middle ear ossicles and the oval window. The large size of the tympanic membrane compared to the oval window and the mechanical properties of the ossicle chain amplify the movements and allow optimal transfer of energy between air and perilymph, from sound waves to vibrations of the tissues and fluid-filled chambers.

FIGURE **23–30** Cochlear duct and spiral ganglion.

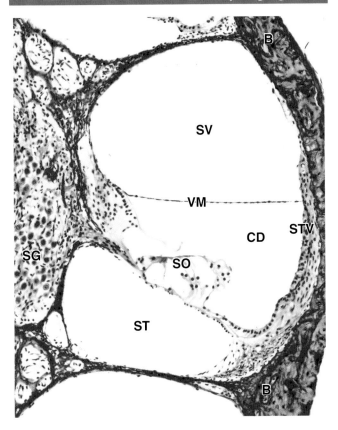

The spiral organ (**SO**) or organ of Corti lies within the cochlear duct (**CD**), or scala media. This duct is filled with endolymph produced in the stria vascularis (**STV**), an unusual association of epithelial cells and the capillaries in the periosteum of the bone (**B**). On either side of the cochlear duct are the scala vestibuli (**SV**) and scala tympani (**ST**), which are filled with perilymph and are continuous at the apex of the cochlea. The vestibular membrane (**VM**) separates perilymph in the scala vestibuli from endolymph in the cochlear duct. Cell bodies of bipolar neurons in the spiral ganglion (**SG**) send dendrites to the hair cells of the spiral organ and axons to the cochlear nuclei of the brain. (X25; H&E)

FIGURE **23–31** Stereocilia hair bundles of cochlear hair cells.

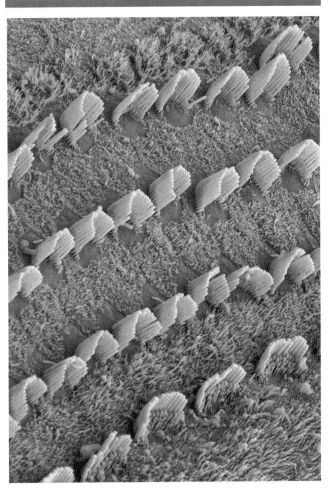

With the tectorial membrane removed, SEM shows the **apical plate** of the rat spiral organ through which rigid **stereocilia bundles** project into endolymph. Shown are hair bundles from three rows of **outer hair cells** and the single row of **inner hair cells** (bottom) in the middle turn of the cochlea. (X3000)

(*Used with permission from Dr Bechara Kachar and Dr Leonardo Andrade, Laboratory of Cell Structure and Dynamics, National Institutes of Health, Bethesda, MD.*)

Pressure waves within the perilymph begin at the oval window and move along the scala vestibuli. Each pressure wave causes momentary displacement of the vestibular and/or basilar membranes and the endolymph surrounding the organ of Corti (Figure 23–32).

The width, rigidity, thickness, and other physical properties of the basilar membrane and its organ of Corti all vary in precise gradients along its length. This allows the region of maximal displacement to vary with the sound waves' frequency, that is, the number of waves moving past a point per unit of time (measured in *hertz*). High-frequency sounds displace the basilar membrane maximally near the oval window. Sounds of progressively lower frequency produce pressure waves that move farther along the scala vestibuli and displace the spiral organ at points farther from the oval window

(Figure 23–33). The sounds of the lowest frequency that can be detected produce movement of the basilar membrane at the apex or helicotrema of the cochlea. After crossing the cochlear duct (scala media) and organ of Corti, pressure waves are transferred to the scala tympani and exit the inner ear at the round window (Figure 23–32).

The main mechanoreceptors for the sense of hearing are the more heavily innervated inner hair cells in the organ of Corti. The outer hair cells, with their stereocilia tips embedded in the tectorial membrane, are depolarized when stereocilia are deformed, as described previously, for vestibular hair cells (Figure 23–27). In the organ of Corti, however, hair cell activities are more complex, allowing greater control on sensory reception.

FIGURE **23–32** Path of sound waves through the ear.

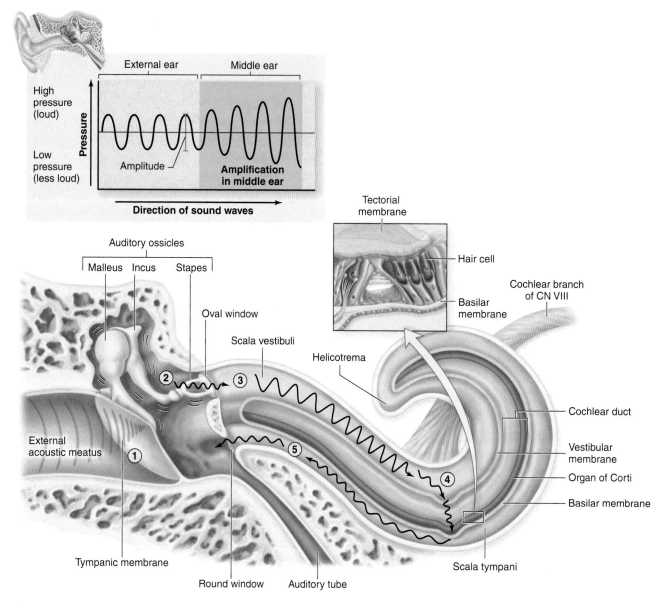

① Sound waves enter ear and cause the tympanic membrane to vibrate.

② Tympanic membrane vibration moves auditory ossicles; sound waves are amplified.

③ The stapes at the oval window generates pressure waves in the perilymph within the scala vestibuli.

④ Pressure waves cause the vestibular membrane to move, resulting in pressure wave formation in the endolymph within the cochlear duct and displacement of a specific region of the basilar membrane. Hair cells in the organ of Corti are distorted, initiating a nerve signal in the cochlear branch of CN VIII.

⑤ Remaining pressure waves are transferred to the scala tympani and exit the inner ear via the round window.

Depolarization of the outer hair cells causes these columnar cells to shorten very rapidly, an effect mediated by an unusual 80-kD transmembrane protein called **prestin** (It. *presto*, very fast) abundant in the lateral cell membranes. Prestin undergoes a voltage-dependent conformational change that affects the cytoskeleton, rapidly shortening the cells when the membrane is depolarized and elongating them again with membrane hyperpolarization. Piston-like movements of the outer hair cells pull down the tectorial membrane against the stereocilia of the inner hair cells (Figure 23–29c), causing depolarization of these cells which then send the signals to the brain for processing as sounds. This sequential role for outer and inner hair cells produces further cochlear amplification of the sound waves.

FIGURE **23–33** Interpretation of sound waves in the cochlea.

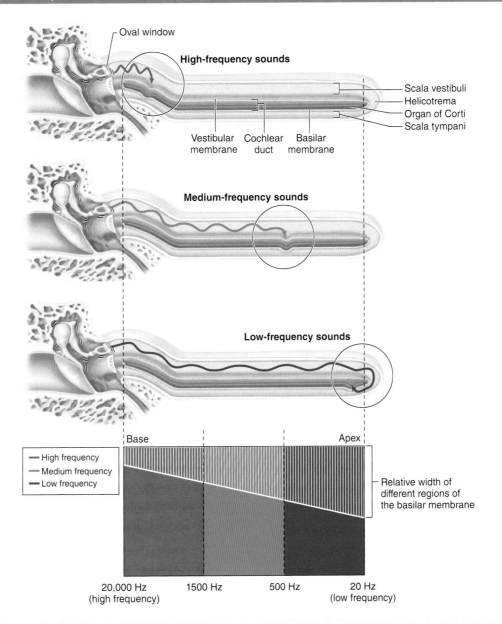

The cochlea is completely straightened in this diagram to more easily show how sound waves are interpreted according to their frequency at specific sites along the organ of Corti. High-frequency sound waves (red arrow) generate pressure waves that displace the basilar membrane near the base of the cochlea, close to the oval window. Medium-frequency sound waves (green arrow) generate pressure waves that displace the membrane at different points along the central region of the cochlea. Low-frequency sound waves (blue arrow) generate waves that displace the membrane near the helicotrema, at the apical end of the cochlea.

❯❯ MEDICAL APPLICATION

Some types of sensorineural deafness can be treated by a **cochlear implant**. A small cable with a series of electrodes is threaded into the scala tympani, with the electrodes along the wall containing branches of the cochlear nerve. A device containing a microphone, a speech processor to filter extraneous sounds, and a transmitter is worn behind the external ear. Sounds of various frequencies transmit signals to a receiver implanted in a bone of the skull and attached to the array of electrodes that stimulate nerve branches appropriate for those frequencies. The neural impulses are interpreted in the brain as sounds. Cochlear implants do not restore normal hearing but can provide a range of sounds that allows understanding of speech.

The Eye & Ear: Special Sense Organs SUMMARY OF KEY POINTS

Eye

- The eye has **three tunics**: the **sclera** and **cornea** form the outer **fibrous tunic**; the middle **vascular layer** (or uvea) consists of the **choroid, ciliary body**, and **iris**; and the **retina** forms the **inner tunic**.
- The transparent **cornea** consists of an anterior **stratified squamous epithelium** on **Bowman membrane**, a thick avascular **stroma**, and an inner **endothelium** on **Descemet membrane**.
- **Aqueous humor** is secreted by **ciliary processes** into the **posterior chamber**, flows through the **pupil** into the **anterior chamber**, and is drained by the **scleral venous sinus** in the **limbus**.
- The **iris** stroma contains **melanocytes** and posteriorly has smooth muscle fibers of the **sphincter pupillae muscle** and the **myoepithelial cells** forming the **dilator pupillae muscle**.
- The **lens** is a unique avascular tissue composed of long **lens fibers**, covered on its anterior side by cuboidal **lens epithelium**, and surrounded by a thick acellular layer called the **lens capsule**.
- The **lens** is suspended behind the **iris** and its central **pupil** by the **ciliary zonule** of fibrillin fibers produced by epithelial cells covering the encircling **ciliary body**.
- The **retina** has the two major parts derived from the embryonic optic cup: the **pigmented epithelium** next to the vascular **choroid** layer and the thicker **neural retina**.
- Cells of the **pigmented epithelium** absorb **scattered light**, form part of a **blood-retina barrier**, regenerate **11-*cis*-retinal**, phagocytose shed discs from rods, and support the rod and cone cells.
- **Rod cells** are **photoreceptors** detecting **light intensity** with short rod-shaped outer segments; less numerous **cone cells**, with conical outer segments, are receptors for the primary **colors** (light of different wavelengths).
- **Rods** have stacked **membrane discs** in which the membranes are densely packed with the protein **rhodopsin** with bound **retinal**.
- Photons of light convert 11-*cis*-retinal to all-*trans*-retinal, causing rhodopsin to release the retinal (**bleaching**), and activate the adjacent G protein **transducin**, which causes a nerve impulse.
- In the **neural retina** the **rod and cone layer** (**RCL**) is nearest to the **retina pigmented epithelium** and near the **outer nuclear layer** (**ONL**) which contains the cell bodies of these photoreceptors.
- An **outer plexiform layer** (OPL) contains the photoreceptor's axons connected in synapses with dendrites of various integrating neurons whose cell bodies form the **INL**.
- Axons from cells in the INL form synapses in the **inner plexiform layer** (IPL) with neurons of **ganglionic layer** (**GL**), which send axons through the **nerve fiber layer** (**NFL**) to the **optic nerve**.
- Eyelids are lined by **conjunctiva**, a stratified columnar epithelium with goblet cells, which also covers the anterior part of the sclera and is continuous with the corneal epithelium.
- **Lacrimal glands** continuously produce the tear film that drains into the nasal cavity via the ducts of the **lacrimal apparatus**.

Ear

- The **acoustic meatus** of the **external ear** ends at the **tympanic membrane** and its mucosa contains sebaceous and **ceruminous glands** that produce an antimicrobial substance, cerumen.
- The **tympanic cavity** of the **middle ear** opens to the nasopharynx via the **auditory** (**eustachian**) **tube**.
- Within the **tympanic cavity**, an articulated series of three small bony **ossicles** (malleus, incus, and stapes) connects the **tympanic membrane** with the **oval window** in the wall of the **internal ear**.
- The **internal ear** consists of a **membranous labyrinth** containing **endolymph**; the membranous labyrinth is enclosed by the temporal bone's **bony labyrinth** which contains perilymph.
- The **membranous labyrinth** has a central **vestibule** with two subdivisions: the **utricle** connects to the three **semicircular ducts** and the **saccule** connects to the **cochlear duct**.
- The walls of the **utricle** and **saccule** each have a thickened area, the **macula**, which contains both sensory **hair cells** with synaptic connections to sensory nerves and **supporting cells**.
- A bundle of rigid **stereocilia** and one rigid **kinocilium** project apically from each **hair cell** and are surrounded by endolymph containing a gel-like matrix with mineralized crystals called **otoliths**.
- Head movements cause **endolymph** and the **otolithic membrane** to move, deforming the rigid apical structures of the **hair cells**, depolarizing them and producing nerve impulses.
- Each of the **semicircular ducts**, oriented 90 degrees from one another, has a terminal **ampulla** region with a thickened **crista ampullaris** containing **hair cells** that contact a gel-like **cupola**.
- Head movements displace endolymph and stereocilia of hair cells in the **utricle, saccule**, and **semicircular ducts** which together produce signals that contribute to the sense of equilibrium.
- The **cochlear duct** is the middle compartment (**scala media**) of the **cochlea** and runs between two other long compartments that contain perilymph: the **scala vestibuli** and the **scala tympani**.
- Along the base of the cochlear duct, the **basilar membrane** supports the **spiral organ of Corti**, which consists largely of **hair cells** connected to sensory fibers of cranial nerve VIII.
- The cochlear hair cells include three to five rows of **outer hair cells** with stereocilia embedded in a gel-like **tectorial membrane** and one row of more heavily innervated **inner hair cells**.
- Sound waves transmitted by the **ossicles** move the oval window and produce pressure waves in the cochlear **perilymph** which deflect the **basilar membrane** and **organ of Corti**, causing nerve impulses which the brain interprets as sounds.

The Eye & Ear: Special Sense Organs ASSESS YOUR KNOWLEDGE

1. Which of the following is the thickest component of the cornea?
   a. Corneal epithelium
   b. Stroma
   c. Descemet membrane
   d. Bowman membrane
   e. Corneal endothelium

2. Which description is accurate for lens fibers?
   a. Are terminally differentiated fibroblasts
   b. Consist of specialized type I collagen
   c. Derived from epithelial cells that produce proteins called crystallins
   d. Consist of type III collagen
   e. Have the same embryonic origin as the neural retina

3. Which structure is the most anterior extension of the eye's vascular layer?
   a. Ciliary body
   b. Cornea
   c. Lens
   d. Iris
   e. Zonule

4. Which cells transmit visual signals from the retina to the brain?
   a. Bipolar cells
   b. Amacrine cells
   c. Ganglion cells
   d. Horizontal cells
   e. Müller cells

5. The epithelial cells within the organ of Corti are supported by which of the following structures?
   a. Spiral limbus
   b. Tectorial membrane
   c. Vestibular membrane
   d. Basilar membrane
   e. Spiral ligament

6. The middle ear contains which of the following structures?
   a. The labyrinth
   b. The modiolus
   c. The perilymph
   d. The vestibular apparatus
   e. The ossicles

7. Which structure in the cochlea is analogous in composition and function to the cupola of each semicircular canal?
   a. Tectorial membrane
   b. Pillar cells
   c. Inner tunnel
   d. Saccule
   e. Spiral ganglion

8. A 47-year-old woman is referred to an ophthalmologist after reporting increased difficulty with tasks at night or in dark places for the past 3 or 4 years. She has trouble walking in dimly lit rooms and the movie theater. She has given up driving at night and describes a prolonged adaptation period going from light to dark. She also describes her daylight vision as "tunneled," as she frequently walks into furniture. A family history indicates that her father had a similar condition. In this disease, a single point mutation in the rhodopsin gene leads to disruption of signal transduction. Visual transduction in the affected cells involves which of the following?
   a. Inactivation of phosphodiesterase
   b. Increase in cGMP levels
   c. Conversion of all-*trans*-retinal to 11-*cis*-retinal
   d. Closing of Na^+ channels
   e. Depolarization of the cell membranes

9. During an eye examination a 55-year-old man is found to show intraocular pressure exceeding 40 mm Hg and blockage of fluid flow within the eye is suspected. Such a diagnosis would involve a blockage at which of the following sites?
   a. Choroid body
   b. Vessels at the optic disk
   c. Pupillary junction of the posterior and anterior chambers
   d. Choroid layer
   e. Trabecular meshwork

10. During a boxing match, a 23-year-old fighter sustains a direct blow to the right ear. He presents with dizziness, vertigo, imbalance, nausea, vomiting, tinnitus, and fullness in the ears. His vertigo increases with activity and is relieved by rest. He has some hearing loss. The symptoms worsen with coughing, sneezing, or blowing his nose, as well as with exertion. He is diagnosed with perilymphatic fistula, which allows leakage of perilymph. In which of the following structures is perilymph normally found?
    a. Scala media
    b. Scala tympani
    c. Semicircular canals
    d. Saccule
    e. Utricle

Appendix | *Light Microscopy Stains*

Hematoxylin and Eosin (H&E)

Hematoxylin stains cellular regions rich in basophilic macromolecules (DNA or RNA) a purplish blue color. It is the most common stain for demonstrating cell nuclei and cytoplasm rich in rough ER. Usually used as the contrasting "counterstain" with hematoxylin, **eosin** is an acidic stain that binds to basic macromolecules such as collagen and most cytoplasmic proteins, especially those of mitochondria. Eosin stains regions rich in such structures a pinkish red color. Tissue sections showing only structures with shades of purple and pink are most likely stained with H&E.

Pararosaniline-Toluidine Blue (PT)

This dye combination stains chromatin shades of purple and cytoplasm and collagen a lighter violet. These stains penetrate plastic sections more readily than H&E and are used in this atlas primarily with acrylic resin-embedded sections to provide better detail of cell and tissue structures. Toluidine blue is also commonly used for differential staining of cellular components, particularly cytoplasmic granules.

Mallory Trichrome

This procedure employs a combination of stains applied in series which results in nuclei staining purple; cytoplasm, keratin, and erythrocytes staining bright red or orange; and collagen bright or light blue. **Mallory trichrome** is particularly useful in demonstrating cells and small blood vessels of connective tissue. Similar stains, such as **Masson trichrome** and Gomori trichrome, yield comparable results except that collagen stains blue-green or green.

Picro-Sirius-Hematoxylin (PSH)

The dye Sirius red in a solution of picric acid stains collagen red and cytoplasm a lighter violet or pink, with nuclei purple if first stained with hematoxylin. Under the polarizing microscope, collagen stained with **picro-sirius red** is birefringent and can be detected specifically.

Periodic Acid–Schiff Reaction (PAS)

This histochemical procedure stains complex carbohydrate-containing cell components, which become magenta (shades of purplish pink). **PAS** is used most commonly to demonstrate cells filled with mucin granules, glycogen deposits, or the glycocalyx.

Wright-Giemsa Stain

These are two similar combinations of stains that are widely used on fixed cells of blood or bone marrow smears to demonstrate types of blood cells. Granules in leukocytes are seen to have differential affinity for the stain components. Nuclei stain purple and erythrocytes stain uniformly pink or pinkish orange.

Silver or Gold Stains

Various procedures employing solutions of **silver or gold salts** have been developed to demonstrate filamentous structures in neurons and fibers of reticulin (type III collagen). By these "metal impregnation" techniques these filaments stain dark brown or black. Such stains have been largely replaced now by immunohistochemical procedures.

Stains for Elastin

Several staining methods have been developed to distinguish elastic structures from collagen, most of which stain the elastin-rich structures brown or shades of purple. Examples of such stains are Weigert's **resorcin fuchsin**, **aldehyde fuchsin**, and **orcein**.

Stains for Lipid

When special preparation techniques are used to retain lipids of cells, such as in frozen sections, lipophilic dyes are used to demonstrate lipid droplets and myelin. **Oil red O** and **Sudan black** stains lipid-rich structures as their names suggest. **Osmium tetroxide** (osmic acid), which is used as a fixative for TEM, is reduced to a black substance by unsaturated fatty acids and is also used to demonstrate lipids, including those of myelin and cell membranes.

Other Common Stains

Many basic aniline dyes, including azures, cresyl violet, brilliant cresyl blue, luxol fast blue, and light green, are used because of the permanence and brightness of the colors they impart to cellular and extracellular structures in paraffin sections. Many such stains were initially developed for use in the textile industry.

525

Figure Credits

Figure numbers in boldface indicate those appearing for the first time in this text; figure numbers in lightface indicate those taken from other sources.

Berman I. *Color Atlas of Basic Histology.* 3rd ed. New York, NY: McGraw-Hill; 2003.

Eckel CM. *Human Anatomy Lab Manual.* New York, NY: McGraw-Hill; 2008.

Fitzpatrick TB, et al. *Dermatology in General Medicine.* New York, NY: McGraw-Hill; 1971.

Hartwell L, Hood L, Goldberg M., et al. *Genetics: From Genes to Genomes.* 4th ed. New York, NY: McGraw-Hill; 2010.

Kaushansky K, Lichtman M, Beutler E, et al. *Williams Hematology.* 8th ed. New York, NY: McGraw-Hill; 2010.

Lewis R, Gaffin D, Hoefnagels M, et al. *Life.* 5th ed. New York, NY: McGraw-Hill; 2004.

Lichtman MA, Shafer MS, Felgar RE, Wang N: *Lichtman's Atlas of Hematology.* New York, NY: 2007. http://www.accessmedicine.com.

McKinley M, O'Loughlin VD. *Human Anatomy.* 2nd ed. New York, NY: McGraw-Hill; 2008.

McKinley M, O'Loughlin VD. *Human Anatomy.* 3rd ed. New York, NY: McGraw-Hill; 2012.

McKinley MP, O'Loughlin VD, Bidle TS. *Anatomy & Physiology: An Integrative Approach.* New York, NY: McGraw-Hill; 2013.

McKinley MP, O'Loughlin VD, Bidle TS. *Anatomy & Physiology: An Integrative Approach, 2nd Edition.* New York, NY: McGraw-Hill; 2016.

Murray RK, Bender DA, Botham KM, et al. *Harper's Illustrated Biochemistry.* 28th ed. New York, NY: McGraw-Hill; 2009.

Raven P, Johnson GB, Losos JB, et al. *Biology.* 7th ed. New York, NY: McGraw-Hill; 2005.

Weiss L, Greep RO. *Histology.* 4th ed. New York, NY: McGraw-Hill; 1977.

Widmaier EP, Raff H, Strang KT. *Vander's Human Physiology.* 11th ed. New York, NY: McGraw-Hill; 2008.

Chapter 1

1-14: McKinley et al 1-5.

Chapter 2

2-3: McKinley et al 4-5a; **2-6:** McKinley, O'Loughlin (2nd ed) 2-7; **2-8:** McKinley et al 4-19; **2-10a:** McKinley, O'Loughlin 2-8; **2-10b:** McKinley, O'Loughlin 2-8; **2-13a (left side):** McKinley, O'Loughlin 2-9; **2-16b:** McKinley, O'Loughlin 2-10; **2-16c:** McKinley, O'Loughlin 2-10; **2-20 (top part):** McKinley, O'Loughlin (2nd ed) 2-12; **2-21b:** McKinley, O'Loughlin 2-11; **2-24:** McKinley, O'Loughlin (2nd ed) 2-35.

Chapter 3

3-2: McKinley, O'Loughlin 2-17; **3-10:** Hartwell et al 17-22b; **3-12 (right):** McKinley, O'Loughlin 2-19; **3-18:** Lewis et al 9-10.

Chapter 4

4-4: Raven 7-13; **4-5:** Weiss 3-12; **4-20:** McKinley, O'Loughlin (2nd ed) 4-4; **4-15d:** Berman 1-16; **4-21a:** McKinley, O'Loughlin 4-6; **4-21b:** McKinley, O'Loughlin 4-6; **4-21c:** McKinley, O'Loughlin 4-6c.

Chapter 5

5-2: McKinley et al 16-3; **5-3a:** Berman 2-6; **5-8a:** Berman 2-7; **5-12b:** Berman 2-24.

Chapter 6

6-1c: Berman 2-20; **6-1d:** Berman 2-19.

Chapter 7

7-1: McKinley, O'Loughlin (2nd ed) 6-1; **7-5a:** Berman 3-4.

Chapter 8

8-1: McKinley et al 7-7; **8-9:** Berman 4-4; **8-13a:** Berman 5-7; **8-14:** McKinley, O'Loughlin (2nd ed) 6-11; **8-16:** McKinley, O'Loughlin (2nd ed) 6-12a,b; **8-17a:** Berman 5-3; **8-17b:** Berman 5-4; **8-18:** McKinley, O'Loughlin (2nd ed) 7-13; **8-19:** McKinley, O'Loughlin (2nd ed) 7-17.

Chapter 9

9-1: McKinley, O'Loughlin (2nd ed) 14-1; **9-2:** McKinley, O'Loughlin (2nd ed) 14-16; **9-3:** McKinley et al 12-2; **9-4:** McKinley et al 12-1 (table); **9-5:** Berman 6-8; **9-6a:** McKinley, O'Loughlin (2nd ed) 14-14b; **9-7:** McKinley, O'Loughlin (2nd ed) 14-13c; **9-8b:** Eckel 4-28b; **9-9:** McKinley, O'Loughlin (2nd ed) 12-5, 12-6; **9-10a:** Berman 9-11a; **9-17:** Eckel 16-1c; **9-18a:** McKinley, O'Loughlin (2nd ed) 16-2b; **9-19:** McKinley, O'Loughlin 15-4; **9-20c:** McKinley, O'Loughlin (2nd ed) 15-7a; **9-21a:** McKinley, O'Loughlin (2nd ed) 14-8(1); **9-21b:** McKinley, O'Loughlin (2nd ed) 14-8(2); **9-21c:** McKinley, O'Loughlin (2nd ed) 14-8(3); **9-21d:** McKinley, O'Loughlin (2nd ed) 14-8(4); **9-22:** Berman 6-21; **9-25:** McKinley, O'Loughlin (2nd ed) 14-10a; **9-26a:** McKinley, O'Loughlin (2nd ed) 14-12a; **9-26b:** Berman 6-15; **9-26d:** McKinley, O'Loughlin (2nd ed) 14-12b; **9-28b:** Berman 6-19; **9-28d:** Berman 6-18; **9-29a:** Berman 6-10; **9-29c:** Berman 6-12.

Chapter 10

10-1: Widmaier 9-1; **10-2:** McKinley, O'Loughlin (2nd ed) 10-4; **10-3:** McKinley et al 10-1; **10-6a:** Berman 7-2; **10-6c:** Berman 7-4; **10-8:** McKinley, O'Loughlin (2nd ed) 10-6; **10-9:** McKinley et al 10-3a; **10-12:** McKinley et al 10-12; **10-13:** Widmaier 9-14; **10-14a:** Widmaier 10-4; Berman 7-6; **10-14b:** Berman 7-7; **10-15:** McKinley, O'Loughlin (2nd ed) 10-12; **10-16:** McKinley, O'Loughlin (2nd ed) 22-10a; **10-17a:** Berman 7-10; **10-17b:** Berman 7-11; **10-18a:** Berman 7-12; **10-20a:** McKinley, O'Loughlin (2nd ed) 10-16.

Chapter 11

11-1: McKinley, O'Loughlin (2nd ed) 19-3; **11-2:** McKinley, O'Loughlin (2nd ed) 22-11; **11-5:** Berman 11-2; **11-6:** McKinley, O'Loughlin (2nd ed) 23-1; **11-8a:** Berman 11-11; **11-8b:** Berman 11-12; **11-13:** McKinley et al 20-5; **11-14a:** Berman 11-20; **11-14b:** Berman 11-22; **11-15:** McKinley et al 20-8; **11-16:** Berman 11-25; **11-21b:** Berman 11-21; **11-22b:** Berman 11-18; **11-22c:** Berman 11-13; **11-22d:** Berman 11-19; **11-24b:** McKinley, O'Loughlin (2nd ed) 24-2b.

Chapter 12

12-1: McKinley, O'Loughlin (2nd ed) 18-1; **12-3:** McKinley et al 18-2; **12-4a:** Widmaier 12-67; **12-4b,c:** McKinley, O'Loughlin (2nd ed) 21-4; **12-8:** Lichtman II.A.4; **12-10d:** Lichtman II.E.7; **12-11b:** Berman 8-5; **12-12c:** Berman 8-6; **12-12d:** Berman 8-1; **12-13a:** Berman 8-9; **12-14:** McKinley, O'Loughlin (2nd ed) 21-10.

Chapter 13

13-1: Kaushansky et al 4-1; **13-2:** McKinley, O'Loughlin (2nd ed) 18-3; **13-5:** McKinley et al 18-4b; **13-7a:** Berman 9-6 through 9-9; **13-7b:** Berman 8-8; **13-10 top, bottom, insets:** Berman 9-2, 9-1; 9-4, 9-5; **13-13a:** Berman 9-11; **13-13b:** Berman 9-13; **13-14:** Berman 9-14.

Chapter 14

14-1: McKinley et al 21-1; **14-2:** McKinley et al 22-17; **14-3:** McKinley et al 22-18a; **14-5:** McKinley et al 22-9; **14-6:** McKinley et al 22-18; **14-8a:** McKinley et al 21-5; **14-8c:** McKinley et al 21-5; **14-11:** McKinley et al 22-14; **14-13a:** Berman 10-5; **14-16:** McKinley et al 21-6.

Chapter 15

15-1: McKinley, O'Loughlin (2nd ed) 26-1; **15-2:** McKinley, O'Loughlin (2nd ed) 26-9; **15-4:** McKinley et al 16-7; **15-5a:** Berman 12-10; **15-5b:** Berman 12-12; **15-6a:** McKinley, O'Loughlin 26-6c; **15-6b:** McKinley/O'Loughlin (2nd ed) 26-5; **15-10a:** Berman 12-1; **15-10b:** Berman 12-4; **15-12:** McKinley, O'Loughlin 26-10; **15-13a:** Berman 12-16; **15-14a:** McKinley et al 26-9a; **15-14b:** McKinley et al 26-10; **15-15:** Berman 12-22; **15-17d:** McKinley et al 26-10c; **15-19:** McKinley, O'Loughlin (2nd ed) 26-11; **15-22:** McKinley, O'Loughlin (2nd ed) 26-15; **15-31:** McKinley, O'Loughlin (2nd ed) 26-26; **15-32a:** McKinley et al 26-23a; **15-32b:** McKinley et al 26-23b; **15-33a:** Berman 12-41; **15-33b:** Berman 12-43.

Chapter 16

16-1: McKinley, O'Loughlin (2nd ed) 26-4a; **16-3b:** Berman 13-26; **16-5a:** Berman 13-29; **16-5b:** Berman 13-32; **16-7:** McKinley, O'Loughlin 26-20; **16-8:** Berman 13-17; **16-9a:** Berman 13-21; **16-11a,b:** McKinley, O'Loughlin 26-19; **16-12a:** Berman 13-3; **16-12b:** Berman 13-4; **16-13b:** Berman 13-7; **16-19:** McKinley et al 26-17; **16-20a:** Berman 13-15.

Chapter 17

17-1: McKinley, O'Loughlin (2nd ed) 25-1; **17-3:** McKinley et al 16-6; **17-4:** Berman 14-1; **17-6:** McKinley, O'Loughlin (2nd ed) 25-8; **17-7:** Berman 14-10; **17-8a:** Berman 14-11; **17-8b:** Berman 14-12; **17-9a:** Berman 14-13; **17-9c:** Berman 14-14; **17-11a:** McKinley et al 26-11; **17-11b,c:** McKinley, O'Loughlin (2nd ed) 25-9; **17-12:** Berman 14-18; **17-13a:** McKinley et al 23-12; **17-13b:** McKinley, O'Loughlin (2nd ed) 25-10; **17-14:** Berman 14-20; **17-18a:** McKinley, O'Loughlin (2nd ed) 25-11.

Chapter 18

18-1: McKinley et al 6-6; **18-2:** McKinley, O'Loughlin (2nd ed) 5-2; **18-3:** Berman 15-4; **18-5:** Berman 15-3; **18-6a:** Berman 15-2; **18-6b:** McKinley, O'Loughlin (2nd ed) 5-4a; **18-7b:** Fitzpatrick 70-9; **18-9:** Fitzpatrick 7-6; **18-10:** McKinley, O'Loughlin (2nd ed) 19-5; **18-11:** McKinley et al 16-3; **18-12:** Eckel 17-2; **18-13a:** McKinley et al 6-9; **18-13c:** McKinley, O'Loughlin (2nd ed) 5-9; **18-14a:** Berman 15-15; **18-14b:** Berman 15-14; **18-14c:** Berman 15-13; **18-15a,b:** McKinley, O'Loughlin (2nd ed) 5-8; **18-16:** McKinley et al 6-10a; **18-17a:** Berman 15-10; **18-19:** Fitzpatrick 81-2; **18-20:** McKinley et al 6-12.

Chapter 19

19-1: McKinley et al 24-3 (right side); **19-2:** McKinley et al 24-4; **19-3:** McKinley et al 24-8; **19-4:** Berman 16-4; **19-5a:** McKinley, O'Loughlin (2nd ed) 27-6; **19-5c:** McKinley et al 24-11a; **19-5d:** McKinley, O'Loughlin (2nd ed) 27-6; **19-6b:** Berman 16-11; **19-6c:** McKinley et al 24-11b; **19-8a:** Berman 16-8; **19-9a,b,c:** McKinley, O'Loughlin (2nd ed) 27-7; **19-13:** McKinley et al 24-9; **19-16:** McKinley, O'Loughlin (2nd ed) 27-8; **19-17a:** Berman 16-18.

Chapter 20

20-1: McKinley et al 17-3; **20-2:** McKinley, O'Loughlin (2nd ed) 20-4; **20-3:** McKinley, O'Loughlin (2nd ed) 20-15; **20-4:** Berman 17-1; **20-5a:** McKinley, O'Loughlin 20-8; **20-5b:** McKinley, O'Loughlin 20-6; **20-6:** Berman 17-3; **20-8:** McKinley et al 17-4; **20-9:** Berman 17-4, McKinley, O'Loughlin (2nd ed) 20-10; **20-12:** McKinley, O'Loughlin (2nd ed) 20-13a; **20-14:** McKinley, O'Loughlin (2nd ed) 20-13c,d; **20-17c:** Berman 17-13; **20-17e:** McKinley, O'Loughlin (2nd ed) rt qt; **20-18:** McKinley, O'Loughlin (2nd ed) 20-9; **20-19:** Berman 17-15; **20-21 (upper right):** McKinley et al 17-18; **20-22:** McKinley, O'Loughlin (2nd ed) 20-11a; **20-23:** Berman 17-17.

Chapter 21

21-1: McKinley, O'Loughlin (2nd ed) 28-11; **21-2:** McKinley, O'Loughlin (2nd ed) 28-13; **21-3:** Berman 18-2; **21-4a:** Berman 18-5; **21-5:** McKinley et al 28-18; **21-6a:** Berman 18-7; **21-6b:** Berman 18-8; **21-9a:** Berman 18-10; **21-9b:** Berman 18-11; **21-10b:** Berman 18-13; **21-11a:** Berman 18-14; **21-12a:** Berman 18-16; **21-13a:** McKinley et al 28-19; **21-14a:** Berman 18-18; **21-16a:** Berman 18-20; **21-16b:** Berman 18-21; **21-17:** McKinley, O'Loughlin (2nd ed) 18-17b; **21-18:** Berman 18-23.

Chapter 22

22-1a,b: McKinley, O'Loughlin (2nd ed) 28-11; **22-2:** McKinley, O'Loughlin (2nd ed) 28-4; **22-9:** McKinley, O'Loughlin (2nd ed) 3-7; **22-10:** McKinley et al 28-6; **22-11:** McKinley et al 28-8; **22-13:** Berman 19-8; **22-14:** McKinley, O'Loughlin (2nd ed) 28-7; **22-15a:** Berman 19-16; **22-17:** McKinley, O'Loughlin (2nd ed) 28-6; **22-19a:** Berman 19-19; **22-19b:** Berman 19-20; **22-19c:** Berman 19-21; **22-20:** McKinley, O'Loughlin (2nd ed) 2-6; **22-21c:** McKinley et al 29-7 **22-23a:** Berman 19-22; **22-24a:** Berman 19-23; **22-26a:** Berman 19-24; **22-26b:** Berman 19-25; **22-26c:** Berman 19-26; **22-27a:** Berman 19-27.

Chapter 23

23-1: McKinley 19-12b; **23-2:** McKinley, O'Loughlin (2nd ed) 19-19; **23-8:** McKinley et al 16-16; **23-10a:** Berman 20-4; **23-14:** McKinley, O'Loughlin (2nd ed) 19-14a,b; **23-15:** Berman 20-9; **23-21:** McKinley, O'Loughlin (2nd ed) 19-20; **23-23:** McKinley et al 16-25; **23-24:** McKinley et al 26-26; **23-25:** McKinley et al 16-32; **23-28:** McKinley, O'Loughlin (2nd ed) 19-25; **23-29a,b:** McKinley et al 16-27a,b; **23-29c:** McKinley 19-27; **23-29d:** McKinley et al 16-27d; **23-32:** McKinley et al 16-28; **23-33:** McKinley et al 16-29.

Index

Note: Page numbers followed by *f* indicate figures; and page numbers followed by *t* indicate tables

orthochromatophilic erythroblast, 259

osmium tetroxide, 3

osmosis, 24*t*

osmotic pressure, 114

ossification
endochondral, 149–151, 150*f*–151*f*
intramembranous, 148, 149*f*

ossification centers, 149, 151*f*

ossification zone, epiphyseal plate, 151, 152*f*

osteitis fibrosa cystica, 148

osteoarthritis, 131

osteocalcin, 141, 143

osteoclasts, 99, 100*t*, 138, 139*f*–140*f*, 143, 144*f*

osteogenesis
endochondral ossification, 149–151
intramembranous ossification, 148
overview, 148–149

osteogenesis imperfecta, 30, 108*t*, 149

osteoid, 140*f*, 141

osteomalacia, 148, 151

osteonectin, 143

osteons, 145, 146*f*–147*f*

osteopetrosis, 143

osteoporosis, 143

osteoprogenitor cells, 143

osteosarcoma, 141

otitis media, 510

otolithic membrane, 513

otoliths, 513, 513*f*

outer hair cells (OHC), 517

outer limiting membrane, retina, 503

outer membrane, mitochondria, 38

outer plexiform layer, 503

outer segments, retina, 504

oval cells, 345

oval window, 510, 517–519

ovarian cycle, 469*f*, 475*f*

ovarian follicles
atresia, 465
growth of, 463–465
overview, 461

ovaries
corpus luteum, 467, 469–470
early development of, 460–461
follicles. *See* ovarian follicles
hormonal regulation of, 469*f*
overview of, 460, 461*f*
ovulation, 466–467

oviducts, 472*f*, 476

ovulation, 466–467

ovum, 471

oxidases, 39

oxygen, in blood, 238*f*

oxyhemoglobin, 241

oxyphil cells, 434

oxytocin, 421, 423*t*, 485

P

pacemaker, 217–218

Pacinian corpuscles, 382*f*, 383

palatine tonsils, 281–282

PALS (periarteriolar lymphoid sheath), 286, 289*f*

pampiniform venous plexus, 439

pancreas
acinar cells, 333, 336*f*–337*f*
cancer of, 333
islets of Langerhans, 332, 427–429, 428*f*, 429*t*
overview, 332–335, 335*f*

pancreatic polypeptide, 429, 429*t*

pancreatitis, 334

Paneth cells, 315, 320*f*

Papanicolaou procedure (Pap smear), 482*f*, 483

papillae
circumvallate, 299–300
definition of, 72
dermal, 371, 372*f*, 374*f*, 378, 380, 382–383, 384*f*–385*f*
filiform, 299
foliate, 299
fungiform, 299
renal, 408*f*
on tongue, 298, 299*f*–300*f*
vallate, 299–300

papillary duct, 406

papillary layer, dermis, 380

paracortex, lymph node, 282, 284, 285*f*, 287*f*, 287*f*

paracrine secretion, 413

paracrine signaling, 23, 87

parafollicular cells, 430, 431*f*–432*f*, 436*t*

paraganglia, 223

paranasal sinuses, 351

parasympathetic divisions, 187

parathyroid glands, 414*f*, 432–434, 434*f*, 436*t*

parathyroid hormone (PTH), 155, 433

paraventricular nuclei, 416

parenchyma, 71

parietal cells, 310, 310*f*–313*f*

parietal layer
of glomerular capsule, 397
of pericardium, 217

parietal pleura, 368, 368*f*

Parkinson disease, 163

parotid glands, 329, 331*f*, 332

pars distalis, 416–418, 418*f*, 420*f*

pars intermedia, 418, 420, 421*f*

pars nervosa, 413, 423*f*

pars tuberalis, 416, 418

parturition, 473

PAS (periodic acid-Schiff reagent), 3, 4*f*

passive diffusion, 24*t*

passive immunity, 270

PCOS (polycystic ovary syndrome), 464

PCR (polymerase chain reaction), 14

PD (pigment deposits), 48*f*

pedicels, 397, 398*f*

peg cells, 471

pemphigus, 380

pemphigus vulgaris, 76

pendrin, 430

penicillar arterioles, 290

penis, 456–457, 456*f*–457*f*

pepsin, 307, 312

pepsinogen, 312

peptidases, 315

perforating canals, 145, 147*f*

perforating fibers, 143

perforin, 271

periarteriolar lymphoid sheath (PALS), 286, 289*f*

pericardium, 217

perichondrium, 129, 133

pericytes, 228*f*, 229

perikaryon, 163, 165–166

perilymph fluid, 511

perimetrium, 473

perimysium, 194, 195*f*

perineurium, 184, 186*f*

perinuclear space, 53

periodic acid-Schiff reagent (PAS), 3, 4*f*

periodontal diseases, 304

periodontal ligament, 301, 304, 306*f*

periodontitis, 304

periodontium, 306*f*

periosteum, 138, 139*f*, 143